R. S. Naylor

The Mantle Sample: Inclusions in Kimberlites and Other Volcanics

SECOND INTERNATIONAL KIMBERLITE CONFERENCE

SPONSORS

American Geophysical Union

The Navajo Tribe, Museum & Research

Carnegie Institution of Washington, Geophysical Laboratory

U. S. Geological Survey

U. S. National Committee for Geology

U. S. National Committee for Geochemistry

U. S. National Committee/IUGG

U. S. Geodynamics Committee

The San Carlos Apache Tribe

International Association of Volanology & Chemistry of the Earth's Interior

International Association of Geochemistry & Cosmochemistry

Inter-Union Commission on Geodynamics/IUGG

Commission on Experimental Petrology/IUGS

Arkansas Geological Commission

Conveners: L. H. Ahrens, R.S.A.; F. R. Boyd, U.S.A.; J. B. Dawson, U.K.

Organizing Committee: O. L. Anderson, A. L. Boettcher, F. R. Boyd (Chmn.), D. H. Eggler, S. R. Hart, T. H. Jordan, T. R. McGetchin, W. G. Melson, H. O. A. Meyer, M. E. McCallum, J. R. Smyth, G. A. Swann, H. G. Wilshire

Associate Editors: N. Z. Boctor, S. R. Hart, I. D. MacGregor, R. H. Mitchell, D. Smith, J. R. Smyth, H. G. Wilshire

The Mantle Sample: Inclusions in Kimberlites and Other Volcanics

F. R. Boyd
Henry O. A. Meyer
editors

Proceedings of the Second International Kimberlite Conference
Volume 2

American Geophysical Union
Washington, D. C. 20006
1979

The Mantle Sample:
Inclusions in Kimberlites and Other Volcanics

Copyright © 1979 by the
American Geophysical Union
1909 K Street, N.W.
Washington, D.C. 20006

Library of Congress Catalog No. 78-72026
ISBN 0-87590-213-8

Printed by
LithoCrafters, Inc.
Chelsea, Michigan 48118

PREFACE

A dedicated student of xenoliths once remarked (privately) that when confronted by a bus full of people, what interested him was the people and not the bus. This is a book about the people, not the bus. Those who (rightfully) turn on to the genesis and mineralogic complexities of kimberlites, the mechanisms of diatreme formation, and the parageneses of diamond will find much to excite them in Volume I of these Proceedings--Kimberlites, Diatremes and Diamonds. Volume I is primarily a book about the bus.

Papers contained in these volumes were presented at the Second International Kimberlite Conference, held in Santa Fe, New Mexico, October 3-7, 1977. An aim of this Conference was "to provide an opportunity for rationalization of mantle models developed through study of kimberlite xenoliths, basalt xenoliths and through geophysical methods." In field trips associated with the Conference, participants had an opportunity to visit kimberlites in Arkansas, the Colorado Front Range and the Colorado Plateau, together with spectacular basalt nodule localities at Kilbourne Hole, San Carlos and Chino Valley. These trips, in particular, provided marvelous opportunities for communication and discussion between participants of diverse backgrounds and specialties. It is hoped that this communication will have an important impact on progress in mantle petrology and kimberlite-oriented research, but the needed rationalization of mantle models has yet to come. Major problems remain; some have been around for a long time, while the existence of some is only recently appreciated.

For example, much progress has been made in the application of experimental data to determine the temperatures and depths of origin of garnet peridotite xenoliths and megacrysts. Further refinement of these techniques is certainly needed, but of more critical importance is the development of analogous techniques that can be applied to eclogites, spinel peridotites and harzburgites. These latter rock types are abundant as xenoliths in kimberlites as well as basalts. Progress has also been made in determining which inclusions in kimberlite are truly accidental, but debate continues over the relationship between kimberlites and the megacrysts of garnet, pyroxene and ilmenite that they commonly contain.

An outstanding development since the First Kimberlite Conference, held in Cape Town in 1973, has come in the interpretation of the deformation textures found in peridotite xenoliths. It now appears generally accepted that severe deformation has accompanied eruption and is not the product of slow creep at depth. Moreover, it now appears possible to use the degree of recrystallization of certain olivine neoblasts and the growth of exsolution lamellae in pyroxenes to estimate velocities of eruption.

These are exciting developments. The energy and enthusiasm demonstrated by participants in the Second International Kimberlite Conference make it clear that more exciting developments are on the way.--F. R. Boyd

Ken 30 Sept '77

CONTENTS

I. GEOPHYSICS

II. ECLOGITES AND PERIDOTITES FROM KIMBERLITES

III. MEGACRYSTS

IV. XENOLITHS FROM THE COLORADO PLATEAU

V. XENOLITHS FROM BASALTS AND OTHER VOLCANICS

I. GEOPHYSICS

MINERALOGIES, DENSITIES AND SEISMIC VELOCITIES OF GARNET LHERZOLITES AND THEIR GEOPHYSICAL IMPLICATIONS

Thomas H. Jordan

Geological Research Division, Scripps Institution of Oceanography, La Jolla, California 92093

Abstract. A simple algorithm for estimating the mineral composition, density and seismic velocities of a garnet lherzolite from its whole rock oxide composition has been devised and successfully tested against observations. The algorithm has been applied to 78 analyses of garnet lherzolite xenoliths from kimberlite pipes and to 9 model compositions for the oceanic upper mantle. Normative densities and velocities for these samples correlate well with Fe/Fe + Mg. Removal of a basaltic fraction from fertile garnet lherzolite lowers its normative density, in agreement with the conclusions of O'Hara [1975] and Boyd and McCallister [1976]. The average continental garnet lherzolite (ACGL) constructed from the xenolith compositions is significantly depleted in normative garnet and clinopyroxene relative to the range for oceanic upper mantle models, and its normative density is significantly less (3.353 gm/cm^3 vs. 3.397 gm/cm^3 for pyrolite). The data support Jordan's [1975a] hypothesis that the large sub-lithospheric temperature gradients associated with deep continental root zones are dynamically stabilized by compositional gradients. It is suggested that substantial variations in lithospheric density and thickness are generated in regions of anomalous oceanic vulcanism (e.g. Hawaii and Iceland); these may contribute to observed gravity anomalies and residual depth anomalies, and may exert some dynamical control on the subduction process.

Introduction

The great horizontal motions on the surface of the Earth are envisaged to be a consequence of gravitational instability within the mantle. The lateral variations in density implied by this hypothesis depend on the compositions of mantle rocks, as well as their temperatures. Large-scale chemical homogeneity of the sub-lithospheric mantle is usually postulated in most quantitative discussions of convective mass transport, but this assumption may not be adequate for understanding important aspects of mantle dynamics. The vast outpourings of basalt from the interior compel us to believe that processes of chemical differentiation are active within the Earth. The likely source region for these basalts is the garnet lherzolite layer of the mantle [Yoder, 1976]. The removal of basaltic magma from a garnet lherzolite leaves a residuum whose physical properties differ significantly from the parental material.

The variations of density induced by basaltic differentiation have been the subject of several previous papers. Shaw and Jackson [1973; see also Jackson and Wright, 1970] estimated that, at equal pressures and temperatures, the residues from the melting which produces the Hawaiian volcanics are about 3% more dense than the parental composition. This conclusion formed the basis for their hypothesis that residual material will sink into the mantle to form a "gravitational anchor" beneath Hawaii. However, O'Hara [1975], in a discussion of Icelandic vulcanism, pointed out that partial fusion will generally decrease Fe/Fe + Mg in the residuum and deplete the dense garnet component. He concluded that residual material should be about 2.5% *less* dense than undepleted mantle, and, therefore, mantle depleted in a basaltic component will not sink unless large temperature contrasts exist. O'Hara's results were supported and strengthened by Boyd and McCallister's [1976] experimental study of two natural garnet peridotites. Green and Liebermann [1976] computed densities and elastic parameters for several theoretical compositions, including refractory peridotite and garnet pyrolite, which they compared with seismic models of the oceanic upper mantle. Their calculations also indicated that refractory material will be less dense, but the magnitude of their density differences was only half that of O'Hara's.

In this paper, the effects of chemical variations on the densities and seismic velocities of garnet lherzolites are examined in detail. A simple procedure for estimating the mineralogy and elastic parameters of lherzolites in the garnet stability field from whole rock compositions is tested by comparing its results with the experimental data of Boyd and McCallister [1976].

The procedure is then applied to a large number of garnet lherzolite xenoliths from kimberlite pipes and to a variety of model compositions for the oceanic upper mantle.

The conclusions of this study are used to speculate on several topics concerning mantle structure and dynamics.

Estimation of Garnet Lherzolite Mineralogies, Densities and Seismic Velocities

Sufficient data now exist for estimating the bulk elastic properties of a garnet lherzolite if the mineralogy of the rock is known. A simple analytical procedure for calculating the approximate mineral compositions from a whole rock composition has been devised for this purpose.

Let X_A^α be the mole fraction of component A in phase α, and define the apparent distribution coefficient

$$K_{A/B}^{\alpha/\beta} = (X_A^\alpha / X_B^\alpha)/(X_A^\beta / X_B^\beta) \quad (1)$$

Seven oxide components (SiO_2, Al_2O_3, Cr_2O_3, FeO, MgO, CaO, Na_2O; total Fe recomputed as FeO) and four phases (garnet, clinopyroxene, orthopyroxene, olivine) are considered. The problem is to determine the 28 unknown X_A^α 's from the 7 observed X_A^{WR} 's (WR = whole rock). Nine of these are assumed to be zero:

$$X_{Al}^{Opx} = X_{Al}^{Ol} = 0 \quad (2a)$$

$$X_{Cr}^{Opx} = X_{Cr}^{Ol} = 0 \quad (2b)$$

$$X_{Ca}^{Opx} = X_{Ca}^{Ol} = 0 \quad (2c)$$

$$X_{Na}^{Gt} = X_{Na}^{Opx} = X_{Na}^{Ol} = 0 \quad (2d)$$

Five apparent distribution coefficients are specified:

$$K_{Al/Cr}^{Gt/Cpx} = 1.0 \quad (3a)$$

$$K_{Fe/Mg}^{Gt/Ol} = 1.8 \quad (3b)$$

$$K_{Fe/Mg}^{Cpx/Ol} = 0.9 \quad (3c)$$

$$K_{Fe/Mg}^{Opx/Ol} = 0.9 \quad (3d)$$

$$K_{Ca/Mg}^{Gt/Cpx} = 0.3 \quad (3e)$$

The values of these coefficients are chosen to be roughly consistent with the observed apparent distribution coefficients in garnet lherzolite xenoliths from kimberlite pipes [e.g. Nixon and Boyd, 1973; Cox *et al*., 1973; Gurney *et al*., 1975].

The stoichiometric equations and the equations of mass balance reduce the problem to a system of four nonlinear equations in five unknowns. The system can be made determinate by specifying the calcium-magnesium ratio of the clinopyroxene. For the calculations discussed here, it is assumed that

$$X_{Ca}^{Cpx}/(X_{Ca}^{Cpx} + X_{Mg}^{Cpx}) = 0.4, \quad (4)$$

which corresponds to a temperature of 1200°C on the 30 kb diopside-enstatite solvus of Davis and Boyd [1966].

The nonlinearity of the resulting 4 × 4 system of equations is not severe, and the solution of the system can be achieved by a simple iterative algorithm. No multiplicity of solutions satisfying positivity constraints was discovered, and, for the compositions considered in this paper, at least one non-negative solution was available.

Densities were determined from the mineral compositions by assuming the additivity of end-member molar volumes. The seismic velocities were obtained from end-member elastic parameters using the averaging procedure of Hill [1952]. The end-member densities and seismic velocities at P = 1 bar and T = 25°C are listed in Table 1. All calculated densities and velocities correspond to these standard conditions and will be referred to as "normative"; the normative parameters will be denoted by $\hat{\rho}$, $\hat{v}_P$ and $\hat{v}_S$, respectively.

The results of the algorithm are compared with the experimental data of Boyd and McCallister [1976] for two natural garnet lherzolites (PHN1569 and PHN1611) in Tables 2 and 3. The agreement is good. In particular, the whole rock density difference $\hat{\rho}_{1611} - \hat{\rho}_{1569}$ is calculated to be 0.106 ± 0.010 gm/cm^3, compared with an experimentally determined value of 0.09 ± 0.03 gm/cm^3. The error assignment on the former estimate is discussed below.

The densities and seismic velocities of these rocks are primarily sensitive to two parameters: the atomic ratio

$$R \equiv X_{Fe}^{WR}/(X_{Fe}^{WR} + X_{Mg}^{WR}), \quad (5)$$

and the mole fraction of garnet molecule X_{Gt}^{WR}. For PHN1611, the partial derivatives of logarithmic density with respect to these parameters have the (dimensionless) values:

$$(\partial \ln \hat{\rho}/\partial R)_{X_{Gt}^{WR}} = 0.32 \quad (6a)$$

$$(\partial \ln \hat{\rho}/\partial X_{Gt}^{WR})_R = 0.10 \quad (6b)$$

If basaltic differentiation is responsible for chemical heterogeneity in the upper mantle, then the parameters R and X_{Gt}^{WR} will be correlated, and

TABLE 1. Densities and Velocities of End Members at P=1 Bar, T=25°C.

End Member Name	Formula	ρ (gm/cm³)	Source	v_p* (km/s)	v_s* (km/s)	Source
Forsterite	Mg_2SiO_4	3.214	(9)	8.57	5.02	(5)
Fayalite	Fe_2SiO_4	4.393	(9)	6.64	3.49	(2)
Orthoenstatite	$MgSiO_3$	3.198	(9)	8.32	4.98	(6)
Orthoferrosilite	$FeSiO_3$	3.990	(1)	6.90	3.72	(6)
Clinoenstatite	$MgSiO_3$	3.190	(9)	8.32	4.98	(12)
Clinoferrosilite	$FeSiO_3$	4.068	(8)	6.90	3.72	(12)
Diopside	$CaMgSi_2O_6$	3.277	(9)	8.06	4.77	(7)
Jadeite	$NaAlSi_2O_6$	3.315	(9)	9.01	5.35	(3)
Ureyite	$NaCrSi_2O_6$	3.590	(8)	8.62	5.12	(11)
Pyrope	$Mg_3Al_2Si_3O_{12}$	3.559	(9)	8.96	5.05	(4)
Almadine	$Fe_3Al_2Si_3O_{12}$	4.318	(9)	8.42	4.68	(4)
Grossular	$Ca_3Al_2Si_3O_{12}$	3.595	(9)	9.31	5.43	(4)
Knorringite	$Mg_3Cr_2Si_3O_{12}$	3.852	(8)	8.50	4.79	(11)
	$Fe_3Cr_2Si_3O_{12}$	4.578	(10)	7.99	4.44	(11)
Uvarovite	$Ca_3Cr_2Si_3O_{12}$	3.851	(9)	8.88	5.18	(11)

* Voigt-Reuss-Hill averages

References: (1) Bragg *et al.* [1965]; (2) Chung [1970]; (3) Hughes and Nishitake [1963]; (4) Isaak and Graham [1976]; (5) Kumazawa and Anderson [1969]; (6) Liebermann [1974]; (7) Liebermann and Mayson [1976]; (8) Roberts *et al.* [1974]; (9) Robie *et al.* [1966]; (10) Inferred from lattice parameter systematics; (11) Inferred from velocity-density systematics; (12) Assumed to be the same as for orthopyroxenes.

the density variations due to variations in these ratios will add with the same sign. For the seismic velocities, the derivatives computed from PHN1611 are

$$(\partial \ln \hat{v}_p/\partial R)_{X_{Gt}^{WR}} = -0.27 \tag{7a}$$

$$(\partial \ln \hat{v}_p/\partial X_{Gt}^{WR})_R = 0.08 \tag{7b}$$

$$(\partial \ln \hat{v}_s/\partial R)_{X_{Gt}^{WR}} = -0.34 \tag{7c}$$

$$(\partial \ln \hat{v}_s/\partial X_{Gt}^{WR})_R = 0.05 \tag{7d}$$

Thus, increasing the garnet content of a rock increases the seismic velocities, whereas increasing the iron content decreases the velocities; these relations are concordant with well-established velocity-density systematics [e.g. Anderson *et al.*, 1972]. If the factors R and X_{Gt}^{WR} are positively correlated, the effects of their variations on the velocities will tend to cancel. It should be noted, however, that the

TABLE 2. Comparison of Observed and Calculated Parameters for PHN1569.

	Whole Rock		Garnet		Clinopyroxene		Orthopyroxene		Olivine	
	Obs.*	Comp.**	Obs.*	Comp.	Obs.*	Comp.	Obs.*	Comp.	Obs.*	Comp.
Composition (Wt.%)										
SiO_2	48.58	48.58	41.35	41.82	55.51	56.21	57.43	58.71	41.00	41.40
Al_2O_3	0.64	0.64	18.41	18.37	2.04	3.91	0.91	---	0.02	---
Cr_2O_3	0.28	0.28	7.52	7.87	1.72	1.67	0.36	---	0.01	---
FeO	5.76	5.76	6.47	5.09	1.48	2.08	4.29	4.33	7.05	6.96
MgO	44.18	44.18	19.12	20.98	16.96	17.12	36.58	36.96	51.90	51.64
CaO	0.50	0.50	7.11	5.87	20.85	15.96	0.37	---	0.01	---
Na_2O	0.06	0.06	0.01	---	1.44	3.06	0.06	---	0.00	---
Mode & Norm (Wt.%)	---	---	0.8	3.1	1.4	2.0	45.3	39.7	52.5	55.2
$\hat{\rho}$ (gm/cm³)	3.30±.02	3.309	3.67	3.749	3.25	3.295	3.27	3.266	3.33	3.320
$\hat{v}_p$ (km/s)	---	8.300	---	8.824	---	8.098	---	8.181	---	8.364
$\hat{v}_s$ (km/s)	---	4.868	---	4.973	---	4.784	---	4.869	---	4.866

* Observed values from Nixon and Boyd [1973] and Boyd and McCallister [1976].

** Computed whole rock compositions are required to equal the observed values.

TABLE 3. Comparison of Observed and Calculated Parameters for PHN1611.

	Whole Rock		Garnet		Clinopyroxene		Orthopyroxene		Olivine	
	Obs.*	Comp.**	Obs.*	Comp.	Obs.*	Comp.	Obs.*	Comp.	Obs.*	Comp.
Composition (Wt.%)										
SiO_2	44.78	44.78	43.13	41.92	56.09	55.82	56.58	57.78	40.34	40.37
Al_2O_3	2.82	2.82	21.37	22.19	2.47	3.11	1.35	---	0.10	---
Cr_2O_3	0.29	0.29	1.48	2.26	0.50	0.32	0.21	---	0.05	---
FeO	10.30	10.30	8.72	9.11	5.26	4.24	7.08	7.89	11.42	12.46
MgO	38.14	38.14	20.88	19.16	20.80	17.85	32.84	34.33	47.96	47.17
CaO	3.34	3.34	4.35	5.36	13.34	16.63	1.60	---	0.13	---
Na_2O	0.34	0.34	0.07	---	1.54	2.02	0.34	---	0.00	---
Mode & Norm (Wt.%)	---	---	10.4	10.4	19.7	16.7	11.1	9.5	58.8	63.4
$\hat{\rho}$ (gm/cm^3)	3.39±.02	3.415	3.69	3.768	3.30	3.315	3.28	3.321	3.38	3.405
$\hat{v}_p$ (km/s)	---	8.217	---	8.806	---	8.026	---	8.080	---	8.211
$\hat{v}_s$ (km/s)	---	4.763	---	4.975	---	4.718	---	4.782	---	4.746

* Observed values from Nixon and Boyd [1973] and Boyd and McCallister [1976].
** Computed whole rock compositions are required to equal the observed values.

normative shear velocities are much less sensitive to garnet content than to iron content. Also, because of the high compressional velocity of olivine relative to pyroxene (Table 1), the normative compressional velocities are sensitive to a third factor, the degree of silica saturation.

The derivatives given above have been used to estimate roughly the accuracy of the normative calculations. For good wet chemical analyses, the ratio R is generally determined to within ± 0.002 [Peck, 1964, p. 54], and the uncertainties for other methods (atomic absorption, XRF) are comparable [J. Natland, private communication]. A comparison of the calculated norms with the observed modes for PHN1569, PHN1611 and other garnet lherzolites [e.g. Chen, 1971] indicates that X_{Gt}^{WR} is generally estimated to within ± 0.02. A primary source of error is the variable solubility of Al_2O_3 in the pyroxenes not modelled in the normative calculation. To first order, the uncertainties in the calculated density and velocity *differences* do not depend on the uncertainties in the densities and velocities of the end members. Consequently, the expected errors in the calculated differences can be estimated from the errors in R and $\hat{X}_{Gt}^{WR}$ using standard variational analysis. The uncertainties are ± 0.010 gm/cm^3 for $\hat{\rho}$, ± 0.019 km/s for $\hat{v}_p$, and ± 0.008 km/s for $\hat{v}_s$. Thus, the expected error in the calculated difference $\hat{\rho}_{1611} - \hat{\rho}_{1569}$, quoted above, is about 10% of its value. The relative errors for the corresponding $\hat{v}_p$ and $\hat{v}_s$ differences are 23% and 8%, respectively.

TABLE 4. The Distribution of Continental Garnet Lherzolites from Kimberlite Pipes Used in This Study

Pipe	Location	No. of Samples	Symbol*	References
Thaba Putsoa	Lesotho	6	○	Nixon et al. (1963); Nixon and Boyd (1973)
Lighobong	Lesotho	1	◁	Nixon and Boyd (1973)
Mothae	Lesotho	3	△	Nixon and Boyd (1973)
Matsoku	Lesotho	11	▽	Carswell and Dawson (1970); Cox et al. (1973)
Bultfontein	S. Africa	22	+	Carswell and Dawson (1970); Chen (1971)
Jagersfontein	S. Africa	10	×	Carswell and Dawson (1970); Chen (1971)
Wesselton	S. Africa	15	⅄	Carswell and Dawson (1970); Chen (1971)
Dutoitspan	S. Africa	1	Y	Ito and Kennedy (1967)
Lourwencia	S.W. Africa	1	≻	Nixon et al. (1963)
W Diatreme	Montana, U.S.A.	2	□	B.C. Hearn (unpublished)
Udachnaya	Yakutia, U.S.S.R.	6	◇	Sobolev (1977)

* Plotting symbol used in Figures 1-4 and Figure 10.

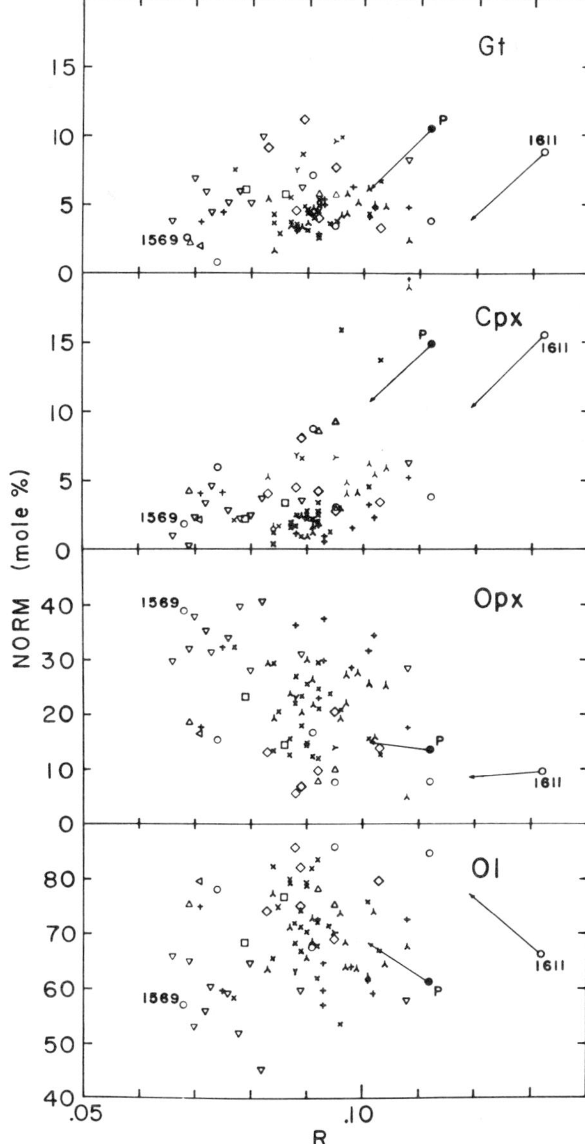

Figure 1. Normative mineral fractions vs. R = $X^{WR}_{Fe}/(X^{WR}_{Fe} + X^{WR}_{Mg})$ for the 78 continental garnet lherzolite analyses identified in Table 4. The points labelled P correspond to Ringwood's [1966] pyrolite. The vectors attached to pyrolite and to PHN1611 are displacements produced by subtraction of 10 mole % basalt [see text and Table 6].

The normative calculations described above provide a basis for deducing the effects of compositional variations on the densities and seismic velocities of volatile-free mantle rocks at standard temperature and pressure. Obviously, corrections for elevated temperatures and pressures, as well as the effects of volatiles, must be included before the normative parameters can be compared directly with the values observed at mantle conditions. Since a direct comparison with seismic data is not attempted here, these corrections are not applied. The primary concern of this paper is to examine the differences in the normative parameters induced by compositional differences. At subsolidus conditions, a change in the reference state to an elevated temperature and pressure will have an effect on the differences in the normative parameters which is second-order in small quantities and can therefore be ignored. However, at supersolidus conditions, the amount and distribution of partial melting, which may depend strongly on composition, especially volatile content, is important, particularly for the calculation of seismic velocities. These effects will be qualitatively considered in the discussions below.

Application to Mantle Compositions

Continental Garnet Lherzolites. The algorithm for computing normative mineralogies, densities and seismic velocities has been applied to 78 whole rock analyses of garnet lherzolite xenoliths from kimberlite pipes. The geographic distribution of the samples and the references for the chemical analyses are given in Table 4. Most of the xenoliths are from South African localities, but pipes in Montana and Siberia have also been sampled.

The mineral norms are plotted against the Fe/Fe + Mg ratio R in Figure 1. Generally, the continental lherzolites are depleted in normative garnet and normative clinopyroxene relative to pyrolite and the fertile mantle composition of PHN1611. This fact, as well as other aspects of the whole rock chemistry, suggests that basaltic differentiation may play an important role in establishing the observed compositional variations. Nixon and Boyd [1973], Nixon et al.

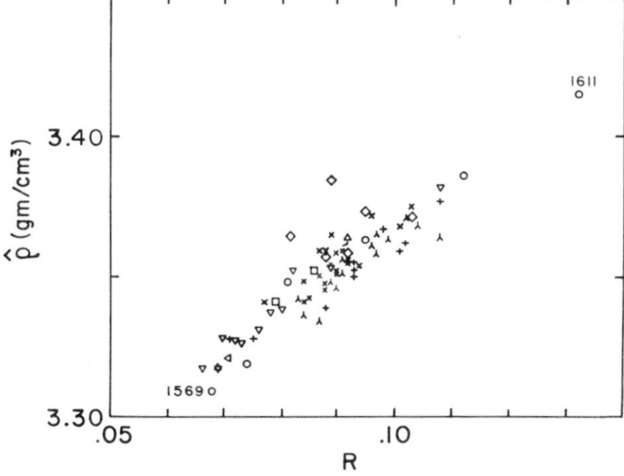

Figure 2. Normative density vs. R for the 78 continental garnet lherzolite analyses identified in Table 4.

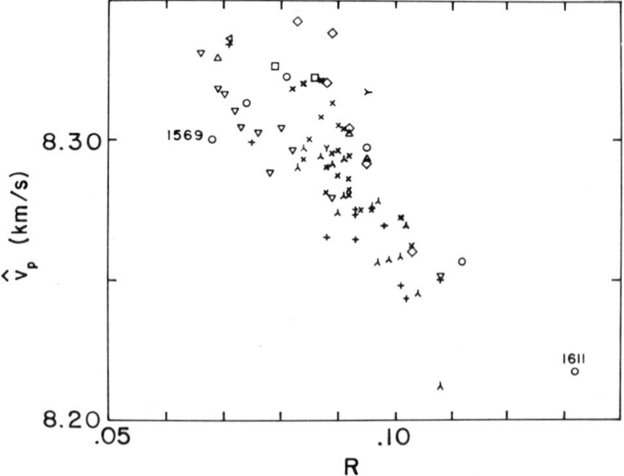

Figure 3. Normative compressional velocity vs. R for the 78 continental garnet lherzolite analyses identified in Table 4.

[1973] and Boyd and Nixon [1975] have discussed this point most thoroughly, although the importance of basaltic fractionation has been recognized by many other authors [e.g. Ringwood, 1966; Kuno and Aoki, 1970; Gurney et al., 1975].

The densities and seismic velocities computed from the normative mineralogies are plotted against R in Figures 2-4; $\hat{\rho}$ correlates positively with R, and $\hat{v}_p$ and $\hat{v}_s$ correlate negatively with R, as expected from velocity-density systematics. For shear velocity, almost the entire variation can be ascribed to the variation in mean atomic weight. The densities are more sensitive than shear velocities to $\hat{X}_{Gt}^{WR}$, and the imperfect correlation between $\hat{X}_{Gt}^{WR}$ and R (Figure 1) introduces some scatter. The scatter is greatest for the compressional velocities, because they are also sensitive to $\hat{X}_{Ol}^{WR}$, a highly variable parameter uncorrelated with R (Figure 1). Generally, the normative variations in the continental garnet lherzolites are nicely bracketed by PHN1569 and PHN1611, which illustrates the extreme compositions of these two samples [Boyd and McCallister, 1976]. $\hat{v}_p$ is lower for PHN1569 than for many other samples because PHN1569 has a low normative olivine content.

The mean oxide composition of the 78 xenoliths is listed with its normative parameters in Table 5. This average continental garnet lherzolite (ACGL) is similar to the mean compositions of other authors [e.g. Ito and Kennedy, 1967; Carswell and Dawson, 1970; Sobolev, 1977], although more samples with a greater geographic distribution have been used here.

I will assume that the ACGL in Table 5 represents the average mantle composition beneath the stable continental shields and platforms in the depth range 100-250 km. This depth range is apparently well sampled by the xenolith population [Boyd and Nixon, 1975; Mercier and Carter, 1975], but the appropriateness of this composition may be questioned for at least two reasons. First, it is not known whether the mantle sampled by the kimberlite pipes is representative of the stable continental environment. The kimberlite pipes listed in Table 4, in fact, predominately occur in, or peripheral to, regions with histories of crustal extension and extrusion of other magmas. Since these processes are indicative of continuing basaltic fractionation, the ACGL may be biased to a more fertile mantle composition than is characteristic of the cratons. However, the conclusions advocated below are robust with respect to this type of bias. Second, the entire xenolith population is not represented by the samples averaged to obtain the ACGL; both highly refractory rocks (e.g. harzbergites, dunites) and supposed melting products or cumulates (e.g. eclogites) have been purposely excluded, as have other mineral associations found as xenoliths (e.g. wehrlites, websterites). However, ignoring these extreme compositions is a valid statistical procedure: they are numerically a small fraction of the xenolith population and, as a whole, tend to average to a typical garnet lherzolite. Mathias et al. [1970] have established these conclusions in a comprehensive study of the South African xenolith distribution. Their estimate of the mean modal composition of the South African upper mantle is: garnet 6%, clinopyroxene 3%, orthopyroxene 27%, olivine 64%, remarkably close to the normative mineralogy for the ACGL listed in Table 5.

Model Compositions for the Oceanic Upper Mantle. In the ocean basins, garnet lherzolites are notably rare; their only documented occurrence is on Oahu, Hawaii [Jackson and Wright, 1970]. All other ultrabasic rocks thus far found in the ocean basins appear to have equili-

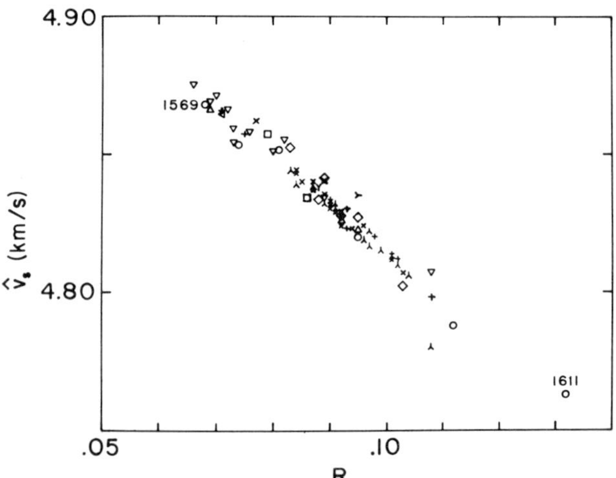

Figure 4. Normative shear velocity vs. R for the 78 continental garnet lherzolite analyses identified in Table 4.

TABLE 5. Parameters of an Average Continental Garnet Lherzolite Compared with Those for Model Compositions of the Oceanic Upper Mantle.

		Average Continental Garnet Lherzolite*	(1)	(2)	(3)	(4)	(5)	(6)	(7)	(8)	(9)
Composition (Wt.%)†											
SiO_2		45.55 ± 1.45	45.30	44.99	46.13	44.63	44.46	43.00	45.32	48.28	43.06
TiO_2		0.11 ± 0.06	0.71	0.24	0.20	0.29	0.34	0.33	0.50	0.22	0.60
Al_2O_3		1.43 ± 0.60	3.55	4.31	4.30	3.78	3.69	7.01	4.12	4.91	4.34
Cr_2O_3		0.34 ± 0.14	0.43	0.40	---	0.52	0.43	0.18	0.30	0.25	0.31
FeO		7.61 ± 0.94	8.48	8.22	8.21	8.15	8.69	9.32	9.75	9.95	12.68
MnO		0.11 ± 0.02	0.14	0.10	---	0.12	0.13	0.14	0.20	0.14	0.18
MgO		43.55 ± 1.98	37.60	38.98	37.63	39.40	38.94	35.18	36.88	32.53	35.02
CaO		1.05 ± 0.77	3.09	2.51	3.10	2.67	2.67	4.38	2.31	2.99	3.06
Na_2O		0.14 ± 0.10	0.57	0.23	0.40	0.34	0.57	0.45	0.60	0.66	0.65
K_2O		0.11 ± 0.10	0.13	0.02	0.03	0.10	0.08	0.00	0.02	0.07	0.10
Norm (Wt.%)	Gt	6.0	12.3	17.6	15.2	15.3	12.8	26.8	14.1	16.9	14.9
	Cpx	4.5	16.3	9.3	14.1	11.3	13.9	14.5	12.1	15.5	16.8
	Opx	22.5	13.5	15.2	17.3	11.9	9.3	0.0	17.4	31.8	3.5
	Ol	67.0	57.9	57.9	53.4	61.5	64.0	58.7	56.4	35.8	64.8
$\hat{\rho}$ (gm/cm^3)		3.353	3.397	3.407	3.396	3.401	3.401	3.463	3.420	3.432	3.474
$\hat{v}_p$ (km/s)		8.290	8.265	8.314	8.289	8.305	8.284	8.341	8.247	8.204	8.186
$\hat{v}_s$ (km/s)		4.833	4.802	4.826	4.817	4.821	4.807	4.810	4.785	4.769	4.720

*Average of the 78 analyses listed in Table 4; oxide composition expressed as sample mean ± sample standard deviation.
†Totals normalized to 100%; total Fe expressed as FeO
Oceanic model compositions: (1) Pyrolite [Ringwood, 1966]; (2) Pyrolite, 99% Lizard peridotite + 1% nephelinite [Ringwood, 1975, table 5-2]; (3) Pyrolite, 83% harzburgite + 17% tholeiite [Ringwood, 1975, table 5-2]; (4) St. Paul's Rocks, weighted average [Melson *et al.*, 1967, table 1]; (5) Parent composition for Honolulu Series lavas [Jackson and Wright, 1970, table 6a]; (6) Mantle beneath Kilbourne Hole, N.M. [Carter, 1970, table 5]; (7) Volatile-depleted mantle [Nicholls, 1967, table 9-1]; (8) Spinel lherzolite nodule used in experiments by Kushiro *et al.* [1968, table 1]; (9) Parent composition for Mauna Loa lavas [Wright, 1971, table 19].

brated in the spinel lherzolite or a lower pressure facies, indicating source depths less than ∿ 70 km [Ringwood, 1975; Yoder, 1976]. Consequently, the average major element chemistry of the garnet lherzolite layer beneath the oceans is based largely on indirect evidence, and estimates usually rely on certain modelling assumptions. Many approaches to this problem have been published, and the subject has been fraught with controversy. In an attempt to avoid dogmatism and overly naïve conclusions, nine representative models are considered in this paper, and these are given in Table 5.

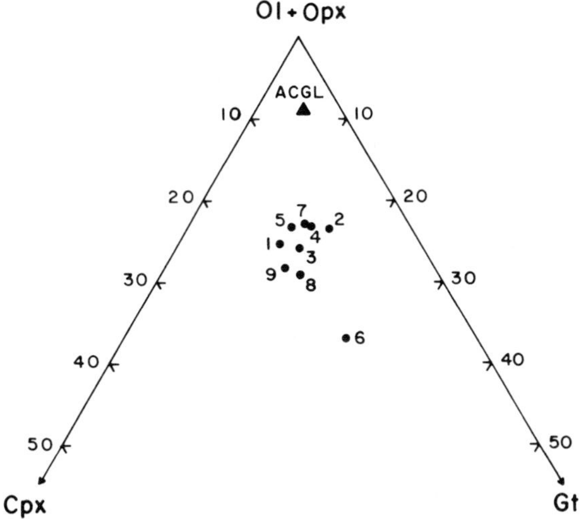

Figure 5. Normative mineral fractions (mole %) of the average continental garnet lherzolite (ACGL) and oceanic model compositions in Table 5 plotted on a Gt - Cpx - (Opx + Ol) diagram.

Two models, (5) and (8), are based directly on the observed compositions of spinel lherzolite xenoliths [66-PAL-3 of Jackson and Wright, 1970, and sample 24 of Kuno and Aoki, 1970, respectively]. The other models are derived by mixing a residual rock with a presumed "primary magma" in proportions chosen to satisfy various geochemical and geophysical criteria. (For details, the reader should consult the original references (Table 5); the general discussion of Ringwood [1975, Chapter 5] is especially recommended.) The substantial range in these model compositions can be attributed to the variety of possible primary magma compositions and the range of mixing proportions allowed by the observational constraints.

Although a particular model for the oceanic upper mantle is not advocated here, an hypothesis basic to the conclusions of this paper is postulated: any function (e.g. $\hat{\rho}$) of the average oceanic upper mantle major element chemistry is assumed to be bounded by the values calculated from models (1)-(9) of Table 5.

The range of normative parameters permitted by

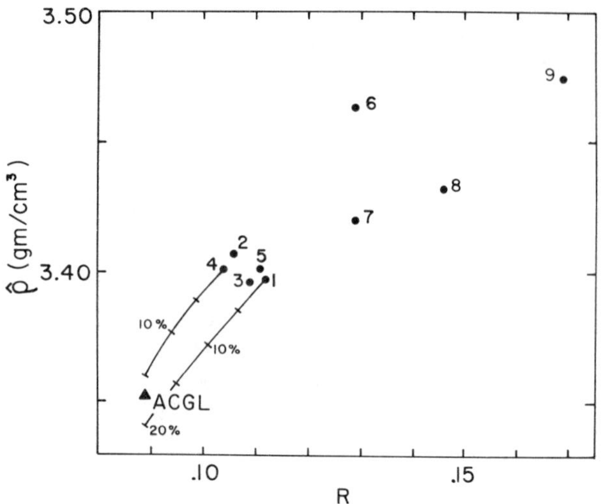

Figure 6. Normative density vs. R for the average continental garnet lherzolite (ACGL) and oceanic model compositions in Table 5. Displacements of models 1 and 4 by subtraction of basaltic components are shown [see text and Table 6].

this hypothesis is considerable (Table 5, Figures 5-8). The normative mineralogies include werhlite (model 6), pyroxenite (model 8) and lherzolite (all other models). The Fe/Fe + Mg ratios vary between 0.104 (model 4) and 0.169 (model 9), and the alumina contents vary between 3.6 wt.% (model 1) and 7.0 wt.% (model 6). These induce large variations in the normative parameters $\hat{\rho}$, $\hat{v}_p$ and $\hat{v}_s$ (Figures 6-8).

It is noted that the normative calculations for the pyrolite model agree reasonably well with those of Green and Liebermann [1976]. The normative density of garnet pyrolite is about 1% greater than the value calculated by these authors and the normative velocities are lower by about the same magnitude, but at least part of the difference is due to differences in assumed end-member properties.

A large uncertainty is allowed in the average oceanic upper mantle composition; the models in Table 5 encompass a wide variety of modelling assumptions. Nevertheless, several normative parameters of the ACGL model lie outside the range permitted for the oceanic mantle. Because the oceanic models are rich in Al_2O_3, CaO and Na_2O, the normative Ol + Opx is less than 80 mole % in every case, whereas the normative Ol + Opx exceeds 90 mole % for the depleted ACGL composition (Figure 5). The Fe/Fe + Mg ratio R is 0.089 for the ACGL model and its normative garnet is 6.0 wt.%, both considerably less than the range for the oceanic models. Consequently, the ACGL normative density falls below the oceanic range (Figure 6). This conclusion may have considerable geophysical importance, and its implications will be explored below.

Basalt Subtraction from Parental Compositions. The effects on the normative parameters of basalt subtraction from parental compositions are demonstrated in Table 6. Three parent-differentiate pairs representing a range of compositions are shown. Their trajectories are illustrated in Figure 1 and Figure 6-8. O'Hara's [1975] and Boyd and McCallister's [1976] conclusion that the removal of a basaltic liquid from a fertile mantle composition will lower the density of the residual rock is substantiated (Figure 6). For example, the subtraction of 20 mole % olivine basalt from Ringwood's [1966] pyrolite composition lowers $\hat{\rho}$ by 1.6%. There is a corresponding increase in both $\hat{v}_p$ and $\hat{v}_s$ (Figures 7 and 8), but the relative magnitudes are much smaller (0.1% and 0.5%, respectively). Similar results are obtained by the subtraction of an average oceanic tholeiite from the weighted mean composition of St. Paul's Rocks.

Geophysical Implications and Speculations

Oceanic Lithospheric Structure. The residuum/basalt ratio implied by the oceanic upper mantle models in Table 5 lies between 4:1 and 2:1. Thus, for every volume of basalt removed from the mantle, a volume of residual rock several times larger is left behind. Shaw and Jackson [1973] recognized the importance of explaining the fate of this residual material, but the "gravitational anchors" envisaged by these authors appear to be dynamically unfeasible. It is the conclusion of O'Hara [1975], Boyd and McCallister [1976] and this study that the nor-

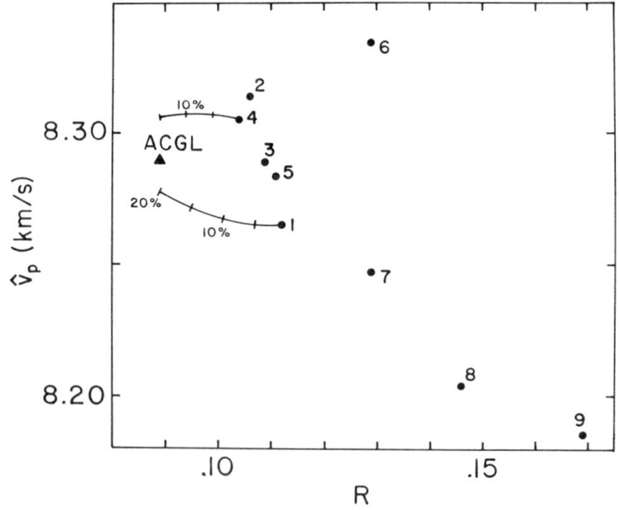

Figure 7. Normative compressional velocity vs. R for the average continental garnet lherzolite (ACGL) and oceanic model compositions in Table 5. Displacements of models 1 and 4 by subtraction of basaltic components are shown [see text and Table 6].

mative density of a residuum is less than the normative density of undepleted mantle. For the models considered here, the density difference between fertile and depleted garnet lherzolites is smallest for parental compositions with low Fe/Fe + Mg, such as Ringwood's [1966] pyrolite (Figure 6). The fractionation of 20 mole % olivine basalt from pyrolite lowers $\hat{\rho}$ by 1.6% (Table 6). For a coefficient of thermal expansion $\alpha = 3 \times 10^{-5}/^{o}C$, the temperature decrease required to maintain constant density is about 530°C. Therefore, the residual material produced by basalt fractionation cannot be easily returned to the deep mantle at the site of differentiation; it will float upon undepleted mantle [O'Hara, 1975; Boyd and McCallister, 1976; Oxburgh and Parmentier, 1977].

The implications of this conclusion deserve more scrutiny. A simple working hypothesis is suggested: Because its density is lower and because its resistance to deformation is increased by elevation of its solidus and by the loss of volatiles, depleted mantle is presumed to attach itself to the lithosphere at a position laterally near the site of differentiation. This hypothesis agrees well with the accepted two-dimensional models of lithospheric development near spreading centers, where a refractory peridotite layer is assumed to underlie the oceanic crust [e.g. Ringwood, 1969; Oxburgh and Parmentier, 1977]. (The question of density relationship in the spinel stability field (depths <70 km) has not been formally treated. However, the densities of most non-hydrated garnet-deficient peridotites are controlled primarily by the Fe/Fe + Mg index, and the quali-

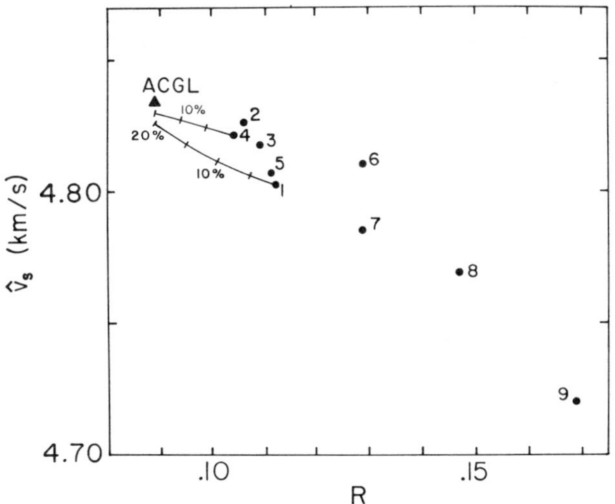

Figure 8. Normative shear velocity vs. R for the average continental garnet lherzolite (ACGL) and oceanic model compositions in Table 5. Displacements of models 1 and 4 by subtraction of basaltic components are shown [see text and Table 6].

tative relationships deduced for garnet lherzolites are applicable.) Furthermore, it implies a negative correlation between crustal thickness and the vertically integrated density of the lithosphere beneath the crust. For example, the large volcanic piles such as Iceland and Hawaii should be characterized by considerable mass deficits in the mantle. The expected mantle mass deficit is approximately $-f\Delta\rho\Delta V$, where ΔV

TABLE 6. Parameters of Residual Compositions Derived by the Removal of Basaltic Fractions from Model Compositions of the Upper Mantle.

Parent - Differentiate	Amount Fractionated (mole %)	Residual Rock							
		$\frac{X_{Fe}^{WR}}{(X_{Fe}^{WR} + X_{Mg}^{WR})}$	Norm (Wt.%) Gt	Cpx	Opx	Ol	$\hat{\rho}$ (gm/cm^3)	$\hat{V}_p$ (km/s)	$\hat{V}_s$ (km/s)
Pyrolite - Olivine Basalt 1)	0	0.112	12.3	16.3	13.5	57.9	3.397	8.265	4.802
	5	0.107	9.8	14.3	14.1	61.7	3.385	8.265	4.806
	10	0.101	7.4	11.9	15.0	65.7	3.372	8.268	4.811
	15	0.095	4.6	9.0	16.3	70.1	3.357	8.272	4.818
	20	0.088	1.7	5.6	17.7	75.0	3.341	8.277	4.826
St. Paul's Rocks - Average Oceanic Tholeiite 2)	0	0.104	15.3	11.3	11.9	61.5	3.401	8.305	4.821
	5	0.099	12.6	8.1	12.2	67.0	3.389	8.307	4.824
	10	0.094	9.8	4.3	13.0	73.0	3.377	8.307	4.827
	15	0.089	5.1	1.6	11.4	82.0	3.360	8.306	4.830
PHN1611 - Average Continental Basalt 3)	0	0.132	10.4	16.7	9.5	63.4	3.415	8.217	4.763
	5	0.125	7.9	13.9	9.3	68.9	3.405	8.218	4.766
	10	0.119	4.8	11.0	8.7	75.5	3.392	8.219	4.769

1) Pyrolite composition from Ringwood [1966]; olivine basalt composition from Ringwood [1975, Table 4-2].
2) Composition of St. Paul's Rocks from Melson et al. [1967, Table 1]; average oceanic tholeiite composition from Engel et al. [1965, Table 2].
3) Composition of PHN1611 from Nixon and Boyd [1973]; average continental basalt composition from Manson [1967, Table III].

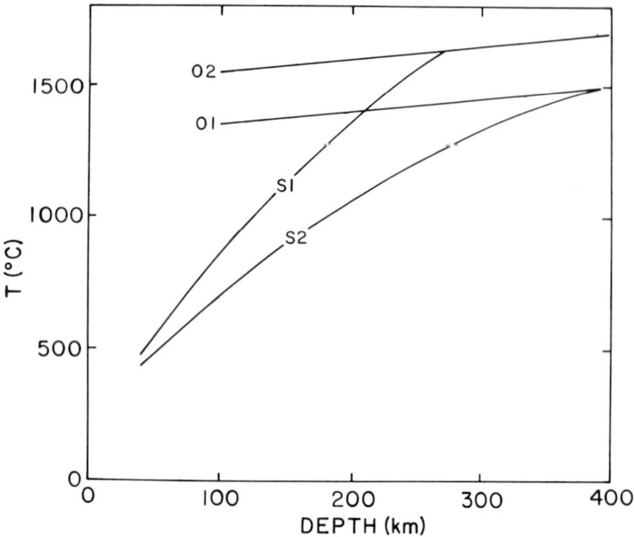

Figure 9. Temperature models for the mantle beneath ocean basins (01 and 02) and shields (S1 and S2). 01 and 02 are adiabats with gradients of 0.5°C/km and potential temperatures of 1300°C and 1500°C, respectively. S1 and S2 are conductive geotherms obtained by varying the heat production in the lower crust; both are characterized by surface heat flows of 1 H.F.U.

is the excess volume of basaltic crust, $\Delta\rho$ is the density difference between parental and residual mantle and f is the residuum/basalt ratio. The product $f\Delta\rho$ is insensitive to the assumed model; for the parent-residuum pairs in Table 6, it ranges from 0.20 to 0.23 gm/cm^3. This value is a large fraction of the crust-mantle density contrast ($\sim$ 0.5 gm/cm^3). Consequently, the mass deficits caused by mantle depletion may contribute significantly to the observed gravity field and topography in the oceans. In particular, this mechanism is expected to yield a positive correlation between gravity anomalies and residual depth anomalies [cf. Watts, 1974]. Such correlations are observed in regions of active ridge and mid-plate vulcanism [Anderson et al., 1973; Sclater et al., 1975; Watts, 1976], but these anomalies and correlations have been ascribed to dynamically maintained stress systems, at least at the low-wavenumber end of the spectrum [Sclater et al., 1975]. Nevertheless, the lateral density variations and lithospheric thickness variations established by basaltic depletion can be significant, and their relevance to these data deserves a more thorough investigation.

The localized formation of oceanic lithosphere anomalously depleted in basaltic components may also exert some dynamical control on the subduction process. Several authors have recognized that lithosphere capped by an unusually thick oceanic crust is hard to subduct and have correlated the arrival of shallow ocean floor at a trench with changes in arc shape and even reversals in subduction polarity [e.g. Vogt et al., 1976; Burke et al., 1978]. It is speculated here that such effects are as much due to the existence of a low-density sub-crustal lithosphere as to the existence of buoyant oceanic crust.

Formal tests of these hypotheses must be deferred to a subsequent publication.

Continental Roots: Stabilization of Temperature Gradients by Chemical Gradients. In terms of its classical definition (which will be retained throughout this discussion) the continental lithosphere is less than 100 km in thickness [Banks et al., 1977; McNutt and Parker, 1977]. Below this depth, any significant density differences maintained over geologically long periods of time (>10 My) must be accompanied by mass flow in the mantle. However, systematic large-scale seismic velocity and attenuation differences between the ancient continental cratons and the ocean basins most certainly extend to depths exceeding 100 km [e.g. Knopoff, 1972; Sacks and Okada, 1974; Sipkin and Jordan, 1975, 1976], and these are accompanied by large temperature differences [Jordan, 1975a]. I have ar-

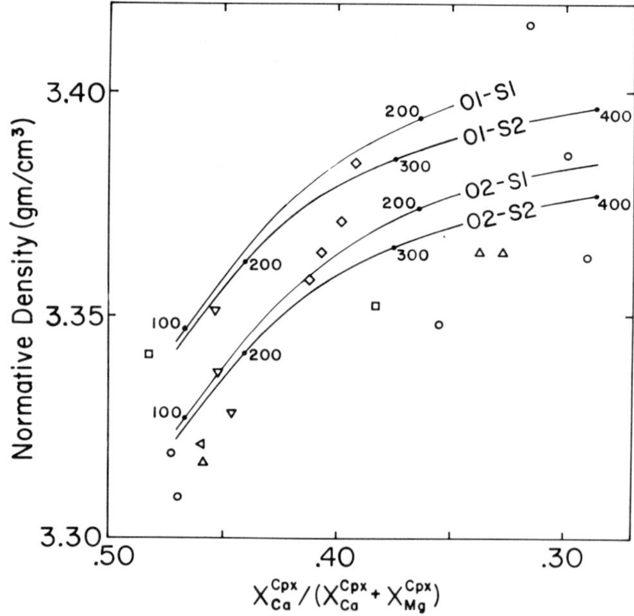

Figure 10. Normative density vs. Ca/Ca + Mg in clinopyroxene. The symbols are the continental garnet lherzolites identified in Table 4 for which clinopyroxene analyses were available. The theoretical curves are based on the temperature-depletion compensation hypothesis discussed in the text and were constructed from the temperature models in Figure 9 and Nehru and Wyllie's [1974] 30 kb diopside-enstatite solvus. The numbers on the curves are depths in km appropriate to the various models.

gued in the last-cited publication that significant super-adiabatic gradients extend beneath the shields to depths of at least 300 km and that the depth interval 100-300 km is characterized by large horizontal temperature gradients in the transition from shield to oceanic structures. Since it is not likely that these gradients are dynamically maintained, a mechanism for stabilizing them against advection must exist.

Based on these considerations, a specific model for sub-lithospheric continental root zones was proposed [Jordan, 1975a, b]. It was hypothesized that, below a depth of 100 km or so, the average *in situ* density differences at a given level beneath continents and oceans are zero, and that the anomalous temperature gradients associated with continental structures are stabilized against flow by chemical gradients. It was further proposed that these chemical gradients have been generated by the removal of a basaltic fraction from the sub-continental upper mantle [Jordan, 1976].

This model receives considerable support from the present study and other studies of continental major element chemistry [Boyd and Nixon, 1975; Boyd and McCallister, 1976]. In the depth interval 100-200 km, the temperature difference between a typical shield geotherm and the average oceanic adiabat is about -400°C [Jordan, 1975a, Figure 4]. In a chemically homogeneous mantle this temperature difference would induce a density difference of about +1.2%. (In this comparison the effects of partial melting have been ignored. If the sub-oceanic mantle contains, say, 1% melt and the sub-continental mantle contains none, then this difference is augmented by about 0.2%.) From Table 5 it is seen that the normative density difference between the ACGL model and Ringwood's [1966] pyrolite is -1.3%. Thus, the effects of temperature differences and chemical differences indeed have the same magnitude, but opposite signs, and will cancel, as predicted by the model.

A more precise test of these hypotheses is shown in Figure 10, which is a graph of normative density versus the Ca/Ca + Mg ratio in the clinopyroxene. All samples from Table 4 for which clinopyroxene analyses were available have been plotted. The correlation between $\hat{\rho}$ and $X_{Ca}^{Cpx}/(X_{Ca}^{Cpx} + X_{Mg}^{Cpx})$ is strongly positive. Since $X_{Ca}^{Cpx}/(X_{Ca}^{Cpx} + X_{Mg}^{Cpx})$ correlates negatively with temperature [Davis and Boyd, 1966] and temperature increases with depth, the general trend of these data is indicative of increasing normative density with depth caused by an increase in Fe/Fe + Mg and normative garnet. The fact that the xenoliths from deeper levels are more fertile than those from shallower depths was discussed by Boyd and Nixon [1975], and the density relationships were recognized by Boyd and McCallister [1976].

Theoretical curves based on the temperature-depletion compensation hypothesis are also plotted in Figure 10. These curves were constructed from the ocean-shield temperature differences derived from Figure 9. Although the temperature distributions in the upper mantle are not well constrained, the pairs of models in Figure 9 probably bracket the actual values [cf. Jordan, 1975a]. Normative density differences were computed assuming a pyrolite composition for the oceanic upper mantle (model 1, $\hat{\rho}$ = 3.397 gm/cm^3) and a thermal expansion coefficient of 3 × 10^{-5}/°C. The Ca/Ca + Mg index for clinopyroxene was calculated from the shield temperatures using the diopside-enstatite solvus of Nehru and Wyllie [1974]. It is seen from Figure 10 that the locations of the theoretical curves are insensitive to the assumed continental geotherm, but they do depend on the potential temperature of the oceanic adiabat. The curves can also be translated vertically by changing the assumed oceanic normative density. An uncertainty in the shapes of the curves is introduced by the uncertainty in the shape of the diopside-enstatite solvus and its probable dependence on pressure [Mori and Green, 1975; Lindsley and Dixon, 1976]. In addition, the effects of other cations, most notably Fe^{2+}, on the location of the solvus have been ignored. For the relatively Fe-poor compositions of the rocks in Figure 10, an approximate correction for natural systems referred to the pure enstatite-diopside solvus can be obtained by adding Fe^{2+} to Mg when forming the cation ratio plotted as the ordinate [Mercier and Carter, 1975], thus translating the points to the right. This correction is largest for the rocks with the greatest normative densities, but its magnitude is small in all cases and does not significantly distort the distribution of points in Figure 10.

None of these uncertainties alters the fact that a general increase in $\hat{\rho}$ as a function of Ca/Ca + Mg in clinopyroxene is observed and the magnitude of the increase is predicted by the geophysical model. Hence, the hypothesis that chemical gradients dynamically stabilize the super-adiabatic gradients below the continental lithosphere is supported by the major-element data.

The model of a basalt-depleted sub-continental root zone, or *tectosphere* in the terminology proposed by Jordan [1975a, b], is in good agreement with the seismic evidence. For example, large lateral variations in vertical shear-wave travel times through the upper mantle have been observed [Sipkin and Jordan, 1975, 1976]: on the average, about a 4 s difference in one-way vertical travel time exists for ocean basins and continental shields, implying higher shear velocities beneath the shields. These higher shear velocities are thought to extend to considerable depth [Sipkin and Jordan, 1975, 1976; Jordan, 1975a, b], although precisely how deep is still the subject of discussion and investigation among seismologists. Temperature is obviously the important factor in controlling the lateral variations in shear velocities, but chemical depletion can also contribute to the increased shear velocities beneath the cratons. The mean atomic weight is lowered by basalt removal, which inincreases the shear velocities (Figures 4 and 8), but, more importantly, basalt depletion raises the mantle solidus, which for temperatures near the solidus can substantially elevate the velocities and reduce attenuation [Goetze, 1977].

Further comments on the composition and evolution of the continental tectosphere can be found in another publication [Jordan, 1978].

Acknowledgments. I thank Dr. B. C. Hearn for providing me with his unpublished analyses of two Montana xenoliths and Drs. F. R. Boyd and J. Natland for helpful discussions. This research was sponsored by the National Science Foundation under grant DES75-20993.

References

Anderson, D.L., C. Sammis, and T.H. Jordan, Composition of the mantle and core, in The Nature of the Solid Earth, edited by E.C. Robertson, McGraw-Hill, New York, 1972.

Anderson, R.N., D.P. McKenzie, and J.G. Sclater, Gravity, bathymetry and convection in the earth, Earth Planet. Sci. Lett., 18, 391-407, 1973.

Banks, R.J., R.L. Parker, and S.P. Huestis, Isostatic compensation on a continental scale: local versus regional mechanisms, Geophys. J. Roy. Astr. Soc., 51, 431-452, 1977.

Boyd, F.R., and R.H. McCallister, Densities of fertile and sterile garnet peridotites, Geophys. Res. Lett., 3, 509-512, 1976.

Boyd, F.R., and P.H. Nixon, Origin of the ultramafic nodules from some kimberlites of northern Lesotho and the Monastary Mine, South Africa, in Physics and Chemistry of the Earth, v.9, edited by L.H. Ahrens, J.B. Dawson, A.R. Dawson, and A.J. Erlank, Pergamon Press, Oxford, 1975.

Bragg, L., G.F. Claringbull, and W.H. Taylor, Crystal Structures of Minerals, Cornell University Press, 1965.

Brooks, C., D.E. James, and S.R. Hart, Ancient lithosphere: its role in young continental vulcanism, Science, 193, 1086-1094, 1976.

Burke, K., P.J. Fox, and A.M.C. Şengör, Buoyant ocean floor and the evolution of the Caribbean, J. Geophys. Res., in press, 1978.

Carswell, D.A., and J.B. Dawson, Garnet peridotite xenoliths in South African kimberlite pipes and their petrogenesis, Contr. Mineral. Petrol., 25, 163-184, 1970.

Carter, J.L., Mineralogy and chemistry of the Earth's upper mantle based on the partial fusion-partial crystallization model, Geol. Soc. Am. Bull., 81, 2021-2034, 1970.

Chen, J., Petrology and chemistry of garnet lherzolite nodules in kimberlite from South Africa, Am. Mineral., 56, 2098-2110, 1971.

Chung, D.H., Effects of Iron/Magnesium ratio on P- and S-wave velocities in olivine, J. Geophys. Res., 75, 7353-7361, 1970.

Cox, K.G., J.J. Gurney, and B. Harte, Xenoliths from the Matsoku pipe, in Lesotho Kimberlites, edited by P.H. Nixon, Lesotho National Development Corp., Maseru, Lesotho, 1973.

Davis, B.T.C., and F.R. Boyd, The join $Mg_2Si_2O_6$ - $CaMgSi_2O_6$ at 30 kb pressure and its application to pyroxenes from kimberlite, J. Geophys. Res., 71, 3567-3576, 1966.

Engel, A.E.J., C.G. Engel, and R.G. Havens, Chemical characteristics of oceanic basalts and the upper mantle, Geol. Soc. Am. Bull., 76, 719-734, 1965.

Erlank, A.J., and N. Shimizu, Strontium and strontium isotope distributions in some kimberlite nodules and minerals, Second International Kimberlite Conference, extended abstracts, 1977.

Goetze, C., A brief summary of our present day understanding of the effect of volatiles and partial melt on the mechanical properties of the upper mantle, in High-Pressure Research, Applications in Geophysics, edited by M.H. Manghnani and S.-I. Akimoto, Academic Press, New York, 1977.

Green, D.H., and R.C. Liebermann, Phase equilibria and elastic properties of a pyrolite model for the oceanic upper mantle, Tectonophysics, 32, 61-92, 1976.

Gurney, J.J., B. Harte, and K.G. Cox, Mantle xenoliths in Matsoku kimberlite pipe, in Physics and Chemistry of the Earth, v.9, edited by L.H. Ahrens, J.B. Dawson, A.R. Dawson, and A.J. Erlank, Pergamon Press, Oxford, 1975.

Hill, R., The elastic behavior of a crystalline aggregate, Proc. Phys. Soc. London, Sec. A., 65, 349-354, 1952.

Hughes, D.S., and T. Nishitake, Measurement of elastic wave velocity in Armco iron and jadeite under high pressures and high temperatures, in Geophysical Papers Dedicated to Professor Kenzo Sasso, Kyoto University Geophysical Institute, 379-385, 1963.

Isaak, D.G., and E.K. Graham, The elastic properties of an almadine-spessartine garnet and elasticity in the garnet solid solution series, J. Geophys. Res., 81, 2483-2489, 1976.

Ito, K., and G.C. Kennedy, Melting and phase relations in a natural peridotite to 40 kilobars, Am. J. Sci., 265, 519-538, 1967.

Jackson, E.D., and T.L. Wright, Xenoliths of the Honolulu volcanic series, Hawaii, J. Petrol., 11, 405-430, 1970.

Jordan, T.H., The continental tectosphere, Rev. Geophys. Space Phys., 13, no.3, 1-12, 1975a.

Jordan, T.H., Lateral heterogeneity and mantle dynamics, Nature, 257, 745-750, 1975b.

Jordan, T.H., Peridotites, eclogites and the evolution of the continental tectosphere, Geol. Soc. Am. Abstracts, 8, 944-945, 1976.

Jordan, T.H., Composition and development of the

continental tectosphere, Nature, in press, 1978.

Knopoff, L., Observation and inversion of surface-wave dispersion, Tectonophysics, 13, 497-519, 1972.

Kumazawa, M., and O.L. Anderson, Elastic moduli, pressure derivatives, and temperature derivatives of single-crystal olivine and single-crystal forsterite, J. Geophys. Res., 74, 5961-5972, 1969.

Kuno, H., and K. Aoki, Chemistry of ultramafic nodules and their bearing on the origin of basaltic magmas, Phys. Earth. Planet. Interiors, 3, 273-301, 1970.

Kushiro, I., Y. Syono, and S. Akimoto, Melting of a peridotite nodule at high pressures and high water pressures, J. Geophys. Res., 73, 6023-6029, 1968.

Liebermann, R.C., Elasticity of pyroxene-garnet and pyroxene-ilmenite phase transformation in germanates, Phys. Earth. Planet. Interiors, 8, 361-374, 1974.

Liebermann, R.C., and D.J. Mayson, Elastic properties of polycrystalline diopside ($CaMgSi_2O_6$), Phys. Earth Planet. Interiors, 11, P1-P4, 1976.

Lindsley, D.H., and S.A. Dixon, Diopside-enstatite equilibria at 850° to 1400°C, 5 to 35 kb, Amer. J. Sci., 276, 1285-1301, 1976.

Manson, V., Geochemistry of basaltic rocks: major elements, in Basalts: the Poldervaart Treatise on Rocks of Basaltic Composition, v.1, edited by H.H. Hess and A. Poldervaart, Interscience, New York, 1967.

Mathais, M., J.C. Siebert, and P.C. Rickwood, Some aspects of the mineralogy and petrology of ultramafic xenoliths in kimberlite, Contr. Mineral. Petrol., 26, 75-123, 1970.

McNutt, M., and R.L. Parker, Isostasy in Australia and the evolution of the compensation mechanism, Science, 199, 773-775, 1978.

Melson, W.G., E. Jarosewich, V.T. Bowen, and G. Thompson, St. Peter and St. Paul Rocks: a high temperature, mantle-derived intrusion, Science, 155, 1532-1535, 1967.

Mercier, J.-C., and N.L. Carter, Pyroxene geotherms, J. Geophys. Res., 80, 3349-3362, 1975.

Mori, T., and D.H. Green, Pyroxenes in the system $Mg_2Si_2O_6$ - $CaMgSi_2O_6$ at high pressure, Earth Planet. Sci. Lett., 26, 277-286, 1975.

Nehru, C.E., and P.J. Wyllie, Electron microprobe measurement of pyroxenes coexisting with H_2O-undersaturated liquid in the join $CaMgSi_2O_6$ - $Mg_2Si_2O_6$ - H_2O at 30 kilobars, with applications to geothermometry, Contrib. Mineral. Petrol., 48, 221-228, 1974.

Nicholls, G.D., Geochemical studies in the ocean as evidence for the composition of the mantle, in Mantles of the Earth and Terrestrial Planets, edited by S.K. Runcorn, Interscience, New York, 1967.

Nixon, P.H., and F.R. Boyd, Petrogenesis of the granular and sheared ultrabasic nodule suite in kimberlites, in Lesotho Kimberlites, edited by P.H. Nixon, Lesotho National Development Corp., Maseru, Lesotho, 1973.

Nixon, P.H., F.R. Boyd, and A.M. Boullier, The evidence of kimberlite and its inclusion on the constitution of the outer part of the earth, in Lesotho Kimberlites, edited by P.H. Nixon, Lesotho National Development Corp., Maseru, Lesotho, 1973.

Nixon, P.H., O. von Knorring and J.M. Rooke, Kimberlites and associated inclusions of Basutoland: a mineralogical and geochemical study, Am. Mineral., 48, 1090-1132, 1963.

O'Hara, M.J., Is there an Icelandic mantle plume?, Nature, 253, 708-710, 1975.

Oxburgh, E.R., and E.M. Parmentier, Compositional and density stratification in oceanic lithosphere--causes and consequences, J. Geol. Soc. London, 133, 343-355, 1977.

Peck, L.C., Systematic analysis of silicates, U.S. Geol. Survey Bull. 1170, U.S. Government Printing Office, 1964.

Ringwood, A.E., The chemical composition and origin of the earth, in Advances in Earth Sciences, edited by P. Hurley, M.I.T. Press, Cambridge, Mass., 1966.

Ringwood, A.E., Composition and evolution of the upper mantle, in The Earth's Crust and Upper Mantle, edited by P.J. Hart, Am. Geophys. Union Monograph 13, 1-17, 1969.

Ringwood, A.E., Composition and Petrology of the Earth's Mantle, McGraw-Hill, New York, 1975.

Roberts. W.L., G.R. Rapp, and J. Weber, Encyclopedia of Minerals, Van Nostrand Reinhold, New York, 1974.

Robie, R.A., P.M. Bethke, M.S. Toulmin, and J.L. Edwards, X-ray crystallographic data densities, and molar volumes of minerals, in Handbook of Physical Constants, edited by S.P. Clark, Geol. Soc. Am. Memoir, 97, 1966.

Sclater, J.G., L.A. Lawver, and B. Parsons, Comparison of long-wavelength residual elevation and free air gravity anomalies in the North Atlantic and possible implications for the thickness of the lithospheric plate, J. Geophys. Res., 80, 1031-1052, 1975.

Sacks, I.S., and H. Okada, A comparison of the anelasticity structure beneath western South America and Japan, Phys. Earth Planet. Interiors, 9, 211-219, 1974.

Shaw, H.R., and E.D. Jackson, Linear island chains in the Pacific: results of thermal plumes or gravitational anchors?, J. Geophys. Res., 78, 8634-8652, 1973.

Sipkin, S.A., and T.H. Jordan, Lateral heterogeneity of the upper mantle determined from the travel times of ScS, J. Geophys. Res., 80, 1474-1484, 1975.

Sipkin, S.A., and T.H. Jordan, Lateral heterogeneity of the upper mantle determined from the travel times of multiple ScS, J. Geophys. Res., 81, 6307-6320, 1976.

Sobolev, N.V., Deep-Seated Inclusions in Kimberlites and the Problem of the Composition of the Upper Mantle, trans. by D.A. Brown, Am. Geophys. Union, Washington, D.C., 1977.

Vogt, P.R., A. Lowrie, D.R. Bracey, and R.N. Hey, Subduction of aseismic oceanic ridges: effects on shape, seismicity, and other characteristics of consuming plate boundaries, Geol. Soc. Amer. Spec. Paper 172, 59 pp., 1976.

Watts, A.B., Gravity and bathymetry in the central Pacific Ocean, J. Geophys. Res., 81, 1533-1553, 1976.

Wright, T.L., Chemistry of Kilauea and Mauna Loa lava in space and time, U.S. Geol. Survey Prof. Paper 735, U.S. Government Printing Office, 1971.

Yoder, H.S., Jr., Generation of Basaltic Magma, National Academy of Sciences, Washington, D.C., 1976.

METAMORPHISM IN A MODEL MANTLE
I. PREDICTIONS OF P-T-X RELATIONS IN $CaO-Al_2O_3-MgO-SiO_2$

Alan Bruce Thompson

Institut für Kristallographie & Petrographie, ETH-Zentrum, CH-8092 Zürich, Switzerland

Abstract. Metamorphic reactions in most ultramafic and related rocks, that are likely to occur in the Lower Crust and Upper Mantle, are largely controlled by net-transfer reactions and exchange equilibria in the $CaSiO_3$-$MgSiO_3$-Al_2O_3 plane of the Model Mantle system CaO-Al_2O_3-MgO-SiO_2. Additional net-transfer reactions, changing the modal proportions of reacting phases, result from the reaction of garnet and pyroxene with phases (such as anorthite, kyanite, quartz or spinel, forsterite) on either side of the $CaSiO_3$-$MgSiO_3$-Al_2O_3 plane. Each continuous reaction in the Model Mantle system includes several net-transfer reactions and exchange equilibria and cannot be satisfactorily calibrated with available data. The techniques presented can be used to obtain maximum thermodynamic information from relevant experimental studies, including those involving silicate melt. Additional components, such as FeO, Fe_2O_3, Na_2O, Cr_2O_3, K_2O and TiO_2, may be considered through calibration of further exchange equilibria together with activity data on multicomponent crystalline solutions. The Model Mantle system may be used to predict compositional and modal changes of reacting minerals as functions of pressure and temperature. These calculations can be extended to determine variations of bulk modulii, seismic velocities and thermal conductivities with changing pressure and temperature. Furthermore, the discontinuous reactions may be examined as the cause of seismic discontinuities.

Introduction

Although geophysical measurements of the physical properties of the Lower Crust and Upper Mantle and petrological examination of nodule material can be used to constrain mineralogical constitution, the boundary values currently available are too large to be able to resolve most of the important details and controversies. In order to understand the mineralogical significance of seismic discontinuities and to be consistent with the constraints of thermal structure and viscosity, electrical conductivity and gravity data we need to be able to predict the resulting mineralogy for a given bulk chemical composition as a function of pressure (P) and temperature (T). While the effect of iron substitution in mineral lattices and the appearance of dense phases (such as garnet) on the above physical properties is known in a qualitative sense, the exchange equilibria controlling the distribution of species between coexisting minerals and the effect of these complex equilibria on the discontinuous reactions producing new minerals in only crudely understood. The difficulties arise from several sources including, (a) a poor knowledge of the behaviour and distribution of the alkali metals (mainly Na and K), the alkali earths (mainly Ca) and most trace elements in the Lower Crust and Upper Mantle, (b) the existence of only a few high-variance mineral assemblages (few phases containing many components) which are stable over a wide range of P-T conditions, (c) large discrepancies in the experimental location of equilibrium boundaries between key mineral phases and significant inconsistencies in the P-T calibration of exchange equilibria of major and minor elements, (d) lack of knowledge concerning the relationships between the random varieties of distinctly different nodule types in adjacent kimberlite pipes and lava flows and whether they represent lateral chemical inhomogeneities over a small range in P-T or vertical inhomogeneities over a large range in P-T, (e) insufficient data both on nodule mineralogy and petrology and on experimental calibration of relevant equilibria.

If we consider the Lower Crust and Upper Mantle as metamorphic rocks then we can apply many of the predictive arguments developed for crustal metamorphism where regional correlation can confirm the predicted effects of P and T on the compositions of coexisting minerals. These predictive methods, when adequately calibrated by means of experiment or by examining the behaviour of likely Mantle compositions involved in progressive crustal metamorphism, can then be applied to the random sample of nodular Mantle material referred to in (d) above. Given the appropriate data it is possible to predict not only how a given bulk chemical composition will be represented by coexisting minerals of complex composition at a given P and T, it will be possible to calculate bulk modulii and seismic velocities for each complex mineralogy. Furthermore, by the techniques outlined below, it is possible to relate distinctly different nodular mineralogies and decide whether these represent a compositional (or facies) variation at similar P-T or whether they represent samples from distinctly different depths in the Mantle.

In order to find the most suitable Model Mantle system by which variations in the proportions and compositions of mantle minerals can be described, it is necessary to distinguish between those chemical components which result in significant modal changes through reactions and those which result primarily in changing the compositions of coexisting phases through cation exchange. The components CaO-Al_2O_3-MgO-SiO_2 have been used by many workers to model the mineral reactions in mafic and ultramafic rocks in the Lower Crust and Upper Mantle.

Reactions in the CaO-Al_2O_3-MgO-SiO_2 Compositional Space

Many of the phases of interest in the Model Mantle system are contained within the $CaSiO_3$-Al_2O_3-MgO-SiO_2 compositional volume (Fig. 1a) excluding for the moment monticellite, merwinite, melilite and other calc-silicates. There are eleven anhydrous and ancarbonous minerals that may be considered as pure phases or as components of crystalline solutions. The pure phases in this system include: corundum (Cor, Al_2O_3); kyanite (Kya, Al_2SiO_5); forsterite (For, Mg_2SiO_4); anorthite (Anr, $CaAl_2Si_2O_8$); quartz or other silica polymorphs (Qtz, SiO_2); and spinel (Spn, $MgAl_2O_4$, assuming no solution towards γ-Al_2O_3).

The garnet [Gar] crystalline solution components grossular (gr, $Ca_3Al_2Si_3O_{12}$) and pyrope (py, $Mg_3Al_2Si_3O_{12}$) and the pyroxene-pyroxenoid [Pyx] crystalline solution components wollastonite (wo, $CaSiO_3$), diopside (di, $CaMgSi_2O_6$) and enstatite (en, $MgSiO_3$) are each related by the Ca-Mg exchange coordinate or, alternatively, the $CaMg_{-1}$ exchange vector. In the $CaSiO_3$-$MgSiO_3$-Al_2O_3 plane both clinopyroxene [Cpx] and orthopyroxene [Opx] are ternary solution phases whose compositions can be expressed in terms of the components di+en+ct ($CaAl_2SiO_6$) or alternatively mt ($MgAl_2SiO_6$ = en+ct+di). It should be noted that <u>phases</u> are denoted by 3-letter abbreviations and components by 2-letter abbreviations. Thus a garnet [Gar] phase, [Pyp] if Mg-rich or [Gro] if Ca-rich, would be a crystalline solution between py($Mg_3Al_2Si_3O_{12}$) and gr($Ca_3Al_2Si_3O_{12}$). All garnet compositions can be expressed in terms of py+$CaMg_{-1}$ exchange vector. The compositions of pyroxene-pyroxenoid [Pyx] phases [Wol], [Cpx] and [Opx] can be expressed in terms of en($MgSiO_3$) + $CaMg_{-1}$ + $Al_2Mg_{-1}Si_{-1}$ exchange vectors. The concepts and applications of exchange vectors are discussed more fully by Thompson and Thompson (1976 and in prep.) and by Laird <u>et al</u>. (1978). The compositions of all ternary pyroxenes can also be expressed by the component $MgSiO_3$ plus the $CaMg_{-1}$ and the $Al_2Mg_{-1}Si_{-1}$ (Tschermak) exchange vectors. The above situation is easily visualised by projecting down the $CaMg_{-1}$ exchange vector (equivalent to combining molar CaO-MgO) as shown in Fig. 1b.

Although several other components are necessary to adequately describe the Model Mantle system (notably FeO, MnO, Cr_2O_3, Fe_2O_3, TiO_2 Na_2O and K_2O) their inclusion is best accomplished by additional exchange vectors relative to the phases in CaO-Al_2O_3-MgO-SiO_2. These additional components rarely decrease the variance of the system as they generally partition between phases already present in CaO-Al_2O_3-MgO-SiO_2. They usually do not result in new saturating phases except for high concentrations of TiO_2 and K_2O or where immiscibility gaps occur in the complex crystalline solutions.

In order to adequately describe the modal and compositional changes of a given assemblage with changing P and T, we must define all independent net-transfer reactions (Thompson and Thompson, 1976, p.255) and exchange equilibria. Such systematisation is necessary for it often reveals unexpected reactions of the net-transfer type which produce changes in the modal proportions of phases. Exchange equilibria involve only compositional changes in coexisting phases with no change in the modal proportions of the phases. The continuous and discontinuous variations in facies are clearly combinations of both net-transfer and exchange equilibria. It is practical to formulate these equilibria such that they are most easily related to their experimental calibration.

It is the crystalline solution of garnet and pyroxene within the plane $CaSiO_3$(wo)-$MgSiO_3$(en) - Al_2O_3(Cor), that causes the complex reactions in the Model Mantle system. The other phases of interest lie on the silica-rich side (Anr, Kya, Qtz) or the silica-poor side (Spn,For) of wo-en-Cor. This means that we can distinguish four compositional <u>Reaction Types</u> in the CaO-Al_2O_3-MgO-SiO_2 system:

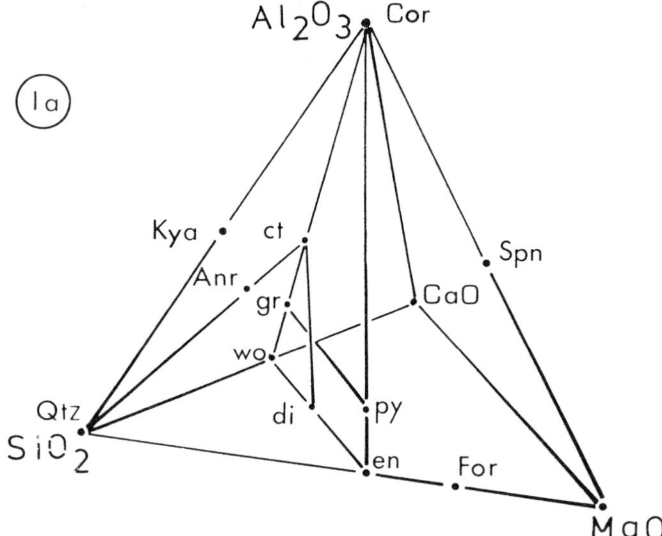

Fig. 1a. The compositional tetrahedron $CaO-Al_2O_3-MgO-SiO_2$ showing the location of end-member components of crystalline solutions in the $CaSiO_3$(wo)-$MgSiO_3$(Cor) plane and the pure phases in the plane and on high-silica and low-silica sides. The abbreviations used are: en($MgMgSi_2O_6$), di($CaMgSi_2O_6$), wo($CaCaSi_2O_6$), py($Mg_3Al_2Si_3O_{12}$), gr($Ca_3Al_2Si_3O_{12}$), ct($CaAl_2SiO_6$), Cor(Al_2O_3), Qtz(SiO_2), Kya(Al_2SiO_5), Anr($CaAl_2Si_2O_8$), For(Mg_2SiO_4) and Spn($MgAl_2O_4$).

(i) those in the wo-en-Cor plane (with Gar, Pyx and Cor)

(ii) those on the SiO_2-rich side (with Kya, Anr, Qtz plus i)

(iii) those on the SiO_2-poor side (with For, Spn, plus ii)

(iv) those crossing the wo-en-Cor plane (combination of the above).

It is important to note that the presence of any one of these phases from outside the wo-en-Cor plane (i.e. For, Spn, Anr, Kya,Qtz) will not result in any new reactions. For example the presence of For with Pyp+Cpx+Opx (garnet lherzolite) will not influence the reactions in the wo-en-Cor plane. The case of two (or more) phases lying outside wo-en-Cor will be discussed further below.

Reactions in the wo($CaSiO_3$)-en($MgSiO_3$)-Cor(Al_2O_3) plane

The extensive crystalline solutions of garnet and pyroxene have received the attention of many authors (see summaries by Boyd, 1970; Wood, 1975; Newton, 1977) largely as a means by which the P-T conditions of origin of ultramafic assemblages can be determined by calibration of exchange equilibria. The garnet and pyroxene crystalline solutions in $CaO-Al_2O_3-MgO-SiO_2$ lie within the plane wo-en-Cor. The topology deduced by Boyd (1970, p.69) for this plane at 1200°C and 30kbar is shown in the central part of Figure 2, and consists of four 3-phase fields separated by bundles of 2-phase tie lines. Very few ultramafic and related rocks are completely represented by the plane wo-en-Cor with the obvious exceptions of rare monomineralic types such as clinopyroxenites, etc. Some 2-phase regions are represented among natural assemblages, for example Pyp+Opx or Pyp+Cpx (garnet clinopyroxenites or eclogites, although many components are neglected in the simple system). Of the four 3-phase fields, Pyp+Cpx+Opx represents garnet websterites, Pyp+Cpx+Opx represents corundum eclogites, Gro+Cpx+Cor represents some grospydites (grospycorites) and Wol+Gro+Cpx may be found in calc-silicates or metarodingites. However, most ultramafic and related rocks consist of one, two or three phase assemblages in wo-en-Cor coexisting with one or more minerals from either side of this plane (for example garnet lherzolites contain For+Pyp+Cpx+Opx). It is necessary to consider which equilibria control the P-T behaviour of the 3-phase triangles and 2-phase regions in wo-en-Cor.

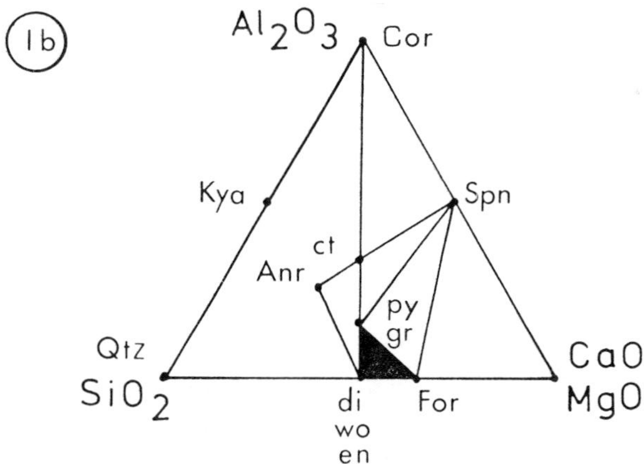

Fig. 1b. The section (CaO+MgO)-Al_2O_3-SiO_2 showing the relation of high-silica and low-silica phases to wo-en-Cor. Pyroxenes and garnet crystalline solutions are related by the Ca-Mg-1 exchange vector. The shaded area shows the compositional range for most ultramafic rock types. The area [(di, en)-Anr-Spn-For] represents the compositional volume involved in the plagioclase to spinel lherzolite transition and the smaller area [(gr,py)-(di,en)]-Spn-For represents the compositional volume involved in the spinel to garnet lherzolite transition.

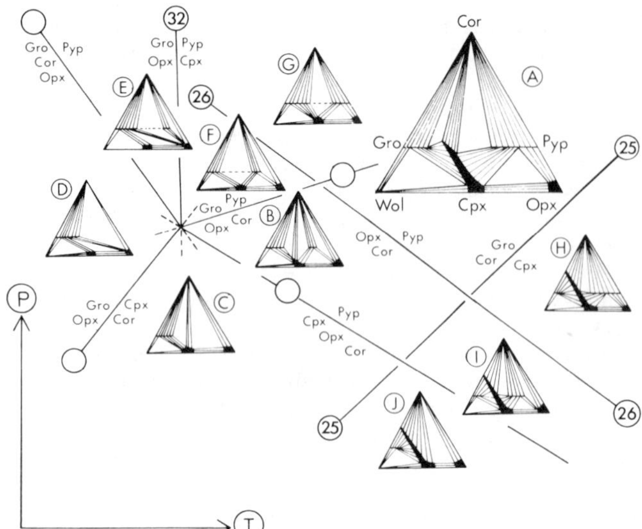

Fig. 2. A schematic P-T diagram showing the relative development of Facies Types from that proposed by Boyd (1970) for 1200°C and 30kbar. The numbers correspond to the reactions discussed in the text.

Exchange equilibria and crystalline solutions in wo-en-Cor

The phase relations in wo-en-Cor (Fig.2) are dominated by $CaMg_{-1}$ exchange equilibria between Pyp, Gro, Cpx, Opx and possibly Wol at high enough temperatures, and by the Tschermak exchange ($Al_2Mg_{-1}Si_{-1}$) in Cpx and Opx. The exchange equilibria only involve exchange of material between the minerals with no change in the modal proportions. Between the five phases Pyp, Gro, Cpx, Opx and Wol there are 10 possible (5!/2!3!) $CaMg_{-1}$ exchanges between pairs of phases :

(1) $CaMg_{-1}$[Gro] = $CaMg_{-1}$[Pyp]
(2) $CaMg_{-1}$[Cpx] = $CaMg_{-1}$[Opx]
(3) $CaMg_{-1}$[Wol] = $CaMg_{-1}$[Cpx]
(4) $CaMg_{-1}$[Gro] = $CaMg_{-1}$[Wol]
(5) $CaMg_{-1}$[Gro] = $CaMg_{-1}$[Opx]
(6) $CaMg_{-1}$[Gro] = $CaMg_{-1}$[Cpx]
(7) $CaMg_{-1}$[Pyp] = $CaMg_{-1}$[Cpx]
(8) $CaMg_{-1}$[Pyp] = $CaMg_{-1}$[Opx]
(9) $CaMg_{-1}$[Pyp] = $CaMg_{-1}$[Wol]
(10) $CaMg_{-1}$[Wol] = $CaMg_{-1}$[Opx]

The condensed notation used above can be understood by writing for example, the exchange equilibria for (2) as

$CaMgSi_2O_6$[Cpx] + $MgMgSi_2O_6$[Opx]
= $CaMgSi_2O_6$[Opx] + $MgMgSi_2O_6$[Cpx] (2a)

or as

($CaMgSi_2O_6$-$MgMgSi_2O_6$)[Cpx]
= ($CaMgSi_2O_6$-$MgMgSi_2O_6$)[Opx] (2b)

All of the above $CaMg_{-1}$ exchange equilibria, with the possible exceptions of (9) and (10), are necessary to describe the phase relations in wo-en-Cor.

The Tschermak exchange equilibria between [Cpx] and [Opx] may be written

$CaAl_2SiO_6$[Cpx] + $MgMgSi_2O_6$[Opx]
= $CaMgSi_2O_6$[Cpx] + $MgAl_2SiO_6$[Opx] (11a)

or as

($CaAl_2SiO_6$-$CaMgSi_2O_6$)[Cpx]
= ($MgAl_2SiO_6$-$MgMgSi_2O_6$)[Opx] (11b)

or in condensed notation as

$Al_2Mg_{-1}Si_{-1}$[Cpx] = $Al_2Mg_{-1}Si_{-1}$[Opx] (11)

The exchange equilibria (11) could have been written for either Ca <u>or</u> Mg components of both phases by combining (11) with (2). The possibility that [Pyp] and [Gro] could involve the Tschermak substitution at high pressure is not considered here.

Relations (2b) and (11b) may be used together with volume and entropy data for end-member pyroxenes to calculate ΔV and ΔS for $CaMg_{-1}$ and $Al_2Mg_{-1}Si_{-1}$ exchange in pyroxene.

Net-transfer reactions in wo-en-Cor

Net-transfer reactions (Thompson and Thompson, 1976, p.255) involve changes in the modal proportions of phases and in the present case can be considered in terms of Ca or Mg components in the respective subsystems $CaO+Al_2O_3+SiO_2$ or $MgO+Al_2O_3+SiO_2$. In order to determine the net-transfer reactions in the wo-en-Cor plane it is useful to express the formulae of phase components in terms of $MgSiO_3$ <u>or</u> Al_2O_3 and the $CaMg_{-1}$ and $Al_2Mg_{-1}Si_{-1}$ exchange vectors, as in Table 1.

Continuous variations in facies result from displacements of the four 3-phase fields in the wo-en-Cor plane, in the central part of Figure 2. These continuous variations may be expressed by a combination of net-transfer reactions with the appropriate $CaMg_{-1}$ and $Al_2Mg_{-1}Si_{-1}$ exchanges. In the wo-en-Cor plane there are only five kinds of net-transfer reactions. Two involve the $Al_2Mg_{-1}Si_{-1}$ variation in pyroxene which is buffered by Cor+Gar or by Gar+Pyx, i.e.

$Mg_3Al_2Si_3O_{12}$[Gar] + $3Al_2Mg_{-1}Si_{-1}$[Pyx]
= $4Al_2O_3$[Cor] (19)

and

$$Mg_3Al_2Si_3O_{12}[Gar] = 4MgSiO_3[Pyx] + Al_2Mg_{-1}Si_{-1}[Pyx] \quad (13)$$

Net-transfer (19) can be written in a more conventional form, i.e.

$$Mg_3Al_2Si_3O_{12}[Pyp] + CaAl_2SiO_6[Cpx] = 4Al_2O_3[Cor] + CaMgSi_2O_6[Cpx] \quad (19a)$$

Net-transfer (13) can be written in a more conventional form, i.e.

$$Mg_3Al_2Si_3O_{12}[Gar] + CaMgSi_2O_6[Cpx] = 4MgSiO_3[Opx] + CaAl_2SiO_6[Cpx] \quad (13a)$$

When net-transfer (13) is written in terms of Pyp+Opx alone, i.e.

$$Mg_3Al_2Si_3O_{12}[Pyp] = 4MgSiO_3[Opx] + Al_2Mg_{-1}Si_{-1}[Opx] \quad (13b)$$

it reveals that garnet increases at the expense of aluminous orthopyroxene alone with increasing P (even in $MgO-Al_2O_3-SiO_2$).

Two other net-transfer reactions involve Cpx+Opx or Cpx+Wol. The net-transfer $MgSiO_3[Cpx] = MgSiO_3[Wol]$ is only likely to be important for high temperature assemblages and can probably be safely ignored for most cases. The net-transfer

$$MgSiO_3[Cpx] = MgSiO_3[Opx] \quad (23)$$

is important for many ultramafic and mafic assemblages. This could be written as

$$CaMgSi_2O_6[Cpx] = CaMgSi_2O_6[Opx] \quad (23a)$$

These reactions express the change in activities and hence composition of coexisting [Cpx] and [Opx]. Compositions between $CaMgSi_2O_6$ and $MgSiO_3$ will exhibit different modal behaviour as a function of changing T or P. Compositions close to $CaMgSi_2O_6$ will show an increase of [Cpx] at the expense of [Opx] and compositions close to $MgSiO_3$ will show an increase of [Opx] at the expense of [Cpx] with increasing T at constant P.

The fifth net-transfer reaction

$$Mg_2Al_2Si_3O_{12}[Gro] = Mg_3Al_2Si_3O_{12}[Pyp] \quad (24)$$

is that representing the Ca-Mg garnet miscibility gap (plus the $CaMg_{-1}$ exchange [Gro] = [Pyp] and the P-T behaviour would be quantitatively the same as for $CaMg_{-1}$ pyroxenes.

Consequently the four net-transfer reactions (13), (19), (23) and (24) in their various forms can be combined with the appropriate $CaMg_{-1}$ and $Al_2Mg_{-1}Si_{-1}$ exchange equilibria to completely define the continuous variations in wo-en-Cor as functions of P and T.

Continuous facies variations in $CaSiO_3$-$MgSiO_3$-Al_2O_3

The net-transfer and exchange equilibria may be used to predict modal and compositional changes as functions of P and T for Pyp+Cpx+Opx (garnet

TABLE 1.

Phase Components in terms of $MgSiO_3$ and exchange vectors.

$CaMgSi_2O_6$	=	$2MgSiO_3 + CaMg_{-1}$	(12)
$Mg_3Al_2Si_3O_{12}$	=	$4MgSiO_3 + Al_2Mg_{-1}Si_{-1}$	(13)
$CaAl_2SiO_6$	=	$2MgSiO_3 + CaMg_{-1} + Al_2Mg_{-1}Si_{-1}$	(14)
$Ca_3Al_2Si_3O_{12}$	=	$4MgSiO_3 + 3CaMg_{-1} + Al_2Mg_{-1}Si_{-1}$	(15)
$CaSiO_3$	=	$MgSiO_3 + CaMg_{-1}$	(16)
Al_2O_3	=	$MgSiO_3 + Al_2Mg_{-1}Si_{-1}$	(17)

Phase Components in terms of Al_2O_3 and exchange vectors

$CaMgSi_2O_6$	=	$2Al_2O_3 - 2Al_2Mg_{-1}Si_{-1} + CaMg_{-1}$	(18)
$Mg_3Al_2Si_3O_{12}$	=	$4Al_2O_3 - 3Al_2Mg_{-1}Si_{-1}$	(19)
$CaAl_2SiO_6$	=	$2Al_2O_3 - Al_2Mg_{-1}Si_{-1} + CaMg_{-1}$	(20)
$Ca_3Al_2Si_3O_{12}$	=	$4Al_2O_3 - 3Al_2Mg_{-1}Si_{-1} + 3CaMg_{-1}$	(21)
$CaSiO_3$	=	$Al_2O_3 - Al_2Mg_{-1}Si_{-1} + CaMg_{-1}$	(22)

pyroxenites or garnet lherzolites with For), Pyp+Cpx+Opx (corundum garnet-clinopyroxenites or eclogites), Gro+Cpx+Cor (grospydites, really grospycorites) and Gro+Cpx+Wol (metarodingites).

P-T behaviour of Pyp+Cpx+Opx

Two independent net-tranfer reactions of the three possible ones

$$Mg_3Al_2Si_3O_{12}[Pyp] = 4MgSiO_3[Opx] + Al_2Mg_{-1}Si_{-1}[Opx] \quad (13b)$$

$$Mg_3Al_2Si_3O_{12}[Pyp] = 4MgSiO_3[Cpx] + Al_2Mg_{-1}Si_{-1}[Cpx] \quad (13c)$$

$$MgSiO_3[Cpx] = MgSiO_3[Opx] \quad (23)$$

plus two of the three possible $CaMg_{-1}$ exchanges (2), (7), (8), i.e.

$$CaMg_{-1}[Pyp] = CaMg_{-1}[Cpx] = CaMg_{-1}[Opx]$$

plus the Tschermak exchange

$$Al_2Mg_{-1}Si_{-1}[Cpx] = Al_2Mg_{-1}Si_{-1}[Opx] \quad (11)$$

are needed to define the P-T behaviour of Pyp+Cpx+Opx.

Increasing temperature increases ternary solution of both Cpx and Opx, while increasing pressure does the reverse. On the basis of available volume and entropy data (Robie and Waldbaum, 1968; Newton et al, 1977; Thompson et al, 1978) we may obtain dP/dT = +3.3bar deg^{-1} for $Al_2Mg_{-1}Si_{-1}$[Cpx] isopleths coexisting with [Pyp] and [Opx] using equation (13a). [Pyp] and diopsidic [Cpx] form the low T-high P assemblage. Consequently, in the 3-phase assemblage Pyp+Cpx+Opx, [Cpx] will become more tschermakitic and [Pyp] less pyropic (more grossularitic!) with increasing T. Conversely, increasing P makes [Pyp] more pyropic and [Cpx] more diopsidic when coexisting with [Opx]. With this information it is possible to predict the relative modal and compositional changes of Pyp+Cpx+Opx assemblages with P and T. With increasing T the three phase triangle dinimishes because of increased $CaMg_{-1}$ exchange and $Al_2Mg_{-1}Si_{-1}$ exchange in both Cpx and Opx. Thus with increasing T a garnet websterite (Pyp+Cpx+Opx) with a composition close to Cpx+Oxp would lose [Pyp], a composition close to Cpx+Pyp would lose [Opx] and become an "eclogite" and a composition close to Opx+Pyp would lose [Cpx] and become a garnet-orthopyroxenite.

Furthermore, a garnet lherzolite (For+Pyp+Cpx+Opx) could be replaced by a garnet wehrlite (For+Pyp+Cpx) for compositions low in [Opx], a garnet harzburgite (For+Pyp+Opx) for compositions low in [Cpx], or lose [Pyp] for compositions close to (Cpx+Opx+For), with increasing T.

As increasing pressure decreases ternary solution of both Cpx and Opx, the consequent modal and compositional changes would be the reverse of the above.

P-T behaviour of Pyp+Cpx+Cor

The 2-phase region Pyp+Cpx can serve as a simple model for garnet-clinopyroxenites or even "eclogites". These minerals are often saturated with alumina and coexist with corundum at low $aSiO_2$. To describe the continuous variations in Pyp+Cpx+Cor, two independent net-transfer reactions of the three possible ones :

$$Mg_3Al_2Si_3O_{12}[Pyp] + 3Al_2Mg_{-1}Si_{-1}[Cpx] = 4Al_2O_3[Cor] \quad (19b)$$

$$Mg_3Al_2Si_3O_{12}[Pyp] = 4MgSiO_3[Cpx] + Al_2Mg_{-1}Si_{-1}[Cpx] \quad (13c)$$

$$MgSiO_3[Cpx] + Al_2Mg_{-1}Si_{-1}[Cpx] = Al_2O_3[Cor] \quad (17')$$

plus the $CaMg_{-1}$ exchange

$$CaMg_{-1}[Pyp] = CaMg_{-1}[Cpx] \quad (7)$$

are required. On the basis of available volume and entropy data (loc.cit.) we may obtain dP/dT = +197bar deg^{-1} for $Al_2Mg_{-1}Si_{-1}$[Cpx] isopleths coexisting with [Pyp] and [Cor] using equation (19a). Diopsidic [Cpx] and [Cor] form the low T-high P assemblage.

Consequently in the 3-phase assemblage Pyp+Cpx+Cor, [Cpx] will become more tschermakitic and [Pyp] more pyropic and increase in amount with a decrease in the amount of [Cor], with increasing T. (This can be viewed as the Pyp+Cpx tie-line rotation). Conversely, increasing P will cause [Cpx] to become more diopsidic and [Pyp] more grossularitic, with a decrease in the amount of garnet and an increase in the amount of corundum. The calculated $Al_2Mg_{-1}Si_{-1}$ isoplethic dP/dT (= 197bar deg^{-1}) for (19a) indicates that the temperature effect is much greater than the pressure effect. However, if the dP/dT (= 3.3bar deg^{-1}) for $Al_2Mg_{-1}Si_{-1}$ for [Cpx] in (13c) is the same as that for Opx in (13a), then this net-transfer with increasing T will cause [Pyp] to become more grossularitic and decrease in amount as the [Cpx] becomes more enstatic and tschermakitic. For this case increasing P causes [Pyp] to increase in amount and become pyropic, while the [Opx] becomes less enstatitic and tschermakitic. Because of the opposing effects of these competing net-transfer reactions it is difficult to predict exact modal changes of

Pyp+Cpx+Cor assemblages, as functions of P and T, without better experimental data.

P-T behaviour of Gro+Cpx+Cor

This assemblage is that of grospycorites and when with kyanite of grospydites. The net-transfer reactions can be considered to be the same as those for Pyp+Cpx+Cor (i.e. 19, 13, 17) plus the $CaMg_{-1}$ exchange [Gro] = [Cpx]. However, if the [Cpx] is considered only to be binary between di and ct then only one net-transfer reaction

$$Ca_3Al_2Si_3O_{12}[Gro] + 2Al_2O_3[Cor] = 3CaAl_2SiO_6[Cpx] \quad (25)$$

plus the $CaMg_{-1}$ exchange [Gro] = [Cpx] is required to describe the facies variation. Reaction (25) has a positive dP/dT (= +50bar deg^{-1}) with $CaAl_2SiO_6$ as the high T-low P phase (Hays, 1967). With increasing T, the [Cpx] will become more tschermakitic and [Gro] more grossularitic, and corundum (or kyanite) clinopyroxenites will form at the expense of grospycorites (grospydites). With increasing P, the [Cpx] will become more diopsidic and [Gro] more pyropic.

P-T behaviour of Wol+Gro+Cpx

Although this assemblage would require unusual compositions in mafic rocks it may occur in metasomatised equivalents (rodingites + serpentinite) and in calc-silicates. The net-transfer reactions can be considered to be the same as those for Pyp+Cpx+Opx, and if $Al_2Ca_{-1}Si_{-1}$ (= $Al_2Mg_{-1}Si_{-1}$ - $CaMg_{-1}$) substitution may be neglected in [Wol] we need only consider

$$Ca_3Al_2Si_3O_{12}[Gro] = 4CaSiO_3[Wol] + Al_2Ca_{-1}Si_{-1}[Cpx] \quad (13d)$$

plus $CaMg_{-1}$[Gro] = [Cpx]. The $Al_2Ca_{-1}Si_{-1}$ isopleths have dP/dT = 28bar deg^{-1} and indicate that with increasing T, the [Cpx] becomes more tschermakitic and [Gro] more pyropic. With increasing P, the [Cpx] becomes more diopsidic and [Gro] more grossularitic.

Mineral facies in $CaSiO_3$-$MgSiO_3$-Al_2O_3

From the techniques outlined above we may predict how the mineral facies in wo-en-Cor, occurring over a whole range of bulk compositions, can be used to define a petrogenetic grid. Though simplified, this system completely contains the crystalline solution variation of garnet and pyroxene and consequently is easily extended to include other phases (and components).

The phase diagram for $CaSiO_3$-$MgSiO_3$-Al_2O_3 from Boyd (1970, p.69, Fig.11) for 1200°C at 30kbar is shown in the central part of Figure 2 (Facies Type A). The approach outlined above has been used to determine the P-T displacements for other possible 3-phase assemblages likely to occur in $CaSiO_3$-$MgSiO_3$-Al_2O_3. The systematic treatment of the additional 3-phase assemblages may be simplified by considering them in terms of topological types.

Topological types and continuous reactions in $CaSiO_3$-$MgSiO_3$-Al_2O_3

It was noted above that Pyp+Cpx+Opx and Gro+Cpx+Wol could be treated the same way with regard to net-transfer and exchange equilibria and belong to the topological type [Gar-Pyx-Pyx]. Similarly Pyp+Cpx+Cor and Gro+Cpx+Cor belong to the topological type [Gar+Pyx+Cor]. Between the 6-phases [Gro], [Pyp], [Cpx], [Opx], [Wol], [Cor], there is a possibility of 6!/3!3! = 20 such 3-phase regions. If the assemblages [Pyp] + [Wol] + (Cor,Gro,Cpx,Opx) and [Opx] + [Wol] + (Pyp,Gro,Cpx,Cor) and Gro+Wol+Cor are not considered, then eleven such 3-phase regions are relevant. These may be reduced to 5 topological types in terms of the net-transfer reactions (Table 2).

Some information on the continuous variation of the 1200°C 30kbar Facies Type (A) determined by Boyd (1970) may be obtained from additional experimental studies. The CaO-Al_2O_3-SiO_2 reaction

$$Ca_3Al_2Si_3O_{12}[Gro] + 2Al_2O_3[Cor] = 3CaAl_2SiO_6[Cpx] \quad (25)$$

investigated by Hays (1967) indicates that the 3-phase triangle Gro+Cpx+Cor disappears at about 1450°C at 30kbar where a complete di-ct crystalline solution could exist. The absence of a 3-phase triangle Pyp+Opx+Cor in Facies Type (A) indicates that the MgO-Al_2O_3-SiO_2 reaction

$$3MgSiO_3[Opx] + Al_2O_3[Cor] = Mg_3Al_2Si_3O_{12}[Pyp] \quad (26)$$

involving an aluminous [Opx], occurs below 1200°C at 30kbar. This has a negative dP/dT on the basis of volume-entropy data mentioned above. Further transformations of Facies Type (A) occur through discontinuous variations in facies in wo-en-Cor.

Discontinuities in facies in $CaSiO_3$-$MgSiO_3$-Al_2O_3

Discontinuous reactions producing abrupt facies variations in large ranges of bulk composition occur when the compositions of phases changing through two continuous reactions coincide. In this ternary plane four continuous reactions, each

TABLE 2. Continuous Reactions in $CaSiO_3$ - $MgSiO_3$ - Al_2O_3.

	Topological Types	Net-transfer Reactions	
I	$[Gar^* - Pyx^{**} - Cor]$	Two of	
	Gro + Cpx + Cor	$Mg_3Al_2Si_3O_{12}$[Gar]+$3Al_2Mg_{-1}Si_{-1}$[Pyx] = $4Al_2O_3$[Cor]	(19)
	Pyp + Cpx + Cor	$Mg_3Al_2Si_3O_{12}$[Gar] = $4MgSiO_3$[Pyx]+$Al_2Mg_{-1}Si_{-1}$[Pyx]	(13)
	Pyp + Opx + Cor	$MgSiO_3$[Pyx]+$Al_2Mg_{-1}Si_{-1}$[Pyx] = Al_2O_3[Cor]	(17)
	Gro + Opx + Cor		
II	[Gar - Pyx - Pyx]		
	Pyp + Cpx + Opx	$Mg_3Al_2Si_3O_{12}$[Gar] = $4MgSiO_3$[Pyx]+$Al_2Mg_{-1}Si_{-1}$[Pyx]	(13)
	Gro + Cpx + Wol	$MgSiO_3$[Pyx] = $MgSiO_3$[Pyx]	(23)
	Gro + Cpx + Opx		
III	[Gar - Gar - Pyx]	$Mg_3Al_2Si_3O_{12}$[Gar] = $4MgSiO_3$[Pyx]+$Al_2Mg_{-1}Si_{-1}$[Pyx]	(13)
	Gro + Pyp + Opx	$Mg_3Al_2Si_3O_{12}$[Gar] = $Mg_3Al_2Si_3O_{12}$[Gar]	(24)
	Gro + Pyp + Cpx		
IV	[Pyx - Pyx - Cor]	$MgSiO_3$[Pyx]+$Al_2Mg_{-1}Si_{-1}$[Pyx] = Al_2O_3[Cor]	(17)
	Cpx + Opx + Cor	$MgSiO_3$[Pyx] = $MgSiO_3$[Pyx]	(23)
V	[Gar - Gar - Cor]	$Mg_3Al_2Si_3O_{12}$[Gar] = $Mg_3Al_2Si_3O_{12}$[Gar]	(24)
	Gro + Pyp + Cor		

* [Gar] refers either to [Pyp] or [Gro].

** [Pyx] refers either to [Cpx],[Opx] or [Wol].

involving 3-phases are involved in each discontinuity in facies. The continuous variations discussed above have been combined in Figure 2 to show the relative stabilities of Facies Types in wo-en-Cor, together with the discontinuities separating them. The entropy and volume data discussed above have been used to predict the P-T displacements of continuous reactions and dP/dT for discontinuous reactions.

Some of the higher pressure Facies Types show the coexistence of two garnets with the critical composition displaced towards pyrope. This two-phase region is only stable at high P. With increasing T, the critical line will be reached permitting continuous solution in Ca-Mg garnet. The details concerning the critical phenomena and the coincidence of [Cpx] + [Cor] with [Gar] will be discussed elsewhere (Thompson and Thompson, in prep.).

The continuous and discontinuous variations in facies involving [Gar] + [Pyx] are not modified by the presence of one phase [eg.For] lying outside the plane wo-en-Cor in $CaO-Al_2O_3-MgO-SiO_2$. The additional reactions occurring between [Gar] + [Pyx] and two phases lying outside wo-en-Cor can be considered in terms of Reaction Types (ii), (iii) and (iv).

Reactions in the Model Mantle system : $CaO-Al_2O_3-MgO-SiO_2$

The additional continuous and discontinuous reactions involving [Anr,Kya,Qtz] or [Spn,For] are better understood by remembering that,

$$CaAl_2Si_2O_8[Anr] = CaAl_2SiO_6[Pyx] + SiO_2[Qtz] \quad (27)$$

$$Al_2SiO_5[Kya] = Al_2O_3[Cor] + SiO_2[Qtz] \quad (28)$$

$$Mg_2SiO_4[For] = 2MgSiO_3[Pyx] - SiO_2[Qtz] \quad (29)$$

$$MgAl_2O_4[Spn] = Mg_2SiO_4[For] + Al_2Mg_{-1}Si_{-1}[Pyx] \quad (30)$$

These formulations can be combined with those for wo-en-Cor phase components written in terms of $MgSiO_3$ or Al_2O_3 plus $CaMg_{-1}$ and $Al_2Mg_{-1}Si_{-1}$ exchange vectors. An important example of this concerns the buffering of $\mu Al_2Mg_{-1}Si_{-1}$ in [Pyx] by assemblages on either side of the wo-en-Cor in CaO-Al_2O_3-MgO-SiO_2.

Alumina content of pyroxene in CaO-Al_2O_3-MgO-SiO_2

Although the $Al_2Mg_{-1}Si_{-1}$ variation in [Pyx] occurs within the wo-en-Cor plane and can be buffered by [Gar] + [Cor] or [Gar] + [Pyx], the Tschermak variation can be buffered by other pairs of phases. Reaction (30) shows that [Spn] + [For] can buffer $Al_2Mg_{-1}Si_{-1}$ in either [Cpx] or [Opx], although to lower values than with [Gar] through reaction (13). Consequently [Spn] + [For] saturation will correspond to one of the 2-phase tie-lines between aluminous [Cpx] and [Opx] below that bounding the 3-phase field Pyp+Cpx+Opx in Figure 2. Combining (28) with (17) we obtain

$$Al_2SiO_5[Kya] = MgSiO_3[Pyx] + SiO_2[Qtz] + Al_2Mg_{-1}Si_{-1}[Pyx] \quad (31)$$

indicating that the pair [Kya] + [Qtz] can buffer $Al_2Mg_{-1}Si_{-1}$ in either [Cpx] or [Opx]. To show how a pair of CaO-Al_2O_3-MgO-SiO_2 phases lying outside wo-en-Cor generate reactions that are controlled by the equilibria in the plane we will consider those with [Kya] + [Qtz].

Reactions in wo-en-Cor-SiO_2

Although quartz (or other SiO_2 polymorph) is certainly not expected in many ultramafic and related associations, it may be more common than previously thought in high pressure parageneses (see Smyth and Hatton, 1977). Moreover, reactions such as (27) release SiO_2 with pressure as high-silica minerals such as [Anr] are replaced by denser lower silica minerals such as [Cpx].

Many of the possible reactions in wo-en-Cor-Qtz have been discussed by Hensen (1976) involving the phases [Anr],[Kya] or [Sil] and [Qtz] in addition to [Gar] and [Pyx]. Most of the reactions encountered in wo-en-Cor-Qtz merely permit assemblages already stable in wo-en-Cor to coexist with [Anr],[Kya] or [Qtz]. To illustrate how the equilibria in wo-en-Cor influence the reactions in the larger compositional volume, we will consider reactions leading to the formation of kyanite-corundum grospydites and eclogites. In Figure 3, the phase relations involving [Gro], [Pyp],[Cpx],[Opx],[Kya],[Qtz] (the Anr-absent invariant point discussed by Hensen, 1976, p.282, Fig.1) have been combined with those involving [Gro],[Pyp],[Cpx],[Opx],[Cor] (the Wol-absent invariant point in Fig. 2 above). The two invariant points are related through the discontinuous reaction

$$Gro + Opx = Pyp + Cpx \quad (32)$$

which is defined by the intersection of the lower temperature continuous reactions Gro+Pyp+Opx and Gro+Cpx+Opx (or the coincidence of [Gro] or [Opx] compositions in the different assemblages) and the two higher temperature continuous reactions Gro+Pyp+Cpx and Pyp+Cpx+Opx. Each of the discontinuities in facies around the lower pressure invariant point involving the five phases [Gro+Pyp+Cpx+Opx+Cor] result from the intersection of the continuous reactions between four sets of 3-phases in the wo-en-Cor plane as discussed above. In contrast, the discontinuities in facies around higher pressure invariant point involving the six phases [Gro+Pyp+Cpx+Opx+Cor] result from the intersection of the continuous reactions between five sets of 4-phases in the wo-en-Cor-Qtz volume. For example, the discontinuity

$$Gro + Pyp + Qtz = Kya + Cpx \quad (33)$$

occurs at the intersection of the five reaction volumes

Kya - Gro - Pyp - Cpx (Qtz-absent)
Kya - Gro - Cpx - Qtz (Pyp-absent)
Kya - Pyp - Cpx - Qtz (Gro-absent)

Kya - Gro - Pyp - Qtz (Cpx-absent)
Gro - Pyp - Cpx - Qtz (Kya-absent)

The volume Pyp-Cpx-Qtz-Opx occurs on both sides of the discontinuity (33). Although [Kya] and [Qtz] are pure phases in this system their modal amounts can vary in the assemblage as functions of P and T. For example reaction (31)

$$Al_2SiO_5[Kya] = MgSiO_3[Pyx] + SiO_2[Qtz] + Al_2Mg_{-1}Si_{-1}[Pyx] \quad (31)$$

is a net-transfer reaction also buffering $\mu Al_2Mg_{-1}Si_{-1}$ in [Opx] or [Cpx]. Because kyanite is the low volume side, kyanite will increase and quartz will decrease in modal amount with increasing P as [Pyx] becomes less tschermakitic. The discontinuity (33) will occur when the values of $\mu Al_2Mg_{-1}Si_{-1}[Pyx]$ as buffered by (31) and (13) become identical through changing P and T. The other discontinuities

$$Cpx + Opx + Kya = Pyp + Qtz \quad (34)$$
$$Gro + Kya + Opx = Pyp + Qtz \quad (35)$$
$$Gro + Opx + Qtz = Kya + Cpx \quad (36)$$

may be described in a similar fashion when the intersection of reaction volumes representing the

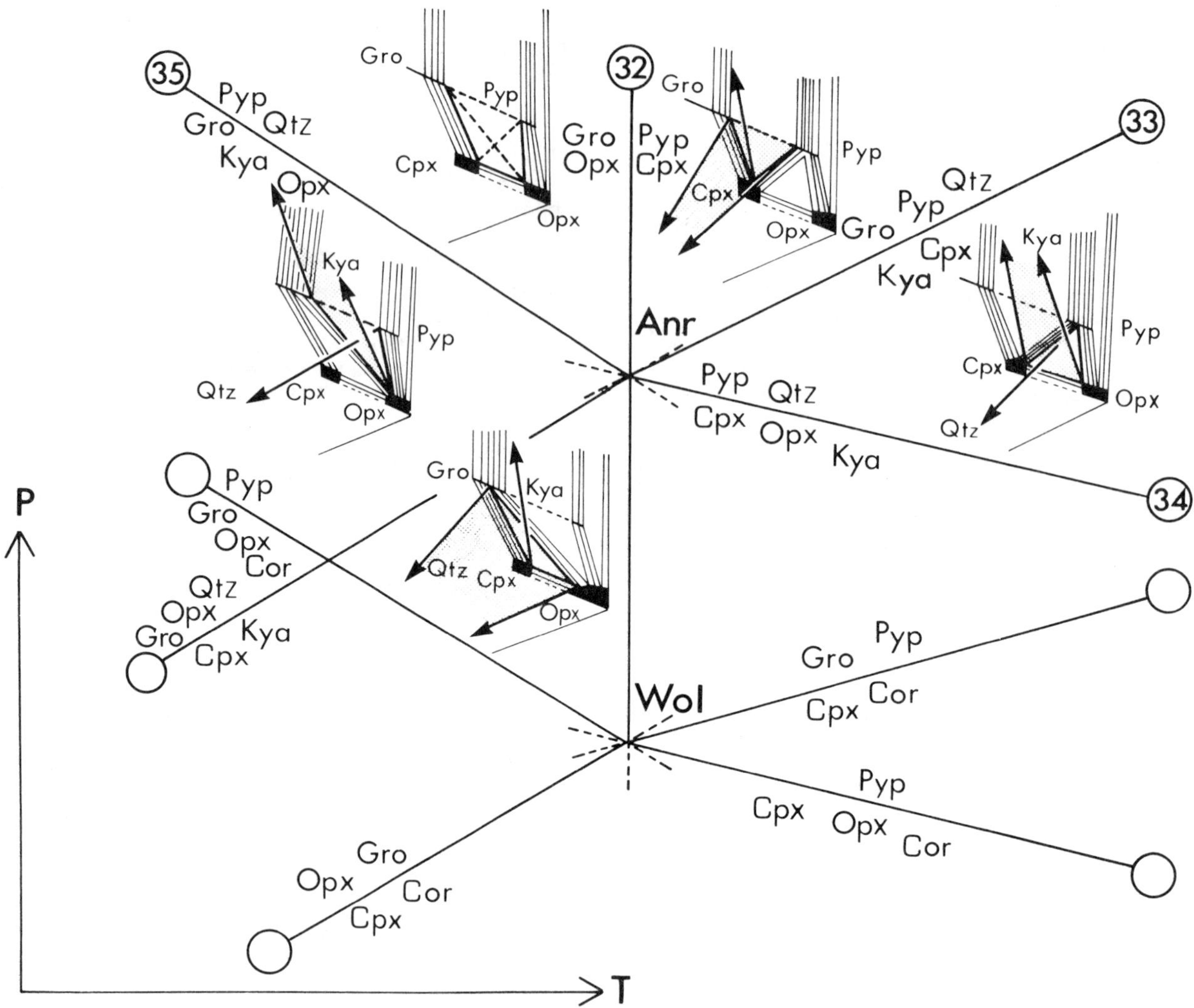

Fig. 3. Schematic P-T diagram showing the relative stability of Kya+Qtz+Gro+Pyp+Cpx+Opx (Hensen, 1976, p.282, invariant point [Anr]) and Cor+Gro+Pyp+Cpx+Opx (Figure 2 above, invariant point [Wol]). The compositional diagrams illustrate the topological changes for the univariant reactions involving Kya+Qtz. At these univariant reactions the compositions of three-phase assemblages of Gro+Pyp+Cpx+Opx are defined by the reactions in the plane wo-en-Cor. Numbers correspond to the reactions discussed in the text.

appropriate continuous reactions coincide. Again these discontinuities represent equalisation of $\mu Al_2Mg_{-1}Si_{-1}$[Pyx] buffered by the various forms of (13) and (31). This situation is illustrated in Figure 3 by the heavy lines representing the intersection of the appropriate compositional volumes. The relative pressure locations of [Cor] versus [Kya] + [Cor] and [Kya] + [Qtz] bearing assemblages indicate that corundum eclogites and grospycorites (with or without kyanite) can form at lower pressures than quartz-kyanite eclogites and quartz-grospydites, at least with reference to assemblages in $CaO-Al_2O_3-MgO-SiO_2$.

Anorthite may also be regarded as a pure phase in $CaO-Al_2O_3-MgO-SiO_2$ and with [Qtz] can buffer $\mu CaAl_2SiO_6$[Pyx] through reaction (27). Similarly [Anr] with [Kya] can buffer $\mu CaSiO_3$[Pyx] or $\mu MgSiO_3$[Pyx] through

$$CaAl_2Si_2O_8[Anr] - Al_2SiO_5[Kya] = CaSiO_3[Pyx] \quad (37)$$

$$CaAl_2Si_2O_8[Anr] - Al_2SiO_5[Kya] = MgSiO_3[Pyx] + CaMg_{-1}[Pyx] \quad (38)$$

When combined with net-transfer reactions in wo-en-Cor these net-transfer reactions will also produce appropriate continuous and discontinuous reactions. The modal and compositional changes can also be predicted using the methods outlined above.

Reactions in wo-en-Cor-MgO

The addition of either forsterite or spinel to garnet + pyroxene assemblages in wo-en-Cor produces some important Upper Mantle parageneses (notably lherzolites, Cpx+Opx+For±Pxp±Spn) but does not generate reactions additional to those in wo-en-Cor. The presence of both [For] and [Spn] with [Gar] + [Pyx] adds new net-transfer reactions through

$$MgAl_2O_4[Spn] = Mg_2SiO_4[For] + Al_2Mg_{-1}Si_{-1}[Cpx\ or\ Opx] \quad (30)$$

The calculated dP/dT = +96bar deg^{-1} for $Al_2Mg_{-1}Si_{-1}$[Cpx or Opx] for (30) indicates that with increasing P, [Spn] will increase and [For] will decrease in modal amount as the pyroxene [Cpx or Opx] becomes less tschermakitic. The reverse will occur with increasing T.

The discontinuous reaction

$$Pyp + For = Cpx + Opx + Spn \quad (39)$$

is often considered to represent the transition from spinel to garnet lherzolite. The same reaction also affects a larger compositional volume than the lherzolites (Fig. 1b) and includes the following five reaction volumes :

Pyp-For-Spn-Opx (Spn-Gar-harzburgite) (40)

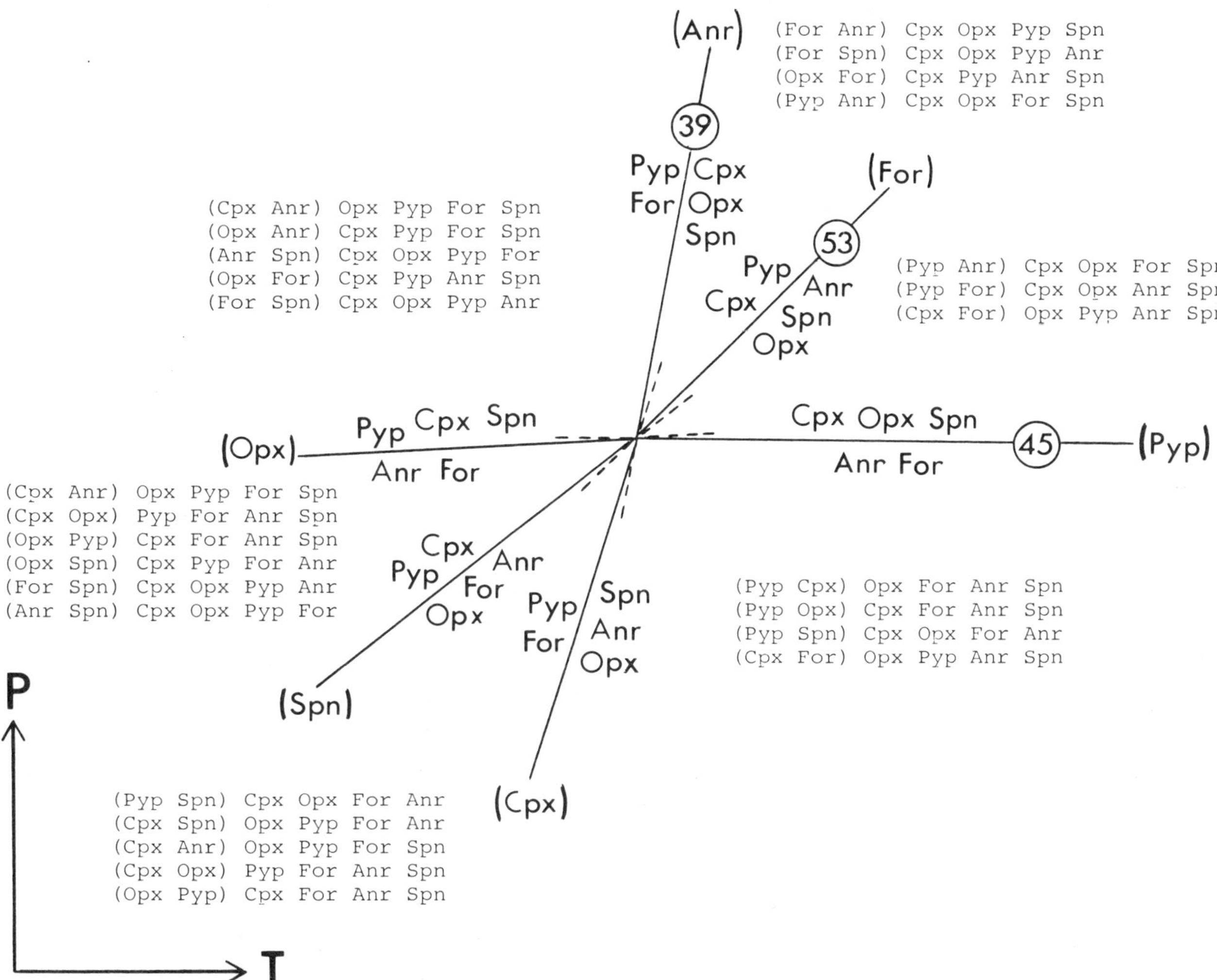

Fig. 4. Phase relations around the "lherzolite" invariant point [Cpx+Opx+Pyp+For+Anr+Spn] with garnet more pyrope-rich than Opx+Anr+Spn (1/6gr5/6py). The divariant assemblages are denoted by absent phases in parentheses. Some divariant assemblages and univariant reactions alter with changing garnet composition. Numbers correspond to reactions discussed in the text.

Pyp-For-Spn-Cpx	(Spn-Gar-wehrlite)	(41)
Cpx-Opx-For-Pyp	(Gar-lherzolite)	(42)
Cpx-Oxp-Spn-For	(Spn-lherzolite)	(43)
Cpx-Opx-Spn-Pyp	(Spn-Gar-websterite/ Ariégite)	(44)

The total net-transfer reactions influencing the P-T development of (40), (41) and (43) must include reaction (30) in addition to those in wo-en-Cor. At lower pressures, the presence of plagioclase in lherzolite and other mafic assemblages means that additional reactions involving phase relations that cross the wo-en-Cor plane must be considered.

Reactions crossing the wo-en-Cor plane involving anorthite, spinel or forsterite

The stabilities of the aluminous phases [Anr], [Spn],[Pyp] are often used to describe the P-T conditions of origin of various lherzolitic and related rocks. The compositional volumes in $CaO-Al_2O_3-MgO-SiO_2$ affected by the various mineral reactions are restricted such that a particular reaction may not occur for a given ultramafic rock composition. The example of the spinel to garnet lherzolite reaction was discussed above. Also the transition from plagioclase to spinel lherzolite is often represented by the discontinuous reaction

$$\text{Anr + For = Cpx + Opx + Spn} \quad (45)$$

and includes the following five reaction volumes :

Cpx-Opx-Spn-Anr	(Spn-Pla-websterite/ Seilandite)	(46)
Cpx-Opx-Spn-For	(Spn-lherzolite)	(47)
Anr-For-Cpx-Opx	(Pla-lherzolite)	(48)
Anr-For-Cpx-Spn	(Spn-Pla-wehrlite)	(49)
Anr-For-Opx-Spn	(Spn-Pla-harzburgite)	(50)

The association [Anr] + [For] or [Anr] + [Spn] introduces new net-transfer reactions that must be considered in addition to the above. By considering reactions (27), (28) and (30) together with those in wo-en-Cor we obtain

$$CaAl_2Si_2O_8[Anr] + Mg_2SiO_4[For] = 4MgSiO_3[Pyx] + Al_2Mg_{-1}Si_{-1}[Pyx] + CaMg_{-1}[Pyx] \quad (51)$$

and

$$CaAl_2Si_2O_8[Anr] + MgAl_2O_4[Spn] = 4MgSiO_3[Pyx] + 2Al_2Mg_{-1}Si_{-1}[Pyx] + CaMg_{-1}[Pyx] \quad (52)$$

These new net-transfer reactions (51) and (52) are related through (30) and in some assemblages will not be independent. Both (51) and (52) could influence the $CaMg_{-1}$ composition of [Pyx] or [Gar] if present.

The volume and entropy data referred to above have been used to calculate dP/dT for reactions (51) and (52) involving [Pyx]. For (51) the calculated dP/dT = +11bar deg^{-1} indicates that with increasing P, [Anr] and [For] decrease and [Pyx] increases in amount while [Pyx] becomes more tschermakitic and less enstatitic. The reverse will occur with increasing T. For (52) the calculated dP/dT = +3bar deg^{-1} indicates that with increasing P, [Anr] and [Spn] decrease and [Pyx] increase in amount while [Pyx] becomes more tschermakitic and less enstatitic. The reverse will occur with increasing T.

Any consideration of the continuous reactions involved in the discontinuity in facies (45) must include (51) and (52) in addition to the appropriate net-transfer reactions discussed above.

Relative stability of Lherzolite and other assemblages in the Model Mantle system $CaO-Al_2O_3-MgO-SiO_2$

The continuous reactions used to model the transition from plagioclase to spinel lherzolite in the system $CaO-Al_2O_3-MgO-SiO_2$

$$\text{Anr + For = Cpx + Opx + Spn} \quad (45)$$

and from spinel to garnet lherzolite

$$\text{Cpx + Opx + Spn = Pyp + For} \quad (39)$$

could intersect at an invariant point at low T and P (see Kushiro and Yoder, 1966, p.344). Four additional discontinuous reactions are generated at this invariant point (see Fig.4), each involving continuous reactions between four phases. Between the six phases [Cpx+Opx+For+Pyp+Anr+Spn] there are (6!/4!2!) = 15 four-phase reaction volumes. These four-phase assemblages are shown in the appropriate divariant fields in Fig.4, determined by the absent phases in parentheses. One important feature descernible from the relative stability of the four-phase assemblages is that mineral facies cannot be specified by the presence of one mineral alone, simply because every one of the six phases is stable in every divariant field although not in the same assemblage. Many rare or unknown rock types are indicated in Fig.4 but they are stable in the compositional space defined above, in the system $CaO-Al_2O_3-MgO-SiO_2$, if the invariant point exists. All of the net-transfer reactions and exchange equilibria contributing to the various continuous reactions must be considered in detail to determine not only the stability of the invariant but also the reactant or product nature or each phase involved in the continuous reactions. Whether a phase is a reactant or product depends upon the relative compositions of the phases involved. For example, consider the continuous reactions involved in the $CaO-Al_2O_3-MgO-SiO_2$ discontinuous reaction

$$Opx + Spn + Anr = Pyp + Cpx \quad (53)$$

The situation shown by the labelling of (53) in Figure 4 involves the continuous reaction volumes

Pyp-Cpx-Spn-Anr	(Spn-Pla-Gar-clinopyroxenite!)	(54)
Pyp-Cpx-Opx-Anr	(Gar-Pla-websterite)	(55)
Pyp-Cpx-Opx-Spn	(Spn-Gar-websterite/Ariégite)	(44)
Cpx-Opx-Spn-Anr	(Spn-Pla-websterite/Seilandite)	(46)
Opx-Pyp-Spn-Anr	(Spn-Pla-Gar orthopyroxenite!)	(56)

This labelling holds for garnet that is more pyropic than defined by the plane Opx+Anr+Spn. Although the garnet composition defined by $4MgSiO_3+CaAl_2Si_2O_8+MgAl_2O_4$ is 1/6gr5/6py, the actual garnet will be slightly more calcic because of $CaMg_{-1}$[Opx] = [Pyp]. However, the experimental data of Kushiro and Yoder (1966, p.343) indicate that garnet (py 74-78 gr26-22) is growing at the expense of [Anr],[Spn],[For] and [Cpx]. Because the garnet composition lies between that defined by the plane Opx+Anr+Spn (1/6gr5/6py) and that defined by the join [Anr] + [For] ($CaAl_2Si_2O_8$ + Mg_2SiO_4 = 1/3gr2/3py) the data of Kushiro and Yoder suggest that the assemblage [Pyp+Anr+Spn+Opx] is stable at lower pressures than the discontinuity (53) when written in the form

$$Opx + Cpx + Spn + Anr = Pyp \quad (53a)$$

Consequently, the form of the discontinuous reaction is a function of the relative stability of five-phase reaction volumes which is in turn determined by the relative compositions of reacting phases. Thus along some discontinuities, singularities in composition of reacting phases can change dP/dT, reaction stoichiometry and the relative stability of four-phase assemblages. Obviously equilibria determined at high P and T cannot be extrapolated to lower P and T without considering the effects of the net-transfer reactions and exchange equilibria on the particular discontinuity. Much of the disagreement in the P-T location of discontinuous reactions may be due to such unjustified extrapolations (as also discussed by Obata, 1976 and Newton, 1977). Furthermore, calculations with the limited available experimental and thermochemical data by Obata (1976) suggest that the invariant point shown in Figure 4 may not be stable in $CaO-Al_2O_3-MgO-SiO_2$. The complex reactions discussed above and the effects of additional components must be considered in much greater detail to determine if such phase relations exist in more complex compositional space.

Metamorphism in the Upper Mantle

As only small volumes of Upper Mantle are available for direct petrographic examination, the predictive methods outlined above can only be successfully calibrated by careful experiments. However, petrological study of ultramafic nodules and crystalline massifs can provide significant insight into the reactions occurring between the chemically complex minerals and indicate the effects of additional components on the P-T displacements of the continuous and discontinuous equilibria in $CaO-Al_2O_3-MgO-SiO_2$. The components Na_2O, FeO, Fe_2O_3 and Cr_2O_3 generally only enter into crystalline solution and through an increase in the variance of a particular association permit additional phases to coexist. The components K_2O and TiO_2 are often present in saturating phases, such as biotite (when hydrous) and rutile. Although these components can be added through further binary exchange equilibria (eg. $Fe^{2+}Mg_{-1}$, $NaSiCa_{-1}Al_{-1}$, $Fe^{3+}Al_{-1}$, $CrAl_{-1}$, KNa_{-1}, $TiSi_{-1}$), the application to natural assemblages requires specification of the appropriate exchange components for the multicomponent crystalline solutions. The problem is not only one of determination of multicomponent activities but also of schemes by which mineral analyses may be recalculated into practical components.

The techniques discussed here provide means by which coexisting nodule assemblages may be compared in such a way as to determine if they are related by compositional differences or by P and T. This includes that possibility that the compositions of coexisting minerals in grospydites or eclogites may be used to calibrate those in lherzolites, etc. Conversely, a given bulk composition can be transformed into different mineral assemblages as functions of changing P and T (see also Nicholls, 1977).

The net-transfer and exchange equilibria between crystalline solutions still apply when these phases are in equilibrium with a silicate metl and consequently in part determine the likely compositions of magmas in the Upper Mantle.

Most of the reactions occurring in the Upper Mantle will be of the continuous type and will only result in changing proportions and compositions of the phases. Discontinuous reactions producing dense phases such as garnet could be responsible for seismic discontinuities. However, the width of the transition interval is dependent upon the compositions of the minerals involved, which vary with P and T as discussed above, in relation to the bulk composition of the Upper Mantle material. Because discontinuous reactions affect large ranges of bulk composition with different mineral assemblages, seismic discontinuities could result from reaction discontinuities in a heterogeneous Mantle. It is possible to assess the width of reaction intervals and calculate the bulk modulii using the techniques presented here given more experimental and thermochemical data.

Acknowledgements

Discussions with B.W.Evans, T.Finnerty, S.E.Kesson, J.Laird and R.J.Tracy significantly modified the manuscript and collaboration with J.B.Thompson greatly advanced the concepts and applications of exchange vectors. Support by the Schweizerische Nationalfonds (Grant 2.772-0.77) is greatly acknowledged.

References

Boyd, F.R., Garnet peridotites and the system $CaSiO_3$-$MgSiO_3$-Al_2O_3, Mineral. Soc. Amer. Spec. Pap. 3, 63-75, 1970.

Hays, J.F., Lime-alumina-silica, Carnegie Inst. Wash. Year Book 65, 234-239, 1966.

Hensen, B.J., The stability of pyrope-grossular garnet with excess silica, Contrib. Mineral. Petrol. 55, 279-292, 1976.

Kushiro, I. and H.S. Yoder, jr., Anorthite-forsterite and anorthite-enstatite reactions and their bearing on the basalt-eclogite transformation, Jour. Petrol. 7, 337-362, 1966.

Laird, J., A.B. Thompson, and J.B. Thompson, Jr., Amphibole reactions in mafic schist, Trans. Amer. Geophys. Un. 59, 408, 1978.

Newton, R.C., Thermochemistry of garnet and aluminous pyroxenes in the CMAS system, in D.G. Fraser (ed.), Thermodynamics in Geology, 29-55, 1977.

Newton, R.C., A.B. Thompson, and K.M. Krupka, Heat capacity of synthetic $Mg_3Al_2Si_3O_{12}$ from 350 to 1000K and the entropy of pyrope, Trans. Amer. Geophys. Un. 58, 523, 1977.

Nicholls, J., The calculation of mineral compositions and modes of olivine-two pyroxene-spinel assemblages, Contr. Mineral. Petrol. 60, 119-142, 1977.

Obata, M., The solubility of Al_2O_3 in orthopyroxenes in spinel and plagioclase peridotites and spinel pyroxenite, Amer. Min. 61, 804-816, 1976.

Robie, R.A., and D.R. Waldbaum, Thermodynamic properties of minerals and related substances at 298.15K (25.0°C) and one atmosphere (1.013 bars) pressure at high temperatures, U.S. Geol. Surv. Bull. 1259, 1968.

Smyth, J.R., and C.J. Hatton, A coesite-sanidine grospydite from the Roberts Victor Kimberlite, Earth and Planet. Sci. Letts. 34, 284-290, 1977.

Thompson, A.B., D. Perkins, U. Sonderegger, and R.C. Newton, Heat capacities and thermodynamic properties of $CaAl_2SiO_6$-$CaMgSi_2O_6$ pyroxenes, Trans. Amer. Geophys. Un. 59, 395, 1978.

Thompson, J.B., and A.B. Thompson, A model system for mineral reactions in pelitic schists. Contr. Mineral. Petrol. 58, 243-277, 1976.

Thompson, J.B., and A.B. Thompson, Metamorphism in the Earth's Crust and Mantle, in prep.

Wood, B.J., The application of thermodynamics to some subsolidus equilibria involving solid solutions, Fortschr. Miner. 52, 21-45, 1975.

II. ECLOGITES AND PERIDOTITES FROM KIMBERLITES

A DIAMOND-GRAPHITE ECLOGITE FROM THE ROBERTS VICTOR MINE

C.J. Hatton and J.J. Gurney

Department of Geochemistry, University of Cape Town, Rondebosch, 7700. South Africa.

Abstract. The coexistence of diamond and graphite in a Roberts Victor eclogite indicate that the eclogite formed under pressure and temperature conditions close to the diamond-graphite reaction curve. The silicates have crystallised over a range of temperature and pressure conditions which transect the diamond-graphite equilibrium curve at 1020 to 1140°C and 42 to 45 kbar. In this pressure-temperature range the carbon-gas (CCO) buffer curve defines oxygen fugacities close to those considered reasonable approximations of the oxygen fugacity of the mantle. Diamond and graphite therefore may have formed by reduction of carbon oxides during cooling of the silicate magma. The eclogite is believed to have formed by the crystallisation of a partial melt produced by disequilibrium melting of garnet lherzolite under volatile-rich conditions.

Introduction

Diamonds are most commonly recovered from the fine-grained matrix of kimberlite, but although a genetic relationship between diamond and kimberlite has been suggested (Wagner, 1914, pp159-163; Williams, 1932, pp 410-420), diamond does not necessarily crystallise from a kimberlite magma. The majority of syngenetic mineral inclusions in diamond may be related to either a peridotite or eclogite paragenesis (Meyer and Svisero, 1975; Prinz et al., 1975). It is possible therefore that since diamond is related to the xenolith suites, diamond is itself a xenocrystal mineral. However, some mineral inclusions in diamonds, notably high-iron chromite (Meyer and Boyd, 1972), high-iron ilmenite and zircon (Meyer and Svisero, 1975) cannot be clearly related to the eclogite or peridotite parageneses and may represent minerals crystallising at depth from kimberlite or protokimberlite. The evidence of the inclusions in diamonds is therefore equivocal.

Another approach in the determination of the origin of diamond is a study of those xenoliths in which diamond apparently occurs as a primary component. Eclogite is the most commonly reported host rock (e.g. Reid et al., 1976), but diamonds have also been found in peridotite xenoliths (Sobolev et al., 1969; Dawson and Smith, 1975; Eggler and McCallum, 1975; Pokhilenko et al., 1977). In most kimberlite pipes peridotite is more abundant than eclogite. Similarly, peridotitic minerals are more commonly encountered in diamond inclusions than are eclogitic minerals (Meyer and Tsai, 1976; Sobolev, 1977; Harris and Gurney, 1977). It is therefore surprising that eclogite xenoliths most commonly contain diamonds. Diamond has been found in biminerallic eclogite and in eclogite containing kyanite, corundum, rutile and graphite (Rickwood et al., 1969). The coexistence of diamond and graphite is of considerable interest.

The diamond-graphite reaction curve has been determined by a number of workers, and the theoretical and experimental curves are in reasonable agreement (Bundy et al., 1961; Kennedy and Kennedy, 1976). There is at present no evidence that diamond or graphite will nucleate outside their stability fields, although metastable growth of diamond on pre-existing nuclei can occur (Evans, 1976). Thus, if it can be shown that diamond and graphite formed together with the other minerals in a rock, the conditions of formation of that rock may be constrained to temperatures and pressures on or near the diamond-graphite reaction curve. It might be expected that such conditions are only rarely encountered in the crystallisation of eclogites from a melt, and that in consequence diamond-graphite eclogites are relatively uncommon. However, of some 70 known diamondiferous eclogites (Rickwood et al., 1969; Reid et al., 1976; Sobolev, 1977; Shee and Gurney, 1977; Hatton and Gurney, unpublished data), 8 (11%) contain both diamond and graphite (Beck, 1907; Bobrievich et al., 1960; Gurney et al., 1969; Sobolev, 1977, p 98; Robinson, 1977). Conditions necessary for simultaneous crystallisation of both diamond and graphite appear to be more common than anticipated; an observation which must be taken into account when discussing possible models for the origins of these unusual rocks.

In this study particular attention is paid to

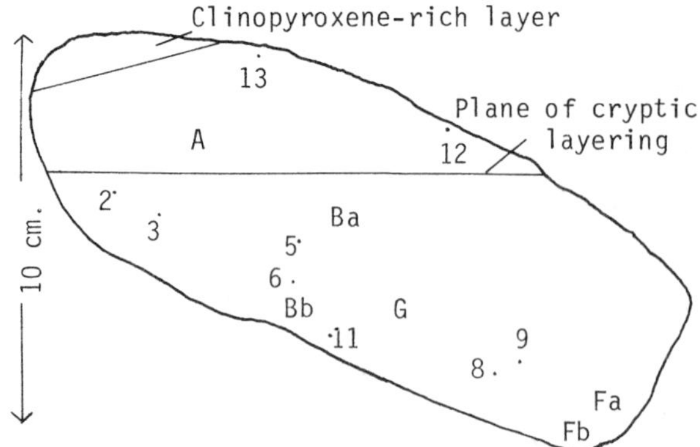

Fig. 1. Section through central part of nodule perpendicular to layering. Alphabetical characters indicate approximate locations of slides cut. Numbered dots locate individual grains removed.

the silicates in diamond-graphite eclogite HRV 247 from the Roberts Victor mine. The morphological features of diamond and graphite in this rock (Robinson, 1976, 1977) are also taken into account.

Comparison of the chemistry of diamondiferous and non-diamondiferous eclogites from the Roberts Victor mine indicates that diamonds tend to occur in the more evolved compositions. It is convenient to recognise within the Roberts Victor eclogites extremes of composition represented by iron-rich members with greater than 18 wt% FeO in garnet, and calcium-rich members with greater than 6 wt% CaO in garnet. Calcium-rich eclogites have the highest Na 0 contents and on this account are termed the most evolved members (Hatton, 1978). Of 136 Roberts Victor eclogites in which garnet and clinopyroxene have been analysed (Kushiro and Aoki, 1968; Switzer and Melson, 1969; Whitfield, 1971; O'Hara et al., 1975; Sobolev, 1977; Hatton, 1978) 42 are calcium-rich, while 10 of the 12 diamondiferous eclogites analysed are calcium-rich. Thus while calcium-rich eclogites comprise 31% of the total Roberts Victor eclogite population, 83% of the diamondiferous eclogites are calcium-rich. The reasons for this bias are discussed later.

Description of Eclogite Xenolith HRV247

The nodule was collected from the Roberts Victor kimberlite pipe and had original dimensions of 20 x 14 x 7cm. A number of graphite crystals and fifteen diamonds were observed on broken surfaces. On further examination and solution of three portions of the nodule, 49 diamond crystals and 200 graphite crystals were recovered (Robinson, 1977). Six slides were cut and nine garnet grains were separated from portions of the nodule (Figure 1). The modal proportions of garnet and clinopyroxene are constant throughout most of the rock but mineral layering is defined by a thin layer of clinopyroxene-rich eclogite at one edge of the nodule. Cryptic layering (Wager and Brown, 1967, p 548) is defined by consistent changes in garnet and clinopyroxene composition across the rock, and the planes of cryptic layering broadly coincide with the mineral layering (Figure 1).

Clear rounded garnet of 5mm mean grain size, and fresh, grass-green clinopyroxene comprise over 99% of the nodule. Besides diamond and graphite, pyrrhotite and rutile are present as minor constituents which appear to have formed penecontemporaneously with garnet and clinopyroxene. The inter-granular mineralogy is dominantly phlogopite with amphibole and minor opaque phases. Some of the phlogopite may have crystallised together with garnet and clinopyroxene. The graphite crystals are euhedral, and because of the low crystallising power of graphite it is probable that the graphite crystals grew in a liquid medium and hence that the rock formed by igneous processes. There is no evidence of formation of diamond from graphite or vice versa, (Robinson, 1976) and we consider that diamond, like graphite, is a primary igneous constituent of the eclogite.

Inhomogeneities within the Eclogite

Zoning of the Garnets

All the garnets analysed have MgO-rich rims. Cores are enriched in CaO and TiO_2 and to a

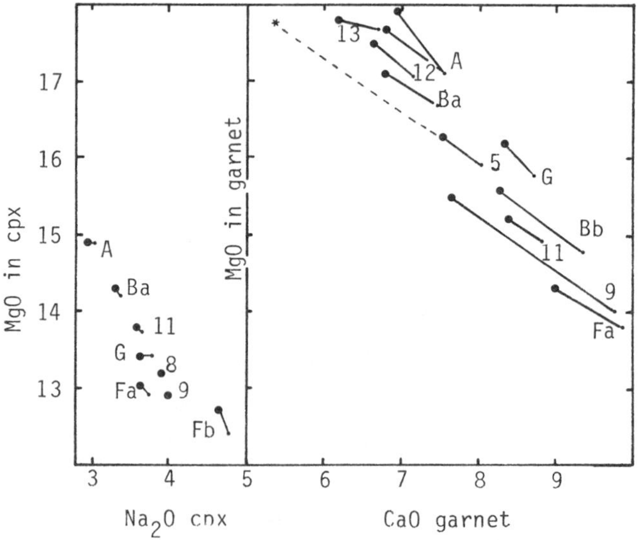

Fig. 2. Plot of wt% MgO in clinopyroxene against wt% Na_2O in clinopyroxene and of wt% MgO in garnet against wt% CaO in garnet. Small dots represent cores of grains, large dots represent rims. Some data have been omitted for clarity. * represents garnet inclusion in garnet 5.

lesser extent FeO. Other elements analysed show little variation in concentration. In terms of end members the variation in garnet composition can be described as enrichment of the rims in pyrope content with concomitant depletion of grossular and, to a lesser extent, of almandine. Zoning proceeds smoothly from core to rim with no sudden fluctuations in composition. The enrichment in MgO from core to rim is matched by an enrichment in MgO from one end of the nodule, F, to the opposite end, A. The other elements also show a correspondence between changes in concentration from core to rim and changes in concentration from F to A. Since rims cannot have crystallised before cores, garnets in A cannot have crystallized before garnets in F.

Zoning of the clinopyroxenes

The clinopyroxenes are weakly zoned with CaO, MgO-rich rims and Na_2O, Al_2O_3, TiO_2-rich cores. The enrichment of the rims in the diopside end member at the expense of the jadeite end member is matched by the enrichment in the diopside end member from portion F to portion A (Figure 2). The Ca:Mg:Fe ratio of cores and rims and of clinopyroxenes throughout the rock remains relatively constant.

Garnet inclusion in garnet

In the centre of garnet crystal 5 a garnet inclusion of a composition quite dissimilar to the surrounding garnet is present. The CaO content of the garnet inclusion is lower than that of any other garnet in HRV247 (Figure 2, Table 1). Garnet crystal 5 is 5 x 3mm in size and the included garnet is 3 x 1mm. The origin of the garnet inclusion is problematical.

Chemical Variation across the layers

The bulk composition of the sample, with the exception of the clinopyroxene-rich layer, can

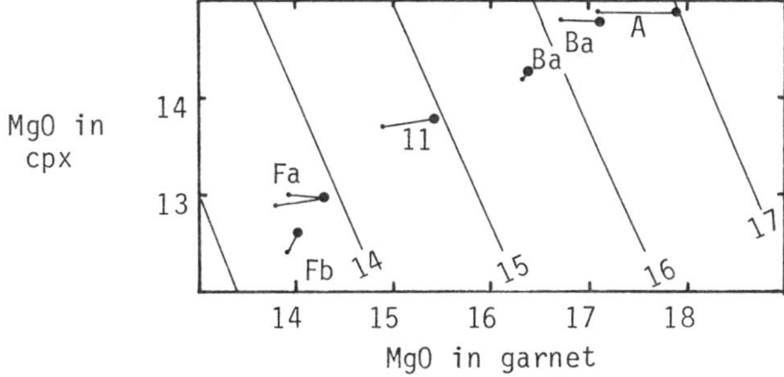

Fig. 3. Plot of wt% MgO in clinopyroxene against wt% MgO in garnet. Small dots represent cores of grains, large dots represent rims. Oblique lines represent bulk compositions calculated from the ratio 0.7 garnet to 0.3 clinopyroxene.

be calculated by combining garnet and clinopyroxene in the proportions indicated by the modal abundances, i.e. 0.7 garnet and 0.3 clinopyroxene. In a plot of MgO in clinopyroxene against MgO in garnet, bulk MgO is represented by lines along which garnet and clinopyroxene are combined in the appropriate ratio. Bulk MgO increases from approximately 13 wt% in F to 17 wt% in A, and the sequence core-to rim follows the pattern of increasing MgO.

Variation in diamond and graphite contents

Graphite and diamond contents were determined in three portions of the rock by dissolution in HF of 416g of A, 122g of G (central portion) and 405g of F (Robinson, 1976). Virtually no diamonds were recovered from G but three diamonds with a total mass of approximately 1/10 carat or .02g were removed prior to dissolution. The diamond content of G can then be estimated as approximately 200ppm. Diamond content decreases from 1590ppm in A through 200ppm in G to 2ppm in F. Graphite content is 150ppm in A, 1464 ppm in G and 980ppm in F. Thus graphite content is highest in the central portion of the nodule.

Range in temperature and pressure

The ratio K_D = (FeO/MgO gt)/(FeO/MgO cpx) is a function of temperature and pressure (Raheim and Green, 1974). The ln K_D ratio, assuming all iron as FeO, shows a consistent decrease from 1.44 in F to 1.19 in A (Figure 4). Although the inhomogeneity of the rock demonstrates that equilibrium of the bulk rock system was not attained, the regularity of the compositional change and of the K_D ratios suggests that local equilibrium between edges of adjacent garnet and clinopyroxene may have been attained. If local equilibrium was attained, and if the silicates crystallised under conditions close to the theoretical diamond-graphite reaction curve (Bundy et al., 1961), the $\ln K_D$ ratios indicate that the temperature and pressure of equilibration of the rock lies in the range 1020 to 1140°C at 42 to 45 kbar.

The Origin of Diamond and Graphite

The elemental carbon content of subcratonic upper mantle is not well known but probably is extremely low. Carbon necessary for the formation of diamond or graphite may be introduced in a mobile phase as CO_2, or combined in carbonate or as elemental carbon dissolved in subphide (Brett et al., 1977). Irrespective of the original source of carbon, the oxygen fugacity of the eclogite rock system has an important bearing on the formation of diamond and graphite, for these phases may be oxidised to carbon dioxide in an environment where the prevailing oxygen fugacity is too high. The

TABLE 1. Representative analyses of garnet and clinopyroxene compositions.

	Garnet compositions							Clinopyroxene compositions			
	Slide Ba		Garnet 5			Slide F		Slide Ba		Slide F	
	Centre	Edge	Inclusion	Centre	Edge	Centre	Edge	Centre	Edge	Centre	Edge
SiO_2	41.6	41.6	41.5	41.3	41.4	40.6	40.6	55.0	55.4	55.2	55.5
TiO_2	.27	.22	.36	.30	.27	.38	.32	.24	.26	.31	.29
Al_2O_3	23.6	23.6	23.3	23.3	23.4	23.0	23.2	5.13	5.15	7.80	7.76
Cr_2O_3	.06	.06	.10	.05	.05	.05	.09	.07	.08	.07	.07
FeO	10.4	10.4	11.1	10.6	10.6	11.8	12.2	2.51	2.49	2.61	2.62
MnO	.26	.26	.33	.22	.24	.22	.23	.03	.04	.05	.01
MgO	16.7	17.1	17.8	15.9	16.3	13.9	14.2	14.8	14.8	12.6	12.6
CaO	7.45	6.80	5.34	8.04	7.55	9.80	9.25	19.2	19.3	17.5	17.3
Na_2O	.09	.10	.06	.09	.08	.11	.12	3.01	2.91	4.65	4.57
K_2O								.05	.06	.06	.06
Total	100.4	100.1	99.9	99.8	99.9	99.9	100.2	100.0	100.5	100.9	100.8
	Number of cations for 12 oxygens							Number of cations for 6 oxygens			
Si	2.984	2.987	2.987	2.990	2.990	2.973	2.964	1.973	1.977	1.959	1.968
Ti	.015	.012	.019	.016	.015	.021	.018	.006	.007	.008	.008
Al	1.995	1.997	1.977	1.988	1.992	1.985	1.996	.217	.217	.326	.324
Cr	.003	.003	.006	.003	.003	.003	.005	.002	.002	.002	.002
Fe	.624	.625	.668	.642	.640	.723	.745	.075	.074	.077	.078
Mn	.016	.016	.020	.013	.015	.014	.014	.001	.001	.002	.000
Mg	1.786	1.831	1.909	1.716	1.754	1.517	1.545	.792	.787	.667	.666
Ca	.573	.523	.412	.624	.584	.769	.723	.738	.738	.666	.657
Na	.013	.014	.008	.013	.011	.016	.017	.209	.201	.320	.314
K								.002	.003	.003	.003
Total	8.008	8.008	8.007	8.004	8.004	8.020	8.027	4.016	4.008	4.030	4.020

graphite -CO-CO_2 (CCO) buffer may be used as a model for the oxygen fugacity at which graphite or diamond are oxidised. If carbon was introduced as an oxide or carbonate the fO_2 prevailing during melting was probably higher than the CCO buffer, so that carbon could be transported as an oxide or carbonate. During the formation of diamond and graphite the oxygen fugacity decreased to values below the CCO buffer. Thus the eclogite probably formed in a pressure-temperature range where diamond and graphite could coexist and where the oxygen fugacity was close to the CCO buffer curve. Oxygen fugacities of different buffers at 44 kbar are illustrated in Figure 5. At this pressure the diamond-graphite equilibrium curve falls within the temperature range suggested by the K_D ratios, and the CCO buffer (Woermann et al., 1977) defines oxygen fugacities between the FMQ and WM buffers, (Huebner, 1971) which may represent the range of oxygen fugacities encountered in the mantle (Rosenhauer et al., 1977). Thus graphite, diamond, and CO_2 could have coexisted temporarily in the pressure-temperature range suggested by the silicate exchange reactions, so that carbon dioxide may have been reduced to both graphite and diamond during cooling of the magma.

Possible error in pressure-temperature determination

The pressures and temperatures suggested depend on the accuracy with which the K_D ratios have been determined in this study, and on the accuracy with which the diamond-graphite reaction curve has been determined. The K_D ratios were calculated on the assumption that all iron occurs as FeO. If ferric iron is present in garnet or clinopyroxene the cation sum will be higher than the stoichiometric cation sum, and the amount of ferric iron will be proportional to the cation sum excess. However, only small overestimates of the cation sum lead to relatively large estimates of the proportion of ferric iron present, particularly in clinopyroxenes analysed here (2.5 wt% FeO). A 0.25% overestimate of the cation sum can be compensated by conversion of 10% of ferrous iron to ferric iron. The reduced amount of ferrous iron lowers the temperature calculated from the K_D ratio by approximately 40°C at 40 to 50 kbar. The theoretical diamond-graphite equilibrium curve (Bundy et al., 1961) is approximately 80°C higher than the curve determined by Kennedy and Kennedy (1976) at 40 to 50 kbar.

In view of the possible errors a pressure of

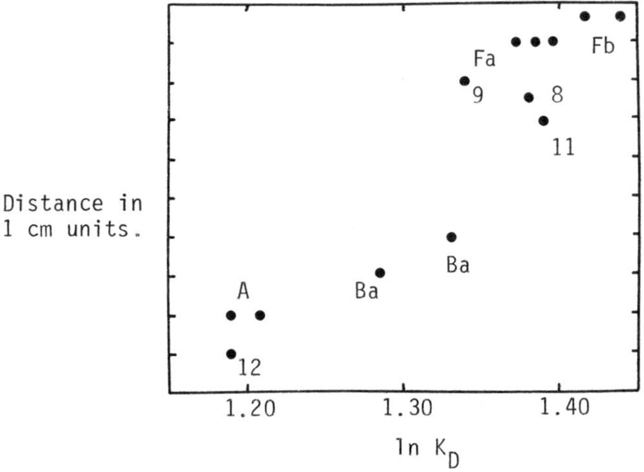

Fig. 4. Plot of distance across nodule against ln K_D ratio.

44 kbar may be 4 to 6 kbar too low, and the diamond-graphite reaction curve and the K_D estimates will overlap at slightly lower temperatures at these higher pressures. However, the pressure-temperature range continues to overlap with the temperature interval over which the CCO buffer lies between the FMQ and WM buffers , and reduction of CO_2 to both diamond and graphite over a small temperature range remains a feasible process at these higher pressures.

Formation of the Silicates

The rims of the silicate grains have higher MgO contents than the cores. MgO is usually depleted in a magma during crystallisation; thus if the silicates formed, as we believe, in an igneous environment, the coexisting magma must have been continuously enriched in MgO during crystallisation. Mysen and Kushiro (1977) have shown that the MgO content of the liquid produced by partial melting of lherzolite increases as melting proceeds. Liquids are produced over a relatively large temperature interval at the very beginning of melting as a result of the presence of small amounts of depolymerising ions, Na, K, and H (Mysen and Kushiro, 1977). We suggest that HRV247 formed by disequilibrium crystallisation of liquids produced by volatile-induced partial melting of garnet lherzolite. Access of volatiles, including CO_2, to peridotite, at temperatures on or above the geotherm would induce partial melting at temperatures well below the anhydrous solidus (Kushiro et al., 1968). When a few percent of melt is present the magma will migrate upwards as rapidly as it is produced (Turcotte and Ahern, 1978), with the consequent production of a continuous range of magma compositions with alkali-rich low-temperature magmas rising first, followed by progressively higher-temperature, MgO-rich compositions. On cooling at a higher level the alkali-rich magma will crystallise. Although the successive additions of MgO-rich magmas will be at higher temperatures, crystallisation will continue provided the rate of heat loss remains constant, since the higher temperature of the successive increments of magma will be counterbalanced by the higher liquidus temperatures of these more MgO-rich compositions. The first formed melts crystallise to produce the silicates in F, and the successive increments with higher liquidus temperatures collect below the earlier liquids and crystallise to produce the silicates in portions G and A.

Formation of Diamond and Graphite

The temperature at which diamond and graphite crystallise may not depend on the liquidus temperature of the silicate phases, but rather on the temperature at which the oxygen fugacity of the CCO buffer rises above the oxygen fugacity of the silicate system. This temperature may be somewhat higher than the liquidus temperature of the silicate phases, so that diamond and graphite may crystallise for a period prior to crystallisation of the silicates.

In the melting sequence outlined above diamond may crystallise from the first-produced, low-temperature melts while graphite crystallises

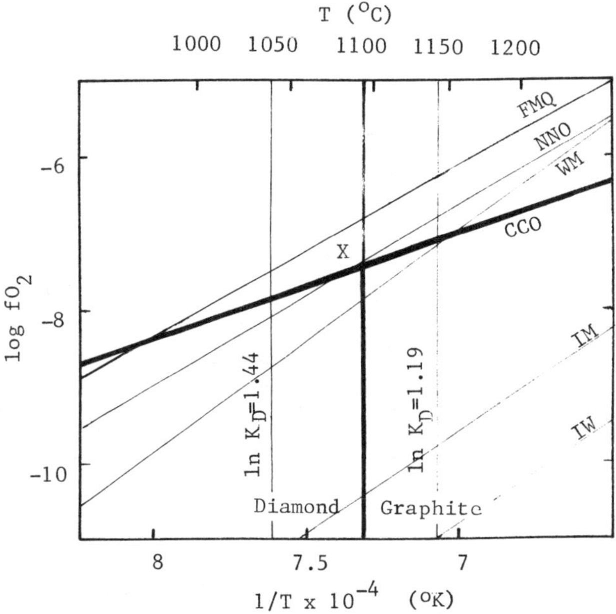

Fig. 5. Oxygen fugacities of different buffers, temperature of diamond-graphite equilibrium curve and temperature range indicated by silicate exchange reactions at 44 kbar. At point X CO_2 may be reduced to either diamond or graphite over a narrow temperature range if the oxygen fugacity of the eclogite system was approximated by the NNO buffer. X lies within the temperature range suggested by the silicate exchange reactions.

from the higher-temperature melt. However, because of the significant density difference between diamond and graphite, even at high pressure and temperature (Berman, 1965; Robinson, 1977), there is likely to be gravitational separation of diamond and graphite, with graphite rising into the low-temperature liquid and diamond settling into the higher-temperature liquid. The proposition that diamond and graphite leave their stability fields is consistent with the observation that both diamond and graphite exhibit etch features (Robinson, 1976). The endothermic reduction of carbon oxides to diamond and graphite (Rosenhauer et al., 1977) hastens the crystallisation of the silicate liquid.

Provided that the conduits through which the melt rises are wider than the maximum grain size of diamond or graphite, for example if the conduits are narrow dykes, gravitational separation of diamond and graphite may take place while the magma rises. On crystallisation of the silicates graphite will then be most abundant in the upper portion (F) while diamond will be most abundant in the lower portion (A).

Discussion

In the model outlined above it was supposed that a small volume of magma, stratified with respect to composition and temperature, was formed in the upper mantle. The source rock from which this magma was derived is most likely to be garnet lherzolite, since this rock type is generally held to be the dominant component of the continental upper mantle and is present at Roberts Victor. The first liquid formed by equilibrium melting of olivine, orthopyroxene, clinopyroxene and garnet will crystallise these phases unless some of these phases bear reaction relations with the liquid. In order to crystallise garnet and clinopyroxene only, there must exist reaction relationships by which olivine and orthopyroxene are eliminated from the melt. The reaction of orthopyroxene with liquid has been demonstrated at 30 kbar by O'Hara and Yoder (1967). In the anhydrous system a reaction relation between olivine and liquid ceases at 26 kbar (Kushiro and Yoder, 1974), but persists to greater pressure in the presence of water. Thus water and other volatiles probably were present during melting if the eclogitic liquid was derived by partial melting of garnet lherzolite at 40 to 50 kbar. In the model proposed it is implied that liquid crystallising garnet and clinopyroxene only can form by melting of garnet lherzolite. This implication is supported by nodule HRV310 from Roberts Victor in which a layer of eclogite occurs on the margin of an altered stratified garnet lherzolite (Hatton, 1978). Similarly eclogite pods have been recorded within mantle-derived garnet lherzolite (Carswell, 1968).

The presence of CO_2 in the volatile phase is an essential feature of the model for the formation of diamond outlined here. Since CO_2 is not concentrated in the solid phases of garnet lherzolite or eclogite, high concentrations of CO_2 are likely to be encountered during low-temperature, volatile-induced partial melting or in the late stages of fractional crystallisation. In both cases the silicates which crystallise will be evolved compositions with Na_2O-rich clinopyroxenes and CaO-rich garnets, hence the observed tendency for diamond and graphite to be most commonly encountered in the evolved calcium-rich eclogites, kyanite eclogites, and corundum eclogites. Another reason why diamonds tend to be most common in eclogites of evolved compositions may be that these rocks have a lower-temperature origin than more primitive eclogites, at whose higher temperature of origin the CCO buffer may have defined a lower oxygen fugacity than the oxygen fugacity of the silicate system so that carbon dioxide was stable relative to diamond or graphite.

Conclusion

The presence of graphite in 11% of the currently known diamondiferous eclogites suggests that diamondiferous eclogites may form in close proximity to the diamond-graphite equilibrium curve at pressures greater than 40 kbar but probably less than 50 kbar, in the temperature range 1000 to 1150°C. We believe that diamond-graphite eclogites form from small, rapidly crystallised volumes of magma where gravitational separation of diamond and graphite is ineffective. Small volumes of eclogitic magma may be common products of volatile-induced partial melting of garnet lherzolite.

In the proposed pressure-temperature range the CCO buffer defines oxygen fugacities close to those of three buffers, FMQ, NNO, and WM, which may be reasonable approximations of the oxygen fugacity of the mantle (Rosenhauer et al., 1977). On cooling, the CCO buffer crosses each of these buffers in turn, so that carbon oxides are reduced to diamond or graphite (Figure 5). If the oxygen fugacity of the eclogite system can be approximated by the NNO buffer, carbon oxides will be reduced to diamond or graphite in the temperature-pressure range suggested by the silicates in the nodule.

The overlap between the temperature-pressure range suggested by the silicates in the diamondiferous eclogite and the temperature-pressure range at which carbon oxides are reduced to diamond indicates that reduction of volatile phases in the mantle is a reasonable model for diamond formation. An upper limit to diamond stability is reached at temperatures on the geotherm when the oxygen fugacity of the mantle is higher than the CCO buffer (Rosenhauer et al., 1977) while a lower limit is reached at the diamond-graphite equilibrium curve, so that diamond may be stable only in a restricted zone of the mantle.

We have proposed that diamond and graphite formed together with eclogite. The PreCambrian age of the Roberts Victor eclogites (Kramers, 1977) and the Cretaceous age of the Roberts Victor kimberlite (Davis, 1977) implies that eclogite and kimberlite are unrelated, and suggests that diamonds of the eclogite paragenesis are not directly related to kimberlite formation.

Acknowledgements. We would like to thank R. Fudali, S.R. Shee, J.R. Smyth and anonymous reviewers for critical comments and suggestions. Financial support during the study was provided by the C.S.I.R. and the Anglo American Corporation. The organising committee of the Second International Kimberlite Conference helped defray travel expenses.

References

Beck, R., Untersuchungen uber einige Sudafrikanische Diamantlagerstatten, Zeit, Deut. Geol. Gesell., 59, 275-307, 1907.

Berman, R., Thermal properties, in Physical Properties of Diamond, edited by R. Berman, pp. 371-393, Oxford University Press, 1965.

Bishop, F.C., J.V. Smith, and J.V. Dawson, Na, P and Ti and coordination of Si in garnet from peridotite and eclogite xenoliths, Nature, 260, 696-697.

Bishop, F.C., J.V. Smith, and J.B. Dawson, Na, K, P and Ti in garnet, pyroxene and olivine from peridotite and eclogite xenoliths from African kimberlites, in preparation.

Bobrievich, A.P., G.I. Smirnov, and V.S. Sobolev, A xenolith of diamond-bearing eclogite, Dokl. Acad. Sci. USSR, Earth Sci. Sect., 126, 581-583, 1960.

Brett, R., J.S. Huebner, and M. Sato, Measured oxygen fugacities of the Angra dos Reis achondrite as a function of temperature, Earth Planet. Sci. Letters, 35, 363-368, 1977.

Bundy, F.P., H.P. Bovenkerk, H.M. Strong, and R.H. Wentorf, Diamond-graphite equilibrium line from growth and graphitisation of diamond, J. Chem. Phys., 35, 383-391, 1961.

Carswell, D.A., Picritic magma-residual dunite relationships in garnet peridotite at Kalskaret near Tajford, South Norway, Contr. Mineral. Petrol., 19, 97-124, 1968.

Davis, G.L., The age and uranium contents of zircons from kimberlites and associated rocks, Extended Abstracts, Second International Kimberlite Conference, Sante Fe, New Mexico, 1977.

Dawson, J.B., and J.V. Smith, Occurrence of diamond in a mica-garnet lherzolite xenolith from kimberlite, Nature, 254, 580-581, 1975.

Eggler, D.H., and M.E. McCullum, Diamond-bearing peridotite in a Wyoming kimberlite pipe, Annual meeting Geol. Soc. Amer. Abstracts, 7, 1065, 1975.

Evans, T., Diamonds, Contemp. Physics, 17, 45-70, 1976.

Gurney, J.J., J.C. Siebert, and G.G. White-Cooper, A diamondiferous eclogite from the Roberts Victor mine, Geol. Soc. S. Africa, Spec. Publ. No. 2, 351-357, 1969.

Harris, J.W., and J.J. Gurney, Inclusions in diamond, in Physical Properties of the Diamond, edited by J.G. Field, in press, 1978.

Hatton, C.J., Geochemistry and origin of eclogite xenoliths from the Roberts Victor mine, unpublished Ph.D. thesis, University of Cape Town, 1978.

Huebner, J.S., Buffering techniques for hydrostatic systems at elevated pressures, in Research Techniques for High Pressure and High Temperature, edited by G.C. Ulmer, pp. 123-177, Springer-Verlag, 1971.

Kennedy, C.S., and G.C. Kennedy, The equilibrium boundary between graphite and diamond, J. Geophys. Res., 81, 2467-2470, 1976.

Kramers, J.D., Lead and strontium isotopes in inclusions in diamonds and in mantle-derived xenoliths from Southern Africa, Extended Abstracts, Second International Kimberlite Conference, Sante Fe, New Mexico, 1977.

Kushiro, I., and K. Aoki, Origin of some eclogite inclusions in kimberlite, Amer. Mineral., 53, 1347-1367, 1968.

Kushiro, I., and H.S. Yoder Jr., Formation of eclogite from garnet lherzolite: liquidus relations in a portion of the system $MgSiO_3$-$CaSiO_3$ -Al_2O_3 at high pressures, Carnegie Inst. Wash. Yearbk. 73, 266-269, 1974.

Kushiro, I., Y. Syono, and S. Akimoto, Melting of peridotite at high pressures and high water pressures, J. Geophys. Res., 73, 6023-6029, 1968.

Meyer, H.O.A., and F.R. Boyd Jr, Composition and origin of crystalline inclusions in natural diamonds, Geochim. Cosmochim. Acta, 36, 1255-1273, 1972.

Meyer, H.O.A., and D.P. Svisero, Mineral inclusions in Brazilian diamonds, Phys. Chem. Earth, 9, 785-796, 1975.

Meyer, H.O.A., and H. Tsai, The nature and significance of mineral inclusions in natural diamond: a review, Minerals Sci. Engng., 8, 242-261, 1976.

Mysen, B.O., and I. Kushiro, Compositional variation of coexisting phases with degree of melting of peridotite in the upper mantle, Amer. Mineral., 62, 843-856, 1977.

O'Hara, M.J., M.J. Saunders, and E.L.P. Mercy, Garnet peridotite, primary ultrabasic magma and eclogite; interpretation of upper mantle processes in kimberlite, Phys. Chem. Earth, 9, 571-604, 1975.

Pokhilenko, N.P., N.V. Sobolev, and Yu.G. Lavrent'ev, Xenoliths of diamondiferous ultramafic rocks from Yakutian kimberlites, Extended Abstracts, Second International

Kimberlite Conference, Santa Fe, New Mexico, 1977.

Prinz, M., D.V. Manson, P.F. Hlava, and K.Keil, Inclusions in diamonds: Garnet lherzolite and eclogite assemblages, Phys. Chem. Earth, 9, 797-816, 1975.

Raheim, A., and D.H. Green, Experimental determination of the temperature and pressure dependence of the Fe-Mg partition coefficient for coexisting garnet and clinopyroxene, Contrib.Mineral. Petrol., 48, 179-203, 1974.

Reid, A.M., R.W. Brown, J.B. Dawson, G.G. Whitfield and J.C. Siebert, Garnet and pyroxene compositions in some diamondiferous eclogites, Contrib. Mineral. Petrol., 58, 203-220, 1976.

Rickwood, P.C., J.J. Gurney, and D.R. White-Cooper, The nature and occurrence of eclogitic xenoliths in the kimberlites of South Africa, Geol. Soc. S.Africa Spec. Publ. No 2, 371-393, 1969.

Robinson, D.N., The characteristics of diamond and graphite in the Roberts Victor eclogite xenolith HRV247, unpublished report, Anglo American Research Laboratories, Project No. R. 19, Report No. 99, 1976.

Robinson, D.N., Diamond and graphite in eclogite xenoliths from kimberlite, Extended Abstracts, Second International Kimberlite Conference, Sante Fe, New Mexico, 1977.

Rosenhauer, M., E. Woermann, B. Knecht, and G.C. Ulmer, The stability of graphite and diamond as a function of the oxygen fugacity in the mantle, Extended Abstracts, Second International Kimberlite Conference, Sante Fe, New Mexico, 1977.

Shee, S.R., and J.J. Gurney, The mineralogy of xenoliths from Orapa, Botswana, Extended Abstracts, Second International Kimberlite Conference, Sante Fe, New Mexico, 1977.

Sobolev, N.V., Deep-seated inclusions in kimberlites and the problem of the composition of the upper mantle, Amer. Geophy. Union Washington, D.C., 1977.

Sobolev, N.V., and I.K. Kuznetsova, The mineralogy of diamond-bearing eclogites, Dokl. Acad. Sci. USSR, Earth Sci. Sect., 167, 168-170, 1966.

Sobolev, V.S., B.S. Nai, N.V. Sobolev, Yu. G. Lavrentiev, and L.N. Pospelova, Xenoliths of diamond-bearing pyrope sepentinite from the Aykhal pipe, Yakutia, Dokl. Acad. Sci. USSR, Earth Sci. Sect., 188, 168-170, 1969.

Switzer, G., and W.G. Melson, Partially melted kyanite eclogite from the Roberts Victor mine, South Africa, Smithsonian Contr. Earth Sci., 1, 1-9, 1969.

Wager, L.R., and G.M. Brown, Layered Igneous Rocks, Oliver and Boyd, Edinburgh, 1967.

Wagner, P.A., The Diamond Fields of Southern C. Struik, Cape Town, 1914.

Whitfield, G.G., A petrological and mineralogical study of peridotite and eclogite xenoliths from certain kimberlite pipes, Unpublished MSc. Thesis, Rhodes University, Grahamstown, 1971.

Williams, A.F., The Genesis of the Diamond, 2 vols., Benn, London, 1932.

Woermann, E., B. Knecht, M. Rosenhauer, and G.C. Ulmer, The stability of graphite in the system C-O, Extended Abstracts, Second International Kimberlite Conference, Santa Fe, New Mexico, 1977.

THE MINERALOGY OF XENOLITHS FROM ORAPA, BOTSWANA

Simon R. Shee

De Beers Consolidated Mines Limited, Box 616, 8300 Kimberley, S. Africa

John J. Gurney

Department of Geochemistry, University of Cape Town, 7700 Rondebosch, S. Africa

Abstract. Xenoliths of eclogite and megacrysts from the upper mantle have been found at Orapa, Botswana. Peridotite has been found in kimberlites in the same cluster, but not at Orapa.

Minor phases in the eclogitic rocks include orthopyroxene, chromite, amphibole, phlogopite, rutile, kyanite, corundum, graphite and diamond.

The eclogites are divided into two main groups on textural and chemical grounds. A wide range in mineral chemistry which is very similar to that reported for Roberts Victor has been observed. The coefficient for Fe/Mg distribution between coexisting garnet-cpx pairs shows a wide range which has been interpreted as being partly due to differences in equilibration temperature and partly due to the effect of increased calcium in garnet. Garnet megacrysts are high titanium, low chrome pyropes similar to those from Monastery Mine. Cpx megacrysts fall into two groups: high chrome, calcic and low chrome, sub-calcic as in the Colorado/Wyoming kimberlites. Ilmenites are chrome rich picroilmenites.

Garnet xenocrysts appear to be derived from disaggregated eclogite and peridotite xenoliths, and from megacrysts.

The inclusions in the Orapa diamonds are predominantly eclogitic, with important sulphide and minor peridotitic components.

Introduction

The Orapa (2125 A/K1) kimberlite pipe is the largest in a cluster of 23 pipes found in north eastern Botswana in the last decade. It is the second biggest in the world (1560 x 950 metres). The pipe is a major diamond producer (2,4 million carats in 1977). Proven reserves to a depth of only 37 metres exceed 85 million carats.

The near surface filling of the pipe is sedimentary (epiclastic) kimberlite indicating that the pipe has been only slightly eroded since emplacement (Hawthorne, 1975). Serpentinised primary kimberlite is intersected in borehole cores below 90 m. The pipe is intrusive into basaltic lavas of Stormberg age. A zircon age of 93,1 m.y. (Davis, 1977) is close to ages obtained by the same method for kimberlites from South Africa and Northern Lesotho.

Small samples (1½ - 5 cm in longest dimension) of mantle material are recovered during the mining process. We have found abundant eclogite xenoliths and garnet, clinopyroxene and ilmenite megacrysts. No olivine-bearing rocks have been found in the mine concentrates nor in many hundreds of metres of borehole core. Peridotite xenoliths have been found at the nearby Bk9 and Dk1 kimberlite pipes.

It has been established by visual observation (J.W. Harris; pers. comm.) and electron microprobe analysis (R.S. Rickard; pers. comm.) that the inclusions in diamonds from Orapa are predominantly eclogitic with common sulphides and rare peridotite minerals. This contrasts with Roberts Victor (O.F.S.) and Bellsbank (C.P.) kimberlites where, although the xenoliths are also eclogitic, the inclusions have been found to be peridotitic by the above sources (Harris and Rickard).

Mineral analyses were carried out by electron microprobe using close standards where possible and the correction procedure used was that of Bence and Albee (1968) as modified by Albee and Ray (1970).

Eclogite Xenoliths

Eclogites are the commonest form of mantle material found at Orapa. All of the samples recovered from the mining processes are small (1½ - 5 cm in longest dimension). This means that samples showing layering and gross inhomogeneities similar to those exhibited by the Roberts Victor (O.F.S.) eclogites and described by Hatton and Gurney (1977) and Hatton (1978) are not known.

Garnet and clinopyroxene are always the principal minerals in the eclogites but variants with minor orthopyroxene, chromite, amphibole, phlogopite, rutile, kyanite, corundum, diamond and graphite have been found. The eclogite suite have been studied by petrographic and electron microprobe techniques and representative analyses are presented in Table 1.

Bimineralic eclogites

Approximately 84% of all the eclogites found at Orapa contain the simple bimineralic assemblage, garnet and clinopyroxene. The eclogites can be divided into two major groups on textural and chemical grounds using the scheme proposed by MacGregor and Carter (1970) for the Roberts Victor eclogites.

Group I eclogites are characterized by altered clinopyroxene, dusky garnet and rounded grains. Corundum, kyanite, rutile, diamond and graphite bearing varieties are included. Na_2O in the garnet and K_2O in the clinopyroxene are also used to classify this group but these elements are not always detected by electron probe.

Group II eclogites are recognised by the fresh appearance of minerals, straight edged grains and the occasional appearance of tiny oriented inclusions of what is probably rutile in both garnet and clinopyroxene. Some of the Group II eclogites are characterized by a fabric with a roughly planar arrangement of elongated garnet and clinopyroxene grains. Sodium in the garnet and potassium in clinopyroxene are not detectable by electron probe in Group II eclogites. Chromite, orthopyroxene, graphite, amphibole, corundum, kyanite and rutile -bearing varieties are found.

The Orapa eclogites display a wide range in mineral compositions. Figure 1 is a Ca: Mg: Fe projection for the eclogite garnets. The garnets show two main trends with coincident magnesian garnet compositons. In the one case there is major enrichment in iron and minor enrichment in calcium and in the second the reverse. Eclogites from Groups I and II display both trends.

Figure 2 is a plot of diopside versus jadeite content of the clinopyroxenes. Most of the clinopyroxenes are composed of 80%-90% diopside plus jadeite with major variation in composition occurring between these end members. The few samples that deviate from the 80%-90% diopside-jadeite join contain either more enstatite, ureyite, pseudojadeite or acmite.

In general the most magnesian garnets co-exist with the most diopside-rich clinopyroxenes whilst the most grossular-rich garnets co-exist with the most jadeitic pyroxenes. All intermediate compositions occur.

Garnet Websterites (orthopyroxene eclogites)

Orthopyroxene has been observed in two samples. Strictly speaking these rocks should not be termed eclogites but should rather be called garnet websterites. However, they are clearly related to other rocks in the eclogite suite. The orthopyroxene is not present in large amounts, it makes up $<$ 1 volume percent of the rock. It is pale-yellow and presents a fresh appearance. The grains are anhedral and small (1-1½ mm in diameter). The orthopyroxene coexists with pale-purple, coarse grained ($>$ 2 mm), anhedral garnet and bright-green, coarse ($>$ 2 mm) tabular clinopyroxene. On textural and chemical grounds the garnet websterites are Group II eclogites.

Mineral analyses of garnet websterite AK1/132 are presented in Table 1. The garnets are magnesian and contain more chrome (0,56 wt. % Cr_2O_3) than other eclogite types except for the high chrome eclogites. The garnets plot in the same field as the chrome-rich eclogite garnets in Figure 1. The clinopyroxenes are diopsides with 0,32 wt. % Cr_2O_3. The orthopyroxenes contain 0,64 wt. % Al_2O_3 and are similar to enstatites from mantle garnet lherzolites.

Chromite Eclogites and Chrome-rich Eclogites

Five xenoliths similar to those described as high-chrome "eclogites" by Stephens and Dawson (1977) have been found at Orapa. These rocks are characterized by high chrome contents in the garnets and clinopyroxenes (both $>$ 1,00 wt. % Cr_2O_3) which are higher than those commonly found in eclogites. Chrome spinel occurs in some of these xenoliths.

The modal proportions of the minerals in these eclogites vary considerably. The garnets vary in colour from mauve to pale-purple and are coarse grained ($>$ 3 mm). Clinopyroxenes are bright-green and very fresh. In two of the samples the clinopyroxenes occur as a narrow band ($<$ 1 x 6 mm) in garnet whilst in two other specimens the clinopyroxene grains are coarse ($>$ 2 mm) with anhedral to rounded shapes. The clinopyroxene in the fifth specimen occurs as small (1 x ½ mm) inclusions in the garnet. Chromite occurs as small inclusions ($<$ ¼ mm) in garnet or as slightly coarser grains ($<$ ½ mm) at the boundaries between garnet and clinopyroxene. Pale-green amphibole (subcalcic edenite) occurs in one of the chromite eclogites. The amphibole appears to be a replacement mineral. These eclogites are all Group II.

Mineral analyses of AK1/3, a chromite-bearing eclogite are presented in Table 1. The garnets of the five xenoliths are magnesian and plot in field 1 on Figure 1. Chrome contents of the garnets range from 1,95 to 5,72 wt. % Cr_2O_3. The clinopyroxenes are ureyitic diopsides (terminology of Stephens and Dawson, 1977) with significant chrome contents (1,26 - 3,49 wt. % Cr_2O_3) and have low sodium compared with other Orapa clinopyroxenes. The chromites have 52,1 - 57,0 wt. % Cr_2O_3 and Cr/Cr + Al ratios from 0.656 - 0,813.

Amphibole Eclogites

Amphiboles have been found in a number of eclogites where the amphibole contents range from 2 to 45 volume percent. The textures exhibited by the garnet and clinopyroxene grains in the

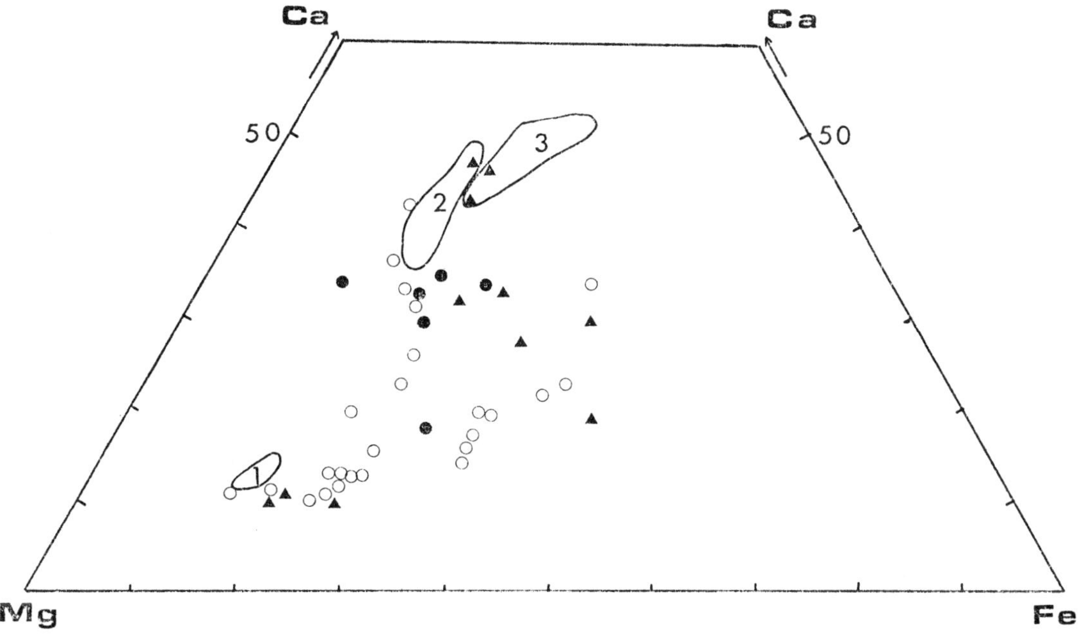

Figure 1. Ca, Mg, Fe plot for eclogite garnets from Orapa. Solid circles are diamond-bearing eclogites, solid triangles are graphite-bearing eclogites, open circles are other Orapa eclogites. Garnets from chrome-rich eclogites and garnet websterites plot in field 1, garnets from corundum eclogites and kyanite eclogites plot in fields 2 and 3 respectively.

amphibole eclogites are similar to those in other Orapa Group II eclogites.

The amphiboles which range in colour from brown to light-green, poikilitically enclose both garnet and clinopyroxene grains and in some instances clinopyroxene inclusions in amphibole are seen to be in optical continuity. Garnet and clinopyroxene grains in contact with amphibole sometimes display corroded borders. These features indicate that the amphibole could be a replacement mineral.

Analyses of AK1/20, an amphibole-bearing eclogite are listed in Table 1. The garnets in amphibole-bearing eclogites show a range in composition i.e. four analysed garnets fall on the iron enrichment trend, one lies on the calcium enrichment trend and one occurs in a chromite eclogite. The amphiboles are brown edenites e.g. Ak1/20, brown pargasitic hornblende and pale-green, sub-calcic edenite (terminology of Leake, 1968).

Kyanite Eclogites and Grospydites

Kyanite eclogites are found at many kimberlite localities, the best known being the Roberts Victor pipe in South Africa (Rickwood et al., 1969) and the Zagadochnaya pipe in Yakutia, U.S.S.R. (Sobolev et al., 1968). Kyanite eclogites have also been found at Orapa.

The specimens recovered were all small (< 5 cm long). The garnets are dark-brown ([illegible] 2 mm diameter), anhedral, rounded grains with smoothly curving boundaries. The garnets are cracked and fine grained brown alteration products and tiny, secondary, dusky-green, aluminous spinels occur

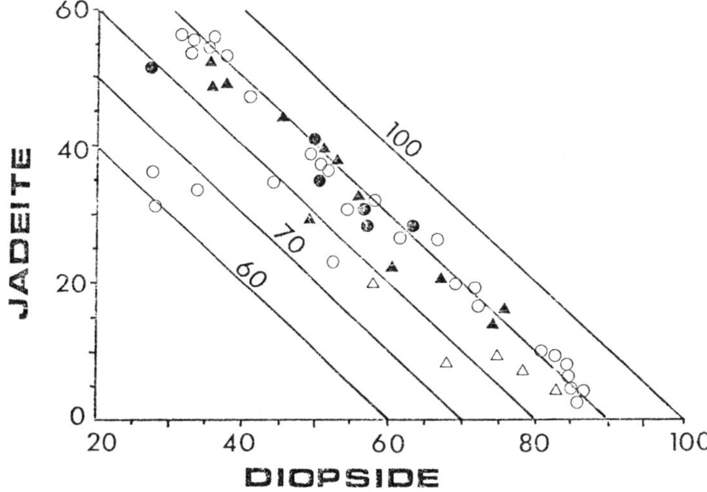

Figure 2. Plot of percent jadeite content versus percent diopside content for eclogite pyroxenes from Orapa. The lines in the diagram represent jadeite plus diopside contents. Solid circles are diamond-bearing eclogites, solid triangles are graphite-bearing eclogites, open circles are clinopyroxenes from eclogites without a carbon polymorph and open triangles are chrome-rich eclogites.

TABLE 1. Representative Mineral Analyses for Orapa Eclogites including Diamond and Graphite bearing rocks.

	Group 1		Group 1		Group 1		Group 1		Group 1		Group 1		Group 1	
	Graph. Ecl. AK1/26		Dia. Ecl. AK1/9		Dia. Ecl. Ak1/10		Dia.+Graph.Ecl. AK1/25		Dia.+Rut.Ecl. AK1/111		Dia. Ecl. AK1/204		Dia. Ecl. No. 18 *	
	Gt	Cpx	Gt	Cpx	Gt	Cpx	Gt	Cpx	Gt3	Cpx	Gt2	Cpx1	Gt	Cpx
SiO_2	39.3	54.6	40.4	55.3	41.6	54.8	40.3	54.7	40.5	55.2	39.2	55.0	40.2	56.0
TiO_2	0.25	0.36	0.45	0.48	0.19	0.09	0.51	0.39	0.61	0.60	0.55	0.62	0.32	0.32
Al_2O_3	22.6	8.75	22.4	10.6	23.3	9.90	22.9	9.27	22.9	7.80	22.1	10.7	22.1	18.1
Cr_2O_3	0.05	0.06	0.05	0.04	0.12	0.08	0.12	0.09	0.07	0.09	0.05	ND	0.07	0.08
FeO	21.2	5.87	11.6	2.53	6.97	2.15	10.3	3.52	14.7	4.96	17.6	5.41	11.0	1.86
MnO	0.37	0.05	0.17	ND	0.15	ND	0.20	ND	0.31	0.05	0.31	0.04	0.23	ND
MgO	9.66	10.2	13.0	10.8	14.7	12.5	12.5	11.8	14.5	12.0	8.01	8.90	11.6	6.65
CaO	7.12	15.1	11.1	15.2	12.7	16.6	13.0	15.5	6.57	14.8	12.2	13.2	12.9	10.5
Na_2O	0.07	5.28	0.16	5.51	0.12	4.18	0.08	4.81	0.16	5.21	0.20	6.49	0.51	7.75
K_2O	ND	0.07	ND	0.12	ND	0.11	ND	0.12		0.04		0.08		
Total	100.7	100.3	99.9	100.6	99.9	100.5	100.0	100.3	100.3	100.7	100.2	100.5	98.6	101.3
							Ionic Proportions							
Si	2.963	1.964	2.967	1.951	2.997	1.937	2.960	1.948	2.966	1.969	2.967	1.962	3.002	1.922
Ti	0.014	0.010	0.025	0.013	0.010	0.002	0.028	0.010	0.034	0.016	0.031	0.017	0.018	0.008
Al	2.009	0.371	1.986	0.441	1.982	0.413	1.983	0.389	1.977	0.328	1.972	0.450	1.946	0.732
Cr	0.003	0.002	0.003	0.001	0.007	0.002	0.007	0.003	0.004	0.003	0.003		0.004	0.002
Fe	1.339	1.177	1.713	0.075	0.420	0.064	0.632	0.105	0.900	0.148	1.114	0.161	0.687	0.053
Mn	0.024	0.002	0.011		0.009		0.012		0.019	0.002	0.020	0.001	0.015	
Mg	1.085	0.547	1.423	0.568	1.581	0.660	1.367	0.627	1.583	0.638	0.903	0.473	1.291	0.340
Ca	0.575	0.582	0.874	0.575	0.984	0.629	1.022	0.593	0.516	0.566	0.989	0.505	1.032	0.386
Na	0.010	0.368	0.023	0.377	0.017	0.286	0.011	0.332	0.023	0.360	0.029	0.449	0.022	0.516
K		0.003		0.005		0.005		0.005		0.002		0.004		
Total	8.023	8.025	8.025	4.006	8.007	3.999	8.024	4.014	8.021	4.031	8.029	4.022	8.016	3.960
Oxygen	12	6	12	6	12	6	12	6	12	6	12	6	12	6
K_D	3.81		3.81		2.76		2.76		2.45		3.61		3.39	

* - A.M. Reid, *et al.*, (1976)

TABLE 1. (contd.).

	Group II			Group II			Group II			Group I		Group I	
	Garnet Websterite AK1/132			Chromite + Eclogite AK1/3			Amphibole + Eclogite AK1/20			Kya.+ Rut. Ecl. AK1/131		Corundum Ecl. AK1/127	
	Gt	Cpx	Opx	Gt	Cpx	Chr	Gt	Cpx	Amp	Gt	Cpx	Gt	Cpx
SiO_2	41.6	54.9	57.8	40.9	54.1		40.4	54.9	47.1	40.2	56.0	40.0	54.9
TiO_2	0.08	0.15	0.09	0.20	0.21	1.23	0.10	0.22	0.83	0.27	0.36	0.32	0.41
Al_2O_3	23.6	1.85	0.85	19.8	3.28	8.80	23.2	4.21	11.8	23.1	16.5	23.2	18.0
Cr_2O_3	0.54	0.32	0.07	5.72	2.95	57.1	0.05	0.08	0.08	ND	ND	0.08	0.04
FeO	8.32	1.59	4.98	8.20	2.32	21.1	16.9	3.94	6.38	11.6	1.95	9.06	1.50
MnO	0.31	0.05	0.08	0.46	0.09	0.98	0.42	0.04	ND	0.17	ND	0.20	ND
MgO	19.9	17.6	35.6	19.2	15.1	11.5	14.2	14.5	18.2	8.59	8.79	8.94	6.74
CaO	4.60	22.9	0.27	5.62	18.2		5.34	20.2	9.77	16.4	12.4	17.9	11.6
Na_2O	ND	1.02	ND	ND	3.11		ND	2.73	4.19	ND	4.46	0.12	8.13
K_2O	ND	ND	ND	ND	ND		ND	ND	1.17	ND	0.37	ND	ND
Total	99.0	100.4	99.8	100.2	99.4	100.7	100.6	100.8	99.5	100.3	100.8	99.8	101.5
					Ionic Proportions								
Si	2.980	1.975	1.983	2.959	1.969		2.967	1.974	6.886	2.980	1.928	2.964	1.892
Ti	0.004	0.004	0.002	0.011	0.006	0.031	0.006	0.006	0.091	0.015	0.009	0.018	0.011
Al	1.993	0.078	0.034	1.690	0.141	0.344	2.008	0.178	2.034	2.019	0.670	2.026	0.731
Cr	0.031	0.009	0.002	0.327	0.085	1.495	0.003	0.002	0.009			0.005	0.001
Fe	0.498	0.048	0.143	0.496	0.071	0.584	1.038	0.118	0.780	0.719	0.056	0.561	0.043
Mn	0.019	0.002	0.002	0.028	0.003	0.027	0.026	0.001		0.011		0.013	
Mg	2.125	0.943	1.820	2.070	0.819	0.569	1.554	0.777	3.966	0.949	0.451	0.987	0.346
Ca	0.353	0.883	0.010	0.436	0.710		0.420	0.778	1.531	1.303	0.457	1.421	0.428
Na		0.071			0.219			0.190	1.188		0.298	0.017	0.555
K									0.218		0.016		
Total	8.004	4.013	3.997	8.018	4.022	3.050	8.023	4.026	16.704	7.996	3.885	8.012	4.009
Oxygen	12	6	6	12	6	4	12	6	24	12	6	12	6
K_D		4.63			2.78			4.17		6.09		4.55	

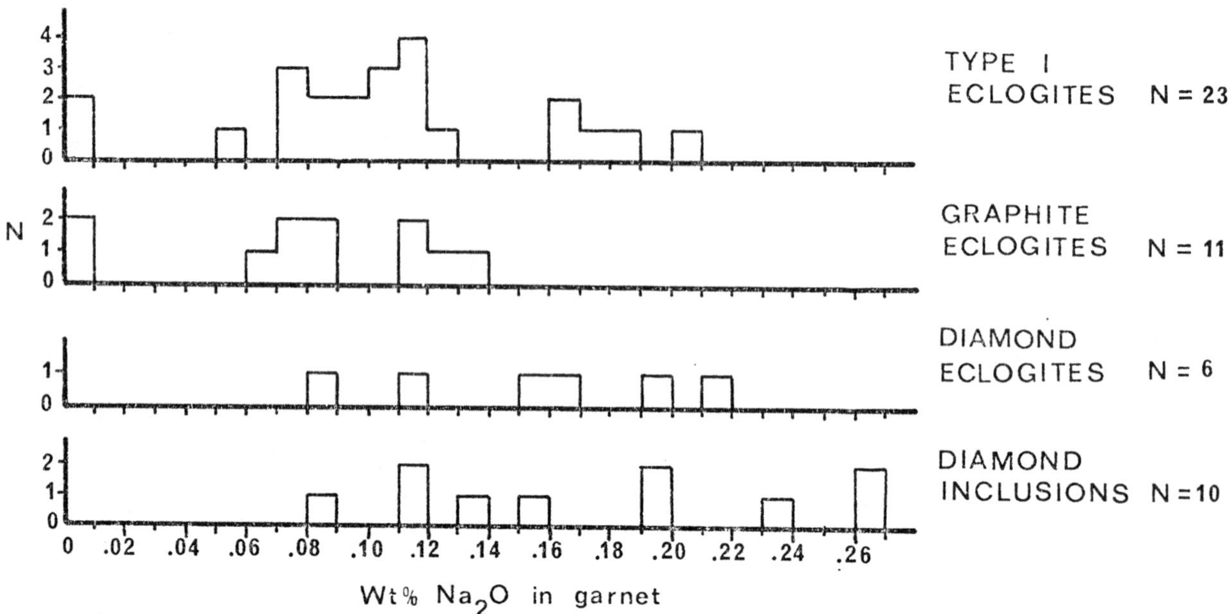

Figure 3. Histogram of sodium contents in garnets from Type 1 eclogites, graphite eclogites, diamond eclogites and diamond inclusions from Orapa. The diamond inclusion data is from R.S. Rickard (unpublished analyses).

in the cracks and around the rims of the grains. The clinopyroxenes are coarse-grained, anhedral crystals which are usually partially altered along the rims to a white, opaque, irresolvable product. Some of the pyroxenes have been completely altered. The pale-blue kyanite occurs as elliptical grains ranging from 2 - 6 mm in length. Kyanite may be enclosed in both garnet and clinopyroxene grains. Accessory rutile, graphite and corundum occur. Most of the kyanite eclogites recovered were Group I but Group II kyanite eclogites do occur at Orapa.

Mineral analyses for a typical Orapa, Group I kyanite eclogite (AK1/131) can be found in Table 1. The garnets are grossular-rich (see Figure 1) and in some samples the grossular content of the garnets exceeds 50%. These rocks can be termed grospydites after Sobolev et al. (1968). The garnets in all the Group I kyanite eclogites are characterized by significant amounts of sodium (0,07 - 0,18 wt. % Na_2O) and low chrome ($<$ 0,05 wt. % Cr_2O_3). The xenolith with the most grossular-rich garnets found at Orapa (51,8% grossular and 18,7 wt. % CaO) contains a garnet inclusion of a less calcic composition. The inclusion is zoned with the core more magnesian and less calcic than the rim (core : 8,39 wt.% CaO and 22,3% grossular; rim : 9,58 wt. % CaO and 25,5% grossular). A similar feature has been recorded in inhomogeneous kyanite eclogite from the Roberts Victor Mine (Hatton and Gurney, 1977).

The clinopyroxenes are diopside jadeites (23,1% - 53% jadeite) using the terminology of Stephens and Dawson (1977). They are characterized by variable potassium contents (N.D. - 0,46 wt. % K_2O). The latter is the highest found at Orapa except for some diamond inclusions (0,25 - 0,55 wt. % K_2O). (Rickard pers. comm. 1978). The kyanite grains have low chrome contents ($<$ 0.06 wt. % Cr_2O_3) unlike the kyanites described by Sobolev et al., (1968) which have not yet been found in southern Africa.

Corundum Eclogites

Corundum eclogites have been reported at several localities e.g. Bellsbank/Bobbejaan, Frank Smith, Newlands, Crown, Jagersfontein, Roberts Victor (Rickwood et al., 1969), Zagadochnaya in Yakutia (Sobolev et al., 1968) and Colorado/Wyoming (Eggler and McCallum, 1974) but few mineral analyses are available in the literature. Corundum eclogites have also been found at Orapa.

In hand specimen the garnets and clinopyroxenes of the corundum-bearing nodules are similar to those of the kyanite eclogites. The corundum is colourless to pale-purple and occurs as small (½ mm - 3 mm), anhedral grains with irregular shapes in both garnet and clinopyroxene.

Garnet and clinopyroxene compositions of a corundum eclogite (AK1/127) are listed in Table 1. The garnets are similar to those of kyanite eclogites in that they are grossular-rich (Figure 1), have low chrome ($<$ 0,08 wt. % Cr_2O_3) and detectable sodium contents (0,07 - 0,13 wt. % Na_2O). Garnets from corundum eclogites are less iron-rich (9,07 - 9,90 wt % FeO) than those from kyanite eclogites (9,84 - 12,9 wt. % FeO), (see

Figure 1), a fact first pointed out by Sobolev et al., (1968).

The clinopyroxenes are diopsidic jadeites with ~55% jadeite content. The clinopyroxenes from corundum eclogites are less iron-rich (1,30 - 1,50 wt. % FeO) than those from kyanite eclogites (1,40 - 3,58 wt. % FeO). Potassium was not detected in any clinopyroxenes from the corundum eclogites.

Diamond and Graphite Eclogites

Diamond and graphite eclogites have been found at Orapa and it has been established by visual observation (J.W. Harris, pers. comm.) that all of the diamond inclusions are of eclogite affinity. Orapa is the first known kimberlite occurrence where the xenolith and diamond inclusion suites are both predominantly eclogitic. This contrasts with the Roberts Victor and Bobbejaan Mines where the xenoliths are nearly all eclogite but the diamond inclusions are peridotitic (Harris and Rickard, pers. comm.)

This study includes a total of six diamond-bearing eclogites and ten graphite-bearing eclogites. The mineral assemblages found are :

diamond, garnet, clinopyroxene (4)
diamond, rutile, garnet, clinopyroxene (1)
diamond, graphite, garnet, clinopyroxene (1)
graphite, garnet, clinopyroxene (5)
graphite, rutile, garnet, clinopyroxene (1)
graphite, kyanite, garnet, clinopyroxene (2)
graphite, corundum, garnet, clinopyroxene (1)

Diamond bearing kyanite or corundum eclogites have not been found at Orapa. This is surprising because the occurrence of diamond in kyanite eclo-

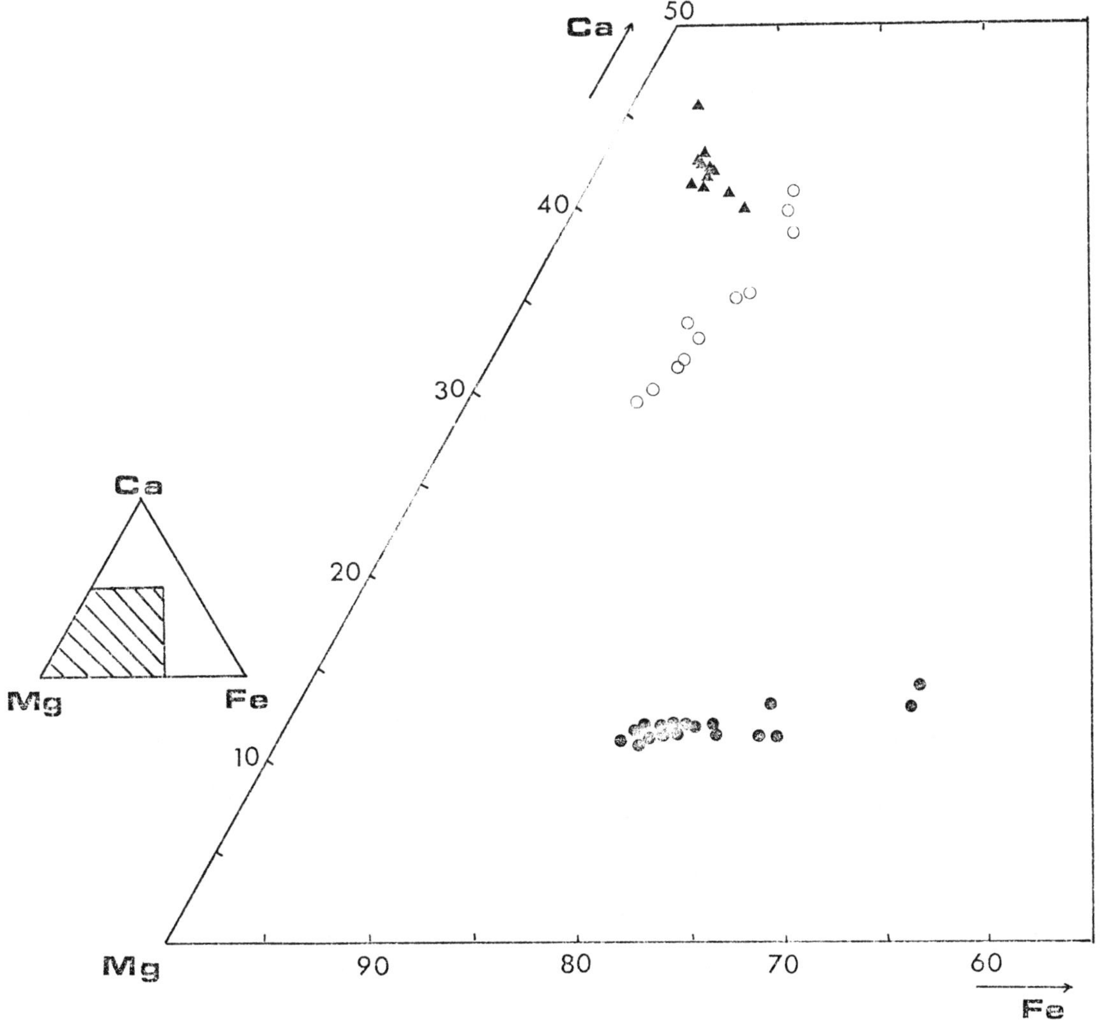

Figure 4. Part of a Ca, Mg, Fe diagram for garnet and clinopyroxene megacrysts from Orapa. Solid circles are garnet megacrysts, open circles are low-chrome, subcalcic clinopyroxene megacrysts and solid triangles are high-chrome, calcic clinopyroxene megacrysts.

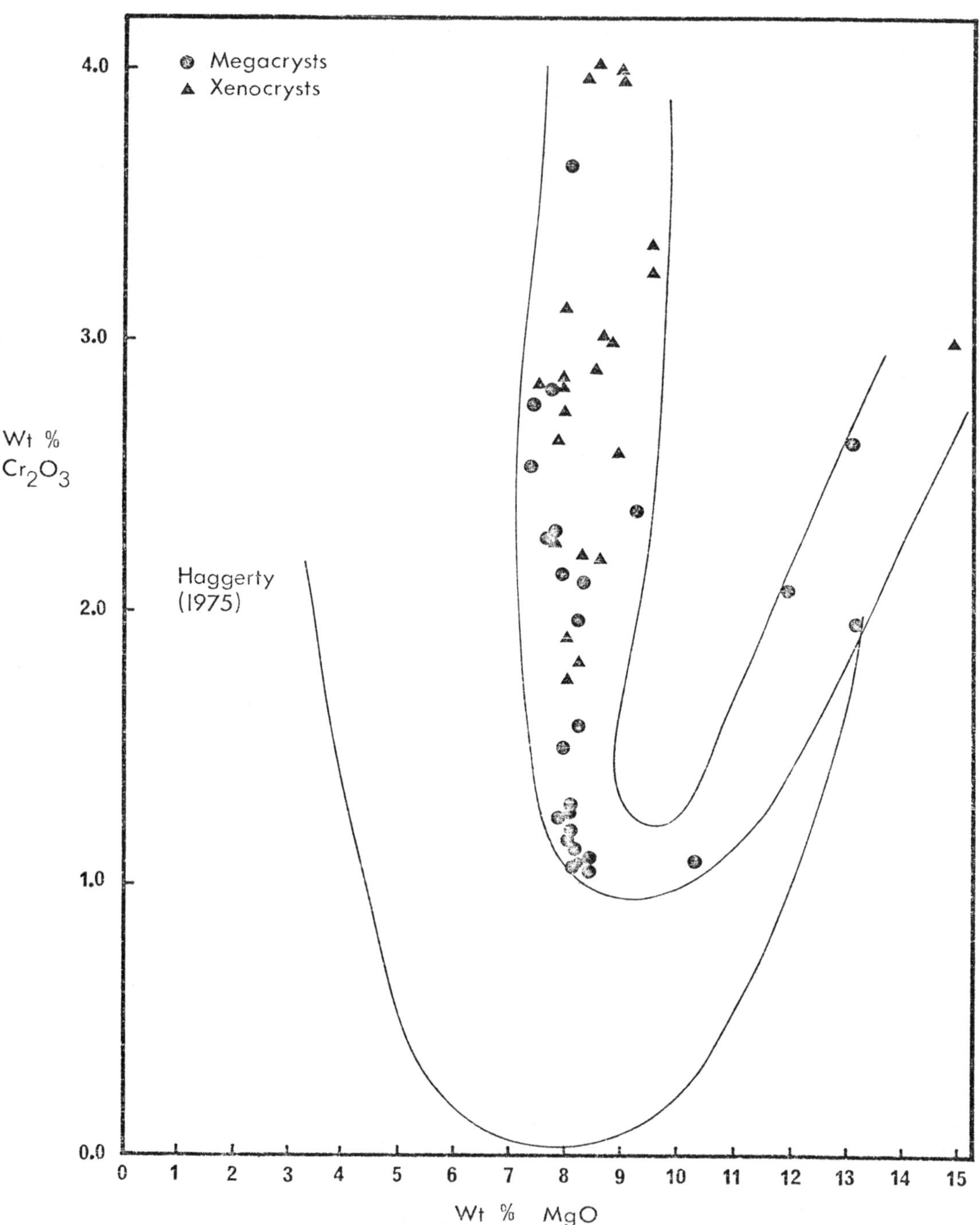

Figure 5. Plot of wt. % Cr_2O_3 versus wt. % MgO for ilmenites from Orapa. Solid circles are ilmenite megacrysts and solid triangles are ilmenite xenocrysts. Also represented is the parabolic curve for kimberlitic ilmenites (Haggerty, 1975).

gites from other localities is relatively common, and because kyanite and corundum eclogites including graphite-bearing varieties do occur at Orapa.

The modal percentages of garnet and clinopyroxene vary widely, from 20 to 80 volume percent garnet. When present, kyanite and corundum account for 5 to 10 volume % of the eclogite. All of the samples studied have coarse (> 3 mm), anhedral, rounded, orange garnets and pale green, coarse (> 2 mm), anhedral clinopyroxenes which are turbid in parts with a white alteration product.

The number of diamonds in each specimen ranges from 1 to 11. The diamonds are mostly octahedral-dodecahedral combined forms and range from 0,003 to 0,32 carats (Robinson, 1977). The number of graphite crystals exposed at the surface of the nodules varies from 2 to 28. These exposed graphite crystals are invariably damaged owing to the softness of the mineral. Dissolution of some of the graphite eclogites reveals discrete graphite crystals up to 2,5 mm in diameter, which are predominantly tabular, hexagonal prisms with slightly rounded edges. The graphite crystals in some specimens have embayed and irregular edges, which can be interpreted either to have grown in a confined space or to have experienced partial dissolution (Robinson, 1977).

Mineral analyses of all six diamond eclogites and one graphite eclogite are presented in Table 1 and the garnet compositions of all the diamond and graphite eclogites are illustrated in Figure 1. These garnets show a wide range in composition but are all characterized by low chrome contents ($<$ 0,12 wt. % Cr_2O_3). Diamond inclusion garnets (Rickard, unpublished data), diamond eclogite garnets and all but two of the graphite eclogite garnets contain small but significant amounts of sodium. The theoretical detection limit for sodium on the Cambridge Microscan microprobe at the University of Cape Town is 0,05 wt.% Na_2O and the sodium contents of these garnets range from 0,06 to 0,26 wt. % Na_2O. The sodium contents of garnets from diamond inclusions, diamond eclogites, graphite eclogites and Type 1 eclogites from Orapa are displayed in a histogram in Figure 3 which shows that these garnets cannot be distinguished from each other on the basis of sodium content.

Clinopyroxenes from diamond and graphite eclogites are similar to those in other Group I eclogites and range in composition from jadeitic diopsides to diopsidic jadeites. Small but significant amounts of potassium have been detected in the clinopyroxenes from four of the diamond eclogites (0,05 - 0,12 wt. % K_2O), from five of the graphite eclogites (0,05 - 0,17 wt. % K_2O) and from all of the diamond inclusions (0,25 - 0,55 wt. % K_2O) (Rickard, pers. comm.). Thus clinopyroxenes from diamond and graphite-bearing eclogites cannot be distinguished from each other on the basis of K_2O content but both contain less potassium than do clinopyroxene inclusions in diamond.

Other workers, Switzer and Melson (1969), Sobolev et al., (1972), Prinz et al., (1975), Reid et al., (1976) and R.S. Rickard (pers. comm. 1978) all report variable potassium contents in clinopyroxenes from diamond inclusions and diamond eclogites. Erlank (1970) suggested that the K_2O in clinopyroxenes could be due to the presence of submicroscopic intergrowths of amphibole in the omphacite. Reid et al., (1976) reported the presence of K-rich blebs of $\sim$ 20 microns diameter in clinopyroxenes from diamond eclogites which approximated the composition of K felspar. They suggested that K_2O is incorporated in the clinopyroxene lattice at high pressures and that as pressure decreases on emplacement some of the K_2O is lost. A possible explanation for the high K_2O contents in the diamond inclusion clinopyroxenes is that loss of the potassium was prevented by the armouring effect of the diamond.

Discussion

The parameters which define the physical conditions of formation of the diamond and graphite-bearing eclogites are pressure, temperature and oxygen fugacity. The presence of diamond or graphite in the eclogites places constraints upon the pressures and temperatures of formation of the xenoliths. The diamond-graphite stability curve has been well determined by Bundy et al., (1961). If it is assumed that the diamond or graphite crystallized in equilibrium with the garnet and clinopyroxene then the eclogite must have equilibrated in the diamond or graphite stability fields.

If it is assumed that diamond or graphite formed by reduction of carbon/oxide gases, knowledge of the oxygen fugacity at the time of formation of the carbon polymorph would further constrain the conditions of diamond or graphite genesis. The stability of graphite in the system C - O has been studied by Rosenhauer et al., (197 (1977) and by Woerman et al., (1977) who postulated that an oxygen fugacity slightly below that of the nickel-nickel oxide (NNO) buffer is applicable to mantle conditions. Extrapolations of the relevant equations to high temperatures and pressures cannot be made with confidence at present but oxygen fugacities at high pressures are currently being determined (Ulmer pers. comm. to C.J. Hatton, 1978) and hopefully this new data will clarify the situation with regard to oxygen fugacities likely to prevail under upper mantle pressure and temperature conditions.

Use of the Raheim and Green (1974) geothermometer provides a further constraint on the conditions of equilibration of the xenoliths. This geothermometer is discussed below.

Raheim and Green (1974) measured the empirical distribution coefficient of $K_D = (FeO/MgO)^{gt} / (FeO/MgO)^{cpx}$ for garnet and clinopyroxene pairs crystallized from glasses of basaltic compositions at 30 kb over a wide range of temperatures. They concluded that the effect of bulk chemistry for basaltic compositions between $6,2 < 100\ Mg^{2+} / Mg^{2+} + Fe^{2+} < 85$, does not perceptibly affect the K_D value. Raheim and Green (1974) did not attempt to test the effect of changing the amount of CaO, Al_2O_3 and Na_2O in bulk chemical systems.

They proposed that the equation -

$$T^oK = \frac{3686 + 28,35\ (Pkb)}{\ln K_D + 2,33} \qquad (1)$$

could be used to determine the temperature of

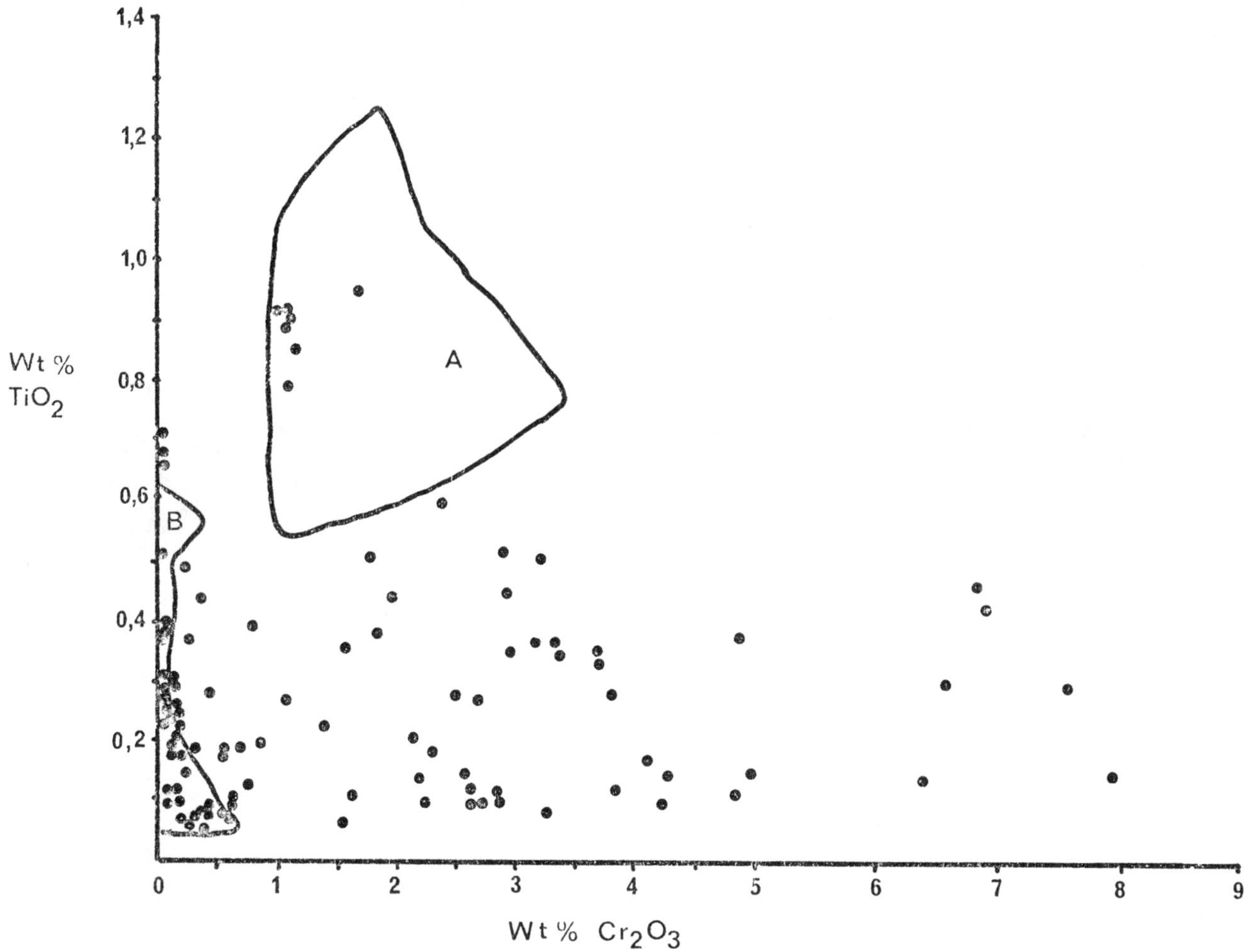

Figure 6. Plot of wt. % TiO_2 versus wt. % Cr_2O_3 for garnet xenocrysts from Orapa. Solid circles are garnet xenocrysts, A is the compositional field for garnet megacrysts and B is the compositional field for garnets from eclogites.

equilibration of natural eclogites provided that the K_D is known and that a pressure can be given.

Wood (1975) investigated the pressure dependence of K_D for garnet-clinopyroxene pairs in which the grossular content of garnet is not a function of pressure. He concluded that the pressure dependence of K_D is smaller for ultramafic bulk compositions than for basaltic compositions and attributed the compositional dependence of K_D to varying activity coefficients of one or more of the components of interest. Although Wood could not provide positive proof he was able to suggest that the calcium content of garnet was probably of greatest importance.

From the above discussion it can be concluded that the Raheim and Green (1974) equation (1 above) can be applied to eclogites of basaltic composition but that caution should be exercised when dealing with eclogites with high Ca contents in garnets i.e. kyanite eclogites, corundum eclogites, grospydites and high-calcium eclogites.

The Raheim and Green (1974) equation (1 above) can be combined with the equation defining the graphite-diamond stability curve (Bundy et al., 1961)

$$Pkb = 7,1 + 0,027\ T^{o}K \qquad (2)$$

to estimate the conditions of equilibration of the diamond and graphite eclogites assuming that the diamond and graphite crystallized in equilibrium with the garnet and clinopyroxene.

K_D values have been calculated for coexisting garnet clinopyroxene pairs in the diamond and graphite eclogites and are used in equations (1) and (2) to yield minimum and maximum estimates of pressure and temperature of formation for the diamond eclogites and the graphite eclogites respectively.

The diamond eclogites give minimum estimates ranging from 1091° - 1285° and 44 - 49 kb.

AK1/25 which contains both diamond and graphite and probably equilibrated very close to the diamond-graphite stability curve yields values of 1251° and 48 kb. The Orapa graphite eclogites indicate a range in maximum temperatures and pressures of equilibration from 894° - 1251° and 39 - 48 kb. The lowest values are given by graphite-bearing kyanite eclogite AK1/12 which contains a very grossular-rich garnet (45% grossular). As discussed above the effect of Ca on the distribution of Fe and Mg between garnet and clinopyroxene may be considerable and the Raheim and Green (1974) geothermometer cannot be applied to AK1/12 with any confidence. The large apparent range of pressures of equilibration is due in part at least, to the effect of the grossular component of the garnets but the large range in mineral chemistry observed in these rocks is not likely to have been generated at one temperature and pressure. Since the eclogites form a related suite of rocks, a range in equilibration temperatures for rocks sampled from a restricted zone in the mantle is preferred.

The presence of Mg rich garnets as inclusions in Ca-rich garnets in some eclogites indicates that the Ca-rich eclogites formed in close proximity to Mg-rich eclogites and suggests that the whole eclogite suite is genetically related. The origin of the eclogites is considered to most probably be related to a partial melting event in the mantle e.g. a partial melt of garnet lherzolite (O'Hara et al., 1975). Diamonds contain inclusions of eclogitic minerals and are found in eclogite.

They also contain significant quantities of H_2O, CO_2, CH_4, and dropplets of liquid (Fesq et al., 1975), which supports the contention that on crystallographic grounds they grew in a liquid (Giardini and Tydings, 1962). This by inference connects eclogite with a magmatic process.

Although the small sample size at Orapa has prevented the recognition of certain large scale textural and chemical relationships found at Roberts Victor by Hatton (1977), the similarity in mineral assemblages and the observed ranges of mineral compositions is most striking.

Megacrysts

Megacrysts of garnet, clinopyroxene and ilmenite occur at Orapa.

Garnet Megacrysts are red to orange in colour and are between 1 and 3 cm long. The garnets are usually fractured but are very fresh.

The garnets are chemically homogeneous, unzoned, low-chrome, high titanium pyropes which show increasing Fe/Mg with gradually increasing calcium (4,3 - 5,87 wt. % CaO), (see Figure 4). Titanium and chrome range from 0,5 - 1,26 wt. % TiO_2 (mean 0,97 wt. %) and 0,36 - 3,41 wt. % Cr_2O_3 (mean 1,68 wt. %) respectively. Sodium is detected in all the megacryst garnets, 0,09 - 0,18 wt. % Na_2O (mean 0,12 wt. %).

These megacrysts are similar to garnets found at the Monastery Mine, O.F.S. (Jakob, 1977; Gurney et al., 1977) and to the low-chrome ($<$4,0 wt. % Cr_2O_3) megacrysts from the Colorado-Wyoming State Line and Iron Mountain district kimberlites, U.S.A. (Eggler and McCallum, 1974; McCallum et al., 1977; Eggler et al., 1977).

Clinopyroxene megacrysts are commonly 1 - 2 cm long. The colour varies from bright-green to grey-green. Intergrowths with ilmenite are not found. Chemically they fall into two groups similar to the high-chrome calcic and the low-chrome subcalcic megacrysts from the Colorado-Wyoming State Line kimberlites.

The high-chrome, calcic clinopyroxene megacrysts have Ca/Ca + Mg ratios between 0,436 and 0,469; Mg/Mg + Fe ratios of between 0,861 - 0,938 and contain between 0,71 and 2,88 wt. % Cr_2O_3. This type of megacryst has not been recorded in Southern African kimberlites before. The low-chrome, calcic clinopyroxene megacrysts are similar to those found at the Monastery Mine (Gurney et al., 1977). They are characterized by low-chrome contents between 0.05 and 0.56 wt. % Cr_2O_3 and show a wide variation in Ca/Ca + Mg ratios from 0,321 - 0,456. Mg/Mg + Fe ratios vary from 0.813 - 0,881.

The two groups of clinopyroxene megacrysts are clearly separated in the Ca-Mg-Fe diagram (Figure 4).

Ilmenite megacrysts are between 1 and 3 cm long. They are the least abundant megacryst type found at Orapa whereas at Monastery they are the most abundant.

The ilmenite megacrysts studied were all discrete nodules, no intergrowths with other phases were found. The study was further extended by the analysis of a number of ilmenite xenocrysts from the heavy concentrate.

The ilmenites are typically picroilmenite with a geikielite content of 28 - 51 per cent. The Fe_2O_3 content of the ilmenites tends to increase with increasing Fe TiO_3 content (4,8 - 12,5 percent hematite).

Figure 5 shows the relationship between Cr_2O_3 and MgO in the ilmenites analysed in this study. Haggerty (1975) obtained a parabolic relationship for kimberlitic ilmenites from Southern and West African localities in a similar plot with most of the analyses lying along the arms of the parabola. He did note, however, that the lowest and highest Cr_2O_3 values determined lay between 6 and 10 wt. % MgO. The majority of the Orapa ilmenites lie in this region where they show a trend of increasing Cr_2O_3 from 1,0 to 4 wt. % within the narrow range of 7,5 to 9,5 wt. % MgO. The remainder of the Orapa ilmenites show a trend of increasing Cr_2O_3 with increasing MgO which merges into Haggerty's parabola at high MgO values of about 12 wt. %.

The solubility of Cr_2O_3 in picroilmenite is temperature dependent (Wyatt 1977) and the chrome distribution pattern in the Orapa ilmenites could therefore be due to crystallization from a cooling magma with decreasing Cr linked to decreasing Mg/Fe and increasing Fe^{3+}.

It is suggested that the chemical features displayed by the garnet, ilmenite and low chrome, subcalcic clinopyroxene megacrysts could best be explained by a process of igneous differentiation from a liquid as postulated by Boyd and Nixon (1973) and Gurney et al., (1977) for the megacrysts from the Monastery Mine. The crystallisation temperatures of the subcalcic clinopyroxene megacrysts range from 1375° to 1023°C using the diopside - enstatite solvus of Davis and Boyd (1966). The high chrome, calcic clinopyroxene megacrysts are chemically similar to clinopyroxenes from peridotitic xenoliths but their large grain size precludes the possibility that they could be peridotite disaggregation products. They appear to have equilibrated at much lower temperatures (970° - 1095°C) than the subcalcic clinopyroxenes, which is incompatible with their higher Mg/Mg + Fe and higher Cr_2O_3.

Kimberlite Xenocrysts

The study of heavy concentrates from the mining process is a simple way of extending the range of sampling and yields useful information on the xenolith suite because many of the xenocrysts are considered to have been derived from disaggregated mantle rocks sampled by the kimberlite (Dawson, 1972)

Examination of the heavy concentrate from Orapa reveals that the most abundant mineral is ilmenite (~ 45%) followed by garnet (~ 40%) and clinopyroxene (~ 5%). Rutile, kyanite, corundum, and diamond have also been identified. Olivine and orthopyroxene xenocrysts have not been found.

The ilmenite xenocrysts have similar compositions to the ilmenite megacrysts and are obviously derived from them. Garnet xenocrysts range in colour from orange to mauve and a plot of wt. % TiO_2 versus wt. % Cr_2O_3 (Figure 6) for these xenocrysts reveals that a large range in titanium and chrome content exists. The compositional fields shown by garnet megacrysts and by eclogitic garnets from Orapa are also delineated on Figure 6. High titanium ($>$ 0,70 wt. % TiO_2) and moderate chrome (~ 1,00 wt. % Cr_2O_3) garnet xenocrysts plot in the garnet megacryst field whilst low chrome ($<$ 1,00 wt. % Cr_2O_3) garnet xenocrysts plot within and around the field of garnets from eclogites. These xenocrysts could represent disaggregated garnet megacrysts and eclogites respectively.

Many of the garnet xenocrysts are characterized by moderate titanium ($<$ 0,5 wt. % TiO_2) and high chrome (1,00 - 7,95 wt. % Cr_2O_3) and do not plot in the above fields. These garnets are pyropes with high magnesium (18,9 - 21,8 wt. % MgO) and moderate calcium (3,23 - 5,99 wt. % CaO) and resemble garnets from garnet lherzolites. It is reasonable to postulate that olivine and orthopyroxene crystals in garnet lherzolite xenoliths would alter in the hydrous environment of the sedimentary kimberlite and that the friable nodules would disaggregate during the mining process leaving the more resistant garnets as xenocrysts. Thus it appears likely that the garnet xenocrysts at Orapa are derived from disaggregated eclogites, megacrysts and garnet lherzolites.

Conclusions

Eclogite is a common upper mantle xenolith in the Orapa Kimberlite. These eclogites form a genetically related suite of rocks, probably sampled from a restricted mantle zone. An igneous origin is favoured. The wide range in mineral chemistry is probably due to changes in melt composition and to changes in equilibration temperature.

Diamonds and graphite co-exist in some eclogites which have a range in mineral compositions. It is concluded that equilibration conditions were close to the diamond graphite reaction curve boundary for these rocks.

The presence of peridotite xenoliths is inferred by the occurrence of peridotitic garnets in Kimberlite concentrate.

Megacryst minerals are quite abundant and the clinopyroxenes display a wide range in apparent equilibration temperatures.

The mantle rocks sampled by the Kimberlite from beneath Orapa are very similar to those sampled by the Roberts Victor Kimberlite some 1 000 km to the south. This similarity does not extend to the inclusions in the diamonds, nor to the megacrysts.

Acknowledgements. The authors wish to thank J.B. Hawthorne (De Beers Consolidated Mines Limited) and J. Gibson (De Beers Consolidated Mines, Botswana) for making this study possible and for help in sample collection.

S.R.S. gratefully acknowledges financial assistance from De Beers Consolidated Mines Limited; the University of Cape Town and from the Convenors of this conference. Financial support was provided in part by a Newman Trust Scholarship, Rhodesia.

References

Albee, A.L., and L. Ray, Correction factors for the electron probe microanalysis of silicates, oxides, carbonates, phosphates and sulphates, Analytical Chemistry, 42, 1408 - 1414, 1970.

Bence, A.L., and A.L. Albee, Empirical corrections for the electron microanalysis of silicates and oxides, J. Geol., 76, 382 - 403, 1968.

Boyd, F.R., and P.H. Nixon, Origin of the discrete nodules in the kimberlites of Northern Lesotho, Extended Abstracts, First International Kimberlite Conference, Cape Town, South Africa, 1973.

Bundy, F.P., H.P. Bovenkirk, H.M. Strong, and R.H. Wentorf, Diamond-graphite equilibrium line from growth and graphitization of diamond,

J.Chem. Phys., 35, 383 - 391, 1961.

Davis, G.L., The ages and uranium contents of zircons from kimberlites and associated rocks, Extended Abstracts, Second International Kimberlite Conference, Santa Fé, New Mexico, 1977.

Davis, B.T.L., and F.R. Boyd, The join $Mg_2 Si_2O_6$-Ca Mg Si_2O_6 at 30 kb pressure and its application to pyroxenes from kimberlite, Jour. Geophys. Res., 71, 3567 - 3576, 1966.

Dawson, J.B., Kimberlites and their relation to the mantle, Phil. Trans. Roy. Soc. London, Sect. A., 271, 297 - 311 1972.

Eggler, D.H. and M.E. McCallum, Xenoliths in diatremes of the Western United States, Carniegie Inst, Washington Yearb., 73, 294 - 300, 1974.

Eggler, D.H., M.E. McCallum, and C.B. Smith, Discrete nodule assemblages in kimberlites from Northern Colorado and Southern Wyoming : Evidence for a diapiric origin of kimberlite, Extended Abstracts, Second International Kimberlite Conference, Santa Fé, New Mexico, 1977.

Erlank, A.J., Distribution of potassium in mafic and ultramafic nodules, Carniegie Inst. Washington Yearb., 68, 433 - 439, 1970.

Fesq, H.W., D.M. Bibby, C.S. Erasmus, E.J.D. Kable, and I.P.F. Sellschop, A comparative trace element study of diamonds from Premier, Finsch and Jagersfontein Mines, South Africa, Phys. Chem. Earth. 9, 817 - 836, 1975.

Giardini, A.A., and J.E. Tydings, Diamond Synthetics : observations on the mechanism of formation, Amer. Min., 47, 1393 - 1421, 1962.

Gurney, J.J., W.R.O. Jakob, and J.B. Dawson, Megacrysts from the Monastery Mine, Extended Abstracts, Second International Kimberlite Conference, Santa Fé, New Mexico, 1977.

Haggerty, S.E., The chemistry and genesis of opaque minerals in kimberlites, Phys. Chem. Earth, 9, 295 - 307, 1975.

Hatton, C.J., The geochemistry and origin of xenoliths from the Roberts Victor Mine, Unpublished PhD. thesis, University of Cape Town, South Africa, 1978.

Hatton, C.J., and J.J. Gurney, Kyanite eclogites from the Roberts Victor Mine, Extended Abstracts, Second International Kimberlite Conference, Santa Fé, New Mexico, 1977.

Hatton, C.J. and J.J. Gurney, Igneous fractionation trends in Roberts Victor eclogites, Extended Abstracts, Second International Kimberlite Conference, Santa Fé, New Mexico, 1977.

Hawthorne, J.B., Model of a kimberlite pipe, Phys. Chem. Earth, 9, 1 - 15, 1975.

Leake, B.E., A catalogue of analysed calciferous and subcalciferous amphiboles together with their nomenclature and associated minerals, Geol. Soc. Am. Special Paper, 98, 210 p., 1968.

MacGregor, I.D., and J.L. Carter, The chemistry of clinopyroxenes and garnets of eclogites and peridotite xenoliths from the Roberts Victor mine, South Africa, Phys. Earth Planet. Interiors, 3, 391 - 397, 1970.

McCallum, M.E., D.H. Eggler, and C.B. Smith, Discrete nodule assemblages in kimberlite from Northern Colorado and Southern Wyoming, Extended Abstracts, Second International Kimberlite Conference, Santa Fé, New Mexico, 1977.

O'Hara, M.J., M.J. Saunders, and E.L.P. Mercy, Garnet Peridotite, primary ultrabasic magma and eclogite; interpretation of upper mantle processes in kimberlite, Phys. Chem. Earth, 9, 571 - 604, 1975.

Prinz, M., D.B. Manson, P.F. Hlava, and K. Keil, Inclusions in diamond : garnet lherzolite and eclogite assemblages, Phys. Chem. Earth, 9, 797 - 815, 1975.

Raheim, A., and D.H. Green, Experimental determination of the temperature and pressure dependence of the Fe-Mg partition coefficient for coexisting garnet and clinopyroxene, Contrib. Mineral. and Petrol., 48, 179 - 283, 1974.

Reid, A.M., R.W. Brown, J.B. Dawson, G.G. Whitfield and J.C. Siebert, Garnet and pyroxene compositions in some diamondiferous eclogites, Contrib. Mineral, and Petrol,, 58, 203 - 220, 1976.

Rickwood, P.C., J.J. Gurney, and M.White-Cooper, The nature and occurrences of eclogite xenoliths in the kimberlites of Southern Africa, Geol. Soc. South Africa Special Publication, 2, Upper Mantle Project, 1969.

Robinson, D.N., Diamond and graphite in eclogite xenoliths from kimberlite, Extended Abstracts Second International Kimberlite Conference, Santa Fé, New Mexico, 1977.

Rosenhauer, M., E. Woerman, B. Knecht, and C.G. Ulmer, The stability of graphite and diamond as a function of the oxygen fugacity in the mantle, Extended Abstracts, Second International Kimberlite Conference, Santa Fé, New Mexico,1977.

Sobolev, N.V., I.K. Kuznetsova, and N.I. Zyuzin, The petrology of grospydite xenoliths from the Zagadochnaya Kimberlite pipe in Yakutia, J. Petrol., 9, 253 - 280, 1968.

Sobolev, V.S., N.V. Sobolev, and Yu.G. Lavrent'ev, Inclusions in the diamonds in diamondiferous eclogite, Dokl. Acad. Sci. U.S.S.R., Earth. Sci. Sect., 207, 121 - 123, 1972.

Stephens, W.E. and J.B. Dawson, Statistical comparison between pyroxenes from kimberlites and their associated xenoliths, Jour. Geol., 85, 433 - 449, 1977.

Woerman, E., B.Knecht, M. Rosenhauer, and G.C. Ulmer, The stability of graphite in the system C-O, Extended Abstracts, Second International Kimberlite Conference, Santa Fé, New Mexico, 1977.

Wood, B.J., The partitioning of iron and magnesium between garnet and clinopyroxene, Carniegie Inst. Washington Yearb., 75, 571 - 574, 1975.

DIAMOND AND GRAPHITE IN ECLOGITE XENOLITHS FROM KIMBERLITE

Derek N. Robinson

Anglo American Research Laboratories,
P.O. Box 106, Crown Mines, 2025,
Republic of South Africa

Abstract. The characteristics of diamond and graphite in a number of diamond- and/or graphite-bearing eclogite xenoliths are described. In most cases of eclogite specimens containing diamond, the proportions of diamond present exceed that in kimberlite by more than a factor of three. The diamonds are similar to those in kimberlite and very small proportions of disaggregated diamond-bearing eclogite can account for much, even all, of the diamond in some kimberlite. Morphological features of diamond and graphite crystals, and spatial relationships between them in diamond-graphite eclogite, indicate that both minerals are primary and favour an igneous origin for the rocks. In one specimen there is evidence that partial resorption and etching of diamond crystals was effected by the liquid from which diamond crystallized, and not by kimberlite. In another specimen, however, kimberlite volatiles were evidently responsible for resorbing and etching diamond. Considering the low density of graphite, graphite-bearing eclogite is unlikely to represent crystal cumulate derived by gravitational settling. At least some of the diamond-graphite eclogite xenoliths are probably samples of abundant, very small bodies of carbon-bearing eclogite. Simple crystallization paths can be inferred for most specimens studied. In one xenolith, however, two generations of diamond are clearly evident and it is apparent that an increase in temperature occurred during crystallization of the rock.

Introduction

The characteristics and modes of occurrence of diamond and graphite in eclogite xenoliths from kimberlite are relevant to both the origin of diamond in kimberlite and the petrogenesis of eclogite. This paper deals with the diamond and graphite in nine diamond eclogite xenoliths, five diamond-graphite eclogite xenoliths and ten graphite eclogite xenoliths. Kyanite- and corundum-bearing varieties are represented. Particulars of each xenolith are summarized in Table 1.

By 1976 at least 63 individual diamond-bearing eclogite xenoliths had been recorded in the literature. These include 33 specimens from ten Southern African kimberlite localities (29 in Rickwood et al., 1969 and at least four additional specimens in Reid et al., 1976) and 30 from three kimberlites in the U.S.S.R. (Sobolev, 1970; Sobolev, 1973; Ponomarenko et al., 1976; Ponomarenko and Spetzius, 1976). At least six newly-discovered or previously unreported diamond-bearing xenoliths are included in the present study (see Table 1). More than 20 graphite eclogite xenoliths are currently known (Rickwood et al., 1969; this paper). One specimen (AK1/25) initially classified as graphite eclogite in the present study yielded a diamond upon dissolution of the eclogite.

Diamond

Small xenoliths containing a large diamond which constitutes a few per cent of the specimen (e.g. DB 1, DB 2, DB 4 and DB 7) cannot be considered as representative samples. More meaningful indications of the diamond content of possibly substantial bodies of eclogite can, however, probably be obtained from relatively large xenoliths containing a number of diamonds. Diamond crystal size distribution in each xenolith is represented in Figure 1. Specimens AK1/9, XRV 22 and HRV 247 yield normal distributions so are likely to more closely reflect parent rock diamond contents than do the other specimens studied. Computed diamond grades for these three, and two less representative, xenoliths are recorded in Table 2. These data should be compared with the less than 0.6 carat per tonne grade of most economically viable South African kimberlites (after Wagner, 1914, p.137). Although the commercial recovery of small diamonds from kimberlite may be inefficient it is, nevertheless, apparent that some eclogite xenoliths contain more than 10,000 times as much diamond as does kimberlite while most diamond-bearing eclogite specimens contain at least 1,000 times as much. Only a few

TABLE 1. Xenolith Details.

Specimen Number	Kimberlite locality	Weight examined gm	Approx. modal composition⁺ cpx ga other	No. of diamond crystals	No. of graphite crystals
AK1/9	Orapa	7.4	60 40	14*	
AK1/10	Orapa	14.7	60 40	2*	
DB 1	Doornkloof	15.2	50 50	1*	
DB 2	Doornkloof	13.3	30 70	1*	
DB 3	Doornkloof	3.1	30 70	1*	
DB 4	Bobbejaan	9.1	60 40	1*	
DB 5	Jaggersfontein	1.5	65 35	5*	
DB 6	Excelsior	24.2	75 20 ky 5	1*	
DB 7	Excelsior	10.8	50 45 ky 5	1*	
HRV 247	Roberts Victor	943.0	30 70	49	±200
XRV 22	Roberts Victor	248.1	70 30	119	±200
PJL 18	Roberts Victor	115.7	75 25	3	6
JJG 531	Jagersfontein	13.1	85 15	6	±50
AK1/25	Orapa	12.0	55 45	1	16
AK1/12	Orapa	11.6	60 30 ky 10		3
AK1/13	Orapa	8.5	30 70		3
AK1/24	Orapa	8.4	50 50 cor tr		25
AK1/26	Orapa	2.1	65 35		3*
AK1/27	Orapa	4.9	50 50 ky tr		5
AK1/28	Orapa	1.8	80 20		28*
AK1/29	Orapa	8.8	20 80		11
AK1/30	Orapa	8.5	15 85		11*
AK1/31	Orapa	24.4	35 65		3
AK1/34	Orapa	21.2	60 35 cor 5		2*

⁺ Diamond and graphite excluded. Accessory rutile and secondary (?) phlogopite commonly also present. cpx = clinopyroxene, ga = garnet, ky = kyanite, cor = corundum.
* Crystals exposed at xenolith surface only.
Discoverers of new xenoliths as follows: AK1 series S.R. Shee. HRV 247 C.J. Hatton. PJL 18 P.J. Lawless. JJG 531 J.J. Gurney. Either DB 6 or DB 7 may also not previously have been reported.

hundred ppm of disaggregated diamond-bearing eclogite can, therefore, account for all of the diamond in some, even economically viable, kimberlite. In this context it must be noted, however, that studies of inclusions in diamond indicate that eclogite-derived diamonds are subordinate in most (but not all) kimberlites for which data are available (Harris and Gurney, in press; Sobolev, 1977).

It is also apparent from Figure 1 that much of the range in diamond crystal size present in kimberlite is represented among the xenoliths studied.

Most diamonds in the xenoliths appear to be colourless. Brownish colouration becomes apparent with increasing crystal size among XRV 22 diamonds, however, while two very light green crystals and a very light brown crystal were recovered, with colourless diamonds of similar size, from specimen HRV 247. According to Fesq et al., (1975), brown and green diamonds commonly contain higher levels of impurities, probably in the form of submicroscopic inclusions, than colourless diamonds. Variation in growth conditions during the crystallization of HRV 247 diamonds may, therefore, be indicated. Local, green surface staining attributable to epigenetic radiation damage (Vance et al., 1973) is also apparent on some colourless HRV 247 diamonds. Thus, insofar as colour is concerned, the xenolith diamonds are not unlike diamonds in kimberlite.

A variety of diamond crystal forms is represented in the xenoliths and some contain more than one form (see Table 2). Octahedral and dodecahedral forms are most common, but cubes are also represented in the XRV 22 diamond population (Figure 2) and most HRV 247 diamonds exhibit subordinate cubic surfaces. Where octahedral and dodecahedral crystals are associated in a xenolith, the largest crystals tend to be dodecahedra and the smallest octahedra (e.g. see Figure 2).

In diamond synthesis experiments octahedral diamond crystallizes at relatively high temperatures and the cubic form at relatively low temperatures (Giardini and Tydings, 1962). Thus, a range in diamond crystallization

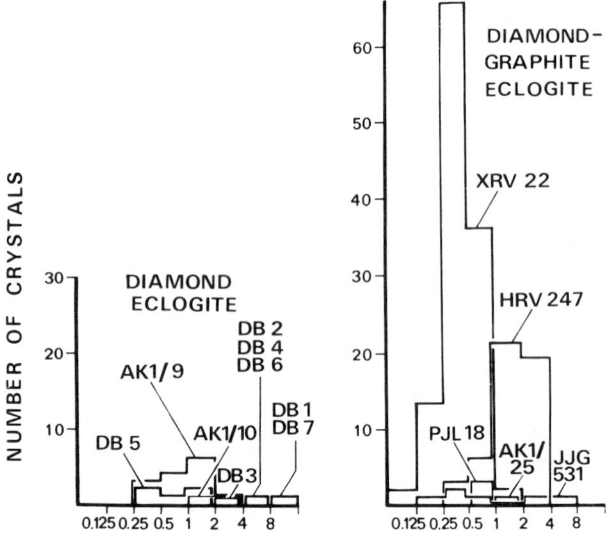

Fig. 1. Size frequency distribution of diamond crystals in individual xenoliths. (Approximate weights for octahedra : 0.125mm = 5 x 10^{-6} carat, 0.25mm = 45 x 10^{-6} carat, 0.5mm = 350 x 10^{-6} carats, 1mm = 0.003 carat, 2mm = 0.02 carat, 4mm = 0.2 carat, 8mm = 1.5 carats. 1 carat = 0.2gm).

temperature is indicated for specimen XRV 22 in which octahedral and cubic diamond crystals are associated.

It has been demonstrated (e.g. Seal, 1963; Moore and Lang, 1974) that the dodecahedron is not a growth form of diamond but results from partial resorption of the octahedron or cube. The tendency for a growth form to be more frequently preserved as small, rather than large, crystals in some of the xenoliths can be explained by small crystals having been more frequently enclosed in other mineral grains while diamond was being resorbed. Alternatively, more than one generation of diamond may be represented. Among cases in which diamonds were recovered from the interior of xenoliths (see Table 1), only in specimen JJG 531 does it appear that dodecahedral morphology is best developed on diamonds nearest the xenolith exterior. In this specimen all of the diamonds exhibit local, black coats of amorphous carbon and this material is thickest over dodecahedral surfaces which are restricted to a large, partly exposed diamond. According to Phaal (1965), amorphous carbon forms as an intermediate stage in the oxidation of diamond and is preserved if the rate of its formation from diamond exceeds the rate of its own oxidation. Resorption of the JJG 531 diamond was, therefore, probably a consequence of oxidation and is likely to have occurred after the xenolith was incorporated in kimberlite. Such is not the case, however, for

TABLE 2. Diamond Crystal Forms, Diamond Content and Graphite Crystal Habits.

Diamond-bearing specimen	Diamond crystal forms+	Diamond grade cpmt*	Graphite crystal habits**	Graphite-bearing specimen
AK1/9	(od),(do)	18,000	etp	AK1/12
AK1/10	(od)	-	i,e,etch	AK1/13
DB 1	fragment	-	etp,e,etch	AK1/24
DB 2	(od)	-	etp	AK1/26
DB 3	d	-	etp	AK1/27
DB 4	o	-	etp,r	AK1/28
DB 5	(od)	-	etp,r	AK1/29
DB 6	o	-	etp	AK1/30
DB 7	o	-	(damaged)	AK1/31
			etp	AK1/34
HRV 247	(oda),(doa), (od),o	3,500	re,etch	HRV 247
XRV 22	(od),(do),d, o,(ad)	650	etp,i,r	XRV 22
PJL 18	o,(od)	25	etp,i	PJL 18
JJG 531	o,(od)	19,000	etp,r,i	JJG 531
AK1/25	o	-	e,r,etp	AK1/25

+ Given in order of decreasing abundance : o = octahedron, d = dodecahedron, a = cube; combined forms bracketed with predominant form first.
* cpmt = carats per metric tonne.
** Given in order of decreasing abundance :- etp = euhedral tabular prism, r = rounded crystal, e = embayed crystal, i = irregular crystal, etch = pitted crystal.

Fig. 2. Some diamonds recovered from specimen XRV 22. Most are octahedra with, or without, minor dodecahedral surfaces but the dodecahedral form predominates in the two largest crystals and the incomplete crystal (at upper left) while a cube is also represented (second in top row). Two octahedra are contact twins (third and fourth in second row).

the partially resorbed diamond crystals at least of specimen HRV 247. In that specimen, nearly perfect moulds were revealed opposite diamond crystals when the xenolith was broken. X-ray diffraction analysis of particles scratched from such a mould surface showed that dodecahedral diamond had been in contact with clinopyroxene, indicating that diamond resorption preceeded eclogite crystallization.

Contact- and interpenetrant-twinned crystals of diamond occur together with single crystals in some xenoliths (e.g. in specimen HRV 247 two contact twins and four interpenetrant-twinned aggregates are among the diamonds recovered; also see Figure 2). Exposed diamonds are commonly freshly broken (e.g. specimens DB 1, DB 4, DB 5 and DB 7) while numerous small fragments, possibly resulting from crystal breakage during laboratory treatment, were recovered from xenoliths HRV 247 and XRV 22.

It is apparent that the morphologies of most of the xenolith diamonds are no different from those of diamonds common in kimberlite. Some unusual features were, however, noted. For example: (i) Some XRV 22 dodecahedral crystals exhibit large cavities or appear to be incomplete (see Figure 2). Such crystals are likely to have grown about pre-existing crystals of another mineral. (ii) A group of parallel, small diamond forms is exposed at an irregular cubic surface of a relatively large, broken crystal of specimen HRV 247 (Figure 3a). These small forms can be interpretated either as inclusions or as overgrowths. Like their host, they exhibit dodecahedral surfaces showing that they were partially resorbed. (iii) Tiny octahedral forms of diamond are evident on pitted, cubic surfaces of many HRV 247 diamonds (Figure 3b). These octahedral forms commonly occur within or astride etch pits in their host, but they display no evidence of having been resorbed or etched (Figure 3c). They are epitaxial with respect to their hosts and are interpreted as overgrowths which grew subsequently to the resorption and etching of hosts. Considering their octahedral form and the partly cubic form of their hosts, the overgrowths are likely to have crystallized at a higher temperature. Overgrowth crystallites are not included in the diamond size distribution data for xenolith HRV 247 shown in Figure 1.

Most of the diamonds exhibit surface textures such as trigonal pits on octahedral surfaces, tetragonal pits on cubic surfaces and elongate hillocks on dodecahedral surfaces. These features are invariably in the negative orientation (terminology of Frank et al., 1958), as is usually the case for diamonds recovered from kimberlite. Flat-bottomed hexagonal pits (which are fairly common features on octahedral diamond from some kimberlites) are developed, however, on the diamonds of specimens DB 6 and DB 7. Experimental investigations (cf. Phaal, 1965) indicate that negatively-orientated, diamond surface textures result from etching (e.g. by dissolution in magma or oxidation by water vapour or wet carbon dioxide gas) at temperatures above 1,000^{o}C, while hexagonal pits are likely to result from the combined effect of two etchants, one of which is oxygen gas, at temperatures between 950 and 1,000^{o}C. Negatively-orientated textures are developed beneath the amorphous carbon noted on diamond of specimen JJG 531, while amorphous carbon also coats the diamond exhibiting hexagonal pits in specimen DB 6. Etching in these cases must have involved oxidation and can be ascribed to kimberlitic fluid. At least in the case of specimen HRV 247, with its unetched diamond overgrowths, however, all etching of diamond is likely to have preceded incorporation of the xenolith into kimberlite.

No attempt was made to determine the spectral properties of the diamonds. However, the good crystal shape of the majority and a lack of yellow colouration are compatible with their being of the most common spectral type, viz. type Ia (see Bruton, 1970, pp. 309-311).

Graphite

The morphological characteristics of graphite crystals in pertinent specimens are summarized in Table 2. In both diamond-graphite eclogite

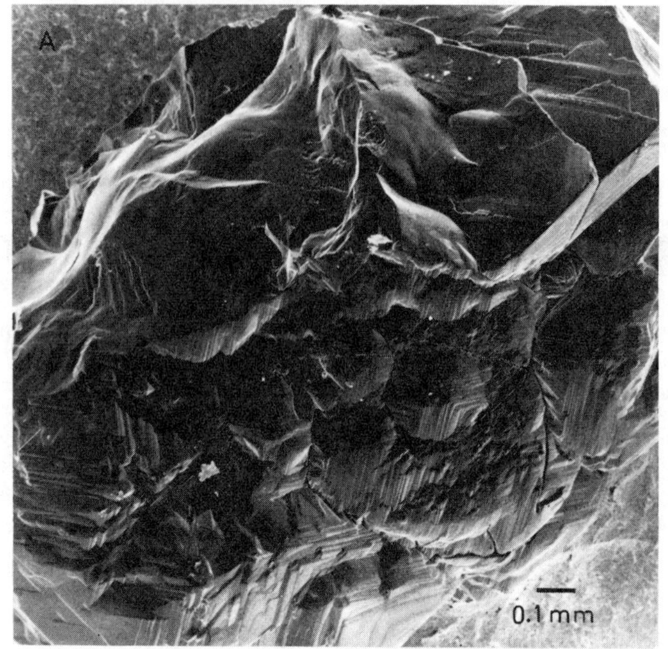

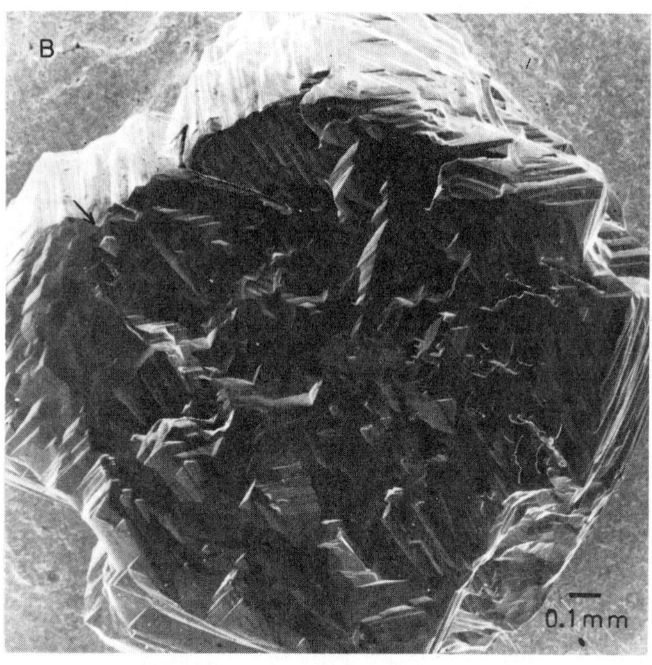

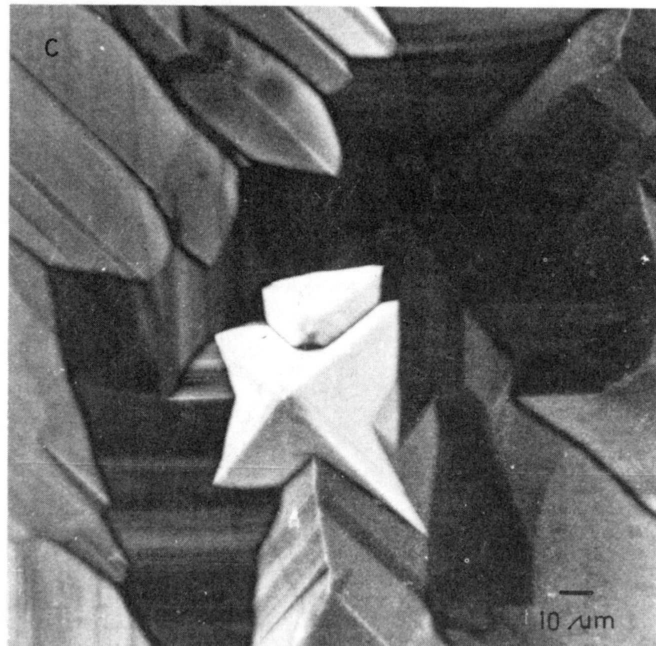

Fig. 3. Examples of unusual features of HRV 247 diamonds:-
a. Parallel diamond forms protruding from an irregular cubic surface of a large broken crystal. Both the host and protruding forms exhibit minor dodecahedral surfaces.
b. Irregular cubic surface of diamond crystal with very small, epitaxial overgrowths of octahedral form (some examples arrowed).
c. Parallel growth octahedral overgrowth astride edges to tetragonal etch pits. Note the sharp edges and smooth crystal surfaces of the overgrowth.

and graphite eclogite, graphite occurs as discrete grains which are frequently euhedral (Figure 4a), suggesting crystallization in a fluid medium. Rounded crystals (Figure 4b) are conspicuous in the diamond-graphite eclogite specimens HRV 247 and AK1/25 and in specimen AK1/29, while embayed crystals (Figure 4c) are common in the diamond-graphite eclogite specimens JJG 531 and AK1/25 and in specimens AK1/13 and AK1/30. Rounded or embayed crystals are

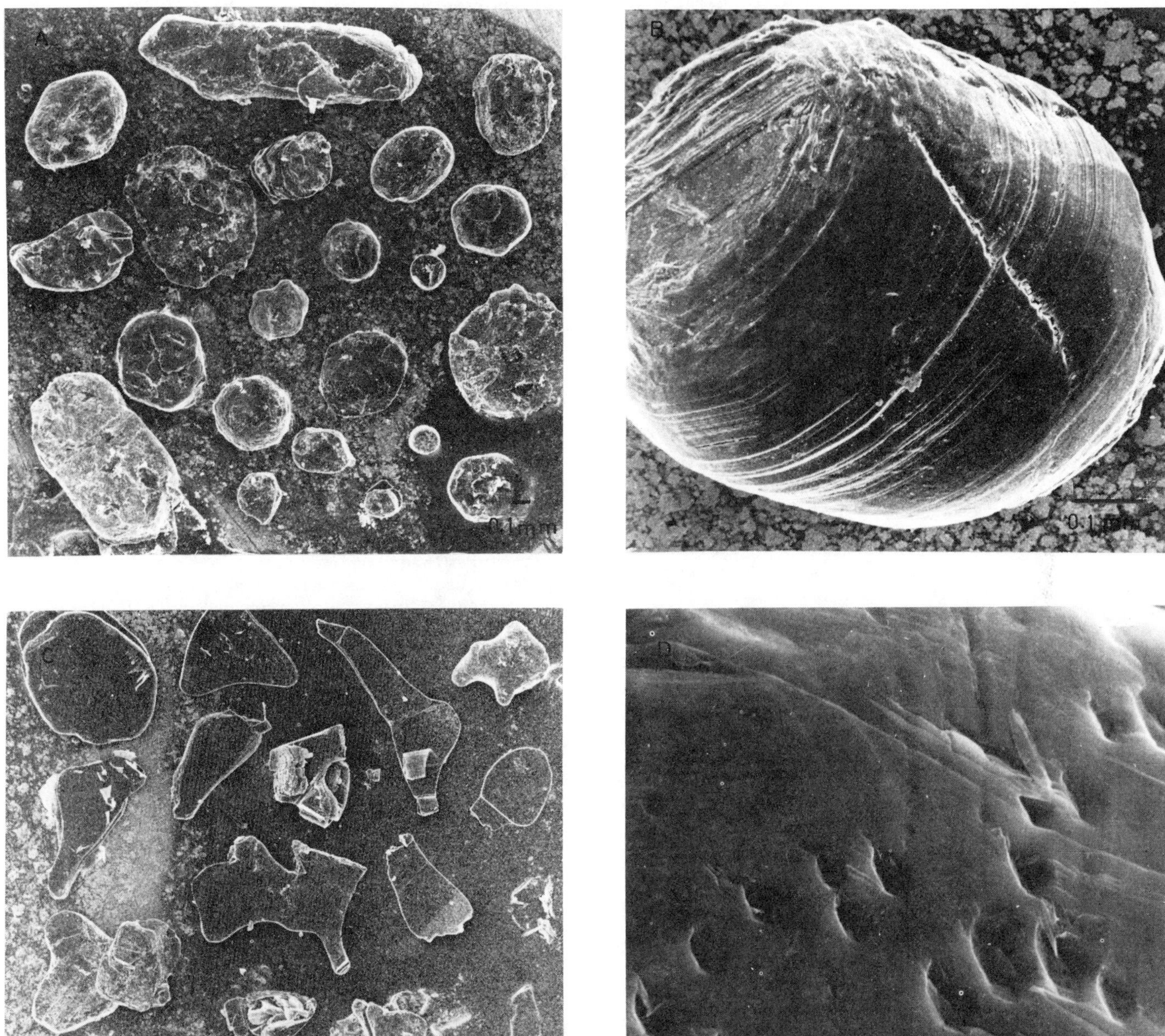

Fig. 4. Examples of graphite crystal characteristics:-
a. Euhedral tabular prisms and some rounded and slightly embayed crystals from specimen XRV 22.
b. Rounded prismatic crystal of specimen HRV 247.
c. Rounded and conspicuously embayed crystals of specimen AK1/25.
d. Pits in the prismatic surface of a rounded crystal of specimen HRV 247.

likely to have been partially resorbed. Some are finely pitted (Figure 4d). Occasional, irregular graphite crystals are also present in some of the xenoliths.

Some xenoliths (e.g. HRV 247, JJG 531 and AK1/24) contain both euhedral graphite crystals and others which evidently experienced resorption. It is possible that more than one

Fig. 5. Specimen HRV 247 reassembled. Diamond was discovered only after the xenolith had been broken and the missing portion (portion G) consisted of small fragments. The dashed line approximately marks the boundary between diamond-rich (D*) and graphite-rich (G*) eclogite. Portion A yielded 0.66gm of diamond and 0.06gm of graphite. Portion F2 yielded 0.0008gm of diamond and 0.40gm of graphite. Portion G yielded 0.18gm of graphite.

generation of graphite is present in such cases.

Because of the low density of graphite (even at very high pressure if temperature is high, e.g. 2.2 gm/cc at approximately 47 kb and 1200°C; after Berman, 1965), graphite-bearing eclogite is unlikely to be a gravitationally-induced, igneous crystal cumulate unless flotation, rather than sinking, of crystals is involved.

Relationships Between Diamond and Graphite

The direct derivation of one form of carbon from another is seen only in the amorphous coatings noted on the diamonds in two xenoliths. There is no evidence to suggest that either discrete diamond or discrete graphite crystals are not primary.

The diamond-graphite eclogite xenolith HRV 247 consists of two parts, one containing most of the diamond and few graphite crystals and the other containing most of the graphite and few diamond crystals (Figure 5). Specimen XRV 22 had been crushed prior to examination by the author but was also noted (J.J. Gurney, personal communication) to consist of a diamond-rich portion and a graphite-rich portion. Likely processes whereby spatial separation between primary diamond and primary graphite crystals could have been accomplished include:- (i) Gravitational separation within liquid as a consequence of the contrasting densities (e.g. approximately 3.5 gm/cc for diamond and 2.2 gm/cc for graphite at 47 kb and 1,200°C; after Berman, 1965) of the two minerals. The degree of separation accomplished by such a process should depend on factors such as the viscosity of the liquid. (ii) Crystallization at the wall of a conduit or magma chamber from fluid undergoing a change in temperature so as to precipitate first one, and then the other, allotrope of carbon with eclogite. A clean separation is to be anticipated from such a process.

Suggestion (i) is preferred to suggestion (ii) because separation between diamond and graphite is imperfect in the two xenoliths. In addition, in the case of specimen HRV 247 Hatton and Gurney (1977) detected geochemical differences between the silicate minerals of the diamond-enriched and graphite-enriched portions which indicate that the diamond-rich portion crystallized at a higher temperature than the graphite-rich portion. Since graphite is the relatively high temperature allotrope of carbon (Berman, 1965), for suggestion (ii) to be compatible with the findings of Hatton and Gurney would require that an increase in pressure be also involved.

In the other diamond-graphite eclogite xenoliths examined diamond and graphite appear to be interspersed. This relationship can be explained by viscous or crystal mush conditions during carbon crystallization.

It is unlikely that samples of single interfaces between diamond-enriched and graphite-enriched portions of large eclogite bodies would often be found. Should separation between diamond and graphite have been effected by gravity, therefore, xenoliths consisting of a diamond-enriched portion and a graphite-enriched portion are likely to be samples of abundant, very small bodies.

Petrogenesis

An igneous origin for the xenoliths is strongly suggested by the following:- (i) Evidence that partial resorption of diamond preceded eclogite crystallization at least in specimen HRV 247. (ii) Common euhedralism among graphite (and diamond) crystals. (iii) The concentration of diamond and graphite in separate portions of two xenoliths.

An igneous model also allows for certain features, such as the presence of incomplete diamond crystals in specimen XRV 22 and irregular graphite crystals in some xenoliths, by late growth of these crystals.

It is unnecessary to invoke metastable nucleation of either graphite or diamond in the xenoliths studied. Instead, diamond resorption can be related to graphite crystallization and, likewise, graphite resorption to diamond crystallization. This accords with the principle (upon which commercial diamond synthesis depends; see Wentorf, 1965) that if one carbon allotrope in equilibrium with liquid is subjected to conditions at which the other

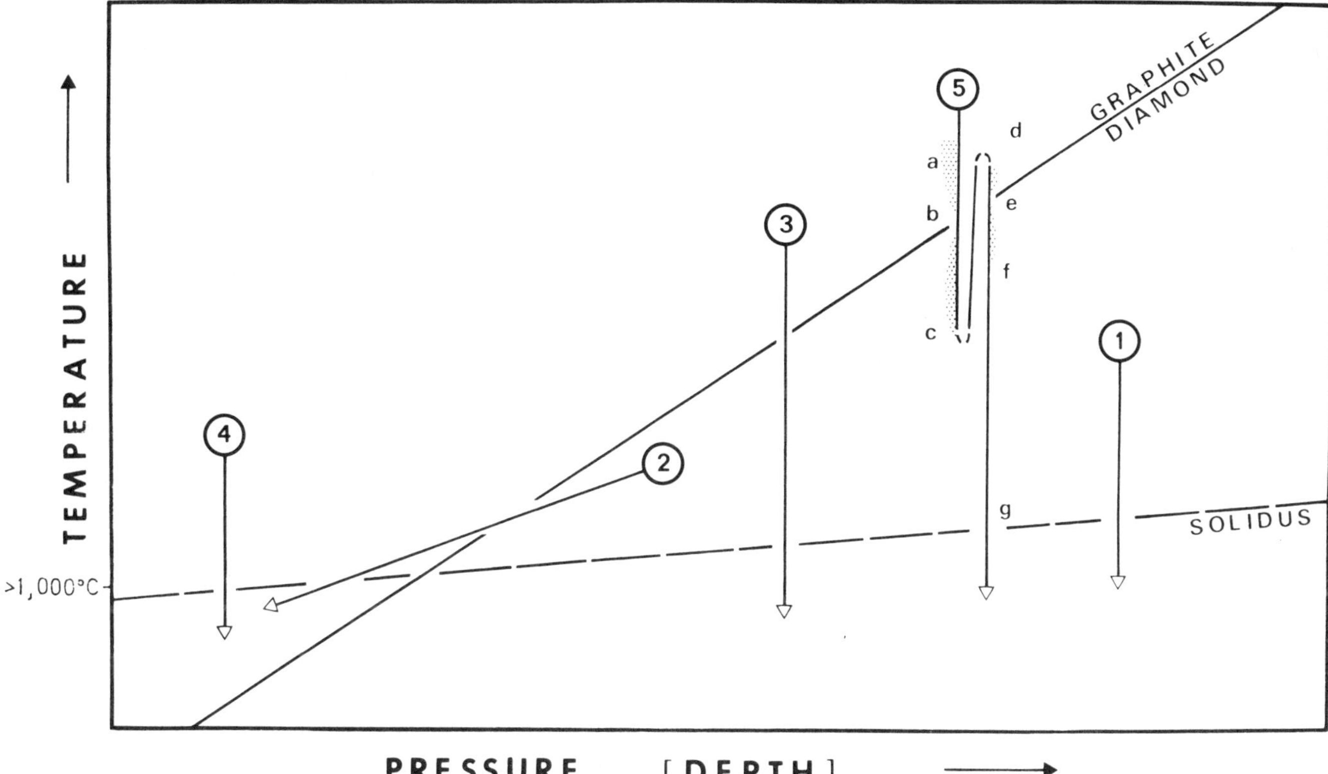

Fig. 6. Proposed thermodynamic crystallization paths for various carbon-bearing eclogites.
1. Diamond eclogite with unresorbed diamond crystals (e.g. DB 4, DB 6, DB 7).
2. Diamond eclogite with partially resorbed diamond crystals (e.g. AK1/9, AK1/10, DB 2, DB 3, DB 5) and graphite floated off to form graphite eclogite, with unresorbed graphite crystals, elsewhere (e.g. AK1/26, AK1/34). Also eclogite with diamond-rich and graphite-rich halves if very small volume of magma involved (e.g. XRV 22), and interspersed diamond-graphite eclogite if viscous magma or late crystallization of carbon (e.g. PJL 18).
3. Diamond-graphite eclogite with unresorbed diamond crystals and partially resorbed graphite crystals (e.g. JJG 531, AK1/25). Also, separate diamond eclogite (e.g. DB 4, DB 6, DB 7) and graphite eclogite (e.g. AK1/13, AK1/24).
4. Graphite eclogite with unresorbed graphite crystals (e.g. AK1/12, AK1/26, AK1/27, AK1/30, AK1/34).

allotrope is stable it will dissolve, due to an increase in its free energy, while the other allotrope will crystallize once the liquid becomes over-saturated with respect to it. Oxidation of diamond can be relegated to a late-stage process affecting free carbon in some of the xenoliths only after their incorporation by kimberlite magma.

Likely crystallization paths for the xenoliths studied are illustrated in Figure 6. It must be noted that in many cases other paths also fit the observations and interpretations made. In such cases the simplest alternative involving decreasing temperature has been selected. Also, only one generation of diamond or graphite is assumed unless evidence to the contrary is convincing. If it be accepted that some negatively-orientated surface textures on the diamonds resulted from etching by eclogite fluid, as is evident in the case of specimen HRV 247, then a minimum temperature of 1,000^{o}C can be fixed in Figure 6.

The crystallization history of specimen HRV 247 is relatively complicated in that there is compelling evidence for at least two periods of diamond crystal growth separated by a period during which diamond was resorbed and etched, in the xenolith. In addition, there is evidence that the diamonds of the earlier generation crystallized at a lower temperature than those crystallizing later, so that an increase in temperature during crystallization of the rock is indicated. Hatton and Gurney (1977) discerned geochemical zoning in the garnet and clinopyroxene of xenolith HRV 247 which also indicates that temperature increased during crystallization. As previously mentioned, these investigators also determined that the minerals of the diamond-rich portion of the rock crystallized at higher temperatures than the minerals of the graphite-rich portion. A model which allows for two periods of diamond (and graphite) crystallization and supposes that graphite crystals would tend to float in magma, is offered for the genesis of HRV 247. This model requires that the xenolith represents part of a very small body of original eclogitic liquid and is not a crystal cumulate derived from some other liquid. Contrary to the model offered by Hatton and Gurney, (1977), no increase in pressure during crystallization is necessary.

Acknowledgements. The author is grateful to his colleagues for criticism of a draft manuscript of this paper, and to the Management of Anglo American Research Laboratories for permission to publish.

References

Berman, R., Thermal properties, in Physical Properties of Diamond, edited by R. Berman, pp. 371-395, Clarendon Press, Oxford, 1965.

Bruton, E., Diamonds, N.A.G. Press, London, 1970.

Fesq, H.W., D.M. Bibby., C.S. Erasmus., E.J.D. Kable, and J.P.F. Sellschop, A comparative trace element study of diamonds from Premier, Finsch and Jagersfontein Mines, South Africa, in Physics and Chemistry of the Earth, 9, edited by L.H. Ahrens, J.B. Dawson, A.R. Duncan, and E.J. Erlank, pp. 817-836, Pergamon Press, Oxford, 1975.

Frank, F.C., K.E. Puttick, and E.M. Wilks, Etch pits and trigons on diamond, Phil. Mag., 3, 1273-1279, 1958.

Giardini, A.A., and J.E. Tydings, Diamond synthesis: observations on the mechanism of formation, Amer. Mineral., 47, 1393-1421, 1962.

Harris, J.W., and J.J. Gurney, Inclusions in diamond, in Properties of Diamond, edited by J.E. Field, Academic Press, London, (in press).

Hatton, C.J., and J.J. Gurney, A diamond graphite eclogite from the Roberts Victor Mine, in Extended Abstracts, Second International Kimberlite Conference, Santa Fe, New Mexico, 1977.

Moore, M., and A.R. Lang, On the origin of the rounded dodecahedral habit of natural diamond, J. Cryst. Growth, 26, 133-139, 1974.

Phaal, C., Surface studies of diamond, Ind. Diamond Rev., 25, 486-489 and 591-595, 1965.

Ponomarenko, A.I., N.V. Sobolev, N.P. Pokhilenko, Yu.G. Lavrentyev, and V.S. Sobolev, Diamond-bearing grosspidite and diamond-bearing disthene eclogites from the kimberlite pipe "Udachnaya", Yakutia, Dokl. Akad. Nauk SSSR, 226, No. 4 (Petrography), 927-930, 1976, (in Russian).

Ponomarenko, A.I., and Z.V. Spetzius, Diamond-bearing eclogites from kimberlite pipe "Sitikanskaya", Geologiya i Geofizika, No. 6, 103-106, 1976, (in Russian).

Reid, A.M., R.W. Brown, J.B. Dawson, G.G. Whitfield, and J.C. Siebert, Garnet and pyroxene compositions in some diamondiferous eclogites, Contrib. Mineral. Petrol., 58, 203-220, 1976.

Rickwood, P.C., J.J. Gurney, and D.R. White-Cooper, The nature and occurrences of eclogite xenoliths in the kimberlites of Southern Africa, in Upper Mantle Project, Geol. Soc. S.Afr. Special Publication No. 2, pp. 371-394, 1969.

Seal, M., The growth history of natural diamonds as revealed by etching experiments, in Proceedings of the First International Congress on Diamonds in Industry, Paris, 1962, pp. 361-376, Industrial Diamond Information Bureau, London, 1963.

Sobolev, N.V., On the origin of diamond-bearing eclogites, Geologiya i Geofizika, No. 12, (Criticism and Discussions), 1970, (in Russian).

Sobolev, V.S., First finds of diamond-bearing eclogite in kimberlite pipe "Udachnaya", Dokl. Akad. Nauk SSSR, 209, (Petrography) 188-189, 1973.

Sobolev, N.V., Mineral paragenesis of natural diamonds, Extended Abstracts, Second International Kimberlite Conference, Santa, Fe, New Mexico, 1977.

Vance, E.R., J.W. Harris, and H.J. Milledge, Possible origins of α-damage in diamonds from kimberlite and alluvial sources, Mineral. Mag., 39, 349-360, 1973.

Wagner, P.A., The Diamond Fields of Southern Africa, 1914, (Second impression Struik, Cape Town, 1971).

Wentorf, R.H., Diamond synthesis, in Advances in Chemical Physics, 9, edited by I. Prigogine, pp. 365-404, Interscience, New York, 1965.

LOWER-CRUSTAL GRANULITES AND ECLOGITES FROM LESOTHO, SOUTHERN AFRICA

W. L. Griffin

Geologisk Museum, Oslo, Norway

D. A. Carswell

Department of Geology, Sheffield University, Sheffield, England

P. H. Nixon

Geology Department, Box 4820 University P.O., Papua, New Guinea

Abstract. Nodules of garnet granulites and "crustal" eclogites occur in many kimberlites around the edge of the Kaapvaal craton, but are apparently absent from pipes within the craton. In most Lesotho pipes this suite is dominated by basic cpx+plag+gnt±opx granulites, with smaller numbers of eclogites, garnet websterites and intermediate/acid granulites. Two-pyroxene granulites are common at Monastery Mine.

The basic granulites are essentially olivine-basaltic, with minor normative Hy or Ne. The absence of plagioclase in eclogites and garnet websterites is controlled mainly by bulk chemistry; only rocks with >45% normative An+Ab contain modal feldspar. Microprobe data for 50 rocks demonstrate the consanguinity of the garnet granulite-eclogite-websterite suite. Clinopyroxenes from all garnetiferous samples are eclogitic (Jd/Ts > ½). Jd/Jd+Ts is proportional to Ab/Ab+An in the coexisting plagioclase (An 5-77%); Jd ranges from <5 to 30%. Garnets in all rock types resemble those from eclogites in gneiss and blueschist. High-sulfur (3.5-6.8% SO_3) scapolite occurs in six samples.

Most samples of the garnet granulite suite yield T-P estimates between 550-700°C and 5-13 Kb, implying an origin in the lower crust. Seismic data on adjacent areas, measured densities, and data on nodule abundances suggest that the lower crust beneath Lesotho contains 50-70% basic granulite+eclogite+websterite, and 50-30% intermediate to acid granulite. Fe-rich garnet websterites give P [illegible]timates in the range 13-19 Kb, and may represent an important rock type in the uppermost mantle.

Our crustal P-T estimates lie closer to predicted oceanic geotherms than to the "Shield" geotherm, suggesting that the latter is only valid for the deeper parts of the upper mantle. The high T at lower-crustal depths during mid-Cretaceous time presumably was a consequence of the Karroo volcanic activity.

Introduction

Petrological studies of supposed deep-crustal rocks (granulites, eclogites, etc.) usually contain an inherent ambiguity. The exposure of these rocks on the earth's surface implies that they have been brought up from depth by large-scale tectonic movements. Unless this uplift is, geologically speaking, very rapid, the mineralogy and chemistry of the rocks will probably be modified by partial or complete reequilibration at shallower crustal depths. This modification may prevent study, or even recognition, of the original nature of the rocks.

Kimberlite eruption, on the other hand, is an extremely rapid process. McGetchin and Ullrich (1973) estimate that an ascending kimberlite magma would traverse the thickness of the continental crust in less than 15 minutes. Magma temperatures have been estimated to be around 500°C (Sheppard and Dawson, 1975). The circulation of meteoric water, besides causing serpentinization, probably reduces the temperature of emplaced diatreme-facies kimberlite to <100°C in a very short time. Nodules of deep-seated crustal rocks in these diatremes may therefore be regarded as quenched samples, at least by comparison with normal surface outcrops. If they escape hydrothermal alteration during the hydration of the kimberlite, these nodules offer a unique chance to study deep-crustal chemistry and mineral equilibria.

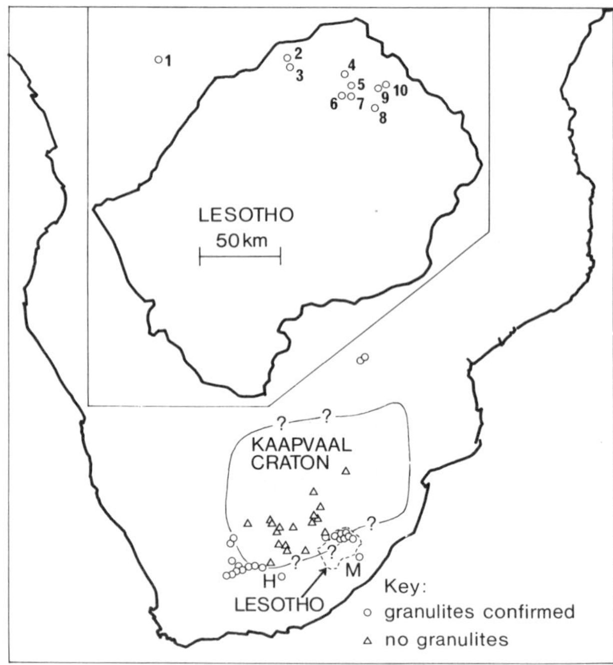

Fig. 1. Distribution of granulite nodules in the kimberlites of Southern Africa, based largely on survey of Rickwood et al. (1969). H, Hanover pipe; M, Melkfontein. Inset: Lesotho, showing sample localities: 1, Monastery Mine; 2, Lipelenang; 3, Khabos; 4, Lemphane; 5, Lighobong; 6, Kao; 7, Pipe 200; 8, Matsoku; 9, Mothae; 10, Letseng-la-Terae.

This paper presents a study of 55 granulite, eclogite and garnet websterite nodules from 11 kimberlites in Lesotho and adjacent South Africa. We have also studied a suite of spinel lherzolites, presumably from the shallow mantle or deep crust, in the Lipelenang and Ngopetseu kimberlites; primary data on these will be presented elsewhere.

Occurrence of granulite-suite nodules

The "garnet granulite suite" described here includes pyroxene-garnet granulites (cpx+gnt+plag±opx), eclogites (cpx+gnt) and garnet websterite (cpx+opx+gnt). The essentially continuous variation in modal composition and mineral chemistry among these three groups (see below) suggests that most, if not all, of these rocks formed through metamorphism of a dominantly mafic suite under a rather restricted range of P and T. Two-pyroxene, garnet-free granulites occur at Monastery Mine.

Our collection includes nodules from nine kimberlites in Lesotho and three (Monastery Mine; Hanover; and Melkfontein, E. Griqualand) in adjoining South Africa. A survey of the literature and available collections shows that garnet granulites ("plagioclase eclogites" of Ringwood et al., 1969) have been found only in kimberlites on, or outside of, the edge of the Kaapvaal craton (Fig. 1). Pipes within the craton apparently lack granulite nodules, and contain eclogite nodules only of the "griquaite" type (Nixon, 1973). Basement xenoliths in the pipes of the craton interior are usually intermediate to acid, amphibolite-facies gneisses that may or may not represent deep-crustal material.

The Kaapvaal craton consists largely of Early Archaean gneisses and greenstone belts. It is bordered by younger mobile belts characterized by high-grade regional metamorphism, commonly in granulite facies. Kröner (1977) argues that these belts were formed primarily by intracratonic reworking of the older rocks, with little addition of new material. The restriction of the granulite nodules to kimberlites lying along the craton-mobile belt boundary may indicate that the granulite suite is a product of the reworking processes. This suggestion is strengthened by the Proterozoic U/Pb ages obtained on zircons from two granulite nodules from Lesotho kimberlites (Davis, 1977). If the granulite suite is present beneath the craton, something must have prevented the kimberlites from sampling it there.

Petrography

Most of our samples are 1-3 cm in size and relatively homogeneous. The larger ones, however, commonly display compositional layering on a scale of mm to cm. Interlayering of eclogite and granulite on these scales is illustrated in "Lesotho Kimberlites" (P. Nixon, ed. 1973) (pl. 43A, our sample KN-195), and by our sample L-12 (Table 1). Many other samples show a foliation defined by thin, discontinuous trains of mafic minerals in plagioclase, or of garnet in clinopyroxene (see "Lesotho Kimberlites", pl. 33A, our PHN-2438).

The dominant texture in all rock types is polygonal granoblastic (see "Lesotho Kimberlites" Pl. 15A, 29B, 34, 40A). The rocks are typically equigranular, with average grain sizes from 0.5-1.5 mm. Despite the compositional layering, dimensional orientation of grains is rare, as is undulatory extinction or other evidence of strain. The textures are interpreted as the result of annealing in a static environment (Nixon, 1973), perhaps reflecting heating during the Karroo volcanic episode (see below).

Clinopyroxene, orthopyroxene, garnet and plagioclase account for >95% of the mode in most samples (Table 1). Within the garnet granulite suite there is essentially continuous modal variation from granulites to eclogites and garnet pyroxenites (as plagioclase ranges from 50% to <1%) and from eclogites to garnet websterites (opx from 0 to 30%) (Fig. 2). Hydrous minerals are rare; where present they

Table 1. Modal Data for Granulites, Eclogites and Pyroxenites

Sample	type	dens.	gr. size	cpx	opx	gnt	plag	%An	scap.	amp.	bio.	ap.	rut.	ilm.	other phases	comments
HANOVER																
PHN-2317	GG	-	F	40	-	38	22	21	-	-	-	-	-	-	Qtz	Foliated
KAO																
K-1	GG*	2.39	M	+		+	+	-	-	-	-	-	-	+	Natrol., seric.	Compl. alt.
K-2	GG	2.94	M/C	26	-	31	40	20	3	-	-	-	+	-	Kspar	Anom. rutile
K-3	GG	3.27	M/C	48	-	34	18	13	-	+	+	-	+	-	Pyrite	Sec. amph.
K-4	GG	-	C	26	-	69	5	77	-	+	-	-	-	-		- " -
K-5	Ec	3.66	C	52	-	42	-	-	-	-	-	-	-	6		
KAO-1	Ec	-	M	46	-	51	-	-	-	-	3	-	+	-		Anom. rutile
KAO-2	Ec	-	M/C	67	-	33	-	-	-	-	+	-	+	-		- " -, sec. bio.
KC-22	Ec	-	M	36	-	64	-	-	-	-	-	-	-	-		
KC-23	Ec	-	M	61	-	39	-	-	-	-	-	-	-	-		
KC-28	GG	-	M	34	-	44	1	50	-	-	-	-	-	21		
KC-29	GG	-	M/C	78	-	20	2	alt.	-	-	-	-	-	-		
KC-30	Ec	-	M/C	49	-	40	-	-	-	2	-	-	-	9		
KC-31	Ec	-	M	44	-	53	-	-	-	-	3	-	+	-		
PHN-2438	GG	2.69	F/M	33	-	21	46	8	-	-	-	-	-	-		"Lesotho Kimb." pl.33A
KHABOS																
PHN-1442j	GG	-	M	8	-	46	44	44	-	-	-	-	-	-	2% Kyan.	Al_2SiO_5/cpx sympl.
LEMPHANE																
PHN-2508	GW	-	M	47	15	15	-	-	-	8	-	+	-	-	15% Mgt.	Gnt. exsol. in pyrox., sec. poik. amph.
LETSENG-LA-TERAE																
LT-2	GG*	2.74	M/C	27	+	23	28	27	+[1]	-	3	+	+	+	18% Qtz.	Opx incl. in cpx.
LT-6	GG	-	M	34	-	22	41	30	-	-	-	-	+	3		
PHN-2003/1	GG	-	M	13	-	20	67	alt.	-	-	+	-	-	-		Heav. alt.
PHN-2567/5	GW	-	M/C	69	9	22	-	-	-	-	-	-	+	+		
LIPELENANG																
L-5	GG*	3.03	M/C	28	4	21	47	52	-	-	-	-	-	-	Sillim., corun., spinel	Coronite. Al_2SiO_5 in plagioclase.
LIQHOBONG																
LQ-1	GW	3.35	M	52	5	40	-	-	-	3	+	-	-	-	Fe, Ni sulf.	Anom. rutile
LQ-2	GG	2.94	M	31	-	43	23	32	3	-	-	-	+	-	Qtz	Altered
LQ-4	GG*	3.24	M	39	-	30	10	5	-	-	-	+	8	3	3% Qtz, 7% Ksp.	- " -
PHN-2494	GG	3.08	F/M	26	-	28	46	12	-	-	-	-	-	-	Ksp. (Ab_{32})	Plag = $Or_{4.5}$
PHN-2495	GG*	3.37	C	52	6	41	tr.	16	-	-	+	+	1	-		Anom. rutile
KN-195	GG	3.16	F/M	26	-	43	31	12	-	-	-	-	+	-	Ksp. (Ab_{35})	"Lesotho Kimb."pl.43A
KN-206	GW	3.50	C	39	30	26	-	-	-	+	-	+	5	+		Anom. rut. (SE Blow)
MATSOKU																
M-1	GG*	3.13	M	32	-	34	34	23	-	-	-	-	-	-	Kyan. (sec.)	Finely layered
M-3	GW	3.37	C	58	23	19	-	-	-	-	-	-	+	+		Anom. rutile
M-5	GW	3.39	M	60	15	25	-	-	-	-	-	-	+	+		- " -
L-9	GW*	3.38	C	75	6	19	-	-	-	-	-	-	+	+		- " -, opx alt.

Table 1, Cont'd.

{ L-12A	GG*	3.15	M	47	-	23	30	19	-	-	-	-	-	-	Qtz.	
L-12B	Ec*	3.15	M/C	70	-	26	-	-	+	-	-	-	+	+	4% Qtz, pectol.	Layered granul./eclog.
L-13	GG*	3.10	M	37	-	29	34	23	-	-	+	-	+	-	Kyan.(sec.)	
L-16	GW*	3.31	M	60	19	21	-	-	-	-	-	-	+	+		Anom. rut.,opx alt.
L-17	GW*	3.40	M	58	5	37	-	-	-	-	-	-	+	+		
L-20	GG*	3.17	M	26	5	46	20	17	-	-	1	-	+	-	Ksp.	
PHN-1646A	GG	3.21	M	50	-	29	21	19	-	-	+	-	+	-		Anom. rut., (cf.Rogers,
PHN-1670	GG*	3.05	M	20	-	33	47	23	+	-	-	-	+	-		Altered 1977)
PHN-2852	GG*	3.03	M	30	-	32	35	19	3	-	-	-	+	-		Anom. rut.(cf.Rogers,
OVK-F 10303	GG*	3.00	M	19	-	37	44	19	-	-	-	-	+	+	Kyan. (sec.)	Sec. carb. 1977)
																pect.(cf.Rogers,1977)
MONASTERY																
MO-4	PG	3.15	F/M	48	22	-	21	47	-	8	1	-	-	-	Spinel	
MO-6	PG	2.97	M	43	6	-	34	55	-	17	-	-	-	-		
MO-9	SP	-	F	-	40	-	-	-	-	45	-	-	-	-	15% spinel	
PHN-2630/1	GW*	3.52	F/M	33	2	53	-	-	-	-	12	-	-	-		
PHN-2630/2	GG*	3.14	M	38	8	32	18	50	-	-	+	-	-	4		Kelyphite
PHN-2645	PG*	2.94	M	39	10	-	45	35	-	-	4	-	-	-	Ti-mgt	
MOTHAE																
PHN-1919	GG*	-	M	36	-	33	29	18	-	-	2	-	+	+		Heavy alt. "Les.Kimb." pl. 15A(cf.Rogers,1977)
PIPE 200																
PHN-2450	Ec	-	C	37	-	61	-	-	-	+	1	-	1	-		Anom. rutile
PHN-2496	Ec	3.34	C	71	-	23	-	-	-	5	1	-	+	-		- " -
PHN-2532	Ec*	3.42	M/C	55	-	36	-	-	-	-	-	+	9	+		Sec. carb.
PHN-2533	GG*	3.13	M	45	-	27	27	12	-	-	1	+	+	+		Kelyphite(cf.Rogers, 1977)
PHN-2685/1A	Ec*	3.13	C	45	-	50	-	-	-	-	5	-	+	-	Carbonate	Anom.rut.Altered.
MELKFONTEIN																
PHN-3017	GG	-	M/C	45	-	45	5	27	5	-	-	-	-	-		

Rock types: GG, garnet granulite; PG, pyroxene granulite; Ec, eclogite; GP, garnet pyroxenite; SP, spinel pyroxenite.
*Sample analyzed (Table 2).

Grain size: F, fine-grained (<0.5 mm average diameter); M, medium-grained (0.5-1.5 mm); C, coarse-grained (>1.5 mm).

1: One thin section contains no scapolite, another ca. 5%.

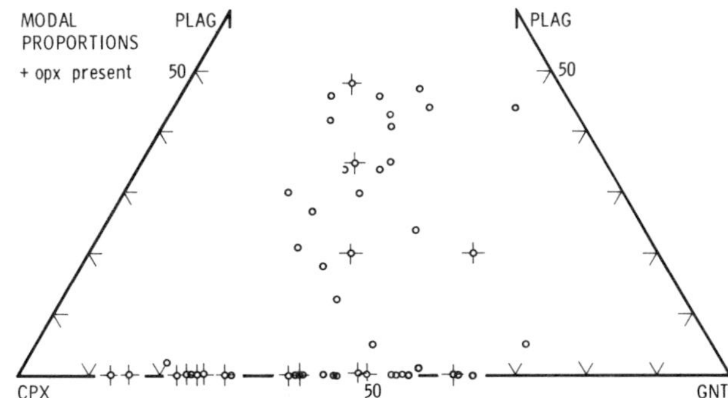

Fig. 2. Modal proportions of pyroxenes, garnet and plagioclase. See also Table 1.

usually appear to be in textural equilibrium with the anhydrous phases. The main exceptions to this rule are the pyroxene granulites from Monastery Mine, in which amphibole surrounds and replaces pyroxenes, and the garnet websterite PHN-2508, which contains abundant poikilitic amphibole. Other cases of disequilibrium features are noted in Table 1.

The most important accessory mineral is rutile, which occurs both as discrete grains and commonly as acicular inclusions in garnet and pyroxene. These needles always show inclined extinction, the result of an unusual habit that appears to be characteristic of rutile inclusions in minerals formed at high P and T (Griffin et al., 1971). Ilmenite, zircon and apatite are also common accessory minerals. Quartz and scapolite are less common, but are abundant ($\lessapprox$15%) in some specimens, as are magnetite, orthoclase, kyanite and sillimanite (Table 1).

Many samples show hydrothermal alteration, presumably by fluids from the kimberlite. Natrolite, pectolite, sericite and carbonate replace feldspars. Orthopyroxene is commonly altered to a brown serpentine-like material, especially in the garnet websterites, and green "kelyphite" rims garnets in a few samples. Scapolite is commonly altered to an unidentified light brown material; nearly complete loss of SO_3 is associated with the early stages of this alteration. Grain boundaries and cracks in the clinopyroxenes of some eclogites and garnet websterites show wide, sharply bounded zones filled with vermicular, apparently empty, channels in a symplectite-like pattern. Similar rims have been observed in garnet lherzolite nodules (Carswell, 1975).

Although equilibrium metamorphic textures are the rule in this nodule suite, some important exceptions should be noted.

(1) Opx in LT-2 occurs only as rounded inclusions in clinopyroxene, suggesting that the reaction opx+plag $\rightarrow$ cpx+gnt+qtz has taken place.

(2) L-6 is a coronite, showing "primary" Al-rich pyroxenes surrounded by zones of granular garnet with inclusions of spinel and corundum. The following reactions have probably occurred:

(a) $3MgAl_2SiO_6 \rightarrow Mg_3Al_2Si_3O_{12} + 2Al_2O_3$

(b) $MgAl_2O_4 + 3MgSiO_3 + CaAl_2Si_2O_6 \rightarrow$

$CaMgSi_2O_6 + Mg_3Al_2Si_3O_{12} + Al_2O_3$

The plagioclase is full of randomly oriented needles of an Al_2SiO_5 phase (sillimanite?).

(3) In PHN-1442j rinds of coarse-grained garnet surround areas in which poikilitic clinopyroxene encloses tiny garnet euhedra and laths of kyanite. Elsewhere, sheaves of kyanite needles (X-ray identification; B. Harte, pers. comm. 1977) are intergrown with clinopyroxene (cf. Meyer and Brookins, 1976, Fig. 2). A coarser-grained kyanite/cpx intergrowth was seen in a coronite anorthosite from Thaba Putsoa (not analyzed). The reactions involve the breakdown of plagioclase and a primary clinopyroxene rich in Tschermak's molecule, to produce kyanite, a less aluminous secondary clinopyroxene and a Ca, Fe-rich garnet (Table 3). The reactions responsible for the texture may be of the type:

(c) 3Mg-Ts + 3An + Ab $\rightarrow$ Gros + Py + Jd + 4Kyan

(d) 3Mg-Ts + 3An + 3Ab $\rightarrow$ 3Di + 3Jd + 6Kyan

The higher Jd content and lower Ts content of the secondary pyroxene and the higher K_D (Fe/Mg) of the secondary gnt + cpx pair indicate that these reactions proceeded with decreasing T.

The intergrowths of kyanite needles with clinopyroxene are also present, though rare, in samples M-1, L-13 and OVK-F10303, all from Matsoku.

(4) In PHN-2508, garnet occurs as small rods, apparently exsolved, in clinopyroxene and orthopyroxene.

(5) The pyroxene granulites from Monastery Mine show extensive exsolution of orthopyroxene lamellae in clinopyroxene, a feature that is rare in the garnet-granulite suite.

Whole-rock Chemistry

Whole-rock analyses and CIPW norms for 25 nodules are presented in Table 2. One strikingly banded nodule (L-12) was split into eclogite and granulite portions for analytical purposes. Analyses were done at the University of Sheffield, mainly by XRF; major elements were analyzed on fusion pellets (1:9 lithium tetraborate), trace elements on pressed powders. Contents of FeO, Na_2O, H_2O and CO_2 were determined separately by standard wet-chemical techniques.

Four of our analyzed samples have also been analyzed by Rogers (1977). The large differences in major-element values between our analyses

Table 2. Whole Rock Analyses and CIPW Norms

Rock Type	Garnet Granulites											Banded Specimen	
Sample Number	PHN 2630/2	LQ 4	PHN 2852	L 20	L 6	PHN 1670	PHN 2533	L 13	M 1	OVKF 10303	LT 2	L 12A	L 12B
SiO_2	42.97	43.97	44.98	46.82	47.23	49.18	49.69	49.89	50.37	50.89	59.93	49.50	49.65
TiO_2	1.86	3.30	1.39	0.06	0.06	0.53	0.67	0.36	0.35	0.26	0.87	0.43	0.69
Al_2O_3	12.25	11.89	16.87	13.81	23.93	18.24	14.69	18.55	18.89	19.12	13.07	16.23	12.80
Fe_2O_3	3.71	5.03	1.35	2.07	0.72	0.61	2.16	0.60	0.71	0.90	0.60	1.07	0.88
FeO	12.40	13.22	8.33	10.24	3.82	6.44	9.09	5.57	5.52	5.73	4.66	6.70	8.22
MnO	0.28	0.27	0.15	0.18	0.07	0.11	0.21	0.13	0.10	0.12	0.10	0.14	0.17
MgO	7.52	5.16	8.27	9.04	8.18	7.75	6.93	8.11	8.05	6.80	2.42	8.70	10.82
CaO	12.46	9.87	10.85	10.91	11.15	10.07	9.30	10.44	10.38	8.59	9.77	11.44	13.33
Na_2O	1.30	2.39	3.16	2.16	2.54	3.09	3.26	3.45	3.46	4.14	4.69	3.41	2.25
K_2O	1.05	1.13	0.39	1.13	0.18	0.96	1.88	0.72	0.88	1.03	0.92	0.47	0.42
P_2O_5	0.44	0.74	0.14	0.25	0.02	0.11	0.03	0.04	0.06	0.05	0.21	0.03	0.03
S	0.08	0.15	0.06	0.05	0.03	0.04	0.04	0.05	0.05	0.04	0.06	0.07	0.16
H_2O	1.65	1.69	3.41	2.05	0.80	1.94	1.19	0.89	1.08	1.20	1.62	0.75	0.43
CO_2	0.92	NIL	NIL	NIL	NIL	NIL	NIL	NIL	NIL	0.36	NIL	0.18	TRACE
Total+	99.39	99.00	99.77	100.55	98.94	99.77	99.58	98.98	100.11	99.65	99.00	99.43	100.25
$\frac{100\ Mg}{Mg+Fe}$	46.0	34.1	60.7	57.1	76.5	66.4	52.8	70.3	70.0	64.9	45.3	66.9	68.1
$\frac{100\ Na}{Na+Ca}$	15.9	30.5	34.5	26.4	29.2	35.7	38.8	37.4	46.6	46.6	46.5	35.0	23.4
PPM													
Ni	156	60	56	114	32	69	39	144	83	59	23	141	221
Co	78	114	80	94	48	60	84	83	90	52	60	88	82
V	425	348	131	225	12	68	309	138	139	96	129	158	268
Cr	223	36	427	261	228	347	174	284	297	254	98	315	890
Zn	121	190	96	107	27	64	235	46	48	50	40	59	65
Cu	200	59	10	51	11	18	18	21	13	21	28	45	127
Rb	52	21	12	21	2	30	24	10	13	15	3	11	5
Sr	396	283	1419	1070	232	3556	652	882	1081	2415	294	812	698
Zr	56	137	29	106	3	29	5	14	19	16	76	18	22
Ba	3611	535	2451	2428	1002	1854	2949	183	237	682	388	1217	358
K/Rb	168	447	268	448	751	266	650	599	561	572	2557	352	689

+ Total includes wt.% Cr_2O_3, NiO, SrO and BaO

Table 2. Cont.

Rock Type	2 Pyroxene Granulite	Garnet Websterites							Eclogite		Highly Altered Xenoliths		
Sample Number	PHN 2645	L 17	PHN 2495	KN 206	L 9	M 5	L 16	M 3	PHN 2630/1	PHN 2532	PHN 2685/1A	PHN 1919	K 1
SiO_2	47.34	45.87	46.98	47.15	47.88	48.07	48.50	48.50	41.06	42.92	32.75	44.57	46.21
TiO_2	0.76	0.71	1.85	1.58	1.24	0.49	0.71	0.51	0.51	2.22	1.36	1.01	0.27
Al_2O_3	13.45	10.54	11.05	7.91	7.75	10.24	8.36	7.79	12.72	12.45	12.23	15.69	19.15
Fe_2O_3	3.20	3.23	2.58	4.23	4.71	4.34	3.01	3.55	1.93	2.76	1.86	3.33	0.45
FeO	7.86	9.55	10.31	16.74	9.47	5.45	6.89	10.50	19.45	12.54	13.88	7.13	1.42
MnO	0.18	0.22	0.21	0.29	0.22	0.20	0.19	0.25	0.58	0.21	0.33	0.19	0.05
MgO	8.59	13.52	13.38	16.48	13.17	16.77	15.70	14.26	9.02	9.52	14.57	9.23	0.20
CaO	8.95	12.46	10.21	4.07	12.39	12.50	13.28	11.21	10.60	12.36	10.12	7.92	8.37
Na_2O	2.60	1.47	1.68	0.79	1.84	0.78	1.56	1.24	0.21	1.74	0.11	5.19	14.07
K_2O	2.42	0.04	0.26	0.03	0.02	0.06	0.06	0.03	0.84	0.02	0.21	0.28	0.15
P_2O_5	0.10	0.11	0.48	0.06	0.04	0.02	0.04	0.03	0.04	0.04	0.20	0.04	0.02
S	0.04	0.00	0.03	NIL	0.09	0.01	0.04	0.08	0.04	0.03	0.40	0.04	0.03
H_2O	1.97	1.36	0.74	1.09	0.98	1.20	1.60	1.53	1.30	1.26	4.76	5.67	7.80
CO_2	0.73	TRACE	NIL	NIL	NIL	NIL	TRACE	NIL	0.42	0.50	4.65	NIL	NIL
Total+	99.74	99.25	100.12	100.42	99.44	100.13	100.57	99.48	98.90	98.66	99.15	100.42	98.22
$\frac{100\ Mg}{Mg+Fe}$	58.8	65.9	65.4	59.2	64.0	76.3	74.5	65.1	43.1	53.0	62.5	61.9	16.3
$\frac{100\ Na}{Na+Ca}$	35.4	17.6	22.9	26.0	21.2	10.1	17.5	16.7	3.5	20.3	1.9	54.3	75.3
PPM													
Ni	87	380	357		264		581		257	114	168	173	8
Co	51	104	111		87		78		77	110	81	84	4
V	268	358	253		418		288		169	410	384	176	19
Cr	157	771	1536		903		3717		679	311	356	149	51
Zn	83	90	113		114		73		81	118	340	97	13
Cu	19	39	41		89		94		31	18	177	99	14
Rb	51	0	1		5		6		77	1	11	22	7
Sr	891	113	163		182		111		67	147	610	252	200
Zr	47	19	146		33		41		399	30	103	16	96
Ba	12611	70	637		64		77		407	123	14172	556	80
K/Rb	395	-	2158		33		83		91	125	162	105	183

+ Total includes wt.% Cr_2O_3, NiO, SrO and BaO

Table 3. Pyroxenes and Garnets from Granulite-suite Nodules

Clinopyroxenes

	PHN 2317	K-2	K-3	K-4	K-5	KAO 2	KC 22	KC 23	KC 28	KC 29	KC 30	KC 31	PHN 2438
Rk type	GG	GG	GG	Ec	Ec	Ec	Ec	Ec	GG	GG	Ec	Ec	GG
SiO_2	51.6	51.9	52.6	51.4	52.3	52.9	53.5	53.4	53.1	51.6	51.8	53.8	53.4
TiO_2	0.4	0.6	0.4	0.6	0.3	0.8	0.5	0.3	0.4	0.3	0.2	0.3	0.8
Al_2O_3	5.3	8.3	6.4	6.6	4.3	8.4	8.6	8.2	7.3	3.8	3.6	6.6	8.3
Fe_2O_3	4.5	1.3	4.7	0.0	5.0	2.2	1.4	3.2	3.9	5.1	3.8	2.6	1.8
FeO	4.1	2.9	4.0	3.6	4.5	1.6	2.2	3.9	1.2	4.2	5.7	0.6	3.0
MnO	-	-	0.1	0.09	0.1	0.15	-	-	0.15	0.15	0.2	0.15	-
MgO	11.4	12.2	10.6	13.9	11.6	12.0	12.6	9.7	12.0	12.4	12.0	13.6	11.5
CaO	20.0	20.6	17.7	22.7	18.8	17.5	19.5	16.5	17.5	21.3	20.5	19.6	18.4
Na_2O	2.6	2.5	3.8	0.94	2.9	4.0	3.2	4.7	4.0	1.8	1.8	3.1	3.8
Σ	99.9	100.3	100.3	99.83	99.8	99.55	101.5	99.9	99.55	100.65	99.6	100.35	101.0
Si	1.904	1.877	1.921	1.878	1.932	1.909	1.898	1.938	1.921	1.904	1.928	1.926	1.910
Al^{IV}	0.096	0.123	0.079	0.122	0.068	0.091	0.102	0.062	0.079	0.096	0.072	0.074	0.090
Al^{VI}	0.135	0.231	0.197	0.160	0.120	0.267	0.258	0.289	0.233	0.067	0.086	0.206	0.262
Ti	0.012	0.016	0.011	0.015	0.008	0.022	0.014	0.008	0.010	0.008	0.005	0.008	0.020
Fe^{3+}	0.124	0.034	0.129	-	0.139	0.059	0.036	0.088	0.107	0.141	0.106	0.070	0.048
Fe^{2+}	0.126	0.087	0.121	0.111	0.139	0.047	0.064	0.118	0.038	0.129	0.178	0.017	0.089
Mn	-	-	0.003	0.003	0.003	0.005	-	-	0.005	0.005	0.006	0.005	-
Mg	0.627	0.658	0.577	0.758	0.639	0.645	0.666	0.525	0.649	0.679	0.668	0.726	0.614
Ca	0.791	0.798	0.693	0.886	0.744	0.677	0.741	0.642	0.678	0.843	0.819	0.750	0.706
Na	0.186	0.175	0.269	0.067	0.208	0.279	0.220	0.331	0.281	0.128	0.131	0.218	0.261

Table 3 (cont.)

Clinopyroxenes	PHN 1442j prim.	PHN 1442j sec.	PHN 2508	LT 2	LT 6	PHN 2567/5	L 6	LQ 1	LQ 2	LQ 4	PHN 2494	PHN 2495	KN 195
Rk type	GG	GG	GP	GG	GG	GP	GG	GP	GG	GG	GG	GG	GGG
SiO_2	50.5	52.2	52.5	52.0	51.1	54.1	50.3	53.0	52.6	53.6	53.6	53.2	53.7
TiO_2	0.0	-	0.1	0.6	0.7	0.2	0.8	0.4	0.5	0.2	0.5	0.5	0.5
Al_2O_3	9.5	7.0	2.5	4.2	5.7	4.1	7.1	6.9	6.9	7.2	7.7	6.4	7.3
Fe_2O_3	2.7	1.5	3.7	3.2	2.1	1.7	3.5	3.7	0.0	6.5	3.2	3.7	4.0
FeO	0.9	1.5	4.3	5.0	6.1	4.1	1.2	1.6	2.8	5.7	2.4	2.9	1.7
MnO	-	-	0.1	-	0.06	-	-	0.0	0.04	0.1	-	-	-
MgO	12.5	13.4	13.7	12.8	11.7	13.9	14.0	12.3	13.8	7.5	11.4	11.9	11.6
CaO	21.8	22.0	21.2	20.7	20.8	19.8	21.7	19.1	21.5	14.2	17.4	17.8	17.4
Na_2O	2.0	1.5	1.5	1.8	1.8	2.3	1.5	3.4	1.8	5.8	4.2	3.7	4.3
Σ	99.9	99.5	99.6	100.3	100.06	100.2	100.1	100.4	99.94	100.8	100.4	100.1	100.5
Si	1.831	1.896	1.944	1.913	1.890	1.964	1.831	1.911	1.903	1.953	1.927	1.930	1.929
Al^{IV}	0.169	0.104	0.056	0.087	0.110	0.036	0.169	0.089	0.097	0.047	0.073	0.070	0.071
Al^{VI}	0.237	0.196	0.053	0.095	0.139	0.140	0.135	0.205	0.195	0.263	0.254	0.204	0.238
Ti	-	-	0.004	0.015	0.019	0.005	0.022	0.011	0.014	0.008	0.013	0.013	0.013
Fe^{3+}	0.073	0.041	0.103	0.089	0.059	0.047	0.096	0.100	-	0.177	0.086	0.101	0.107
Fe^{2+}	0.027	0.047	0.132	0.154	0.189	0.123	0.035	0.048	0.084	0.174	0.073	0.087	0.052
Mn	-	-	0.003	-	0.002	-	-	-	0.001	0.003	-	-	-
Mg	0.676	0.726	0.756	0.702	0.643	0.752	0.760	0.661	0.747	0.408	0.611	0.643	0.621
Ca	0.847	0.856	0.841	0.816	0.822	0.770	0.846	0.738	0.835	0.556	0.670	0.692	0.670
Na	0.141	0.134	0.108	0.128	0.127	0.162	0.106	0.238	0.123	0.411	0.293	0.260	0.299

Table 3 (cont.)

Clinopyroxenes

	KN 206	M 1	M 3	M 5	L 9	L 12A	L 12B	L 13	L 16	L 17	L 20	PHN 1646A	PHN 1670
Rk type	GP	GG	GP	GP	GP	GG	Ec	GG	GP	GP	GG	GG	GG
SiO_2	53.8	52.2	53.5	53.0	54.0	52.4	53.0	52.7	53.4	53.4	52.8	52.3	52.5
TiO_2	0.26	0.6	0.17	0.2	0.1	0.3	0.3	0.6	0.3	0.2	0.3	0.2	0.7
Al_2O_3	5.9	10.6	4.8	4.2	4.7	9.9	8.7	10.8	5.5	5.1	5.4	4.8	7.4
Fe_2O_3	5.3	1.2	4.0	1.3	3.2	1.3	1.5	1.6	2.6	2.6	2.9	3.7	2.1
FeO	4.2	1.5	3.7	2.9	3.9	2.6	2.9	1.3	2.2	3.2	4.2	2.8	1.8
MnO	-	-	-	-	0.1	0.1	0.1	-	0.1	0.1	0.1	-	-
MgO	10.3	11.5	12.3	14.6	12.2	10.8	11.5	11.3	13.6	12.9	12.1	13.0	12.9
CaO	15.8	19.3	18.5	21.6	18.4	18.2	18.1	19.5	19.9	19.3	19.9	20.0	20.0
Na_2O	4.7	3.5	3.2	1.5	3.3	3.8	3.7	3.7	2.6	2.8	2.6	2.4	2.8
Σ	100.26	100.4	100.17	99.3	99.9	99.4	99.8	101.5	100.2	99.6	100.3	99.2	100.2
Si	1.960	1.866	1.950	1.939	1.969	1.896	1.915	1.865	1.930	1.948	1.928	1.925	1.894
Al^{IV}	0.040	0.134	0.050	0.061	0.031	0.104	0.085	0.135	0.070	0.052	0.072	0.075	0.106
Al^{VI}	0.214	0.312	0.156	0.120	0.171	0.319	0.285	0.315	0.164	0.167	0.160	0.133	0.208
Ti	0.007	0.016	0.005	0.006	0.003	0.008	0.008	0.016	0.008	0.005	0.008	0.006	0.018
Fe^{3+}	0.144	0.033	0.111	0.037	0.087	0.036	0.042	0.042	0.072	0.072	0.080	0.102	0.058
Fe^{2+}	0.127	0.045	0.112	0.089	0.120	0.078	0.088	0.038	0.067	0.099	0.128	0.086	0.053
Mn	-	-	-	-	0.003	0.003	0.003	-	0.003	0.003	0.003	-	-
Mg	0.559	0.613	0.668	0.796	0.663	0.585	0.618	0.596	0.733	0.701	0.658	0.713	0.694
Ca	0.617	0.739	0.722	0.847	0.719	0.705	0.699	0.739	0.771	0.754	0.778	0.789	0.773
Na	0.332	0.243	0.226	0.106	0.233	0.267	0.258	0.254	0.182	0.198	0.184	0.171	0.196

Table 3 (cont.)

Clinopyroxenes	PHN 2852	OVKF 10303	MO 4	MO 6	PHN 2630/1	PHN 2630/2	PHN 2645	PHN 1919	PHN 2450	PHN 2496	PHN 2532	PHN 2533	PHN 2685/1A	PHN 3017
Rk type	GG	GG	PG	PG	GP	GG	PG	GG	Ec	Ec	Ec	GG	Ec	GG
SiO_2	52.9	53.0	53.0	51.9	52.0	53.4	52.5	52.2	53.5	53.9	53.3	55.0	52.9	49.2
TiO_2	0.6	0.6	0.4	0.3	0.2	0.3	0.2	0.5	0.3	0.3	0.4	0.2	0.5	1.2
Al_2O_3	7.4	9.8	2.6	3.3	2.3	3.8	2.5	7.2	4.7	4.8	7.8	7.9	7.7	8.2
Fe_2O_3	2.4	1.2	1.0	0.6	1.7	0.0	1.4	3.4	1.2	0.0	3.7	0.0	3.3	4.1
FeO	2.9	2.2	4.9	9.8	8.4	8.9	7.8	2.5	3.0	3.9	4.0	6.0	4.6	4.3
MnO	-	-	-	-	-	-	-	-	0.10	0.11	-	-	-	-
MgO	12.4	11.4	15.2	12.7	12.8	11.8	13.5	11.9	14.3	14.1	10.1	10.4	10.3	10.6
CaO	19.4	17.9	22.6	21.2	21.8	20.6	21.4	19.0	21.3	20.8	17.3	16.2	17.0	20.0
Na_2O	3.0	4.0	0.6	0.6	0.7	0.9	0.8	3.2	1.8	1.9	4.3	4.1	4.1	2.4
Σ	101.0	100.1	100.3	100.4	99.9	99.7	100.1	99.9	100.2	99.81	100.9	99.8	100.4	100.0
Si	1.899	1.899	1.942	1.932	1.945	1.996	1.949	1.897	1.938	1.958	1.923	1.995	1.920	1.816
Al^{IV}	0.101	0.101	0.058	0.068	0.055	0.004	0.051	0.103	0.062	0.042	0.077	0.005	0.080	0.184
Al^{VI}	0.212	0.313	0.056	0.076	0.047	0.163	0.059	0.205	0.139	0.164	0.255	0.333	0.251	0.174
Ti	0.016	0.017	0.010	0.009	0.006	0.008	0.006	0.014	0.008	0.008	0.011	0.005	0.013	0.034
Fe^{3+}	0.065	0.031	0.027	0.017	0.048	-	0.038	0.094	0.032	-	0.099	-	0.090	0.114
Fe^{2+}	0.088	0.064	0.150	0.304	0.262	0.278	0.241	0.076	0.092	0.118	0.121	0.182	0.140	0.132
Mn	-	-	-	-	-	-	-	-	0.003	0.003	-	-	-	-
Mg	0.664	0.609	0.830	0.704	0.714	0.657	0.747	0.645	0.772	0.763	0.543	0.562	0.556	0.581
Ca	0.746	0.687	0.885	0.847	0.874	0.825	0.851	0.740	0.827	0.810	0.669	0.630	0.661	0.792
Na	0.209	0.278	0.043	0.043	0.051	0.068	0.058	0.225	0.126	0.134	0.301	0.288	0.288	0.172

Table 3 (cont.)

Orthopyroxenes

	PHN 2508	LT 2	PHN 2567/5	L 6	LQ 1	PHN 2495	KN 206	M 3	M 5	L 9	L 16	L 17	L 20	MO 4	MO 6	PHN 2630/1	PHN 2630/2	PHN 2645
Rk type	GP	GG	GP	GG	GP	GG	GP	GP	GP	GP	GP	GP	GG	PG	PG	GP	GG	PG
SiO_2	53.5	53.1	55.5	54.5	54.8	54.2	53.7	53.8	55.6	54.5	55.0	54.5	54.2	54.1	51.8	52.0	52.6	53.1
TiO_2	0.0	0.5	0.0	0.0	-	0.0	0.0	0.0	0.0	0.0	0.0	-	0.0	0.0	0.0	0.1	0.06	0.0
Al_2O_3	0.75	1.6	1.0	4.0	2.1	1.6	0.83	0.76	2.1	0.85	1.8	1.2	1.7	1.3	1.5	0.94	2.1	1.1
Fe_2O_3	1.1	1.5	0.0	1.9	1.8	0.58	0.0	0.0	0.6	0.0	0.59	1.0	1.1	0.0	0.0	0.78	0.0	0.0
FeO	19.1	19.3	14.3	7.0	10.7	14.7	20.3	18.9	10.5	17.9	11.6	14.2	15.9	17.0	27.1	27.0	26.4	24.0
MnO	0.22	-	-	-	0.1	-	-	-	-	-	0.10	0.1	0.0	-	-	-	-	-
MgO	24.9	24.8	28.8	32.3	30.5	28.0	24.0	25.2	31.1	26.2	30.1	28.3	27.2	26.3	18.6	19.5	17.4	21.3
CaO	0.25	0.36	0.20	0.45	0.28	0.18	0.30	0.35	0.41	0.32	0.35	0.35	0.36	0.43	0.58	0.40	0.40	0.47
Na_2O	0.0	-	-	0.0	0.0	0.0	0.0	0.0	0.0	0.0	0.0	0.0	0.0	0.0	0.0	0.0	0.0	0.0
Σ	99.82	101.16	99.80	100.15	100.28	99.26	99.13	99.01	100.31	99.77	99.54	99.65	100.46	99.13	99.58	100.72	98.96	99.97
Si	1.969	1.930	1.984	1.894	1.933	1.958	1.994	1.985	1.948	1.984	1.954	1.961	1.949	1.976	1.980	1.966	2.032	1.990
Al^{IV}	0.031	0.069	0.016	0.106	0.067	0.042	0.006	0.015	0.052	0.016	0.046	0.039	0.051	0.024	0.020	0.034	-	0.010
Al^{VI}	0.002	-	0.030	0.058	0.020	0.026	0.030	0.018	0.035	0.021	0.030	0.012	0.021	0.033	0.049	0.006	0.096	0.039
Ti	-	0.013	-	-	-	-	-	-	-	-	-	-	-	-	-	0.003	0.002	-
Fe^{3+}	0.029	0.042	-	0.049	0.047	0.016	-	-	0.017	-	0.016	0.028	0.031	-	-	0.022	-	-
Fe^{2+}	0.586	0.587	0.428	0.204	0.316	0.443	0.630	0.583	0.309	0.545	0.344	0.427	0.477	0.519	0.866	0.854	0.853	0.752
Mn	0.007	-	-	-	0.003	-	-	-	-	-	0.003	0.003	-	-	-	-	-	-
Mg	1.366	1.344	1.535	1.673	1.603	1.508	1.328	1.386	1.624	1.422	1.594	1.518	1.458	1.431	1.061	1.099	1.002	1.190
Ca	0.010	0.014	0.008	0.017	0.011	0.007	0.012	0.014	0.015	0.012	0.013	0.013	0.014	0.017	0.024	0.016	0.017	0.019
Na	-	-	-	-	-	-	-	-	-	-	-	-	-	-	-	-	-	-

Table 3 (cont.)

Garnets	KAO 2	KC 22	KC 23	KC 28	KC 29	KC 30	KC 31	PHN 2438	PHN 1442j prim.	PHN 1442j sec.	PHN 2508	LT 2	LT 6
Rk type	Ec	Ec	Ec	GG	GG	Ec	Ec	GG	GG	GG	GP	GG	GG
SiO_2	40.3	41.4	39.5	39.6	38.1	38.4	41.0	40.1	40.3	40.2	38.9	39.5	39.1
TiO_2	0.0	0.0	0.0	0.0	0.0	0.0	0.0	0.0	-	0.0	0.0	0.0	0.0
Al_2O_3	22.4	23.6	22.2	23.7	22.2	21.3	23.5	22.7	23.0	23.2	22.0	21.6	21.6
FeO	15.3	15.0	23.6	18.8	26.9	27.3	13.8	19.1	14.3	15.3	25.1	24.6	23.4
MnO	0.49	-	-	0.71	0.83	1.1	0.57	-	-	-	0.84	-	0.57
MgO	15.3	14.1	7.0	13.7	5.7	6.0	17.3	11.5	11.2	10.4	8.1	8.9	7.4
CaO	5.4	7.1	8.4	4.1	7.5	6.7	4.5	6.0	11.2	11.1	5.6	6.0	8.3
Σ	99.19	101.2	100.7	100.61	101.23	100.8	100.67	99.4	100.0	100.2	100.54	100.6	100.37
Si	2.984	2.998	3.011	2.927	2.948	2.984	2.960	3.009	2.987	2.985	2.984	3.010	3.001
Al^{IV}	0.016	0.002	-	0.073	0.052	0.016	0.040	-	0.013	0.015	0.016	-	-
Al^{VI}	1.942	2.013	1.995	1.995	1.972	1.938	1.960	2.010	1.996	2.015	1.972	1.940	1.955
Ti	-	-	-	-	-	-	-	-	-	-	-	-	-
Fe	0.947	0.909	1.505	1.162	1.737	1.773	0.831	1.198	0.886	0.950	1.610	1.568	1.499
Mn	0.031	-	-	0.044	0.054	0.072	0.035	-	-	-	0.055	-	0.037
Mg	1.686	1.522	0.795	1.514	0.652	0.699	1.864	1.290	1.237	1.151	0.926	1.011	0.851
Ca	0.431	0.551	0.686	0.325	0.624	0.557	0.350	0.481	0.889	0.883	0.460	0.490	0.679

Table 3 (cont.)

Garnets	PHN 2567/5	L 6 inner	LQ 11	LQ 21	LQ 4	PHN 2494	PHN 2495	KN 195	KN 206	M 1	M 3	M 5
Rk type	GP	GG	GP	GG	GG	GG	GG	GG	GP	GG	GP	GP
SiO_2	40.4	40.3	40.2	41.1	37.9	40.1	40.0	40.1	39.2	40.7	39.1	40.6
TiO_2	0.0	0.0	0.0	0.0	0.0	0.0	0.0	0.0	0.0	0.0	0.05	0.0
Al_2O_3	22.2	23.5	22.7	23.3	21.6	22.7	22.3	22.8	21.9	23.7	22.0	22.8
FeO	21.1	14.9	18.9	13.6	32.3	20.3	21.0	20.7	25.2	14.2	23.8	16.4
MnO	-	-	0.6	0.41	0.6	-	-	-	-	0.2	-	-
MgO	11.3	15.6	13.6	15.3	4.3	12.1	11.7	12.0	8.7	13.1	9.4	13.9
CaO	5.0	5.5	4.6	6.7	5.2	4.9	4.5	4.8	4.1	7.9	4.7	5.5
Σ	100.0	99.8	100.6	100.41	101.9	100.1	99.5	100.4	99.1	99.8	99.05	99.2
Si	3.032	2.953	2.977	2.992	2.961	2.996	3.014	2.992	3.024	2.991	3.005	3.010
Al^{IV}	-	0.047	0.023	0.008	0.039	0.004	-	0.008	-	0.009	-	-
Al^{VI}	1.964	1.983	1.958	1.988	1.951	1.996	1.981	1.996	1.991	2.041	1.993	1.993
Ti	-	-	-	-	-	-	-	-	-	-	0.003	-
Fe	1.324	0.913	1.170	0.824	2.113	1.269	1.323	1.292	1.626	0.871	1.530	1.017
Mn	-	-	0.038	0.025	0.040	-	-	-	-	0.012	-	-
Mg	1.264	1.704	1.501	1.653	0.501	1.348	1.314	1.334	1.000	1.438	1.077	1.536
Ca	0.402	0.432	0.365	0.519	0.438	0.392	0.363	0.384	0.339	0.621	0.387	0.437

Table 3 (cont.)

Garnets

	PHN 2317	K 2	K 3	K 4	K 5	L 9	L 12A	L 12B	L 13	L 16	L 17	L 20
Rk type	GG	GG	GG	Ec	Ec	GP	GG	Ec	GG	GP	GP	GG
SiO_2	38.6	40.5	39.0	41.1	38.7	40.0	40.4	40.4	40.9	40.7	39.8	40.1
TiO_2	0.0	0.0	0.0	0.0	0.0	0.0	0.0	0.0	0.0	0.0	0.0	0.0
Al_2O_3	21.7	23.1	21.7	22.8	22.2	21.2	23.8	23.6	22.2	22.3	21.6	21.9
FeO	24.1	16.8	24.8	12.9	25.7	23.0	17.2	18.0	15.1	17.7	20.6	22.5
MnO	-	0.3	0.7	0.56	0.6	0.6	0.4	0.4	0.3	0.5	0.5	0.3
MgO	6.7	12.1	8.4	15.4	8.5	11.0	11.8	12.5	13.3	13.8	12.0	10.7
CaO	8.4	8.1	6.0	8.2	5.3	4.6	7.3	5.9	8.4	5.0	4.8	5.3
Σ	99.5	100.9	100.6	100.96	101.0	100.4	100.9	100.8	100.2	100.0	99.3	100.8
Si	2.994	2.984	2.988	2.981	2.958	3.028	2.976	2.974	3.016	3.015	3.014	3.015
Al^{IV}	0.006	0.016	0.012	0.019	0.042	-	0.024	0.026	-	-	-	-
Al^{VI}	1.978	1.990	1.947	1.929	1.958	1.891	2.043	2.025	1.929	1.947	1.928	1.941
Ti	-	-	-	-	-	-	-	-	-	-	-	-
Fe	1.563	1.035	1.589	0.781	1.643	1.456	1.056	1.107	0.931	1.097	1.305	1.415
Mn	-	0.019	0.045	0.034	0.039	0.038	0.025	0.025	0.019	0.031	0.032	0.019
Mg	0.775	1.329	0.959	1.663	0.968	1.241	1.289	1.376	1.462	1.524	1.354	1.199
Ca	0.698	0.639	0.429	0.638	0.434	0.373	0.578	0.468	0.664	0.397	0.389	0.427

Table 3 (cont.)

Garnets

	PHN 1646A	PHN 1670	PHN 2852	OVKF 10303	PHN 2630/1	PHN 2630/2	PHN 1919	PHN 2450	PHN 2496	PHN 2532	PHN 2533	PHN 2685/1A	PHN 3017
Rk type	GG	GG	GG	GG	GP	GG	GG	Ec	Ec	Ec	Ec	Ec	GG
SiO_2	39.1	40.5	40.6	40.2	38.2	39.7	39.5	40.2	39.8	39.4	40.7	39.2	39.4
TiO_2	0.0	0.0	0.0	0.0	0.0	0.0	0.0	0.0	0.0	0.0	0.0	0.0	0.1
Al_2O_3	22.4	23.0	23.1	23.7	21.7	22.5	22.5	23.0	22.7	22.4	23.8	22.1	21.9
FeO	21.6	16.0	19.3	15.9	28.1	26.0	20.5	19.0	19.9	24.2	21.5	23.6	22.3
MnO	-	-	-	-	-	-	-	0.79	0.84	-	-	-	-
MgO	10.7	14.3	11.7	13.2	4.9	4.7	10.9	12.6	11.8	7.2	8.0	8.6	7.9
CaO	5.2	6.3	6.2	7.3	6.7	7.0	6.5	5.2	5.1	7.8	6.1	6.2	8.6
Σ	99.0	100.1	100.9	100.3	99.6	99.9	99.9	100.79	100.14	101.0	100.1	99.7	100.2
Si	2.981	2.980	3.003	2.958	3.001	3.065	2.979	2.979	2.983	2.998	3.054	3.004	3.005
Al^{IV}	0.019	0.020	-	0.042	-	-	0.021	0.021	0.017	0.002	-	-	-
Al^{VI}	1.994	1.974	2.013	2.013	2.009	2.047	1.978	1.987	1.988	2.006	2.105	1.997	1.974
Ti	-	-	-	-	-	-	-	-	-	-	-	-	0.006
Fe	1.377	0.984	1.194	0.978	1.846	1.679	1.293	1.177	1.247	1.540	1.349	1.509	1.421
Mn	-	-	-	-	-	-	-	0.050	0.054	-	-	-	-
Mg	1.216	1.568	1.290	1.448	0.574	0.541	1.225	1.392	1.318	0.816	0.895	0.980	0.894
Ca	0.425	0.497	0.491	0.575	0.564	0.579	0.515	0.413	0.410	0.636	0.490	0.507	0.704

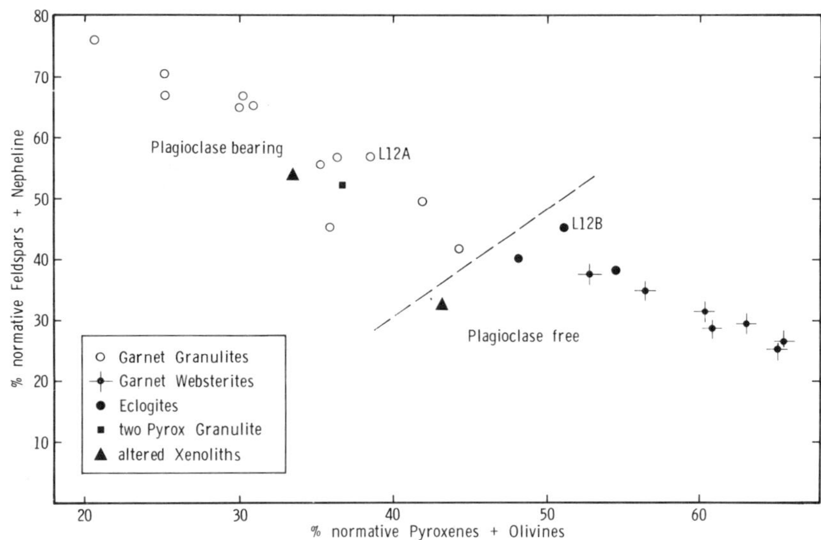

Fig. 3. Chemical controls on the presence of modal plagioclase in the garnet granulite suite.

and his imply that the original samples were quite inhomogeneous (especially OVKF10303).

The basic granulites and eclogite are essentially olivine-basaltic in composition, with minor normative Ne (0-5%) or Hy (0-2%, one with 9%). In contrast, the garnet websterites are more strongly Hy-normative (6-25%, one with 59%). The felsic granulite LT-2 contains significant normative and modal quartz, and is of intermediate composition.

The absence of plagioclase in the eclogites and garnet pyroxenites is largely controlled by rock chemistry. Only rocks with >35% normative plagioclase and <45% normative pyroxenes + olivine contain modal plagioclase (Fig. 3). Plagioclase-bearing nodules also have generally higher Na/Na+Ca than those lacking plagioclase. This compositional control on the presence of plagioclase, and the interlayering of eclogite and granulite on the cm scale (L12A, B) makes it unnecessary to postulate that the eclogites originated at greater depths than the granulites.

The garnet websterites are lower in Al, and higher in Mg/Mg+Fe, than the granulites, as well as being more mafic. This trend is compatible with the websterites being mafic cumulates from a basic magma now represented by the granulites. However, within the granulites there is a well-defined trend toward higher Mg/Mg+Fe in the less mafic samples. This is difficult to explain in terms of crystal fractionation, and argues against a simple genetic relation among the various granulites, or between these rocks and the garnet websterites.

Extensive hydration of primary mafic phases is accompanied in some cases by carbonation (PHN-2685/1A) or natrolitization (PHN-1919, K-1). Many of the fresher samples show similar effects on a smaller scale, especially partial alteration of primary feldspars and scapolite to secondary minerals such as natrolite, pectolite, sericite and carbonate. The fluids responsible for the post-granulite-facies introduction of H_2O, CO_2 and Na_2O presumably come from the enclosing kimberlite, and the alteration may be regarded as analogous to the fenitization that commonly accompanies alkaline-carbonatite complexes.

The contents of LIL elements such as K, Rb, Sr, Ba, Zr, Ti and P are highly variable, and the possibility of contamination from the kimberlite must be considered. In some samples (PHN-2685/1A, PHN-2645) there appears to be a correlation between high Ba and Sr contents and the presence of secondary carbonate, but this does not hold in general. Furthermore, microprobe studies of PHN-2645 show that the Ba mostly resides in tiny (<10μm) K-feldspar inclusions in the plagioclase. The K in most nodules resides in biotite and/or K-feldspar (Table 1), as does, presumably, most of the Rb. Since these phases appear on textural grounds to be in equilibrium with the other members of the granulite-facies or eclogite assemblages, the present levels of K, Rb, Sr and Ba were probably established before or during the granulite-facies metamorphism. The same is probably true of Ti and P, which reside in rutile (or ilmenite) and apatite, respectively; there are no obvious grounds for regarding either of these phases as secondary with respect to the primary metamorphic assemblages.

Analyses of REE in Lesotho granulite nodules (including our samples PHN-1646, -1919, -2852, OVK-F10303) show a wide range in both ΣREE and in the degree of enrichment of the light REE (Rogers, 1977). Samples with low REE have

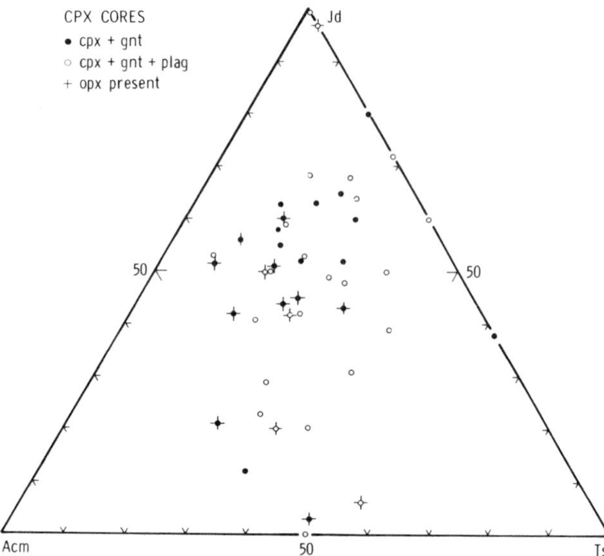

Fig. 4. Proportions of jadeite (Jd), Ca-Tschermak's molecule (Ts) and acmite (Acm) end-members in clinopyroxenes of the garnet granulite suite.

flat, unfractionated or light-REE-depleted patterns relative to chondrites, while increasing REE contents are accompanied by greater enrichment in the light REE. Most samples, however, show pronounced positive Eu anomalies. Both garnet and clinopyroxene also show this positive Eu anomaly (Rogers, 1977; and pers. comm.). This implies that the REE patterns were established, and the Eu^{2+} partly oxidized to Eu^{3+}, prior to the formation of the high-P granulite/eclogite mineral assemblages, which in turn argues against significant contamination from the kimberlite. Rogers (1977) has also shown that other incompatible elements (Nb, Zr, Hf, Ta) vary coherently with the REE, suggesting that the contents of these elements were also relatively little affected by alteration.

Thus, even after discounting kimberlite-contamination effects, the trace-element and minor-element contents of these nodules show wider variations than the major-element concentrations. Many samples show the high K/Rb ratios and low K/Ba ratios characteristic of medium- and high-pressure granulite-facies rocks (Heier, 1973). The very low contents of Ti, K, P and Zr in some nodules are anomalous for basaltic compositions and also suggest depletion. On the Ti-Zr-P discrimination diagram (Floyd and Winchester, 1975) 3 nodules plot as alkali basalts, 10 as tholeiitic basalts, and the rest fall outside any defined fields. Thus the minor and trace-element compositions do not appear to reflect original magmatic distributions. The REE patterns discussed above show very large variations in abundances and degree of fractionation among nodules of similar bulk composition. These variations are difficult to reconcile with an igneous origin, but could be explained by variable removal of incompatible elements during an anatectic event. We conclude that the trace-element and minor-element contents of these nodules reflect modification of some primary distribution(s) by high-grade metamorphic processes, perhaps including anatexis (cf. Rogers, 1977).

Mineral Chemistry

Methods: Minerals were analyzed using a manual ARL-EMX electron microprobe at the Central Institute for Industrial Research, Oslo. Accelerating voltages were 15 kV and sample currents 0.01-0.03 μ amps. Natural and synthetic mineral standards were used, and the data were reduced using the method of Bence and Albee (1968). Fe^{3+} was calculated in the pyroxenes on the assumption of charge balance (program PYROX, by E. R. Neumann). Our experience suggests that this procedure tends to produce maximum values for Fe^{3+}, since analytical errors and non-stoichiometry in the pyroxenes usually produce low values for SiO_2. Analyses with unusually high Fe^{3+}/Fe^{2+} ratios have been run in duplicate to confirm the analyses. Calculated Fe^{3+} values are not given for the garnets, since these values are usually so small (<0.5%) that we interpret them as resulting

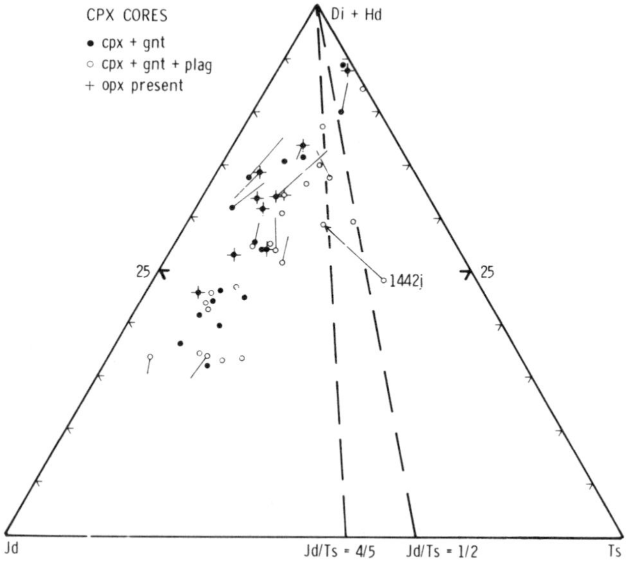

Fig. 5. Proportions of Jd, Ts and diopside + hedenbergite (Di + Hd) end-members in the clinopyroxenes. Short lines connect analyses of the cores of the grains (symbols) with analyses of rims, where zoning was observed. For sample PHN-1442j an arrow runs from the primary to the secondary pyroxene.

from analytical error. This is typical of granulite and eclogite garnets, in our experience.

Clinopyroxenes: Most of the analyzed clinopyroxenes, regardless of rock type, have high contents of Jd+Acm+Ts (Table 3). The relative proportions of these components vary widely, but Jd is usually dominant (Fig. 4). These pyroxenes thus resemble those from metamorphic eclogites. Preliminary single-crystal X-ray studies of pyroxenes from M-1, LQ-1 and LQ-4 show that they have disordered structures (C2/c space group), suggesting crystallization at T $>600^oC$ (A. Mottana, pers. comm.). Further studies of pyroxenes are in progress.

The ratio Jd/Ts $>\frac{1}{2}$ usually distinguishes granulite-facies from eclogite pyroxenes (omphacites) (White, 1964). According to this criterion, the analyzed clinopyroxenes of both eclogites and garnet granulites in the Lesotho suite are eclogitic (Fig. 5). The four exceptions are all low in both Jd and Ts, and the errors on both components may be large relative to the absolute contents. Clinopyroxenes from the Monastery pyroxene granulites have lower Jd/Ts and higher Acm. There is a reasonably good correlation of the Jd/Ts ratio of the pyroxenes with the Ab content of the coexisting plagioclase (Fig. 6) in the garnet granulites.

Some of the clinopyroxenes show zoning, usually in the form of lower Jd content within about 100 μm of the rims. The marked zones with tabular channels, described above, also have much lower Jd and Ts contents than their host pyroxenes (Fig. 5). The Jd-depleted rims

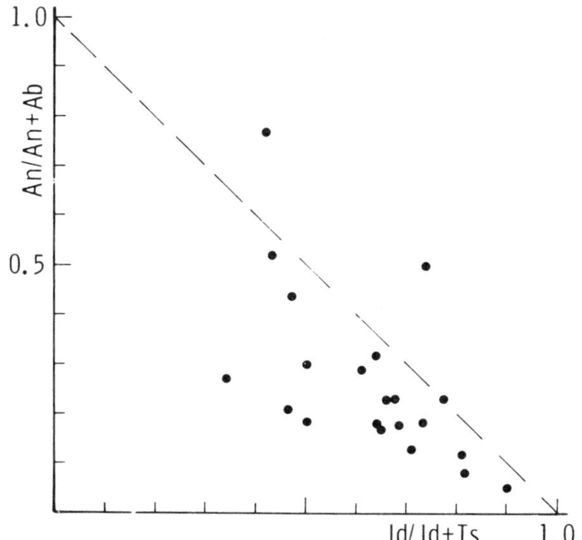

Fig. 6. Relation between plagioclase composition and the Jd/Jd+Ts ratio of coexisting clinopyroxene.

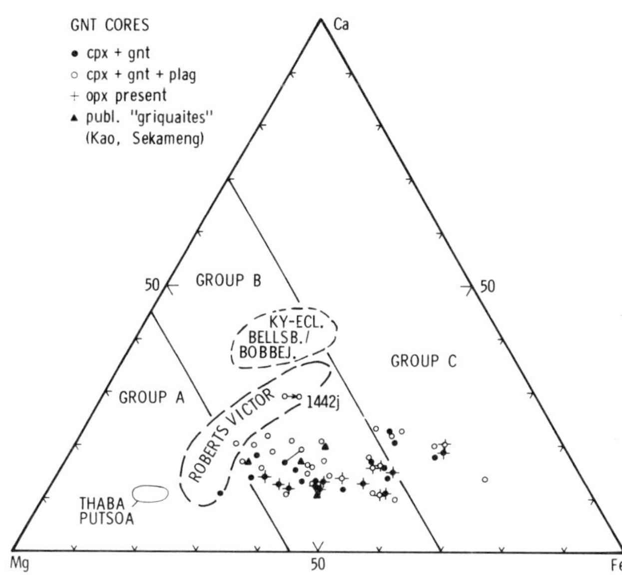

Fig. 7. Atomic proportions of Fe, Mg and Ca in garnets. Group A (eclogites in kimberlites), B (eclogites in gneiss) and C (eclogites in blueschist) are adapted from Coleman et al. (1965). Data on kyanite-eclogites from Bellsbank Fissure and Bobbejaan are by Carswell (unpublished).

are interpreted as the combined result of metasomatism and decompression during the transport of the nodules in the kimberlite, and give an idea of the amount of reequilibration of the mineral assemblages that may have occurred since the nodules were picked up by the kimberlite.

Orthopyroxenes: The orthopyroxenes of the garnet granulite suite show uniformly low CaO contents consist with their relatively low-temperature metamorphic equilibration (see below). The Al_2O_3 contents vary considerably. In the garnet websterites, lower Al_2O_3 contents are systematically related to higher FeO contents, as is characteristic of granulite-facies pyroxenes (Griffin and Heier, 1969). The exception to this trend is PHN-2630/1 from Monastery Mine. The orthopyroxenes from the garnet granulites and pyroxene granulites do not show this Al_2O_3-FeO correlation. No other clear differences in mineral chemistry are observed between the different rock types in the garnet granulite suite.

Garnets: The garnets show a wide range in Fe/Mg, reflecting variations in bulk-rock composition and in the P/T-controlled partitioning between garnet and pyroxenes. The grossular content ranges only between 10 and 23% except for PHN-1442j. The observed ranges in Fe/Mg/Ca are approximately the same for granulites, eclogites and garnet pyroxenites

Table 4. Analyses of Mica, Amphibole and Scapolite

Micas					Amphiboles			
	K-3	LQ-1	LT-2	PHN 2630/1	PHN 2508	206	MO-6	MO-9
SiO_2	40.5	38.1	37.3	36.9	44.0	46.4	42.4	44.3
TiO_2	4.0	3.8	6.0	5.7	0.7	1.1	2.3	0.3
Al_2O_3	11.8	13.7	11.8	13.7	10.7	9.0	12.9	14.6
FeO	8.7	6.9	10.9	17.5	10.1	10.9	14.8	5.2
MnO	-	0.0	-	-	0.1	-	-	-
MgO	19.5	18.7	16.7	12.5	16.1	15.5	10.1	17.0
CaO	0.14	0.06	0.17	-	10.8	8.3	11.4	11.5
Na_2O	0.34	0.38	0.16	0.1	3.4	4.6	1.8	3.0
K_2O	9.7	9.6	9.5	9.4	0.9	0.5	1.6	0.5
BaO	-			-				
Σ	94.7	91.2	92.5	95.8	96.8	98.3	97.3	96.4

Scapolites	K-2	PHN 2852	L-12	LQ-2	PHN 1670	PHN 3017	LT-2
SiO_2	47.3	49.2	51.7	48.2	50.1	48.2	48.6
Al_2O_3	27.5	26.9	23.4	25.4	24.4	22.2	24.9
CaO	15.7	14.3	13.7	16.3	15.0	14.7	14.9
Na_2O	4.3	4.9	5.2	4.2	4.8	4.4	4.8
K_2O	0.0	0.1	0.1	0.0	0.1	0.1	0.1
SO_3	3.5	5.0	5.7	3.9	5.0	6.8	4.7
Cl	0.0	0.0	0.0	0.0	0.0	0.0	0.05
Σ	98.3	100.5	99.8	98.0	99.4	97.0	98.05

PHN-3017 also contains MgO 0.3, FeO 0.3.

(Fig. 7). All of the garnets plot in the fields for Group B (eclogites in gneiss) and Group C (eclogites and blueschists) of Coleman et al. (1965), except for K-31 from Kao. Published analyses of "griquaite" garnets from Kao and Sekameng plot in the Group B field together with our analyses, whereas the garnets of type "Griquaites" are generally distinct in composition from ours (Fig. 7). About 30% of our analyzed garnets classify in Group 5 (magnesian almandine) of Dawson and Stevens (1975); the rest fall in Group 3 (calcic pyrope-almandine). This separation does not correspond to differences in rock type.

Micas and amphiboles: The analyzed micas, all of which appear to be in textural equilibrium with pyroxenes and garnet, include both phlogopites and biotites. Some biotites are unusually rich in TiO_2 (Table 4). The amphiboles are Na-rich hornblendes.

Feldspars: Plagioclase compositions range from An_5-An_{77} (Table 1), but cluster in the range An_{10}-An_{35}. K_2O contents are low (Or <5%) and antiperthite is very rare. Zoning (both "normal" and "reversed") of up to 5% An from core to rim of single grains has been observed in a few samples. Homogeneous grains of orthoclase (Ab_{30-35}) occur rarely (Table 1).

Scapolites: High-sulfer scapolites have been recognized in several samples (Table 4). Similar scapolites have been reported from high-P, high-T nodules in the Delegate pipes (Lovering and White, 1964) and in anorthosite interlayered with eclogite in the Bergen Arcs, Norway (Griffin, 1972). The distribution of Na and Ca between these scapolites and the coexisting plagioclase, if compared with experimental work (Goldsmith and Newton, 1977), would imply temperatures >1300°C. Other mineral equilibria (see below) suggest much lower temperatures; the experimental data probably are not applicable to these rocks.

P/T Estimates

The analyzed rocks contain mineral assemblages that allow estimates of the pressure and temperature of equilibration through the use of published geothermometers and geobarometers. Wood (1976) has shown that the P dependence of Fe/Mg partitioning between coexisting garnet and clinopyroxene in the Ca-free system is less than that found by Råheim and Green (1974)

Table 5. P-T Estimates

Sample	Type	K_D (gnt/cpx)	T°, C	P, Kb	Method	Comments
PHN-2317	GG	10.0	560	5.5	1+5	Qtz present
K-2	GG	5.9	690	10	1+5	
K-3	GG	7.9	630	10	1+5	
K-4	GG	3.2	860	10	1+5	Plag An_{77}, no Fe^{3+}
KAO-2	Ec	7.7	590	5	1+4	No plag, minimum P
			630	10.5	1+5	- " -
KC-23	Ec	8.4	580	5	1+4	- " -
			625	11.5	1+5	- " -
KC-28	GG	13.1	550	12	1+5	
PHN-2438	GG	6.4	670	9.5	1+5	
PHN-1442j	GG	12.7	<500	<0	1+5	Disequil. textures (sec. cpx/gnt)
PHN-2508	GW	10.0	590	10.8	1+2	gnt exsolved from pyroxenes
			795	-	7	
LT-2	GG	7.1	630	6.6	1+2	opx incl. in cpx, maximum P
			680	14	1+3	
			<600	~ 0	1+5	
			810	-	7	
LT-6	GG	6.0	660	5	1+5	
L-6	GG	11.6	<650	<0	1+5	Coronite, plag An_{52}
PHN-2567/5	GW	6.4	715	15.8	1+2	
			843	-	7	
LQ-1	GW	10.7	515	1.1	1+2	575°, 10 Kb w/ form. fr / analysis
			550	1	1+5	No plag, minimum P
			700	-	7	
LQ-2	GG	4.4	750	7.5	1+5	Qtz present
LQ-4	GG	9.9	560	5.5	1+4	- " -
			610	10.5	1+5	
PHN-2494	GG	7.9	600	5.5	1+4	
			625	10	1+5	
			650	at 5	6	
			700	at 10	6	
PHN-2495	GG	7.4	650	9	1+2	
			635	9	1+5	
			770	-	7	
KN-195	GG	11.5	550	9.5	1+5	
			700	at 10	6	
KN-206	GW	7.2	690	16.5	1+2	
			750	-	7	
M-1	GG	8.2	575	8	1+4	
			625	10.5	1+5	
M-3	GW	8.5	640	13.6	1+2	
			780	-	7	
M-5	GW	5.9	680	7.9	1+2	
			790	-	7	
L-9	GW	6.5	740	20.5	1+2	
			765	-	7	
L-12A	GG	6.1	680	7.5	1+4	
			710	13.5	1+5	
L-13	GG	10.0	560	4.5	1+4	
			580	11	1+5	
L-16	GW	7.9	605	6.0	1+2	
			790	-	7	
L-17	GW	6.8	695	14.5	1+2	
			775	-	7	
L-20	GG	6.1	680	9.6	1+2	
			660	6.5	1+5	
			720	15.5	1+3	No qtz, maximum P
			730	-	7	

Table 5, (cont.)

Sample	Type	K_D (gnt/cpx)	T°, C	P, Kb	Method	Comments
PHN-1646A	GG	9.4	<550	<0	1+5	
PHN-1670	GG	8.2	600	7	1+5	
PHN-2852	GG	7.0	650	9	1+5	
OVK-F 10303	GG	6.4	645	6.5	1+4	
			695	13	1+5	
PHN-2630/1	GW	8.8	580	5.8	1+2	
			850	-	7	
PHN-2630/2	GG	7.3	615	5.5	1+2	
			660	12	1+3	
			<600	< 0	1+5	Plag An_{50}. Unreliable
			875	-	7	
PHN-1919	GG	9.0	580	8.5	1+5	
PHN-2532	Ec	8.5	570	4.5	1+4	No plag, minimum P
			615	10.5	1+5	
PHN-2533	GG	4.7	760	11	1+4	
			770	11.5	1+5	
PHN-3017	GG	7.0	625	6.5	1+5	

Methods: 1: gnt/cpx; Råheim and Green (1974). 2: opx/gnt; Wood (1974).
3: opx+plag+gnt+qtz; Wood (1975). 4: Di/Ab; Kushiro (1969).
5: cpx/plag; Currie and Curtis (1976). 6: Ab/Or; Stormer (1975).
7: cpx/opx (Fe/Mg); Wells (1977).

in tholeiitic compositions. However, the rocks studied here are basaltic in composition, and the garnets thus contain significant amounts of Ca. We will therefore use the P-T calibration of Råheim and Green (1974). As mentioned above, our procedure for estimating the Fe^{3+} content of the pyroxenes probably biases our estimates toward higher K_D (lower T) and lower Jd (lower P). On the basis of other studies (Mysen and Griffin, 1973) we estimate the probable error resulting from analytical uncertainty to be ± 75°C and 2 Kb.

The Jd content of diopside coexisting with albite can be used to define the P at given T, using the data of Kushiro (1969); in the absence of quartz this will be a minimum P. Currie and Curtis (1976) have recalibrated this barometer, taking into account the presence of other components in the clinopyroxene and plagioclase. By assuming that the last plagioclase to disappear was albite, a minimum P can also be calculated for eclogites having >20% Jd in the pyroxene.

The Al content of orthopyroxene coexisting with garnet can be used to calculate a P/T relation using the method of Wood (1974), which allows for the presence of Fe in both phases and of Ca in garnet. Wood (1975) has also proposed a geobarometer for opx+gnt+plag+qtz assemblages, but the calibration of this must be regarded as preliminary.

The only available geothermometer for the two-pyroxene granulites from Monastery is that of Wood and Banno (1973) recently recalibrated by Wells (1977). Unfortunately this method appears in our experience to be of dubious reliability at T <800°C (see below). The clinopyroxenes in these rocks contain too little Jd, and the plagioclases too much An, to allow reliable use of the Currie and Curtis (1976) method.

The P-T estimates presented in Table 5 are calculated using the analyses in Table 3, all of which represent the cores of the grains. We regard the marginal zoning observed in some pyroxenes and plagioclase as a result of modification during transport in the kimberlite. The core analyses should therefore give the closest approach to equilibrium conditions.

Five samples give P $\lesssim$0 Kb; two of these have calcic plagioclase and three show disequilibrium textures. In general the Currie and Curtis method gives much higher P than would be estimated from Kushiro's data. As might be expected, the discrepancy is greatest for samples with Fe-rich pyroxenes and calcic plagioclases. There is reasonably good agreement between the Currie and Curtis values and those calculated from opx/gnt partitioning (Wood, 1974) for two samples (PHN-2495, L-20) with sodic plagioclase, but not for PHN-2630/2, in which the plagioclase is An_{50}. We conclude that the Currie and Curtis values are reasonably good for samples with plagioclase less calcic than An_{35}.

Several garnet websterites yield pressures $\gtrsim$15 Kb, and appear anomalous relative to the other samples (Fig. 8). These are all cases in which the orthopyroxene is Fe-rich and Al-poor. These samples otherwise appear, as noted above, to form part of a continuous variation series with the websterites that give much

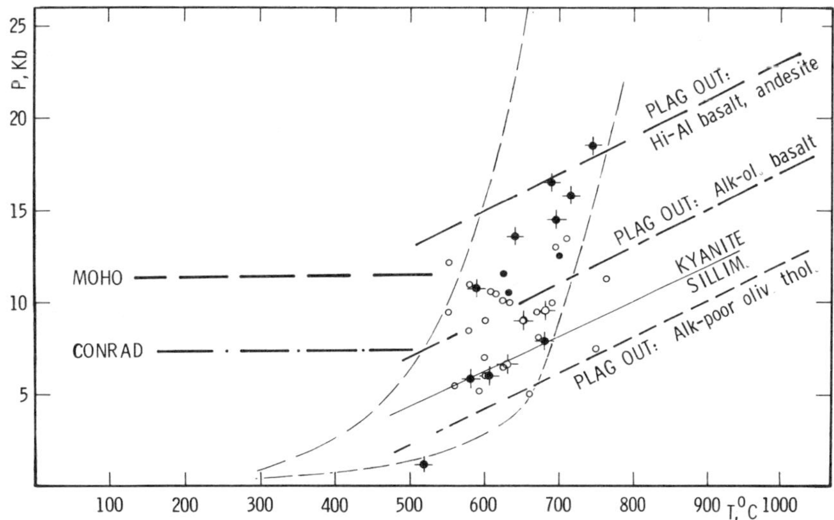

Fig. 8. P-T determinations on nodules of the garnet granulite suite (Table 5). Symbols as in Fig. 3-7. Positions of Moho and Conrad discontinuities are from the Transvaal (Hales and Sacks, 1959). Plag-out curves based on data for relevant compositions at 1100°C, extrapolated at 20 bars/°C (Ringwood, 1975).

lower P, and there is also a correlation of calculated P with whole-rock Fe/Mg.

We therefore prefer to regard these high pressures with caution. The geobarometer of Wood (1974) is calibrated essentially on Mg-rich compositions, and the extrapolation to higher Fe/Mg values may involve a systematic error. The P estimates for these low-Al samples are also very sensitive to K_D (gnt/cpx) and thus to errors in the calculated Fe^{3+} contents of the clinopyroxenes. The effect of overestimating Fe^{3+} is, however, to lower both P and T. For example, calculation of sample L-17 without Fe^{3+} raises P from 14.5 Kb to 28.5 Kb, and T from 695° to 945°C.

The calculated P-T values for the garnet granulites spread mostly between 5-13 Kb and 550-770°C (Fig. 8). There is no obvious correlation of high T with low values of Fe^{3+} in the pyroxenes; both the upper and lower envelopes enclosing the data points on Figure 8 are closely limited by samples with Acm-free clinopyroxenes. Thus the indicated lower-T envelope cannot be moved to significantly higher T. The upper envelope may in fact lie at too high temperatures, but our experience with the calculation of Fe^{3+} in pyroxenes suggests that this is probably not so. Higher T estimates would lead to higher P estimates, and would imply that most of the plagioclase-bearing granulites come from below the Moho; we regard this as unlikely.

The P/T boundaries for the disappearance of plagioclase from garnet-granulites of basaltic composition depend heavily on the details of bulk composition. The boundaries suggested by Green and Ringwood (1972) at 1100°C are plotted in Figure 8, extrapolated to lower T along a gradient of 20 bars/°C as suggested by Ringwood (1975). Most garnet websterites lie above the plag-out line for alkali-poor olivine tholeiite, the closest relevant, experimentally studied composition. All eclogites with >20% Jd in the clinopyroxene plot above the plag-out line for alkali-olivine basalt (cf. analysis of PHN-2532, Table 2). Most of the garnet granulites are equivalent to high-Al olivine basalts, and the experimental boundaries in Figure 8 imply that they should retain plagioclase up to P = 15-17 Kb at their calculated temperatures. The few analyzed granulites with lower Al contents, which should lose their plagioclase at P >10 Kb, all plot below the plag-out line for alkali-olivine basalts.

The internal consistency between the calculated P-T conditions and the extrapolated experimental boundaries suggests that the P-T estimates are reasonable, though it does not place narrow limits on the possible errors. We conclude that the bulk of the garnet granulite suite probably equilibrated as metamorphic rocks at temperatures between 550-700°C and pressures of 5-13 Kb, and that these nodules come from various levels in the lower crust.

The pyroxene granulites from Monastery Mine must have equilibrated below the garnet-in line for alkali-olivine basalt, which essentially overlaps the lowest plag-out curve in Figure 8. The three analyzed samples (as well as a garnet granulite and a garnet pyroxenite from Monastery) give cpx/opx temperatures ranging from 800 to

900°C by the method of Wood and Banno (1973). This method gives temperatures for the rest of the garnet granulite suite from 780-900° (100-200°C higher than the T given by the other methods). We consider these temperatures to be unrealistically high, but the complete overlap may indicate that the pyroxene granulites from Monastery Mine crystallized at similar T, but lower P than the Lesotho garnet granulite suite.

Three spinel lherzolites from Lipelenang (Carswell and Griffin, unpublished) give temperatures of 820-860°C by both the cpx/opx thermometer of Wood and Banno (1973) and the opx/spinel geothermometer of Obata (1976), and 715-790°C by the thermometer of Wells (1977). We assume that these rocks come from below the base of the crust, but that they crystallized at pressures below the spinel lherzolite/garnet lherzolite boundary. The lower T estimates are similar to those obtained on the garnet granulite suite by the same method. This is consistent with equilibration at P = 12-15 Kb according to the boundary of O'Hara et al. (1971), roughly along the same geotherm as the granulites.

Discussion

Composition of the lower crust: No seismic data are available from Lesotho, so inferences about the crustal structure there must be drawn from geophysical studies of the neighboring regions. Hales and Sacks (1959) used earth tremors in the Witwatersrand to study crustal structure along a traverse from Johannesburg eastward to the coast, across the edge of the Kaapvaal craton. They found the depth to Moho along the traverse to be about 37 km, in agreement with earlier studies. The sub-Moho V_p (7.96 km/sec) is lower than in other parts of the Transvaal. The seismic data define a two-layer crust. The upper layer has V_p = 6.0 km/sec, indicating a low density and probably a "granitic" composition. Well within the craton, the lower layer gives V_p = 6.7 km/sec, and the depth to the Conrad Discontinuity is about 22 km. The V_p increases both with depth within the layer, and along the traverse toward the coast, reaching V_p = 7.2 km/sec near the edge of the craton, as the depth to Conrad increases to about 28 km. Data for both P and S waves at stations outside the edge of the craton suggest a depth to Conrad of ca. 25 km. The seismic data thus indicate the presence of an intermediate-velocity lower-crustal layer beneath both the craton and the bordering mobile belt; this layer appears to be both thicker and denser near the edge of the craton.

If these data can be applied to Lesotho, then our P-T estimates (Table 5, Fig. 8) indicate that the nodules of the garnet-granulite suite come largely, if not entirely, from the sub-Conrad layer. This is not unreasonable, since many lines of evidence suggest that the deep continental crust must consist mainly of granulite-facies rocks (Heier, 1973).

The overall composition of the lower-crustal layer will depend on the proportions of basic granulites, eclogites/garnet websterites and intermediate/acidic granulites. Our sampling was not designed to be representative, and is clearly biased toward the mafic rock types. This is mainly because the more felsic granulites collected by us are typically heavily altered, especially in the smaller nodules. About 12% of the "eclogites" examined by Mathias et al. (1970) contained plagioclase; this is clearly not a representative proportion either.

The most reliable estimate is probably that of Bloomer and Nixon (1973), who counted the proportions of different rock types in a large sample from the Letseng-la-Terae concentrates. Garnet-bearing rocks of presumed lower-crustal origin made up 62% of all nodules and 85% of the crustal ones. Of these, 36% were garnet granulite, 6% eclogites and the rest (58%) acid garnet gneisses, garnet amphibolites, etc. The separation of lower-crustal from upper-crustal lithologies in the last group is difficult, but many of these probably correspond to our sample LT-2, an intermediate granulite. The relative proportions of rock types clearly vary from pipe to pipe: for example, eclogites are quite abundant at Kao. However, a 1:1 ratio of basic granulites + eclogites + garnet websterites to acid + intermediate granulites may be roughly correct for the nodule suite. Is it also correct for the lower crust?

We have measured the density of our samples where they were large enough and fresh enough to give reliable results (Table 1). The average density of the basic garnet granulites is 3.1 g/cm^3; the eclogites and garnet websterites average 3.4 g/cm^3. If these values are combined in the 6:1 proportion found by Bloomer and Nixon (1973), the average density of the garnet-granulite suite is about 3.15 g/cm^3. The only fresh intermediate granulite (LT-2) has a much lower density (2.74 g/cm^3).

The lower-crustal seismic velocities measured by Hales and Sacks (1959) beneath the craton (6.7 km/sec) and beneath the mobile belt (7.2 km/sec) may be converted to average densities using the empirical relation found for granulite-facies rocks by Christensen and Fountain (1975). According to this relation ($V_p = 0.33 + 2.27\rho$) the cratonic lower crust has an average density of 2.81 g/cm^3, while that beneath the mobile belt has ρ = 3.03 g/cm^3. If the crust is a mixture of the basic granulite suite and intermediate rocks like LT-2, then the proportion of basic rocks must increase from 15-20% beneath the craton to 70-75% beneath the mobile belt. If more acidic rocks are present, or if the very dense eclogites are regarded as mantle rocks, the proportion of mafic rocks must be somewhat larger in both cases. The lower proportion of basic rocks beneath the craton is consistent with the apparent absence (or great rarity?) of such nodules in the extensively studied cratonic kimberlites.

The estimates based on seismic velocities are only approximate, and assume that seismic data from the Transvaal are relevant to Lesotho. Nevertheless, these data and those of Bloomer and Nixon (1973) imply that the lower crust beneath Lesotho is a roughly equal mixture of basic granulites and intermediate/acid granulites, with smaller proportions of eclogites and garnet websterites. The layered nature of our samples suggests that this part of the crust is compositionally layered on scales ranging from cm to meters. The overall composition may be roughly dioritic, and was proposed for the deep continental crust in general by den Tex (1965). This inference is also consistent with the conclusion of Windom and Boettcher (1976), based on experimental studies of plagioclase-garnet reactions, that a Conrad discontinuity should be observable only in areas where the lower crust contains large amounts of basic rocks. This lower-crustal layer is essentially anhydrous, though small amounts of SO_3 and CO_2 may be stored in scapolites, which are present in at least 15% of our samples.

Origin of the Granulite Complex

The lower crust in continental shield areas is generally believed to be intermediate in average composition, as shown by numerous studies of medium-and high-pressure granulite-facies terranes (Heier, 1973). Our interpretation of the seismic data suggests a larger proportion of basic rocks and a more basic average composition. How has this complex formed?

The basic rocks of the garnet-granulite suite are unlikely to represent recrystallized Karroo-period igneous rocks. The Karroo volcanics are tholeiitic, and lower in Al_2O_3 than the basic granulites. Furthermore, the zircon ages from two nodules indicate that the granulites are Precambrian rocks. Whole-rock chemical data discussed above indicate that the basic rocks are not genetically related to one another as members of a single igneous suite. The trace-element data further suggest that the garnet granulite suite has been depleted in LIL elements, probably during high-grade metamorphism and anatexis.

We therefore envisage the lower-crustal suite as a polygenetic complex with a prolonged history of intrusion, metamorphism and anatexis. These processes have produced a residuum enriched in refractory basic lithologies, and depleted in acidic rocks, relative to the normal continental lower crust. It may be comparable to, for example, the Fraser Range block of the Australian shield, estimated to contain ca. 60% basic granulite-facies rocks (Lambert and Heier, 1968).

The mobile belt bordering the Kaapvaal craton is believed to have formed through multistage reworking of the cratonic rocks (Kröner, 1977). The reworking processes were apparently responsible for both the formation of granulite-facies rocks and for the accumulation of basic rocks in deep crust. We envisage three possible mechanisms: (1) anatexis and subsequent removal of the more acidic components, (2) intrusion of basic rocks into the lower crust during and after orogenic episodes, (3) lateral emplacement of basic material (? oceanic crust) beneath the craton margin by subduction (cf. the apparent existence of subducted oceanic rocks beneath the Colorado Plateau; Helmstaedt and Doig, 1975). The absence of the granulite suite within the craton suggests that these mechanisms were specific to the mobile belt.

Comparison with other suites

(1) Surface outcrops of granulite and eclogite: The Lesotho nodules differ markedly in one respect from most granulite suites; the clinopyroxenes have the high Jd/Ts ratios characteristic of eclogites (Fig. 5). White's (1964) distinction between granulite and eclogite pyroxenes was drawn on the basis of samples from surface outcrops, and has proven to hold true for many granulites and eclogites analyzed since 1964. We interpret the chemistry of the clinopyroxenes in the Lesotho rocks as reflecting a quenched high-pressure pyroxene/plagioclase equilibrium consistent with that predicted from experimental studies (Green and Ringwood, 1967). To our knowledge this relation has not been demonstrated previously.

Under uniform P-T conditions, one would expect a positive correlation of Jd/Jd+Ts in the clinopyroxenes with Ab/Ab+An in the plagioclase over a wide range of bulk compositions. The scatter observed in Figure 6 may be attributed partly to the equilibration of these samples under a range of P-T conditions. Analytical errors, especially in the estimation of Fe^{3+} in clinopyroxene, will also contribute to scatter in the calculated Jd/Jd+Ts. The low Jd/Ts ratios of the pyroxenes in granulites from surface outcrops are probably an effect of reequilibration of the pyroxene-plagioclase pairs during uplift and cooling (Griffin and Carswell, in prep.).

The polygonal granoblastic textures so characteristic of these nodules are apparently not typical of eclogites in gneisses nor of most granulite terranes. These annealed textures may be typical of rocks in the lower crust, where recrystallization has occurred under static conditions over long periods of time (Nixon, 1973; cf. Padovani and Carter, 1977). However, the high heat flow during the Karroo volcanic period may also have been important in promoting textural equilibrium.

(2) Griquaites: Nixon (1973) has suggested reserving the name "eclogite" for crustal rocks and calling mantle-derived garnet-clinopyroxene rocks "griquaites". Crustal types were to be recognized on the basis of their finer grain size and granoblastic textures. The problem of making this distinction in practice is shown by the fact that four "griquaites" from Kao and Sekameng (Nixon and Boyd, 1973a, b) are indistinguishable in garnet and clinopyroxene chemistry from the granulites and fine-to-medium-grained eclogites analyzed by us (Fig. 7). On the other hand, the type griquaites of Roberts Victor Mine and the similar rocks from Newlands (both diamondiferous), as well as the Thaba Putsoa type with subcalcic diopside, can all be separated from our garnet-granulite suite on the basis of garnet chemistry (Fig. 7). Some of these griquaite garnets have moderate TiO_2 and Cr_2O_3 contents, while our samples are essentially Ti- and Cr-free. The classic griquaites also appear to have lower values of K_D^{Fe-Mg} (gnt-cpx) than the eclogites of the garnet-granulite suite. The use of the term "griquaite" for these mantle-derived eclogites may therefore be useful if the distinction is made on the basis of mineral chemistry.

(3) Other crustal nodule suites: Few comparable nodule suites have been described in detail previously. Irving (1974) has described garnet clinopyroxenites ($\pm$plag), garnet websterites and two-pyroxene granulites from the basaltic pipes at Delegate, New S. Wales, Australia. The granulites were inferred to have come from the lower crust, but equilibration temperatures could only be fixed to $<1150^oC$. The garnet pyroxenites and garnet websterites were interpreted as upper mantle (13-17 Kb) cumulates.

Padovani and Carter (1977) describe nodules of two-pyroxene basic granulites and aluminous garnet granulites from the Kilbourne Hole maar, which lies in a region of high heat flow.

McCallum and Eggler (1976) have inferred pressures of equilibration of 12-25 Kb (T = 700-850°C) for garnet websterites from the Sloan diatreme, Wyoming; they also suggested that the lower crust here consists of granulites and pyroxenites, but presented no analyses of these rocks. McGetchin and Silver (1972) presented size/abundance data on the nodule suite from Moses Rock dike, Utah, and concluded that basic garnet granulite ("garnetiferous metagabbro") was the dominant rock type in the lower crust, grading downward to eclogitic rocks and pyroxenites. The Lesotho data suggest that these rocks may in fact "coexist" with granulites, but data on the mineral chemistry of the Moses Rock nodules would be necessary to confirm this.

Meyer and Brookins (1976) described aluminous basic granulites from the Stockdale pipes, Kansas. Their estimated P-T conditions (850°C, 12 Kb) are based on treating all Fe as FeO. Recalculation of the analyses for sample 1128d (the one that satisfies charge-balance requirements) on the basis used here yields P = 5-10 Kb, T = 625-690°, compatible with the estimates for Lesotho. Eclogites from this pipe (Meyer and Brookins, 1971) are similar in mineral chemistry to those described here, and probably come from the lower crust rather than from the mantle.

The available data thus suggest that the lower crust in most areas sampled by kimberlites or basaltic diatremes is dominantly basic in composition. Where heat flow is unusually high, as in areas undergoing basaltic volcanism, these rocks are represented mainly by two pyroxene granulites. In areas of moderate heat flow, garnet granulites, pyroxenites and eclogites dominate the lower-crustal nodule suite.

The Mohorovic Discontinuity beneath Lesotho

The ultramafic nodules from the Lesotho kimberlites have been intensively studied in several laboratories, and their equilibrium P and T have been estimated by reference to experimental work on the Di-En and En-Py systems (Boyd, 1973; Boyd and Nixon, 1975). These estimates define a "pyroxene geotherm" in the depth range 85-150 km, which closely approximates the generalized "Shield geotherm" of Clark and Ringwood (1964). However, these studies have not identified any nodules that originate from the upper mantle at depths between ca. 35 and 85 km. Spinel lherzolites do occur in the kimberlites at Lipelenang and Ngopetseu in Lesotho. However, the rarity of such materials in the Lesotho and South African kimberlites suggests that they do not represent the normal upper mantle beneath this area.

It is possible that the kimberlites simply do not sample the uppermost mantle, for mechanical reasons. Alternatively, the uppermost mantle may be represented by dunites, harzburgites or other lithologies for which P-T estimates are not yet available, but these nodules are also rare in the Lesotho pipes. Carswell et al. (this volume), have argued that the apparent gap in the sampling may be an artifact produced by the calculation procedure. The smaller T dependence of the Ca-rich arm of the Di-En solvus at low temperatures may result in an overestimation of temperatures in the 700-950°C range, and displace all determinations to higher T (and thus higher P) (cf. Mori and Green, 1976; Lindsley and Dixon, 1976). In this case the uppermost mantle may in fact be represented by numerous garnet-lherzolite nodules.

Another possibility is that the upper mantle from 35-85 km depth consists largely of eclogites and garnet websterites, with smaller amounts of garnet granulites of gabbroic-anorthosite or andesitic composition (Fig. 8). We interpret the eclogites studied here as crustal rocks, because of their similarities to the granulites, but it is important to note that the pressure estimates on these rocks are only minimum values. Several garnet websterites from our suite yield P-T estimates that would place them at depths of 45-60 km (Fig. 8). These high pressures may be spurious, but pending better calibration of the opx/gnt barometer, we must consider the possibility that these rocks are in fact fragments of the uppermost mantle. The eclogites and websterites have densities corresponding to the measured sub-Moho seismic velocities in the Transvsaal (Vp = 7.9-8.0, Hales and Sacks (1959)).

The arguments against an eclogitic upper mantle (summarized by Ringwood, 1975) are based heavily on large-scale geophysical and geochemical models. They may not be relevant to the existence of a relatively thin (40-50 km) eclogitic layer immediately underlying the continental crust in shield areas. Until some ultrabasic rocks are definitely identified as coming from the uppermost mantle, the possibility of an eclogite- or pyroxenite-rich zone must remain open. This model would imply that the Moho beneath Lesotho represents a narrow zone beneath which eclogite and garnet websterite become the dominant rock types. We envisage such a zone as being produced by an underplating process. Ascending magmas would tend to pond up, at least temporarily, at the base of the crust due to the density contrast, which would reduce their buoyancy. Cumulates from these magmas would be relatively mafic, and on cooling would be converted to eclogites and garnet websterites. Eclogites formed from large cumulate masses in the lower crust would also, with time, tend to sink through the crust and accumulate near the Moho. Eventually, these processes could build up a thick layer of basic rocks with eclogite-facies mineral assemblages. The shallower parts of the garnet-granulite suite, representing most of our samples, may also be the result of these processes, as suggested above.

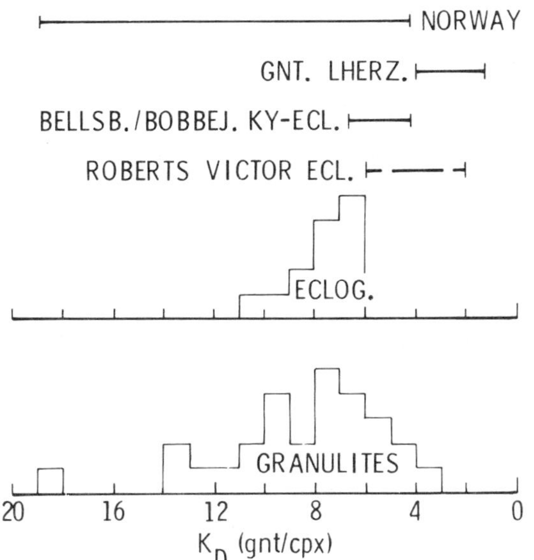

Fig. 9. Ranges of K_D (Fe(Mg)gnt/Fe(Mg)cpx) in the garnet granulite suite and in some comparable suites of eclogites. The Norwegian eclogites are "crustal" (Krogh, 1977), the others are mantle-derived (in part diamondiferous).

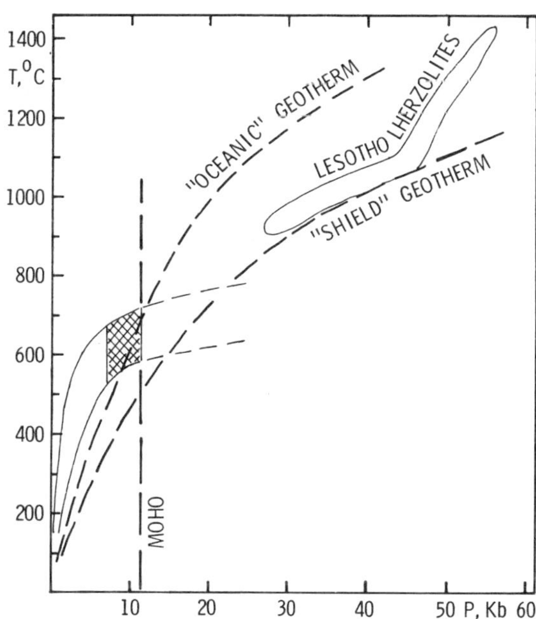

Fig. 10. Relation between the P-T field of the garnet granulite suite (crosshatched) and various proposed geotherms. Crossed dots are the high-Fe, low-Al garnet pyroxenites; these P-T estimates may be spurious.

The Mid-Cretaceous geotherm beneath Lesotho

The P-T estimates in Figure 8 are mostly contained between two curves, which are sharply concave due to the low pressures estimated for a few samples. Moving these samples to higher pressures (and thus higher T) would allow the high-T curve to be smoothed out somewhat, but the positions of these curves in the 6-10 Kb range seem reasonably well-established and imply average temperatures of 600-650°C in this depth interval.

These temperatures are higher by 200-250°C than those predicted by the "Shield geotherm" of Clark and Ringwood (1964) and the low-P extension of the "pyroxene geotherm" of Boyd and Nixon (1975) (Fig. 10). It might be argued that these temperatures are relics after a granulite-facies regional metamorphic event, and that the mineral assemblages have failed to completely equilibrate during post-metamorphic cooling to a "Shield geotherm". However, the present-day heat flow in the Karroo basin is very high (1.41 h.f.u., Clark and Ringwood, 1964), and presumably was still higher immediately after the Karroo volcanism had ceased ca. 90 m.y. ago. The crustal geotherm defined by the nodule suite is consistent with this high heat flow; this suggests that the equilibration temperatures of the nodules do record the actual crustal temperatures at the time of kimberlite eruption. Such a conclusion is also consistent with the annealed textures of most nodules, which may be ascribed to Karroo-period heating in a static environment.

Gabbroic intrusions of varying size, intruded during the Karroo volcanic activity, would perturb the local geotherm to varying degrees. This effect may explain the wide range in the T estimates, at each P, shown by the garnet granulite suite. The two-pyroxene granulites from Monastery Mine appear to imply somewhat elevated T at shallow P. Since gabbro is one of the most common nodule types at Monastery Mine, the pyroxene granulite nodules may be samples of a widespread contact-metamorphic aureole.

Our data suggest that the "pyroxene geotherm" of Boyd and Nixon (1975) is only valid for the deeper parts of the upper mantle, while temperatures in the uppermost mantle and lower crust lay well above those predicted by the "Shield geotherm". This in turn implies a complicated thermal structure, with several inflections, in the 0-250 km depth region during Cretaceous time.

Acknowledgements. The authors are grateful to the Norwegian Scientific Research Council Nansenfondet, the Natural Evnironmental Research Council and The Royal Society for financial support. P. Kretsen kindly provided important extra samples and useful discussions. The considerable assistance of Dr. R. Kanaris-Sotiriou and V. A. Somogyi with the whole-rock analyses is gratefully acknowledged. We would also like to thank the organizers and guides of the 1973 Lesotho excursions for making this work possible.

References

Bence, A. E., and A. L. Albee, Empirical correction factors for the electron microanalysis of silicates and oxides, J. Geol., 78, 382-403, 1968

Bloomer, A. G., and P. H. Nixon, The geology of the Letseng-la-terae kimberlite pipes, in Lesotho Kimberlites, edited by P. H. Nixon, Lesotho National Development Cooporation, Masern, Lesotho, pp. 20-32, 1973.

Boyd, F. R., A pyroxene geotherm, Geochim. Cosmochim. Acta, 37, 2533-2546, 1973.

Boyd, F. R., and P. H. Nixon, Origins of the ultramafic nodules from some kimberlites of northern Lesotho and the Monastery Mine, South Africa, Phys. Chem. Earth, 9, 431-454, 1975.

Carswell, D. A., Primary and secondary phlogopites and clinopyroxenes in garnet lherzolite xenoliths, Phys. Chem. Earth, 9, 417-429, 1975.

Carswell, D. A., D. B. Clarke, and R. H. Mitchell, The geochemistry of ultramafic nodules from Pipe 200 and their bearing on the nature of the upper mantle beneath Lesotho, Extended Abstracts, Second International Kimberlite Conference, Santa Fe, New Mexico, 1977.

Christensen, N. I., and D. M. Fountain, Constitution of the lower continental crust based on experimental studies of seismic velocities in granulite, Bull. Geol. Soc. Amer., 86, 227-236, 1975.

Clark, S. P., Jr., and A. E. Ringwood, Density distribution and constitution of the mantle, Rev. Geophys., 2, 35-88, 1964.

Coleman, R. G., D. E. Lee, L. B. Beatty, and W. W. Brannock, Eclogites and eclogites: their differences and similarities, Bull. Geol. Soc. Amer., 76, 483-508, 1965.

Currie, K. L., and L. W. Curtis, An application of multicomponent solution theory to jadeitic pyroxenes, J. Geol., 84, 179-194, 1976.

Davis, G. L., The ages and uranium contents of zircons from kimberlites and associated rocks, Extended Abstracts, Second International Kimberlite Conference, Santa Fe, New Mexico, 1977.

Dawson, J. B., and W. E. Stevens, Statistical classification of garnets from kimberlite and associated xenoliths, J. Geol., 83, 589-607, 1975.

den Tex, E., Metamorphic lineages of orogenic plutonism, Geol. Mijnbouw, 44, 103-132, 1965.

Floyd, P. A., and J. A. Winchester, Magma type and tectonic setting discrimination using immobile elements, Earth Planet, Sci. Letters, 27, 211-218, 1975.

Goldsmith, J. R. and R. C. Newton, Scapolite-plagioclase stability relations at high pressures and temperatures in the system $NaAlSi_3O_8$-$CaAl_2Si_2O_8$-$CaCO_3$-$CaSO_4$, Amer. Mineral., 62, 1063-1081, 1977.

Green, D. H., and A. E. Ringwood, An experimental investigation of the gabbro to eclogite transformation and its petrological applications, Geochim. Cosmochim. Acta, 31, 767-833, 1967.

Green, D. H., and A. E. Ringwood, A comparison of recent experimental data on the gabbro-garnet granulite-eclogite transition, J. Geol., 80, 272-288, 1972.

Griffin, W. L., Formation of eclogites and the coronas in anorthosite, Bergen area, Norway, Geol. Soc. Amer. Memoir, 35, 37-63, 1972.

Griffin, W. L., and K. S. Heier, Paragenesis of garnet in granulite-facies rocks,

Lofoten-Vesterålen, Norway, Contr. Mineral. Petrol., 23, 89-116, 1969.

Griffin, W. L., and K. S., Heier, Petrological implications of some corona structures, Lithos, 6, 315-335, 1973.

Griffin, W. L., B. B. Jensen, and S. N. Misra, Anomalously elongated rutile in eclogite-facies pyroxene and garnet, Norsk Geol. Tidsskr., 51, 177-185, 1971

Hales, A. L. and I. S. Sacks, Evidence for an intermediate layer from crustal seismic studies in the Eastern Transvaal, Geophys. J. Roy. Astr. Soc., 2, 15-33, 1959.

Heier, K. S., Geochemistry of granulite facies and problems of their origin, Phil. Trans. R. Soc. Lond., A 273, 429-442, 1973.

Helmstaedt, H., and R. Doig, Eclogite nodules from kimberlite pipes of the Colorado Plateau-samples of subducted Franciscan-type oceanic lithosphere, Phys. Chem. Earth, 9, 95-111, 1975.

Irving, A. J. Geochemical and high pressure experimental studies of garnet pyroxenite and pyroxene granulite xenoliths from the Delegate basaltic pipes, Australia, J. Petrol., 15, 1-40, 1974.

Krogh, E., Evidence of Precambrian continent-continent collision in western Norway, Nature, 267, 17-19, 1977.

Kröner, A., The Precambrian geotectonic evolution of Africa: plate accretion versus plate destruction, Precambrian Res., 4, 163-213, 1977.

Kushiro, I., Clinopyroxene solid solutions formed by reactions between diopside and plagioclase at high pressures, Mineral. Soc. Amer., Spec. Paper, 2, 179-191, 1969.

Lambert, I. S., and K. S. Heier, Geochemical investigations of deep seated rocks in the Australian Shield, Lithos, 1, 30-53, 1968.

Lindsley, D. H., and S. Dixon, Diopside-enstatite equilibria at 850°C to 1400°C, 5 to 35 Kb, Amer. J. Sci., 276, 1285-1301, 1976.

Lovering, J. F., and A. J. R. White, The significance of primary scapolite in granulitic inclusions from deep seated pipes, J. Petrol., 5, 195-218, 1964.

Mathias, M., J. C. Siebert, and P. C. Rickwood, Some aspects of the mineralogy and petrology of ultramafic xenoliths in kimberlites, Contr. Mineral. Petrol., 26, 75-123, 1970.

McCallum, M. E., and D. H. Eggler, Diamonds in an upper mantle peridotite nodule from kimberlite in southern Wyoming, Science, 192, 253-256, 1976.

McGetchin, T. R., and L. T. Silver, A crustal-upper mantle model for the Colorado Plateau based on observations of crystalline rock fragments in the Moses Rock dike, J. Geophys. Res., 77, 7022-7037, 1972.

McGetchin, T. R., and G. W. Ullrich, Xenoliths in maars and diatremes with inferences for the moon, Mars and Venus, J. Geophys. Res., 78, 1832-1853, 1973.

Meyer, H. O. A., and D. G. Brookins, Eclogite xenoliths from the Stockdale kimberlite, Kansas, Contr. Mineral. Petrol., 34, 60-72, 1971

Meyer, H. O. A. and D. C. Brookins, Sapphirine, sillimanite and garnet in granulite xenoliths from Stockdale kimberlite, Kansas, Amer. Mineral., 61, 1194-1202, 1976.

Mori, T., and D. H. Green, Subsolidus equilibria between pyroxenes in the $CaO-MgO-SiO_2$ system at high pressure and temperature, Amer. Mineral., 61, 616-625, 1976.

Mysen, B. O., and W. L. Griffin, Pyroxene stoichiometry and the breakdown of omphacite, Amer. Mineral., 56, 60-63, 1973.

Nixon, P. H. (ed.), Lesotho Kimberlites, Lesotho Nat. Devel. Corp. 350 pp. 1973.

Nixon, P. H., Perspective, in Lesotho Kimberlites, edited by P. H. Nixon, Lesotho Nat. Devel. Corp. 1973, 350 pp. 1973.

Nixon, P. H. and F. R. Boyd, Deep-seated nodules in Lesotho Kimberlites, edited by P. H. Nixon, Lesotho Nat. Devel. Corp. 1973, 350 pp. 1973a.

Nixon, P. H. and F. R. Boyd, Carbonated ultrabasic nodules from Sekameng, in Lesotho Kimberlites, edited by P. H. Nixon, Lesotho Nat. Devel. Corp. 1973, 350 pp. 1973b.

Obata, M., The solubility of Al_2O_3 in orthopyroxenes in spinel and plagioclase peridotites and spinel pyroxenite, Amer. Mineral., 61, 804-816, 1976.

O'Hara, M. J., S. W. Richardson, and G. Wilson, Garnet-peridotite stability, Contr. Mineral. Petrol., 32, 48-68, 1971.

Padovani, E. R. and J. L. Carter, Granulite facies xenoliths from Kilbourne Hole maar, New Mexico, and their bearing on deep crustal evolution, Extended Abstracts, Second International Kimberlite Conference, Santa Fe, New Mexico, 1977.

Råheim, A., and D. H. Green, Experimental determination of the temperature and pressure dependence of the Fe-Mg partition coefficient for coexisting garnet and clinopyroxene, Contr. Mineral. Petrol., 48, 179-203, 1974.

Rickwood, P. C., J. J. Gurney, and D. R. White-Cooper, The nature and occurrence of eclogite xenoliths in the kimberlites of southern Africa. Upper Mantle Project, Geol. Soc. S. Africa Spec. Publ., 2, 371-393, 1969.

Ringwood, A. E., Composition and Petrology of the Earth's Mantle, McGraw Hill, New York, 618 pp. 1975.

Rogers, N. W., Granulite xenoliths from Lesotho kimberlites and the composition of the lower continental crust, Nature, 270, 681-684, 1977.

Sheppard, S. M. F. and J. B. Dawson, Hydrogen, carbon and oxygen isotope studies of megacryst and matrix minerals from Lesothan and South African kimberlites, Phys. Chem. of the Earth, 9, 747, 764, 1975.

Stormer, J. C., Jr., A practical two-feldspar geothermometer, Amer. Mineral., 60, 667-674, 1975.

Wells, P. R. A., Pyroxene thermometry in simple and complex systems, Contr. Mineral. Petrol., 62, 129-139, 1977.

White, A. J. R., Clinopyroxenes from eclogites and basic granulites, Amer. Mineral., 49, 883-886, 1964.

Windom, K. E., and A. L. Boettcher, The effect of reduced activity of anorthite on the reaction grossular + quartz = anorthite + wollastonite: a model for plagioclase in the lower crust and upper mantle, Amer. Mineral., 61, 889-896, 1976.

Wood, B. J., Solubility of alumina in orthopyroxene coexisting with garnet, Contr. Mineral. Petrol., 46, 1-5, 1974.

Wood, B. J., The influence of pressure, temperature and bulk composition on the appearance of garnet in orthogneisses - an example from South Harris, Scotland, Earth Planet. Sci. Let., 26, 299-311, 1975.

Wood, B. J., The partitioning of iron and magnesium between garnet and clinopyroxene, Geophys. Lab. Wash. Yearbook, 75, 571-574, 1976.

Wood, B. J., and S. Banno, Garnet-orthopyroxene and orthopyroxene-clinopyroxene relationships in simple and complex systems, Contr. Mineral. Petrol., 42, 109-124, 1973.

MINERALOGY AND S^{34}/S^{32} RATIOS OF SULFIDES ASSOCIATED WITH KIMBERLITE, XENOLITHS AND DIAMONDS

Hsiao-ming Tsai, Yuch-ning Shieh and Henry O.A. Meyer

Department of Geosciences, Purdue University, West Lafayette, Indiana 47907

Abstract. The mineralogy of sulfides from eclogite xenoliths and inclusions in diamond, and the S^{34}/S^{32} ratios of sulfides from several eclogites and other crustal xenoliths in kimberlite have been studied:
1) Sulfides in thirteen eclogites (griquaites) from various localities, including Premier, Roberts Victor, Jagersfontein, Bobbejaan and Obnazhennaya (Siberia) have been examined. The most common assemblage observed is an intergrowth of pyrrhotite-pentlandite with minor chalcopyrite. Nickeliferous pyrrhotite is common, specially in the sulfides from Roberts Victor kimberlite; Ni-bearing pyrite (or Ni-bearing marcasite) has also been found in association with this phase. Textural relations indicate the presence of an original immiscible sulfide melt, and subsequent re-equilibration at lower temperatures from probably an initial M_{ss} phase to form the present intergrowths. K-bearing Fe-Ni-sulfide (8 wt% K) is present in one sample from Obnazhannaya, and is probably due to a later stage metasomatic reaction. A unique assemblage in a garnet xenocryst from Premier consists of granular aggregate of pentlandite, chalcopyrite, magnetite and silicates. Secondary pyrite was observed in samples from Bobbejaan and Roberts Victor.
2) Examination of sulfide inclusions in diamonds from Finsch, Koffiefontein, and Premier kimberlite mines reveals the presence of a variety of sulfide assemblages, including pyrrhotite-pentlandite intergrowth with or without minor chalcopyrite; pyrrhotite-chalcopyrite intergrowth with minor pentlandite; and Ni-rich pentlandite-magnetite with possible minor chalcopyrite. Pyrite is present in one diamond from Koffiefontein.
3) S^{34}/S^{32} ratios have been determined for sulfides in crustal and eclogitic xenoliths and discrete crystals of sulfides from Premier, Finsch, Koffiefontein, Roberts Victor and Bobbejaan kimberlites. Most pyrite-bearing crustal xenoliths as well as secondary pyrite in a Bobbejaan eclogite give results suggestive of sedimentary origin of the sulfur (δS^{34} ranging from -37.3 to +15.1). Discrete pyrite from the Premier kimberlite has a δS^{34} of +1.3, suggesting the sulfur may be of magmatic origin. Chalcopyrite/pentlandite and pyrrhotite/pentlandite assemblages from Premier and Roberts Victor give δS^{34} values of +0.2 and +2.1, respectively. This supports the inference that sulfur of presumed mantle origin has a δ-value close to zero.

Introduction

In the last few years, several investigations concerning sulfide mineralogy and geochemistry in eclogite, lherzolite and associated xenoliths from kimberlite have been undertaken (Vakhrushev and Sobolev, 1971; Frick, 1973; Desborough and Czamanske, 1973; Meyer and Boctor, 1975; Bishop et al., 1975; Tzepin and Lazko, 1976; Clarke et al., 1977). Since it is believed that the eclogite and ultramafic xenoliths represent fragments of the upper mantle, transported to the earth's surface by the ascending movement of the host kimberlite, these studies have provided us with knowledge of the possible composition of sulfide material in the upper mantle. However, compared to the silicate minerals the amount of detailed chemical data pertinent to sulfides in such xenoliths is relatively scarce. We have therefore examined sulfides in eclogite and associated xenoliths in order to both enhance our knowledge of these phases in mantle rocks and also to study variations in chemistry of sulfides in xenoliths from different localities.

Furthermore, we have extended the study to include the sulfide mineral inclusions in diamond. If one assumes that diamond grows stably in nature then presumably the primary sulfide inclusions within diamond also represent stable mantle phases. In theory these sulfides should be better preserved than those in the xenoliths since they have been armored by diamond and unable to react chemically with other phases. Unfortunately, as will be noted later, there is often ambiguity as to whether the sulfide is a primary or secondary inclusion.

Prior to this study, sulfides in diamond have received scant attention. Sharp (1966)

TABLE 1. Localities and assemblages of sulfides from xenoliths in kimberlites.

Group	No.	Locality	Assemblage
Ia	JAG 35	Jagersfontein	Po+Pn+Cp intergrowth plus Cp rim
	JAG 356	Jagersfontein	
	WM 7	Roberts Victor	
Ib	XRV 10	Roberts Victor	Ni-Po+Pn intergrowth plus Cp rim
	KRV 7	Roberts Victor	
	SRV 1	Roberts Victor	
II	JJG 548	Obnazhennaya	Pn+Po+Cp intergrowth plus Cp rim and K-Fe-Ni sulfide adjacent to Cp rim
III	XRV 7	Roberts Victor	Ni-Po+Ni-Py (or Ni-marcasite) plus Cp rim and an adjacent high Ni phase
IV	*173/53	Premier	Granular aggregates of sulfides (Cp+Pn), silicates (andradite + ilvaite + chlorite) and oxide (magnetite)
V	AMRV 1	Roberts Victor	Massive Py (secondary)
V	173/33	Bobbejaan	Massive Py (secondary)
VI	PJL 18	Roberts Victor	Ni-Py+ violarite + magnetite and Cp rim
VI	JJG 4	Roberts Victor	Ni-Po+Ni-Py and Cp rim

*This sulfide aggregate is found in a garnet xenocryst; all others are from eclogite xenoliths.

Po = pyrrhotite, Pn = pentlandite, Cp = chalcopyrite, Py = pyrite

first identified the presence of pyrrhotite and pentlandite as diamond inclusions by X-ray powder diffraction. Harris (1968, 1972) reported these same minerals, and analyses of pyrrhotite from Brazilian diamonds were presented by Meyer and Svisero (1975). Prinz et al. (1975) have also commented on the occurrence of pentlandite and chalcopyrite as diamond inclusions.

A further problem that has to be addressed is the origin and source of the sulfur forming the sulfides, not only in the mantle xenoliths and inclusions in diamond, but also in crustal xenoliths and in kimberlite itself. At this time it is with some difficulty that isotopic data can be obtained from the minute inclusions in diamond. Accordingly, in this phase of our study we have determined the S^{34}/S^{32} values for predominantly crustal xenoliths from a number of South African kimberlites, a garnet xenocryst from Premier Mine, South African and an eclogite xenolith from Roberts Victor.

Sulfides in Xenoliths

Sample Description and Techniques

Sulfides in twelve eclogites (griquaites) and one garnet xenocryst from various localities, including Premier, Roberts Victor, Jagersfontein, Bobbejaan (all South Africa) and Obnazhennaya (Siberia) have been examined. The sulfides are generally rounded in shape, varying from less than 0.5 mm to several mm in diameter. The sulfides usually occupy interstitial positions relative to the silicate minerals, and may also be observed along silicate grain boundaries as discontinuous strings of tiny grains. Most sulfides occur as polymineralic assemblages.

Quantitative analyses were obtained using an automated M.A.C. 500 electron microprobe with on-line computer reduction of data by a modified Bence-Albee (1968) and Albee and Ray (1970) matrix refinement which includes background, deadtime, drift, fluorescence and atomic number effects. FeS and pure metal elements were used as standards. A low specimen current (0.005 μa) was used due to the small grain size and close intergrowth of most sulfide minerals. Unfortunately in extreme cases the fineness of the intergrowth was beyond the resolution of electron microprobe beam and therefore, only bulk (see note, Table II) analyses were obtained using broad beam techniques.

Mineralogy

Several different assemblages consisting of a limited number of sulfide phases were observed in the samples examined (Table I). The most common assemblage, including two from Jagersfontein (JAG 35 and JAG 356), and four from Roberts Victor (WN7, XRV10, SRV1 and KRV7) is an intergrowth of pyrrhotite and pentlandite surrounded by a narrow chalcopyrite rim. Chalcopyrite may also occur as minor, very fine exsolution lamellae in the pyrrhotite. However, differences in relative abundances of the individual phases do exist among the various samples. Another assemblage observed in only one sample from Roberts Victor (XRV-7) consists mainly of Ni-pyrrhotite and Ni-pyrite (or Ni-marcasite; the term Ni-pyrite or Ni-pyrrhotite, etc. is used here to indicate that Ni is present in quantitities greater than normal but does not mean the Ni end-member).

A unique assemblage in the garnet xenocryst from Premier (173/53) consists of granular aggregates of pentlandite, chalcopyrite, magnetite and silicates. Two eclogites, one from Roberts Victor (AMRV1) and one from Bobbejaan (173/33) contained only pyrite, and this mineral is believed to be of secondary origin. Finally, two other specimens from Roberts Victor (PJL18 and JJG4) appear to be highly altered and some of the sulfide phases cannot be identified with certainty. Of special interest is the occurrence of K-sulfide (Clarke et al., 1977) at the margin of sulfides in JJG548, the specimen from Obnazhennaya, Siberia.

The silicate mineralogy of these samples has been independantly investigated by Hatton and Gurney (pers. comm.) except for SRV-1 which is a coesite-bearing grospydite described by Smyth and Hatton (1977), and Smyth (1977).

JAG35. Four sulfide globules are present in this sample with size ranging from 2.5 mm (JAG35-A) to less than 1 mm in diameter (JAG35-B, C and D). Among these JAG35-A is especially interesting in that it contains unusually thick (about 0.05 mm) exsolution lamellae of Co-bearing pentlandite (Co $\sim$ 3.5 wt%) in the pyrrhotite host (Fig. 1a and Table II). These exsolution lamellae appear to be oriented in one direction and tend to be concentrated in the inner part of the sulfide globule. Also observed in JAG35-A are the presence of numerous fine lamellae, several microns in width, oriented in two general directions (Fig. 1a), unfortunately these lamellae are too fine to be individually analyzed. Recalculation of the bulk composition (Table II) obtained by broad beam probe analyses indicates that approximately 5% chalcopyrite is present in the host pyrrhotite as sub-micron exsolution lamellae.

One typical feature in all sulfide assemblages is that chalcopyrite forms a narrow rim around the major central phases. Often between this chalcopyrite and the central assemblage there are thin zones of pentlandite. The composition of this pentlandite is identical to that of the pentlandite lamellae $(Fe_{4.7}Ni_{3.6}Co_{0.4})S_8$. Similar features have also been observed along cracks with chalcopyrite occurring on both sides of the cracks and pentlandite separating the chalcopyrite from the main assemblage (Fig. 1b).

The other three sulfide globules in this eclogite show similar characters except the thick pentlandite lamellae are absent.

WM7 and JAG356. Sulfide globules in WM7 are rounded with maximum dimension of 1.4 mm in diameter, and contain extremely high abundances of fine exsolution lamellae (Fig. 1c). Bulk analyses (Table II) indicate the presence of two types of exsolution lamellae; pentlandite and chalcopyrite. Although pentlandite is exceedingly abundant the host phase appears to be pyrrhotite; however, different cross-sections may result in changes in the proportion of each mineral phase. Minor chalcopyrite occurs at the margins of the globules and along cracks. Pentlandite is also observed adjacent to the chalcopyrite as in JAG35 (Fig. 1c).

JAG356 is generally similar to WM7 except the pentlandite and chalcopyrite exsolution lamellae are less abundant. Only bulk analyses were obtained (Table II).

XRV10 and SRV1. Sample SRV1 is of particular interest as this eclogite (or grospydite) contains crystals of coesite. The silicate and oxide minerals of this specimen have been previously described by Smyth and Hatton (1977). Sulfides from SRV1 and XRV10 are basically similar to those described above with only small differences. For instance, the host pyrrhotite in these two samples is slightly nickeliferous. Chalcopyrite lamellae are absent and the pentlandite lamellae are much less abundant. However, chalcopyrite does occur as a discontinuous rim, and the discrete pentlandite is more common at the margin and along cracks (Fig. 1d). Analyses of all individual mineral phases in SRV1 were obtained (Table II), but only a bulk analysis for XRV10 (Table II).

KRV7. Five sulfide globules are present in this section, but unfortunately most have undergone various degrees of alteration (Fig. 2a and 2b). The least altered (Fig. 2a) has a mineralogy similar to that in SRV1 and XRV10. However, of particular note is the presence of pentlandite lamellae in the chalcopyrite rim. This feature appears to be unique to this one specimen. Another feature in the sulfide globules is the occurrence of presumably secondary pyrite as massive discrete crystals and as small, somewhat skeletal grains outside the margin of the sulfide globule (Fig. 2b).

Altered sulfide aggregates are particularly difficult to analyze, but it appears that granular Ni-pyrrhotite and Ni-pyrite are the two major alteration phases replacing the central sulfides. The outer rims of chalcopyrite often remain unaltered as a reef surrounding the embayed and corroded central altered area. Chalcopyrite also

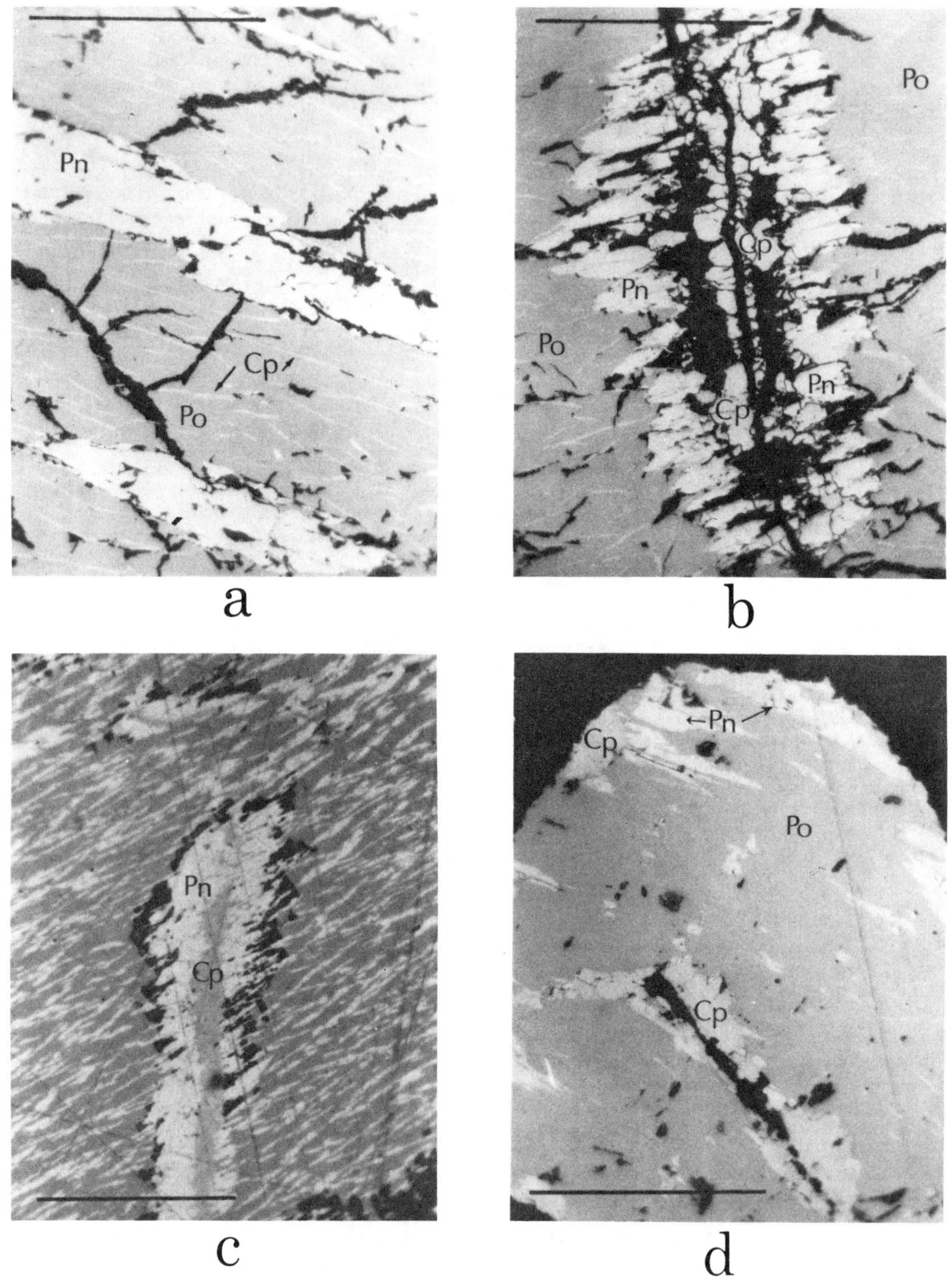

Fig. 1a) Coarse pentlandite (Pn) lamellae and fine chalcopyrite (Cp) lamellae in pyrrhotite (Po) (JAG35). Bar is 0.1 mm. b) Chalcopyrite (Cp) along crack in pyrrhotite (Po) with pentlandite (Pn) intervening between Cp and Po (JAG35). Bar is 0.1 mm. c) Abundant pentlandite (Pn) lamellae (light gray) in pyrrhotite (Po) (dark gray). Also shown in a vein of chalcopyrite (Cp) and pentlandite (Pn) (WM7). Bar is 0.03 mm. d) Pyrrhotite (Po) containing few flame-like exsolution lamellae of pentlandite (Pn). Chalcopyrite (Cp) and pentlandite also occur at the margin and along cracks. (SRV1). Bar is 0.1 mm.

TABLE 2. Representative analyses of sulfide minerals in xenoliths from kimberlite.

	JAG 35A				JAG 356	WM7	SRV 1		
	Pn	Po	Cp	Bulk+	Bulk+	Bulk+	Po	Cp	Pn
Fe	33.9	61.6	30.5	59.8	51.4	56.8	58.1	31.0	28.4
Ni	28.3	0.00	0.00	0.05	9.48	2.31	0.93	0.02	34.6
Co	3.36	0.01	0.09	0.14	0.47	0.32	0.14	0.14	2.15
Cu	0.00	0.07	33.6	1.69	1.52	1.37	0.07	33.7	0.00
S	34.3	37.5	36.3	38.0	35.6	37.8	40.3	35.5	34.6
Total	99.9	99.2	100.5	99.6	98.4	98.6	99.5	100.4	99.8

	XRV 10	JJG 548			XRV 7			
	Bulk+	Pn	K-S	Bulk+	Po	Ni-Py	Cp	High Ni-phase
Fe	54.0	32.0	37.5	36.3	56.1	44.7	30.9	28.19 - 33.53
Ni	3.86	32.7	12.2	23.4	2.15	1.61	0.00	20.54 - 26.38
Co	0.36	0.96	0.08	0.47	0.13	0.50	0.05	0.39 - 0.53
Cu	0.02	0.00	4.59	1.87	0.00	0.18	33.8	0.00
S	40.9	33.7	32.7	34.5	41.3	53.1	35.2	42.96 - 46.27
K			8.31					
Total	99.1	99.6	95.5*	96.5	99.7	100.1	100.0	

	173/53			AMRV 1	173/33/ K64/xm1	PJL 18	
	Gran. Pn	Lam. Pn	Cp	Py	Py	Violarite	Ni-Py
Fe	27.2	27.2	30.5	46.4	46.9	18.4	43.5
Ni	38.7	38.1	0.00	0.00	0.00	37.8	3.13
Co	0.83	0.82	0.01	-	0.04	0.26	0.37
Cu	0.00	0.00	33.9	0.07	0.04	0.11	0.11
S	33.6	33.7	35.7	53.4	52.9	42.8	52.9
Total	100.3	99.8	100.1	99.9	99.9	99.4	100.0

* Mg and Al < 1.0 wt% respectively

+Note : The term bulk is used here (and in text and Table V) for analyses obtained by broad-beam microprobe techniques. Although several phases were included within beam area these analyses should not be considered as representing true overall composition. They are used as indications of elemental abundances, especially in thin lamellae beyond resolution of 1 μm probe beam.

occurs as veins cutting through the sulfide aggregate.

JJG548. Two of the three sulfide assemblages observed in this specimen from Siberia are more irregular in shape than is usually observed (Fig. 2c). Also unique to this sample is the presence of a brownish-gray phase identified to be K-bearing sulfide, which occurs on the outer margin of the sulfide assemblage. This phase is closely associated with chalcopyrite and also contains numerous tiny veinlets which can only be seen under very high magnification (x1500) (Fig. 2d). Consequently; because of the small size, close association and the tiny lamellae in the K-sulfide, it was not possible to obtain good analyses for this phase (Table II) or for the neighboring chalcopyrite. In contrast to other specimens examined, the predominant phase in JJG548 appears to be pentlandite containing fine lamellae of pyrrhotite and less frequently, chalcopyrite. Recalculation of bulk analyses (Table II) gives 75 wt% Pn, 19 wt% Po and 6 wt% Cp. Occurrences of pyrrhotite and chalcopyrite lamellae seem to decrease gradually from the center to the margin. Pentlandite devoid of lamellae near the margin has a composition of $(Fe_{4.4}Ni_{4.4}Co_{0.1})S_8$ (Table II). Mackinawite was reported

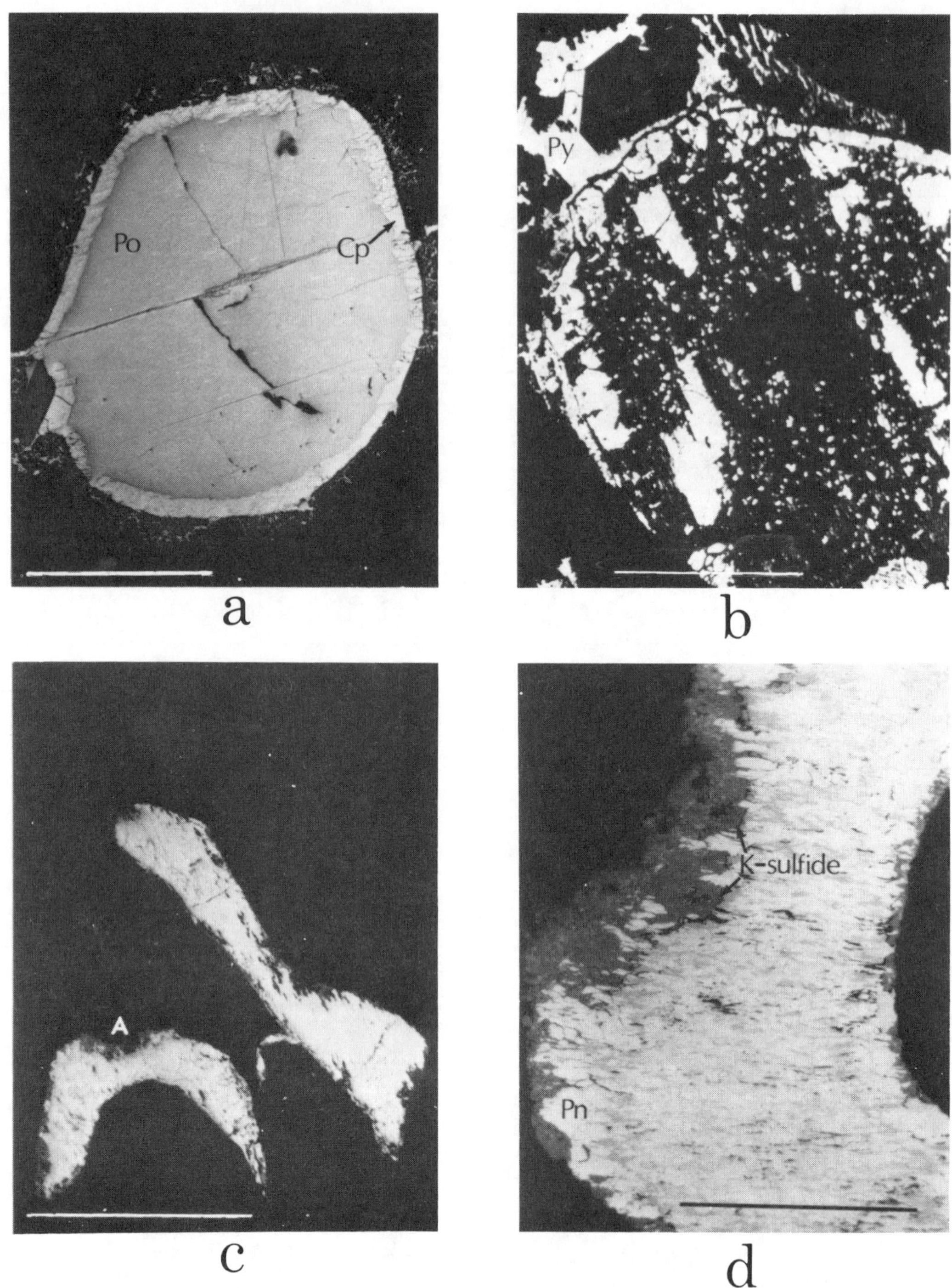

Fig. 2a) Sulfide globule consisting of pyrrhotite host (Po) with very fine pentlandite (Pn) lamellae surrounded by well-developed chalcopyrite (Cp) rim. (KRV7-E). Bar is 0.2 mm. b) Skeletal pyrite (Py) at margin of altered sulfide globule (KRV7-C). Bar is 0.2 mm. c) Irregular shaped sulfide grains in JJG548. Area A is shown enlarged in Fig. 2d. Bar is 0.5 mm. d) K-sulfide occurring at margin of sulfide grain. Pentlandite (Pn) is host to numerous pyrrhotite (Po) lamellae (JJG548). Bar is 0.1 mm.

by Vakhrushev and Sobolev (1971) as thin plates intergrown with pentlandite and chalcopyrite. Attempts to determine the presence of this phase in our sample by X-ray diffraction were unsuccessful.

XRV7. This specimen contains five sulfide grains and is interesting in that it most closely resembles the sample described by Desborough and Czamanske (1973). The major host phase is nickeliferous pyrrhotite (Table II) containing

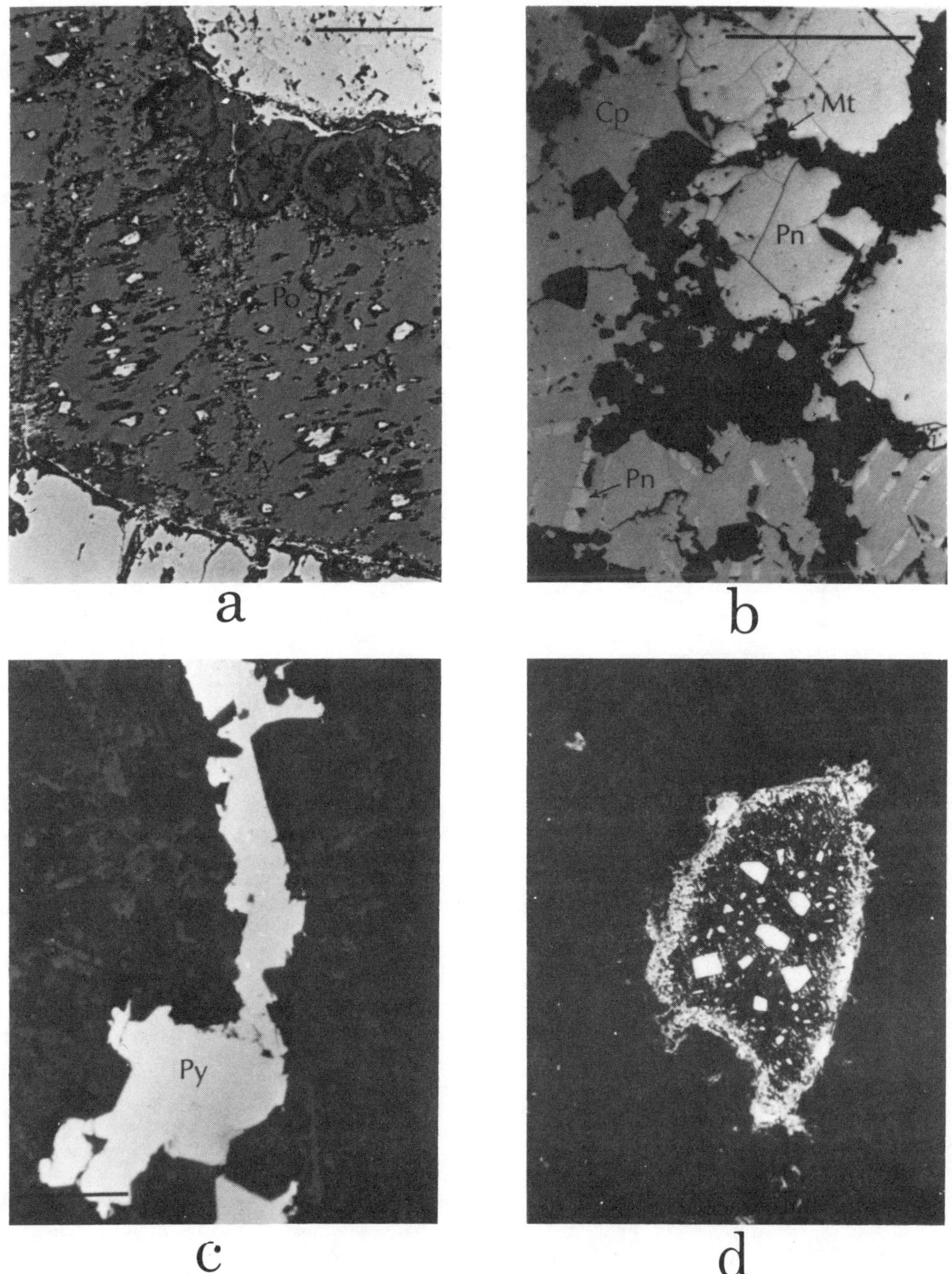

Fig. 3a) Nickeliferous pyrrhotite (Po) containing pyrite (Py). Chalcopyrite at margin (XRV7). (Color contrast due to carbon coat). Bar is 0.1 mm. b) Granular aggregate of chalcopyrite (Cp) and pentlandite (Pn). Pn also occurs as lamellae in Cp. Small grains of magnetite (Mt, dark gray) occurs associated with sulfides and silicates (black). (173/53). Bar is 0.2 mm. c) Secondary subhedral pyrite is AMRV1. Bar is 0.2 mm. d) Isolated angular grains of pyrite in altered sulfide assemblage (JJG4). Length of sulfide mass approximately 0.4mm.

1.7 to 3.5 wt% Ni. Included inside the Ni-pyrrhotite are small subhedral to euhedral Ni-bearing pyrite (or marcasite) (Fig. 3a and Table II) with Ni contents ranging from 0.32 to 2.11 wt%. A thin rim of stoichiometric chalcopyrite (Table II) surrounds the Ni-pyrrhotite + Ni-pyrite assemblage. Also found close to the margin are minute grains of a high-Ni phase containing approximately 20-26 wt% Ni (Table II). Due to the small size of these grains, positive

TABLE 3. Representative analyses of a garnet xenocryst and silicates in the included sulfide aggregate.

Oxide	Garnet	Andradite	Ilvaite
SiO_2	41.3	37.2	30.1
TiO_2	1.17	0.02	0.00
Al_2O_3	20.8	5.47	0.09
Cr_2O_3	0.41	0.00	0.00
FeO*	12.6	21.8	51.8
MgO	18.8	0.09	0.20
CaO	4.46	34.4	13.9
MnO	0.34	0.21	1.64
NiO	0.00	0.13	0.00
Na_2O	0.21	0.00	0.00
K_2O	0.00	0.00	0.00
	100.1	99.3	97.7 (+OH)

* All Fe reported as FeO.

identification is difficult, and probe analyses are ambiguous; however, these analyses are similar to those for the high-Ni phase of Desborough and Czamanske (1973).

173/53. This sample, a garnet xenocryst from Premier Mine, is unique in that it contains a fairly large (∿3.5 mm) granular aggregate of sulfide, oxide and silicate minerals (Fig. 3b). Sulfides are the most abundant phases in this aggregate, and include pentlandite and chalcopyrite in about equal proportions. However, the chalcopyrite contains minor lamellae of pentlandite that are identical in composition to the larger granular pentlandite ($Fe_{3.7}Ni_{5.0}Co_{0.1})S_8$ (Table II). The chalcopyrite is close to stoichiometry in composition (Table II). Small amounts of a weakly anisotropic, pinkish phase, possibly cubanite, as well as a bluish phase, possibly chalcocite occur but unfortunately these are too small to be positively identified with the microprobe and optical techniques.

Other minerals occurring in this aggregate are magnetite (Fe_3O_4), andradite ($Ca_3Fe_2Si_3O_{12}$), ilvaite $\{CaFe_2(Fe,OH)(SiO_4)_2\}$ and secondary needle-shaped chlorite. Analyses of the garnet constituting the xenocryst, plus the andradite and ilvaite associated with the sulfides are presented in Table III. The magnetite is virtually free of other minor elements. A similar assemblage of pentlandite, chalcopyrite, magnetite, ilvaite, andradite, chlorite, plus a number of other minerals has been reported in fissured zones of serpentinites of the Oberhalbstein, Grisons, Switzerland (Dietrich, 1972). However, this is the first occurrence of such assemblage found associated with a kimberlitic environment.

AMRV1 and 173/33. Sulfides from both these eclogites were identified to be homogeneous pyrite (Table II) by microprobe analysis and X-ray diffraction. The pyrite occurs in angular grains showing well-formed crystal outlines (Fig. 3c) which contrast the globular or rounded shapes of most of the other sulfide assemblage observed in this study. The mode of occurrence of this pyrite, in one instance along a fracture surface, is believed to indicate a secondary origin for this mineral.

PJL18 and JJG4. These two eclogites appear to contain the most altered sulfide assemblages. Ni-bearing pyrite is the major phase in the center of PJL18, together with small amounts of magnetite and violarite $(Fe,Co,Ni)_{2.9}S_4$ (Table II). A narrow rim of chalcopyrite remains around the margin. Four sulfide globules in JJG4 display various degrees of alteration. Chalcopyrite always occurs at the edge, and Ni-pyrrhotite appears to be the predominant phase for the less altered assemblages, while Ni-pyrite, occurring in isolated areas among a fine-grained dark alteration matrix (Fig. 3d), is the major phase in the more altered assemblages. The excessive alteration in these samples results in only a few areas where respectable analyses can be obtained.

Sulfide Mineral Paragenesis

The majority of sulfide samples consist generally of a pyrrhotite host in which flame like exsolutions of pentlandite occur, with the whole surrounded by a marginal rim of chalcopyrite. Minor variations in the amounts and compositions of these phases occur and account for the subgroups in Table I. For example, in sub-group Ib the pyrrhotite contains Ni (∿ 2 wt%); and JAG35-A (subgroup Ia) is notable for the very wide parallel exsolution lamellae of pentlandite. Also noted in subgroup Ia is the presence of chalcopyrite as fine lamellae.

This assemblage Po-Pn-Cp in eclogite has been recognized previously (e.g., Vakhrushev and Sobolev, 1971; Frick, 1973; Meyer and Boctor, 1975). It is also common in magmatic sulfide ores, such as Sudbury (Graterol and Naldrett, 1971). The paragenesis of the assemblage probably closely follows the crystallization and subsolidus changes determined from experimental studies in the systems Cu-Fe-Ni-S, (Craig and Kullerud, 1969), Fe-Ni-S and Cu-Fe-S (Kullerud, et al., 1969; Craig and Scott, 1974). Unfortunately, the pressure effect on these sulfide equilibria is not well known, although Bell et al. (1964) showed drastic lowering of pentlandite stability at high pressure. Such knowledge is important, as the eclogite and the included primary sulfide minerals have presumably all formed at high pressures and temperatures.

Compositions of the various coexisting sulfide phases in the eclogites are presented in the 850°C isotherm (Fig. 4) of the Cu-Fe-Ni-S

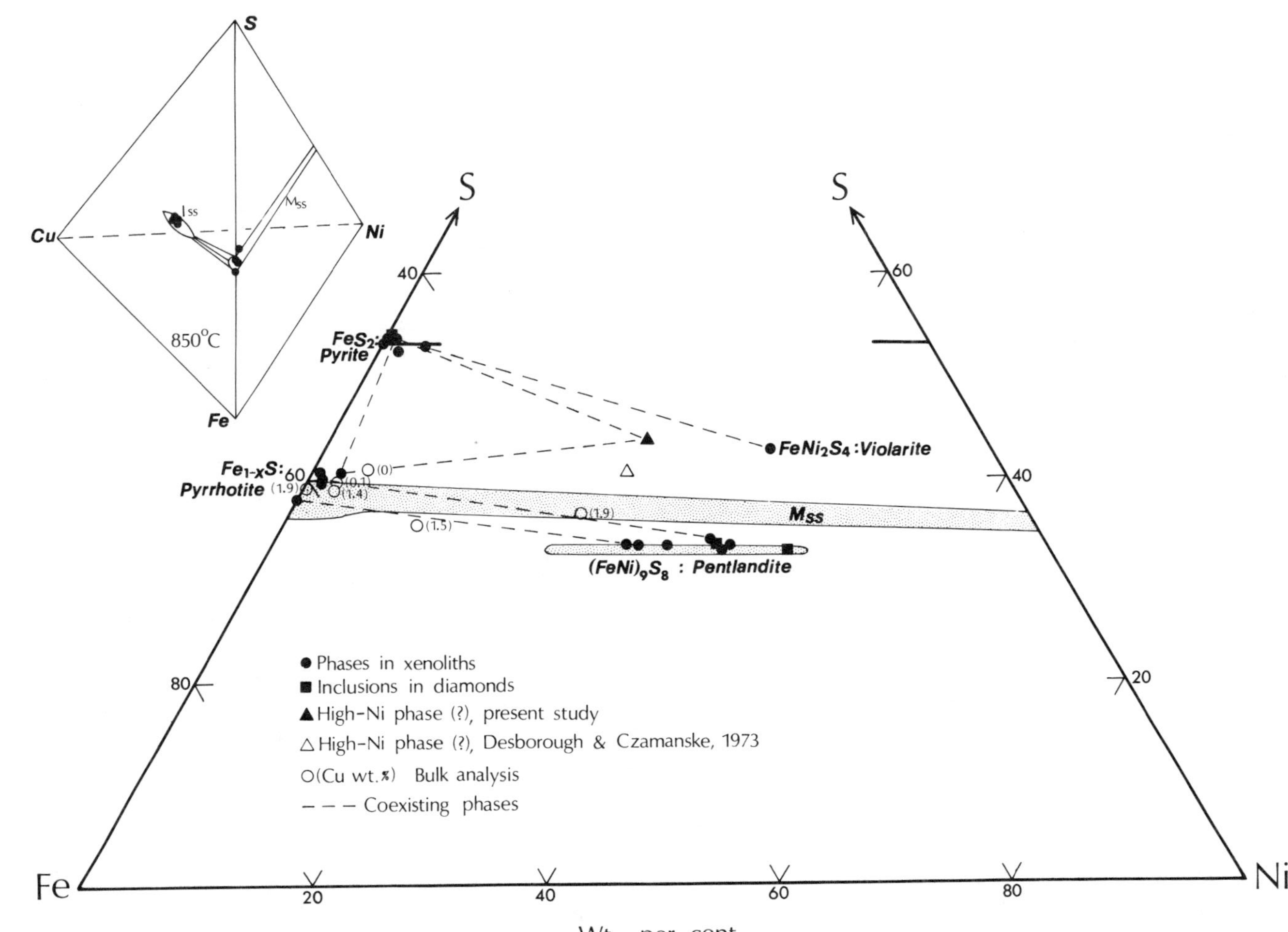

Fig. 4. Sulfide minerals and bulk analyses (broad beam microprobe) plotted in the Fe-Ni-S system. The fields for monosulfide$_{ss}$ (M_{ss}), pentlandite and pyrite are taken from the 500°C isotherm of Graterol and Naldrett (1971).

system (Craig and Kullerud, 1969) and in the 500°C isotherm (Fig. 4) of the Fe-Ni-S system (Graterol and Naldrett, 1971). In general chalcopyrite shows very limited solid solution and lies within the I_{ss} (Intermediate$_{ss}$) field of Cabri (1973) and Barton (1973) (Fig. 4), whereas pyrrhotite, in the various samples, ranges from Ni-free to nickeliferous (Ni ∿ 2 wt%) (Fig. 4). Pentlandite displays a range in Fe/Ni+Co ratio but all lie within the pentlandite stability field at 500°C (Graterol and Naldrett, 1971) (Fig. 4). Pyrite, similar to pyrrhotite, also shows a variation in Ni content (0 ∿ 3.5 wt%).

In discussing the mineral paragenesis of the various phases it is necessary to know the initial bulk composition of the presumably immiscible sulfide melt that existed in association with the silicate magma. However, due to the sample size and other factors it was not possible to obtain serial sections and thus average the overall bulk composition. Nevertheless the results of broad beam probe analyses (Table II, Fig. 4) suggest that the original composition of the sulfide liquid was lying within the Cu-Fe-Ni-S system close to the region of the monosulfide stability field in the Fe-Ni-S face.

From this sulfide liquid, the first phase to crystallize would be an M_{ss}, coexisting with Cu-rich liquid. This liquid probably has a composition close to, if not in, the region later occupied by the intermediate$_{ss}$ (I_{ss}) (Cabri, 1973; Barton, 1973), but unfortunately at this time no experimental data are available that define the composition of this Cu-rich liquid (Craig, pers. comm.). The separation of Cu-rich liquid probably occurs at about 850°C with the liquid migrating to the outer portion of the mass and along cracks. Subsequently, this Cu-rich liquid crystallized to form the present rim of chalcopyrite and veins along cracks throughout the pyrrhotite. Further cooling resulted in the exsolution of pentlandite from the monosulfide and the associated change in composition to produce pyrrhotite or Ni-pyrrhotite. The inclusions of minor amounts of Cu in the original M_{ss} is probably denoted by the rare exsolution lamellae of chalcopyrite in the Po host. Also to be explained is the occurrence of Pn between the chalcopyrite and pyrrhotite (Fig. 1b and 1c). It is believed that I_{ss} can contain a small percentage of Ni, and thus on sub-solidus re-equilibration pentlandite is likely to form from pre-existing phases produced from the Cu-rich liquid close to the Iss field. Interestingly pentlandite either as flame-like exsolution or as discrete grains may contain up to 3.5 wt% Co. Vaasjoki et al. (1974) have shown that Co can raise the thermal stability of pentlandite; however, the amount of cobalt present in our sample is probably too small to have any significant effect.

Of particular interest is the sulfide assemblage in sample XRV-7 from Roberts Victor. This assemblage is similar to that reported by Desborough and Czamanske (1973). However, the low Ni phase proposed by Desborough and Czamanske appears in our sample to be Ni-Po (Table II, Fig. 9). We also observe a high Ni phase similar to that described by Desborough and Czamanske (1973) (Fig. 4). This phase is not so common as reported by these authors and unfortunately due to sample size we were unable to obtain totally satisfactory analyses (Table II). The association of Ni-Po, Ni-Py, Cp plus an unknown high Ni phase is not common, and appears to have not been reported in any experimental studies.

The unique assemblage of sulfide, oxide and silicate in a garnet xenocryst (173/53) from Premier is probably derived from a sulfide melt compositionally different from that which produced the sulfide assemblages discussed above or reported elsewhere. It has been recognized that minor, but distinct differences in chemistry exist between xenocrysts and ultramafic xenoliths. For example, most xenocrysts are more Fe-rich than their counterparts in ultramafic xenoliths (Nixon and Boyd, 1973; Meyer and Tsai, 1977). It has been suggested (Nixon and Boyd, 1973; Gurney et al., 1977) that these xenocrysts represent cumulate minerals which precipitated from a magma produced by a partial melting episode in the mantle. Accordingly it is possible that the composition of this sulfide melt may be the result of the same episode of partial melting.

The textural relation of the sulfide phases is also unusual in that it consists of equigranular chalcopyrite and pentlandite, and less abundantly, intergranular magnetite and silicate minerals. In the Cu-Fe-Ni-S system Iss and pentlandite can exist together at 575°C. In most ore deposits pentlandite and chalcopyrite are believed to be the result of exsolution from a higher temperature M_{ss} phase. Thus when chalcopyrite and pentlandite occur as aggregates instead of exsolution lamellae, it seems natural to envisage some form of recrystallization of preexisting exsolved phases (Naldrett and Kullerud, 1967). However, textural evidence in sulfides is not always equivocal and the possibility of coherent and noncoherent exsolution (Brett, 1969) should also be given consideration.

The absence of pyrrhotite from the assemblage in the Premier xenocryst, as well as the textural relationships of the pentlandite and chalcopyrite may be regarded as evidence that these latter phases did not form by the process noted above by Naldrett and Kullerud (1967). Bishop et al. (1975) also had difficulty in explaining their observation of pentlandite-magnetite intergrowths in a spinel-lherzolite from De Beers Mine, Kimberly. Although formation from M_{ss} appears necessary based on phase equilibria in the system Cu-Fe-Ni-S the role of magnetite must be considered. The system Cu-Fe-Ni-S-O is thus more applicable, but knowledge pertinent to this system is minimal. Bishop et al. (1975) invoked the M_{ss}-magnetite cotectic (Naldrett, 1969) to partly explain the origin of their pentlandite-magnetite intergrowth. Such a cotectic could also be invoked here provided the M_{ss} was compositionally suitable to later produce both chalcopyrite and pentlandite.

The presence of the magnetite could also be the result of reaction between the sulfide liquid and the Fe-bearing garnet xenocryst. At the present time information concerning the interaction between sulfide liquid and mafic silicates is sparse, but is essential to understanding many of the phase relations observed in ore deposits. A further problem is the presence of the silicates andradite, ilvaite and chlorite in the same assemblage with sulfides and magnetite.

The occurrence of a K-Fe-Ni sulfide in JJG548 from an Obnazhennaya eclogite is interesting as it has only recently been recognized in ilmenite-pyroxene xenoliths in the Frank Smith kimberlite, South Africa (Clarke et al., 1977). Meyer et al. (1977) have also found this K-bearing sulfide in an enstatite xenocryst from Weltevreden Floors near Frank Smith. The textural association appears similar in that it is a "reaction rim" on the central assemblage. Clarke (1977) has shown experimentally that this K-sulfide phase (dubbed FSKS by Clarke et al., 1977) is not stable at high pressure and temperature and therefore it is believed that this phase results from a metasomatic reaction between K-rich liquid and the original Fe-Ni sulfide. Observations on the texture and composition of the K-sulfide in JJG548 are in agreement with such an origin.

Sulfides in Diamonds

Sample Description

Several hundred diamonds were selected from Koffiefontein, Finsch and Premier kimberlites on the basis of visibly having black included matter. It was hoped that a large percentage of these diamonds would contain sulfide inclusions. However, after detailed optical examination it was found that very few contained discrete crystals of sulfide suitable for microprobe analyses. For example, out of about 100 diamonds we cracked, only 7 yielded inclusions. Nevertheless, the large majority of diamonds examined did contain black material but this is present as extremely thin films on fracture planes within the diamond. When thick enough for analysis, either by X-ray diffraction or microprobe, these films are graphite or compositionally ambiguous sulfide material.

The majority of the diamonds investigated in this study are less than 2 mm in size. A variety of crystal shapes occurred, including dodecahedron, octahedron, macle, flattened, aggregate and irregular (Harris et al., 1975). The color of these diamonds varies from predominantly colorless, less commonly brown to yellow, and rarely green. No correlation between sulfide inclusion and shape or color can be established due to the scarcity of sulfide inclusions. In all instances, inclusions were extracted by cracking the host diamonds in a sealed optical cell.

A major problem concerning these sulfide inclusions is the decision as to whether the inclusions are primary or secondary. Most black inclusions seem to occur associated with rosette or disc-shaped fractures in the diamonds (Fig. 5a). Whether these fractures reach the surface of diamond is not always easy to determine, and therefore, there is some uncertainty with regard to the primary or secondary nature of some of these sulfide inclusions. However, one particularly interesting observation, also noted by Harris (1972), is that several instances of sulfide occurring with undeniably primary olivine inclusions have been observed. In terms of the number of sulfide inclusions found, this association is sufficiently common as to suggest a genetic association.

Besides occurring as small discrete crystals sulfide material also occurs, as mentioned previously, as a thin veneer on cleavage planes within the diamond. Possibly, this material represents remobilized and reequilibrated earlier sulfide phases. Such material has been partly described by Harris (1972).

Mineralogy

A variety of sulfide assemblages, predominantly polyphase aggregates, were observed among sulfide inclusions in diamond. In general, prior to microprobe analysis the samples were identified by X-ray diffraction using a Gandolfi camera. The interpretation of most X-ray results indicates the presence of one or more phases of which the most common are pyrrhotite and pentlandite. This observation was further confirmed by electron microprobe analyses, although, only very few satisfactory results were obtained because of the minute size and intimate association of constituent sulfide phases in most samples. Besides pyrrhotite and pentlandite, chalcopyrite was also noted, both X-ray diffraction and electron microprobe, as well as magnetite in some specimens.

The few sulfide inclusions found may be divided into two types (Table IV) in terms of assemblages, a) pyrrhotite + pentlandite $\pm$ chalcopyrite; b) pentlandite + magnetite $\pm$ pyrrhotite. One inclusion of pyrite was also observed.

Pyrrhotite + Pentlandite $\pm$ Chalcopyrite. Texturally the association of pyrrhotite and pentlandite is somewhat similar to the same assemblage previously described for the sulfides in xenoliths. The major phase is pyrrhotite in which fine exsolution lamellae, as well as discrete grains of pentlandite occur.

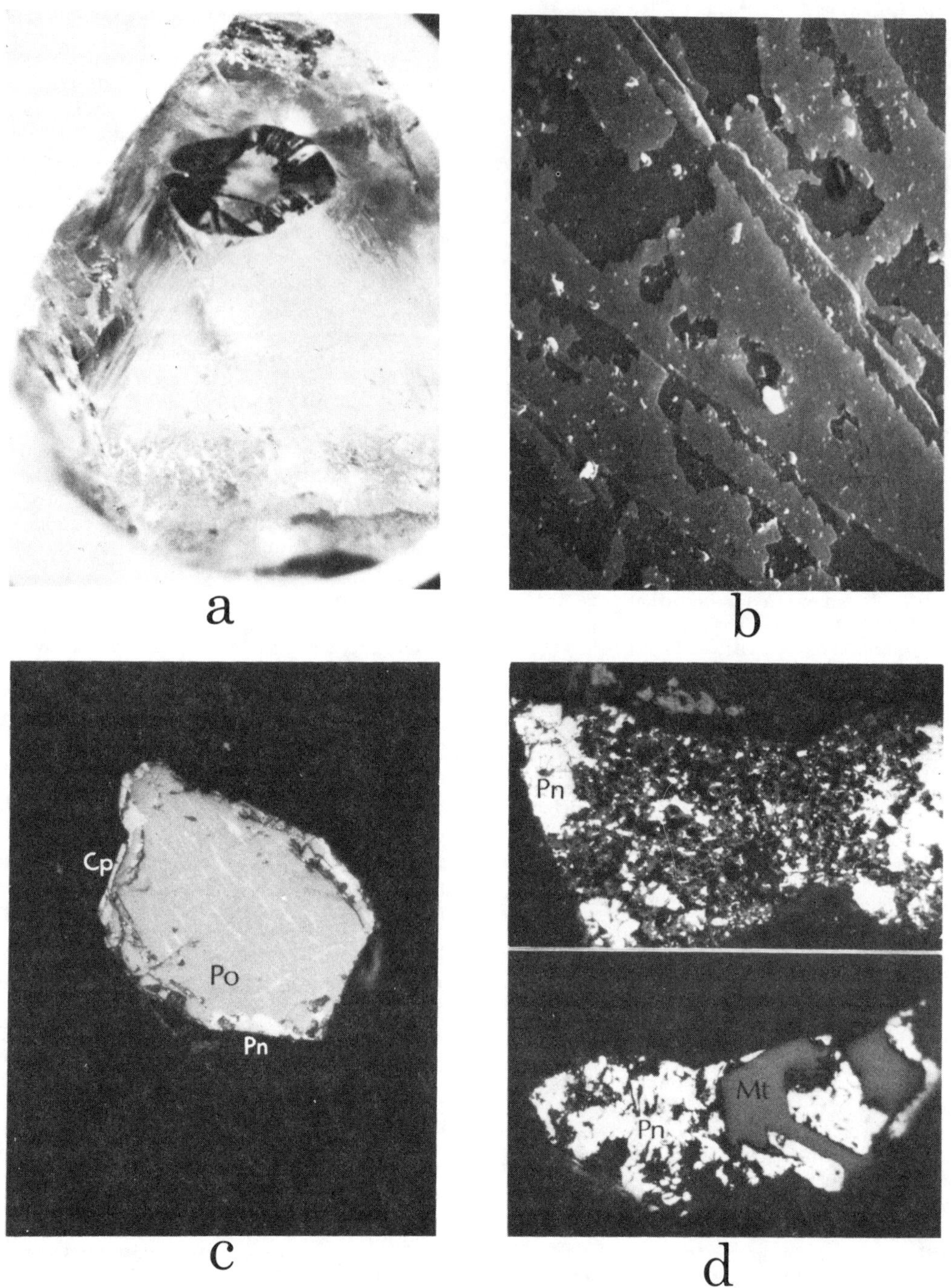

Fig. 5a) Black disc-shaped inclusions shown on broken surface of diamond (FS1). Length of the diamond approximately 2 mm. b) SEM photograph of sulfide material on cleavage plane of diamond in 5a. c) Sulfide inclusion (0.13 mm) consisting of pyrrhotite (Po) host with chalcopyrite lamellae. Minor chalcopyrite and pentlandite occur at edge (FS25). d) Top: Mosaic intergrowth of pentlandite (Pn) and magnetite (dark gray) (K19). Bottom: Discrete grains of Pn and magnetite (Mt). (K19) Width of field in both cases 0.18 mm.

A thin film on a cleavage plane of one diamond (FS1) (Fig. 5b) was analyzed semiquantitatively using an S.E.M. plus energy dispersive system. The result showed the presence of Fe, Ni and S, suggesting either pentlandite, or pyrrhotite + pentlandite are the constituents of the film. Interestingly, a more discrete sulfide from which this film radiated has a bulk composition containing $\sim$ 6 wt% Cu, presumably indicating the presence of chalcopyrite.

TABLE 4. Sulfide assemblages in diamonds

Sample No.	Locality	Assemblage	Identification
FS1	Finsch	a. Po-Pn (film)*	X-ray/SEM
		b. Po-Pn-Cp (?)	Microprobe
FS15	Finsch	Po-Pn ϕ	X-ray/microprobe
FS25	Finsch	Po-Pn-Cp	Microprobe
KS10	Koffiefontein	Po-Pn-Cp	Microprobe
KS5	Koffiefontein	Po-Pn-Cp ϕ	Microprobe
F83B	Finsch	Po-Pn-Cp	Microprobe
K19(A,B,C and D)	Koffiefontein	Po-mt	X-ray/Microprobe
K35B	Koffiefontein	Pn-mt-Po	X-ray
KS8	Koffiefontein	Py	Microprobe

* Coexists with enstatite
ϕ Coexists with olivine

Although this bulk composition may not be concise due to the method of analysis it is possible to recalculate it into pyrrhotite + pentlandite + chalcopyrite.

The occurrence of chalcopyrite, both as fine exsolution lamellae in pyrrhotite (Fig. 5c) and as coarse grains at the pyrrhotite margins, is rare, compared to the ubiquitous occurrence of chalcopyrite in the sulfides from the xenoliths. Due to the very narrow width of both the chalcopyrite and pentlandite in pyrrhotite it was not possible to obtain reasonable analyses of these two phases in this type of assemblage. However, in one instance we were able to obtain an analysis of discrete pentlandite (KS10) at the edge (Table V). Recalculation of broad beam analyses supports the optical evidence for the presence of these three phases. In two cases (KS5 and F83B) the identification is based entirely on the interpretation of the bulk analyses.

Pentlandite + Magnetite. Several fragments of an inclusion from a Koffiefontein diamond (K19) consisted of pentlandite and magnetite. In some instances it is an irregular mosaic intergrowth (Fig. 5d) whereas in others pentlandite and magnetite occur as discrete grains (Fig. 5d). Both pentlandites are Ni-rich (~43 wt% Ni; Table V and Fig. 4) and close to the limit of Ni solubility in the pentlandite structure. Analyses of discrete pentlandite show variable amounts of Cu (0 ~ 6 wt%), which would indicate the possible presence of chalcopyrite. The size of the grains preclude positive optical and probe identifications.

Another assemblage of pentlandite-pyrrhotite-magnetite (K35B) is based on X-ray identification only.

Discussion. The paragenesis of the most common assemblage, Po-Pn-Cp, is probably similar to that previously described for similar sulfide assemblages in the eclogite xenoliths. The rare occurrence of chalcopyrite as an inclusion in diamond seems to indicate a deficiency in Cu content for the original sulfide melt. Previously,

TABLE 5. Representative analyses of sulfide inclusions in diamond

	KS10		FS25	KS5	F83B	K19	KS8
	Pentlandite	Bulk+	Bulk+	Bulk+	Bulk+	Pentlandite	Pyrite
Fe	27.9	56.0	57.2	52.1	52.2	23.3	46.6
Ni	35.7	2.24	0.59	6.53	7.46	43.9	0.00
Co	1.59	0.25	0.06	0.46	0.28	0.87	0.07
Cu	0.00	1.08	2.89	1.51	0.29	0.00	0.00
S	34.3	40.1	39.5	40.2	38.9	32.2	54.5
Total	99.5	99.7	100.2	100.8	99.1	100.3	101.2

\+ See note Table II

chalcopyrite has been identified as an inclusion by Prinz et al. (1975) and recently by Mitchell and Giardini (1977). Since various differences in mineral chemistry exist for silicates between diamond inclusions and xenoliths, it is not surprising to find some mineralogical or chemical differences in sulfides between the two categories.

The association of pentlandite and magnetite resembles somewhat the pentlandite-magnetite intergrowth in a spinel lherzolite reported by Bishop et al. (1975). The texture is more complex in our sample, because pentlandite and magnetite occur both as a mosaic intergrowth and as discrete grains. As discussed earlier the occurrence of these two phases together is not readily explained, especially so in the absence of pertinent experimental data. Interestingly, the possible intergrowth of sulfides and magnetite as inclusions has been noted by Prinz et al. (1975) and both these authors as well as Harris (1968) have commented on the occurrence of magnetite alone in diamond.

These sulfide inclusions, if primary, represent original mantle material that has been chemically unaltered since incorporation within the diamond. The bulk composition of these sulfide assemblages probably represent the original compositions of the immiscible sulfide melts associated with diamond genesis in the mantle. Unfortunately due to the limited number of sulfide inclusions available and the difficulty in determining the primary or secondary nature of these inclusions, it may be presumptious to specify conclusions at this stage. However, it is unlikely that only one single composition was present for the original sulfide melt since various assemblages are observed with noticeably different bulk compositions.

Although a genetic relation maybe suggested for the sulfide-olivine association; the significance and origin of such an association is not understood at this stage.

Sulfur Isotope Studies

The S^{34}/S^{32} ratios were determined in sulfides from fourteen samples which included massive pyrite in kimberlite, pyrite and other sulfides in crustal and eclogitic xenoliths and sulfides in the garnet xenocryst from Premier. Mineral separates were obtained by hand picking or by heavy liquid technique. Purity of pyrite is generally greater than 90% but it is not possible to separate the individual phases in other sulfide assemblages. Sulfides were converted to SO_2 by combustion with CuO at 900°C as described by Fritz et al. (1974). The purified SO_2 was then measured in a Nuclide 3-60-RMS isotope ratio mass spectrometer. The S^{34}/S^{32} ratios are presented in the δ-notation:

$$\delta S^{34} = \left\{ \frac{(S^{34}/S^{32})\ \text{sample}}{(S^{34}/S^{32})\ \text{CDT}} - 1 \right\} \times 1000,$$

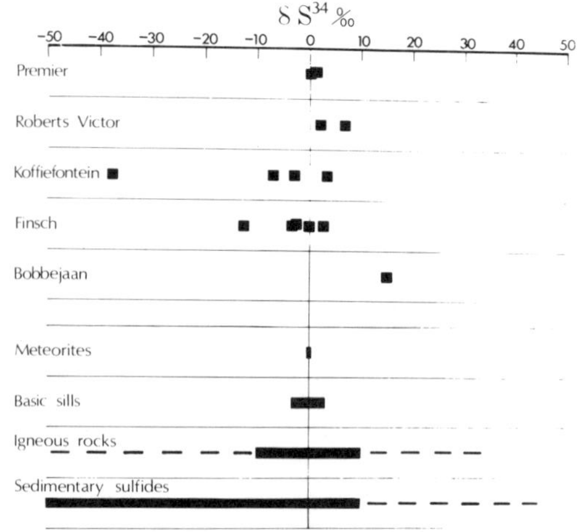

Fig. 6. δS^{34}-values of sulfides from kimberlites and associated xenoliths. Also included are the variation of δS^{34} for common rock types in nature (after Thode et al., 1961).

where CDT is Canyon Diablo troilite. The experimental error is generally $\pm$ 0.2°/oo. The results of the analyses are presented in Table VI and Fig. 6.

As can be seen the samples analyzed show a large variation in δS^{34} from -37.3 to +15.1°/oo. However, this is not unexpected because of the heterogeneity of the rock types from which the sulfides were obtained. The chalcopyrite-pentlandite assemblage in the garnet xenocryst (173/53) from Premier has a δ value of +0.2. Inasmuch as the garnet xenocryst is undoubtedly of mantle origin, it is, therefore, in agreement with the inference that the Earth's mantle sulfur has a δ value close to zero. A second sulfide sample of mantle origin (WM7, a eclogite from Roberts Victor kimberlite) gives a δ value of +2.1. It should be pointed out that this sample produced a very small amount of SO_2 (30 μ moles) and contamination could be a serious problem. Nevertheless, the δ value is sufficiently close to what one would expect from the upper mantle.

The δS^{34} of massive pyrite in Premier kimberlite is +1.3, within the range of basic igneous rocks which suggests that the sulfur in kimberlite is of magmatic origin. The δS^{34} for all the pyrite-bearing crustal xenoliths, massive or disseminated, as well as a secondary pyrite in Bobbejaan kimberlite show a wide range. It is obviously difficult and hazardous to assign an origin to this sulfur on the basis of the limited data available. Perhaps all that can be said at this time is that the wide range of δS^{34} values for the crustal xenoliths is suggestive, as are some of the textures, of a sedimentary source for sulfur. Recent studies (e.g. Ohmoto, 1972; Rye and Ohmoto, 1974) have indicated possible complexities with regard to the ranges of δS^{34} in terms of fS_2 fO_2 and pH.

TABLE 6. S isotope (δS^{34}) values for sulfides in selected xenoliths from South African kimberlites

Sample No.	Locality	Sulfide type	Host rock	δS^{34} $^o/oo$
173/53	Premier	Cp+Pn	garnet xenocryst	+0.2
PS1	Premier	Massive py	Kimberlite	+1.3
173/33	Bobbejaan	Discrete py	Eclogite (xenolith)	+15.1
WM7	Roberts Victor	Po+Pn+Cp	Eclogite (xenolith)	+2.1
RVS1	Roberts Victor	Massive pyrite	Kimberlite concentrate	+6.6
KS1	Koffiefontein	Disseminated py	Shale (xenolith)	-37.3
KS2	Koffiefontein	Disseminated py	Shale (xenolith)	-6.9
KS4	Koffiefontein	Massive py	Shale (xenolith)	-2.6
KS5	Koffiefontein	Disseminated py	Amphibolite (xenolith)	+3.1
FS1	Finsch	Disseminated py	Shale (xenolith)	-3.1
FS3	Finsch	Massive py	Shale (xenolith)	+2.4
FS4	Finsch	Disseminated py	Cherty shale (xenolith)	0.0
FS5	Finsch	Disseminated py	Shale (xenolith)	-2.1
FS6	Finsch	Disseminated py	Shale (xenolith)	-12.2

The suggestion that δS^{34} value of mantle sulfur is at or near zero has been inferred from studies of basic sills and meteorites (Thode et al., 1961). The analyses of sulfur in these samples of unequivocal mantle origin provided the most direct evidence concerning the sulfur isotope composition of the upper mantle. Obviously more sulfide specimens of undoubted mantle origin are needed before a definite conclusion can be reached. Even more significant would be the studies of sulfur in the pre- or syngenetic sulfide inclusions in diamond. Such inclusions represent true mantle material which due to the armoring effect of diamonds should not have undergone any isotopic exchange with the surrounding environment outside of the host diamond. Unfortunately, at the present time the minute amount of sulfides present as inclusions in any one diamond is too small to be analyzed confidently in a conventional mass spectrometer.

Summary

In the previous sections an attempt has been made to characterize, both texturally and chemically, the various sulfides that occur in predominantly eclogitic xenoliths, and as inclusions in diamond. In general the sulfides consist of polyminerallic assemblages in which pyrrhotite, pentlandite and chalcopyrite are the major constituents. These polyphase assemblages usually occur as a somewhat rounded aggregate which have also been noted for sulfides in eclogitic and ultramafic xenoliths in both Siberian (Vakhrushev and Sobolev, 1971) and South African kimberlites (Frick, 1973). Frick (1973) considered the sulfide in griquaite (eclogite) xenoliths to be the result of an immiscible sulfide melt derived from partial melting of the griquaite. Inasmuch as the sulfides examined in the present study are texturally similar to those discussed by Frick, including the presence of sulfide "stringers" or "veins", we agree on the presence of an immiscible sulfide melt. However, we believe it is unnecessary to postulate a partial melt episode of the griquaite to produce such a melt and prefer to consider the sulfide melt having been originally in association with the cumulus phases and related silicate magma now represented by the silicate mineral constituents of the griquaite.

Perhaps of more importance is the tantalizing hint that sulfides from individual pipes are texturally and chemically distinct. For example, the Roberts Victor kimberlite is characterized by the ubiquitous presence of nickeliferous pyrrhotite (Table II). This phase has also been found in association with Ni-pyrite in two samples from Roberts Victor (This study and Desborough and Czamanske, 1973). Such association has not been recognized elsewhere. In comparison, both samples from Jagersfontein, although Ni-pyrrhotite is absent, contain a pyrrhotite + pentlandite + chalcopyrite assemblage. Probably no comment is warranted for the single Premier sample, nevertheless it is interesting that the only Russian sample (from Obnazhennaya) is identical in texture and mineralogy to previous samples from this same locality described by Vakhrushev and Sobolev (1971), and are different from most sulfides found in South African xenoliths. Further support for the idea of sulfides being texturally and mineralogically distinct from various localities is provided by the occurrence of pentlandite-magnetite intergrowth in lherzolite from De Beers Mine (Bishop et al., 1975) and pyrrhotite + pentlandite + chalcopyrite + $monosulfide_{ss}$ assemblage in eclogite from Stockdale kimberlite, Kansas (Meyer and Boctor, 1975). It is distinctly probable that the variation in sulfide chemistry, due to the usually rapid reequilibration sulfide minerals undergo, is a direct reflection of the thermal history of individual kimberlite pipes. Furthermore, it is conceivable that the phases and textures observed in these sulfides are a result of the rapid ascent (and quenching) of kimberlite. For this reason they are not directly comparable with the more slowly cooled phases of most magmatic ore deposits.

The results of the examination of sulfide inclusions in diamond have been extremely disappointing, especially in the paucity of suitable inclusions obtained from several hundred diamonds. Nevertheless there is a possibility that sulfide inclusions vary in abundance from pipe to pipe and that by chance the three kimberlite pipes studied, Finsch, Koffiefontein and Premier, all have low frequencies of sulfide inclusions in diamonds.

Some ambiguity in the relative abundance is caused by the composition of the less than micron thick films of black material on cleavage planes within diamond. In this study, graphite and sulfide material of unknown composition have been identified as being responsible for these films, and similar observations have been made by Harris (1972). Probably, if one were able to assess the exact nature of all black films it is likely that sulfides may be abundant as inclusions as previously suggested by Sharp (1966). Until such films can be accurately characterized such a conclusion can only be surmized.

The origin of sulfur and sulfides in diamonds is also important in view of the recent age determination (Welke et al., 1974; Kramers, 1977) using Pb isotopes from sulfide inclusions. Especially problematic is the syngenetic or epigenetic origin of the sulfides, however, Kramers (1977) indicates there is isotopic disequilibrium between the diamond inclusions and host kimberlite. Such data would suggest a xenocrystic nature for some diamonds, a conclusion heralded by the early work of Meyer and Boyd (1972) on silicate and oxide inclusions.

It is unfortunate that the present detectability limits of most mass spectrometers are such that the small amounts of isotopic elements provided by the inclusions in diamond are insufficient for accurate analysis. However, it is reassuring to note that the S isotope study undertaken as part of this project does give results consistent with a mantle origin for sulfur from sulfide in a garnet xenocryst, as well as the probable magmatic origin for sulfur in pyrite from kimberlite. These results, although preliminary, demonstrate the usefulness of such studies. Undoubtedly with a larger suite of samples and improved detectibility of mass spectrometer, or the ion microprobe, much exciting and significant information of mantle isotopic composition can be obtained.

Acknowledgments. We thank Drs. G. Kullerud, R.H. McCallister and N.Z. Boctor for reviewing an early version of the manuscript, as well as Drs. Craig, and Barton whose critical and discerning comments improved immensely the final version. Samples were kindly supplied by Drs. J.B. Hawthorne, J.J. Gurney, J.W. Harris, C.J. Hatton, and J.R. Smyth. This research was supported by National Science Foundation - Earth Science Section EAR76-22698 (HOAM); EAR75-19999 (YNS). HMT is grateful for financial support from David Ross Fellowship, Purdue Research Foundation.

References

Albee, A.L. and L. Ray, Correction factors for electron probe microanalysis of silicates, oxides, carbonates, phosphates and sulphates. Anal. Chem., 42, 1408-1414, 1970.

Barton, P.B., Jr., Solid solutions in the system Cu-Fe-S. Part I: The Cu-S and Cu-Fe-S joins. Econ. Geol., 68, 433-465, 1973.

Bence, A.E. and A.L. Albee, Empirical correction factors for the electron microanalysis of silicates and oxides. Jour. Geol., 76, 382-403, 1968.

Bell, P.M., J.L. England, and G. Kullerud, Pressure effect on breakdown of pentlandite. Carnegie Inst. Wash. Yearb., 76, 206-207, 1964.

Bishop, F.C., J.V. Smith, and J.B. Dawson, Pentlandite-magnetite intergrowth in DeBeers spinel lherzolite: review of sulfides in nodules. Phys. Chem. Earth, 9, 323-337, 1975.

Brett, R., Experimental data from the system Cu-Fe-S and their bearing on exsolution textures in ores. Econ. Geol., 59, 1241-1269, 1964

Cabri, L.J., New data on phase relations in the Cu-Fe-S system. Econ. Geol., 68, 443-454, 1973.

Clarke, D.B., Synthesis of K-Fe-Ni sulphide from Frank Smith Mine. (Abstr.) 2nd Int. Kimberlite Conf., Santa Fe, New Mexico, 1977.

Clarke, D.B., G.G. Pe, R.M. MacKay, K.R. Gill, M.J. O'Hara, and J.A. Gard, A new potassium-iron-nickel sulfide from a nodule in kimberlite. Earth Planet. Sci. Lett., 35, 421-428, 1977.

Craig, J.R. and G. Kullerud, Phase relations in the Cu-Fe-Ni-S system and their application to magmatic ore deposits. Econ. Geol. Mono., 4, 344-358, 1969.

Craig, J.R. and S.D. Scott, Sulfide phase equilibria. In Sulfide Mineralogy, Ed. P.H. Ribbe, Min. Soc. Am. short course note, V.1., 1974.

Desborough, G.A. and G.K. Czamanske, Sulfides in eclogite nodules from a kimberlite pipe, South Africa, with comments on violarite stoichiometry. Amer. Mineral., 58, 195-202, 1973.

Dietrich, V., Ferroantigorit and Greenalith als Begleiter oxidisch-sulfidischer Vererzungen in den Oberhalbsteiner Serpentiniten. Schweiz. Min. Petr. Mitt., 52, 57-74, 1972.

Frick, C., The sulfides in griquaite and garnet-peridotite xenoliths in kimberlite. Contr. Mineral. Petrol., 39, 1-16, 1973.

Fritz, P., R.J. Drimmie, and V.K. Nowicki, Preparation of sulfur dioxide for mass spectrometer analyses by combustion of sulfides with copper oxide. Anal. Chem., 46, 164-166, 1974.

Graterol, M. and A.J. Naldrett, Mineralogy of the Marbridge No. 3 and No. 4 nickel-iron sulfide deposits with some comments on low temperature equilibration in the Fe-Ni-S system. Econ. Geol., 66, 886-900, 1971.

Gurney, J.J., W.R.O. Jakob, and J.B. Dawson, Megacrysts from the Monastery Mine. (Abstr.) 2nd Int. Kimberlite Conf., Santa Fe, New Mexico, 1977.

Harris, J.W., The recognition of diamond inclusions. Pt. 1: Syngenetic mineral inclusions. Pt. 2: Epigenetic mineral inclusions. Indust. Dia. Rev., London, 402-410, 458-461, 1968.

Harris, J.W., Black material on mineral inclusions and in internal fracture planes in diamond. Contr. Mineral. Petrol., 35, 22-33, 1972.

Harris, J.W., J.B. Hawthorne, M.M. Oosterveld, and E. Wehmeyer, A classification scheme for diamond and a comparative study of South African diamond characteristics. Phys. Chem. Earth, 9, 765-784, 1975.

Kramers, J.D., Lead and strontium isotopes in inclusions in diamonds and in mantle-derived xenoliths from Southern Africa. (Abstr.) 2nd Int. Kimberlite Conf., Santa Fe, New Mexico, 1977.

Kullerud, G., R.A. Yund, and G. Moh, Phase relations in the Fe-Ni-S, Cu-Fe-S and Cu-Ni-S system. Econ. Geol. Mono. 4, 323-343, 1969.

Meyer, H.O.A. and N.Z. Boctor, Sulfide-oxide minerals in eclogite from Stockdale kimberlite, Kansas. Contr. Mineral. Petrol., 52, 57-68, 1975.

Meyer, H.O.A. and F.R. Boyd, Composition and origin of crystalline inclusions in natural diamond. Geochim. Cosmochim. Acta, 36, 1255-1273, 1972.

Meyer, H.O.A. and D.P. Svisero, Mineral inclusions in Brazilian diamonds. Phys. Chem. Earth, 9, 785-796, 1975.

Meyer, H.O.A. and H.-M. Tsai, Inclusions in diamond and the mineral chemistry of the upper mantle. Proc. 2nd Symp. Origin and Distribution of the Elements. UNESCO, Paris, (in press), 1977.

Meyer, H.O.A., H.-M. Tsai, and J.J. Gurney, Enstatite xenocryst containing coexisting Cr-poor and Cr-rich garnet, Weltevreden Floors, South Africa. (Abstr.) 2nd Int. Kimberlite Conf., Santa Fe, New Mexico, 1977.

Mitchell, R.S. and A.A. Giardini, Some mineral inclusions from African and Brazilian diamond: their nature and significance. Amer. Mineral., 62, 756-762, 1977.

Naldrett, A.J., A portion of the system Fe-S-O between 900 and 1080°C and its application to sulfide ore magmas. Jour. Petrol., 10, 171-201, 1969.

Naldrett, A.J. and G. Kullerud, A study of the Strathcona Mine and its bearing on the origin of the nickel-copper ores of the Sudbury District, Ontario. Jour. Petrol., 8, 453-531, 1967.

Nixon, P.H. and F.R. Boyd, The discrete nodule (megacryst) association in kimberlites from Northern Lesotho. In: Lesotho Kimberlite, Ed. P.H. Nixon, Lesotho Nat. Dev. Corp. Maseru, 67-75, 1973.

Ohmoto, H., Systematics of sulfur and carbon isotopes in hydrothermal ore deposits. Econ. Geol., 67, 551-578. 1972.

Prinz, M., D.V. Manson, P.F. Hlava, and K. Keil, Inclusions in diamonds: garnet lherzolite and eclogite assemblages. Phys. Chem. Earth, 9, 797-815, 1975.

Rye, R.O. and H. Ohmoto, Sulfide and carbon isotopes and ore genesis: A review. Econ. Geol., 69, 826-842, 1974.

Sharp, W.E., Pyrrhotite: a common inclusion in South African diamonds. Nature, 211, 402, 1966.

Smyth, J.R. and C.J. Hatton, A coesite-sanidine grospydite from the Roberts Victor kimberlite. Earth Planet. Sci. Lett., 34, 284-290, 1977.

Smyth, J.R., Quartz pseudomorphs after coesite. Amer. Mineral., 62, 828-830, 1977.

Thode, H.G., J. Monster, and H.B. Dunford, Sulfur isotope geochemistry. Geochim. Cosmochim. Acta, 5, 286-298, 1961.

Tzepin, A.I. and E.E. Lazko, Pyrrhotite-pentlandite-chalcopyrite association from sulfide nodules of Takutian kimberlite rocks. Dokl. Akad. Nauk. SSSR, 231, 195-198, 1976.

Vaasjoki, O., T.A. Hakli, and T. Tonitti, The effect of cobalt on the thermal stability of pentlandite. Econ. Geol., 69, 549-551, 1974.

Vakhrushev, V.A. and N.V. Sobolev, Sulfidic formations in deep xenoliths from kimberlite pipes in Yakutia. Geologiya i Geofizika, 11, 3-11 (in Russian). (Translation: Inter. Geol. Rev., 15, 103-110, 1973), 1971.

Welke, H.J., H.L. Allsopp, and J.W. Harris, Measurements of K, Rb, U, Sr and Pb in diamonds containing inclusions. Nature, 252, 35-37, 1974.

MINERAL AND BULK CHEMISTRY OF GARNET LHERZOLITE AND GARNET HARZBURGITE XENOLITHS FROM THE PREMIER MINE, SOUTH AFRICA

R.V. Danchin

Anglo American Research Laboratories, P.O. Box 106, Crown Mines, 2025, Republic of South Africa

Abstract Mineral and bulk chemical compositions of garnet lherzolite and garnet harzburgite xenoliths from the Precambrian Premier Mine kimberlite are presented and compared with equivalent rocks from Cretaceous African kimberlites. It is shown that in spite of an age difference of the order of 1000 million years, the Premier garnet lherzolites, in common with the majority of those from other pipes in Lesotho and South Africa, can be clearly subdivided into two major groups on the basis of differences in mineral composition and texture. Deformed lherzolites originate from greater depths than those with coarse textures which are in turn more depleted in Fe, Al, Ca and Ti relative to the deformed rocks. Constituent minerals of the deformed xenoliths are extremely titaniferous and also contain the most potassium. Thus a metasomatic origin for Ti and K in the Premier xenoliths is unlikely unless bulk Fe/Mg ratios were also controlled by the metasomatic events.

Garnet harzburgites are also progressively deformed and less depleted with respect to Fe/Mg with increasing depth of origin, and several coarse extremely depleted examples were found to contain minerals equivalent to those found as inclusions in diamond. It is therefore suggested that some diamond now found in kimberlite has its origin at relatively shallow depths in the mantle.

Introduction

The Premier Mine kimberlite pipe is situated some twenty miles east of Pretoria, in the Transvaal, and is the largest of eleven known kimberlite occurrences in the area. The pipe is roughly elliptical, having a major axis of one kilometre at the surface. It is the oldest known kimberlite in Africa and has a minimum age of 1115 ± 15 m.y. (Allsopp et al., 1967). The pipe is complex and at least three separate intrusions are believed to have taken place. Barrett and Allsopp (1973) concluded that the Type 1 kimberlite is close to 1250 m.y. old, and assigned tentative ages of 1400 m.y. and 1200 m.y. to Types 2 and 3 kimberlite respectively.

It has been known for many years that ultramafic xenoliths from the Premier kimberlite are extremely rare (see for example Williams, 1932, p.304). Moreover, the recovery of xenoliths from Premier has been further complicated by the mining and recovery procedure used at this mine. Consequently, in contrast to the Late Cretaceous kimberlites of the Kimberley area, as well as those from northern Lesotho, whose abundant peridotite xenoliths have provided considerable insights into the geochemical nature of the mantle to depths as great as 200 km, no data have been available for the Precambrian mantle sampled by the Premier kimberlite. A recent, temporary modification of mining methods at Premier has, however, made it possible to assemble a large and varied suite of ultramafic inclusions from this pipe. This suite includes garnet lherzolites, garnet harzburgites, garnet websterites, dunites, pyroxenites and eclogites, as well as chromite peridotites and harzburgites. Discrete nodules, predominantly of garnet, diopside, enstatite and ilmenite were also recovered. Preliminary results for some of these xenoliths have been reported by Danchin and Boyd (1976).

Recent geochemical studies of kimberlite xenoliths have been directed mainly towards the garnet-bearing lherzolites and harzburgites. To some extent this is because these nodules are usually the most abundant and thus most relevant in attempts to interpret mantle processes, but also because the assemblage garnet-orthopyroxene-clinopyroxene is most amenable to pressure and temperature estimation in terms of available experimental data (Boyd, 1973a). Although considerable uncertainty persists in the precise application of the experimental data to the xenoliths themselves (e.g. Howells and O'Hara, 1977), it has become clear that relative estimates of temperature and pressure using these phase studies can provide valuable information concerning the stratigraphy of the upper mantle to depths of about 200 km. The

TABLE 1. Summary of Mean Compositions of Minerals from the Garnet Lherzolites and Harzburgites Included in the Premier Kimberlite.

	Mg/(Mg+Fe) Mol. fraction	TiO_2 wt.%	Cr_2O_3 wt.%	Na_2O wt.%	CaO wt.%	Al_2O_3 wt.%	K ppm
GARNETS							
A. Lherzolites							
Deformed (35)	0,845	1.15	3.89	0.09	4.86	19.47	-
Coarse (19)	0.846	0.23	4.16	0.04	5.16	20.82	-
B. Harzburgites							
Group I (20)	0.845	1.32	6.72	0.10	5.46	17.17	-
Group II (16)	0.872	0.20	6.73	<0.03	4.32	18.57	-
Group III (2)	0.888	0.02	7.24	<0.03	3.81	18.59	-
ENSTATITES							
A. Lherzolites							
Deformed	0.918	0.23	0.31	0.30	1.25	1.01	-
Coarse	0.932	0.05	0.34	0.14	0.43	0.97	-
B. Harzburgites							
Group I	0.924	0.17	0.47	0.31	1.22	0.96	-
Group II	0.936	<0.03	0.40	0.11	0.80	0.85	-
Group III	0.950	<0.03	0.54	0.12	0.44	0.90	-
OLIVINES							
A. Lherzolites							
Deformed	0.907	0.03	0.05	-	0.08	0.07	-
Coarse	0.921	<0.03	0.02	-	0.03	0.03	-
B. Harzburgites							
Group I	0.913	<0.03	0.07	-	0.05	0.04	-
Group II	0.926	<0.03	0.06	-	<0.03	0.04	-
Group III	0.943	<0.03	0.05	-	<0.03	0.03	-
DIOPSIDES							
A. Lherzolites							
Deformed	0.907	0.41	1.03	1.71	15.07	2.23	307
Coarse	0.931	0.15	1.85	2.31	19.27	2.94	99

results of geochemical studies of xenoliths from Late Cretaceous pipes of southern Africa have been reviewed by Wyllie (1975) and Harte (1977a). Two main issues have emerged from these studies, and these have been debated at length in the literature. The first concerns the use of the two pyroxene geothermometer-geobarometer to plot fossil geotherms for the Cretaceous mantle. Boyd and Nixon (1975), for example, have reported that geotherms for certain pipes in northern Lesotho and South Africa have a point of inflection near 150 km depth and 1100°C, and these authors proposed that the inflected limb of the geotherm was produced by stress heating during the dispersal of Gondwanaland. Various workers have subsequently questioned the validity of this hypothesis and, indeed, that of the inflection itself, and these aspects are discussed by Mercier and Carter (1975), and Harte (1977a).

The second issue concerns the relationship of the deformation textures of these xenoliths to their mineral and bulk chemical compositions. See for example Boyd and Nixon (1975), Gurney *et al*., (1975 a,b), Dawson *et al*., (1975), and O'Hara *et al*. (1975).

In this contribution attention is focused on the Premier garnet lherzolites and garnet harzburgites, data for which are compared with those from equivalent rocks from the Late Cretaceous African kimberlites in the light of these two central issues.

The Garnet Lherzolites

The garnet lherzolites can be subdivided into two groups that differ markedly in both texture and mineral chemistry. In terms of Harte's (1977b) textural classification, one group consists of coarse equant and coarse tabular

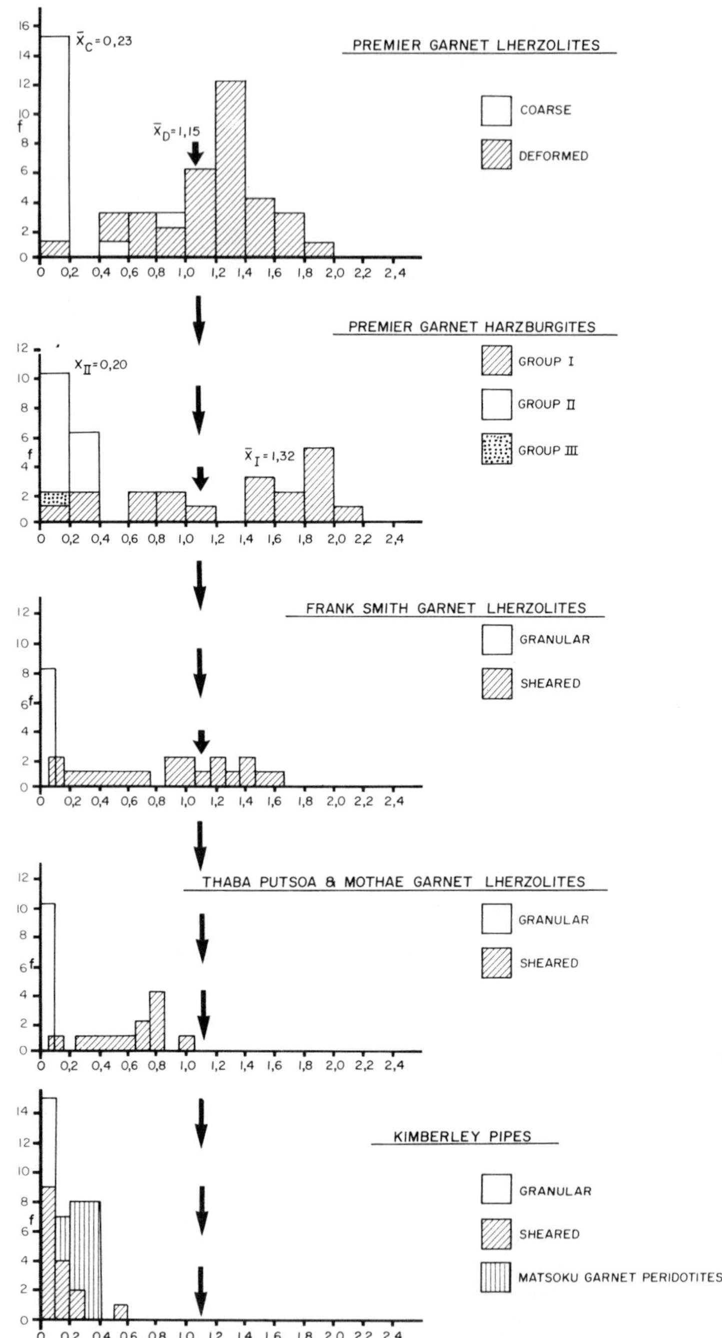

Fig. 1. Frequency distribution of TiO_2 (wt.%) in garnets from Premier Mine garnet lherzolites and garnet harzburgites. $\bar{X}_C$ and $\bar{X}_D$ refer to mean values of the coarse and deformed garnet lherzolites, while $\bar{X}_I$ and $\bar{X}_{II}$ refer to average TiO_2 in the Group I and Group II harzburgite suites. Also shown are histograms for garnet lherzolites from Frank Smith and the Kimberley pipes (Boyd, 1975), Thaba Putsoa and Mothae (Boyd and Nixon, 1975) and Matsoku (Cox et. al., 1973).

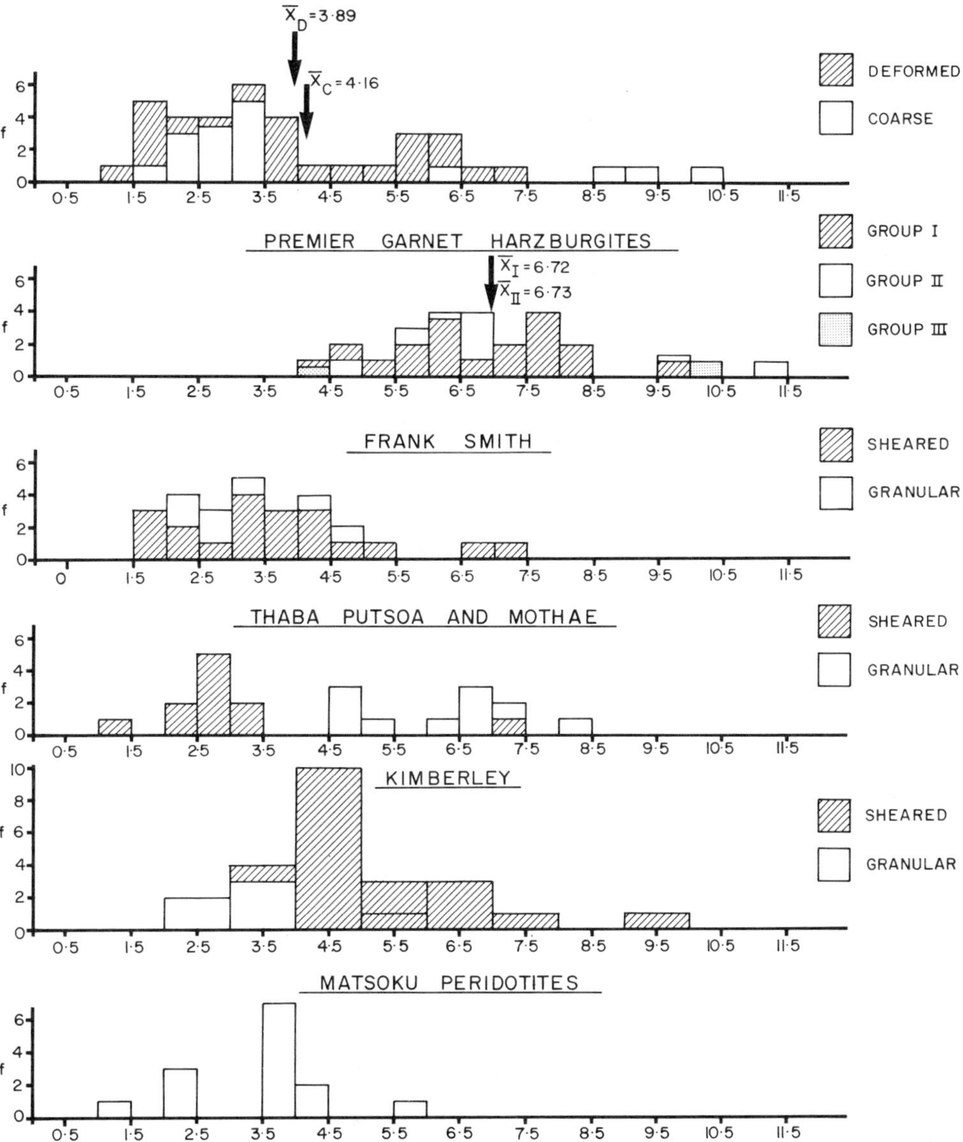

Fig. 2. Frequency distribution of Cr_2O_3 (wt.%) in garnets from Premier Mine garnet lherzolites and garnet harzburgites together with garnet lherzolites from other localities. Ornamentation and sources of data are as for Figure 1.

rocks and the other group is represented by deformed, extensively recrystallized, garnet-rich lherzolites whose textures vary from porphyroclastic to mosaic porphyroclastic. Average mineral compositions for 19 coarse garnet lherzolites and 35 deformed garnet lherzolites are summarised in Table 1, and representative individual analyses for the textural types have been reported by Danchin and Boyd (1976).

A. Mineral Compositions

(i) Garnets. Garnets in the deformed lherzolites are reddish brown and are characterised by unusually high TiO_2 contents (Ave. 1.15 wt.%, Table 1), whereas those from the coarse lherzolites are mauve and titanium poor (Ave. 0.2 wt.%). In Figure 1 the distribution of TiO_2 in these garnets is compared with those from other lherzolite suites in southern Africa, and the marked enrichment of Ti in the garnets of the deformed Premier xenoliths is evident. The Cr_2O_3 contents of the garnets in the deformed and coarse lherzolites, on the other hand, are similar (Table 1), and the bimodal Cr_2O_3 dstribution corresponding to a Cr-poor sheared suite and a Cr-rich granular suite found in nothern Lesotho by Nixon and Boyd (1973) is not repeated at Premier (Fig.2). Within the deformed suite the garnets are increasingly enriched in Cr_2O_3 as the rocks become more magnesian as illustrated in Figure 3. In the coarse lherzolites the distribution of Cr_2O_3 in the garnets is more complex largely because of strong zoning in several instances in which the cores are invariably more Cr-rich than the rims. In one such rock (RVD 153) Cr_2O_3 was found to vary from 4.9 to 9.4% in a single garnet grain. Similar zoning was described by Cox et al. (1973) in peridotitic garnets from Matsoku, and by Boyd and Nixon (1976) in peridotitic garnets from the Kimberley pipes.

(ii) Pyroxenes. The diopsides and enstatites in the deformed xenoliths, in like manner to the coexisting garnets, are markedly enriched in TiO_2 relative to those in the coarse suite (Table 1) and relative to other lherzolite suites (Fig.4). The distribution of Cr_2O_3 in the lherzolite diopsides, unlike the garnets, is similar to that described by Nixon and Boyd (1973) for the clinopyroxenes from Thaba Putsoa lherzolites (Fig.5). In both cases the deformed rocks differ from those with coarse textures in that their diopsides have lower Cr_2O_3 contents. Na_2O and Al_2O_3 are likewise more abundant in the calcic clinopyroxenes of the coarse lherzolites although for the enstatites the situation is

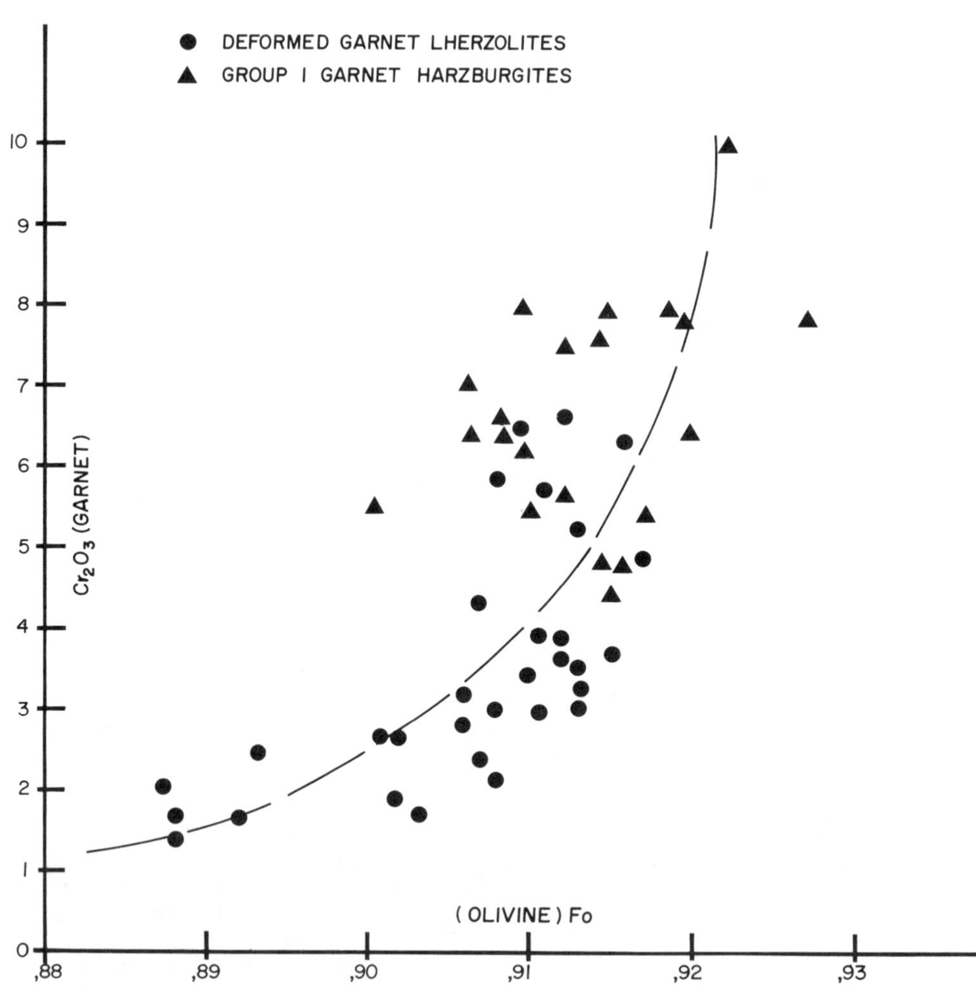

Fig. 3. Plot of Cr_2O_3 contents (wt.%) of garnet versus mole per cent forsterite contents of coexisting olivine for Premier Mine deformed garnet lherzolites and garnet harzburgites.

DIOPSIDES TiO_2

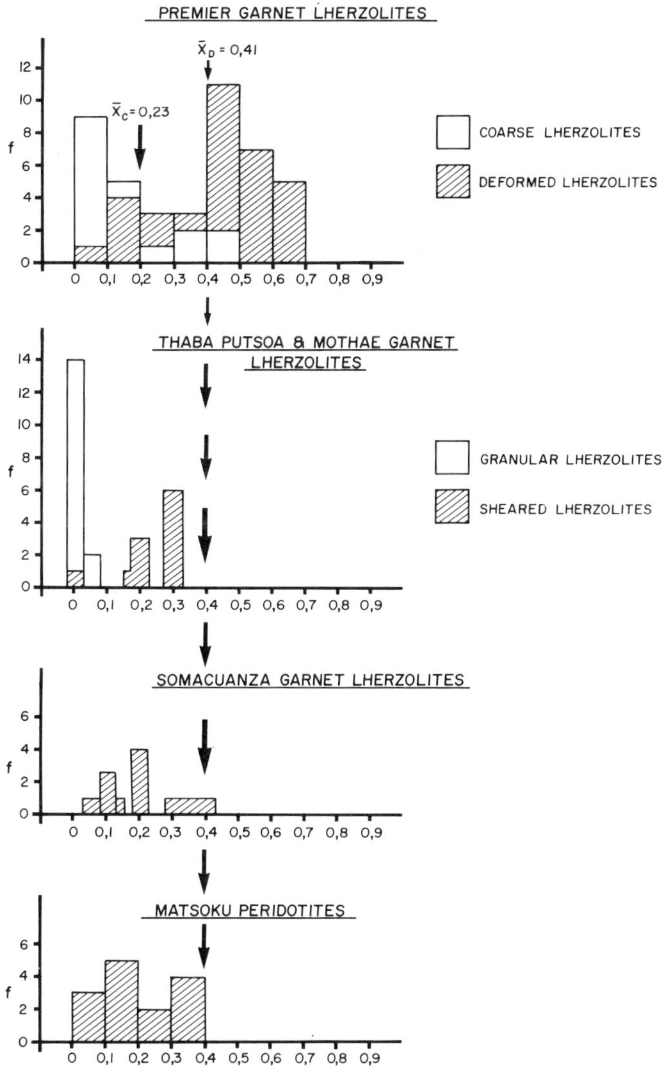

Fig. 4. Frequency distribution of TiO_2 (wt.%) in diopsides from Premier Mine garnet lherzolites. $\bar{X}_C$ and $\bar{X}_D$ refer to the coarse and deformed suites respectively. Also shown are data for granular and sheared garnet lherzolites from Thaba Putsoa and Mothae, Nixon and Boyd (1973), and garnet lherzolites from Somacuanza, Angola (Danchin and Boyd, 1975) and Matsoku (Cox et. al. 1973). No textural distinction has been made for xenoliths from the latter two localities.

reversed (Fig.6). In several instances diopsides from the coarse lherzolites were found to be quite inhomogeneous with respect to Na and Al. This variability was most pronounced in large diopside grains with characteristically curved grain boundaries, and it is possible that these phenomena result from minor partial melting along the grain margins.

Ca/(Ca+Mg) for diopsides in the deformed lherzolites varies from 0.33 to 0.38, and using the diopside solvus of Davis and Boyd (1966) and no other correction procedures, temperatures of equilibration for this suite range from 1355° to 1255°C. In the case of the coarse garnet lherzolites Ca/(Ca+Mg) varies from 0.43 to 0.49 indicating a range of equilibration temperatures between 1115-900°C.

(iii) Olivines. Differences in the olivines in the deformed and coarse xenoliths are pronounced, and these data are also summarised in Table 1. Olivine compositions are compared with those from northern Lesotho, Frank Smith and Kimberley lherzolites (Boyd 1975) in

Figure 7. With the exception of the Kimberley area, where deformed lherzolites with high equilibration temperatures are not found, it is seen that the olivines from the deformed high temperature lherzolites are significantly more iron-rich than those for olivines from the coarse, low temperature varieties. In the Kimberley lherzolites all olivines have Mg/(Mg+Fe) >91 mole per cent, and are thus comparable to coarse, low temperature lherzolites from the other four localities. The effects of these differences on bulk composition are pronounced, and are discussed at some length in the section devoted to major element variations in the Premier rocks.

(iv) Phlogopites. Phlogopite commonly occurs in the coarse Premier lherzolites in at least two generations. Large interstitial flakes of pale phlogopite whose chemical composition (Table 2) corresponds to the "primary" phlogopite of Carswell (1975) are most conspicuous. This mica has characteristically low TiO_2 contents (<1%) as well as low Al_2O_3 contents (<13.5%), and is occasionally mantled by a darker, pleochroic phlogopite richer in both Ti and Fe, as described by Boyd and Nixon (1975). Phlogopite of this type was not found in the deformed lherzolites. Both coarse and deformed lherzolites, however, often contain small irregular patches of dark brown, strongly pleochroic, titaniferous phlogopite whose compositions correspond to the "secondary" phlogopite of Carswell (1975) and examples are given in Table 2. In the Premier lherzolites this mica appears to be of two types. The first, which invariably mantles the kelyphite rims on the garnets, is a high chrome (>1%) variety, and the second, which occurs interstitial to olivine or enstatite grains, contains significantly less Cr_2O_3 (Table 2). A similar distinction between secondary phlogopites was noted by Harte and Gurney (1975) in peridotite nodules from Matsoku.

(v) Spinels. Chromite was found only in the coarse lherzolites and is considerably less abundant than at Thaba Putsoa, where Nixon and

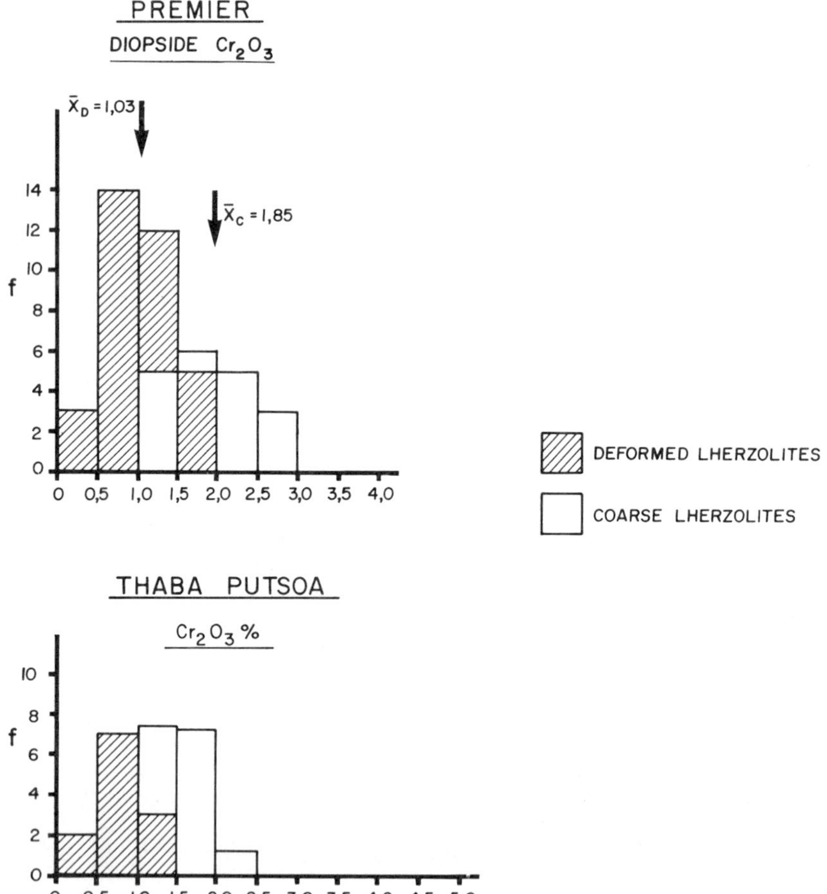

Fig. 5. Frequency distribution of Cr_2O_3 (wt.%) in diopsides from Premier Mine garnet lherzolites compared to equivalent rocks from Thaba Putsoa (Nixon and Boyd, 1973).

PREMIER

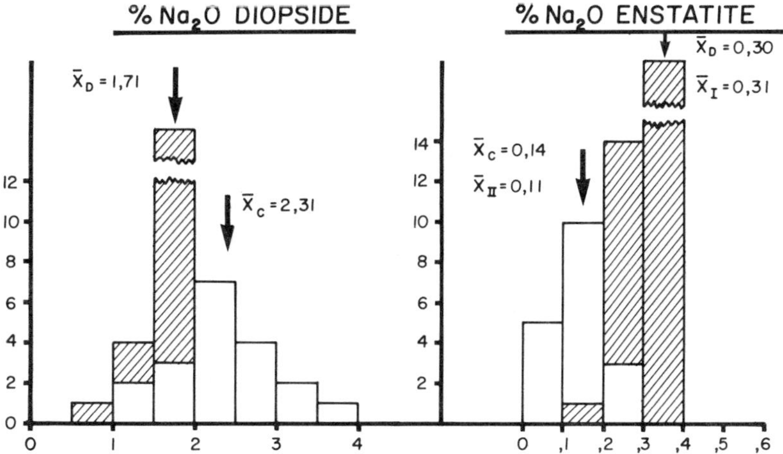

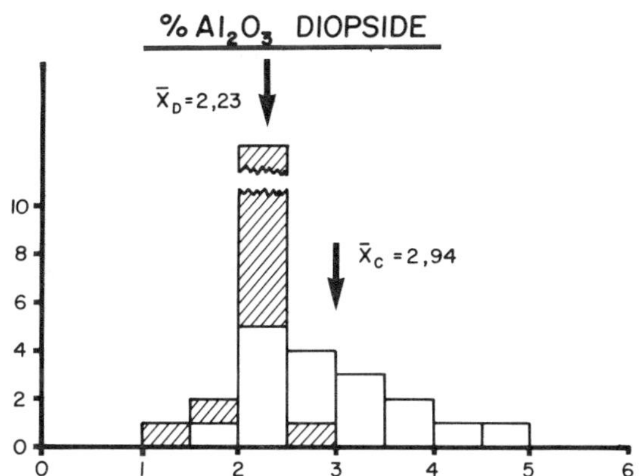

Fig. 6. Frequency distribution of Na_2O and Al_2O_3 (wt.%) in diopsides from Premier Mine coarse and deformed garnet lherzolites, and Na_2O (wt.%) in enstatites from the deformed and coarse garnet lherzolites and garnet harzburgites. Shaded ornamentation refers to deformed suite in each case.

Boyd (1973) reported an average modal concentration of 3 per cent in the granular garnet lherzolites. Like the phlogopite, the chromite occurs as discrete, Ti-poor grains, and as smaller Ti-enriched crystals invariably associated with the garnet kelyphite. Microprobe analyses of these spinel types are given in Table 2.

B. The Premier Pipe Geotherm

Estimates of temperatures and depths of equilibration of both groups of garnet lherzolites are compared with the Lesotho geotherm of Boyd and Nixon (1975) in Figure 8. It is emphasized that the purpose of this plot is comparative, and for this reason temperature has been calculated from the Ca/(Ca+Mg) ratios of the diopsides (Boyd 1973a) using the diopside solvus determined by Davis and Boyd (1966).

Pressure estimates have been made using the raw Al_2O_3 contents of the enstatites (Boyd, 1973a) from isopleths determined for the system $MgSiO_3$ - Al_2O_3 by MacGregor (1974). The temperature-depth relations in Figure 8 show a well developed correlation between the textures and depths of origin of the Premier garnet lherzolites. The coarse suite appear to have originated at depths of 110-170 km, whereas the deformed varieties were derived from depths of the order of 200 km. This is very similar to the pattern reported by Boyd and Nixon (1973, 1975) for Late Cretaceous nodule suites from various localities in southern Africa. It should be stressed, however, that this correlation between depth of origin and degree of deformation is not applicable to all southern Africa xenolith suites, since at other localities wide ranges of textures have been observed for lherzolites which are thought to have

OLIVINES — Mg (Mg + Fe)

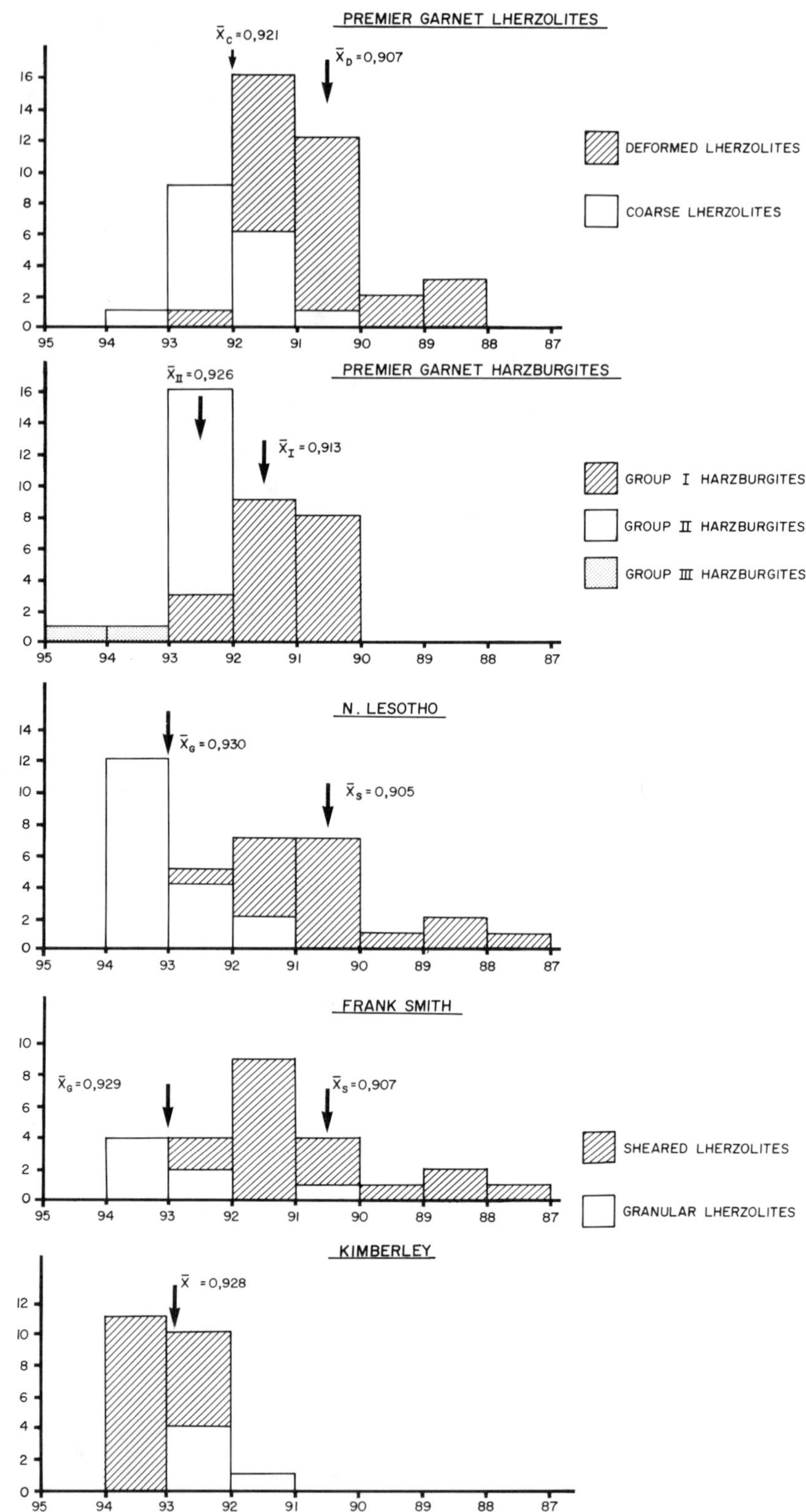

Fig. 7. Frequency distribution of Mg/(Mg + Fe) ratios for olivines from Premier Mine garnet lherzolites and garnet harzburgites as well as olivines from granular and sheared garnet lherzolites from northern Lesotho, Frank Smith and the Kimberley Mines (Boyd, 1975).

TABLE 2. Phlogopite and Spinel Compositions in Premier Mine Garnet Lherzolites.

	Pale "Primary" Phlogopite. Coarse Garnet Lherzolite RVD 175	Phlogopite in Kelyphite. Coarse Garnet Lherzolite RVD 500	Phlogopite in Kelyphite. Deformed Garnet Lherzolite RVD 152	Phlogopite Interstitial to Olivine. Coarse Garnet Lherzolite RVD 500	Phlogopite Intersitital to Olivine. Deformed Garnet Lherzolite RVD 501	Discrete Chromite. Coarse Garnet Lherzolite RVD 417	Chrome Spinel in Kelyphite. Coarse Garnet Lherzolite RVD 417
SiO_2	40.99	38.59	37.66	38.41	37.75	0.07	0.03
TiO_2	0.57	4.50	5.20	4.14	6.44	0.68	4.64
Al_2O_3	13.26	14.57	15.09	16.06	15.82	14.02	15.13
Cr_2O_3	0.78	1.76	1.26	0.27	0.17	52.66	41.90
FeO	3.04	3.65	5.29	4.64	5.06	18.29	22.55
NiO	0.27	0.24	0.16	0.19	0.21	0.11	n.d.
MnO	0.04	<0.03	0.04	0.04	<0.03	0.26	0.39
MgO	26.19	20.77	20.16	22.38	21.29	12.69	13.58
CaO	<0.03	0.08	0.03	<0.03	0.03	n.d.	0.05
K_2O	9.99	9.24	8.91	8.66	8.06	n.d.	n.d.
Na_2O	0.19	0.13	0.69	0.44	0.86	n.d.	n.d.
Total	95.29	93.44	94.49	95.23	95.70	98.78	98.27

originated at a common depth (Dawson *et al.*, 1975; Gurney *et al.*, 1975 a,b, Danchin and Boyd, 1975). In these cases the xenoliths have all equilibrated at relatively low temperatures (950-1050°C) and shallower depths (<175 km). Dawson *et al.*, (1975) and Gurney *et al.*, (1975b) therefore proposed that the undepleted, deformed lherzolites from Thaba Putsoa have high equilibration temperatures as a result of heating in the kimberlite envelope. Mitchell (1976) has argued against this hypothesis on the grounds that high-temperature, coarse xenoliths would also be expected in the Thaba Putsoa suite, but are in fact not found, and also because there is no evidence that kimberlitic liquids ever reach temperatures of the order of 1400°C necessary to achieve such equilibration. Goetze (1975) has suggested that the mosaic and porphyroclastic textures exhibited by the deformed lherzolites were produced in a short time prior to the eruption of the host kimberlite under conditions of high stress and strain rates, and that these textures could not survive under mantle conditions for appreciable periods. Thus the identical association between degree of deformation and depth of origin in mantle rocks contained in kimberlites whose ages differ by more than 1000 m.y. is interpreted as being indicative of similar modes of kimberlite formation and intrusion rather than of any other mantle processes.

The Premier geotherm itself, when calculated in this way, is also very similar to the Late Cretaceous geotherm derived by Boyd and Nixon (1975) for various pipes in northern Lesotho and South Africa. It has already been noted that there has been considerable debate regarding the interpretation of these pyroxene geotherms, notably with respect to the apparent inflection of the high temperature limbs (Fig. 8). Boyd and Nixon (1975) proposed that the inflection was caused by stress heating in the low velocity zone. It has since been argued that the inflected geotherms represent thermal profiles of mantle plumes (Parmentier and Turcotte, 1974) or diapirs (Green and Gueguen, 1974). Although not all geotherms are inflected (Boyd *et al.*, 1976; Eggler and McCallum, 1976), several authors have suggested that the inflection is an artifact of the method of estimation (Mercier and Carter, 1975; Harte, 1977). It has also been suggested that the inflections are spurious as a result of errors in the experimental calibration of the geothermometer (Howells and O'Hara, 1977), and clearly further work is required to resolve these issues. The apparent close similarity of the Precambrian and Cretaceous geotherms tentatively suggests, however, that the thermal evolution of the southern African mantle was essentially complete at least 1115 m.y. ago, and that intervening geological events, such as the dispersal of Gondwanaland, had no lasting effects on prevailing geothermal gradients.

The Garnet Harzburgites

The garnet harzburgite xenoliths from the Premier suite have been subdivided into three distinct groups on the basis of differences in mineral chemistry and texture (Danchin and Boyd, 1976). Mineral data for 38 of these xenoliths are summarised in Table 1. Group I garnet harzburgites have porphyroclastic or mosaic porphyroclastic textures and mineral compositions similar to the deformed garnet lherzolites. For example the garnets in both suites have high TiO_2 contents (Fig. 1) and the enstatites have equivalent Ca/(Ca+Mg) ratios (Fig. 9). The Group III garnet harzburgites, on the other hand, have coarse textures and contain mauve, Ti-poor garnets. Enstatite Ca/(Ca+Mg) ratios correspond to those of the coarse lherzolites. Group II harzburgites are intermediate between these two extremes, particularly with respect to the Ca/(Ca+Mg) ratios of the enstatites (Fig. 9), and the Mg/(Mg+Fe) ratios and Ti and Na contents of the three major mineral phases (Table 1).

Figure 10 is a plot of CaO versus Cr_2O_3 in the garnets for all the Premier lherzolites and harzburgites, and it shows that the lherzolite and Group I harzburgite garnets, all of which contain appreciable amounts of Ti (Fig. 1), exhibit a high degree of positive correlation which is characteristic of lherzolite assemblages (Sobolev et al., 1973). The Group II and Group III harzburgite garnets, which are Ti-poor, however, are enriched in Cr relative to Ca, and appear to define a separate trend (Fig. 10). Moreover, several of these garnets (eg. RVD 183, Danchin and Boyd, 1976) have compositions equivalent to garnets found as diamond inclusions (Meyer, 1968, Sobolev et al., 1969; Meyer and Boyd, 1972; Meyer and Svisero, 1975; Prinz et al., 1975). Sample RVD 183 was also found to contain chromite compositionally equivalent to chromite inclusions in diamond (Danchin and Boyd, 1976).

Temperature-depth estimates for the garnet harzburgites (Fig. 8) depend on the assumption

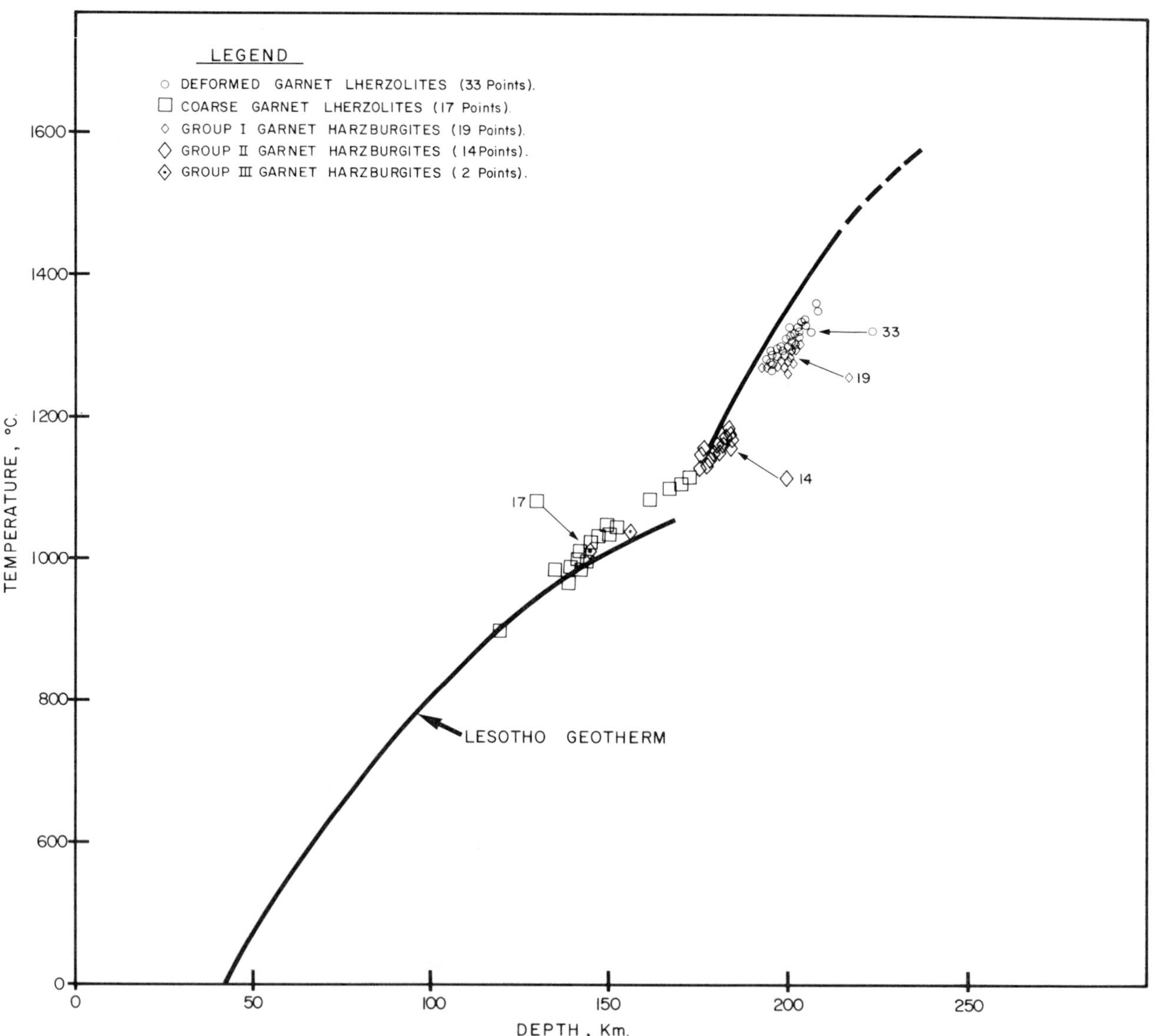

Fig. 8. Estimates of temperatures and depths of equilibration of Premier Mine garnet lherzolites and garnet harzburgites compared with the Lesotho geotherm.

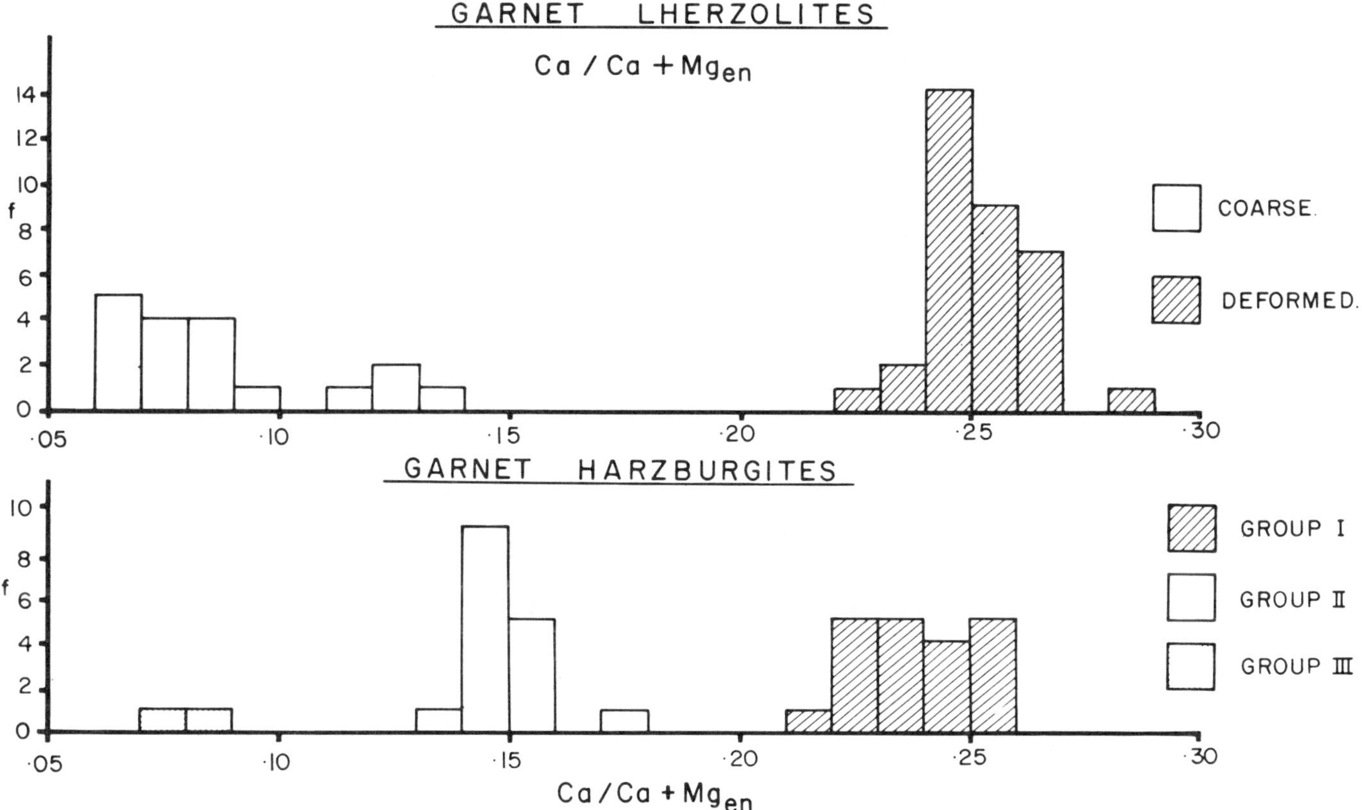

Fig. 9. Frequency distribution of Ca/(Ca + Mg) ratios for enstatites from Premier Mine garnet lherzolites and garnet harzburgites.

that the enstatites are saturated with respect to Ca, despite the apparent absence of diopside. The close textural and chemical correspondence of the Group I harzburgites with the deformed lherzolites, and a similar fit for the coarse members of each suite suggest that this assumption is valid and that the relative temperature sequence is reasonable. The Group I harzburgites thus appear to have equilibrated at depths of about 200 km, and it is noteworthy that the deepest, hottest garnet lherzolites do not appear to have harzburgitic equivalents (Fig.8). The Group II and Group III harzburgites, on the other hand, appear to have equilibrated between 140 and 180 km in the range 1000°-1200°C. Danchin and Boyd (1976) have noted that this range encompasses P-T estimates for the only known diamond-bearing garnet lherzolite (Dawson and Smith, 1975), and a diamondiferous garnet peridotite (McCallum and Eggler, 1976). The compositional similarities between the garnets in harzburgites from this range (and in the case of RVD 183 chromite), and the most common garnet (and chromite) inclusions in diamond suggest that the genesis of these harzburgites and some diamonds in the mantle are related. Also it would appear that the rocks of deepest origin now found as xenoliths in kimberlite are not those in which diamond has most commonly formed. The presence of chromite-rich spinels in these harzburgites is consistent with this suggestion since spinel appears to be replaced in the high P-T mantle suites by Cr-diopside and Cr-garnet (Haggerty, 1976).

Bulk Compositions

Major element compositions have been determined by X-ray fluorescence spectrometry (Appendix 1) for 31 deformed garnet lherzolites, 16 coarse garnet lherzolites and 27 garnet harzburgites, and the results are given in Table 3 in order of decreasing $FeO_t/(FeO_t + MgO)$ for each rock type. These data are summarized in Table 4, where they may be compared with analyses of equivalent xenoliths from some Late Cretaceous, southern African kimberlites reported in the literature, notably PHN 1611, a sheared garnet lherzolite from Thaba Putsoa, adjudged to be the least chemically depleted (particularly in terms of major elements) sample of garnet lherzolite from kimberlite studied thus far (Boyd and Nixon, 1973; Shimizu, 1973), and hence the most likely parental material for the formation of basaltic liquids by partial melting (Kushiro, 1973; Mysen and Kushiro,

1976). Also included in Table 4 is MF 1032, described by O'Hara et al., (1975) as the most fertile-looking garnet lherzolite from Matsoku, but nevertheless rejected by these authors as true source mantle and classified instead as a mantle cumulate. Figure 11 is a plot of $FeO_t/(FeO_t + MgO)$ against several of the major oxide contents for these rocks together with the Premier xenoliths.

In terms of their $FeO_t/(FeO_t + MgO)$ ratios the Premier garnet lherzolites (Category I of O'Hara et al., 1975), show varying degrees of chemical depletion, and, as may be seen from Table 3, relatively undepleted rocks corresponding in chemical composition to PHN 1611 are rare. In general the coarse garnet lherzolites are more depleted than the deformed varieties (Table 4) and mean $FeO_t/(FeO_t + MgO)$ values for the two types are 16.0 and 14.3 respectively. A similar situation pertains for the garnet lherzolites from northern Lesotho analysed by Boyd and Nixon (1973), but whereas in the latter instance the sheared and granular lherzolites could be clearly distinguished using their $FeO_t/(FeO_t + MgO)$ ratios, this is not the case at Premier and overlap between the two categories is extensive. Although additional data are clearly required from localities studied by Boyd and Nixon (1973) it is evident that in both instances, the deformed garnet lherzolites tend to be more iron-rich than those with coarse textures, and notably, that the most iron-rich, least depleted rocks are invariably deformed. The garnet harzburgites are progressively depleted from Group I through Group II to Group III (Table 4) and the respective average $FeO_t/(FeO_t + MgO)$ ratios are 14.4, 12.6 and 11.0.

In Figure 12 bulk $FeO_t/(FeO_t + MgO)$ is plotted against Mg/(Mg + Fe) for major mineral phases in the Premier lherzolites and harzburgites. This figure clearly illustrates that the iron-rich nature of the deformed lherzolites is reflected in their constituent iron-rich minerals and vice versa for the depleted garnet harzburgites.

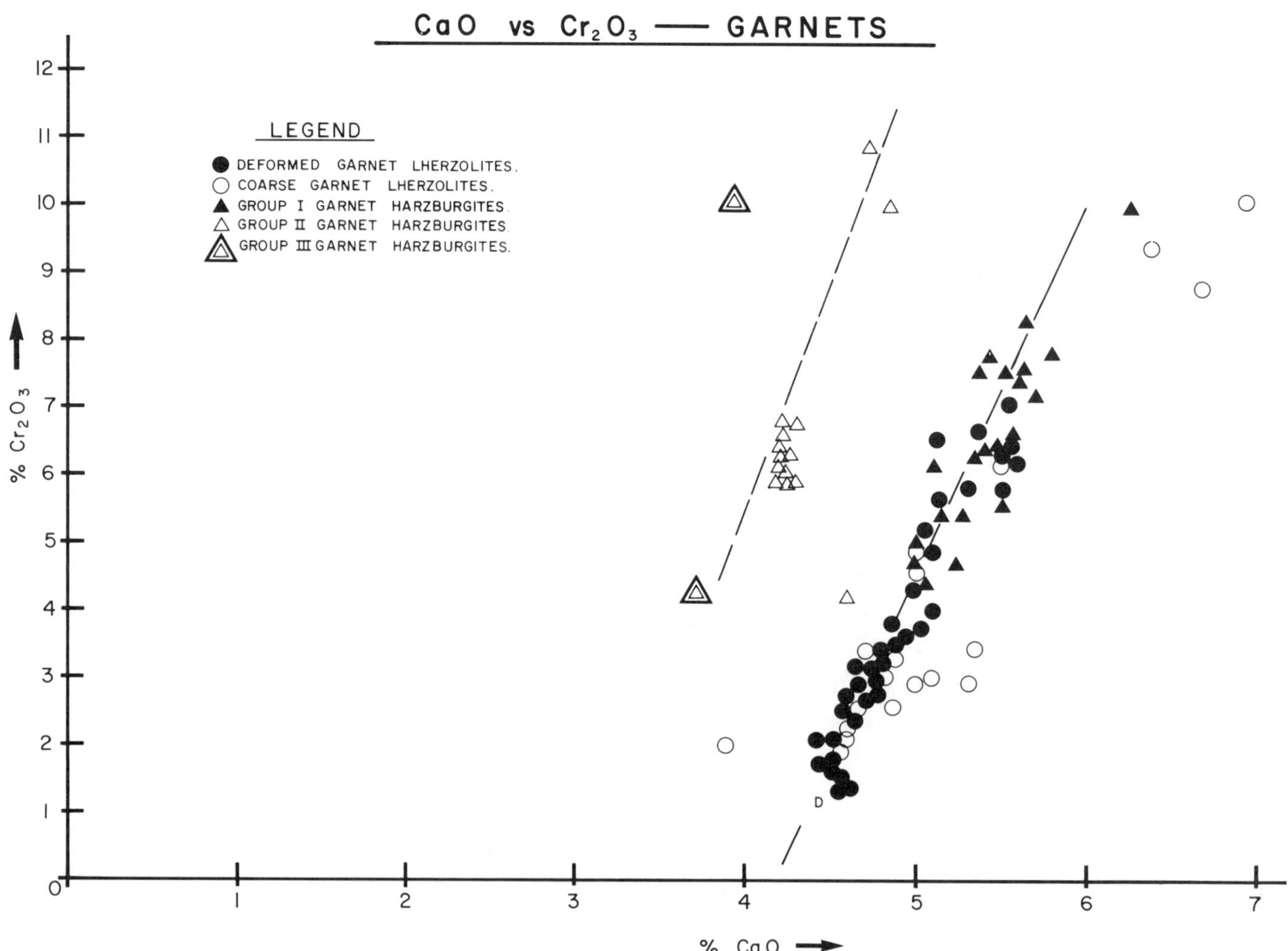

Fig. 10. Plot of CaO versus Cr_2O_3 (wt.%) for lherzolite and harzburgite garnets from Premier Mine.

TABLE 3. Bulk Compositions* of Selected Premier Garnet Lherzolite and Garnet Harzburgite Xenoliths

Deformed Garnet Lherzolites.

SAMPLE NO. RVD	152	121	501	502	158	100	508	112	440	172	507	123	149	413	159
SiO_2	44.76	44.38	43.96	44.14	45.83	43.85	44.68	44.00	45.42	44.55	44.54	43.41	43.49	44.65	43.68
TiO_2	0.28	0.35	0.35	0.32	0.55	0.19	0.19	0.30	0.17	0.23	0.29	0.34	0.17	0.15	0.19
Al_2O_3	3.29	2.96	3.37	3.99	1.83	4.16	3.61	2.72	1.01	3.17	2.90	1.48	1.47	2.30	1.24
Cr_2O_3	0.28	0.34	0.34	0.39	0.22	0.27	0.38	0.28	0.20	0.57	0.51	0.18	0.38	0.18	0.25
Fe_2O_3	5.24	4.80	3.89	3.65	-	3.92	3.90	3.48	4.87	1.90	3.59	3.98	4.03	2.65	4.09
FeO	4.95	5.40	6.03	5.61	8.47+	4.77	4.71	5.24	4.21	6.28	4.86	4.96	4.94	5.65	4.81
MnO	0.12	0.13	0.18	0.17	0.15	0.14	0.17	0.09	0.12	0.19	0.15	0.10	0.11	0.15	0.08
MgO	37.58	38.55	38.96	37.47	39.15	38.57	38.12	41.56	42.42	39.92	40.57	43.27	43.28	41.39	44.28
CaO	2.54	2.25	2.39	3.17	2.05	3.24	3.38	1.41	0.89	2.80	2.12	1.13	1.10	1.83	0.19
Na_2O	0.20	0.17	0.15	0.22	n.d.	0.22	n.d.	n.d.	n.d.	n.d.	n.d.	0.08	n.d.	n.d.	n.d.
K_2O	0.24	0.10	0.18	0.19	0.13	0.24	0.17	0.05	0.07	0.11	0.09	0.30	0.05	0.05	0.04
NiO	0.10	0.25	0.27	0.26	0.30	0.10	0.24	0.24	0.33	0.24	0.26	0.11	0.29	0.30	0.26
Total	99.38	99.78	99.92	99.36	98.68	99.45	99.55	99.37	99.71	99.96	99.88	99.26	99.33	99.30	99.11
FeO_t/FeO_t+MgO	20.5	20.1	19.6	19.2	17.8	17.7	17.7	16.8	16.8	16.6	16.6	16.5	16.5	16.3	16.1

* All analyses calculated on a volatile-free basis; + Total Fe as FeO

SAMPLE NO. RVD	132	161	164	126	128	186	101	120	136	190	119	106	174	402	198	437
SiO_2	45.20	42.40	44.94	43.07	44.78	43.95	43.41	44.32	44.13	43.47	43.74	46.90	44.27	44.08	45.68	44.02
TiO_2	0.21	0.32	0.12	0.15	0.32	0.47	0.23	0.34	0.27	0.11	0.09	0.21	0.16	0.09	0.17	0.09
Al_2O_3	1.51	3.03	3.02	1.11	2.41	2.76	1.42	2.59	1.99	1.54	2.50	1.35	0.94	0.78	1.14	0.64
Cr_2O_3	0.22	0.40	0.24	0.30	0.39	0.32	0.31	0.36	0.29	0.19	0.30	0.30	0.22	0.23	0.31	0.25
Fe_2O_3	3.86	2.37	3.25	3.45	4.32	3.12	3.76	3.80	4.32	2.43	4.03	2.80	2.84	4.22	1.92	4.15
FeO	4.64	5.86	4.73	5.35	3.88	4.95	4.77	4.25	4.05	5.53	4.04	4.67	5.20	3.77	5.29	3.75
MnO	0.09	0.11	0.11	0.10	0.11	0.12	0.09	0.11	0.11	0.12	0.13	0.10	0.12	0.10	0.10	0.11
MgO	42.27	42.29	40.70	45.04	41.58	41.76	44.17	41.50	43.45	44.26	43.71	41.48	45.00	46.00	43.37	46.25
CaO	1.16	2.23	2.41	0.49	1.35	1.80	0.90	1.80	0.99	1.35	1.21	1.15	0.40	0.28	1.01	0.26
Na_2O	n.d.	0.19	0.15	n.d.	n.d.	n.d.	n.d.	n.d.	n.d.	n.d.	n.d.	n.d.	0.12	0.04	n.d.	n.d.
K_2O	0.26	0.09	0.05	0.03	0.07	0.09	0.03	0.07	0.04	0.10	0.03	0.05	0.02	0.03	0.05	0.01
NiO	0.14	0.25	0.24	0.29	0.28	0.24	0.29	0.26	0.30	0.27	0.28	0.26	0.27	0.28	0.24	0.30
Total	99.56	99.35	99.81	99.38	99.49	99.58	99.38	99.40	99.94	99.37	100.06	99.27	99.44	99.86	99.28	99.83
FeO_t/FeO_t+MgO	16.1	15.9	15.8	15.8	15.7	15.7	15.6	15.6	15.4	14.9	14.9	14.8	14.7	14.1	13.9	13.9

Coarse Garnet Lherzolites.

SAMPLE NO. RVD	417	197	189	419	504	500	155	503	424	505	118	509	175	510	506	181
SiO_2	45.29	45.15	45.24	45.15	48.25	45.73	47.00	47.94	45.56	44.95	47.94	45.13	46.71	45.14	44.55	43.57
TiO_2	0.15	0.10	0.31	0.25	0.08	0.09	0.03	0.04	0.23	0.05	0.04	0.17	0.03	0.04	0.03	0.08
Al_2O_3	3.62	3.43	1.44	2.76	1.50	3.02	3.12	5.07	0.72	1.77	2.40	0.40	1.32	1.95	0.90	0.54
Cr_2O_3	0.76	0.31	0.25	0.28	0.38	0.40	0.29	0.69	0.26	0.37	0.38	0.31	0.20	0.38	0.29	0.15
Fe_2O_3	2.56	2.75	2.48	3.74	3.12	2.95	3.38	2.06	3.53	1.97	1.87	3.07	1.81	1.36	2.86	2.86
FeO	5.66	5.25	5.63	4.38	4.46	4.42	3.72	4.31	3.96	5.14	4.28	4.02	4.80	5.03	4.11	3.91
MnO	0.22	0.13	0.13	0.13	0.16	0.15	0.13	0.22	0.11	0.16	0.17	0.13	0.09	0.19	0.11	0.08
MgO	38.28	39.62	41.70	41.09	40.07	39.92	38.55	36.28	43.65	43.59	41.14	44.94	42.55	42.69	45.89	47.60
CaO	2.66	2.11	1.63	1.20	0.87	2.75	2.77	2.32	0.77	1.26	1.02	0.69	1.70	2.09	0.45	0.33
Na_2O	0.30	n.d.	n.d.	0.18	n.d.	0.18	0.30	n.d.	n.d.	0.20	n.d.	n.d.	n.d.	n.d.	0.05	n.d.
K_2O	0.58	0.17	0.39	0.18	0.83	0.15	0.37	0.79	0.07	0.10	0.25	0.04	0.84	0.76	0.03	0.03
NiO	0.23	0.27	0.24	0.25	0.32	0.26	0.24	0.15	0.28	0.26	0.23	0.27	0.23	0.25	0.30	0.32
Total	100.01	99.29	99.44	99.41	100.04	99.84	99.60	99.87	99.14	99.62	99.72	99.17	100.28	99.88	99.52	99.47
FeO_t/FeO_t+MgO	17.2	16.3	15.9	15.9	15.3	15.1	14.9	14.5	14.0	13.7	13.0	13.1	13.1	12.8	12.7	12.0

GROUP I. Harzburgites.

SAMPLE NO. RVD	167	146	404	416	414	438	430	423	401	157	427	400	405	409
SiO_2	42.53	42.89	43.43	46.57	42.51	42.34	45.50	44.57	46.00	42.65	45.40	45.12	45.05	44.43
TiO_2	0.15	0.17	0.21	0.11	0.14	0.10	0.19	0.12	0.21	0.06	0.13	0.09	0.16	0.12
Al_2O_3	0.83	0.87	0.90	0.86	0.66	0.54	0.58	0.90	1.46	0.48	1.68	0.71	0.52	1.25
Cr_2O_3	0.25	0.30	0.23	0.25	0.18	0.13	0.28	0.24	0.27	0.11	0.40	0.15	0.27	0.40
Fe_2O_3	3.72	4.14	3.88	3.90	3.60	3.98	2.26	3.86	3.33	4.00	3.00	3.25	4.01	2.68
FeO	5.29	4.66	4.78	4.14	5.13	4.77	5.62	4.24	4.10	4.23	4.44	4.47	3.60	4.61
MnO	0.10	0.11	0.11	0.12	0.11	0.10	0.13	0.10	0.10	0.06	0.11	0.09	0.08	0.11
MgO	45.93	45.70	45.33	42.33	46.64	47.16	44.07	45.00	42.96	47.56	43.47	45.35	45.20	45.25
CaO	0.40	0.30	0.50	0.48	0.29	0.11	0.51	0.47	0.71	0.12	0.60	0.42	0.32	0.60
Na_2O	n.d.	n.d.	n.d.	n.d.	n.d.	n.d.	n.d.	n.d.	n.d.	n.d.	n.d.	n.d.	n.d.	n.d.
K_2O	0.01	0.01	0.02	0.06	0.01	0.01	0.03	0.02	0.05	0.01	0.03	0.05	0.02	0.05
NiO	0.29	0.27	0.29	0.31	0.30	0.31	0.26	0.30	0.25	0.31	0.25	0.32	0.30	0.27
Total	99.50	99.42	99.68	99.13	99.57	99.55	99.43	99.82	99.44	99.59	99.51	100.02	99.53	99.77
FeO_t/FeO_t+MgO	15.8	15.5	15.4	15.3	15.2	15.1	14.8	14.6	14.2	14.2	14.1	14.0	13.7	13.4

TABLE 3 (Contd.).

SAMPLE NO. RVD	Group II Harzburgites												Group III Garnet Harzburgites	Group III Garnetiferous Dunite
	412	168	435	433	407	410	428	421	406	408	127	425	184	144
SiO_2	45.36	45.97	46.36	45.35	44.50	45.14	45.47	45.40	45.80	45.59	45.87	45.68	44.69	38.88
TiO_2	0.07	0.09	0.09	0.05	0.19	0.05	0.08	0.08	0.04	0.07	0.03	0.08	0.05	0.29
Al_2O_3	1.30	1.14	0.85	1.37	0.71	1.18	0.79	1.37	1.10	1.08	1.08	1.76	1.22	4.16
Cr_2O_3	0.34	0.26	0.20	0.31	0.22	0.30	0.19	0.25	0.24	0.24	0.23	0.25	0.18	0.18
Fe_2O_3	3.41	3.18	2.98	2.53	2.87	3.45	2.85	3.10	3.20	3.05	3.33	2.41	2.44	7.52
FeO	3.65	3.70	3.83	4.24	4.00	3.33	3.83	3.61	3.53	3.53	3.20	3.87	3.58	7.09
MnO	0.11	0.07	0.09	0.10	0.10	0.08	0.09	0.11	0.10	0.11	0.08	0.10	0.09	0.12
MgO	44.84	44.35	44.20	44.77	45.49	45.00	45.00	44.57	44.76	45.08	44.90	44.46	46.26	40.00
CaO	0.43	0.50	0.56	0.52	0.90	0.56	0.43	0.66	0.54	0.53	0.47	0.54	0.55	0.80
Na_2O	n.d.	n.d.	n.d.	n.d.	0.09	n.d.	n.d.	0.04	n.d.	n.d.	n.d.	n.d.	n.d.	n.d.
K_2O	0.06	0.04	0.07	0.04	0.12	0.04	0.06	0.06	0.03	0.04	0.03	0.05	0.10	0.11
NiO	0.27	0.27	0.28	0.25	0.26	0.27	0.26	0.27	0.27	0.26	0.26	0.26	0.26	0.27
Total	99.84	99.57	99.51	99.53	99.36	99.40	99.05	99.48	99.61	99.58	99.48	99.46	99.42	99.42
FeO_t/ FeO_t+MgO	13.0	12.9	12.8	12.7	12.6	12.6	12.5	12.5	12.5	12.2	12.1	11.9	11.0	25.7

Evidently, therefore, these rocks have equilibrated such that their Fe/Mg ratios are a function of bulk composition of the system in which they formed, and not merely of modal permutations of the constituent minerals. Consequently a relatively iron-rich coarse lherzolite such as RVD 197 (Bulk $FeO_t/(FeO_t$ + MgO) = 16.3) has been depleted with respect to iron such that even an extreme modal mixture containing 30% garnet, 30% diopside and 20% each of olivine and enstatite has a bulk $FeO_t/(FeO_t$ + MgO) ratio of only 17.8. It should be noted that although bulk data for lherzolite inclusions in kimberlite are few, the restriction of iron-rich, undepleted compositions to deformed garnet lherzolites at Premier and localities studied by Boyd and Nixon (1973) is not without exception. Gurney et al., (1975 a,b), for example, report that textural variations in lherzolites from the Matsoku kimberlite are totally independant of bulk composition. Similarly at Bultfontein, Gurney et al., (1975b) reported that textural and chemical variations are unrelated. At each of the latter localities, however, high temperature garnet lherzolites containing subcalcic diopsides are not found. The significance of these differences is not presently apparent.

The average modal garnet content of the coarse Premier lherzolites (Fig. 13) is 8 volume per cent compared to a value of 4 per cent for Thaba Putsoa (Nixon and Boyd, 1973) and 4.5 per cent for Matsoku coarse common peridotites (Cox et al., 1973). This difference is strikingly illustrated when bulk Al_2O_3 contents for the suites are compared (Table 4). The mean Al_2O_3 content of the Premier coarse lherzolites is 2.12% compared to 0.5% for the Thaba Putsoa xenoliths. The mean Al_2O_3 content of the Matsoku coarse common peridotites, by comparison, is 1.6%. Unlike the lherzolites studied by Nixon and Boyd (1973) there is extensive overlap in Al_2O_3 content from the Premier coarse lherzolites to the deformed varieties (Fig. 11). The mean Al_2O_3 content of the Premier deformed lherzolites is 2.2% (Max. 4.2%) whereas for the northern Lesotho sheared lherzolites the mean value is 1.84% (2.8% for PHN 1611). Also, it is noteworthy that several of the Premier lherzolites have Cr_2O_3 contents greater than 0.43% (Table 3) - the value assigned to hypothetical pyrolite by Green (1973). It is evident, therefore, that the Premier coarse lherzolites, and to a somewhat lesser extent the deformed lherzolites, are more aluminous (garnetiferous) than equivalent rocks from northern Lesotho studied by Nixon and Boyd (1973) and Cox et al., (1973).

The modal diopside contents of the Premier coarse lherzolites are shown in Figure 13 and the average is 6 volume per cent compared to a value of 5 per cent for granular lherzolites at Thaba Putsoa and 2 per cent for Matsoku coarse common peridotites. Consequently Ca, in like

TABLE 4. Summary of Mean Bulk Compositions* of Premier Garnet Lherzolites and Garnet Harzburgites and Equivalent Rocks from Northern Lesotho.

	Fertile Premier Garnet Lherzolite	PHN 1611	Average Premier Deformed	Average Northern Lesotho Sheared	Average Premier Coarse	Average Northern Lesotho Granular	Average Premier Group I	Average Premier Group II	Premier Group III Garnet	MF 1032 Matsoku
%	RVD 152, 121, 501, 502	Thaba Putsoa[1]	Garnet lherzolite	Garnet lherzolite[1]	Garnet lherzolite	Garnet lherzolite[1]	Garnet harzburgite	Garnet harzburgite	harzburgite	Northern Lesotho[2]
SiO_2	44.31	44.60	44.34	43.86	45.83	45.80	44.21	45.55	44.69	46.62
TiO_2	0.33	0.25	0.24	0.16	0.11	0.03	0.14	0.07	0.05	0.15
Al_2O_3	3.40	2.80	2.20	1.84	2.12	0.50	0.87	1.14	1.22	4.91
Cr_2O_3	0.34	0.28	0.30	0.22	0.36	0.21	0.25	0.25	0.18	0.59
FeO^+	9.46	10.25	8.07	8.58	6.95	6.23	7.56	6.46	7.42	12.04
MnO	0.15	0.13	0.13	0.12	0.14	0.10	0.10	0.09	0.09	0.18
MgO	38.14	37.22	41.87	42.90	41.72	46.06	45.14	44.76	46.26	30.83
CaO	2.59	3.32	1.59	1.86	1.54	0.82	0.42	0.55	0.55	4.33
Na_2O	0.18	0.34	0.15	0.22	0.20	0.06	n.d.	0.07	n.d.	0.35
K_2O	0.18	0.14	0.09	0.04	0.35	0.04	0.02	0.05	0.10	0.04
NiO	0.22	n.d.	0.25	0.23	0.26	0.24	0.29	0.27	0.26	0.22
FeO_t/ FeO_t+MgO	20.0	21.2	16.0	16.1	14.3	11.9	14.4	12.6	11.0	28.1

* All analyses calculated on a volatile-free basis; + Total Fe as FeO; [1]Boyd and Nixon (1973); [2]O'Hara et al. (1975).

manner to Al, but somewhat less strikingly, is relatively enriched in the equivalent Premier rocks. The position in the deformed rocks is reversed, however, and the northern Lesotho lherzolites are slightly more calcic (and sodic) than those from Premier.

Additional analyses of Late Cretaceous xenoliths are clearly needed for bulk data comparisons of the type attempted above, but the available data nevertheless suggest that the Precambrian mantle represented by the coarse Premier garnet lherzolites is significantly less depleted with respect to Fe/Mg, Al and Ca than mantle from northern Lesotho studied by Boyd and Nixon. With the exception of Al, the major elements in the deformed suites are more directly comparable.

The distribution of Ti and K in the Premier xenoliths is complex and problematic. The average TiO_2 content of the deformed lherzolites is 0.24 wt.%, and several xenoliths have TiO_2 contents greater than 0.3 wt.% (Table 3). The mean TiO_2 content of the sheared lherzolites from northern Lesotho studied by Boyd and Nixon (1973) was found to be about 0.16 wt.% (PHN 1611 = 0.25 wt.% TiO_2). The coarse Premier lherzolites are likewise more titaniferous than equivalent xenoliths from northern Lesotho. Furthermore, it is evident from the analyses in Table 3 and Figure 11 that the most iron-rich deformed lherzolites are generally those which contain most TiO_2. A similar trend pertains for the garnet harzburgites as may be seen from Figure 14 where the TiO_2 contents of the constituent garnets are plotted against the forsterite contents of the coexisting olivines.

It has been noted that the TiO_2 contents of the garnets in the deformed Premier lherzolites and harzburgites are amongst the highest yet recorded for such xenoliths, although a similar but less well developed Ti enrichment was found in deformed garnet lherzolites from the Frank Smith Mine (Fig.1) by Boyd (1975), who attributed these high TiO_2 contents to metasomatic processes. Boyd (1975) noted in support of this argument that some fine-grained recrystallized pyroxene in the groundmass of sheared lherzolites from Kimberley was found to be enriched in TiO_2 relative to the porphyroclasts. It is difficult to apply this theory to the deformed Premier lherzolites in view of the fact that similar phenomena were not encountered in these rocks, and particularly because of the well developed correlation of TiO_2 and Fe/Mg noted above. It is suggested that the high TiO_2 contents of the iron-rich xenoliths are consistent with their undepleted major element compositions rather than a preferential process of metasomatism. Also, since the Premier lherzolites are more titaniferous (and aluminous) than those from northern Lesotho, it is concluded that the Precambrian rocks are less depleted with respect to these elements.

In the deformed lherzolites, potassium, like titanium, is concentrated in xenoliths with high Fe/Mg ratios as shown in Figure 15. Relatively undepleted Premier lherzolites with $FeO_t/(FeO_t + MgO)$ >19.0 have a mean K_2O content of 0.18%

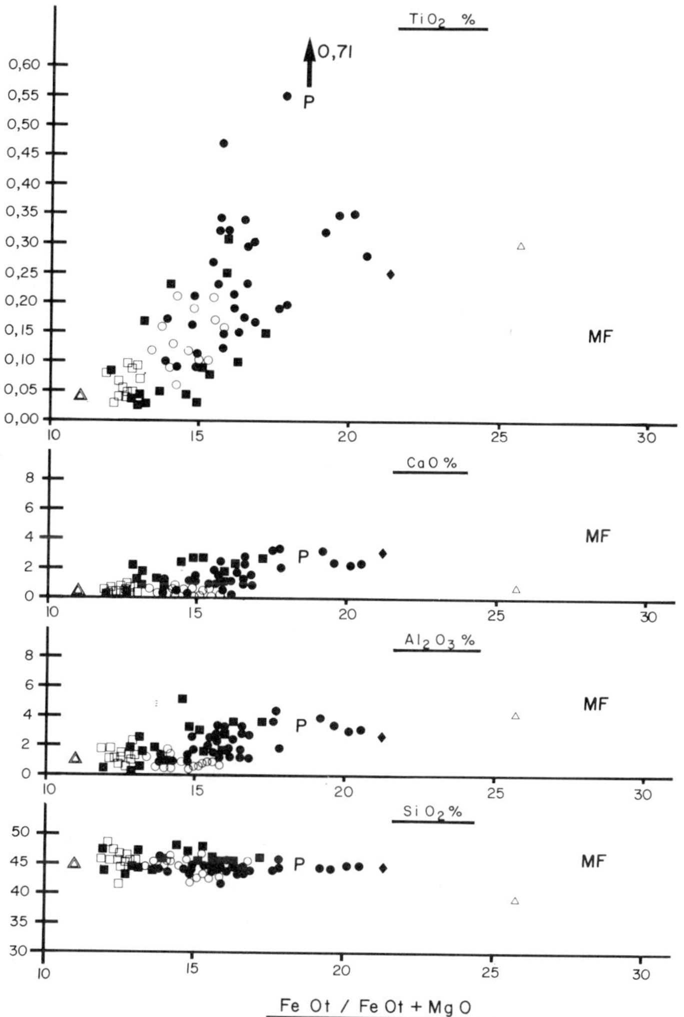

Fig. 11. Plots of FeO_t/FeO_t + MgO versus bulk TiO_2, CaO, Al_2O_3 and SiO_2 (wt.%) for Premier Mine garnet lherzolites and garnet harzburgites. Also shown are a garnetiferous dunite from Premier PHN 1611 (Boyd and Nixon, 1973), hypothetical pyrolite (Green, 1973) and BD 1032 (O'Hara et. al. 1975).

compared to a value of 0.14% for PHN 1611. Unlike Ti, however, K in these rocks is not present in a major mineral phase since careful microprobe analyses for K showed that enstatite, garnet and olivine contain less than 35 ppm, and diopside in the deformed rocks have approximately 300 ppm K (Table 1). The bulk of the K is present in either phlogopite contained in garnet kelyphite or in interstitial serpentine, in a manner first documented by Erlank (1970).

It has been noted that large interstitial flakes of Ti-poor phlogopite are present in the coarse lherzolites whose mean bulk K_2O content is 0.35% (Table 1). The distribution of K in these xenoliths, unlike the deformed suite, is independant of bulk Fe/Mg ratios (Fig. 15).

In terms of these observations, therefore, at least two distinct processes need to be invoked to explain the distribution of K in the lherzolite suites. The presence of "primary" phlogopite in the coarse rocks, where it is texturally in equilibrium with the anhydrous phases has been discussed by Boyd and Nixon (1975). These authors argued for a primary metasomatic origin for this phlogopite since the bulk compositions of the host rocks indicate extensive depletion to have taken place. Harte and Gurney (1975) likewise proposed a primary metasomatic origin to explain the textural and geochemical features of phlogopite present in xenoliths from Matsoku. "Primary" phlogopite is not found in the deformed rocks where the abundance of "secondary" phlogpite is related to the bulk Fe/Mg ratios. The effects of "secondary" phlogopite in the coarse lherzolites are completely masked by the more abundant primary mica. Also, the garnet harzburgites (Table 3) are all K-poor in keeping with their depleted compositions. Thus two separate processes of K introduction (or re-distribution) are indicated. Erlank (1975) considered that the bulk of the phlogopite in kimberlitic nodules equilibrated under upper mantle conditions, but that it has formed either by metasomatic processes or, alternatively, that it is representative of trapped partial melt. A metasomatic origin in the mantle is acceptable for the "primary" phlogopite, but it is not

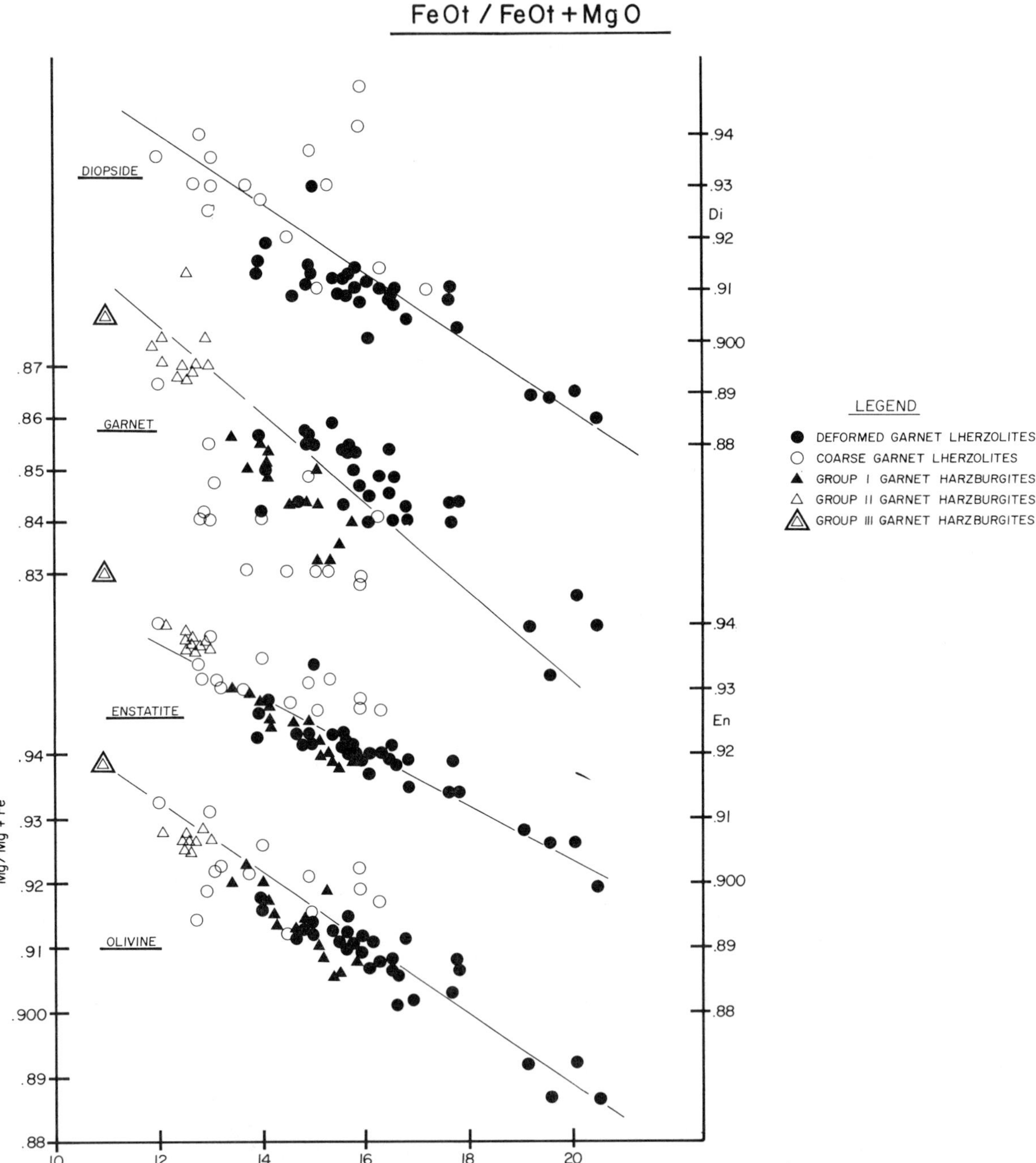

Fig. 12. Plots of bulk FeO_t/FeO_t + MgO versus Mg/(Mg + Fe) ratios for olivine, enstatites and garnet from Premier Mine garnet lherzolites and garnet harzburgites and for diopside from the garnet lherzolites.

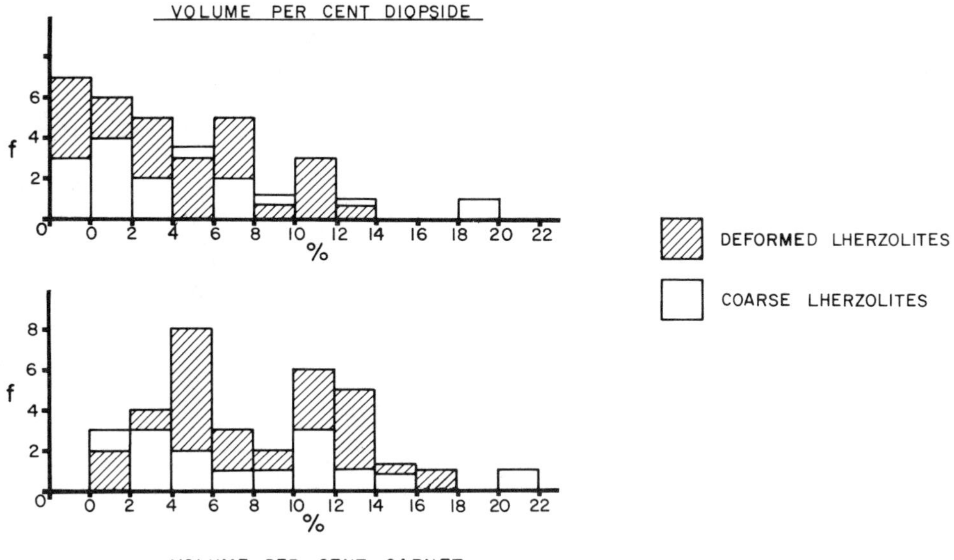

Fig. 13. Frequency distributions of modal diopside and garnet contents (volume per cent) for Premier Mine garnet lherzolites.

clear why the most iron-rich deformed rocks are also the most potassic when one considers the origin of "secondary" phlogopite in these rocks unless the Fe/Mg ratios have also been influenced by a corresponding metasomatic introduction of iron. There is also no textural evidence to support incipient partial melting in the deformed rocks. It is suggested, therefore, that potassium, like titanium in these rocks is more probably a primary feature, and that this

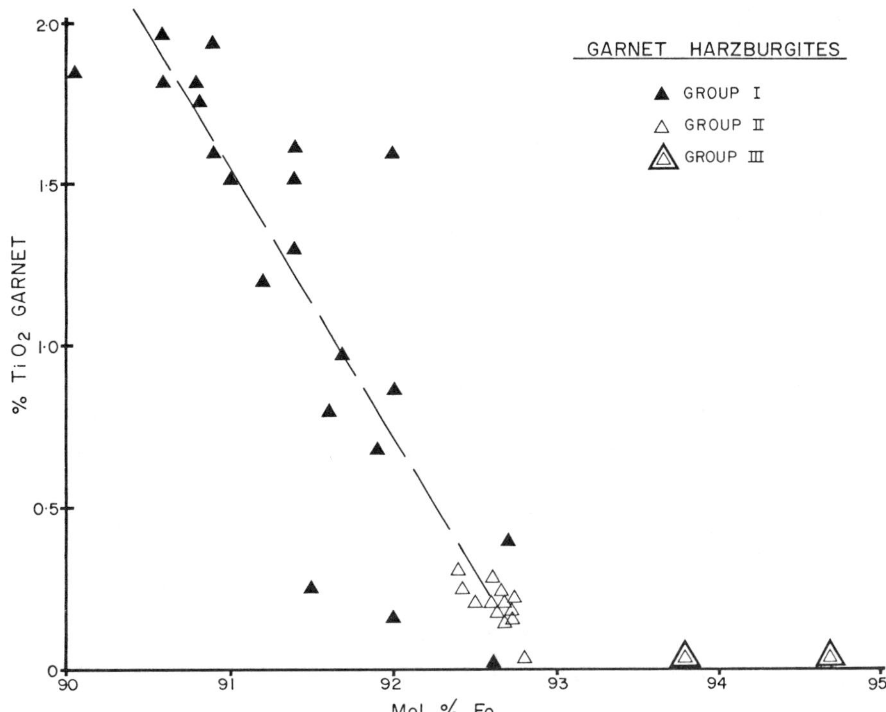

Fig. 14. A plot of TiO_2 (wt.%) in garnet versus forsterite content (mole per cent) in coexisting olivine for Premier Mine garnet harzburgites.

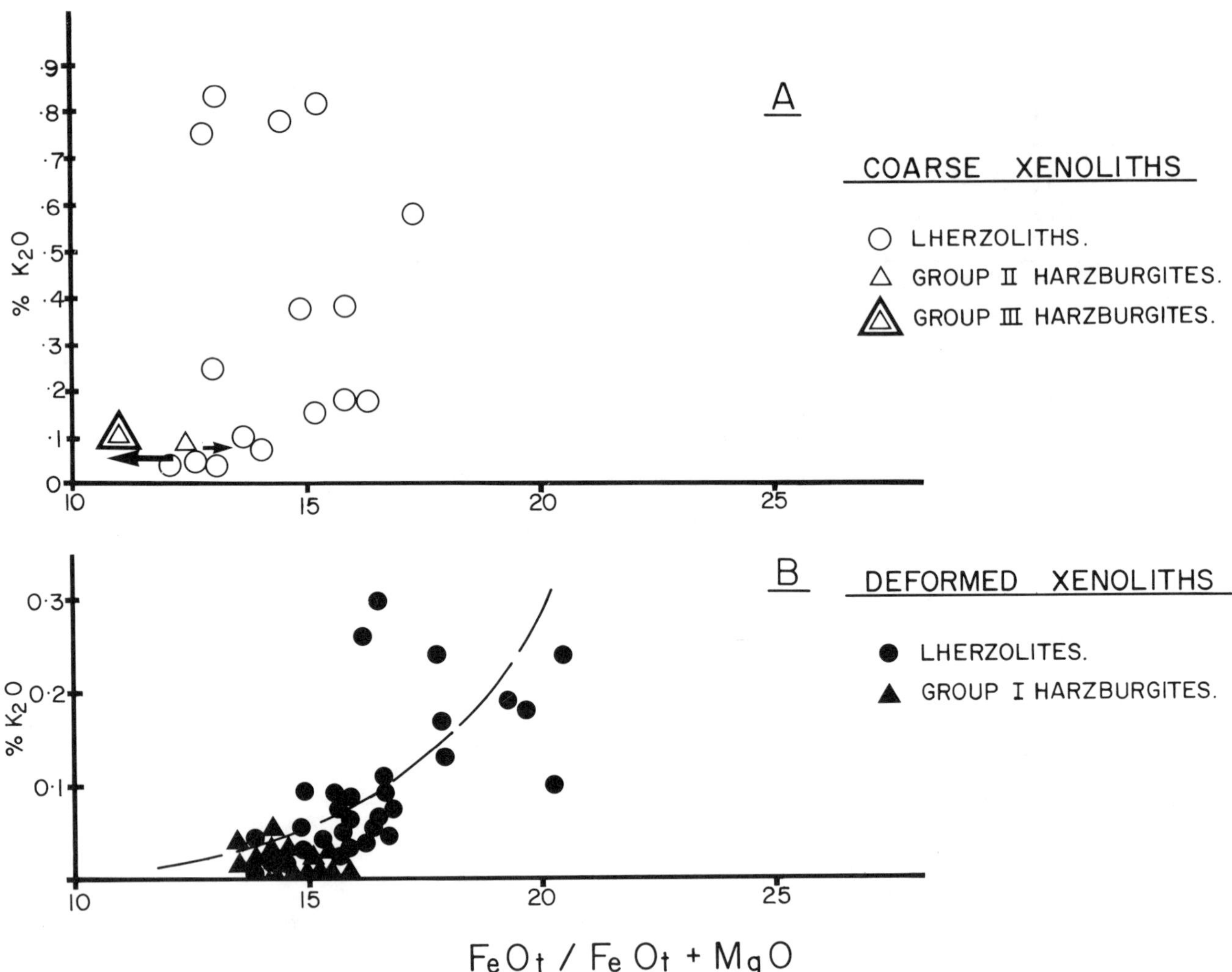

Fig. 15. Plots of bulk FeO_t/FeO_t + MgO versus K_2O (wt.%) for coarse and deformed garnet lherzolites and garnet harzburgites from Premier Mine.

element has not been metasomatically introduced, but rather redistributed in the undepleted Premier garnet lherzolites. That is, it is assumed that K in these rocks is "primary" in the sense that it has always been present, but is not now located in the same site as when these rocks first formed. Erlank (1973) has noted the possibility that entry of K into diopside lattice is pressure dependant, and systematic reduction of pressure could account for the redistribution of K described above.

APPENDIX 1

ANALYTICAL METHODS

Mineral analyses were carried out in part using a Materials Analysis Company Model 400 electron microprobe using techniques described by Boyd (1973b) and also using an ARL SEMQ automated electron microprobe using an almost identical anlytical procedure. At least four analyses at individual points in a grain, or preferably on separate grains were carried out. Individual mineral analyses, not published here due to lack of space, are available from the author on request. Samples for bulk analysis were cleared of weathered edges and reduced in a tungsten-carbide disc mill. Fe, Mn, Ti, Ca, K, P, Si, Al and Mg were determined on a Philips PN 1270 automatic X-ray spectrometer using the fusion method of Norrish and Hutton (1969). Water and Ignition losses were determined at 110° and 1020°C respectively. Na and Ni were determined by atomic absorption spectrophotometry following complete dissolution in acid in platinum crucibles. Cr was determined by colorimetry and ferrous iron by titration with potassium dichromate after acid decomposition in a reducing atmosphere.

Acknowledgements. The author acknowledges with thanks the receipt of a travel grant from the H.O. Wood fund administered by the Carnegie Institution of Washington. I am especially grateful to Dr. F.R. Boyd for his interest, assistance and continued enthusiasm. Thanks are also due to A.A. van Zyl and R. Molyneux for their assistance in collecting the samples and to G. Hutchinson and S.W. Marsh for technical assistance. A.J. Erlank, J.J. Gurney, D. Robinson, B. Scott and B. Wyatt critically read the manuscript. The Anglo American Corporation is thanked for permission to publish the results of this study.

References

Allsopp, H.L., A.J. Burger, and C. van Zyl, A minimum age for the Premier kimberlite pipe yielded by biotite Rb-Sr measurements, with related galena isotopic data, Earth. Planet. Sci., Letts., 3, 161-166, 1967.

Barrett, D.R., and H.L. Allsopp, Rubidium-strontium age determinations on South African kimberlite pipes, in Extended Abstracts, First International Kimberlite Conference, Cape Town, South Africa, 23-25, 1973.

Boyd, F.R., A pyroxene geotherm, Geochimica et Cosmochimica Acta, 37, 2533-2546, 1973a.

Boyd, F.R., Mineral analyses for Lesotho kimberlites, Ed's note, in Lesotho Kimberlites, Edited by P.H. Nixon, Lesotho National Development Corporation, Maseru, 1973b.

Boyd, F.R., Ultramafic xenoliths from the Frank Smith Mine and the Kimberley area, in Extended Abstracts, Kimberlite Symposium, Cambridge, England, 9-14, 1975.

Boyd, F.R., Inflected and non-inflected geotherms, Carnegie Inst. Washington Geophys. Lab. Yrbk., 75, 521-523, 1975.

Boyd, F.R., and P.H. Nixon, Structure of the upper mantle beneath Lesotho, Carnegie Inst. Washington Geophys. Lab. Yrbk., 72, 431-445, 1973.

Boyd, F.R., and P.H. Nixon, Origins of ultramafic nodules from some kimberlites of northern Lesotho and the Monastery mine, South Africa, Phys. Chem. Earth, 9, 431-454, 1975.

Boyd, F.R., and P.H. Nixon, Ultramafic nodules from the Kimberley pipes, South Africa, preprint, 1977.

Boyd, F.R., T. Fujii, and R.V. Danchin, A non-inflected geotherm for the Udachnaya Kimberlite pipe, U.S.S.R., Carnegie Inst. Washington Geophys. Lab. Yrbk., 75, 523-531, 1976.

Carswell, D.A., Primary and secondary phlogopites and clinopyroxenes in garnet lherzolite xenoliths, Phys. Chem. Earth. 9, 417-430, 1975.

Cox, K.G., J.J. Gurney, and B. Harte, Xenoliths from the Matsoku pipe, in Lesotho Kimberlites, edited by P.H. Nixon, Lesotho National Development Corporation, Maseru, 76-98, 1973.

Danchin, R.V., and F.R. Boyd, The geochemistry of ultramafic nodules from three Angolan kimberlites, in Extended Abstracts, Kimberlite Symposium, Cambridge, England, 13-21, 1975.

Danchin, R.V., and F.R. Boyd, Ultramafic nodules from the Premier kimberlite pipe, South Africa, Carnegie Inst. Washington Geophys. Lab. Yrbk., 75, 531-538, 1976.

Davis, B.T.C., and F.R. Boyd., The join $Mg_2Si_2O_6$-$CaMgSi_2O_6$ at 30 kb pressure and its application to pyroxenes from kimberlites, Jour. Geophys. Res., 71, 3567-3576, 1966.

Dawson, J.B., and J.V. Smith, Occurrence of diamond in a mica-garnet lherzolite xenolith from kimberlite, Nature, 254, 580-581, 1975.

Dawson, J.B., J.J. Gurney, and P.J. Lawless, Paleogeothermal gradients derived from xenoliths in kimberlite, Nature, 257, 299-300, 1975.

Eggler, D.H., and M.E. McCallum, A geotherm from megacrysts in the Sloan kimberlite pipes, Colorado, Carnegie Inst. Washington Geophys. Lab. Yrbk., 75, 538-541, 1976.

Erlank, A.J., Distribution of potassium in mafic and ultramafic nodules, in Ann. Rept. of the Dept. of Terrestrial Magnetism, Carnegie Inst. of Washington, 68, 433-438, 1970.

Erlank, A.J., Kimberlite potassic richterite and the distribution of potassium in the upper mantle, in Extended Abstracts, First International Kimberlite Conference, Cape Town, South Africa, 103-106, 1973.

Goetze, C., Sheared lherzolites: From the point of view of rock mechanics, Geology, 3, 172-173, 1975.

Green, D.H., Conditions of melting basanite magma from garnet peridotite, Earth. Planet. Sci. Letts., 17, 456-465, 1973.

Green, H.W., and Y. Guenguen, Origin of kimberlite pipes by diapiric upwelling in the upper mantle, Nature, 249, 617-620, 1974.

Gurney, J.J., B. Harte, and K.G. Cox, Mantle xenoliths in the Matsoku kimberlite pipe, Phys. Chem. Earth, 9, 507-523, 1975a.

Gurney, J.J., J.B. Dawson, B. Harte, and P.J. Lawless, The bulk chemical composition of peridotite-facies rocks from the Matsoku and Bultfontein pipes, in Extended Abstracts, Kimberlite Symposium, Cambridge England, 3-5, 1975b.

Haggerty, S.E., Opaque mineral oxides in terrestrial igneous rocks, Mineral. Soc. Amer., Oxide Minerals Short Course Notes, 3, Hg 101-Hg 300, 1976.

Harte, B., Kimberlite nodules, upper mantle petrology and geotherms, Trans. Royal Soc. London Series A (in press) 1977a.

Harte, B., Rock nomenclature with particular relation to deformation and recrystallisation textures in olivine bearing xenoliths, Jour. Geol., 85, 279-288, 1977b.

Harte, B., K.G. Cox, and J.J. Gurney, Petrography and geological history of upper mantle xenoliths from the Matsoku kimberlite pipe, Phys. Chem. Earth, 9, 477-507, 1975.

Harte, B., and J.J. Gurney, Ore mineral and phlogopite mineralisation within ultramafic nodules from the Matsoku kimberlite pipe, Lesotho, Carnegie Inst. Washington Yrbk., 74, 528-536, 1975.

Howells, S., and M.J. O'Hara, Low solubility of alumina in enstatite and uncertainties in estimated palaeogeotherms, Proc. Royal. Soc. London Series A (In Press), 1977.

Kushiro, I., Partial melting of garnet lherzolites in kimberlites at high pressures, in Lesotho Kimberlites, edited by P.H. Nixon, Lesotho National Development Corporation, Maseru, 1973.

MacGregor, I.D., The system $MgO-Al_2O_3-SiO_2$: solubility of Al_2O_3 in enstatite for spinel and garnet peridotite compositions, Amer. Mineral, 59, 110-119, 1974.

McCallum, M.E., and D.H. Eggler, Diamonds in an upper mantle peridotite nodule from kimberlite in Southern Wyoming, Science, 192, 253-256, 1976.

Mercier, J.C., and N.L. Carter, Pyroxene geotherms, Jour. Geophys. Res., 80, 3349-3362, 1975.

Meyer, H.O.A., Chrome Pyrope: An inclusion in natural diamond, Science, 160, 1446-1447, 1968.

Meyer, H.O.A., and F.R. Boyd, Composition and origin of crystalline inclusions in natural diamonds, Geochemica Cosmochimica Acta, 36, 1255-1273, 1972.

Meyer, H.O.A., and D.P. Svisero, Mineral inclusions in Brazilian diamonds, Phys. Chem. Earth, 9, 785-795, 1975.

Mitchell, R.H., Ultramafic xenoliths from the Elwin Bay kimberlite, the first Canadian palaeogeotherm, Canad. Jour. Earth. Sci., 14, 1202-1210, 1977.

Mysen, B.O., and I. Kushiro, Compositional variation of coexisting phases with degree of melting of peridotite under upper mantle conditions, Carnegie Inst. Washington Geophys. Lab. Yrbk., 75, 546-555, 1976.

Nixon, P.H., and F.R. Boyd, Petrogenesis of the granular and sheared ultrabasic nodule suite in kimberlites, in Lesotho Kimberlites, edited by P.H. Nixon, Lesotho National Development Corporation, Maseru, 48-56, 1973.

Norrish, K., and J.T. Hutton, An accurate X-ray spectrographic method for the analysis of a wide range of geological samples, Geochimica Cosmochimica Acta, 33, 431-453, 1969.

O'Hara, M.J., M.J. Saunders, and E.L.P. Mercy, Garnet peridotite, primary ultrabasic magma and eclogite; interpretation of upper mantle processes in kimberlite, Phys. Chem. Earth, 9, 571-604, 1975.

Parmentier, E.M., and D.L. Turcotte, An explanation of the pyroxene geotherm based on plume convection in the upper mantle, Earth Planet. Sci. Letts. 24, 209-212, 1974.

Prinz, M., D.V. Manson, P.F. Hlava, and K. Keil, Inclusions in diamonds: garnet lherzolite and eclogite assemblages, Phys. Chem. Earth, 9, 797-816, 1975.

Shimizu, N., Rare earth elements (REE) in garnets and clinopyroxenes from garnet lherzolite nodules in kimberlites, in Ann. Rept. of the Dept. of Terrestrial Magnetism, Carnegie Inst. Washington, 73, 954-961,1974.

Sobolev, N.V., Yu.G. Lavrent'ev., L.I. Pospelova, and V.S. Sobolev, Chrome pyrope in diamonds from Yakutia, Dokl. Akad. Nauk SSSR, 189, 162-165, 1969.

Williams, A.F., The Genesis of the Diamond, Ernest Benn. Ltd., London, vol. 1. 1932.

Wyllie, P.J., Chairmans Summary in Extended Abstracts, Kimberlite Symposium, Cambridge, England, 28-31, 1975.

THE PETROLOGY AND GEOCHEMISTRY OF ULTRAMAFIC NODULES FROM PIPE 200, NORTHERN LESOTHO

D. A. Carswell

Department of Geology, University of Sheffield, Sheffield, U.K.

D. B. Clarke

Department of Geology, Dalhousie University, Halifax, Canada

R. H. Mitchell

Department of Geology, Lakehead University, Thunder Bay, Canada

Abstract. The Pipe 200 ultramafic nodule suite is dominated by relatively depleted garnet- and/or chromite-bearing lherzolites (whole rock 100 Mg/(Mg+Fe) 91.0-93.5). These contain olivines (Fa 7.0-8.6), orthopyroxenes (Fs 6.3-7.7) with 0.54-1.04 wt.% Al_2O_3, clinopyroxenes with consistent 100 Ca/(Ca+Mg) (46.0-48.4) but variable Al_2O_3, (1.64-3.99 wt.%), Cr_2O_3 (1.59-4.25 wt.%) and Na_2O (0.81-3.14 wt.%), chrome pyrope garnets and/or primary magnesiochromites. By contrast four ultradepleted chromite harzburgites/lherzolite (100 Mg/(Mg+Fe) 95.0-95.4) contain more magnesian olivines (Fa 4.9-5.0) and orthopyroxenes (Fs 4.1-4.3).

Provisionally preferred P/T estimates for the garnet-bearing lherzolites are 906-949°C and 24.5-30.8 kb, the temperature range being somewhat wider (873-962°C) if one also includes lherzolites in which garnet is absent but is considered likely to have been originally present. An ultradepleted chromite lherzolite nodule, which is unlikely to have ever contained garnet, yields an equivalent equilibration temperature of 950°C. It is concluded that these various ultramafic nodules may all have been derived from much the same depth zone in the mantle (roughly 70-100 km below surface on preferred estimates), the presence or absence of primary chromite relative to garnet being related to shifts in the position of the spinel lherzolite/garnet lherzolite transition boundary for compositions with various $Cr_2O_3/(Cr_2O_3+Al_2O_3)$ ratios. The observed nodule suite does, however, contain one harzburgite nodule with more aluminous primary spinel and orthopyroxene which seems likely to have equilibrated under significantly lower P/T conditions and hence have been derived from depths < 50 km.

Mineralogical and chemical comparison between the Pipe 200 ultramafic nodules and those from other pipes in northern Lesotho and in the Kimberley area, as well as in the Lashaine volcano, indicates that the mantle beneath Pipe 200 and much of northern Lesotho may be extensively depleted down to greater depths, perhaps reflecting the intensive Karroo volcanism in that area.

Introduction

Pipe 200 in northern Lesotho is situated close to the Letele Pass road about 8 km west of the better known kimberlites of the Kao area. The main intrusion of the pipe outcrops on the western bank of the Malibamatso River (see Kresten and Dempster, 1973; including Plate 41B). In addition to numerous nodules of obvious crustal origin (basalts, sandstones, gneisses, etc.) this kimberlite pipe also contains common ultramafic nodules of likely mantle derivation. These are mainly lherzolites with or without small amounts of purple garnet.

We decided to study the whole rock and mineral chemistry of these particular nodules, in order to compare them with the contrasting granular and sheared ultramafic nodule types described by Nixon and Boyd (1973) from other kimberlite pipes in northern Lesotho, notably Thaba Putsoa and Mothae. We also wished to determine whether the presence or absence of garnet in these nodules is related to varying whole rock chemistry or to different depths of derivation in the mantle.

Petrography

All thirty-four ultramafic nodules which we have examined from Pipe 200 are of the common 'coarse grained' or 'granular' textured types as defined by Boullier and Nicholas (1973) and Boyd and Nixon (1972), respectively. Olivines and

TABLE 1. Whole Rock Analyses of Pipe

Sample Number	PTH 107	PHT 108	PTH 201	PTH 202	PTH 203	PTH 205	PTH 207	PTH 208	PTH 210	PTH 400	PTH 401
Rock Type	CH	(G)CL	SH	(G)CL	CL	(G)CL	GCL	GL	GP	GL	GCL
SiO_2	40.49	43.69	44.10	46.07	43.42	43.68	45.69	46.29	52.52	45.55	44.94
TiO_2	0.01	0.04	0.17	0.17	0.35	<0.01	0.02	0.16	0.13	0.04	0.15
Al_2O_3	0.24	0.51	0.47	0.54	0.14	0.24	0.66	1.18	5.10	0.41	0.35
Cr_2O_3	0.09	0.18	0.20	0.23	0.20	0.19	0.34	0.28	0.33	0.16	0.15
Fe_2O_3	2.45	2.58	1.73	1.74	1.49	2.91	1.88	1.97	1.53	2.75	2.45
FeO	2.27	4.54	4.09	4.57	2.80	3.45	4.20	3.95	2.52	4.93	3.37
MnO	0.06	0.12	0.10	0.11	0.07	0.11	0.11	0.11	0.14	0.10	0.09
NiO	0.28	0.29	0.29	0.31	0.34	0.33	0.28	0.29	0.07	0.29	0.31
MgO	47.48	44.25	45.66	43.50	48.17	44.46	42.83	42.17	22.84	42.10	45.06
CaO	0.07	0.92	0.59	0.58	0.18	0.29	0.71	0.73	13.11	1.32	0.39
Na_2O	0.02	0.14	0.03	0.12	0.06	0.06	0.09	0.07	0.43	0.26	0.06
K_2O	0.02	0.20	0.07	0.12	0.04	0.05	0.05	0.03	0.09	0.13	0.04
P_2O_5	0.01	0.03	0.02	0.04	0.04	0.03	0.03	0.04	0.03	0.02	0.05
$H_2O^{\pm}$	5.94	2.52	2.26	2.41	2.11	4.04	3.05	2.47	1.11	1.93	2.53
CO_2	NIL	NIL	0.48	NIL	NIL	NIL	NIL	NIL	0.27	NIL	0.22
TOTAL	99.43	100.01	100.26	100.51	99.41	100.04	99.94	99.74	100.22	99.99	100.16
100 Mg/(Mg+Fe)	95.0	92.0	93.5	92.7	95.4	92.7	92.8	92.9	91.3	91.0	93.5
100 Cr/(Cr+Al)	20.1	19.1	22.2	22.2	48.9	34.7	25.7	13.7	4.2	20.7	22.3

Rock Types: CL - Chromite Lherzolite CH - Chromite Harzburgite SH - Spinel Harzburgite
GCL - Garnet Chromite Lherzolite GL - Garnet Lherzolite GP - Garnet Pyroxenite
(G) - In parentheses indicates that garnet thought likely to be a member of the original primary assemblage, although now absent

enstatites do, however, show evidence of slight deformation with blurred kink band development and more rarely grain flattening giving a crude tabular texture.

Mineralogically most of the Pipe 200 ultramafic nodules are lherzolites with forsteritic olivine (65-75 vol.%) and enstatite (20-30 vol.%) the two dominant minerals. In the analytical tables peridotite nodules are classified as lherzolites if there was any chrome diopside visible in thin section or hand specimen. Amounts of the bright green chrome diopside phase are always small (< 5 vol.%) more typically only 1-2 vol.%. Likewise the amounts of the other characterising phases garnet and/or chromite, are also small (1-2 vol.% on average).

The chrome pyrope garnets invariably have extensive semi-opaque kelyphite alteration rims often coarsening and clearing somewhat towards their outer margin so that one can discern numerous small yellowish or reddish brown spinel grains associated with pale coloured amphibole and/or phlogopite. PTH 202 and 205 now lack garnet but are designated as '(garnet) chromite lherzolites', since characteristic kelyphite alteration patches indicate its previous presence.

Where present, the large discrete deep red brown primary chrome spinels (chromites) are petrographically distinct from the smaller secondary spinels associated with the breakdown of the garnets. PTH 201 is designated a spinel harzburgite on analytical evidence that the primary spinels are more aluminous than in the other nodules. Of the thirty-four ultramafic nodules studied, sixteen contained only garnet (including analysed samples PTH 102, 210, 400, 403, 404, 405, 406 and 407), eight contained both primary chromite and garnet (including analysed samples PTH 207, 401, 409 and 410), two have primary chromite and scarce kelyphite patches indicating the previous presence of garnets (PTH 202 and 205), and eight apparently contained only chromite (analysed samples PTH 107, 108, 201, 203, 301, 302, 303 and 304), although of those PTH 108 and 304 may arguably have also originally contained garnet.

The garnet pyroxene nodule, PTH 210, is exceptional in that it lacks olivine and has high

200 Ultramafic Nodules.

PTH 403	PTH 404	PTH 405	PTH 406	PTH 407	PTH 409
GL	GL	GL	GL	GL	GCL
45.73	45.11	46.22	45.82	45.06	44.82
0.04	0.04	0.02	0.03	0.10	0.17
1.16	0.73	1.06	1.64	1.09	0.58
0.40	0.32	0.32	0.32	0.35	0.28
2.98	2.09	2.62	2.27	2.13	2.64
4.30	3.93	4.16	4.59	4.29	3.41
0.12	0.11	0.11	0.12	0.11	0.11
0.26	0.29	0.28	0.29	0.29	0.30
41.13	43.45	42.81	41.54	43.32	43.01
0.98	0.95	0.57	1.22	0.85	0.43
0.13	0.12	0.10	0.12	0.11	0.10
0.12	0.04	0.07	0.09	0.12	0.09
0.04	0.06	0.03	0.03	0.05	0.03
2.63	2.36	1.65	1.66	2.32	4.05
NIL	NIL	NIL	NIL	0.23	NIL
100.02	99.60	100.02	99.74	100.42	100.02
91.3	93.0	92.1	91.8	92.6	93.0
18.8	22.7	16.8	11.6	17.7	24.4

contents of both garnet and clinopyroxene, approximately 15 and 50 vol.% respectively. A garnet granulite nodule and four eclogite nodules from Pipe 200 are described elsewhere by Griffin et al. (1978).

All ultramafic nodules are relatively fresh but do show some serpentinisation and additional infiltration metasomatism effects reflected in the growth of small amounts of secondary phlogopite, carbonate, apatite, ilmenite and/or rutile.

Rock Chemistry

Whole rock analyses for seventeen nodule samples are presented in Table 1. Samples were analysed by a combination of 'wet' chemical techniques (spectrophotometric, gravimetric, titration and atomic absorption) at the University of Sheffield.

Excluding for the moment the garnet pyroxenite nodule PTH 210, it is apparent that the suite of Pipe 200 ultramafic nodules taken as a whole is characterised by high Mg/(Mg+Fe) and Cr/(Cr+Al) ratios, and low CaO and Na_2O contents. This reflects their refractory and highly depleted chemical nature in terms of their basalt-yielding potential. The garnet-bearing lherzolites from Pipe 200 are of broadly similar chemistry (see Table 2) to the granular textured garnet lherzolites in the nearby Thaba Putsoa, Mothae and Liqhobong pipes, but contrast strikingly with the relatively fertile, sheared garnet lherzolite nodules in those pipes (Nixon and Boyd, 1973). They are also somewhat more depleted than in the common granular textured garnet lherzolites found in the Kimberley area pipes and the unbanded coarse grained granular or flaser textured garnet lherzolites from the Matsoku pipe (Carswell and Dawson, 1970; Cox et al., 1973). It should be noted that some Na_2O values for Matsoku ultramafic nodules, as given in Table 26 of Cox et al. (1973), are almost certainly in serious error. Calculation of whole rock compositions using the mineral analyses and modes indicates that the Na_2O values quoted are often unrealistically low. Hence the average Na_2O value quoted for the Matsoku 'common peridotites' in Table 2 involves a correction to the Na_2O figures for the analyses taken from Cox et al., (1973).

Consideration of the internal chemical variation within the suite of analysed Pipe 200 ultramafic nodules (Table 1) shows that there are fairly constant differences in whole rock chemistry between the garnet-only lherzolites, garnet + chromite lherzolites, and chromite-only harzburgites/lherzolites. Mg/(Mg+ΣFe) and Cr/(Cr+Al) increase and Al_2O_3, CaO and Na_2O contents decrease for the nodule types in that order, with decreasing amounts of chrome diopside as well as garnet. In fact, of the ultra-depleted chromite harzburgites/lherzolites which lack petrographic evidence suggesting that they originally contained garnet, only PTH 203 contains any visible chrome diopside and even then as confirmed by the low CaO and Na_2O contents, only trace amounts. There is also a general increase in MgO/SiO_2 ratio from the garnet lherzolites to the chromite harzburgites/lherzolites reflecting increased modal contents of olivine relative to enstatite. Thus, as illustrated on Figure 1, the chromite harzburgites/lherzolites have more depleted compositions than the garnet lherzolites, indicating that modal mineralogy may be mainly controlled by whole rock chemistry.

The fact that on Figure 1 the composition points for PTH 202 and 108 plot close to the boundary between the garnet-only and garnet + chromite fields supports our petrographic interpretation that these nodules originally contained garnet. PTH 202 (and likewise 205) contains kelyphite patches which indicate the original presence of garnet, whilst PTH 108 contains clusters of small chromite grains associated with coarser phlogopite and chrome diopside which arguably may have formed at the expense of original garnet. On the other hand, the spinel harzburgite PTH 201, which likewise plots in the garnet + chromite field, lacks any petrographic evidence to suggest that it may originally have contained garnet.

As expected, the garnet pyroxenite nodule PTH 210 contrasts strikingly in its chemical composi-

TABLE 2. Comparison of Garnet Lherzolites from Pipe 200 with Those from Other Localities.

Rock Type	A	B	C	D	E	F
SiO_2	46.83	45.80	43.89	47.44	46.56	44.61
TiO_2	0.08	0.03	0.16	0.07	0.08	0.08
Al_2O_3	0.91	0.51	1.85	1.62	1.90	1.85
Cr_2O_3	0.30	0.21	0.22	0.41	0.31	0.44
*FeO	6.43	6.23	8.58	6.48	6.60	6.95
MnO	0.11	0.10	0.12	0.11	0.11	0.12
NiO	0.30	0.18	0.23	0.33	0.29	0.37
MgO	43.97	46.06	42.94	42.39	42.67	44.20
CaO	0.84	0.82	1.86	1.12	1.14	1.22
Na_2O	0.12	0.06	0.21	0.14	0.18	0.17
K_2O	0.08	0.03	0.03	0.11	0.14	0.08
P_2O_5	0.04	0.00	0.00	0.03	0.04	0.08

* Total Fe expressed as FeO and analyses recalculated anhydrous to 100% in all cases.

	A	B	C	D	E	F
100 Mg/(Mg+Fe)	92.4	92.9	89.9	92.1	92.0	92.0
100 Cr/(Cr+Al)	19.4	23.3	7.9	14.8	10.1	14.2

A - Garnet lherzolites from Pipe 200 (mean of 10).

B - Granular garnet lherzolites from Thaba Putsoa, Mothae and Liqhobong (mean of 4) - Nixon and Boyd (1973).

C - Sheared garnet lherzolites from Thaba Putsoa and Mothae (mean of 6) - Nixon and Boyd (1973).

D - Unbanded garnet lherzolites (common peridotites) from Matsoku (mean of 5) - Cox *et al.* (1973); Carswell and Dawson (1970).

E - Common type garnet lherzolites from Kimberley Mines (mean of 23) - Analyses from Holmes (1936), Ito and Kennedy (1967), Whitfield (1971), Berg and O'Hara (1973), Carswell and Dawson (1970), and unpublished data. The data presented by Chen (1971) are deliberately excluded on the grounds of incompatibility with data from other sources, in particular Mg/(Mg+Fe) values are consistently lower at equivalent CaO, Al_2O_3 and Cr/(Cr+Al) values suggesting that the analytical MgO values are too low.

F - Garnet lherzolites from Lashaine volcano, Tanzania (mean of 5) - Rhodes and Dawson (1975), Ridley and Dawson (1975).

tion with the peridotite nodule suite, especially in its much higher Al_2O_3, CaO and Na_2O contents and lower Cr/(Cr+Al) ratio. In terms of rock chemistry, as well as in modal mineralogy, this nodule resembles certain of the Matsoku pyroxenite/peridotite nodules (Cox *et al.*, 1973) interpreted as mantle cumulates of moderately evolved magmatic liquids (Gurney *et al.*, 1975). However, whilst the Matsoku nodules show a striking trend of decreasing Mg/(Mg+Fe) with increased Al_2O_3, CaO and Na_2O contents, this is not apparent in the Pipe 200 garnet pyroxenite. Its 100 Mg/(Mg+Fe) ratio (91.3) is in fact within the range of values (91.0-93.5) observed in the common garnet lherzolites from Pipe 200. It also has a significantly lower Na_2O/CaO ratio (0.033) when compared with the proposed Matsoku cumulate nodules (mean 0.093) and does not show the banded texture seen in many of the latter.

The low, but nevertheless variable, contents

of 'incompatible' elements such as Ti, K and P in the Pipe 200 nodules merit some comment. The variability in the contents of these elements (especially of Ti) in the various peridotite nodules and comparison with contents in the various primary mineral phases indicate that the base levels for these elements in these mantle-derived samples prior to incorporation in the kimberlite are likely to have been extremely low, probably in the order of 0.01-0.02 wt.% for the oxides of each of these elements. Concentrations above these low base levels seem likely to have resulted from infiltration metasomatism (Harte et al., 1975) due to contamination by the enclosing kimberlite magma.

Mineral Chemistry

The minerals in these nodules were analysed by electron microprobe mostly at Dalhousie University but some at the University of Edinburgh. Most quoted analyses are averages of several point analyses of different grains. The data were acquired on both conventional crystal spectrometry and energy dispersive systems and as a result of background uncertainty and detector sensitivity, considerable uncertainty is associated with concentrations of less than 0.10 wt.%.

Olivine

Olivines in all nodules are highly forsteritic (Table 3). Those in the ultra-depleted chromite harzburgite/lherzolite nodules have lower fayalite contents (4.9-5.0% Fa) than those in the garnet ± chromite lherzolites (7.0-8.6% Fa). The fayalite contents of olivines in the (garnet) chromite lherzolite xenoliths (7.4-8.6% Fa) which now lack garnet (namely PTH 108, 202, 205 and 304) are in line with the latter rather than the former. As with other aspects of its mineral chemistry, the spinel harzburgite PTH 201 is exceptional as it contains olivine with 7.0% Fa composition yet lacks evidence that it ever contained garnet.

Orthopyroxene

Orthopyroxenes in all nodules are enstatites (Table 4) although again 100 Fe/(Fe+Mg) values vary somewhat between the different nodule types. The group of ultradepleted chromite harzburgites/lherzolite (PTH 107, 203, 301, 302 and 303) contain orthopyroxenes with 4.1-4.3% Fs in contrast with the garnet ± chromite lherzolites in which orthopyroxene compositions are 6.3-7.7% Fs. Orthopyroxene compositions (6.5-7.5% Fs) in those chromite lherzolites which arguably contained garnet originally are in line with the latter rather than the former. All orthopyroxene analyses, except PTH 207(S) in Table 4 are of the common large primary grains. Such orthopyroxenes in garnet ± chromite lherzolites are characterised by low Al_2O_3 (0.70-1.04 wt.%), moderate CaO (0.33-0.53 wt.%) and high Cr_2O_3 (0.25-0.50 wt.%) contents. Small amounts of discernible secondary orthopyroxene occur associated with secondary chrome spinel at the outer margins of the kelyphite rims around some garnets. Such secondary orthopyroxene, as represented by analysis PTH 207(S) is quite distinct in

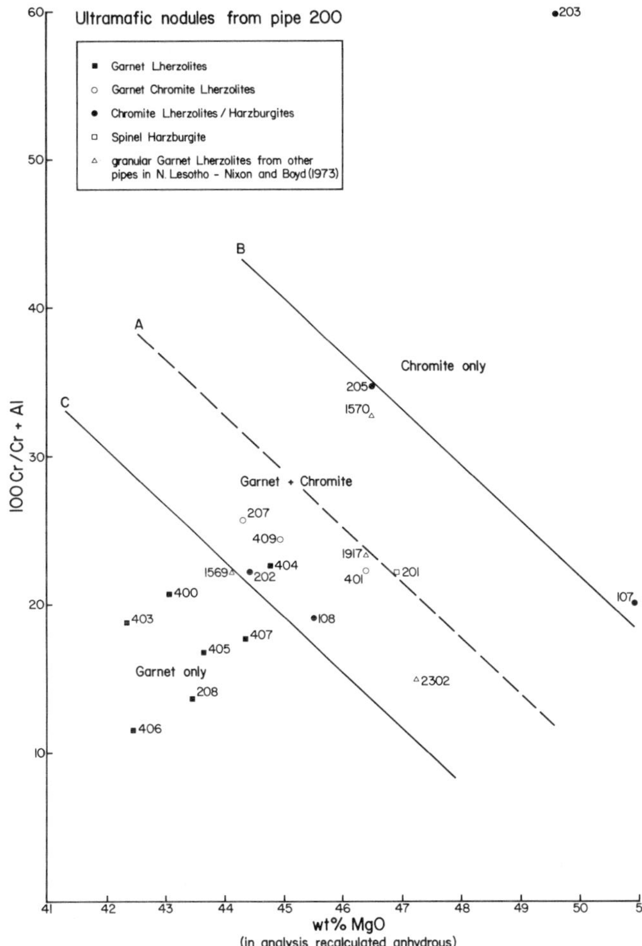

Fig. 1. Whole rock composition plot of wt.% MgO against 100 Cr/(Cr+Al) (in analyses recalculated anhydrous to 100% and with total iron as FeO) for Pipe 200 ultramafic nodules. Compositions for comparable granular textured garnet lherzolites from Thaba Putsoa, Mothae and Liqhobong (Nixon and Boyd, 1973) are also shown. These authors give an analysis for primary chromite in nodule 1570 but do not make it clear whether chromite is absent in the other nodules. Boundaries A and C correspond to those defining the chromite-only, garnet + chromite and garnet-only composition fields for peridotite nodules from the Lashaine volcano (Reid et al., 1973) whilst B represents the approximate position of the chromite + garnet/chromite-only boundary for the Pipe 200 nodules.

TABLE 3. Analyses of Olivines in Pipe

Sample Number	PTH 102	PTH 107	PTH 108	PTH 201	PTH 202	PTH 203	PTH 205	PTH 207	PTH 301	PTH 302	PTH 303	PTH 304	PTH 400
Rock Type	GL	CH	(G)CL	SH	(G)CL	CL	(G)CL	GCL	CH	CH	CH	(G)CL	GL
SiO_2	41.9	41.9	41.1	41.9	41.1	41.8	41.7	42.1	40.7	41.6	41.5	41.0	40.4
TiO_2	0.07	n.d.	0.00	0.00	0.00	0.01	0.08	0.05	n.d.	n.d.	n.d.	n.d.	0.02
Al_2O_3	n.d.	n.d.	0.03	0.04	n.d.	0.04	n.d.	n.d.	n.d.	n.d.	n.d.	n.d.	0.04
Cr_2O_3	0.07	n.d.	0.01	0.10	0.04	0.05	n.d.	0.11	n.d.	n.d.	n.d.	n.d.	0.03
*FeO	7.77	4.97	8.26	6.95	7.45	4.78	7.35	7.75	4.86	5.02	4.97	7.34	8.42
MnO	0.13	n.d.	0.12	0.01	0.02	0.06	n.d.	0.13	n.d.	n.d.	n.d.	n.d.	0.09
NiO	n.d.	0.29	0.44	0.43	n.d.	0.40	n.d.	n.d.	n.d.	n.d.	n.d.	n.d.	0.45
MgO	48.5	52.9	49.6	51.8	50.5	51.9	50.4	50.1	52.8	53.7	53.5	51.5	50.0
CaO	0.06	n.d.	0.01	0.02	n.d.	0.02	0.07	0.05	n.d.	n.d.	n.d.	n.d.	0.03
Total	98.5	100.1	99.6	101.2	99.1	99.0	99.6	100.3	98.4	100.3	100.0	99.8	99.5
100 Fe/(Fe+Mg)	8.3	5.0	8.5	7.0	7.6	4.9	7.6	8.0	4.9	5.0	5.0	7.4	8.6

* All Fe taken as FeO
n.d. Not determined
Sample nomenclature as in Table 1

composition from the primary enstatites with a higher 100 Fe/(Fe+Mg) content (11.4% Fs), an appreciably higher Al_2O_3 content (2.81 wt.%) and much lower CaO (0.16 wt.%) and Cr_2O_3 (0.09 wt.%) contents. Primary orthopyroxenes in PTH 201 and 210 are exceptional in their appreciably higher Al_2O_3 contents, 1.45 and 2.28 wt.% respectively.

Clinopyroxene

Clinopyroxene analyses (Table 5) show variable contents of Al_2O_3 (1.64-3.99 wt.%), Cr_2O_3 (1.59-4.25 wt.%) and Na_2O (0.81-3.14 wt.%). It is clear from Figure 2 that the Al_2O_3 contents vary sympathetically with Na_2O in individual clinopyroxene point analyses, reflecting variable jadeite contents. In most lherzolite nodules individual analyses both of different grains and different points within individual grains show little variation and in these cases the single clinopyroxene analysis given in Table 5 represents the mean value. However, in both PTH 202 and 204 although grain cores are usually of fairly uniform composition, striking jadeite depletion was observed in certain point analyses, notably those from grain rims. For both nodules, the rim analysis given in Table 5 corresponds to the individual point analysis indicating the most extreme jadeite depletion noted in clinopyroxene grains in that rock. The clinopyroxene grains in these nodules have clear cores, but distinctly cloudy margins. The marginal clouding often extends along curved fractures into or even right across individual grains and appears to be due to large numbers of small tubular shaped fluid inclusions. Similar petrographic features and associated jadeite depletion have previously been observed in garnet lherzolite nodules from the Kimberley mines (Carswell, 1975).

The observed jadeite depletion is not accompanied by kosmochlor ($NaCrSi_2O_6$) depletion. The chromite lherzolite PTH 203 contains clinopyroxene with the highest Cr_2O_3 content and this correlates with highest whole rock Cr/(Cr+Al) and Na_2O/CaO ratios. However, plots of Cr/(Cr+Al) Cpx against Cr/(Cr+Al) rock and of wt.% Al_2O_3 and wt.% Cr_2O_3 in clinopyroxene against Na_2O/CaO in the rock for other nodules do not show consistent correlations between rock and clinopyroxene compositions. We conclude that the overall variation in Al_2O_3, Cr_2O_3 and Na_2O contents observed in the clinopyroxenes in these Pipe 200 ultramafic nodules largely reflects the effects of secondary alteration (notably jadeite loss) superimposed on primary chemical and physical controls.

There is only limited variation in 100 Ca/(Ca+Mg) (46.0-48.4 excluding rim analyses) amongst the analysed primary clinopyroxenes in the lherzolite nodules. Rim analyses do not have consistently lower or higher Ca/(Ca+Mg) ratios than corresponding cores. This, together with the fact that there is no significant correlation between Ca/(Ca+Mg) and Na_2O content in the clinopyroxenes in the various lherzolite nodules, points to broadly similar initial equilibration temperatures for the nodule suite and the non-equilibration of the jadeite depleted clinopyroxenes.

In the lherzolite nodules much of what little clinopyroxene is present tends to be closely associated spatially with garnets and its kelyphite

200 Ultramafic Nodules.

PTH 401	PTH 403	PTH 404	PTH 405	PTH 406	PTH 407	PTH 409	PTH 410
GCL	GL	GL	GL	GL	GL	GCL	GCL
41.6	41.6	41.5	41.2	41.5	40.8	41.2	41.2
0.01	0.04	0.07	0.01	0.03	n.d.	0.03	0.03
0.04	0.03	0.04	0.03	0.03	0.03	0.03	0.03
n.d.	0.04	0.09	0.04	n.d.	n.d.	n.d.	n.d.
6.88	7.73	7.36	7.37	7.14	7.44	7.16	7.32
0.06	n.d.	0.12	0.11	0.08	0.09	0.11	0.08
n.d.	n.d.	n.d.	0.45	n.d.	n.d.	n.d.	n.d.
51.6	49.5	50.2	50.8	50.1	50.9	49.8	51.3
0.01	0.05	0.05	0.02	0.04	0.04	0.04	0.03
100.2	99.0	99.4	100.0	98.9	99.2	98.4	99.8
7.0	8.1	7.6	7.5	7.4	7.6	7.5	7.4

alteration products or with clusters of phlogopite and spinel grains which arguably may be secondary after garnet. Such clinopyroxene may actually occur within kelyphite or more typically may form a narrow collar around it. This suggested that at least some of the clinopyroxene might be secondary and related to garnet breakdown. However, only in the case of nodules PTH 207 and 403 were we able to detect clinopyroxenes of secondary aluminous types--analyses 207 (S1 and S2) and 403 (S1 and S2) in Table 5. Such clinopyroxenes occur as very small grains on the outermost margins of the semi-opaque kelyphite alteration zones around garnets and have variable compositions reflecting their unequilibrated nature. However, most clinopyroxene grains immediately adjacent to kelyphite zones and also those in clusters with spinels and phlogopites (as in PTH 108) are chrome diopsides essentially indistinguishable in composition from discrete primary grains. Either such clinopyroxene has re-equilibrated with the lherzolite assemblage following garnet breakdown, or poikiloblastic habit of clinopyroxene towards garnet is a feature which pre-dates kelyphite formation in many nodules. It is possible that the marginal clouding and jadeite depletion of clinopyroxene grains as observed in nodules with extensive kelyphite development, such as PTH 202 and 204, are adjustments in composition of the primary clinopyroxene grains which are related to garnet breakdown, and that both features are a consequence of the combined effects of metasomatism and decompression resulting from incorporation in kimberlite and subsequent diatreme emplacement.

The clinopyroxene in the garnet pyroxenite nodule PTH 210 is distinctly different in composition from clinopyroxenes in lherzolite nodules. It has low Al_2O_3, Cr_2O_3 and Na_2O contents coupled with high 100 Ca/(Ca+Mg) = 50.9. These compositional differences, as with those observed in coexisting orthopyroxene and garnet, may simply reflect the different paragenesis and whole rock composition, rather than different equilibration conditions.

Garnet

Garnet analyses (Table 6) for the lherzolite nodules fall within the range of the common chrome pyropes (Group 9 - Dawson and Stevens, 1975) found elsewhere in garnet lherzolite nodules and as common xenocrysts within kimberlites. However, with the exception of PTH 400 they have somewhat higher Cr_2O_3 and CaO contents (see Figure 3) than the most frequent garnets found within the common garnet lherzolite nodules in the Kimberley area and Matsoku pipes (Carswell and Dawson,1970; Cox et al., 1973; Gurney and Switzer, 1973) and in Yakutian pipes (Sobolev et al., 1973). On the other hand the garnets from the granular garnet lherzolite nodules of the Thaba Putsoa and Mothae pipes (Nixon and Boyd, 1973) have broadly comparable Cr_2O_3 and CaO contents.

It is apparent that the more depleted garnet lherzolites with higher Cr/(Cr+Al) and Mg/(Mg+Fe) ratios have garnets which also have higher Cr/(Cr+Al) ratios and in addition show a sympathetic increase in CaO. This garnet composition trend is perhaps also reflected, although in a more extreme fashion, by the analysed chrome pyrope garnet xenocrysts from the nearby Kao kimberlite (Hornung and Nixon, 1973). We suggest that all the Kao garnets plotted on Figure 3 are from disaggregated lherzolite nodules, those with lowest Cr_2O_3 and CaO contents being from nodules of comparable composition to the common type in the Kimberley and Matsoku pipes, intermediate ones from comparatively depleted nodules of the common Pipe 200 type, whilst the extreme chrome-rich green garnets are likely to be from nodules with extremely high Cr/(Cr+Al) ratios. Garnets showing a comparable range in chemistry and thought to be related to similar lherzolite parageneses have been reported as xenocrysts in Yakutian pipes, notably Udachnaya (Sobolev et al., 1973). In addition garnets, intermediate in composition between those in the Pipe 200 nodules and the extreme Kao green garnet types, have been recorded by Gurney and Switzer (1973) as xenocrysts in the Finsch kimberlite pipe, South Africa. These observations indicate that the garnets of Groups 9 and 12 (Dawson and Stevens, 1975) should best be considered as one group, related to lherzolite parageneses with varying Cr/(Cr+Al) ratios.

The garnet in the garnet pyroxene nodule PTH 210 has an appreciably lower Cr_2O_3 content as well as somewhat higher FeO and MnO contents.

TABLE 4. Analyses of Orthopyroxenes

Sample Number	PTH 102	PTH 107	PTH 108	PTH 201	PTH 202	PTH 203	PTH 205	PTH 207(P)	PTH 207(S)	PTH 210	PTH 301
Rock Type	GL	CH	(G)CL	SH	(G)CL	CL	(G)CL	GCL	GCL	GP	CH
SiO_2	58.0	58.4	58.0	57.9	57.2	58.4	58.2	57.8	57.6	56.8	58.4
TiO_2	0.10	n.d.	0.05	0.00	0.01	0.02	0.08	0.08	0.05	n.d.	n.d.
Al_2O_3	0.86	0.95	0.59	1.45	1.04	0.79	0.74	0.70	2.81	2.28	0.76
Cr_2O_3	0.50	0.48	0.40	0.43	0.33	0.55	0.44	0.46	0.09	0.35	0.36
*FeO	4.83	2.86	5.06	4.52	4.59	2.90	4.77	4.76	7.05	5.94	2.87
MnO	0.14	n.d.	0.14	0.01	0.01	0.07	0.00	0.15	0.13	0.22	n.d.
NiO	n.d.	n.d.	n.d.	0.10	n.d.	0.13	n.d.	n.d.	n.d.	n.d.	n.d.
MgO	34.0	36.6	34.9	35.9	37.1	35.8	35.3	35.4	30.9	34.4	36.4
CaO	0.50	0.34	0.34	0.18	0.29	0.41	0.48	0.47	0.16	0.22	0.11
Na_2O	n.d.	n.d.	0.10	0.02	n.d.	0.16	n.d.	n.d.	n.d.	n.d.	n.d.
Total	98.9	99.6	99.6	100.5	100.6	99.2	100.0	99.8	98.8	100.2	98.8
100 Fe/(Fe+Mg)	7.4	4.2	7.5	6.6	6.5	4.3	7.0	7.0	11.4	8.8	4.2
100 Ca/(Ca+Mg)	1.0	0.6	0.7	0.4	0.6	0.8	1.0	0.9	0.4	0.5	0.2
Structural Formulae on the Basis of 6.000 Oxygens											
Si	2.005	1.987	1.995	1.969	1.948	1.997	1.989	1.984	2.003	1.950	1.998
Al^{IV}	0.000	0.013	0.005	0.031	0.042	0.003	0.011	0.016	0.000	0.050	0.002
Al^{VI}	0.035	0.025	0.019	0.027	0.000	0.029	0.019	0.012	0.115	0.042	0.029
Ti	0.003	--	0.001	0.000	0.000	0.001	0.002	0.002	0.001	--	--
Cr	0.014	0.013	0.011	0.012	0.010	0.015	0.012	0.012	0.002	0.009	0.010
Fe	0.140	0.081	0.145	0.129	0.131	0.083	0.136	0.137	0.205	0.171	0.082
Mn	0.004	--	0.004	0.000	0.000	0.002	0.000	0.004	0.004	0.006	--
Ni	--	--	--	0.003	--	0.004	--	--	--	--	--
Mg	1.750	1.857	1.789	1.818	1.884	1.827	1.799	1.809	1.598	1.762	1.856
Ca	0.019	0.012	0.013	0.007	0.011	0.015	0.018	0.017	0.006	0.008	0.004
Na	--	--	0.007	0.001	--	0.011	--	--	--	--	--
Total	3.968	3.988	3.989	3.996	4.026	3.985	3.986	3.994	3.935	3.999	3.981

* All Fe as FeO
n.d. Not determined

Spinel

Spinel analyses for various nodules are given in Table 7, and are plotted in a wt.% Cr_2O_3 vs. wt.% MgO variation diagram (Figure 4). Discrete primary spinels are low Al magnesiochromites, i.e. 64.9-81.4% M^{2+} Cr_2O_4, except in the spinel harzburgite PTH 201 where they are aluminous chrome spinels with a much lower 100 Cr/(Cr+Al) ratio. The primary chromites in the ultra-depleted harzburgite/lherzolite nodules plot in a slightly different composition field to those in garnet lherzolites (Figure 4) because of their higher absolute Cr_2O_3 and MgO contents. However primary chromites in Pipe 200 garnet lherzolites are similar in composition to those described from comparable granular textured garnet lherzolites in the nearby Thaba Putsoa and Mothae pipes (Nixon and Boyd, 1973) and to those in garnet and/or spinel lherzolite nodules from the Lashaine volcano in northern Tanzania (Reid et al., 1975). Spinels associated with kelyphitic alteration zones around garnets show a wide range of compositions. Those from the innermost parts of kelyphites (PTH 207A, 404, 405A, 407 and 410 in Table 7) are high Al chrome spinels (72.9-77.6% M^{2+} Al_2O_4), fairly analogous in composition to the primary spinels in the common lherzolite inclusions in basalts (Ross et al., 1954). These undoubtedly secondary spinels thus appear to be related to partial breakdown of the high pressure garnet lherzolite assemblage to a lower pressure spinel lherzolite assemblage. However, spinels from the outermost parts of kelyphite zones or immediately adjacent to them (PTH 207B, 403B and 405B in Table 7) often have intermediate 100 Cr/

in Pipe 200 Ultramafic Nodules.

PTH 302	PTH 303	PTH 304	PTH 400	PTH 401	PTH 403	PTH 404	PTH 405	PTH 406	PTH 407	PTH 409	PTH 410
CH	CH	(G)CL	GL	GCL	GL	GL	GL	GL	GL	GCL	GCL
58.9	59.3	58.7	57.8	58.1	58.0	58.9	57.8	57.8	57.3	57.7	58.0
n.d.	n.d.	n.d.	0.06	0.03	0.06	0.04	0.03	0.03	0.02	0.05	0.04
0.36	0.41	0.54	0.75	1.04	0.92	0.82	0.88	0.88	0.87	0.82	0.96
0.23	0.76	0.18	0.25	0.37	0.46	0.42	0.48	0.43	0.45	0.40	0.50
2.79	2.92	4.39	5.13	4.28	4.76	4.39	4.46	4.38	4.56	4.57	4.53
n.d.	n.d.	n.d.	0.11	0.11	0.00	0.11	0.09	0.11	0.10	0.12	0.10
n.d.	n.d.	n.d.	0.12	n.d.	n.d.	n.d.	0.11	n.d.	n.d.	n.d.	n.d.
36.7	36.8	35.7	34.7	35.8	34.6	35.0	35.2	35.0	35.6	34.8	35.7
0.22	0.18	0.34	0.48	0.33	0.47	0.53	0.44	0.43	0.45	0.42	0.46
n.d.	n.d.	n.d.	0.13	0.07	n.d.	0.10	0.16	0.08	0.14	0.15	0.17
99.2	99.9	99.8	99.5	100.2	99.2	100.3	99.6	99.1	99.5	99.0	100.4
4.1	4.3	6.5	7.7	6.3	7.2	6.6	6.6	6.6	6.7	6.9	6.6
0.4	0.4	0.7	1.0	0.7	1.0	1.0	0.9	0.9	0.9	0.9	0.9
2.008	2.009	2.003	1.991	1.980	1.997	2.004	1.984	1.991	1.973	1.992	1.976
0.000	0.000	0.000	0.009	0.020	0.003	0.000	0.016	0.009	0.027	0.008	0.024
0.014	0.016	0.022	0.021	0.022	0.034	0.033	0.020	0.027	0.008	0.025	0.015
--	--	--	0.002	0.001	0.002	0.001	0.001	0.001	0.001	0.001	0.001
0.006	0.007	0.005	0.007	0.010	0.013	0.011	0.013	0.012	0.012	0.011	0.013
0.080	0.083	0.125	0.148	0.122	0.137	0.125	0.128	0.126	0.131	0.132	0.129
--	--	--	0.003	0.003	0.000	0.003	0.003	0.003	0.003	0.004	0.003
--	--	--	0.003	--	--	--	0.003	--	--	--	--
1.865	1.858	1.816	1.782	1.821	1.775	1.774	1.802	1.798	1.826	1.791	1.813
0.008	0.007	0.012	0.018	0.012	0.017	0.019	0.016	0.016	0.017	0.016	0.017
--	--	--	0.009	0.005	--	0.007	0.011	0.005	0.009	0.010	0.011
3.981	3.979	3.983	3.993	3.995	3.977	3.977	3.996	3.988	4.007	3.990	4.002

(Cr+Al) ratios. Furthermore, it was somewhat surprising to discover that the three petrographically distinct types of spinel grains in PTH 204 (large discrete primary (?) grains, small idioblastic grains associated with clusters of clinopyroxene and phlogopite grains, and small grains associated with kelyphitic alteration zones around garnet) all have similar compositions of the low Al magnesiochromite type. Small spinels associated with 'pools' of phlogopite and clinopyroxene grains in PTH 108 are of similar low Al type. Thus, if as seems likely, at least some of these spinels (PTH 108, 204, 207B, 403B and 405B) are a secondary generation derived from the breakdown of primary garnet, then it would appear that they have re-equilibrated with the lherzolite assemblage unlike those in the innermost part of kelyphites.

Phlogopite

Phlogopites typically form narrow coronas around the partly kelyphitised garnets but also, as in PTH 108, occur in clusters with spinel and chrome diopside grains. Their general secondary metasomatic character is borne out by partial analyses (not presented here) which indicate that they usually have high, although somewhat variable TiO_2 (0.21-4.24 wt.%) and Cr_2O_3 (0.84-2.48 wt.%) contents together with 100 Fe/(Fe+Mg) ratios of 5.6-10.6.

Discussion

No garnet-bearing lherzolites with P/T equilibration estimates which indicate derivation from

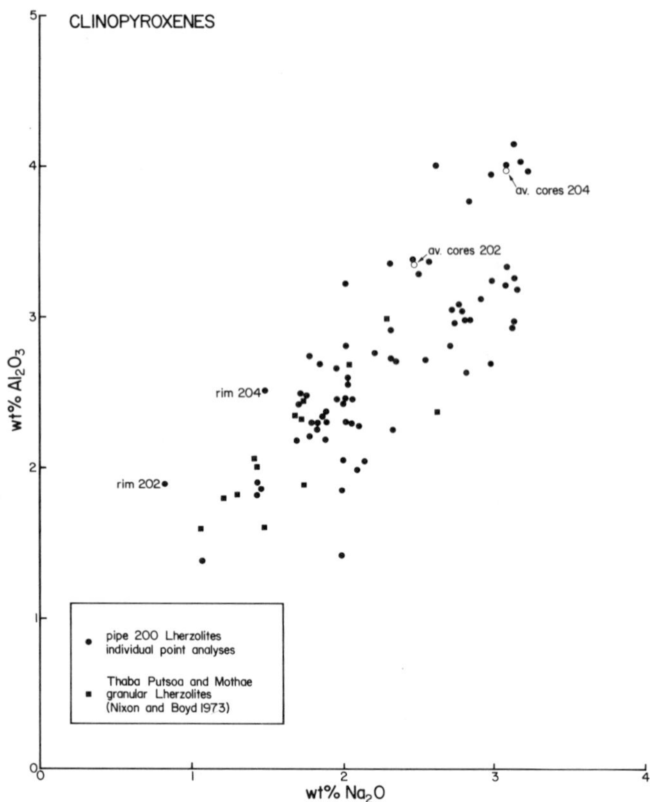

Fig. 2. Wt.% Al_2O_3 against wt.% Na_2O plot for analysed clinopyroxenes from Pipe 200 lherzolite nodules.

TABLE 5: Analyses of Clinopyroxenes

Sample Number	PTH 102	PTH 108	PTH 202 (Cores)	PTH 202 (Rim)	PTH 203	PTH 204 (Cores)
Rock Type	GL	(G)CL	(G)CL	(G)CL	CL	GCL
SiO_2	54.6	54.4	54.4	53.0	54.5	54.2
TiO_2	0.13	0.05	0.06	0.08	0.10	0.06
Al_2O_3	1.90	1.64	3.36	1.90	2.64	3.99
Cr_2O_3	2.22	2.74	2.28	2.44	4.25	2.89
*FeO	2.22	2.31	2.38	2.36	1.47	2.14
MnO	0.04	0.08	0.02	0.02	0.09	0.07
NiO	n.d.	n.d.	0.05	n.d.	0.05	n.d.
MgO	15.8	15.6	15.2	18.2	15.0	14.3
CaO	20.6	20.3	19.4	20.5	17.0	18.1
Na_2O	1.44	1.96	2.47	0.81	2.82	3.08
Total	99.0	99.1	99.6	99.3	98.8	98.8
100 Ca/(Ca+Mg)	48.4	48.3	47.8	44.7	46.2	47.7
Si	1.993	1.991	1.974	1.935	1.987	1.977
Al^{IV}	0.007	0.009	0.026	0.065	0.013	0.023
Al^{VI}	0.075	0.062	0.118	0.017	0.101	0.149
Ti	0.004	0.001	0.002	0.002	0.003	0.002
Cr	0.064	0.079	0.065	0.070	0.123	0.083
Fe	0.068	0.071	0.072	0.072	0.045	0.065
Mn	0.001	0.002	0.001	0.001	0.003	0.002
Ni	--	--	0.001	--	0.001	--
Mg	0.860	0.851	0.821	0.992	0.817	0.777
Ca	0.806	0.796	0.753	0.803	0.701	0.707
Na	0.102	0.139	0.174	0.057	0.199	0.218
Total	3.980	4.002	4.007	4.015	3.992	4.003
$\%NaCrSi_2O_6$	6.6	7.9	6.6	6.0	12.4	8.3
$\%NaAlSi_2O_6$	4.2	5.6	9.5	0.0	7.8	12.9

* All Fe taken as FeO. Stiochiometrically calculated Fe^{3+} values not presented as at these very low total iron values slight analytical errors, especially in SiO_2, result in a widely differing Fe^{3+}/Fe^{2+} ratio.

the top sixty or so kilometres of the mantle have been recognised in previous studies of nodule suites in Lesotho pipes (Boyd, 1973; Boyd and Nixon, 1975; Gurney et al., 1975; Dawson et al., 1975). This may be due to overestimation of P/T values and implied depths of derivation for certain garnet lherzolites by the P/T calculation procedures adopted by these authors. Alternatively the uppermost mantle may be of highly depleted chemical character, and hence largely comprise garnet-free ultramafic rocks, for which it is currently not possible to fix P/T values and depths of origin. Assuming that both our sampling, and that of the kimberlite, were adequate, consideration of the Pipe 200 ultramafic nodule suite may enable us to resolve this problem and provide a more satisfactory characterisation of the nature of the uppermost mantle beneath northern Lesotho at the time of kimberlite emplacement. We intend to present a more detailed discussion elsewhere as for the present we are obliged to restrict ourselves to just a few pertinent remarks on these matters.

We think it reasonable to conclude that the one primary aluminous spinel harzburgite nodule (PTH 201) was derived from shallower mantle depths than any of the other analysed Pipe 200 nodules. Such primary aluminous spinel nodules are apparently rare amongst the nodule suite from pipes in northern Lesotho, although Boyd and Nixon (1975) described one from Liqhobong, and we have observed others from Lipelenang and Ngopetseu. This rarity is in line with the narrow stability zone (probably < 20 km) at the top of the mantle predicted for such assemblages by relevant experimental studies (MacGregor, 1970; O'Hara et al., 1971).

The primary spinels in other garnet free peri-

in Pipe 200 Ultramafic Nodules.

PTH 204 (Rim)	PTH 205	PTH 207(P)	PTH 207(S_1)	PTH 207 S_2	PTH 210	PTH 304	PTH 400	PTH 401	PTH 403(P)	PTH 403(S_1)	PTH 403(S_2)	PTH 404	PTH 405	PTH 407	PTH 409	PTH 410
GCL	(G)CL	GCL	GCL	GCL	GP	(G)CL	GL	GCL	GL	GL	GL	GL	GL	GL	GCL	GCL
52.9	54.4	54.7	49.1	52.6	54.4	54.4	54.2	54.3	54.2	48.6	51.7	55.0	54.2	53.8	54.1	54.5
0.10	0.16	0.10	n.d.	0.17	n.d.	n.d.	0.13	0.01	0.18	0.92	0.40	0.10	0.08	0.01	0.11	0.09
2.52	2.46	2.19	8.12	4.02	1.99	2.95	2.66	2.54	2.84	8.73	4.43	2.34	2.95	2.30	3.26	3.07
3.01	2.88	2.63	3.26	2.19	0.65	3.64	1.59	2.02	3.12	2.25	1.40	2.64	3.20	2.52	3.41	3.33
2.11	2.29	2.34	2.88	3.94	1.59	1.94	2.64	1.79	2.48	4.10	4.45	2.24	2.22	2.17	2.27	2.22
0.08	n.d.	0.06	n.d.	0.48	n.d.	n.d.	0.10	0.07	n.d.	n.d.	0.24	0.05	0.10	0.10	0.08	0.07
n.d.	n.d.	n.d.	n.d.	n.d.	n.d.	n.d.	0.06	n.d.	n.d.	n.d.	n.d.	n.d.	0.04	n.d.	n.d.	n.d.
15.6	15.6	16.1	15.9	21.4	16.5	15.2	16.2	16.1	15.3	15.7	18.6	15.9	15.2	16.1	14.5	15.3
21.1	19.3	19.8	18.8	14.3	23.7	18.2	19.6	20.9	18.6	18.3	16.8	20.0	18.1	19.5	18.0	18.1
1.48	2.01	1.93	0.93	n.d.	0.48	3.14	2.03	1.71	2.49	1.19	1.03	1.88	2.79	2.05	2.94	2.76
98.9	99.1	99.9	99.0	99.6	99.3	99.5	99.2	99.5	99.2	99.7	99.1	100.2	98.9	98.7	98.7	99.5
49.3	47.1	46.9	45.9	32.5	50.9	46.3	46.6	48.2	46.6	45.6	39.4	47.5	46.2	46.5	47.1	46.0
Structural Formulae on the Basis of 6.000 Oxygens																
1.946	1.984	1.980	1.802	1.900	1.981	1.976	1.975	1.973	1.975	1.775	1.890	1.985	1.982	1.974	1.981	1.978
0.054	0.016	0.020	0.198	0.100	0.019	0.024	0.025	0.027	0.025	0.225	0.110	0.015	0.018	0.026	0.019	0.022
0.055	0.090	0.074	0.153	0.071	0.066	0.102	0.089	0.082	0.097	0.151	0.081	0.084	0.109	0.073	0.122	0.109
0.003	0.004	0.003	--	0.005	--	--	0.004	0.000	0.005	0.025	0.011	0.003	0.002	0.000	0.003	0.002
0.088	0.083	0.075	0.095	0.063	0.019	0.105	0.046	0.058	0.090	0.065	0.040	0.075	0.092	0.073	0.099	0.096
0.065	0.070	0.071	0.088	0.119	0.048	0.059	0.080	0.054	0.076	0.125	0.136	0.068	0.068	0.067	0.070	0.067
0.002	--	0.002	--	0.015	--	--	0.003	0.002	--	--	0.007	0.002	0.003	0.003	0.002	0.002
--	--	--	--	--	--	--	0.002	--	--	--	--	--	0.001	--	--	--
0.855	0.846	0.871	0.868	1.152	0.893	0.823	0.879	0.873	0.833	0.855	1.012	0.855	0.825	0.883	0.792	0.828
0.831	0.754	0.769	0.739	0.555	0.925	0.708	0.767	0.814	0.726	0.716	0.658	0.773	0.707	0.767	0.706	0.705
0.106	0.142	0.136	0.066	--	0.034	0.221	0.143	0.120	0.176	0.084	0.073	0.131	0.198	0.146	0.209	0.194
4.006	3.989	4.000	4.008	3.979	3.984	4.019	4.013	4.004	4.002	4.021	4.020	3.990	4.005	4.012	4.001	4.004
8.8	8.4	7.5	6.8	--	1.9	10.6	4.6	5.8	9.0	6.6	4.1	7.6	9.3	7.4	9.9	9.6
0.7	6.1	6.0	0.0	--	1.6	8.0	7.3	5.5	8.2	0.0	0.0	5.8	9.5	4.9	10.8	9.2

% Kosmochlor and % Jadeite calculated in that order, after first calculating $CaTiAl_2O_6$ and $CaAl_2SiO_6$ on the assumption that $Al^{IV} = 2.000 - Si$

n.d. - not determined

dotite nodules from Pipe 200 have 100 Cr/(Cr+Al) ratios which are as high or even higher (Table 7 and Figure 4) than those in the associated garnet lherzolites. This, together with the earlier demonstration of significant differences in whole rock chemistry between these nodule types (Table 1 and Figure 1) negates the necessity to interpret these nodules as having equilibrated at significantly different pressures (and depths). The stability of garnet relative to spinel in lherzolite assemblages is strongly dependent on the 100 Cr/(Cr+Al) ratio in the rock (MacGregor, 1970). In the absence of any way in which we can at present estimate P/T equilibration conditions for the garnet-free peridotites, it is not possible to demonstrate convincingly that they have indeed been derived from depths comparable to those deduced for the associated garnet lherzolites. How-

TABLE 6: Analyses of Garnets in Pipe 200 Ultramafic Nodules.

Sample Number	PTH 102	PTH 204	PTH 207	PTH 210	PTH 400	PTH 403	PTH 404	PTH 405	PTH 406	PTH 407	PTH 409	PTH 410
Rock Type	GL	GCL	GCL	GP	GL	GL	GL	GL	GL	GL	GCL	GCL
SiO_2	41.5	41.8	41.5	41.9	41.7	41.4	41.3	41.4	41.7	40.9	41.4	41.6
TiO_2	0.11	0.02	0.15	n.d.	0.17	0.10	0.10	0.04	0.05	0.04	0.07	0.09
Al_2O_3	18.0	18.1	17.2	23.2	20.3	18.9	17.7	18.9	18.8	18.1	17.9	19.1
Cr_2O_3	7.63	8.00	8.36	1.48	4.29	6.82	7.86	6.29	7.15	7.59	7.68	6.96
$*Fe_2O_3$	0.00	0.00	0.11	0.78	1.91	0.45	0.60	0.46	0.00	1.31	0.03	1.70
FeO	7.04	6.78	6.94	9.06	6.04	6.93	6.43	6.51	6.60	5.62	6.50	5.30
MnO	0.40	0.42	0.40	0.59	0.38	0.25	0.41	0.37	0.39	0.37	0.41	0.35
MgO	18.1	19.0	18.8	18.6	20.6	19.4	19.0	20.1	19.2	19.4	19.2	20.8
CaO	6.72	6.62	6.92	5.72	5.39	6.04	6.82	5.32	6.60	6.50	6.55	5.55
Total	99.5	100.7	100.5	101.3	100.8	100.3	100.2	99.4	100.5	99.8	99.7	101.4
Structural Formulae on the Basis of 24.000 Oxygens												
Si	6.058	6.024	6.024	5.939	5.938	5.970	5.992	5.999	6.001	5.940	6.018	5.911
Al	3.092	3.074	2.947	3.873	3.400	3.211	3.027	3.222	3.192	3.096	3.064	3.197
Ti	0.012	0.002	0.016	--	0.018	0.011	0.011	0.004	0.005	0.004	0.008	0.010
Cr	0.880	0.912	0.959	0.166	0.483	0.778	0.902	0.720	0.814	0.872	0.882	0.781
$*Fe^{3+}$	0.000	0.000	0.013	0.083	0.205	0.049	0.066	0.051	0.000	0.143	0.003	0.181
Fe^{2+}	0.859	0.818	0.842	1.074	0.719	0.837	0.780	0.789	0.795	0.683	0.789	0.630
Mn	0.049	0.051	0.049	0.071	0.046	0.031	0.050	0.045	0.048	0.046	0.050	0.042
Mg	3.942	4.076	4.074	3.926	4.369	4.178	4.113	4.344	4.119	4.204	4.166	4.404
Ca	1.051	1.023	1.076	0.869	0.822	0.934	1.060	0.825	1.018	1.012	1.020	0.844
Total	15.944	15.981	16.000	16.000	16.000	16.000	16.000	16.000	15.991	16.000	16.000	16.000
Percentage of Garnet End Members												
Pyrope	62.2	62.5	63.1	66.1	73.6	66.1	63.8	68.3	65.5	65.8	65.2	69.0
Almandine	14.6	13.7	11.2	18.1	11.5	13.8	11.2	11.7	13.3	11.4	11.5	10.5
Spessartine	0.8	0.9	0.8	1.2	0.8	0.5	0.8	0.8	0.8	0.8	0.9	0.7
Uvarovite	17.8	17.1	18.3	4.2	12.2	15.7	17.7	13.8	17.0	17.0	17.2	14.3
Knorringite	4.6	5.8	6.2	--	--	3.9	4.9	4.3	3.4	5.0	5.1	5.5
Andradite	--	--	--	2.1	1.7	--	--	--	--	--	--	--
Skiagite	--	--	0.3	--	0.3	--	1.7	1.3	--	--	0.1	--
Grossular	--	--	--	8.3	--	--	--	--	--	--	--	--

$*Fe^{3+}$ calculated from anion deficiency in structural formulae based on 16.000 cations

ever, the similarity in clinopyroxene 100 Ca/(Ca+Mg) ratio, and consequently derived temperature estimates for the chromite lherzolite (PTH 203), to the values obtained for the garnet lherzolites suggest that this may well be the case.

It is clear that the Pipe 200 garnet-bearing lherzolite nodules are closely similar in terms of their whole rock and mineral chemistries to the petrographically similar nodules previously described by Nixon and Boyd (1973) in nearby pipes. P/T estimates for all such nodules based on direct application of experimental data for the two-pyroxene solvus (Davis and Boyd, 1966) and garnet-orthopyroxene equilibria (MacGregor, 1974), lie on the non-inflected part of the 'pyroxene' geotherm of Boyd (1973), Boyd and Nixon (1975). It is widely accepted that there is a large measure of uncertainty in the absolute accuracy of such P/T values but it has been assumed (Boyd, 1973; Boyd and Nixon, 1975) that the relative P/T values obtained for different garnet lherzolites should be significant. However, even this may not be the case if there are variations in whole rock chemistry amongst the garnet lherzolite nodules which significantly affect the mineral equilibria.

With this in mind we prefer to use the semi-empirical approaches adopted by Wood and Banno (1973), Wood (1974) and Wells (1977) which attempt to take account of the effect of Fe on the two-pyroxene solvus, and of the complex chemistries of natural garnets and orthopyroxenes. Certainly the low Al_2O_3 contents of the primary orthopyroxenes (0.70-1.04 wt.%) and in their high 100 Cr/(Cr+Al) ratios mean that we cannot justifiably obtain P estimates for the Pipe 200 garnet lherzolites by direct application of MacGregor's (1974) data in the manner followed by Boyd (1973).

The Pipe 200 garnet lherzolites yield T/P estimates of 1034-1071°C/31.4-37.7 kb using the combination of equation (27) of Wood and Banno (1973), and equation (12) of Wood (1974), and 906-949°C/24.5-30.8 kbs. using equation (5) of Wells (1977) and equation (12) of Wood (1974). For both sets of P estimates Cr contents in garnet and orthopyroxene were taken into account in the manner proposed by Wood (1974). Garnet-free lherzolites yield T estimates of 999-1093°C (Wood and Banno, 1973) and 873-962°C (Wells, 1977), but P estimates are unobtainable. These various P/T estimates imply derivation of these nodules from a surprisingly narrow depth zone in the mantle. Whilst it is possible that kimberlite eruptions may be habitually selective in their depth sampling, we consider that the failure of the adopted calculation procedures to adequately take account of the opposing effects of T and P on the clinopyroxene-orthopyroxene and garnet-orthopyroxene equilibria may be responsible for telescoping the derived P/T estimates into an artifically narrow range.

In view of the relative insensitivity of the two-pyroxene thermometer at temperatures below about 1100°C and doubts over the corrections necessary for the effect of P and the presence of Al on the two-pyroxene solvus (Mori and Green, 1975, 1976; Lindsley and Dixon, 1976; Dixon and Presnall, 1977) as well as major uncertainties introduced in the extrapolation of experimental data on simple systems to the interpretation of mineral equilibria in more chemically complex natural minerals, we feel that we cannot unquestionably accept any of the so-derived P/T estimates. Derivation of unequivocal P/T values for the Pipe 200 nodules will probably have to await further experimental data on the effect of Cr_2O_3 on the garnet-orthopyroxene equilibria and experimental calibration of the effects of P and T on the Fe-Mg or Mn-Mg distribution coefficients in these ultramafic assemblages. In the meantime our inclination is towards accepting the relatively lower P/T estimates provided by the combined use of equation (5) of Wells (1977) and equation (12) of Wood (1974) as the most realistic. Such values as shown in Figure 5, are the most compatible with preferred P/T estimates for the lower crustal nodules (Griffin et al., 1978) but necessitate substantial revision to the palaeogeothem of Boyd (1973), Boyd and Nixon (1975) in the region of the lower crust and uppermost mantle.

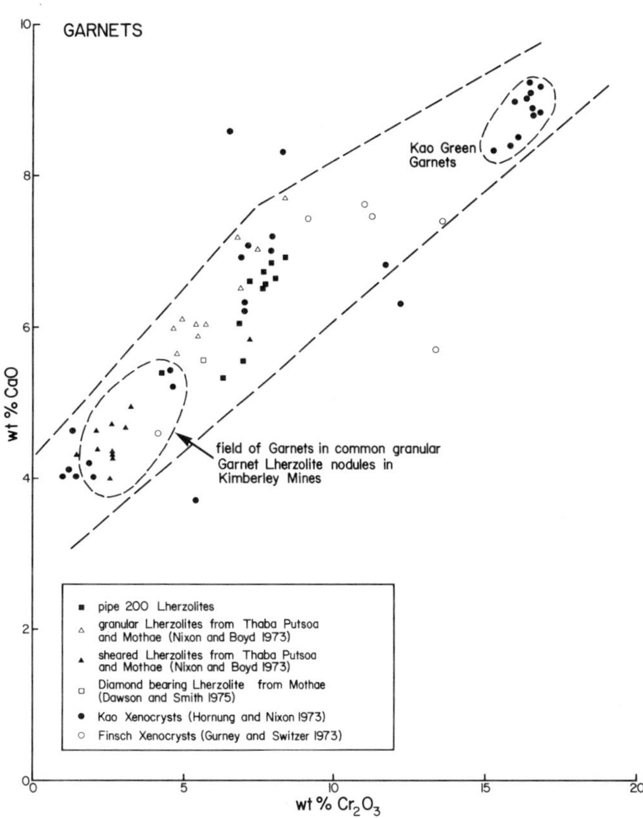

Fig. 3. Wt.% CaO against wt.% Cr_2O_3 plot for analysed garnets from Pipe 200 lherzolite nodules as well as for xenolithic and xenocrystal garnets of likely lherzolite paragenesis from other kimberlites in Lesotho and South Africa.

TABLE 7: Analyses of Spinels

Sample Number	PTH 108	PTH 201	PTH 202	PTH 203	PTH 204A	PTH 204B	PTH 204C	PTH 207A	PTH 207B
Rock Type	(G)CL	SH	(G)CL	CL	GCL	GCL	GCL	GCL	
SiO_2	0.05	0.13	0.04	0.05	n.d.	n.d.	n.d.	n.d.	n.d.
TiO_2	0.73	0.01	0.27	0.34	n.d.	n.d.	n.d.	0.21	0.61
Al_2O_3	5.35	29.4	14.2	8.75	13.9 ±0.7	14.8 ±0.3	15.0 ±0.2	44.9	12.5
Cr_2O_3	59.9	41.5	51.7	63.1	55.0 ±0.7	53.7 ±0.4	53.1 ±0.6	23.0	55.9
$*Fe_2O_3$	6.04	0.53	7.04	1.91	3.99 ±0.4	4.09 ±0.2	4.30 ±0.4	2.01	4.08
FeO	15.9	12.2	12.0	10.8	13.7	13.4	13.7	10.09	12.3
MnO	0.55	0.07	0.06	0.38	n.d.	n.d.	n.d.	0.57	n.d.
NiO	n.d.	0.10	n.d.	0.10	n.d.	n.d.	n.d.	n.d.	n.d.
MgO	11.1	16.0	14.7	14.7	13.4 ±0.3	13.6 ±0.3	13.4 ±0.4	18.6	14.5
Total	99.1	99.9	19.4	100.0	99.6	99.2	99.1	99.2	99.5
100 Cr/(Cr+Al)	88.2	48.6	70.9	82.9	72.6	70.9	70.4	25.6	75.0
					Structural Formulae on the Basis of 4.000 Oxygens				
Si	0.002	0.004	0.001	0.002	--	--	--	--	--
Ti	0.018	0.000	0.06	0.008	--	--	--	0.005	0.015
Al	0.212	1.017	0.529	0.331	0.521	0.554	0.562	1.450	0.468
Cr	1.594	0.963	1.288	1.603	1.383	1.348	1.335	0.498	1.405
$*Fe^{3+}$	0.153	0.012	0.167	0.046	0.096	0.098	0.101	0.041	0.098
Fe^{2+}	0.449	0.300	0.317	0.291	0.365	0.356	0.365	0.231	0.328
Mn	0.016	0.002	0.002	0.010	--	--	--	0.013	--
Ni	--	0.002	--	0.003	--	--	--	--	--
Mg	0.556	0.700	0.689	0.706	0.635	0.644	0.635	0.760	0.687
$\%M^{2+}Cr_2O_4$	81.4	48.4	64.9	81.0	69.2	67.4	66.8	25.0	71.3
$\%M^{2+}Al_2O_4$	10.8	51.0	26.7	16.7	26.0	27.7	28.1	72.9	23.8
$\%M^{2+}Fe_2{}^{3+}O_4$	7.8	0.6	8.4	2.3	4.8	4.9	5.1	2.1	4.9

$*Fe^{3+}$ Calculated by Charge Balance in the Structural Formulae

n.d. Not determined

In most rocks different spinel grains show little chemical variation and are represented by a single mean analysis. However, spinels in PTH 207, 403 and 405 show a range in chemistry as represented by the two sets of analyses presented. 207A, 403A and 405A, represent spinels in the innermost parts of kelyphites around garnets, whilst 207B, 403B, and 405B represent spinels

---

The stability of primary spinel relative to garnet in all Pipe 200 nodules other than PTH 201 appears to have been largely controlled by the level of depletion of Al_2O_3, CaO and FeO relative to Cr_2O_3 and MgO in these rocks (Table 1 and Figure 1). It is of interest to consider whether ultramafic nodules from pipes in other areas have similar assemblages at equivalent levels of Cr/(Cr+Al) and Mg/(Mg+Fe) in the rock. Stability of garnet in rocks with higher Cr/(Cr+Al) ratios would imply higher pressures (MacGregor, 1970) and hence deeper levels of origin, thus indicating that the mantle in that area was highly depleted to greater depths if the depth range of sampling was comparable.

For comparison the garnet-only, garnet and chromite and chromite-only composition fields defined by the ultramafic nodules in the Lashaine volcano, Tanzania (Reid et al., 1975) are shown on Figure 1. Unfortunately, the garnet and chromite/chromite-only boundary (A) is not closely defined, but the apparent lack of overlap in the composition

in Pipe 200 Ultramafic Nodules.

PTH 301	PTH 302	PTH 303	PTH 304	PTH 401	PTH 403A	PTH 403B	PTH 404	PTH 405A	PTH 405B	PTH 407	PTH 409	PTH 410
CH	CH	CH	(G)CL	GCL	GL		GL	GL		GL	GL	GL
n.d.	n.d.	n.d.	n.d.	0.10	n.d.	n.d.	0.12	0.14	0.16	0.19	0.12	0.22
n.d.	n.d.	n.d.	n.d.	0.11	1.39	2.03	0.83	0.29	1.96	0.07	0.49	0.59
11.2	8.03	8.20	9.0	13.8	28.8	15.5	46.3	45.1	33.3	46.7	10.1	48.6
60.1	62.3	61.7	58.6	55.4	38.4	47.8	18.7	22.0	31.4	21.9	57.5	18.7
3.56	4.89	4.91	5.87	3.37	4.11	6.14	2.88	3.07	3.36	1.49	3.80	2.38
9.39	8.79	9.17	13.5	13.5	12.7	15.3	11.0	8.13	12.1	8.64	15.0	9.57
n.d.	n.d.	n.d.	n.d.	0.38	0.24	n.d.	0.31	0.27	0.33	0.23	0.35	0.24
n.d.	n.d.	n.d.	n.d.	n.d.	0.28	0.26	n.d.	0.05	0.21	n.d.	n.d.	n.d.
16.0	16.0	15.7	13.0	13.4	16.6	13.7	18.7	20.2	17.4	20.0	12.2	20.2
99.9	99.5	99.2	99.4	99.7	100.5	100.1	98.6	99.0	99.9	99.0	99.2	100.2
78.3	83.9	83.5	81.4	72.9	46.1	67.4	21.3	24.6	38.7	23.9	79.2	20.5
--	--	--	--	0.003	--	--	0.003	0.004	0.005	0.005	0.004	0.006
--	--	--	--	0.003	0.030	0.048	0.017	0.006	0.042	0.001	0.012	0.012
0.416	0.303	0.311	0.346	0.515	0.977	0.573	1.495	1.444	1.122	1.488	0.389	1.523
1.499	1.579	1.570	1.510	1.392	0.874	1.186	0.405	0.474	0.712	0.468	1.485	0.394
0.085	0.118	0.119	0.144	0.081	0.089	0.145	0.059	0.063	0.072	0.030	0.093	0.048
0.248	0.236	0.247	0.348	0.359	0.306	0.401	0.252	0.185	0.289	0.196	0.410	0.213
--	--	--	--	0.010	0.006	--	0.007	0.006	0.008	0.005	0.010	0.005
--	--	--	--	--	0.006	0.007	--	0.001	0.005	--	--	--
0.752	0.764	0.753	0.632	0.637	0.712	0.641	0.762	0.817	0.745	0.806	0.596	0.799
75.0	78.9	78.5	75.5	70.0	45.0	62.3	20.7	23.9	37.3	23.6	75.5	20.0
20.8	15.2	15.6	17.3	25.9	50.4	30.1	76.3	72.9	58.9	74.9	19.8	77.6
4.2	5.9	5.9	7.2	4.1	4.6	7.6	3.0	3.2	3.8	1.5	4.7	2.4

typically outside, although still adjacent to, turbid kelyphite alteration zones. PTH 204 contains three petrographically distinct types of spinels: (A) large discrete apparently primary grains (B) small idioblastic grains associated with clusters of clinopyroxene and phlogopite grains (C) small grains associated with kelyphitic breakdown of garnet.

Mean and standard deviation are presented for each of these types of spinel grains. In other garnet chromite lherzolites, analyses for PTH 401 and 409 represent primary chromites whilst that for PTH 410 is for secondary spinels associated with garnet breakdown.

fields suggests either derivation from a fairly restricted depth zone or a systematic decrease in the level of depletion with depth. It is apparent that in the ultramafic nodules from Pipe 200 and certain other pipes in northern Lesotho, garnet is stable to significantly higher whole rock Cr/(Cr+Al) values than in the Lashaine nodules (compare also Table 2, columns A, B and F). It therefore seems likely that the mantle is in general more highly depleted to greater depths beneath these northern Lesotho pipes, although in both areas there may well be an overall general decrease in the level of depletion in basalt-yielding constituents with depth. Moreover, as the garnet stable in lherzolites with increased Cr/(Cr+Al) values itself becomes increasing chromiferous, the presence of the exceptionally chrome-rich green garnet xenocrysts in the nearby Kao kimberlite (Figure 3) implies that the mantle beneath northern Lesotho may in places be exceptionally depleted to considerable depths with extremely high Cr/(Cr+Al) ratios. This may well reflect the extensive Karroo volcan-

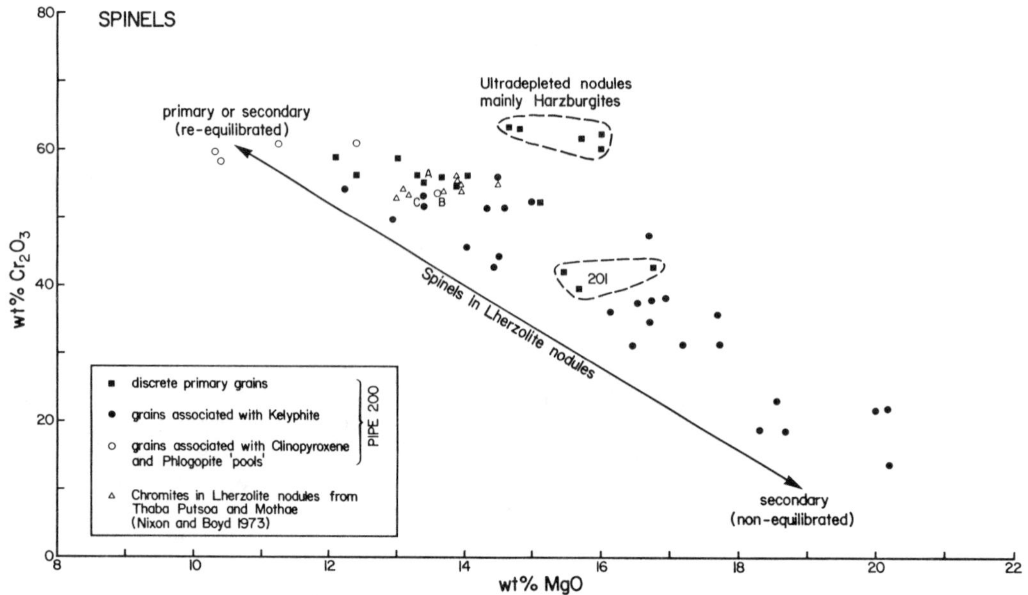

Fig. 4. Wt.% Cr_2O_3 against wt.% MgO plot for analysed spinels from Pipe 200 ultramafic nodules.

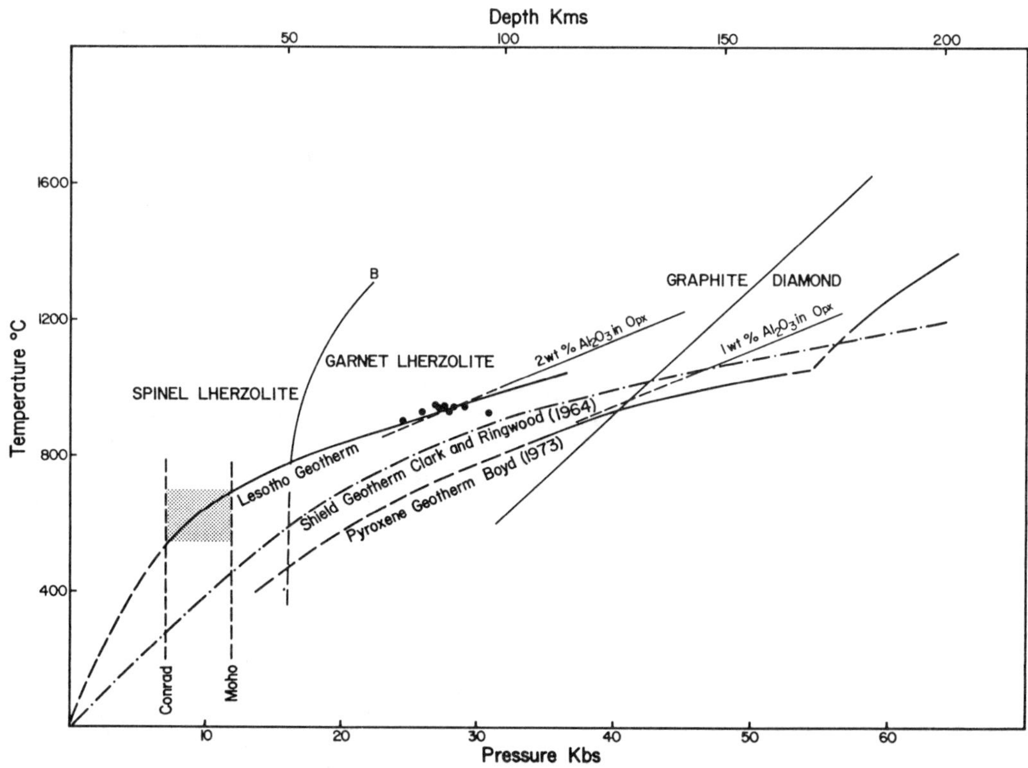

Fig. 5. P/T diagram showing revised Lesotho Geotherm compatible with provisionally preferred P/T estimates for Pipe 200 garnet lherzolite nodules (dots, see text) and P/T estimates for lower crustal granulite/eclogite nodules from Lesotho pipes (stippled field, Griffin et al., 1978). Curve B is the experimentally determined spinel lherzolite/garnet lherzolite boundary for natural materials from O'Hara et al. (1971). The diamond/graphite inversion curve is from Bundy et al. (1961) and the wt.% Al_2O_3 isopleths for orthopyroxene in equilibrium with garnet are from MacGregor (1974). Positions of the Conrad and Moho discontinuities are based on the seismic data for the Transvaal (Hales and Sacks, 1959).

ism which preceded kimberlite emplacement in this area.

Acknowledgements: D. A. Carswell wishes to acknowledge the financial support of The Royal Society, University of Sheffield and Natural Environmental Research Council, D. B. Clarke funding from Dalhousie University and the National Research Council of Canada, and R. H. Mitchell funding from the National Research Council of Canada. J. R. Andrews, M. J. O'Hara and P. Kresten kindly provided additional samples. The considerable assistance of V. A. Somogyi with the whole rock analyses and of P. G. Hill and R. M. MacKay with the probe analyses is gratefully acknowledged. We also wish to thank the organisers and guides of the 1973 Lesotho kimberlite excursions.

References

Berg, W. and M. J. O'Hara, Source mantle, residuum and partial melt compositions deduced from the kimberlite record, Extended Abstracts, First International Kimberlite Conference, Cape Town, 31-33, 1973.

Boullier, A. M., and A. Nicholas, Texture and fabric of peridotite nodules from kimberlite at Mothae, Thaba Putsoa and Kimberley, in Lesotho Kimberlites, edited by P. H. Nixon, pp. 57-66, Lesotho National Development Corp., Maseru, Lesotho, 1973.

Boyd, F. R., A pyroxene geotherm, Geochim. Cosmochim. Acta, 37, 2533-2546, 1973.

Boyd, F. R., and P. H. Nixon, Ultramafic nodules from the Thaba Putsoa kimberlite pipe, Carnegie Inst. Wash. Yearb., 71, 362-373, 1972.

Boyd, F. R., and P. H. Nixon, Origins of the ultramafic nodules from some kimberlites of northern Lesotho and the Monastery Mine, South Africa, Phys. Chem. Earth, 9, 431-454, 1975.

Bundy, F. R., H. P. Bovenkerk, H. M. Strong, and R. H. Wentorf, Jr., Diamond graphite equilibrium line from growth and graphitisation of diamond, J. Chem. Phys., 35, 383-391, 1961.

Carswell, D. A., Primary and secondary phlogopites and clinopyroxenes in garnet lherzolite xenoliths. Phys. Chem. Earth., 9, 417-429, 1975.

Carswell, D. A., and J. B. Dawson, Garnet peridotite xenoliths in South African kimberlite pipes and their petrogenesis, Contr. Mineral. Petrol., 25, 163-184, 1970.

Chen, J. C., Petrology and chemistry of garnet lherzolite nodules in kimberlite from South Africa, Am. Mineral., 56, 2098-2110, 1971.

Clark, S. P., Jr., and A. E. Ringwood, Density distribution and constitution of the mantle. Rev. Geophys., 2, 35-88, 1964.

Cox, K. G., J. J. Gurney, and B. Harte, Xenoliths from the Matsoku Pipe, in Lesotho Kimberlites, edited by P. H. Nixon, pp. 76-92, Lesotho National Development Corp., Maseru, Lesotho, 1973.

Davis, B. T. C., and F. R. Boyd, The join $Mg_2Si_2O_6$-$CaMgSi_2O_6$ at 30 kilobars pressure and its application to pyroxenes from kimberlites. J. Geophys. Res., 71, 3507-3576, 1966.

Dawson, J. B., J. J. Gurney, and P. J. Lawless, Palaeogeothermal gradients derived from xenoliths in kimberlite, Nature, 257, 299-300, 1975.

Dawson, J. B., and J. V. Smith, Occurrence of diamond in a mica-garnet lherzolite xenolith from kimberlite, Nature, 254, 580-581, 1975.

Dawson, J. B., and W. E. Stevens, Statistical classification of garnets from kimberlite and associated xenoliths, J. Geol., 83, 589-607, 1975.

Dixon, J. R., and D. C. Presnall, Geothermometry and geobarometry of synthetic spinel lherzolite in the system CaO-MgO-Al_2O_3-SiO_2, Extended Abstracts, Second International Kimberlite Conference, Santa Fe, New Mexico, 1977.

Griffin, W. L., D. A. Carswell, and P. H. Nixon, Lower crustal granulites and eclogites from Lesotho and South Africa, Proceedings of the Second International Kimberlite Conference, in press.

Gurney, J. J., B. Harte, and K. G. Cox, Mantle xenoliths in the Matsoku kimberlite pipe. Phys. Chem. Earth, 9, 507-523, 1975.

Gurney, J. J., and G. S. Seitzer, The discovery of garnets closely related to diamonds in the Finsch Pipe, South Africa, Contr. Mineral. Petrol., 39, 103-116, 1973.

Hales, A. L., and I. S. Sacks, Evidence for an intermediate layer from crustal seismic studies in the East Transvaal, Geophys. J. Roy. Astr. Soc., 2, 15-23, 1957.

Harte, B., K. G. Cox, and J. J. Gurney, Petrography and geological history of upper mantle xenoliths from the Matsoku kimberlite pipe, Phys. Chem. Earth, 9, 477-506, 1975.

Holmes, A., A contribution to the petrology of kimberlite and its inclusions, Trans. Geol. Soc. S. Africa, 39, 379-428, 1936.

Hornung, G., and P. H. Nixon, Chemical variations in the knorringite-rich garnets, in Lesotho Kimberlites, edited by P. H. Nixon, pp. 122-127, Lesotho National Development Corp., Maseru, Lesotho, 1973.

Ito, K., and G. C. Kennedy, The melting and phase relations in a natural peridotite to 40 kilobars, Am. J. Sci., 265, 519-538, 1967.

Kresten, P., and A. N. Dempster, The geology of Pipe 200 and the Malibamatso dyke swarm, in Lesotho Kimberlites, edited by P. H. Nixon, pp. 172-179, Lesotho National Development Corp., Maseru, Lesotho, 1973.

Lindsley, D. H., and S. A. Dixon, Diopside-enstatite equilibria at 850° to 1400°C, 5 to 35 kb., Am. J. Sci., 276, 1285-1301, 1976.

MacGregor, I. D., The effect of CaO, Cr_2O_3, Fe_2O_3 and Al_2O_3 on the stability of spinel and garnet peridotites, Phys. Earth Planet Inter., 3, 372-377, 1970.

MacGregor, I. D., The system MgO-Al_2O_3-SiO_2: solubility of Al_2O_3 in enstatite for spinel and garnet peridotite compositions, Am. Mineral., 59, 110-119, 1974.

Mori, T., and D. H. Green, Pyroxenes in the system $Mg_2Si_2O_6$-$CaMgSi_2O_6$ at high pressure, Earth Planet. Sci. Letters, 26, 277-286, 1975.

Mori, T., and D. H. Green, Subsolidus equilibria between pyroxenes in the CaO-MgO-SiO_2 systems at high pressures and temperatures, Am. Mineral. 61, 616-625, 1976.

Nixon, P. H., and F. R. Boyd, Petrogenesis of the granular and sheared ultrabasic nodule suite in kimberlites, in Lesotho Kimberlites, edited by P. H. Nixon, pp. 48-56, Lesotho National Development Corp., Maseru, Lesotho, 1973.

O'Hara, M. J., S. W. Richardson, and G. Wilson, Garnet peridotite stability and occurrence in crust and mantle, Contr. Mineral. Petrol., 32, 48-68, 1971.

Reid, A. M., C. H. Donaldson, R. W. Brown, W. I. Ridley, and J. B. Dawson, Mineral chemistry of peridotite xenoliths from the Lashaine volcano, Tanzania, Phys. Chem. Earth, 9, 525-543, 1975.

Rhodes, J. M., and J. B. Dawson, Major and trace element chemistry of peridotite inclusions from the Lashaine volcano, Tanzania, Phys. Chem. Earth, 9, 545-557, 1975.

Ridley, W. I., and J. B. Dawson, Lithophile trace element data bearing on the origin of peridotite xenoliths, ankaramite and carbonatite from Lashaine volcano, North Tanzania, Phys. Chem. Earth, 9, 559-596, 1975.

Ross, C. S., M. D. Foster, and A. T. Myers, Origin of dunites and of olivine-rich inclusions in basaltic rocks, Am. Mineral., 39, 693-737, 1954.

Sobolev, N. V., Yu G. Lavrentev, N. P. Pokhilenko, and L. V. Usova, Chrome-rich garnets from the kimberlites of Yakutia and their paragenesis, Contr. Mineral. Petrol., 40, 39-52, 1973.

Wells, P. R. A., Pyroxene thermometry in simple and complex systems, Contr. Mineral. Petrol., 62, 129-139, 1977.

Whitfield, G. G., A petrological and mineralogical study of peridotite and eclogite xenoliths from certain kimberlite pipes, Unpublished M.Sc. Thesis, Rhodes University, South Africa, 1971.

Wood, B. J., The solubility of alumina in orthopyroxene coexisting with garnet, Contr. Mineral. Petrol., 46, 1-15, 1974.

Wood, B. J., and S. Banno, Garnet-orthopyroxene and orthopyroxene-clinopyroxene relationships in simple and complex systems, Contr. Mineral. Petrol., 42, 109-124, 1973.

POLYMICT PERIDOTITES FROM THE BULTFONTEIN AND DE BEERS MINES, KIMBERLEY, SOUTH AFRICA

P.J. Lawless

Anglo American Research Laboratory, P.O. Box 106, Crown Mines, Transvaal, 2025, South Africa.

J.J. Gurney

Department of Geochemistry, University of Cape Town, Private Bag, Rondebosch, 7700 Cape, South Africa

J.B. Dawson

Department of Geology, Purdie Building, University of St. Andrews Fife, Scotland KY 16 9 ST. United Kingdom.

Abstract. Four samples of hitherto undescribed xenoliths from the Bultfontein and De Beers kimberlites have been studied. The rocks consist of a wide variety of peridotitic xenoliths, eclogite fragments and megacryst minerals cemented together by ilmenite, phlogopite, rutile and sulphides. The rocks have been called polymict peridotites. They are interpreted as remnant conduit filling left behind in the mantle after the migration of a kimberlitic fluid upwards. Subsequent to consolidation the polymict rocks were sampled by one of the Bultfontein kimberlite intrusions.

A zircon age of 82.8 m.y. is thought to apply to the sampling event.

Introduction

The xenolith suites in the Bultfontein and De Beers Mines, Kimberley, consist mainly of peridotites (lherzolites, harzburgites, dunites and wehrlites together with more scarce mica rich rocks (MARID-suite) and rare eclogites.

Here we describe four xenoliths which are unlike any others previously described. The samples are tectonic mixtures of upper mantle rock clasts and minerals that are cemented by variable amounts of picro-ilmenite, phlogopite, rutile and sulphides. Clasts of harzburgite, lherzolite and pyroxenite have been identified. Individual mineral phases include garnet, clinopyroxene, orthopyroxene, olivine and chromite. Zircon has been found in one sample. No crustal rock fragments have been recognised.

The rocks are disequilibrium assemblages and the rock components are derived from many sources. We propose to call these rocks "polymict peridotites" though it must be noted that within them we have recognised megacrysts and phases of eclogitic origin.

Description of the Polymict Peridotites

Using the terminology of Harte (1977) the textures of two of the nodules are phorphyroclastic, one is laminated disrupted porphyroclastic and the fourth is disrupted mosaic porclastic.

The largest of the four rocks measures 18 x 16 x 11 cms.

Olivine. The most abundant silicate constituent in all four polymict rocks is olivine which occurs as porphyroclasts up to 5 mm in diameter; they are usually anhedral and display undulose extinction. Most have been partially serpentinised and many have been partially recrystallised into neoblasts or tabular crystals ($<$ 0.05 mm). Some porphyroclasts have almost completely recrystallised. In JJG 1414 neoblastic olivine makes up $\sim$ 90% of the rock, forming a dark green matrix in which olivine porphyroclasts occur.

The ranges of composition of olivine porphyroclasts vary from rock to rock. Individual porphyroclasts have zoned margins, though all are essentially magnesium-rich forsterites, in which MgO, FeO and SiO_2 are the major constituents. Often, the amounts of CaO, Al_2O_3, Cr_2O_3

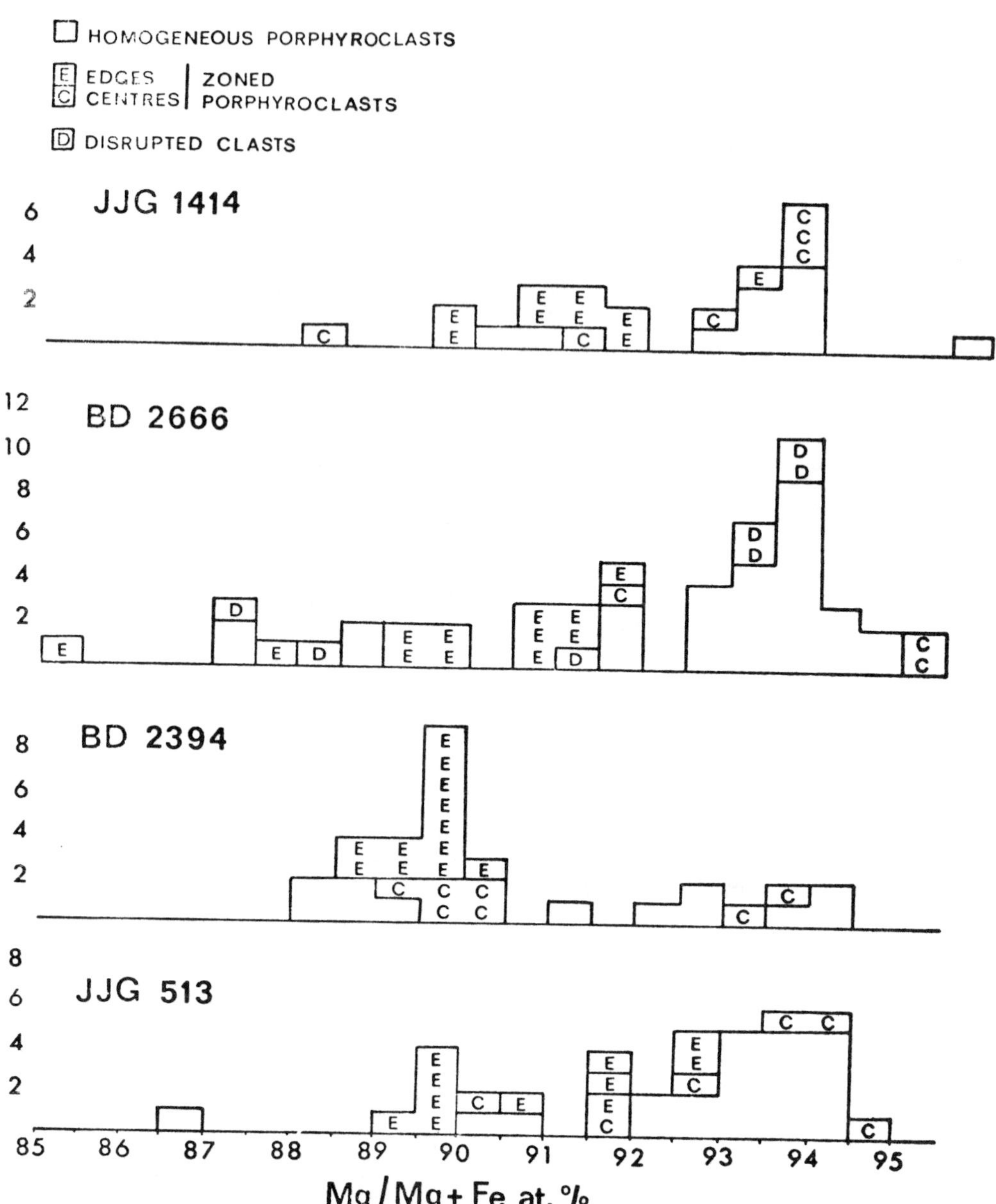

Figure 1. Histogram showing Mg/(Mg+Fe) x 100 for orthopyroxenes, not associated with clinopyroxene grains or megacrysts in JJG 1414, BD 2666, BD 2394 and JJG 513.

and TiO_2 are below detection limits for routine analysis. The Mg/Mg + Fe ratio ranges from 0.86 to 0.96.

In BD 2666 small, clear anhedral unstrained neoblasts occasionally penetrate large porphyroclasts and show major chemical differences compared with the host clast. The zoning in the olivines is erratic but generally the cores are more magnesian rich than the edges. The most magnesian olivine occurs in BD 2394. The olivines in JJG 1414 have not been as extensively studied as those in the other three xenoliths.

Orthopyroxene. Orthopyroxene occurs as porphyroclasts in all four rocks and also within and around clinopyroxene megacrysts.

A few porphyroclasts are clear and not serpentinised but many are "cloudy", serpentinised and display undulose extinction with well developed kink banding. In BD 2666 some orthopyroxenes have been disrupted and strung out into discontinuous chains. The orthopyroxenes in JJG 1414 appear in three separate forms (a) as elongated, oval or rounded porphyroclasts which may have been disrupted, and/or recrystallised and/or partially serpentinised: (b) as aggregates of tiny grains in the olivine matrix and (c) surrounding or enclosed within clinopyroxene megacrysts. Orthopyroxene with apparent exsolved rutile platelets has been seen in three of these xenoliths.

The orthopyroxene around clinopyroxene grains and/or within clinopyroxene megacrysts is usually associated with small phlogopite flakes and finely disseminated opaque minerals.

The orthopyroxenes are magnesian enstatites. As with the olivines, the porphyroclasts are not all chemically homogeneous. Many have been either altered or recrystallised along edges with marked changes in composition, particularly increased levels of Al_2O_3, Cr_2O_3, CaO, Na_2O and TiO_2 and generally decreased Mg/(Mg + Fe). The range of Mg(Mg + Fe) values for orthopyroxenes is shown in Figure 1.

Figures 2 and 3 indicate the large range of Al_2O_3 and CaO contents.

Like the olivines, the orthopyroxenes in JJG 1414 are more deformed than in the other three rocks. Discrete small grains of orthopyroxene and aggregates of grains are present in the olivine matrix of this rock. One such aggregate has been analysed and the orthopyroxenes show a range of Al_2O_3 contents from 1.78 wt. % to 5.94 wt. %. The results of these latter analyses have not been displayed in Figures 2 or 3.

Garnet. The garnets are anhedral and vary in shape, size, colour and composition. The largest is ~ 20 mm in longest dimension but it is the only one > 8 mm. The colours range from yellow-orange through red to mauve. A few are alexandritic. All have kelyphitic margins. Some have been disrupted.

In two specimens (JJG 1414 and BD 2394) yellow garnet "overgrowths" occur around mauve garnets as observed in a megacryst by Meyer et al. (this volume). In one example, small (~1 mm)

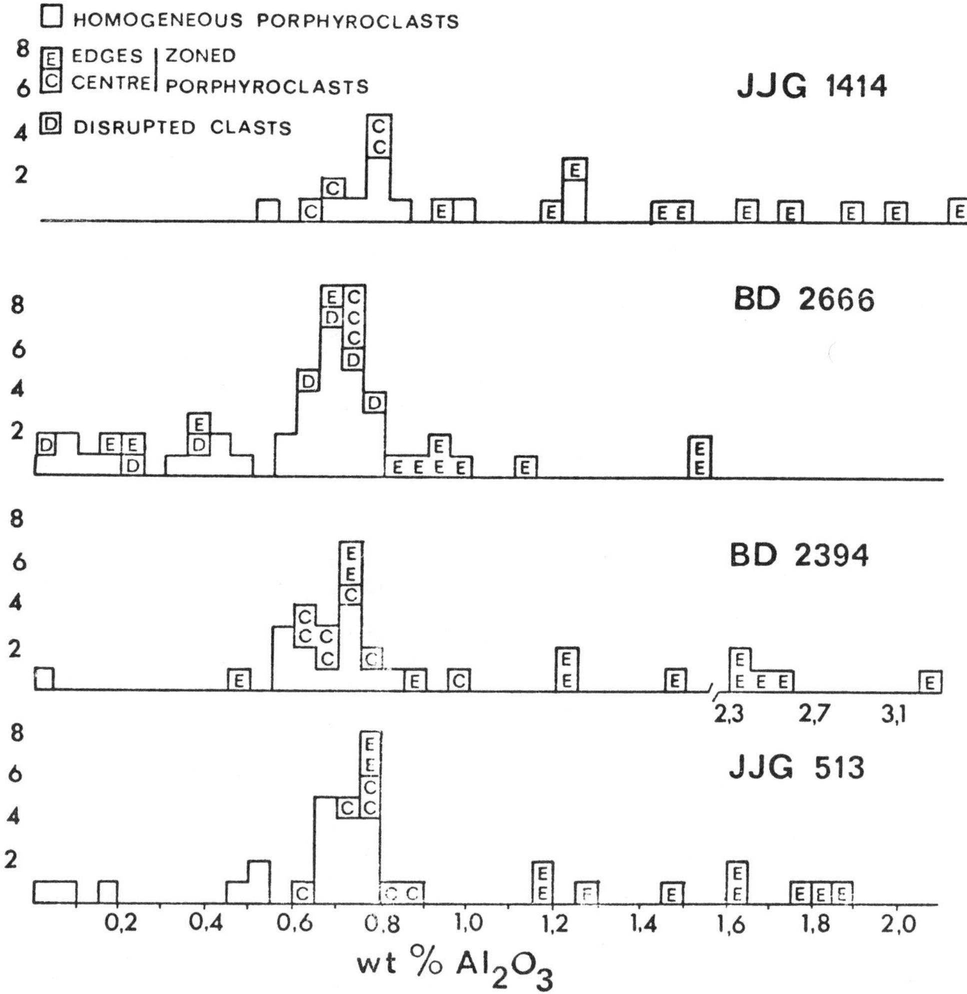

Figure 2. Histogram of Al_2O_3 wt. % for orthopyroxenes, not associated with clinopyroxene, in JJG 1414, BD 2666, JJG 513.

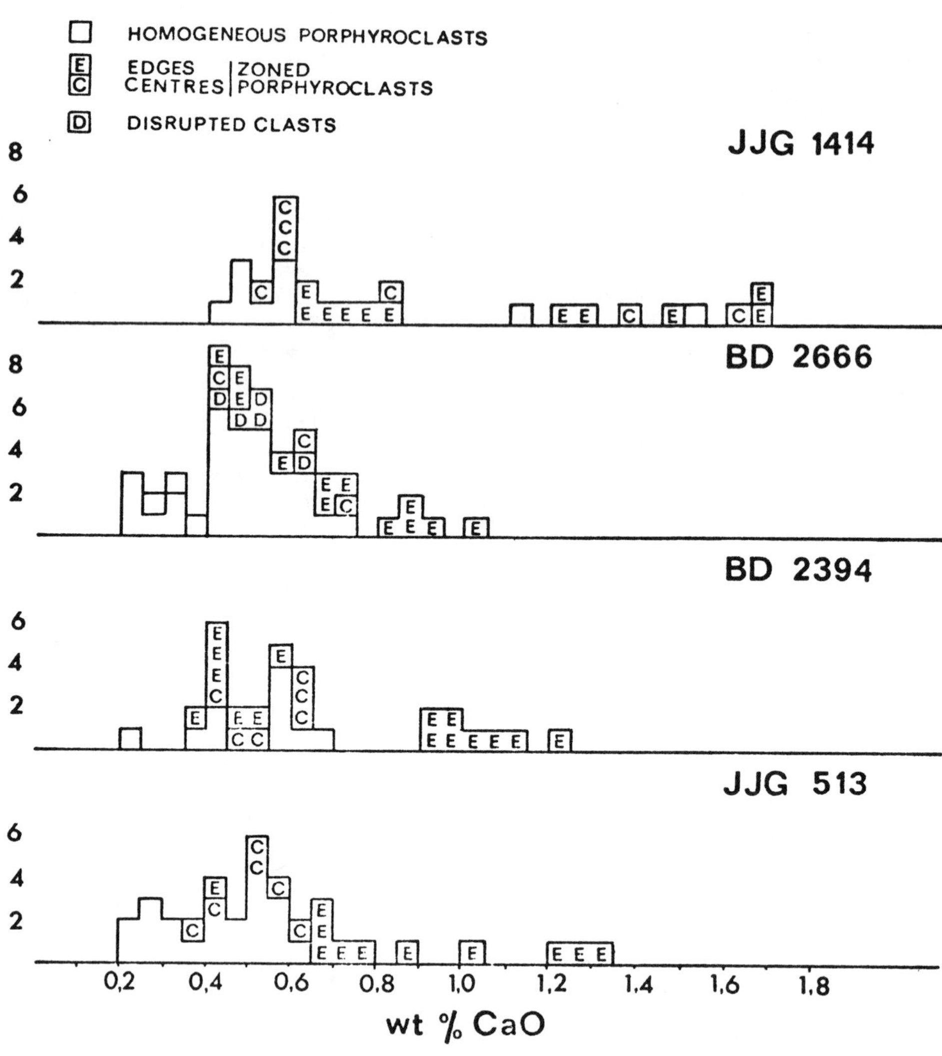

Figure 3. Histogram of CaO wt. % in orthopyroxene not associated with clinopyroxene, JJG 1414, BD 2666, BD 2394 and JJG 513.

yellow garnets on mauve garnets have almost hexagonal shapes. Occasionally garnets are associated with the clinopyroxene-orthopyroxene megacrysts. Clinopyroxene, orthopyroxene, olivine and spinel have been found as inclusions in garnets in all four rocks.

The garnets in the polymict peridotites have been classified into three groups based on the scheme of Dawson and Stephens (1975):

(a) Chrome pyropes similar to those found as inclusions in diamond having less than 3 wt. % CaO, (b) eclogitic garnets having less than 1 wt. % Cr_2O_3 and (c) harzburgite-lherzolite-websterite chromium garnets with more than 3 wt. % Cr_2O_3. This last group has been further subdivided into two smaller groups, one with greater than 0.2 wt. % TiO_2, the other having less than 0.2 wt. % TiO_2. Whilst in four instances, (d), (e), (f) and (g) rare garnets are described separately but are related to group (c).

(a). Five garnets of the "low calcium" variety have been found in BD 2666.

One other "low calcium" garnet was found. This garnet is zoned with respect to calcium and titanium and to a lesser extent chromium. TiO_2 along edges has a maximum value of 0.19 wt. % whereas in the centres TiO_2 is below detection limits. CaO varies from 1.12 wt. % in the centres up to 3.51 wt. % on some edges in this garnet.

(b). Eclogitic garnets have been found in three of the rocks. BD 2394 shows a large range of eclogitic garnets, BD 2666 has four and JJG 513 one.

(c). The "low-TiO_2 harzburgite lherzolite" garnets show variations in CaO and FeO contents and particularly in Cr_2O_3 content. Two very alexandritic garnets have high calcium and chromium levels. (7.06 wt. % CaO and 11.35 wt. % Cr_2O_3 and 10.8 wt. % CaO and 14.7 wt. % Cr_2O_3). "High-TiO_2 harzburgite lherzolite garnets" occur in three rocks and show considerable variation in CaO and Cr_2O_3 and FeO. TiO_2 ranges from 0.2 - 0.7 wt. %. In JJG 1414 three additional types of "harzburgite-lherzolite" garnet have been noted :

(d). The large garnet megacryst is essentially a chromiferous pyrope with the bulk of the garnet having 4.71 wt. % CaO, 8.27 wt. % FeO, 3.32 wt. % Cr_2O_3 and 0.50 wt. % TiO_2. The TiO_2 content, however, increases to 0.78 and 0.98 wt. % along a crack and on an edge with FeO, Cr_2O_3 and CaO varying erratically. The composition of the homogeneous part of this garnet matches that of garnet megacrysts from elsewhere, (eg. Monastery Mine).

(e). Three garnets are strongly but complexly zoned with respect to TiO_2, FeO, CaO and Cr_2O_3, with greatly increased TiO_2 contents on one or more edges compared with the centres. These centres have 0.2 wt % TiO_2 but certain edges may have as much as 1.15 wt. % TiO_2.

(f). Garnets which occur as small disrupted grains or as deformed clasts in the neoblastic olivine matrix and also associated with the clinopyroxene-orthopyroxene megacrysts could be classified into the "high TiO_2 harzburgite-lherzolite" group but are variable in composition with large differences in the relative amounts of CaO, FeO, Cr_2O_3 and TiO_2 from 0.65 to 1.87 wt. %. There is no obvious chemical relationship between these garnets and adjacent grains may have very different compositions.

(g). A most unusual garnet occurs in BD 2394. It has a pink-mauve central core which is compositionally a "low TiO_2 harzburgite-lherzolite" garnet. Forming a corona like overgrowth on this is yellow garnet. The core is homogeneous (0.07 wt. % TiO_2, 5.01 wt. % Ca, 6.58 wt. % FeO and 4,01 wt. % Cr_2O_3). The corona composition is within the ranges: 0.31 - 3.81 wt. % TiO_2, 3.11 - 3.86 wt. % CaO, 9.31 - 10.28 wt. % FeO and 0.91 - 1.93 wt. % Cr_2O_3.

The range of garnet compositions in all four polymict peridotites has been illustrated by the use of Ca-Mg-Cr and Ca-Mg-Fe ternary diagrams (Figure 4 and 5). Not all the garnets analysed have been plotted in these diagrams due to the considerable overlap of compositions in the various rocks.

Clinopyroxene. All the rocks have megacrysts of clinopyroxene completely or partially enclosed

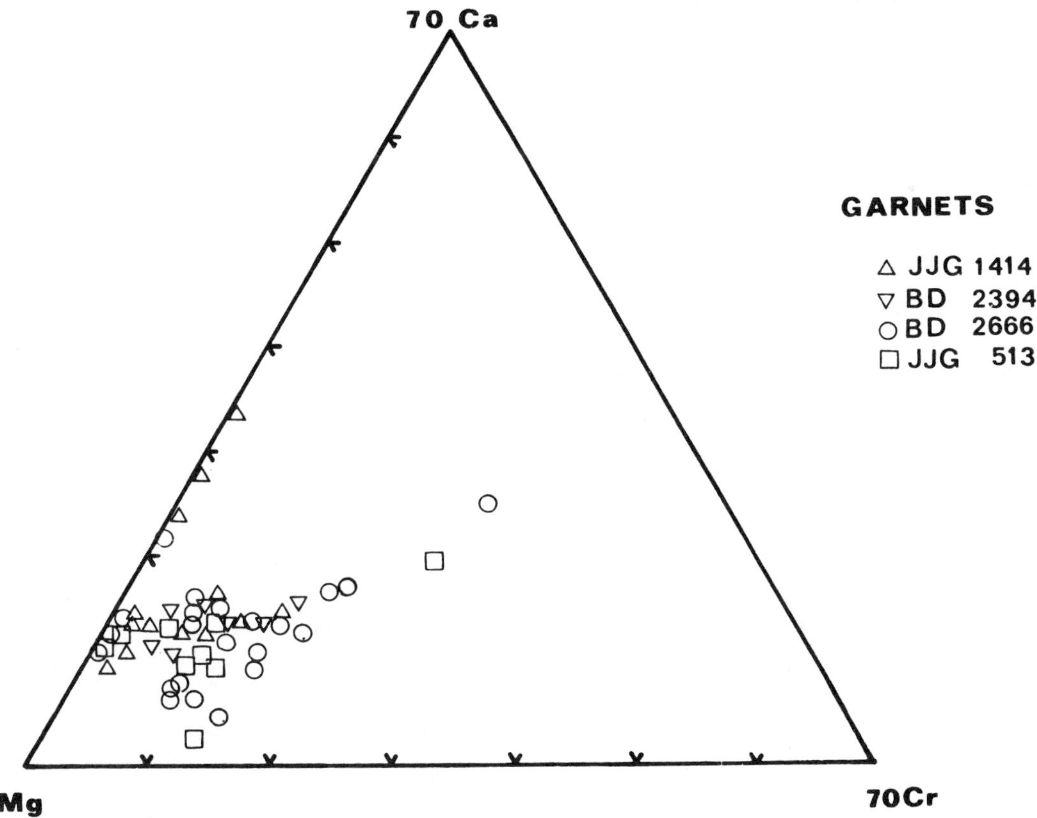

Figure 4. Ca-Mg-Cr ternary diagram for some of the garnets in JJG 513, BD 2394, JJG 1414 and BD 2666.

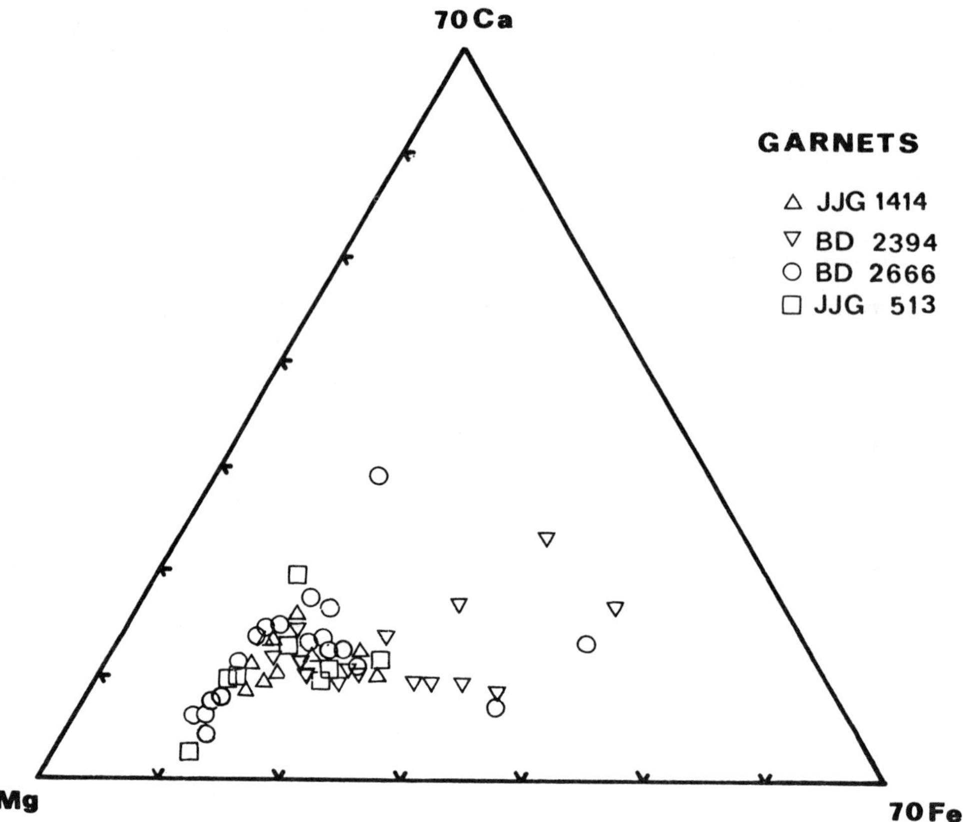

Figure 5. Ca-Mg-Fe ternary diagram for some of the garnets in JJG 513, BD 2394, JJG 1414 and BD 2666.

by randomly orientated orthopyroxene associated with finely disseminated phlogopite and opaque minerals. The clinopyroxene in the megacrysts is anhedral and has irregular poorly defined grain boundaries. Some grains have serrated edges, and the margins may be defined by finely disseminated opaque phases.

Some clinopyroxenes in the clinopyroxene-orthopyroxene association are $<$ 2 cm in maximum dimension and therefore by the terminology of Harte (1977) are not megacrysts. However since these are also rimmed by orthopyroxene and are similar in all other respects to the true megacrysts they are grouped together.

Compositions of the clinopyroxene in megacrysts in BD 2666 and BD 2394 are relatively constant and although there are minor variations between individual grains these differences are small. For the megacrysts in JJG 513 and JJG 1414, however, there are sub-grains within a single megacryst which show chemical variability, particularly for Al_2O_3, Cr_2O_3 and TiO_2.

Tables 1 and 2 illustrate the minimum and maximum values for the oxides, (Mg/(Mg + FeO) and Ca/(Ca + Mg) within single megacrysts for JJG 513 and JJG 1414 respectively. These tables also illustrate minimum and maximum values for the orthopyroxenes associated with these megacrysts.

Within the clinopyroxene-orthopyroxene megacrysts the presence of opaque minerals associated with phlogopite, finely disseminated chiefly within the orthopyroxene rich areas, is ubiquitous. Analysis has confirmed the presence of ilmenite, rutile and chromite. The ilmenites have high chromium (5.3 - 7.7 wt. % Cr_2O_3) and high magnesium ($\pm$ 14 wt. %). The range observed in the chromites is shown in Table 3.

The associated phlogopite usually has $<$ 2 wt. % TiO_2 and $<$ 1.75 wt. % Cr_2O_3. One phlogopite grain is much larger (0.5 cm) than any other and is heavily zoned. (eg. 0.76 - 3.8 wt. % TiO_2).

Apart from the above, rare occurrences of clinopyroxene were noted in rock clasts, as inclusions in olivine, orthopyroxene and garnet and in an ilmenite vain.

<u>Phlogopite</u>. Phlogopite is widely disseminated throughout the rocks. In many cases it is clearly associated with the blebs, veins and stringers in which ilmenite and rutile are also found and which appears to cement the other diverse components together. The phlogopites vary in size from large unzoned relatively undeformed blades in JJG513, to small (0.01 - 0.02 mm) strained grains in BD 2666 where they occur in "stringers" seldom more than 1 mm wide which

TABLE 1. Minimum and Maximum values for clinopyroxenes and orthopyroxenes in megacrysts in JJG 513.

Clinopyroxenes		Megacryst (a)		Megacryst (b)		Megacryst (c)		Megacryst (d)		Megacryst (e)		Megacryst (f)	
		Minimum	Maximum	Minimum	Maximum	Minimum	Maximum	Minimum	Maximum	Minimum	Maximum	Minimum	Maximum
TiO_2	wt.%	0.15	0.49	0.17	0.40	0.30	1.22	0.41	0.46	0.13	0.35	0.30	0.83
Al_2O_3	wt.%	0.63	1.29	1.45	2.05	0.51	2.93	2.08	3.03	1.63	2.57	0.51	2.17
Cr_2O_3	wt.%	0.98	2.09	1.83	3.24	0.44	2.16	1.43	1.74	2.21	2.42	1.52	2.95
FeO	wt.%	2.66	4.02	2.51	3.52	2.87	4.13	3.26	3.69	2.30	3.35	2.87	3.73
MgO	wt.%	16.76	18.12	15.94	17.53	16.86	19.43	15.95	17.43	15.97	17.58	16.51	18.32
CaO	wt.%	19.78	22.24	17.77	20.06	18.52	22.29	17.80	19.64	19.04	19.39	17.70	19.29
Na_2O	wt.%	1.23	1.51	1.44	3.00	0.91	2.55	1.44	2.98	1.68	2.91	1.37	2.66
Mg/Mg+Fe	at.%	88.7	92.0	89.7	92.1	89.3	91.9	89.4	91.3	90.3	92.8	89.8	91.1
Ca/Ca+Mg	at.%	44.0	48.4	43.5	45.1	41.3	46.9	43.0	44.8	42.7	46.5	43.1	43.5
Orthopyroxenes													
TiO_2	wt.%	0.16	0.28	0.12	0.35	0.19	0.31	0.01	0.31	0.16	0.32	0.15	0.28
Al_2O_3	wt.%	0.44	2.68	0.59	1.50	0.83	2.71	0.54	2.95	0.70	1.98	0.81	1.99
Cr_2O_3	wt.%	0.10	1.18	0.25	1.23	0.62	1.68	0.27	1.53	0.43	1.50	0.28	1.62
FeO	wt.%	6.77	8.00	6.77	7.52	6.94	8.08	5.37	7.25	6.81	7.47	5.82	7.16
MgO	wt.%	32.56	35.17	32.61	34.78	32.74	33.86	32.20	36.24	32.91	34.93	32.51	34.39
CaO	wt.%	0.37	1.14	0.46	1.04	0.79	1.61	0.33	1.56	0.67	1.63	0.50	1.24
Na_2O	wt.%	0.05	0.64	0.05	0.19	0.15	0.44	0.08	0.47	0.08	0.25	0.05	0.33
Mg/Mg+Fe	at.%	87.9	90.1	88.6	90.1	88.0	89.4	88.9	92.3	89.2	90.0	88.9	91.2
Ca/Ca+Mg	at.%	0.75	2.41	0.92	2.24	1.65	3.46	0.65	3.37	1.37	3.44	1.03	2.67

TABLE 2. Minimum and Maximum values for clinopyroxenes and orthopyroxenes in megacrysts in JJg 1414.

Clinopyroxenes		Megacryst (a)		Megacryst (b)		Megacryst (c)		Megacryst (d)	
		Minimum	Maximum	Minimum	Maximum	Minimum	Maximum	Minimum	Maximum
TiO_2	wt.%	0.27	0.46	0.33	0.27	0.33	0.35	0.29	0.36
Al_2O_3	wt.%	1.64	2.63	1.91	2.36	2.06	2.48	1.82	2.32
Cr_2O_3	wt.%	2.45	2.70	2.48	2.69	2.48	2.51	2.32	2.52
FeO	wt.%	2.49	2.75	2.61	2.84	2.74	2.71	2.50	2.64
MgO	wt.%	16.62	17.77	16.41	16.97	16.78	17.11	16.57	17.00
CaO	wt.%	18.18	19.95	18.47	18.79	17.94	18.81	18.55	19.23
Na_2O	wt.%	2.37	2.77	2.37	2.62	2.46	2.49	2.38	2.75
Mg/Mg+Fe	at.%	91.7	92.3	91.4	92.0	91.7	91.7	91.8	92.3
Ca/Ca+Fe	at.%	43.6	45.4	44.0	45.1	43.0	44.6	44.4	45.1
Orthopyroxenes									
TiO_2	wt.%	0.15	0.37	0.12	0.23	0.15	0.36	0.12	0.28
Al_2O_3	wt.%	1.33	4.01	1.36	3.12	1.75	4.23	1.93	3.89
Cr_2O_3	wt.%	0.73	1.89	0.79	1.33	0.85	1.69	0.98	1.88
FeO	wt.%	5.11	5.80	5.34	6.10	5.42	5.70	5.30	5.66
MgO	wt.%	32.13	35.45	32.88	34.68	32.52	35.17	32.88	34.49
CaO	wt.%	0.55	1.47	0.81	1.22	0.80	1.54	0.59	1.51
Na_2O	wt.%	0.21	0.46	0.22	0.42	0.20	0.43	0.23	0.44
Mg/Mg+Fe	at.%	90.8	92.3	90.6	91.8	91.1	92.0	91.2	91.8
Ca/Ca+Mg	at.%	1.1	3.1	1.7	2.5	1.6	3.3	1.2	3.2

are generally aligned parallel to the adjacent edges of ilmenite veins and clast grain boundaries.

Phlogopite also occurs, as described earlier, in the megacrysts where it is associated with orthopyroxene and opaque minerals; with altered semi-translucent orthopyroxene and as an alteration product of garnets.

In JJG 1414 there are some small ($\sim$0.5 mm) deformed phlogopite grains completely surrounded by neoblastic olivine.

Zoned phlogopite associated with serpentinisation has pleochroic cores and may show undulose extinction. It may sometimes take the form of pools, up to 15 mm long, of numerous small randomly orientated lathes.

Although the larger blades are usually chemically unzoned individual phlogopites have different Ti, Fe and Cr contents, whilst phlogopites along grain boundaries with serpentine and rutile are often zoned with low Ti and Cr cores relative to edges. One example has variations of 0.58 - 4.33 and 0.38 to 1.09 wt. % TiO_2 and Cr_2O_3 respectively.

Ilmenite. Ilmenite in JJG 513 and BD 2394

TABLE 3. Chromite Compositions in JJG 513.

		Inclusion in OLIVINE JJG513	Inclusion in GARNET JJG513	In rock clast JJG 513	In cpx-opx megacrysts
Cr_2O_3	wt.%	54.5	29.5	62	44.8-54.0
Al_2O_3	wt.%	10.5	36.4	1.5	5.5-10.2
TiO_2	wt.%	0.6	0.1	2.0	2.5-5.85

occurs as blebs up to 4 x 3 cm in size, and in these two rocks is always associated with phlogopite. It is often intimately mixed or rimmed by rutile.

In two rocks ilmenite veins showing a strong parallel lineation are mantled by rutile. These veins often partially enclose silicate minerals and separate adjacent rock clasts.

Ilmenite also occurs in the clinopyroxene - orthopyroxene megacrysts.

All the ilmenites have high magnesium contents (always greater than 12 wt. % MgO) and exhibit compositional differences within individual blebs and veins. These are most evident in the chrome contents. There are only rare examples of blebs which are chemically homogeneous.

In BD 2666 the Cr_2O_3 content varies from 1 to 4.4 wt. % in separate veins and may vary across a single vein. Generally Cr_2O_3 is higher along the edges of veins than in the centres. Two examples from separate veins are :-

a) Centre 1.47; Edges 1.9., 2.18;

b) Centre 2.08; Edges 2.64 and 2.67 wt. % Cr_2O_3.

Some veins in BD 2666 appear to consist of two distinct groups of ilmenite, one with Cr_2O_3 ranging from 1.2 to 2.0 wt. % and the other from 2.5 to 3.1 wt. %.

There is no obvious relationship between the various veins and a few veins within 2 cm of each other can show the whole range of Cr_2O_3 variations. These variations are greatest in the ilmenite veins in JJG 1414 where one vein has Cr_2O_3 levels ranging from 2.20 (centre) to 7.08 wt. % (edge).

BD 2394 has a clast which consists of an intimate mixture of ilmenite and rutile in which the Cr_2O_3 content of the ilmenite varies from 3.97 to 6.75 wt. % with higher values towards clast edges of the ilmenite, and around silicate inclusions in the central portions of the clast. The Cr_2O_3 content of the rutile varies from 4.2 to 6 wt. % and it may have up to 2.5 wt. % Fe_2O_3.

Ilmenites in orthopyroxenes or associated with clinopyroxene megacrysts are not homogeneous.

The range of ilmenite compositions in the polymict peridotites has been diagrammatically illustrated in Figure 6, a Fe_2O_3 - $FeTiO_3$ - $MgTiO_3$ ternary. (The hematite-ilmenite-geikelite components were calculated by the method of Finger (1972)). Figure 6 shows the very restricted range of compositions compared with the field for kimberlitic ilmenites, (See review by Haggerty, 1975). There is no overlap of the ilmenite compositions in these rocks with the ilmenite data taken from Mitchell (1977) for various kimberlites.

Rutile. Rutile appears in all four rocks as inclusions in, and as rims to ilmenite veins and blebs; as discrete veins or blebs; as exsolved tiny oriented platelets in rare orthopyroxene crystals; and associated with serpen-

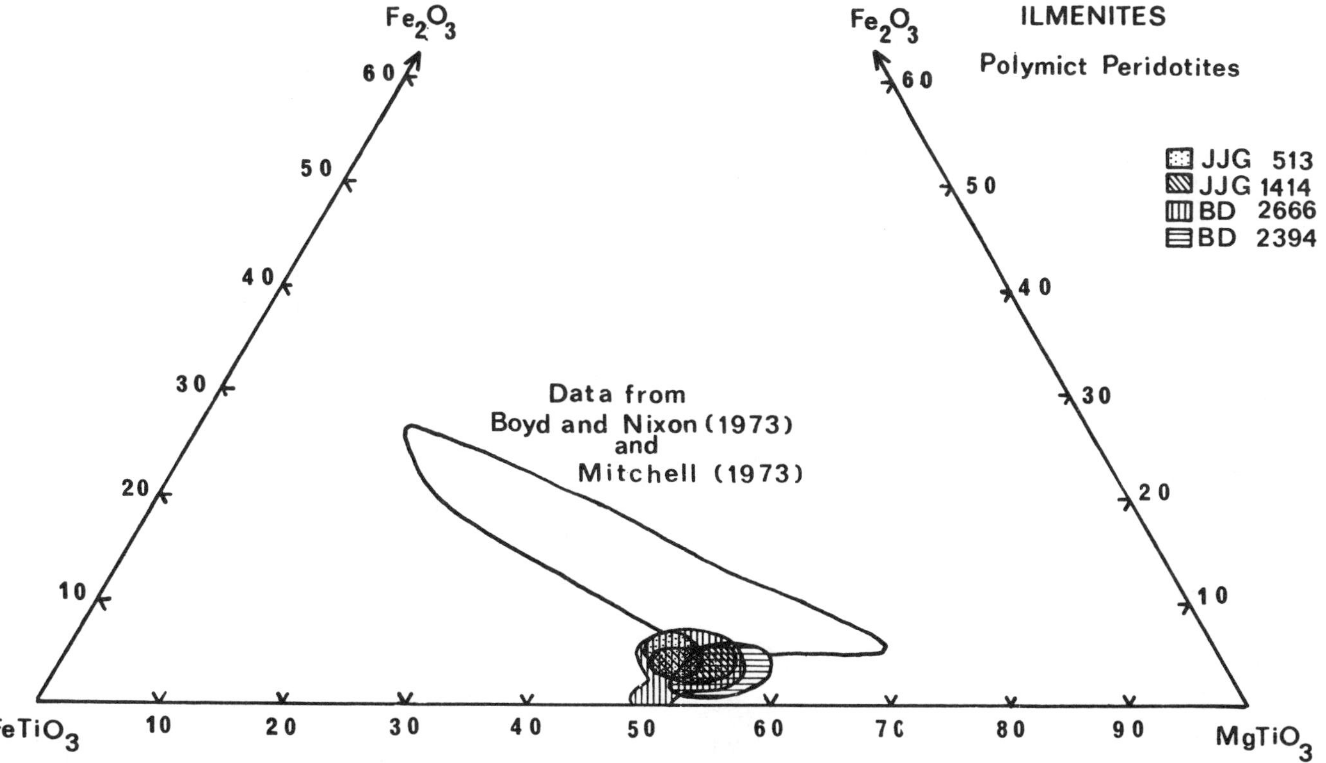

Figure 6. Fe_2O_3-$FeTiO_3$-$MgTiO_3$ ternary diagram for ilmenites in JJG 513, BD 2394, JJG 1414 and BD 2666 (after Haggerty, 1975).

tine veins and phlogopite, occasionally as inclusions in these phlogopites; and as tiny disseminated grains in some of the clinopyroxene-orthopyroxene megacrysts.

Chromite. Chromite occurs as an alteration product of chromiferous garnets in kelyphitic rims; as inclusions in garnets and olivines; in a pyroxenite rock clast; and as the most common opaque mineral associated with the clinopyroxene-orthopyroxene megacrysts.

The chromite compositions vary widely with respect to Cr. Al and Ti. Some of these variations are shown in Table 3.

Sulphides. Sulphide phases occur as irregularly shaped blebs closely associated with ilmenite, phlogopite stringers and serpentine veins. The sulphides have been tentatively identified optically as pentlandite, pyrite, chalcopyrite and possibly pyrrhotite.

Calcite. Calcite is very rare. It has been noted in JJG 513 and BD 2394 associated with serpentine veins and phlogopite.

Rock Clasts. Harzburgite, lherzolite and pyroxenite rock clasts have been identified. The orthopyroxenes in one harzburgite rock clast have been severely altered and show undulose extinction and kink banding. A pyroxenite rock clast in JJG 513 is made up of clinopyroxene and orthopyroxene with chromite inclusions. In BD 2394, one altered rock clast consists of garnet, orthopyroxene, phlogopite and opaque minerals. Nearly all these garnets are yellow and some have overgrowths with a variable amount of alteration of the cores and overgrowths. The mineral compositions of the rock clasts are within the ranges for the constituent minerals discussed in the preceding sections.

Serpentinisation. Serpentinisation, although common in all four xenoliths is not extensive and occurs mainly along cracks in mineral grains, both silicate and opaque, or around grain boundaries and as veins cutting through both porphyroclasts and neoblasts.

Zircon. BD 2666 was found to contain anhedral zircons. Some of these zircons were associated with phlogopite stringers. Two other examples were rounded zircons set into the neoblastic olivine matrix. The zircons gave lead-uranium ages averaging 82.8 million years (Davis, 1977).

Discussion

The four polymict peridotites contain a variety of magnesium rich rock clasts and silicate minerals, together with opaque phases which all appear to have formed in the upper mantle.

Some garnet, orthopyroxene, olivine and chromite compositions closely match the compositions of certain diamond inclusions (Meyer and Boyd, 1972; Sobolev, 1974).

The wide variety of garnets present are commonly of peridotitic and rarely of eclogitic paragenesis. The olivines and orthopyroxenes also show large compositional variations which, even apart from the chemical zoning at grain margins, is so large as to suggest the derivation of all three minerals from the full suite of upper mantle rocks found in the Bultfontein pipe as xenoliths.

Many of the porphyroclasts and the mineral grains in rock clasts have been altered and/or recrystallised giving rise to marked changes in chemical composition between the centres and edges of individual grains. It has previously been observed that there were differences in chemical composition between some orthopyroxene porphyroclasts and associated neoblasts in peridotites (Boyd, 1975). In the polymict peridotites the changes in Al, Cr, Ca, Na and Ti are larger than previously described and there is also an increase in Fe along grain boundaries and/or associated with recrystallisation. These variations are interpreted as being due, in part at least, to the effects of metasomatising fluids, probably a fluid from which the ilmenite, rutile and phlogopite have precipitated.

Clinopyroxene-orthopyroxene megacrysts are quite abundant. The uniqueness of their appearance must be stressed. Typically the clinopyroxene forms a central core surrounded by orthopyroxene, although in larger megacrysts this separation may be irregular and imperfect.

The orthopyroxenes surrounding or enclosed within the clinopyroxene have unusually large variations in Al_2O_3, Cr_2O_3 and CaO contents. Some values are considerably in excess of those found in other upper mantle derived orthopyroxene.

The observed chemistry of the megacrysts is interpreted as being due to metasomatism and partial re-equilibration of clinopyroxene megacrysts to a lower temperature environment within the mantle. A favoured explanation is that megacrysts with sub-calcic compositions were transported from a region of higher temperature and pressure to shallower levels within the mantle by a magmatic event. Re-equilibration to the new P/T regime gave rise to exsolution of orthopyroxene which completely or partially migrated towards crystal boundaries depending on megacryst dimensions. Metasomatic transfer occurred between the transporting medium and the megacrysts to give the observed chemical changes at grain margins, which are probably related to the process in which the large blebs of ilmenite, the large phlogopite blades, the rutile and sulphides crystallised.

The zircon age of 82.8 $\pm$ 2 m.y. (Davis, 1977) is considerably younger than the age of a zircon xenocryst from Bultfontein (91.2 m.y.) and 4.1 m.y. younger than the next youngest zircon age in the Kimberley area of 86.9 m.y. for Kamfersdam (op.cit.).

The 82.8 m.y. age is interpreted as the date of sampling of the rock from the mantle which is considered to have occurred during one of the younger of the several known Kimberlite intrusive events at Bultfontein.

Conclusions

The polymict peridotites consist of xenolithic and xenocrystic minerals of diverse origin cemented by phlogopite ilmenite, rutile and sulphides.

The xenolithic fragments were sampled and transported within the upper mantle prior to their incorporation into the rocks described here. The fragments were derived from a wide range of peridotitic rocks, from eclogite and from megacrysts. They were deformed, partially recrystallised and metasomatised, particularly at grain margins and along cracks, in the sequence of processes during which the cementing minerals were formed and the polymict peridotites were consolidated.

The clinopyroxene megacrysts re-equilibrated imperfectly to more calcic, lower temperature compositions by exsolving orthopyroxene at the same stage, giving rise to the unique rimming with orthopyroxene. The finely disseminated phlogopite and opaque minerals were introduced into the megacrysts by the fluids present at that stage.

The rocks are interpreted to have formed in a two stage process.

1. Migration of a mantle fluid from depth to another shallower location in the mantle. This melt transported the megacrysts with it and sampled the xenolithic components from the mantle traversed during migration. The melt crystallised the ilmenite, phlogopite, sulphides and rutile, and perhaps its last remnants moved on, away and upwards induced by a process such as filter pressing, at the termination of this particular phase of igneous activity.

2. The consolidated polymict rocks were subsequently sampled by one of the Bultfontein Kimberlite intrusions which flushed out the infilling of the old conduit.

It is considered most probably that the initial event was kimberlitic on the basis of the ilmenite and phlogopite compositions. Possibly an earlier Bultfontein kimberlite was involved. The deformation which appears to be a late feature was probably related to the closing of the conduit and the squeezing out of the last remnants of volatile material presumably including carbonates during the first stage mentioned above.

Since this work was completed samples of similar rocks have been found at Du Toits Pan and Monastery Mines.

Acknowledgments. The active support of J.B. Hawthorne and the staff of the Geology Department of De Beers Cons. Mines in Kimberley has been of great assistance in the discovery and study of these rocks.

The financial support of the Anglo American Corporation and De Beers is gratefully acknowledged.

A large portion of the analytical study was completed in the Anglo American Research Laboratories in Johannesburg.

References

Boyd, F.R. (1975). Stress heating and compositional variations in enstatites from sheared lherzolites. Carnegie Inst. Washington Yearb. 74, 525 - 528.

Davis, G.L. (1977), The ages and uranium contents of zircons from Kimberlites and associated rocks. Extended Abstracts, 2nd International Kimberlite Conference, Santa Fé, New Mexico, 1977.

Dawson, J.B. and Stephens, W.E. (1975), Statistical classification of garnets from kimberlite and associated xenoliths. J. Geol., 83, 589-607.

Finger, L.W. (1972), The uncertainty in the calculated ferric iron content of a microprobe analysis. Carnegie Inst. Washington Yearb. 71, 600 - 603.

Haggerty, S.E. (1975), The chemistry and genesis of opaque minerals in kimberlites. Phys. Chem. Earth, 9, 295 - 308.

Harte, B.(1977), Rock nomenclature with particular relation to deformation and recrystallization textures in olivine-bearing xenoliths. J. Geol., 85, 279 - 288.

Meyer, H.O.A. and Boyd, F.R., (1972), Composition and origin of crystalline inclusions in diamond. Geochim. Cosmochim. Acta, 36, 1255 - 1273.

Mitchell, R.H. (1973), Magnesian ilmenite and its role in kimberlite petrogenesis. J. Geol., 81, 301 - 311.

Sobolev, N.V., (1974), Deep seated inclusions in kimberlites and the problem of the composition of the Upper Mantle. A.G.U., 278.

MAFIC AND ULTRAMAFIC XENOLITHS FROM THE KAO KIMBERLITE PIPE

Ian D. MacGregor

Department of Geology, University of California, Davis, California 95616

Abstract. Ultramafic xenoliths from the Kao kimberlite pipe may be subdivided into oxide (± garnet)-bearing and garnet-bearing harzburgites and lherzo-lites. Oxide peridotites are divided into ilmenite and chromite spinel varieties. Both have coarse granular textures; the ilmenite bearing samples show cumulate features while the chromite spinel xenoliths are the product of metamorphic recrystallization. The garnet peridotites occur in three textural types - coarse granular, porphyroclastic and mosaic. Coarse granular and porphyroclastic textured samples are a mixture of harzburgites and lherzolites while the mosaic textured rocks are solely lherzolites. Discrete coarse crystals of the ultramafic minerals also occur.

The coarse grained oxide peridotites occur at the lowest, with the coarse grained, porphyroclastic and mosaic textured garnet peridotites occurring at progressively higher temperatures and pressures. Chromite spinel and coarse grained and porphyroclastic garnet peridotites have compositions indicating that they are refractory residues of earlier partial melting events. Mosaic textured garnet lherzolites have compositions that are more 'primitive'. Ilmenite peridotites are crystal cumulates of mafic magmas in mantle magma chambers. The different rock suites seem to be from texturally and chemically distinct populations and are similar to other previously reported Northern Lesotho xenolith suites.

Eclogites have equilibrated at similar temperatures and pressure to the oxide peridotites and show changes from coarse granular to tabular textures with increasing temperatures and pressures. The textured distinctions also relate to differences in the CaO, Cr_2O_3 and Mg/Mg + Fe of garnet and Cr_2O_3 and Mg/Mg^{2+}Fe of clinopyroxene.

The estimated temperatures and pressures of equilibrium define a curvilinear path in which coarse grained samples are sub-parallel to estimated geothermal gradients for a static shield, while the gradients for the mosaic samples are slightly steeper and support the interpretation that dynamic processes are active at the time of kimberlite formation.

Introduction

A characteristic feature of kimberlite is the presence of xenoliths and xenocrysts whose phase assemblages and compositions indicate that they have been formed at high temperatures and pressures within the earth's upper mantle. These inclusions are prime samples of the mantle and help define the chemical and physical parameters of this region. A number of studies on xenolith suites from a number of continents are now published (Boyd and Nixon, 1975; MacGregor, 1975; Boyd et al, 1977; Danchin and Boyd, 1977, Eggler and McCallum, 1977) and major conclusions are as follows. The estimated temperatures and pressures define a smooth curvilinear curve which is subparallel to calculated geothermal gradients beneath continental shields. At higher temperatures and pressures a number of xenolith suites illustrate an 'inflection' or departure for the theoretical gradient suggesting that the xenoliths and the kimberlite genesis is related to dynamic processes in the mantle. Apart from the Matsoku suite (Gurney et al., 1975), there is a systematic variation in the textures of the xenoliths such that samples formed at progressively higher temperatures and pressure are progressively more intensely deformed. However, the boundary between deformed and undeformed samples is relatively sharp suggestive of a bimodal population. The first is a coarse grained textured suite which characteristically parallels expected stable geotherms, and the second a more highly deformed suite whose temperatures and pressures of equilibration indicate perturbed or apparently steeper gradients than theoretically anticipated. The chemistry of the undeformed suite is characteristic of refractory mantle samples while the deformed xenoliths have unexpectedly high concentrations of those elements normally concentrated in the liquid fraction, and hence appear more 'primitive' or 'fertile'. These generalities are best observed for xenoliths from the South African kimberlites of widely varying age (Boyd and Nixon, 1975; MacGregor, 1975; Danchin and Boyd, 1977), and have been observed in North American (Eggler and McCallum, 1977) and Russian (Boyd et al., 1977) suites. The

TABLE 1. Chemical analyses of olivine.

Sample	137[1]	151[1]	142[1]	209[1]	156[2]	122[2]	127[2]	102[3]	133[3]	157[3]	130[3]	124[4]
SiO_2	41.55	40.64	40.27	42.44	40.27	41.03	39.50	41.90	42.26	41.11	42.45	40.44
TiO_2	.00	.00	.00	.00	.03	.01	---	.02	---	.00	---	---
Al_2O_3	.00	.00	.00	.00	.00	.00	---	.01	---	.00	---	---
Cr_2O_3	.04	.00	.01	.03	.05	.00	---	.02	---	.00	---	---
FeO (total)	7.09	6.89	8.50	6.83	13.17	14.25	20.99	9.46	8.90	7.96	7.97	4.71
MnO	.09	.09	.09	.07	.14	.13	.24	.12	.10	.07	.08	.06
MgO	50.65	51.41	49.17	52.01	46.24	45.40	40.86	49.96	50.32	49.90	50.82	53.53
NiO	---	---	---	.44	---	---	.11	---	.45	---	.45	---
CaO	.01	.00	.00	.00	.02	.02	.06	.03	.04	.01	.04	.01
Na_2O	.02	.01	.00	.02	.02	---	---	---	---	.02	---	.02
Total	99.47	99.04	98.09	101.84	100.00	100.76	101.75	100.52	102.08	99.06	98.80	98.80

TABLE 1. (cont.)

Sample	77[5]	113[5]	125[5]	136[5]	138[5]	147[5]	103[6]	163[6]	203[6]	204[6]	145[6]
SiO_2	41.20	43.97	42.36	40.68	41.16	40.93	41.99	42.05	40.74	39.71	41.17
TiO_2	.00	.01	---	---	.01	.02	.00	.03	.03	.04	.04
Al_2O_3	.00	.00	---	---	.00	.02	.00	.21	.02	.02	.02
Cr_2O_3	.00	.01	---	---	.04	.07	.00	.02	.04	.04	.06
FeO (total)	6.39	8.86	8.14	14.65	7.67	8.75	9.23	9.12	10.97	12.70	8.90
MnO	.04	.10	.09	.17	.09	.10	.10	.14	.12	.13	.10
MgO	51.61	49.27	50.88	45.82	41.16	48.91	49.35	48.73	49.29	46.65	49.38
NiO	.34	---	.42	.40	.02	.04	---	---	.40	---	---
CaO	.00	.05	.04	.06	.03	.04	.05	.05	.06	.05	.07
Na_2O	.00	---	---	---	.03	.03	---	---	.04	.04	.04
Total	99.59	102.27	101.94	101.78	99.82	98.91	100.73	100.36	101.31	99.40	99.79

(1) Coarse granular spinel peridotites, (2) Course granular ilmenite peridotites, (3) Corase granular garnet periodtites, (4) Coarse granular peridotite, (5) Porphyroclastic garnet peridotite.

observations support the conclusions that the petrology and phase chemistry of the mantle beneath continental shields is generally the same and the petrogenesis of kimberlite is governed by similar processes.

The present study reports the results of an investigation of 150 such xenoliths from the Kao kimberlite pipe in Lesotho. The samples were selected to maximize the range of anticipated rock types and temperatures and pressures of equilibration. The range of samples thus does not accurately indicate the population of transported xenoliths but attempts more to characterize the maximum range of the different populations.

Petrology

The general geology of the Kao kimberlite pipe, its inclusions and a description of the diamonds have previously been reported (Rolfe, 1973; Nixon and Boyd, 1973; Clement, 1973; Hornung and Nixon, 1973; Whitelock, 1973). The present report expands the data on the range of mafic and ultramafic xenoliths found in the kimberlite.

The ultramafic xenoliths may be divided into the following phase assemblages:

i) Olivine + orthopyroxene ± clinopyroxene,

ii) a) Olivine + orthopyroxene ± clinopyroxene + chromite spinel ± garnet,

b) Olivine + orthopyroxene ± clinopyroxene + ilmenite ± garnet,

iii) Olivine + orthopyroxene + clinopyroxene + garnet, and

iv) Discrete crystals of clinopyroxene or orthopyroxene.

Mafic assemblages are restricted to eclogite which characteristically have accessory to minor rutile and occassionally ilmenite. Both the garnet and clinopyroxene of the eclogite are often filled with small rod-like rutile inclusions, suggestive of exsolution textures.

The ultramafic xenoliths are readily divided into three textural groups which have been described in detail (Boullier and Nicolas, 1975; Boullier, 1977). They include textures with coarse granular, tabular, porphyroclastic and mosaic textures. Garnet-bearing assemblages show all varieties of textures, while oxide-bearing xenoliths only occur with coarse grained textures. The oxide bearing assemblages which have ilmenite as the oxide phase have textures indicating a

TABLE 2. Chemical analyses of orthopryroxene.

Sample	137[1]	151[1]	142[1]	209[1]	156[2]	112[2]	122[2]	127[2]	120[2]	102[3]	133[3]	153[3]	157[3]	130[3]
SiO_2	57.45	58.96	57.91	59.09	57.93	56.90	55.74	54.11	54.66	58.82	56.02	59.63	58.46	57.39
TiO_2	.00	.00	.03	.01	.14	---	.09	.07	.07	.17	.16	.00	.07	.11
Al_2O_3	.59	1.33	1.94	.80	.96	---	3.26	2.48	2.37	.78	.81	.64	1.51	.81
Cr_2O_3	.27	.47	.39	.43	.42	---	.12	.28	.19	.23	.33	.37	.38	.25
FeO (total)	4.32	4.59	6.06	4.27	5.84	8.51	9.67	9.62	9.82	5.67	5.49	4.34	5.15	5.41
MnO	.10	.09	.11	.08	.12	.15	.14	.18	.14	.13	.11	.09	.10	.12
MgO	36.62	36.13	34.98	36.52	35.06	33.72	31.94	32.43	31.88	34.56	34.57	36.39	35.87	34.94
CaO	.64	.30	.29	.42	1.06	.41	.27	.24	.21	.72	1.01	.45	.19	.94
Na_2O	.13	.05	.01	.09	.19		.00	.04	.01	.13	.25	.08	.04	.31
Total	100.21	101.92	101.74	101.74	100.13	---	101.23	99.49	99.42	101.20	98.72	102.02	101.77	100.30

TABLE 2. (cont.)

Sample	124[4]	77[5]	113[5]	125[5]	136[5]	138[5]	147[5]	103[6]	163[6]	203[6]	204[6]	145[6]	190[7]
SiO_2	58.32	57.90	57.48	58.04	57.80	58.13	58.56	56.93	58.66	58.24	58.04	58.58	56.91
TiO_2	.17	.00	.04	.11	.16	.05	.17	.20	.09	.28	.26	.12	.15
Al_2O_3	.78	.82	.68	1.13	.87	.80		1.06	.98	1.01	1.04	1.36	.81
Cr_2O_3	.29	.23	.32	.37	.22	.28	.43	.12	.32	.24	.17	.35	.00
FeO (total)	7.88	4.52	4.70	5.43	8.79	5.16	5.26	7.21	5.17	6.62	7.66	5.38	10.74
MnO	.14	.07	.10	.12	.16	.13	.11	.12	.12	.12	.13	.09	.18
MgO	33.71	36.07	36.09	34.32	32.93	35.08	34.80	32.60	34.83	34.10	33.39	33.97	31.85
CaO	.62	.18	145	1.55	.71	.99	.92	1.37	1.02	.95	.98	1.50	.73
Na_2O	.19	.04	.14	.50	.20	.24	.27	.46	.18	.26	.27	.39	.18
Total	102.15	99.84	100.00	101.46	101.88	100.89	101.57	100.09	101.38	101.87	102.00	101.80	101.56

(1) Coarse granular spinel peridoties, (2) Coarse granular ilmenite peridotites, (3) Coarse granular garnet peridotites, (4) Coarse granular peridotite, (5) Porphyroclastic garnet peridotite, (6) Mosaic garnet peridotite, (7) Discrete.

TABLE 3. Chemical analyses of ultramafic clinopyroxenes.

	1	2	3	3	3	3	4	5	5	5	5	6
Sample	151	112	102	133	153	157	124	113	125	136	147	103
SiO_2	55.09	54.94	55.99	55.68	55.47	54.88	54.89	55.13	55.23	54.49	55.43	56.11
TiO_2	.02	.35	.34	.05	.01	.28	.35	.20	.19	.34	.32	.17
Al_2O_3	3.19	3.63	2.26	1.99	1.22	3.50	2.37	2.53	2.34	1.11	2.30	1.96
Cr_2O_3	1.95	2.28	1.17	1.79	1.67	1.78	1.80	.97	1.55	1.32	1.70	.74
FeO (total)	1.38	4.27	3.23	2.48	1.89	1.77	4.42	3.39	2.56	4.80	3.17	3.48
MnO	.07	.17	.16	.16	.09	.07	.13	.10	.09	.14	.71	.17
MgO	15.88	14.59	17.98	17.40	18.15	15.94	16.80	16.43	17.79	17.50	18.77	19.38
NiO	.00	---	---	---	.00	.00	.03	.02	.01	.01	.00	---
CaO	21.49	18.13	19.16	20.75	21.66	21.07	17.76	19.16	19.48	19.74	17.22	18.57
Na_2O	2.21	3.34	1.88	1.79	1.15	1.93	2.30	1.88	1.93	.89	2.01	1.41
K_2O	.00	---	---	---	.00	.00	.00	.00	.00	.00	.00	---
Total	101.28	101.71	102.09	101.33	101.23	100.85	99.81	101.16	100.34	101.63	102.00	101.96

TABLE 3. (cont.)

	6	6	6	6	7	7	7	7
Sample	163	203	204	145	189	190	191	196
SiO_2	55.97	55.33	55.50	56.21	55.82	54.96	56.03	56.29
TiO_2	.12	.55	.11	.21	.32	.33	.34	.27
Al_2O_3	2.03	2.59	2.13	2.65	2.66	2.68	2.37	2.51
Cr_2O_3	1.00	.97	.96	.95	.30	.33	.27	.35
FeO (total)	3.26	4.10	3.19	4.09	5.05	5.11	5.77	5.48
MnO	.18	.11	.11	.12	.13	.13	.22	.22
MgO	19.27	18.52	19.21	21.63	20.23	20.61	20.54	21.98
NiO	---	.03	.00	.02	.06	.02	---	---
CaO	18.59	16.98	18.32	.3156	14.75	14.86	11.56	13.07
Na_2O	1.54	2.06	1.54	1.77	1.81	1.90	1.61	1.64
K_2O	---	.00	.00	.00	.00	.00	---	---
Total	101.23		101.06	101.21	101.14	100.94	98.70	101.81

(1) Course granular spinel peridotites, (2) Coarse granular ilmenite peridotites, (3) Coarse granular garnet peridotites, (4) Course granular peridotite, (5) Porphyroclastic garnet peridotite, (6) Mosaic garnet peridotite, (7) Discrete.

cumulate origin and are finer grained than the chromite spinel-bearing xenoliths.

Within the eclogites two textural groups are observed. One is similar to the coarse grained texture of the ultramafic rocks with an equigranular appearance and grain sizes in the range from 2 to 4 mm. The other has a tabular to gneissic texture in which garnet and clinopyroxene crystals are lensoid to tabular in shape. The latter texture is equivalent to the tabular ultramafic texture.

A characteristic feature of the garnet-bearing ultramafic xenoliths is the presence of fine grained reaction rims of intergrown pyroxene and spinel surrounding the garnets. The pyroxenes are often altered to serpentine, micaceous material or amphibole. In the more highly deformed mosaic textured samples the reaction rims are undeformed and appear as tight wormy intergrowths. The spinels are translucent and vary in color from pale green through olive browns to reddish brown.

Mineral Chemistry

The coexisting minerals were analyzed with an ARL-EMXSM microprobe and corrected with the Bence and Albee correction procedures. Representative analyses of all minerals are included in Tables 1 through 7. Relevant discussion of the chemical variables used in this paper are left to subsequent sections.

Estimates of Temperatures and Pressures of Equilibration

The xenoliths brought to the surface in kimberlite pipes represent a jumbled array of samples from a wide range of depths. Crucial to an understanding of their petrogenesis is the interpretation of their depth (pressures) and temperatures of formation (equilibration). The distribution of elements between coexisting phases is dependent on both temperature and pressure and experimental or theoretical calibration of distribution coefficients for simple and more complex chemical systems has allowed reasonable estimates of the intensive parameters.

For ultramafic compositions the mutual solubility of diopside and enstatite has long been used as a geothermometer. A number of experimenters (Davis and Boyd, 1966; Nehru and Wyllie,

1974; Warner and Luth, 1974; Mori and Green, 1975; Lindsley and Dixon, 1976) have defined the dependence of the solvus on both temperature and pressure. A careful analysis of reversed experiments (Lindsley and Dixon, 1976) suggest that the 30 kilobar diopside solvus used by Mori and Green (1975) is preferable for mantle derived ultramafic compositions. However, the natural system contains significant amounts of additional components, the most significant being Fe and Al. The diopside solvus is critically dependent on the Fe content of the pyroxene. However, the ultramafic xenoliths have very low concentrations of Fe, and allowance may be made for the variable Fe contents by using the Ca/Ca + Mg + Fe rather than the Ca/Ca + Mg ratio of the clinopyroxnes for the temperature estimates. The presence of Al_2O_3 widens the diopside solvus (O'Hara and Schairer, 1963; Boyd, 1970; Akella, 1976; Dixon and Presnall, 1977) and would result in anomalously lower temperature estimates based on the Al_2O_3-free systems. Wood and Banno (1973) and more recently Wells (1977) have derived a semi-empirical expression using experimental and theoretical considerations to account for the variable chemistry of natural samples. Estimated temperatures using Well's expression for paired pyroxenes depart significantly from values estimated using Mori and Green's solvus. Well's (1977) expression is a statistical average of experimentally determined values over a wide range of pressures and essentially ignores the demonstrated effect of a widening solvus at higher pressures. Since the affect of Al_2O_3 solubility on the diopside solvus have not been experimentally defined, and the Al_2O_3 contents of the clinopyroxenes is low and falls within a narrow range it is felt that use of Mori and Green's (1975) data gives reasonable results, at least in terms of relative values. The variation of the pyroxene chemistry in the pyroxene quadrilateral is shown in figure 1.

Pressures have been estimated using the temperatures estimated above and the experimentally defined solubility of Al_2O_3 in orthopyroxene coexisting with garnet, clinopyroxene and olivine in the system $CaO-MgO-Al_2O_3-SiO_2$ (Akella, 1976). Estimated temperatures and pressures for garnet peridotites are shown in figure 2. The pressure dependence of Al_2O_3 solubility in orthopyroxenes is not clearly defined. Theoretical (Wood, 1974b; Obata, 1976) and experimental evaluation (Fujii, 1976) in the system

TABLE 4. Chemical analyses of ultramafic garnets.

Sample	209[1]	112[2]	122[2]	102[3]	133[3]	153[3]	157[3]	130[3]	77[5]	113[5]	125[5]	138[5]
SiO_2	42.59	41.95	41.28	43.18	42.28	41.73	42.59	41.97	42.63	42.29	41.38	41.44
TiO_2	.02	.30	.08	.40	.06	.06	.07	.13	.00	.41	.09	.34
Al_2O_3	17.48	20.51	20.30	19.47	17.61	16.59	19.12	19.37	19.87	21.11	17.89	17.94
Cr_2O_3	9.53	3.24	.71	2.99	5.86	7.65	3.08	4.42	2.34	2.46	5.63	5.85
FeO (total)	6.54	12.38	15.36	8.02	7,88	6.48	8.50	7.22	7.25	7.46	6.93	6.82
MnO	.38	.47	.58	.38	.34	.33	.45	.30	.39	.26	.31	.30
MgO	21.79	18.41	16.40	20.90	20.14	20.04	20.25	21.06	20.56	22.10	21.45	21.51
CaO	4.38	4.44	4.90	4.81	6.11	6.81	5.05	5.21	4.89	4.54	5.47	4.73
Na_2O	.01	.08	.02	.01	.04	.02	.04	.04	.01	.08	.05	.07
K_2O	.00	.04	.05	---	.02	.03	.02	.00	.00	.01	.02	.02
Total	102.72	101.82	99.69	100.16	100.32	99.75	99.15	99.72	97.94	10072	99.21	99.02

TABLE 4. (cont.)

Sample	147[5]	103[6]	163[6]	203[6]	204[6]
SiO_2	41.71	43.56	40.90	43.60	43.01
TiO_2	.57	.37	.48	.50	.90
Al_2O_3	16.79	20.25	19.18	21.76	21.06
Cr_2O_3	5.97	2.41	4.06	1.91	2.11
FeO (total)	7.01	7.65	7.10	7.92	9.96
MnO	.27	.31	.28	.29	.33
MgO	21.29	21.39	21.46	21.92	20.39
CaO	5.26	4.80	4.90	4.46	4.58
Na_2O	.08	.00	.07	.06	.11
K_2O	.01	---	.01	.00	.00
Total	98.96	100.74	98.45	102.43	102.43

(1) Corase granular spiel peridotites, (2) Coarse granular ilmenite peridotites, (3) Coarse granular garnet peridotites, (4) Coarse granular peridotite, (5) Porphyroclastic garnet peridotite, (6) Mosaic garnet peridotite.

TABLE 5. Chemical analyses of oxides.

Sample	137[1]	151[1]	142[1]	209[1]	156[2]	112[2]	127[2]	122[2]	122a[2]	122b[2]	122c[2]	127[2]	120[2]
TiO_2	.00	.00	.27	.41	45.61	47.12	43.78	43.60	44.40	12.33	3.04	43.78	51.64
Al_2O_3	9.32	19.95	22.73	10.13	1.04	.54	.19	.54	.68	4.92	7.76	.19	1.42
Cr_2O_3	60.80	46.10	32.55	57.66	7.32	4.55	.55	.80	.65	10.52	10.19	.55	1.04
FeO (total)	14.48	15.36	28.12	17.63	33.80	36.58	45.57	45.88	45.47	66.49	72.11	45.57	32.26
MnO	.34	.30	.28	.08	.21	.25	.25	.23	.14	.18	.91	.25	.24
MgO	18.34	17.98	15.23	13.81	13.17	12.55	9.01	8.47	8.51	5.40	4.15	9.01	15.36
Total	102.55	99.69	99.19	100.16	101.14	101.60	99.35	99.52	99.91	99.92	98.27	99.35	101.97

(1) Corase granular spinel peridotites, (2) Coarse granular ilmenite peridotites.

$MgO-Al_2O_3-SiO_2$ indicate that Al_2O_3 solubility is essentially independent of pressures. However, Dixon and Presnall suggest a negative relationship for the system $CaO-MgO-Al_2O_3-SiO_2$. In this paper Fujii's (1976) data is used. Thus in figure 2 only the range of temperatures for the spinel and spinel plus garnet peridoties is shown.

For those samples where orthopyroxene analyses were solely available an empirical technique equivalent to that used by Boyd and Nixon (1973) was followed. However, the Kao data suggest a systematic relation between the Al_2O_3 content of the orthopyroxene and the width of the orthopyroxene solvus. It was possible to define an empirical relationship between temperatures estimated from Mari and Green's diopside solvus for coexisting clinopyroxenes, and the Al and Ca/Ca + Mg content of the orthopyroxene. It should be noted that in contrast Dixon and Presnall (1977) observe no effect of Al_2O_3 on the orthopyroxene solvus. Thus assuming that clinopyroxene was present temperature estimates have been made for these xenoliths. For clinopyroxene xenocrysts it is assumed that they are in equilibrium with orthopyroxene and only the temperatures are illustrated (figure 2).

Use of the simple four component system (Akella, 1976) to estimate pressure ignores the effect of additional components. As an illustration of the effect of additional components and as an ease of reference to estimates made in previous studies three pressure estimates were made for progressively more complex systems. The first uses the simple three component system $MgO-Al_2O_3-SiO_2$ (MacGregor, 1974), the second the four component system $CaO-MgO-Al_2O_3-SiO_2$ (Akella, 1976) and the third a theoretical expression derived by Wood (1974, equation 12) (Figure 3). It can be seen that there is a significant and progressive decrease in the estimated pressures with an increasing number of components and the scatter in the estimates is greatest for Wood's theoretical expression. Despite the scatter using Wood's expression it is felt that they give the best overall estimate of temperatures and pressures of formation. It is only by using Wood's value that one sample containing graphite (Nixon and Boyd, 1973) falls within the graphite stability field, and a linear least squares fit to the data gives the closest approximation to an estimated geothermal gradient for sub-continental mantle (Clark and Ringwood, 1964) (Figure 3).

Temperatures and pressures for the eclogites were estimated using the distribution of Fe and Mg between coexisting garnet and clinopyroxene defined by Råheim and Green (1975), and assuming that the samples lie on the linear extension of the ultramafic trend (Figure 4).

If the temperatures and pressures shown in figure 3 are reasonable estimates of environmental parameters within the mantle the data may be

TABLE 6. Chemical analyses of eclogitic clinopyroxenes.

Sample	12[1]	14[1]	111[1]	181[1]	179[2]	7[2]	172[2]	4[2]	67[2]
SiO_2	54.23	54.19	53.33	54.67	52.30	54.48	56.09	53.36	54.19
TiO_2	.25	.32	.46	.28	.31	.25	.21	.53	.04
Al_2O_3	5.55	5.35	5.42	5.80	4.08	4.88	6.42	8.46	1.29
Cr_2O_3	.18	.16	.19	.16	.02	.12	.15	.00	.34
FeO (total)	5.80	5.67	3.60	4.25	9.44	3.30	4.14	6.24	2.95
MnO	.08	.07	.04	.06	.12	.03	.08	.05	.11
MgO	12.25	12.66	14.35	.3124	11.84	14.55	12.92	10.20	16.82
CaO	18.31	18.46	21.40	19.33	21.33	19.99	17.67	17.44	24.09
Na_2O	3.54	3.54	1.84	3.24	1.91	2.77	3.94	4.05	.28
K_2O	.03	.03	.00	---	---	---	---	---	.00
Total	100.23	100.45	100.64	101.03	101.36	100.37	101.61	100.33	100.11

(1) Tabular, (2) Coarse granular.

TABLE 7. Chemical analyses of eclogitic garnets.

Sample	12[1]	14[1]	111[1]	181[1]	179[2]	7[2]	172[2]	4[2]	67[2]
SiO_2	40.46	40.51	41.75	41.51	38.73	42.36	40.23	40.04	41.50
TiO_2	.08	.09	.03	.03	.08	.05	.04	.07	.01
Al_2O_3	20.27	20.52	23.14	16.91	19.31	23.38	19.52	21.08	21.61
Cr_2O_3	.22	.24	.21	.14	.04	.15	.17	.02	1.61
FeO (total)	21.68	21.59	15.27	22.04	27.00	14.15	19.68	22.57	14.61
MnO	.57	.57	.32	.37	.77	.24	.33	.49	.90
MgO	11.96	12.05	16.22	15.06	6.16	17.53	13.93	8.84	15.02
CaO	4.55	4.52	5.07	4.35	6.79	4.43	4.17	7.71	6.45
Na_2O	.01	.01	.02	.02	.03	.00	.04	.01	.01
K_2O	---	---	.05	.04	.11	---	.05	---	---
Total	99.80	100.11	102.07	100.46	99.01	102.29	98.16	100.82	101.82

(1) Tabular texture, (2) Coarse granular

used to estimate mantle geothermal gradients. A significant observation of previous studies (Boyd and Nixon, 1975; MacGregor, 1975; Danchin and Boyd, 1977) has been the presence of an inflection or perturbation of the geotherm to steeper gradients at the highest pressures. The perturbation has been variously interpreted as the result of mantle diapirism (Green and Guegen, 1975) or frictional heat in the asthenosphere during continental drift (Boyd and Nixon, 1975). Figure 3 illustrates an apparent inflection for pressures estimated using the simple three component system $MgO-Al_2O_3-SiO_2$, but the inflection is absent for pressures estimated using the four component system $CaO-MgO-Al_2O_3-SiO_2$ and Wood's equation attempting to account

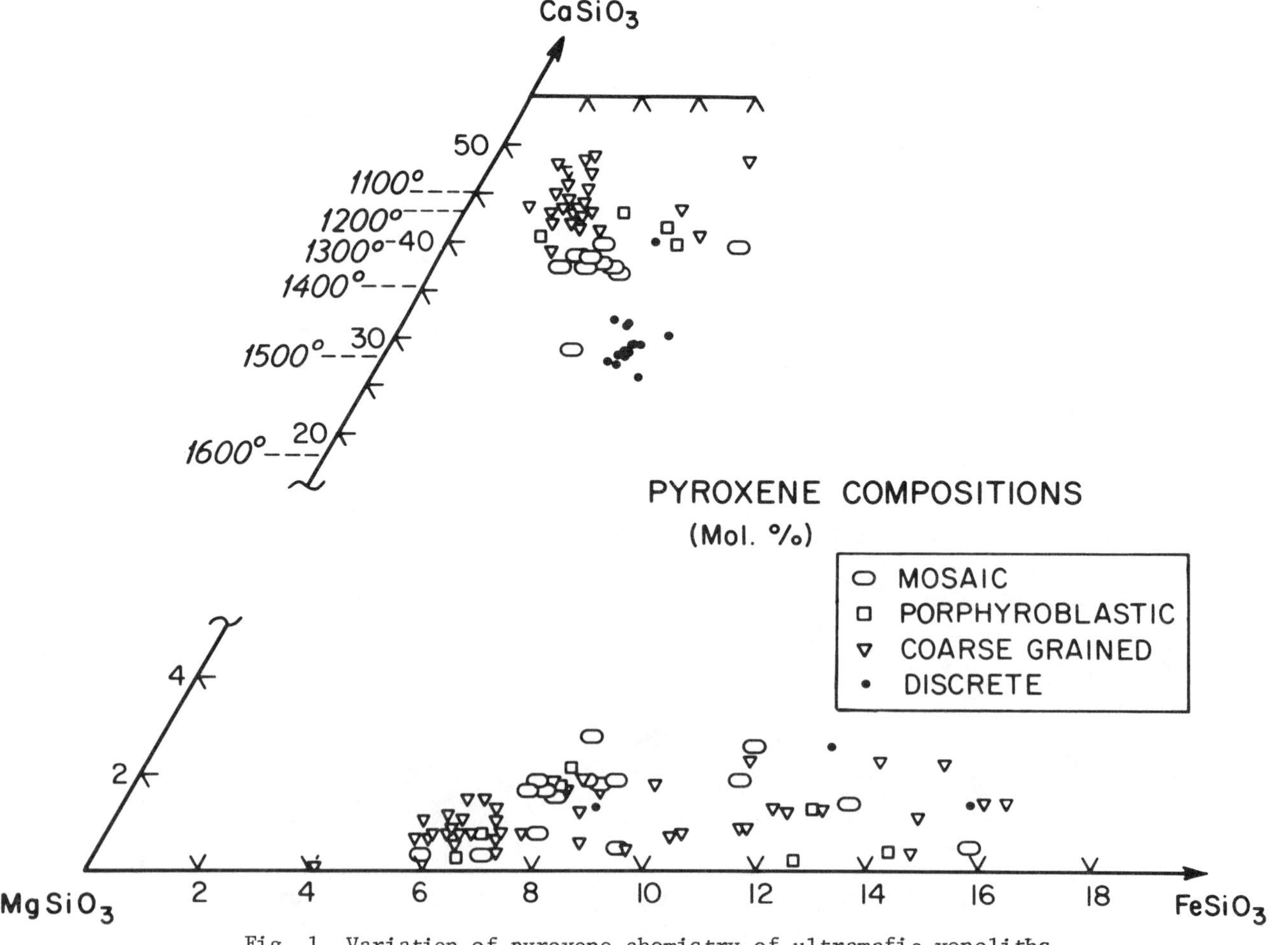

Fig. 1 Variation of pyroxene chemistry of ultramafic xenoliths.

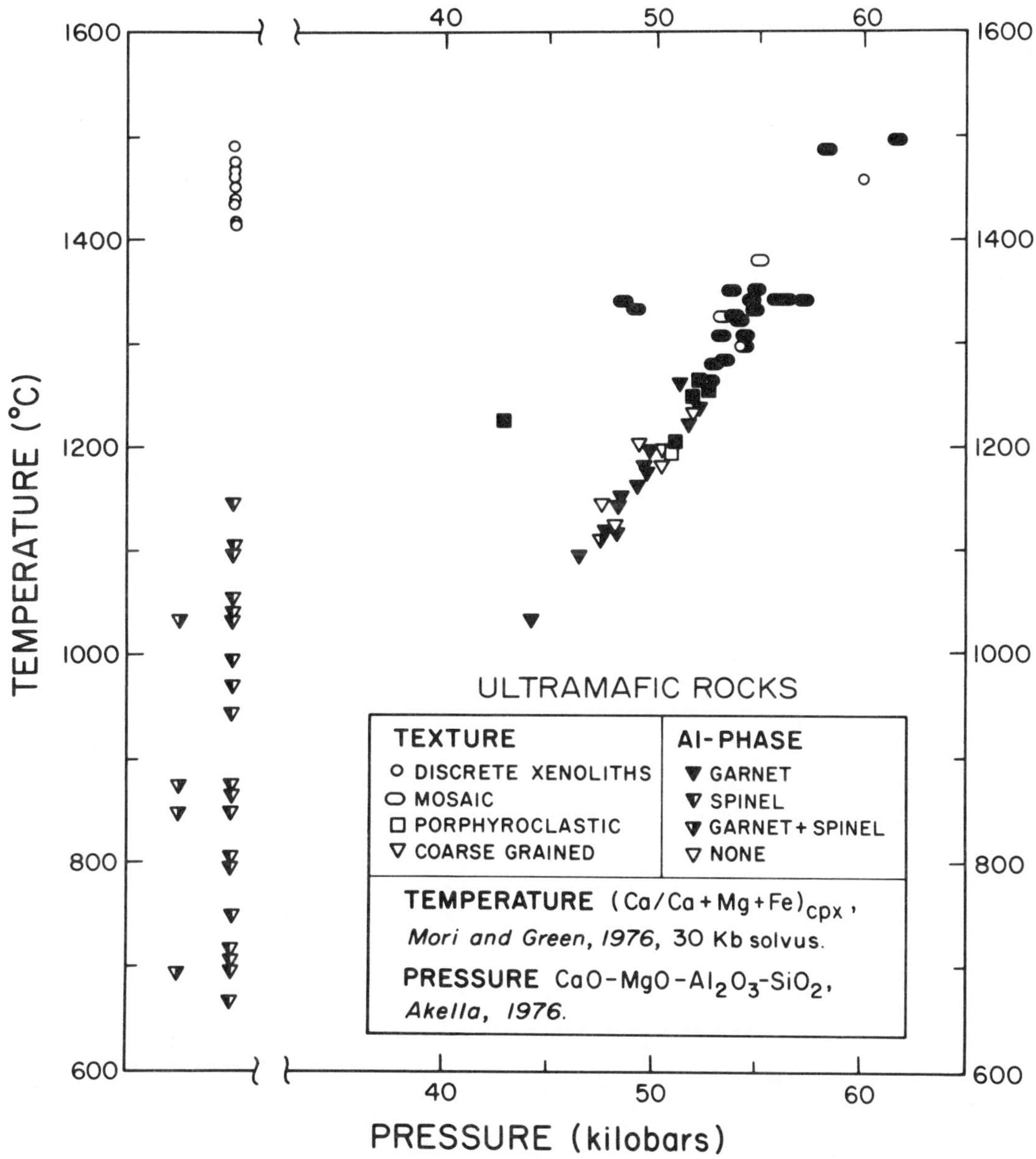

Fig. 2 Variation of estimated temperatures and pressures of equilibration of ultramafic xenoliths.

for additional compositional variables. However, in every case the data at higher temperatures and pressures depart from an expected gradient for a stable static mantle and steeper gradients are observed. Though not as extreme the term 'perturb' may still be considered and suggests that dynamic forces were active in the mantle at the time of kimberlite extrusion.

Since the data for the four component system $CaO-MgO-Al_2O_3-SiO_2$ are more abundant and the results are in rough agreement with Wood's (1974) more complete expression, subsequent discussion on textural and chemical variations use Akella's (1976) base. Although absolute values may vary it is probable that the relative positions of the samples are essentially the same.

Discussion

Phase assemblages

The two most abundant xenolith types are the oxide- and garnet-bearing peridotites. The oxide-bearing peridotites are restricted to lower while the garnet peridotites occur at higher temperatures and pressures (Figure 5). The distribution is as expected but it should

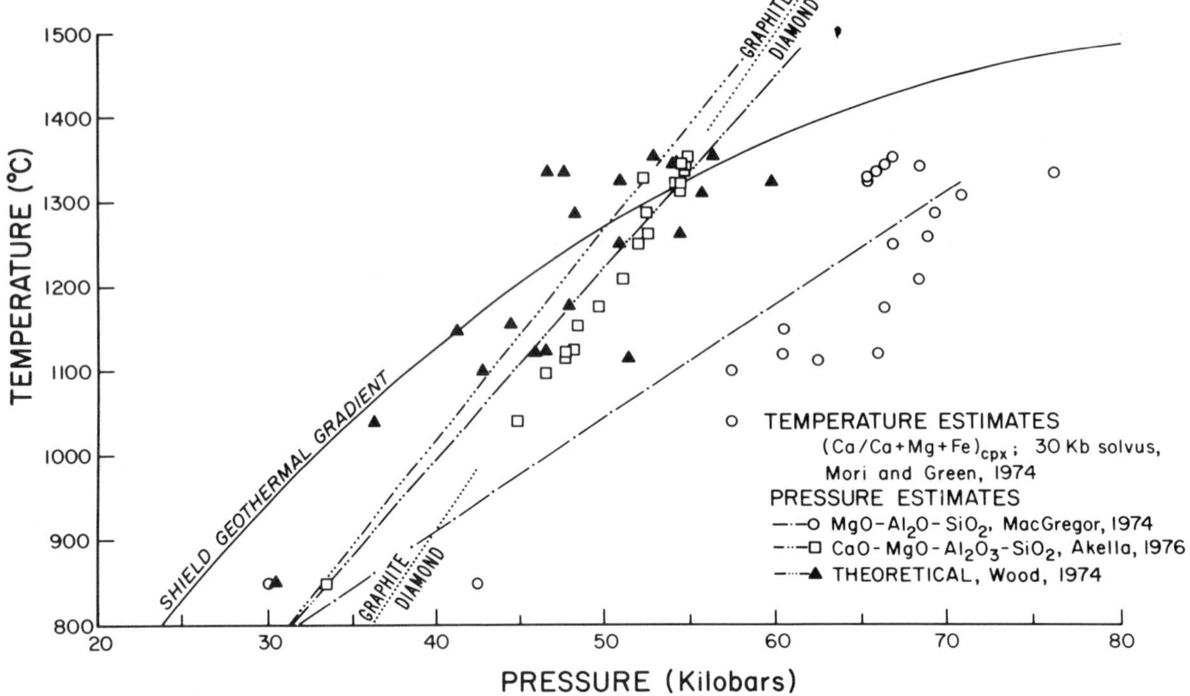

Fig. 3 Comparision of temperature and pressure estimates by different methods.

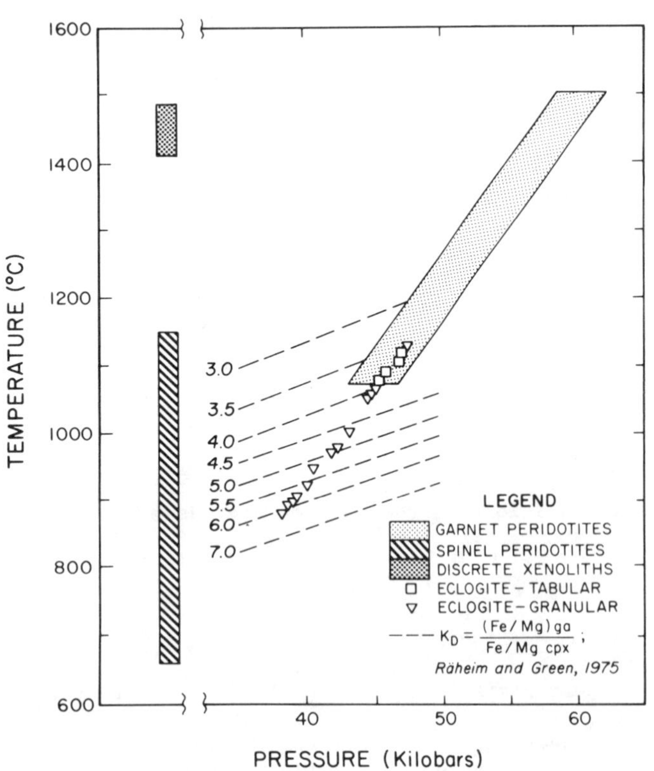

Fig. 4 Estimated temperatures and pressures for ecologite xenoliths.

be noted that the extension of oxide peridotites, and restriction of garnet peridotities, to relatively high temperatures and pressures may suggest that the phase boundary roughly correlates with a compositional boundary (MacGregor, 1969). Specifically, the lower pressure and temperature xenoliths have higher $Cr_2O_3/Cr_2O_3 + Al_2O_3$ ratios. Ilmenite-plus-garnet-bearing peridotites are associated with the oxide peridotites, while peridotites with no spinel or garnet generally overlap the lower temperature range of the garnet peridotite distribution. Orthopyroxenes and clinopyroxenes were the only discrete xenoliths studied; orthopyroxenes occur over a wide range while clinopyroxenes are restricted to temperatures above 1300°C (Figure 3) and are associated with the garnet peridotites.

Mafic compositions are restricted to eclogite phase assemblages and estimated equilibration temperatures (Rä̈heim and Green, 1975) would suggest that they are primarily associated with the oxide-bearing peridotites (Figure 4).

Textures

The textures in the ultramafic xenoliths have been subdivided according to the scheme proposed by Boullier (1977). Their distribution suggests a systematic variation with increasing temperature and pressure such that coarse grained samples occur at lowest temperatures with the porphyroelastic and mosaic textured xenoliths occurring at progressively higher temperatures

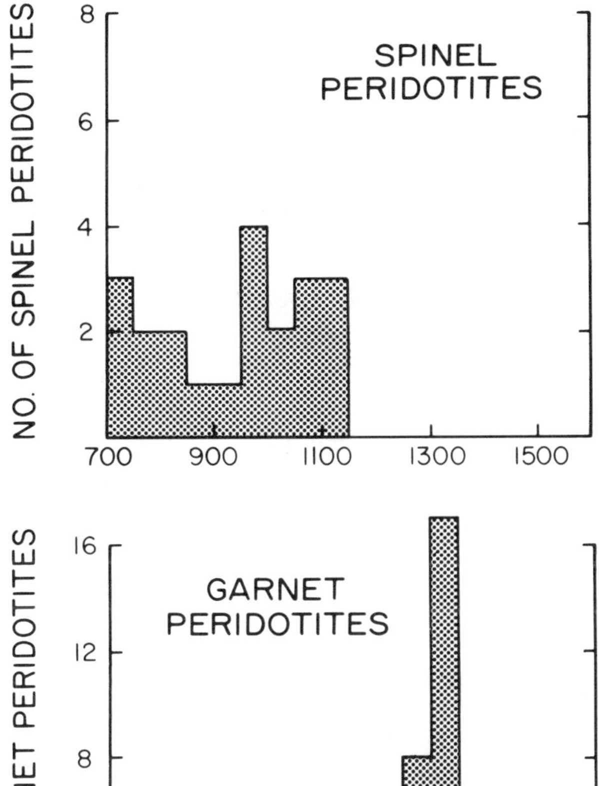

Fig. 5 Distribution of ultramafic phase assemblages with temperature.

(Figure 6). Although the statistical sample is poor it appears that the eclogites with tabular textures are restricted to higher temperatures than the coarse grained varieties suggestive of the textural distribution seen for ultramafic compositions (Figure 4 and 15).

Chemical variations

Mg/Mg + Fe ratio. A measure of systematic chemical variations between the different textural and phase assemblage groups is given by variations in the Mg/Mg + Fe ratio (Figures 7 to 10). Two main features are observed. Firstly, the oxide peridotites have Mg/Mg + Fe ratios for all minerals which cover as wide a range as that observed for garnet peridotites of all textural varieties (Figures 7 through 10). In detail, however, oxide peridotites which have ilmenite (± garnet) as the accessory phase are characterized by low Mg/Mg + Fe ratios while the chromite spinel bearing samples uniformly have the highest Mg/Mg + Fe ratios which correspond to values observed for the coarse grained garnet peridotites (Figures 7 to 10). Though not examined in detail it is felt that the ilmenite-bearing samples which have cumulate textures have a different origin from the coarse granular chromite spinel-bearing xenoliths which represent recrystallized residua of earlier partial melting events.

Secondly, the distributions of the Mg/Mg + Fe ratio for all the garnet-bearing peridotites are generally similar for the coarse granular and porphyroclastic, but greater than that for the mosaic textured samples (Figures 7 through 10).

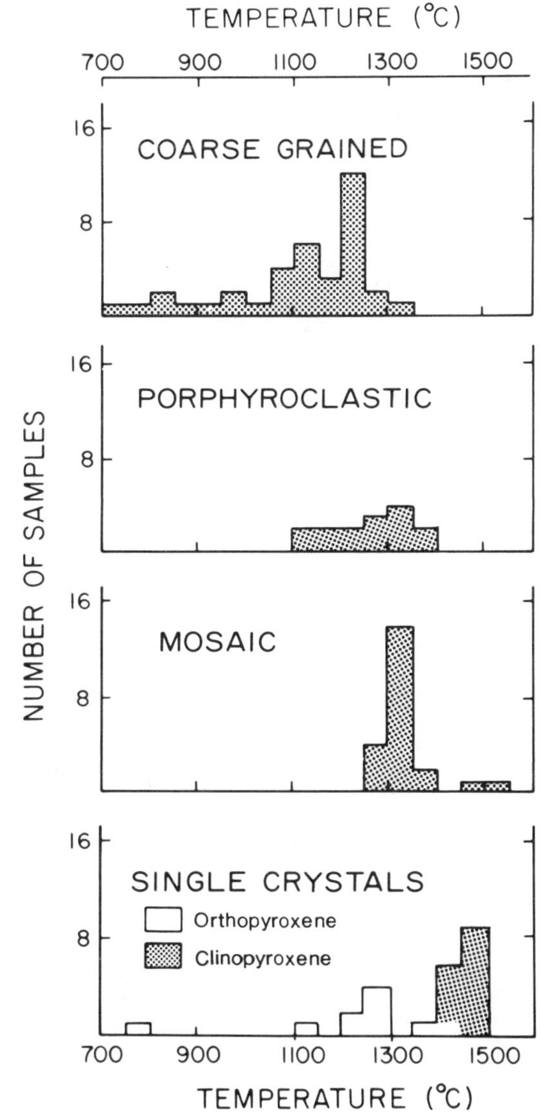

Fig. 6 Distribution of textural types in ultramafic xenoliths with temperature.

Since the same feature is observed for all minerals, it appears that the variations represent comparable variations in the bulk rock chemistry.

The Mg/Mg + Fe ratio is also illustrated as a function of increasing temperature or depth in the mantle (Figure 11). Excluding the ilmenite-bearing assemblages the data suggest two population of samples distributed about a temperature of about 1250°C. At lower temperatures the Mg (Mg+Fe) ratio of the olivine, orthopyroxene clinopyroxene and garnet is distributed over higher values than at higher temperatures. Two separate populations are suggested.

Previous discussion illustrates interesting systematic variations between phase assemblage, texture, a chemical variable and estimated

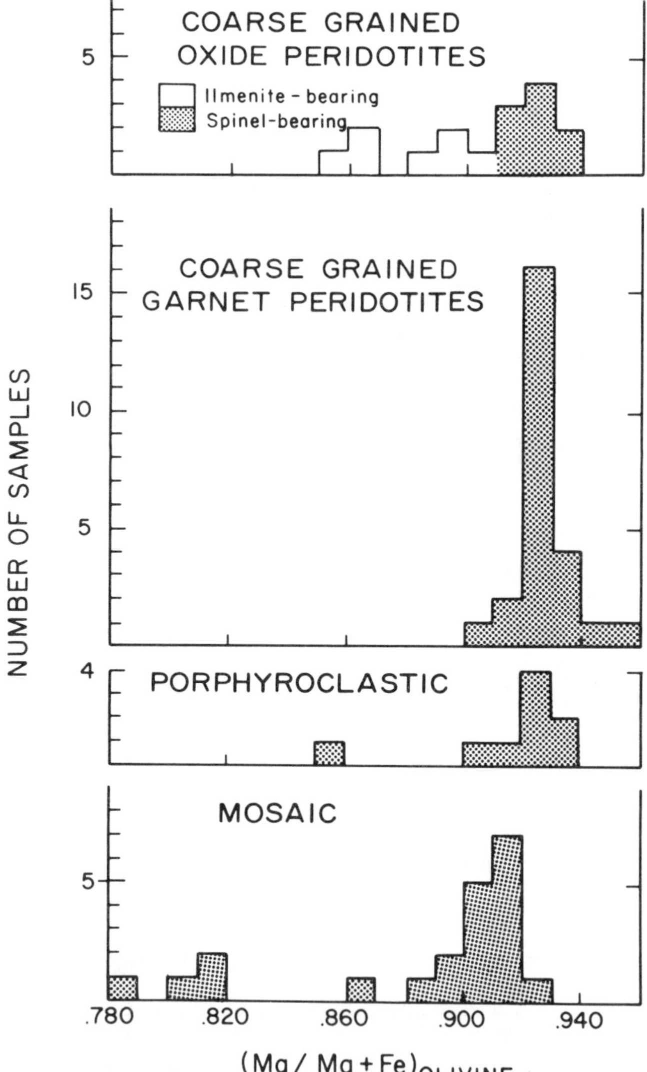

Fig. 7 Distribution of Mg/(Mg+Fe) in olivine from xenoliths with different textures.

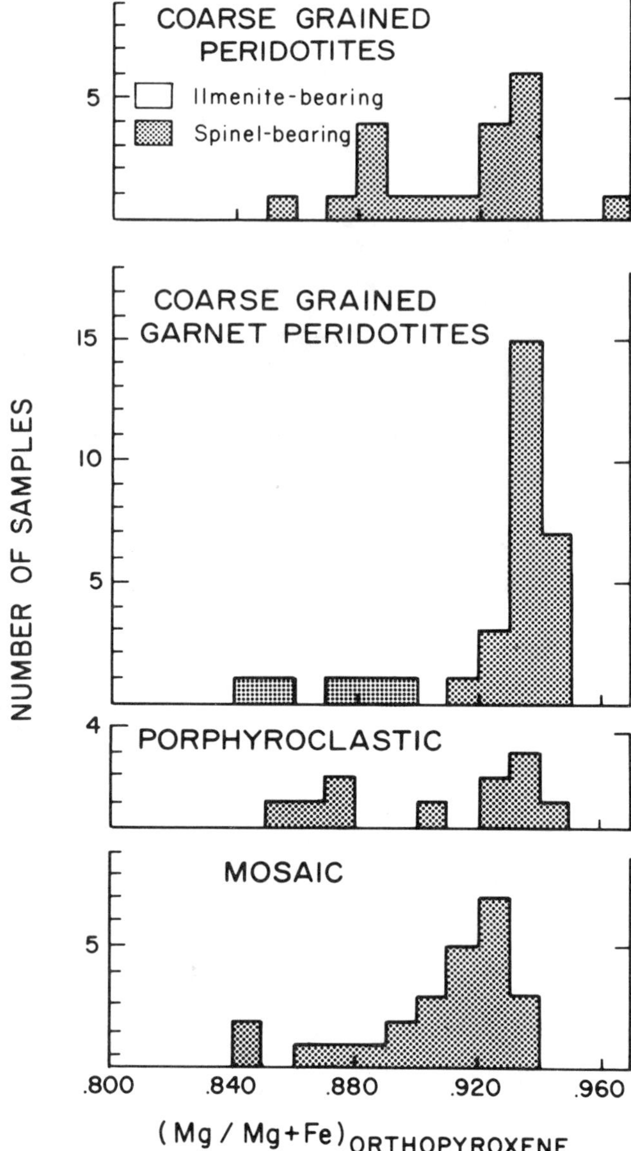

Fig. 8 Distribution of Mg/(Mg+Fe) in orthopyroxene from xenoliths with different textures.

temperatures and pressures. In general, with increasing temperature and pressure, the phase assemblage passes from oxide to garnet peridotites, the textures become more deformed and compositions have a lower Mg/Mg + Fe ratio. There still remains the question as to whether there is a continuous variation of populations or discrete groups are implied, and it is useful to examine other chemical variables.

Mineral chemistry of ultramafic xenoliths. The distribution of the minor elements in olivine as a function of temperature is shown in figure 12. Excluding the ilmenite-bearing xenoliths it is possible to define two separate

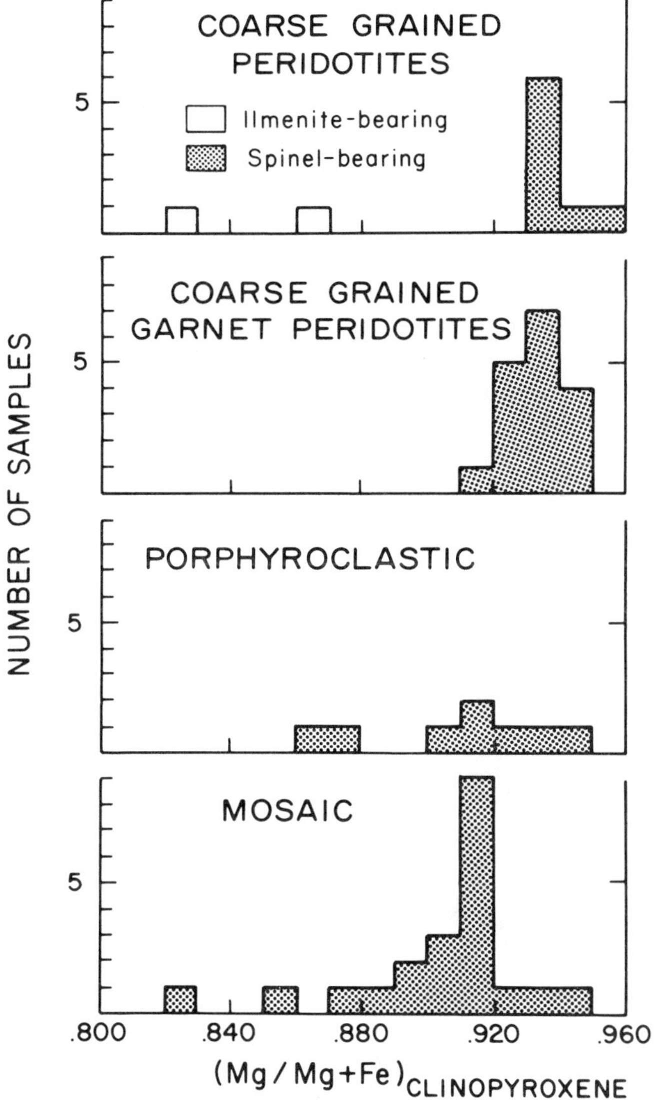

Fig. 9 Distribution of Mg/(Mg+Fe) in clinopyroxene from xenoliths with different textures.

populations as indicated above. Below 1280°C the olivines have CaO, Cr_2O_3, MnO, Na_2O and TiO_2 distributions that are lower than for higher temperatures. The temperature of 1280°C also serves as an approximate boundary for changes in the TiO_2, Na_2O and Cr_2O_3 content of garnets, the TiO_2 and Cr_2O_3 contents of clinopyroxene but illustrates no change in chemical distribution for orthopyroxene (Figures 13 and 14).

Mineral chemistry of eclogites. As previously indicated there is a suggestion that the tabular textured eclogites are restricted to higher temperatures (Figure 4 and 15). A comparable analysis for the chemical variable suggests parallel systematics such that the Mg/Mg + Fe ratio and the CaO content in garnet decreases, while the Cr_2O_3 content generally increases with increasing temperature (Figure 15). The data are not conclusive but included to encourage further study.

Comparison with other kimberlite pipes

It has previously (Nixon et al., 1973; Boyd and Nixon, 1975) been shown that ultramafic xenoliths from Lesotho kimberlites may be divided into different textural varieties such that coarse granular textures are followed at increasing depths by porphyroclastic and mosaic textured varieties. The textural differences are paralleled by variations in the bulk chemistry with mosaic textured rocks having generally higher concentrations of CaO, Al_2O_3, TiO_2, Na_2O and higher Fe/Fe + Mg ratios. Further, chromite spinel peridotites are found at generally shallower depths although only a few samples were examined. Apart from the xenoliths in the Matsoku pipe, the present study is in general agreement with studies on other Northern Lesotho localities. The Kao xenoliths seem to emphasize the distinctions between the different suites and raises the question whether more discrete grouping of populations is not a regional Northern Lesotho phenomena.

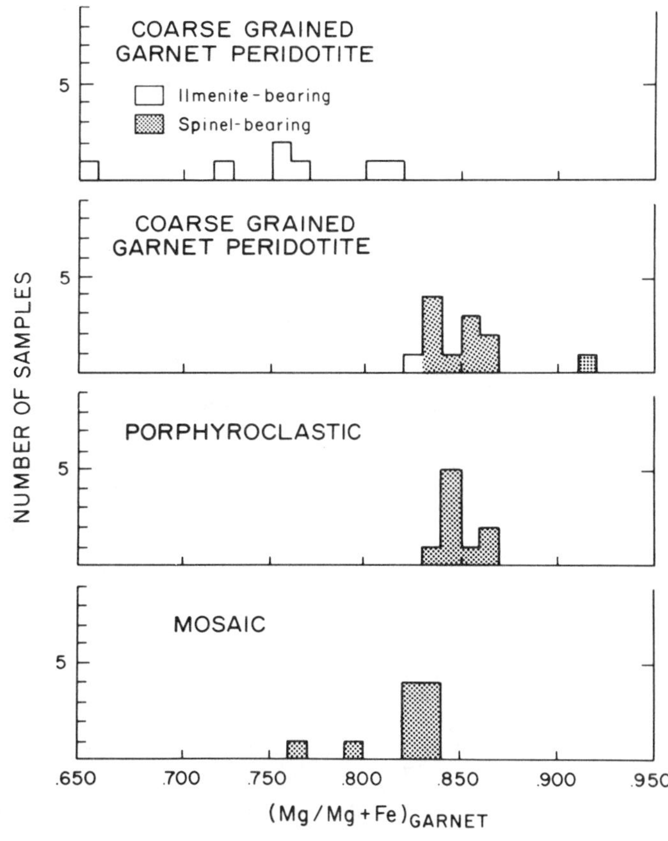

Fig. 10 Distribution of Mg/(Mg+Fe) in garnet from xenoliths with different textures.

The textural variations and systematics have also been observed in other South African (MacGregor, 1975; Boyd and Nixon, 1977; Danchin and Boyd, 1977) and Siberian (Boyd et al., 1977) localities. However, there is not always a corresponding chemical variation with textural type (Boyd and Nixon, 1977) and regional differences in chemistry have been observed (MacGregor, 1975).

Since the most deformed samples are always found at greatest depth it is suggested that kimberlite formation is associated with dynamic mantle processes. However, the lack of a unique correlation with the chemical variables does not define the chemical nature of the source region although there is an indication that more primitive mantle material is often dynamically associated with kimberlite petrogenesis.

Conclusions

The ultramafic xenoliths from the Kao kimberlite appear to come from distinct populations that are distributed within relatively well defined limits of temperature and pressure (depth). With increasing temperature the populations may be defined as follows:

- i) 700 to 1100°C
  - a) Chromite spinel (± garnet) peridotites with coarse grained textures
  - b) Ilmenite (± garnet) peridotites with coarse grained cumulate textures
  - c) Eclogites
- ii) a) 1100 to 1280°C
  Garnet peridotites with coarse

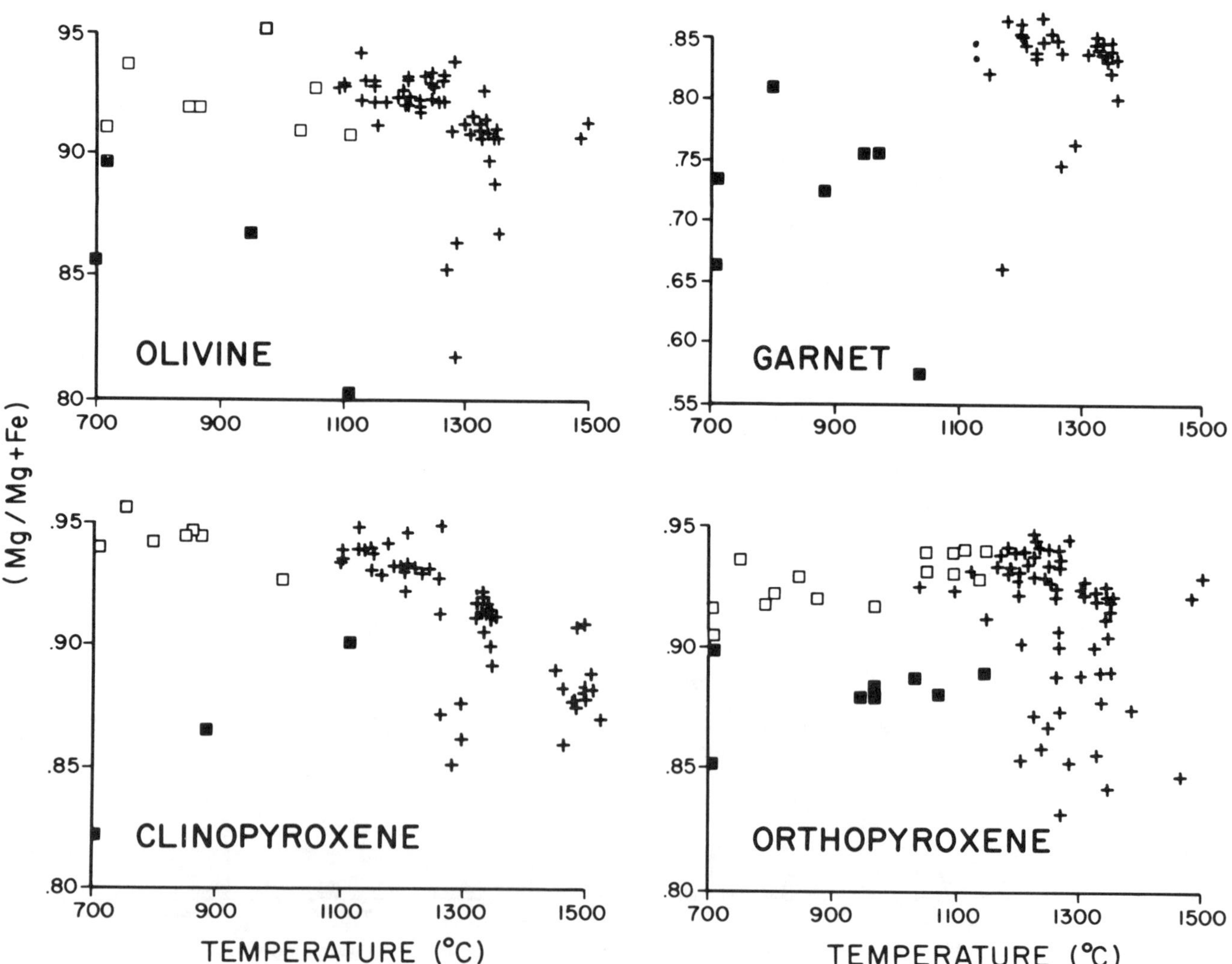

Fig. 11 Variation of Mg/(Mg+Fe) in ultramafic minerals as a function of temperature (open square-spinel peridotite; filled square-ilmenite peridotite; cross-garnet peridotite.)

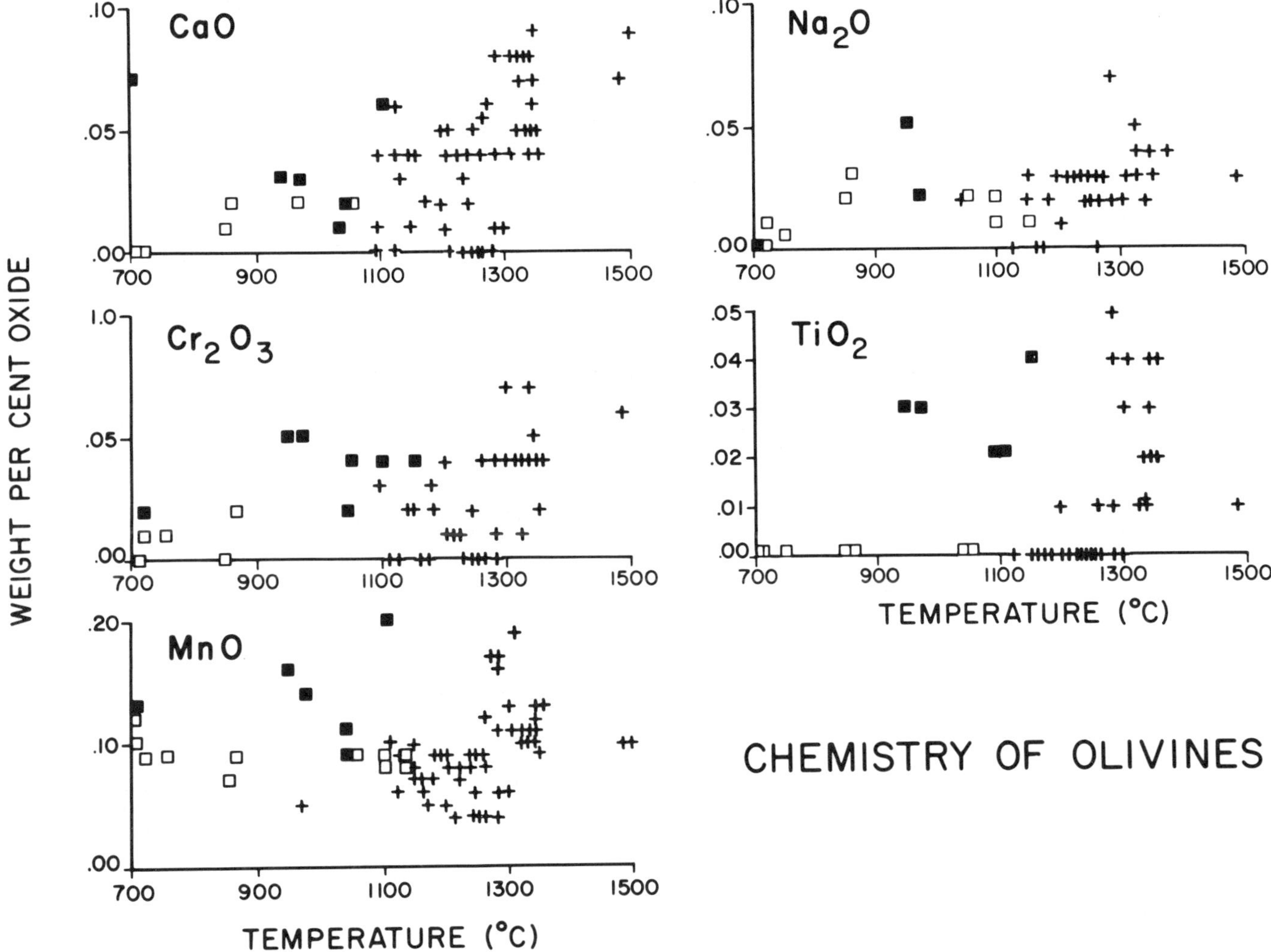

Fig. 12 Variation of olivine chemistry with temperature (open square-spinel peridotite; filled square-ilmenite peridotite; cross-garnet peridotite).

grained textures

b) 1190 to 1280°C
Garnet peridotites with porphyroclastic textures

iii) Greater than 1280°C

a) Garnet peridotites with mosaic textures

b) Discrete xenoliths of clinopyroxene

Chemically the chromite spinel and coarse granular and porphyroclastic garnet peridotites are very similar. They have the highest Mg/Mg + Fe ratios and their minerals have relatively low concentrations of Na_2O, TiO_2 and high contents of Cr_2O_3. The chromite spinel peridotites have higher Cr/Cr + Al ratios and the phase break at 1100°C may be basically related to changes in this chemical parameter. It is felt that these rocks are the refractory residues probably derived from previous episodes of partial melting in the mantle. The ilmenite peridotites have the highest Fe/Fe + Mg ratios and TiO_2 and MnO, and lowest Cr_2O_3 contents. Their cumulate textures suggest an origin by fractional crystallization from deep seated mafic magma chambers.

The mosaic textured xenoliths have low Mg/Mg + Fe ratios, and minerals with relatively lower concentrations of Cr_2O_3 and higher contents of Na_2O, TiO_2 and MnO. Corresponding to the bulk chemistry of other mosaic textured samples (Boyd and Nixon, 1975; Shimizu, 1975) these xenoliths appear to represent less depleted mantle than the coarse grained or porphyroclastic textured xenoliths. The discrete clinopyroxene xenoliths have the chemical characteristics comparable to the mosaic textured suite.

The Kao xenoliths have many features common to the suites of xenoliths from other Northern Lesotho kimberlite pipes (Boyd and Nixon, 1975; MacGregor, 1975). The conclusions of this study focus on the observation that there are distinct

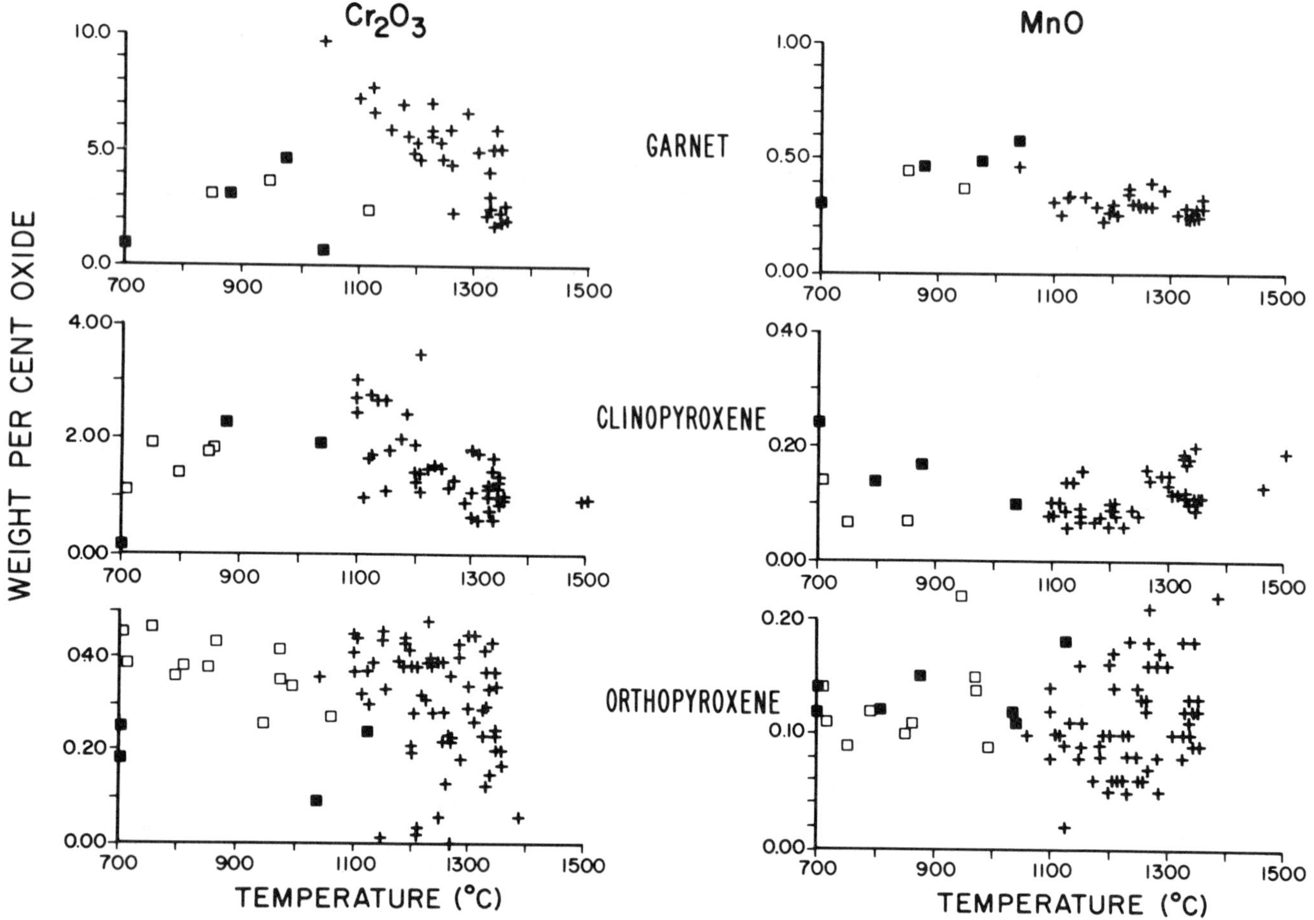

Fig. 13 Variation of Cr_2O_3 and MnO in ultramafic minerals with temperature (open square-spinel peridotite; filled square-ilmenite peridotite; cross-garnet peridotite).

populations of samples despite the fact that textural and chemical distributions show slight overlap, and each population has been formed by different processes not necessarily contemporaneously. Mantle samples from other regions in South Africa, Siberia and North America show the same general characteristics although differences in detail have been observed and the distinct populations noticed at Kao are not so clearly defined. Regional mantle heterogeneity is implied but it appears that within a region such as Northern Lesotho there is local homogeneity.

Using pressure estimates from the simple $MgO-Al_2O_3-SiO_2$ system, a geothermal gradient is identical to that found for other Northern Lesotho pipes. When more complex chemical systems are used in the pressure estimates (Figure 3) the inflection noted previously is less well defined. However, the interpreted temperatures and pressures for the most highly deformed mosaic textured samples still seems to be steeper than that anticipated for a static shield gradient (Figures 2 and 3). Coupled with the observation (Kohlstedt and Goetze, 1974) that the mosaic textured samples were actively being deformed at the time of kimberlite extrusion, the data still favor the interpretation that the highly sheared samples represent a dynamic or perturbed phenomena in the sub continental mantle. In contrast, it seems secure that the coarse granular textures are in static equilibrium with their environment and faithfully record ambient geothermal gradients.

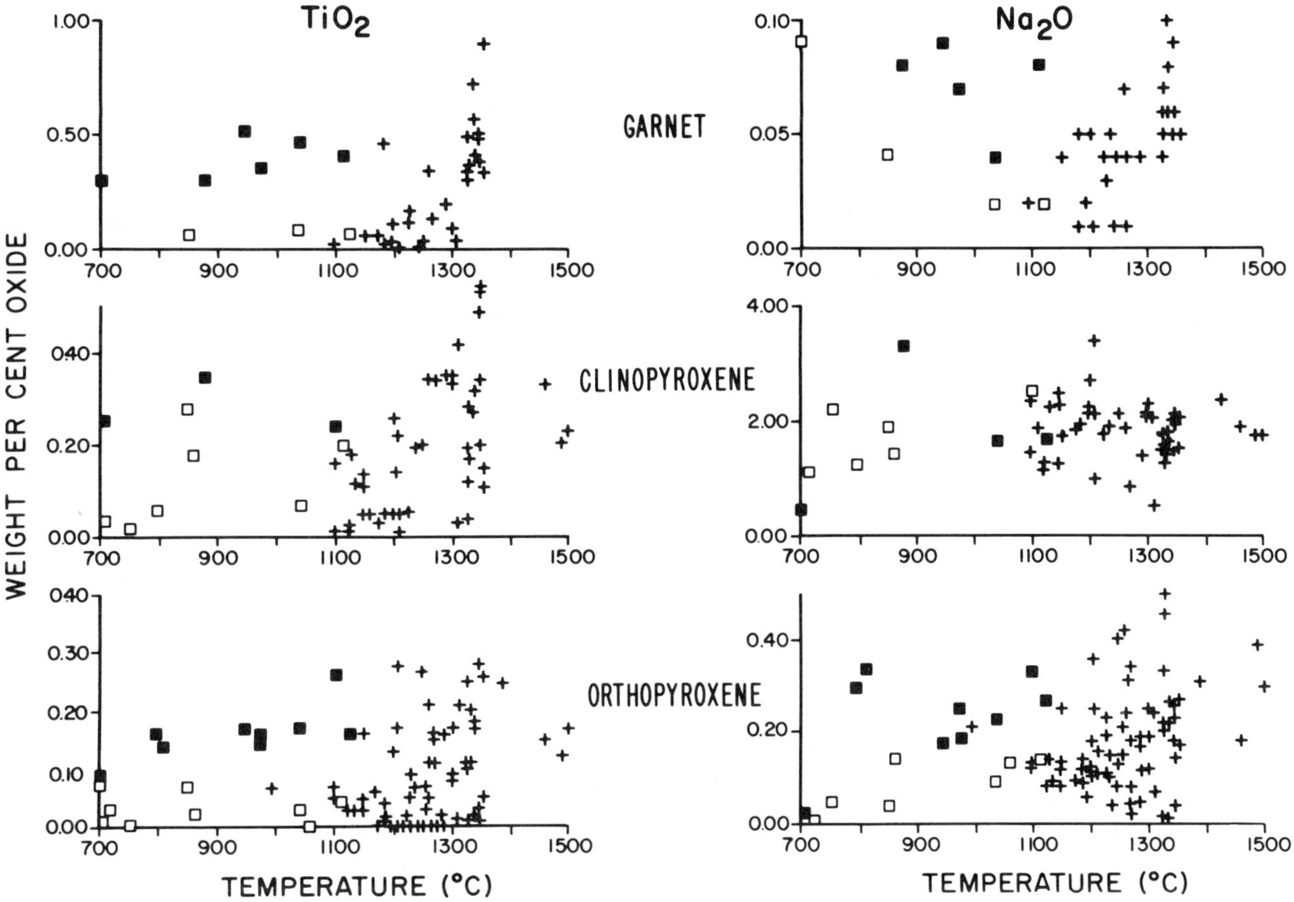

Fig. 14 Variation of TiO_2 and Na_2O in ultramafic minerals with temperature (open square-spinel peridotite; filled square-ilmenite peridotite; cross-garnet peridotite).

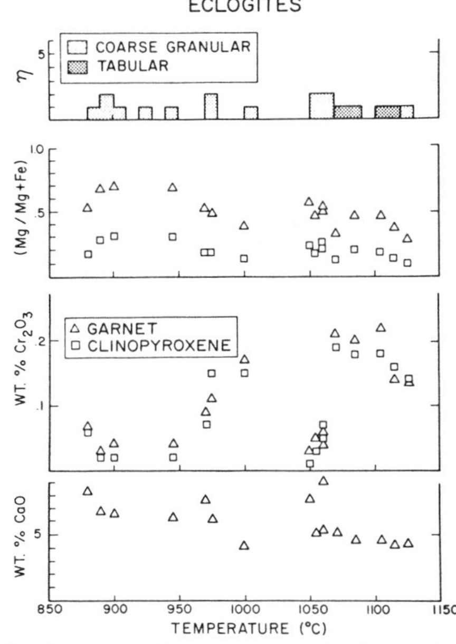

Fig. 15 Variation of texture and mineral chemistry in garnets and clinopyroxenes from cologite xenoliths.

Acknowledgments. The samples used in this study were collected during field trips of the First International Kimberlite Conference in South Africa. Travel funds were supplied by the National Science Foundation (GA-40336). Special thanks are due to B. Hawthorne, his colleagues, and the staff of DeBeers and the Anglo American Corporation of South Africa for organizing the field trips and arranging for the shipment of the samples. R. Wittkopp assisted with the microprobe analyses of the minerals.

References

Akella, J., Garnet pyroxene equilibria in the system $CaSiO_3$-$MgSiO_3$-Al_2O_3 and in a natural mineral mixture. Amer. Mineral., 61, 589-598, 1976.

Boullier, A. M., Structure des peridotites en enclaves dans les kimberlites d'Afrique du Sud. Consequences sur la constitution du manteau superior. Bull. Soc. Fr. Mineral. Cristallogr., 100, 214-219, 1977.

Boullier, A. M. and A. Nicolas, Classification of textures and fabrics of peridotite xenoliths from South African kimberlites. Phys. Chem. Earth, 9, 467-475, 1975.

Boyd, F. R., T. Fujii, and R. V. Danchin, A noninflected geotherm for the Udachnaya kimberlite pipe, USSR. Carnegie Inst. Wash., Yr. Bk. 75, 523-531, 1976.

Boyd, F. R. and P. H. Nixon, Origin of the ilmenite-silicate nodules in kimberlites from Lesotho and South Africa, in Lesotho Kimberlites, ed. P. H. Nixon, Lesotho Nat. Develop. Corp., Maseru, 254-268, 1973.

Boyd, F. R. and P. H. Nixon, Origins of the ultramafic nodules from some kimberlites of northern Lesotho and the Monastery Mine, South Africa. Phys. Chem. Earth, 9, 431-454, 1975.

Bundy, F. P., H. P. Bovenkerk, H. M. Strong, and R. H. Wentorf, Jr., Diamond-graphite equilibrium line from the growth and graphitization of diamond. J. Chem. Phys., 35, 383-391, 1961.

Clement, R., Kimberlites from the Kao pipe, in Lesotho Kimberlite, ed. P. H. Nixon, Lesotho Nat. Develop. Corp., 110-121, 1973.

Danchin, R. V. and Boyd, F. R., Ultramafic nodules from the Premier kimberlite pipe, South Africa. Carnegie Inst. Wash., Yr. Bk. 75, 531-583, 1976.

Davis, B. L. C. and F. R. Boyd, The join $Mg_2Sc_2O_6$ at 30 kilobars pressure and its application to pyroxenes from kimberlites. J. Geophys. Res., 71, 3567-3576, 1966.

Dixon, J. R., and D. C. Presnall, Geothermometry and geobarometry of xynthetic spinel lherzolite in the system $CaO-MgO-Al_2O_3-SiO_2$; 2nd Int. Kimb. Conf., Extended Abs., 1977.

Eggler, D. H. and M. E. McCallum, A geotherm from megacrysts in the Sloan kimberlite pipes, Colorado. Carnegie Inst. Wash., Yr. Bk. 75, 538-541, 1976.

Fujii, T., Solubility of Al_2O_3 in enstatite coexisting with forsterite and spinel. Carnegie Inst. Wash., Yr. Bk. 75, 566-571, 1976.

Green, H. W. and Guegen, Y., Kimberlite pipes: origin by diapiric upwelling in the mantle. Science, 249, 617-620, 1974.

Gurney, J. J., B. Harte and K. G. Cox, Mantle xenoliths in the Matsoku kimberlite pipe; Phys. Chem. Earth., vol. 9, 507-524, 1975.

Hornung, G. and P. H. Nixon, Chemical variations in the knorringite-rich garnets, in Lesotho Kimberlites, ed. P. H. Nixon, Lesotho Nat. Develop. Corp., 122-167, 1973.

Kohlstedt, D. L. and C. Goetze, Low stress high temperature creep in olivine single crystals. J. Geophys. Res., 79, 2045-2051, 1974.

Lindsley, D. H. and S. A. Dixon, Diopside-enstatite equilibria at 850° to 1400°C, 5 to 35 kb. Amer. J. Sci., 276, 1285-1301, 1976.

MacGregor, I. D., The effect of CaO, Cr_2O_3, Fe_2O_3 and Al_2O_3 on the stability of spinel and garnet peridotites. Phys. Earth Planet. Int., 3, 372-377, 1969.

MacGregor, I. D., Petrologic and thermal structure of the upper mantle beneath South Africa in the Cretaceous, Phys. Chem. Earth, 9, 455-466, 1975.

Mori, L. and D. H. Green, Pyroxenes in the system $Mg_2Sc_2O_6-CaMgSi_2O_6$ at high pressure. Earth Planet. Sci. Letters, 26, 277-286, 1975.

Nehru, C. E. and P. J. Wyllie, Electron microprobe measurement of pyroxenes coexisting with H_2O-undersaturated liquid in the join $CaMgSi_2O_6-Mg_2Si_2O_6$ at 30 kilobars, with applications to geothermometry. Contr. Mineral. Petrol., 48, 221-228, 1974.

Nixon, P. H. and F. R. Boyd, Deep seated nodules, in Lesotho Kimberlite, ed. P. H. Nixon, Lesotho Nat. Develop. Corp., 106-109, 1973.

Nixon, P. H., F. R. Boyd, and A. Boullier, The evidence of kimberlite and its inclusions on the constitution of the outer part of the earth, in Lesotho Kimberlites, Ed. P. H. Nixon, Lesotho Nat. Develop. Corp., Maseru, 1973.

Obata, M., The solubility of Al_2O_3 in orthopyroxenes in spinel and plagioclase peridotites and spinel pyroxenite. Amer. Mineral., 61, 804-816, 1976.

Räheim, A. and D. H. Green, P T paths of natural eclogites during metamorphism--a record of subduction. Lithas, 8, 317-328, 1975.

Rolfe, D. G., The geology of the Kao kimberlite pipes, in Lesotho Kimberlite, ed. P. H. Nixon, Lesotho Nat. Develop. Corp., 101-106, 1973.

Warner, R. D. and W. C. Luth, The diopside-orthoenstatite two-phase region in the system $CaMgSi_2O_6-Mg_2Si_2O_6$. Amer. Mineral., 59, 98-109, 1974.

Wells, P. A., Pyroxene thermometry, in simple and complex systems. Contrib. Mineral. Petrol., 62, 129-139, 1977.

Whitelock, T. K., Morphology of the Kao diamonds, in Lesotho Kimberlites, Ed. P. H. Nixon, Lesotho Nat. Develop. Corp., 128-140, 1973.

Wood, B. J., The solubility of alumina in orthopyroxene coexisting with garnet. Contr. Mineral. Petrol., 46, 1-15, 1974a.

Wood, J. B., The application of thermodynamics to some subsolidus equilibria involving solid solutions. Fortschr. Mineral., 52, 21-45, 1974b.

Wood, B. J. and S. Banno, Garnet-orthopyroxene and orthopyroxene-clinopyroxene relationships in simple and complex systems. Contr. Mineral. Petrol., 109-124, 1973.

METASOMATISM OF THE UPPER MANTLE AND THE GENESIS OF KIMBERLITES AND ALKALI BASALTS[†]

Arthur L. Boettcher

Department of Earth and Space Sciences and Institute of Geophysics and Planetary Physics, University of California, Los Angeles, Los Angeles, California 90024

James R. O'Neil

U.S. Geological Survey, Menlo Park, California 94025

Kenneth E. Windom

*Institute of Geophysics and Planetary Physics, University of California, Los Angeles, Los Angeles, California 90024

Dion C. Stewart

Pennsylvania State University, University Park, Pennsylvania 16802

Howard G. Wilshire

U.S. Geological Survey, Menlo Park, California 94025

Abstract. Petrological and geochemical evidence reveals that pervasive metasomatism of upper-mantle lherzolite is precursory to or concomitant with anatexis associated with the genesis of alkali basalts and kimberlites. This introduction of fluids rich in Ti, K, Fe, H_2O and other elements is well displayed in peridotite and eclogite xenoliths in these rocks, particularly in the development of titaniferous phlogopites and amphiboles. Values of δD of phlogopite lie in the restricted range of -60 to $-79^0/_{00}$, which is consonant with other determinations of primordial H_2O and implies constant conditions of formation. On the other hand, amphiboles exhibit a large range in D and H_2O contents, suggesting a more complex history. Values of $\delta^{18}O$ for the phlogopites and amphiboles range from 4.26 to 5.92, also typical of deep-seated materials.

Selective enrichment by aqueous fluids may be, in part, responsible for the chemical heterogeneity of the upper mantle, which is supported by a plethora of other geochemical data. It also provides a ready explanation for the abundance of incompatible elements in kimberlites and alkali basalts as well as a mechanism for localized magma generation within the mantle.

Introduction

Various lines of petrochemical evidence disclose that pervasive metasomatism of mantle lherzolite is precursory to or concomitant with anatexis in the production of many deep-seated alkali basaltic magmas and kimberlites. For example, ultramafic mantle xenoliths in kimberlites and in alkali basalts, basanites, and kindred rocks commonly exhibit evidence of metasomatic enrichment in TiO_2, K_2O, total Fe, H_2O, etc. Evidence for this in many kimberlite xenoliths is abundantly manifest in the development of "secondary" phlogopite, i.e., phlogopite formed in the mantle, but subsequent to the crystallization of the primary lherzolite. In addition to the formation of the phlogopite, this process converts the lherzolite to assemblages rich in clinopyroxene, amphibole, and other minerals (e.g. Lloyd and Bailey, 1975), which are abundant in kimberlites (e.g. Dawson and Smith, 1973) and alkali basalts (e.g. Best and Brimhall, 1974) from many areas of the world.

† Publication No. 1755, Institute of Geophysics and Planetary Physics, University of California at Los Angeles.

* Present address: Department of Earth Sciences, Iowa State University, Ames, Iowa 50011.

Fig. 1. Photomicrographs of phlogopite from DeBeers Mine kimberlite (sample Kb-5-1-A), showing core with reverse pleochroism and rim with high Al_2O_3, Cr_2O_3, and TiO_2 and normal pleochroism (see Table 1). Long dimension of phlogopite grain is 0.9 mm. Plane-polarized light, with vibration direction of the polarizer east-west.

Results

We have selected for chemical and isotopic analysis a number of phlogopites, some of which on the basis of textural evidence appear to be secondary, for comparison with those that have been classified by others as primary. The former occur in veins and overgrowths (Figure 1); the latter as discreet grains (Figure 2), such as the "primary" mica pictured by Dawson et al., (1970, plate 3), and Carswell (1975, Figure 1-A).

A facile examination of Table 1 discloses that the obviously secondary phlogopites, including those that form rims on preexisting mica, are enriched in TiO_2. Carswell (1975) previously pointed out that micas with high TiO_2 contents (> ∿ 1.0%) have textural relationships suggestive of being secondary, and this is in chorus with our findings. Titaniferous phlogopites postulated as primary (see Table 1), such as in the garnet lherzolite xenolith BD 738 from Lashaine Volcano (Dawson et al., 1970) or the alkaline rocks from Jan Mayen (Flower, 1969) and West Kimberly (Prider, 1939), may be secondary.

The low-TiO_2, high-Fe phlogopites in Table 1, some of which are the cores for high-TiO_2 phlogopite rims, themselves commonly appear to be secondary (e.g. Kb-9-5 and B-131). These micas exhibit reverse pleochroic schemes, previously reported in micas from kimberlites and alkalic ultramafic rocks (e.g. Wagner, 1914; Watson, 1955; Hogarth, 1964; Boettcher, 1967; and Suwa and Aoki, 1975). The normal micas contain high totals of Ti, Cr, and Al; the reverse micas are enriched in Fe, a significant proportion of which appears to be Fe^{+3} in the tetrahedral sites.

Eclogite xenoliths also exhibit similar features. Some from the Roberts Victor Mine contain fine-grained intergranular "microphenocrysts" of titaniferous phlogopite, potassic pargasite (Na + K = 1), aluminous augite (Ca-Ts = 14%), green spinel, plagioclase (An_{49}), and analcime in a heterogeneous alkali-rich ($K_2O + Na_2O$ > 12%), carbonate-bearing groundmass (Windom and Boettcher, 1977a; Switzer and Melson, 1969). Phlogopite occurs in veinlets within the garnet, which is $Py_{57}Al_{30}$ (Gr + And_{12}) and homogeneous. By contrast, the clinopyroxene in the eclogite is optically heterogeneous, containing two

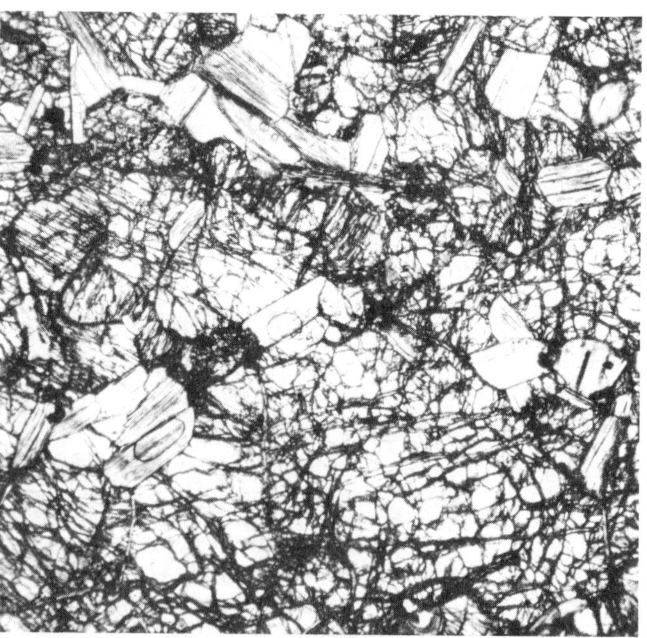

Fig. 2. Photomicrograph of discrete phlogopite crystals in a garnet lherzolite xenolith (sample Kb-9-1) from Bultfontein Floors. Crystals average 0.1 mm diameter. Plane-polarized light.

TABLE 1. Microprobe Analyses of Phlogopites.

	DeBeers Kimberlite Kb-5-1-A		DeBeers Kimberlite Kb-5-1-B		DeBeers Kimberlite Kb-5-1-C	DeBeers Kimberlite Kb-5-13		DeBeers Xenolith Kb-5-10		DeBeers Xenolith Kb-5-14	DeBeers Xenolith Kb-5-2
	core reverse	rim normal	core reverse	rim normal	normal	core reverse	rim normal	core reverse	rim normal	reverse	mildly normal
SiO_2	41.98	40.38	41.72	39.15	40.68	41.81	39.97	41.07	38.69	41.69	42.14
TiO_2	0.83	3.99	0.36	4.46	1.75	0.66	4.58	0.39	3.94	0.68	1.29
Al_2O_3	9.49	14.27	10.55	13.95	10.62	9.45	13.28	10.00	10.84	10.62	9.63
Cr_2O_3	0.29	1.65	0.14	1.23	0.26	0.10	1.59	0.05	0.11	0.11	0.45
FeO*	6.21	4.67	6.86	4.99	6.63	6.28	4.72	6.96	8.52	6.44	6.96
MnO	0.01	0.04	0.02	0.04	0.02	0.02	0.04	0.02	0.04	0.01	0.04
MgO	24.17	19.64	24.69	21.90	23.21	24.66	20.27	24.39	21.71	24.16	23.61
CaO	0.92	0.86	0.26	0.14	0.08	0.11	0.33	0.10	0.11	0.09	0.12
Na_2O	0.26	1.17	0.16	0.29	0.18	0.20	0.25	0.30	0.05	0.22	0.21
K_2O	9.89	9.45	10.19	9.89	9.48	9.69	9.77	9.82	9.68	9.06	9.37
Total	94.05	96.12	94.95	96.03	92.90	92.98	92.81	93.09	93.70	93.08	93.82
					Cations Per 22 Oxygen						
Si	6.112	5.722	6.028	5.567	5.986	6.134	5.598	6.052	5.733	6.083	6.134
Ti	0.091	0.425	0.039	0.477	0.193	0.073	0.507	0.043	0.440	0.074	0.141
Al	1.629	2.384	1.796	2.338	1.841	1.634	2.308	1.737	1.893	1.826	1.653
Cr	0.033	0.185	0.016	0.138	0.030	0.012	0.186	0.005	0.013	0.013	0.051
Fe	0.756	0.553	0.829	0.594	0.816	0.771	0.582	0.857	1.056	0.786	0.848
Mn	0.002	0.005	0.002	0.004	0,002	0.003	0.005	0.003	0.005	0.002	0.004
Mg	5.245	4.148	5.316	4.641	5.089	5.391	4.455	5.356	4.793	5.252	5.122
Ca	0.144	0.130	0.040	0.021	0.012	0.017	0.052	0.016	0.018	0.014	0.019
Na	0.073	0.323	0.044	0.081	0.052	0.058	0.072	0.085	0.015	0.062	0.059
K	1.837	1.708	1.878	1.795	1.780	1.814	1.838	1.846	1.830	1.686	1.740
Total	15.921	15.584	15.988	15.656	15.802	15.906	15.603	15.999	15.796	15.797	15.772

TABLE 1. Microprobe Analyses of Phlogopites (continued)

	Dutoitspan Xenolith Kb-8-8		Dutoitspan Xenolith Kb-8-3		Dutoitspan Xenolith Kb-8-4	Bultfontein Xenolith Kb-9-9-A		Bultfontein Xenolith Kb-9-9-B		Bultfontein Xenolith Kb-9-1	Bultfontein Xenolith Kb-9-5
	core reverse	rim normal	core reverse	rim normal	normal	core reverse	rim normal	core reverse	rim normal	reverse	reverse
SiO_2	41.33	40.57	41.23	40.33	41.85	42.51	41.10	42.73	40.96	42.25	42.92
TiO_2	0.61	3.73	0.43	2.89	2.19	0.35	2.99	0.45	3.66	0.75	0.57
Al_2O_3	9.63	12.69	9.48	11.33	10.13	9.72	12.03	9.86	12.28	10.34	10.24
Cr_2O_3	0.25	1.41	0.07	0.13	0.25	0.07	0.46	0.10	0.24	0.39	0.30
FeO*	6.33	4.58	8.17	6.00	6.99	5.72	5.46	5.81	5.47	4.43	4.64
MnO	0.03	0.02	0.04	0.04	0.04	0.01	0.03	0.03	0.01	0.02	0.01
MgO	24.58	22.85	24.65	23.75	22.08	25.06	22.92	25.22	22.39	25.03	25.61
CaO	0.33	0.14	0.15	0.16	0.14	0.00	0.00	0.00	0.00	0.09	0.12
Na_2O	0.35	0.27	0.15	0.11	0.34	0.07	0.28	0.08	0.29	0.10	0.07
K_2O	9.99	10.13	9.81	9.46	0.74	10.25	9.69	10.33	9.62	9.89	9.89
Total	93.43	96.39	94.18	94.21	93.76	93.76	94.69	94.61	94.92	93.29	94.38
					Cations Per 22 Oxygen						
Si	6.065	5.734	6.042	5.839	6.110	6.170	5.882	6.151	5.859	6.119	6.142
Ti	0.067	0.396	0.047	0.314	0.240	0.039	0.322	0.048	0.394	0.082	0.062
Al	1.666	2.115	1.637	1.934	1.743	1.663	2.030	1.674	2.072	1.765	1.728
Cr	0.029	0.157	0.009	0.015	0.029	0.008	0.052	0.011	0.027	0.045	0.033
Fe	0.777	0.542	1.002	0.727	0.854	0.695	0.654	0.700	0.654	0.537	0.557
Mn	0.004	0.003	0.005	0.005	0.006	0.002	0.003	0.003	0.001	0.002	0.001
Mg	5.376	4.814	5.385	5.125	4.805	5.421	4.889	5.410	4.773	5.402	5.462
Ca	0.052	0.021	0.024	0.024	0.022	0.000	0.000	0.000	0.000	0.013	0.019
Na	0.100	0.074	0.042	0.031	0.097	0.020	0.077	0.023	0.081	0.029	0.019
K	1.870	1.826	1.834	1.748	1.815	1.898	1.769	1.897	1.756	1.828	1.806
Total	16.006	15.684	16.026	15.762	15.720	15.915	15.678	15.918	15.617	15.823	15.828

TABLE 1. Microprobe Analyses of Phlogopites (continued).

	Bultfontein Xenolith Kb-9-4	Bultfontein Xenolith Kb-9-33	Jagersfontein Xenolith Kb-12-33-1	Roberts Victor Xenolith B-16-7	Libby Xenolith B-131
	reverse	intermediate	normal	normal	reverse
SiO_2	41.72	42.15	42.14	38.91	41.97
TiO_2	0.38	1.04	3.56	3.67	0.62
Al_2O_3	8.22	10.11	11.15	14.62	9.83
Cr_2O_3	0.15	0.15	0.54	0.17	0.03
FeO*	8.74	6.73	4.28	9.88	6.76
MnO	0.03	0.01	0.02	0.05	0.09
MgO	24.53	23.40	23.43	17.72	26.26
CaO	0.15	0.00	0.00	0.07	0.15
Na_2O	0.20	0.17	0.44	0.30	0.07
K_2O	10.21	10.16	9.53	9.83	12.03
Total	94.33	93.91	95.09	95.21	97.80
			Cations Per 22 Oxygen		
Si	6.140	6.138	6.172	5.678	5.955
Ti	0.042	0.114	0.041	0.403	0.066
Al	1.427	1.736	1.926	2.516	1.644
Cr	0.017	0.017	0.063	0.020	0.003
Fe	1.075	0.819	0.525	1.206	0.802
Mn	0.004	0.002	0.002	0.006	0.011
Mg	5.379	5.079	5.114	3.854	5.553
Ca	0.023	0.000	0.000	0.011	0.023
Na	0.058	0.047	0.125	0.084	0.020
K	1.917	1.887	1.781	1.831	2.177
Total	16.083	15.839	15.749	15.609	16.254

distinct regions. One contains apparently fresh, homogeneous omphacite with a composition of about 60 (wt)% diopside ($Wo_{43}En_{50}Fs_7$), 39% Jd + Ac, and less than 1.5% Ca-Ts. The patchy areas are chemically very heterogeneous. Harker diagrams suggest that they result from a combination of partial melting (the anatectic liquid being much richer in alkalies, Al_2O_3 and SiO_2 and poorer in Fe, MgO, and CaO than the primary omphacite) combined with the introduction of phlogopite and other components. Our interpretation of these data is that fluids rich in K_2O, TiO_2, and H_2O associated with the formation of the kimberlite permeated the eclogite, causing partial melting of the omphacite.

Xenoliths of spinel lherzolite in alkali basalts and basanites also exhibit evidence of such metasomatic alteration. As many as 50% of these xenoliths from many areas in North America and elsewhere contain pargasitic or kaersutitic amphiboles that have crystallized after the primary lherzolite assemblage but prior to incorporation of the xenoliths into the host magmas (e.g. Wilshire and Trask, 1971; Best, 1974; Francis, 1976; Stewart and Boettcher, 1977). The formation of this amphibole, commonly together with accompanying phlogopite, apatite, magnetite, and other minor phases, is unrelated to contamination by the host magma (see also Lloyd and Bailey, 1975, p. 402), and chemical and textural zonations of these and the primary lherzolite minerals are related to the emplacement of amphibolitic and pyroxenitic veins that predate incorporation of the lherzolite into the host magma. Pike and Schwarzman (1977) and Wilshire and Servals (1975) have presented textural and chemical evidence, respectively, that these features are younger than the primary mantle lherzolite. Our chemical evidence reveals that pargasites in these spinel lherzolites have been metasomatized to kaersutitic amphiboles during emplacement of these veins. For example, minerals in spinel lherzolite xenoliths from Dish Hill, California show variations in the major elements as gradients strongly developed perpendicular to xenolith surfaces covered with rinds of amphibole (the rinds are remnants of veins); no gradients occur parallel to these surfaces. In traverses toward the rinds from within the lherzolites, amphiboles exhibit a relative increase in TiO_2 (> 250%), total Fe (> 75%), and K_2O (> 50%) and a decrease in Cr_2O_3 (> 80%). Concomitant changes in the primary lherzolite minerals are illustrated in Figure 3, using Fe as an example. Spinel, orthopyroxene, and clinopyroxene exhibit decreases in Al_2O_3 (> 35%), and spinel shows a marked increase in TiO_2, all toward the rind. No significant changes in the chemistry of any of the minerals in the lherzolites were detected in traverses from within the xenoliths to the contacts with the lavas. This observation

TABLE 2. Oxygen and Hydrogen Isotope Compositions and H_2O Contents of Hydrous Minerals.

	$\delta^{18}O$ ‰	Yield μ moles O_2/mg	δD ‰	Wt % H_2O
Kb-8-3 (P) glimmerite xenolith, Dutoitspan kimberlite, Africa	5.92	13.5	-60	3.48
Kb-8-4 (P) glimmerite xenolith, Dutoitspan kimberlite, Africa	(-6.93 $\delta^{13}C$)		-79	3.19
Pretorious #2 (P) spinel, lherzolite xenolith, Wesselton kimberlite, Africa	4.68	12.5		
B-8-9 (P) spinel, harzburgite xenolith, Pipe 200 kimberlite, Africa	5.64	13.9	-70	3.08
Kb-9-4 (P) pargasite-rich phlogopite xenolith, Bultfontein Floors kimberlite, Africa	5.64	13.3	-70	3.77
Kb-5-10 (P) Cpx- and Ol-bearing phlogopite xenolith, DeBeers kimberlite, Africa	5.36	14.0	-65	3.90
Kb-5-2 (P) glimmerite xenolith, DeBeers kimberlite, Africa	5.72	13.3	-73	3.41
B-12-S110 (P) lherzolite xenolith, Leghebong kimberlite, Africa	5.80		-68	3.58
B-22-9 (P) megacrysts (45 mm diameter) Monastery kimberlite, Africa	5.30	13.8	-60	3.88
B-22-3 (P) megacrysts (5 mm diameter) Monastery kimberlite, Africa	5.31	13.5		
B-Gold (P) alnoite, Oka, Canada	4.26	12.8		
B-12 (B) biotitite, Libby, Montana	5.68	12.8	-78	
Sp-55 (B) biotitite, Libby, Montana	5.72	13.0	-71	3.20
B-CH-5 (P) megacryst, Crater Hill #32, Australia	5.67		-86	
B-CH-5 (A) megacryst, Crater Hill #32, Australia	5.56			0.06
B-DH-NW-6 (A) megacryst, Dish Hill, California	5.63	7.3	-111	0.11
MK-3 (A) kaersutite megacryst, Kakanui, New Zealand	5.78	13.4		
DL-9 (A) megacryst, Deadman Lake, California	4.65	10.1	-55	0.62
B-BC-2-24 (A) megacryst, Hoover Dam camptonite, Arizona	4.73	13.5		
C-1 (A) megacryst, Gillies Crater, Australia	5.72	14.1	-37	1.15
At-38 (A) megacryst, Mt. Quincan, Australia	5.48	6.5	-59	1.25
H-92 (A) oxidized megacryst, Mt. Emu, Australia	4.85	12.1	-35	0.30
B-VT-3 (A) megacryst, Vulcan's Throne, Arizona	5.40	12.4	-33	0.57

Abbreviations: (A) = kaersutitic amphibole; (P) = phlogopite; (B) = biotite

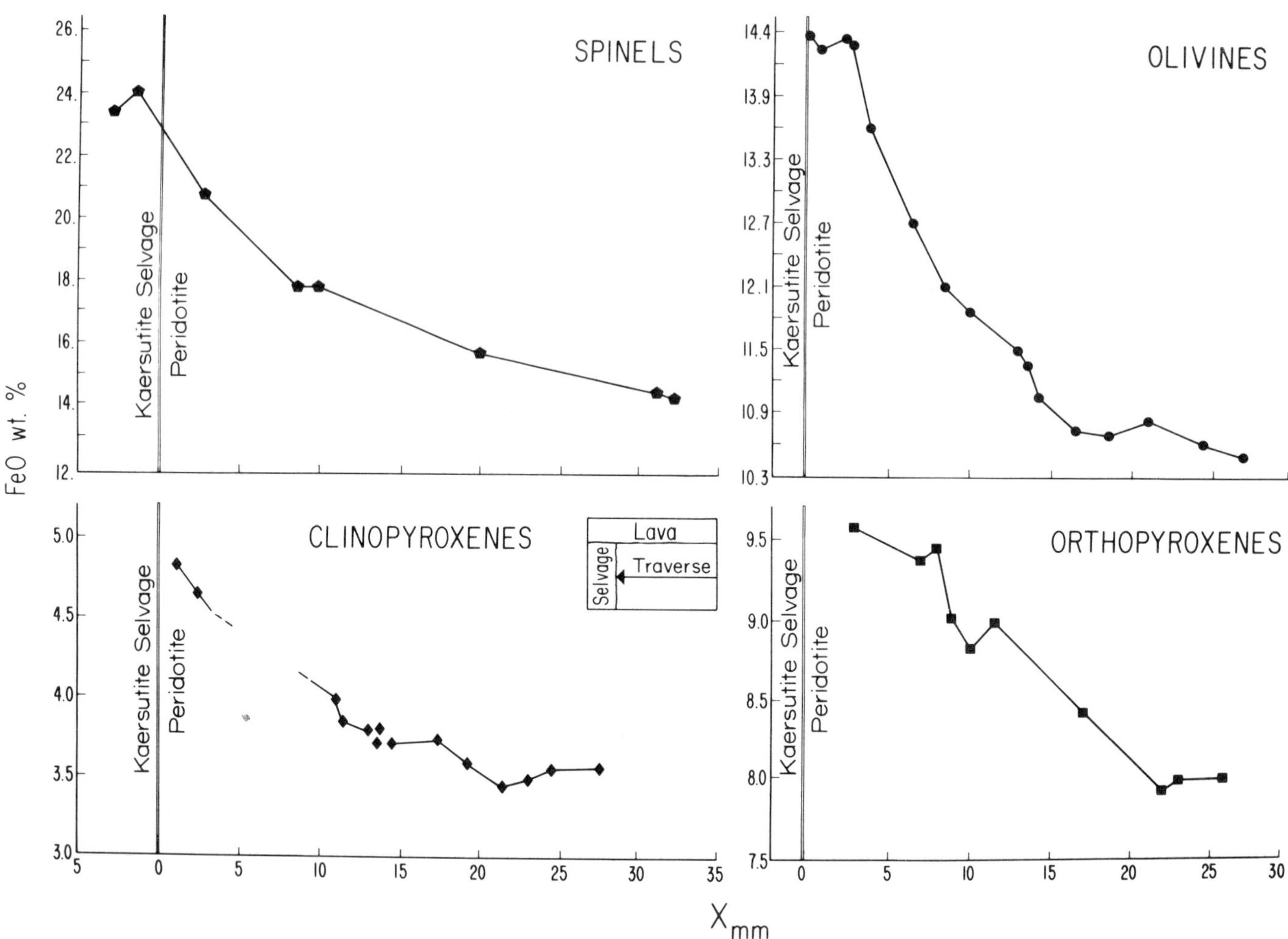

Fig. 3. Variations in total-iron content (expressed as FeO) of minerals in spinel lherzolite xenolith BA-2-9 from Dish Hill, California. The traverses are perpendicular to the xenolith-selvage contact. No significant variations occur in traverses normal to the lava-xenolith contact (spinels not examined in this traverse). Values for spinels and orthopyroxenes are averages of several grains. Those for olivines are for cores of grains; those for clinopyroxenes are each averages of cores and rims. Microprobe determinations by D. C. Stewart.

and those from our studies of xenoliths from other localities support the earlier conclusion of Lloyd and Bailey (1975) and Wilshire and Trask (1971) that the hydrous minerals in xenoliths are not the result of reaction with the host lavas. Caution must be exercised in the application of these mineralogical data to the determinations of temperatures and pressures of crystallization.

These data also support the proposal of Wilshire and Trask (1971) that kaersutite and clinopyroxene (rarely orthopyroxene and spinel) megacrysts in these alkalic lavas are disaggregated remnants of these veins. For example, our kaersutite megacrysts in the Dish Hill basanite have compositions equivalent to those of a rind extrapolated to a distance of 10 mm beyond the inner border of the rind (Stewart and Boettcher, 1977). These data also are consonant with the findings of Basu and Murthy (1977) that the $^{87}Sr/^{86}Sr$ ratios of kaersutite megacrysts from Dish Hill, and probably elsewhere, are the same as those of the kaersutites in the veins. On the other hand, no ready explanation is offered as to why these ratios differ from those of the host basanites (Stuckless and Irving, 1976; Basu and Murthy, 1977), except that the veins predate incorporation of the xenoliths and megacrysts into the magma. This is in contrast to the concept that kaersutite and other megacrysts are near-liquidus phenocrystal phases in the alkaline magmas (see Irving, 1974), although the results of some high-pressure studies are permissive of such an origin (Merrill and Wyllie, 1975; Windom and Boettcher, 1977b; Stewart, in preparation).

$\delta^{18}O$, δD, and H_2O^+ of the amphiboles and phlogopites in the xenoliths and megacrysts were determined for kimberlites, alkali basalts, and basanites. The range of values for δD (relative to SMOW) for the micas is -60 to $-73^0/_{00}$, which is consonant with suggested values of "primordial" H_2O (Sheppard and Epstein, 1970; Craig and Lupton, 1976; Kuroda et al., 1977; Kyser and O'Neil, unpublished data). The narrowness of this range suggests that all H_2O is fractionated into the hydrous minerals, with no vapor present. Values of δD for the amphiboles are more variable, ranging from -30 to $-111^0/_{00}$.

The average $\delta^{18}O$ value for 17 of the 22 samples of amphiboles and micas analyzed is 5.61 ± $0.18^0/_{00}$, which is equal to the value obtained for most fresh tholeiitic basalts (Taylor, 1968; Muehlenbachs and Clayton, 1972). The remaining five samples, again including both amphiboles and phlogopites, yield $\delta^{18}O$ values of 4.63 ± $0.22^0/_{00}$. There is no obvious explanation for the apparent bimodal distribution. These data indicate that most hydrous minerals are in approximate ^{18}O equilibrium with the mantle. At this stage, the stable-isotope data do not enable us to distinguish between a primordial or subducted source of H_2O for these hydrous minerals.

An unusually large quantity of CO_2 (2.67 wt %) was released from a sample of phlogopite (Kb-8-4) during thermal decomposition of the mineral for hydrogen isotope analysis. Based on a high yield of oxygen during fluorination for ^{18}O analyses, the carbon probably exists in the form of CO_2 inclusions. The $\delta^{13}C$ value of -6.93 measured for this gas is typical of carbon of mantle or lower-crustal origin (e.g. Taylor et al., 1967; Pineau et al., 1976).

The range of H_2O contents of the micas analyzed for hydrogen-isotope compositions is narrow (3.08-3.90 wt %) and approximates theoretical proportions. This implies limited substitution by halogens. On the other hand, amphiboles contain only 1.25 to 0.06 wt % H_2O, and at least some are rich in F and Cl. For example, a Dish Hill kaersutite megacryst DH-1A contains 0.40 (wt)% H_2O, 0.13% F, and 0.1% Cl, and a large "oxyamphibole" component (Fe_2O_3 = 13.76%; FeO = 0.29%). These data suggest that the α_{H_2O} is considerably less than unity, even if a discrete vapor may have coexited with the hydrous minerals.

Conclusions

We interpret the above data as evidence that anatexis or mobilization of mantle lherzolite during the genesis of alkali basalts and kimberlites is subsequent to or concomitant with metasomatism by aqueous fluids rich in Ti, Fe, K, and other ions. Additional supportive evidence can be found in papers by Basu and Murthy (1977), Best (1974), Boyd (1973), Edgar, Green, and Hibberson (1976), Ehrenberg (1978), Erlank (1976), Francis (1976), Frey and Green (1974), Lloyd and Bailey (1975), and Varne and Graham (1971). Primary alkali basalt and basanitic magmas unmodified by fractionation, such as those described from Australia (Kesson, 1973), Colorado Plateau (Best and Brimhall, 1974), Mauritius (Indian Ocean) (Baxter, 1976), and the Easter volcanic chain (Pacific Ocean) (Bonatti et al., 1977), are enriched in elements including K, Ti, etc., as well as H_2O. Different degrees of partial melting of the mantle could account for the differences between some enriched primary magmas and others (Baxter, 1976), but marked heterogenity of the source regions is supported by a plethora of geochemical data (Cox et al., 1976; Kesson, 1973; Sun and Hanson, 1975).

Mason (1966) has remarked on the notable observation that Ti, like K, is concentrated in the crust (see Ronov and Yaroshevsky, 1969) relative to the mantle (see Ringwood, 1975, p. 118 and 179), and he later (Mason, 1968) proposed that kaersutite in the upper mantle could be the source of concentration of these elements. Upward migration of H_2O-rich solutions enriched in these and other incompatible elements is in concert with this proposal.

The abundance of these incompatible elements in alkali basalts relative to those in tholeiites has largely been unexplicable in terms of the major proposals for the origins of these magmas (see Ringwood, 1975). The model of O'Hara (1968) requires about 15% anatexis together with extensive eclogite fractionation at depth and as much as 50% olivine extraction at shallower depths. This is not consonant with the common occurrence of megacrysts and xenoliths in these magmas at the surface. Green and Ringwood (1967) proposed that wall-rock reaction would selectively increase the concentration of K, Ti, etc. without significant changes in major-element chemistry, but why would this be restricted to alkaline magmas and not to tholeiites? The indifference of the chemistry of alkali basalts to their environment of eruption, be it continental, oceanic, or island-arc, argues against crustal contamination as a viable explanation (see Schwarzer and Rogers, 1974).

On the basis of the distribution of trace elements, Gast (1968) suggested that alkali basalts are not generated by fractional crystallization but, rather, are the products of small (< 7%) degrees of anatexis of mantle peridotite. This model has much to commend it, but a precursory metasomatic event or events resulting from an aqueous fluid not only provides the requisite elements and eliminates the need for very small fractions of partial melting, it also provides the mechanism for localized magma generation in the mantle. That is, metasomatism will provide the following, which will enhance the likelihood of melting: (1) H_2O, CO_2, and other fluxes, (2) low-melting components such as Fe and alkalies, and (3) heat sources such as ^{40}K.

Migrating fluids associated with, say, mantle diapirism (Wilshire and Pike, 1975; Best and Brimhall, 1974) or plumes (Bonatti et al., 1977) appear to have operated throughout much of the history of the mantle. Consequently, the chemistry of mantle samples has been altered by the very processes that have resulted in bringing them to the surface--Irving's (1976) Heisenberg uncertainty. As a result, xenoliths probably are a poor representation of much of the mantle, and Earth models based on such evidence may portray an upper mantle richer in incompatible elements, Fe, Ti, H_2O, and CO_2 than actually occurs. This line of reasoning is contrary to the generally accepted concept that xenoliths are depleted in these elements relative to most mantle lherzolite!

The samples most representative of unaltered mantle may be inclusions in diamonds. Here the compositional ranges of olivines and orthopyroxenes are narrower (Meyer and Tsai, 1976), and the olivines, clinopyroxenes, garnets, and spinels are much richer in Cr (Meyer, 1975) than they are for these phases in xenoliths.

Even here, however, variations are common. For example, the K_2O content of clinopyroxene inclusions in diamond ranges from $\sim$ 0.0 to 0.3 wt %, perhaps reflecting primary heterogenity in the mantle at the time that they crystallized (Sobolev, 1977, p. 242). However, this range appears to be considerably narrower when only diopsides are considered, neglecting the omphacite group (Sobolev, 1977, p. 195; see also Meyer and Tsai, 1976). Part of this range may also reflect sampling of the mantle at different depths, as the K_2O content of pyroxenes is likely pressure dependent in the depth range where kimberlites originate.

The search for pristine mantle continues.

Acknowledgements

Most of this research was supported by National Science Foundation Grants EAR76-22330 and EAR73-00220-A02 to Boettcher. A few mass spectrometric analyses were performed in Professor Epstein's laboratory at Cal Tech. We are indebted to Denise Maziars, John Perry, Cynthia Ricks, and Robbie Score of U.C.L.A. for their excellent work in sample preparation and to Lanford Adami and Doug White of U.S.G.S. for isotopic analyses. Harriett Arnoff carefully typed and retyped the manuscript. The valuable reviews by Asish Basu and Myron Best are appreciated, as are the editorial comments of Doug Smith.

References

Baird, A. K., D. M. Morton, A. O. Woodford, and K. W. Baird, Transverse Ranges province; a unique structural-petrochemical belt across the San Andreas fault system, Geol. Soc. Am. Bull. 85, 163-74, 1974.

Basu, A. R., and V. M. Murthy, Kaersutites, suboceanic low-velocity zone, and the origin of mid-oceanic ridge basalts, Geology, 5, 365-8, 1977.

Baxter, A. N., Geochemistry and petrogenesis of primitive alkali basalt from Mauritius, Indian Ocean, Geol. Soc. Am. Bull. 87, 1028-34, 1976.

Best, M. G., Mantle-derived amphibole within inclusions in alkalic-basaltic lavas, J. Geophys. Res. 79, 2107-13, 1974.

Best, M. G., and W. H. Brimhall, Late Cenozoic alkalic basaltic magmas in the western Colorado Plateau and the Basin and Range transition zone, U.S.A., and their bearing on mantle dynamics, Geol. Soc. Am. Bull. 85, 1677-90, 1974.

Boettcher, A. L., The Rainy Creek alkaline-ultramafic igneous complex near Libby, Montana. I. Ultramafic rocks and fenite, J. Geol. 75, 526-53, 1967.

Bonatti, E., C. G. A. Harrison, D. E. Fisher, J. Honnorez, J. -G. Schilling, J. J. Stipp, and M. Zentilli, Eastern volcanic chain (southeast Pacific): a mantle hot line, Jour. Geophys. Res. 82, 2457-78, 1977.

Boyd, F. R., A pyroxene geotherm, Geochim. Cosmochim. Acta. 37, 2533-46, 1973.

Campbell, I., and E. T. Schenk, Camptonite dikes near Boulder Dam, Arizona, Amer. Mineral. 35, 671-92, 1950.

Carswell, D. A., Primary and secondary phlogopites and clinopyroxenes in garnet lherzolite xenoliths, Phys. Chem, Earth 9, 417-29, 1975.

Cox, K. G., C. J. Hawkesworth, R. K. O'Nions, and J. D. Appleton, Isotopic evidence for the derivation of some Roman region volcanics from anomalously enriched mantle. Contrib. Mineral. Petrol. 56, 173-180, 1976.

Craig, H., and J. E. Lupton, Primordial neon, helium and hydrogen in oceanic basalts, Earth Planet. Sci. Lett. 31, 369-85, 1976.

Dawson, J. B., D. G. Powell, and A. M. Reid, Ultrabasic xenoliths and lava from the Lashaine volcano, northern Tanzania, J. Petrol. 11, 519-48, 1970.

Dawson, J. B., and J. V. Smith, Alkalic pyroxenite xenoliths from the Lashaine volcano, northern Tanzania, J. Petrol. 14, 113-31, 1973.

Edgar, A. D., D. H. Green, and W. O. Hibberson, Experimental petrology of a highly potassic magma, J. Petrol. 17, 339-356, 1976.

Ehrenberg, S. N., Garnet peridotite xenoliths in minette from the Navajo Volcanic Field, Proceedings of the Second International Kimberlite Conference, in press, 1978.

Erlank, A. J., Upper mantle metasomatism as revealed by potassic richterite bearing peridotite xenoliths from kimberlite, EOS, 57, 597, 1976.

Flower, M. F. J., Phlogopite from Jan Mayen Island (North Atlantic), Earth Planet. Sci. Lett. 6, 461-6, 1969.

Francis, D. M., Amphibole pyroxenite xenoliths: cumulate or replacement phenomena from the upper mantle, Nunivak Island, Alaska, Contrib. Mineral. Petrol. 58, 51-61, 1976.

Frey, F. A., and D. H. Green, The mineralogy, geochemistry, and origin of lherzolite inclusions in Victorian basanites, Geochim. Cosmochim. Acta 38, 1023-59, 1974.

Friedman, I., and J. R. O'Neil, Hydrogen. In: Handbook of Geochemistry II/5 (editor H. Wedepohl). Springer-Verlag, New York, 1978.

Gast, P. W., Trace element fractionation and the origin of tholeiitic and alkaline magma types. Geochim. Cosmochim. Acta 32, 1057-86, 1968.

Green, D. H., and A. E. Ringwood, The genesis of basaltic magmas, Contr. Mineral. Petrol. 15, 103-90, 1967.

Hogarth, D., Normal and reverse pleochroism in biotite, Canadian Mineral. 8, 136, 1964.

Irving, A. J., Megacrysts from the New Basalts and other basaltic rocks of southeastern Australia, Bull. Geol. Soc. Amer. 85, 1503-14, 1974.

Irving, A. J., On the validity of paleogeotherms determined from xenolith suites in basalts and kimberlites, Amer. Mineral. 61, 638-42, 1976.

Kesson, S. E., The primary geochemistry of the Monaro alkaline volcanics, southeastern Australia-evidence for upper mantle heterogeneity, Contrib. Mineral. Petrol. 42, 93-108, 1973.

Kuroda, Y., T. Suzuoki, and S. Matsuo, Hydrogen isotope compositions of deep-seated water, Contrib. Mineral. Petrol. 60, 311-15, 1977.

Kyser, T. K., and J. R. O'Neil, ^{18}O, D, and H_2O contents of basalts and ultramafic nodules from Hawaii, Geol. Soc. Amer. Annual Mtg. 9, 1063-4, 1977.

Lloyd, F. E., and D. K. Bailey, Light element metasomatism of the continental mantle: the evidence and the consequences, Phys. Chem. Earth 9, 389-416, 1975.

Mason, B., Composition of the Earth, Nature 211, 616-18, 1966.

Mason, B., Kaersutite from San Carlos, Arizona, with comments on the paragenesis of this mineral, Mineralog. Mag. 36, 997-1002, 1968.

Merrill, R. B., and P. J. Wyllie, Kaersutite and kaersutite eclogite from Kakanui, New Zealand--water-excess and water-deficient melting to 30 kilobars, Geol. Soc. Am. Bull. 86, 555-70, 1975.

Meyer, H. O. A., Chromium and the genesis of diamond, Geochim. Cosmochim. Acta 39, 929-36, 1975.

Meyer, H. O. A., and H. -M. Tsai, The nature and significance of mineral inclusions in natural diamond: a review, Minerals Sci. Engng. 8, 242-261, 1976.

Muehlenbachs, K., and R. N. Clayton, Oxygen isotope studies of fresh and weathered basalts. Can. J. Earth Sci. 9, 172-84, 1972.

O'Hara, M. J., The bearing of phase equilibria studies in synthetic and natural systems on the origin and evolution of basic and ultrabasic rocks, Earth Science Reviews 4, 69-133, 1968.

Pike, J. E., and E. C. Schwarzman, Classification of textures in ultramafic xenoliths, J. Geology 85, 49-61, 1977.

Pineau, F., M. Javoy, and Y. Bottinga, $^{13}C/^{12}C$ ratios of rocks and inclusions in popping rocks of the Mid-Atlantic Ridge and their bearing on the problems of isotopic compositions of deep-seated carbon, Earth Planet. Sci. Lett. 29, 413-21, 1976.

Prider, R. T., Some minerals from the lucite-rich rocks of the West Kimberly area, Western Australia, Mineralog. Mag. 25, 373-87, 1939.

Ringwood, A. E., Composition and Petrology of the Earth's Mantle, McGraw-Hill, New York,1975.

Ronov, A. B., and A. A. Yaroshevsky, Chemical composition of the earth's crust. In: The Earth's Crust and Upper Mantle, Geophys. Mongr. Ser, v. 13 (editor P. J. Hart), AGU, Washington, 1969.

Savin, S. M., and S. Epstein, The oxygen and hydrogen and shales, Geochim. Cosmochim. Acta 34, 43-63, 1970.

Schwarzer, R. R., and J. J. W. Rogers, A world-wide comparison of alkali olivine basalts and their differentiation trends, Earth Planet. Sci. Lett. 23, 286-96, 1974.

Sheppard, S. M. F., and S. Epstein, D/H and $^{18}O/^{16}O$ ratios of minerals of possible mantle or lower crustal origin, Earth Planet. Sci. Lett. 9, 232-9, 1970.

Sobolev, N. V., Deep-Seated Inclusions in Kimberlites and the Problem of the Composition of the Upper Mantle, American Geophysical Union, Washington, D.C., 1977.

Stewart, D. C., and A. L. Boettcher, Chemical gradients in mantle xenoliths, Geol. Soc. Amer. Annual Mtg. 9, 1191-2, 1977.

Stuckless, J. S., and A. J. Irving, Strontium isotope geochemistry of megacrysts and host basalts from southeastern Australia, Geochim. Cosmochim. Acta 40, 209-13, 1976.

Sun, S. S., and G. N. Hanson, Evolution of the mantle: geochemical evidence from alkali basalt, Geology 3, 297-302, 1975.

Suwa, K., and K. Aoki, Reverse pleochroism of phlogopites in kimberlites and their related rocks from South Africa, 1st Prelim. Rept. Afr. Studies, Nagoya Univ., 60-64, 1975.

Suzuoki, T., and S. Epstein, Hydrogen isotope fractionation between OH-bearing minerals and water, Geochim. Cosmochim. Acta 40, 1229-40, 1976.

Switzer, G., and W. G. Melson, Partially melted kyanite eclogite from the Roberts Victor Mine, South Africa, Smithsonian Contr. Earth Sci., no. 1, 1-9, 1969.

Taylor, H. P., The oxygen isotope geochemistry of igneous rocks, Contr. Mineral. Petrol. 19, 1-71, 1968.

Taylor, H. P., Jr., J. Frechen, and E. T. Degens, Oxygen and carbon isotope studies of carbonatites from the Laacher See District, West Germany and the Alnö District, Sweden, *Geochim. Cosmochim. Acta* 31, 407-30, 1967.

Varne, R., and A. L. Graham, Rare earth abundances in hornblende and clinopyroxene of a hornblende lherzolite xenolith: implications for upper mantle fractionation processes, *Earth Planet. Sci. Lett.* 13, 11-18, 1971.

Wagner, P., *The Diamond Fields of Southern Africa*, Tranwall Leader, 1914.

Watson, K. D., Kimberlite at Bachelor Lake, Quebec, *Amer. Mineral.* 40, 565-79, 1955.

Wilshire, H. G., and N. J. Trask, Structural and textural relationships of amphibole and phlogopite in peridotite inclusions, Dish Hill, California, *Amer. Mineral.* 56, 240-55, 1971.

Wilshire, H. G., and J. E. N. Pike, Upper-mantle diapirism: evidence from analogous features in alpine peridotite and ultramafic inclusions in basalt, *Geology* 3, 467-70, 1975.

Wilshire, H. G., L. C. Calk, and E. C. Schwarzman, Kaersutite--a product of reaction between pargasite and basanite at Dish Hill, California, *Earth Planet. Sci. Lett.* 10, 281-84, 1971.

Wilshire, H. G., and J. W. Shervais, Al-augite and Cr-diopside ultramafic xenoliths in basaltic rocks from western United States: structural and textural relationships, *Phys. Chem. Earth* 9, 257-72, 1975.

Windom, K. E., and A. L. Boettcher, Lamprophyre-kimberlite association exemplified in eclogite from Roberts Victor Mine, South Africa: evidence for metasomatism in the mantle, *Geol. Soc. Amer. Annual Mtg. Program* 9, 1230-1, 1977a.

Windom, K. E., and A. L. Boettcher, Melting relations in the system $NaAlSiO_4$-Mg_2SiO_4-SiO_2, *EOS* 58, 521, 1977b.

SPINELS IN HIGH PRESSURE REGIMES

Stephen E. Haggerty

Department of Geology, University of Massachusetts, Amherst, Massachusetts 01003

Abstract. An empirical study on a large data bank of spinel compositions from high pressure regimes was undertaken with the following objectives: (a) to examine the distribution of spinels in the framework of the lherzolite petrogenetic grid; (b) an attempt to distinguish between spinels that are derived from the lower part of the upper mantle from those that are derived from higher levels in the upper mantle; and (c) to evaluate the disappearance and reappearance of spinels in the P-T interval between garnet lherzolite stability and diamond stability. The results of this study show that spinels in layered intrusive suites and spinels in crustal precipitates are distinct from spinels in high pressure, mantle derived, ultramafic rocks. Spinels in kimberlites exhibit a marked contrast in Cr/Al trends from similar trends in xenolithic spinels from volcanic rocks; the former suggest an increase in Al with decreasing pressure, whereas the latter suggest the reverse. Xenolith suites in volcanic rocks are in accord with the lherzolite grid and these imply that Al/Cr ratios increase with increasing pressure. Based on experimental data for multicomponent spinels, on the dispersion of compositions with respect to cationic site occupancy, and on spinels in diamond inclusions, it is suggested that the differences in Cr and Al spinel pressure trends, and the destabilization and stabilization of spinel in the garnet lherzolite and diamond stability fields respectively, results from a transition in chromium from Cr^{3+} at low pressures to Cr^{2+} at high pressures. This proposal may explain the observation that chromites can be present in low pressure plagioclase lherzolites as well as in high pressure diamond inclusions; it is suggested that the former is entirely in the Cr^{3+} state and the latter is partially Cr^{2+}.

Introduction

Spinels and ilmenites are characteristic constituents of a wide variety of igneous rock types. These minerals have provided significant insights into temperatures of formation and oxidation states of host rock equilibration. Compositional variations among these oxides are controlled by physical and chemical parameters, and by the specific properties of bulk composition, by mineral paragenesis, and by partitioning coefficients of co-equilibrating multivariable assemblages.

Elemental fractionation and thermal stability have long been recognized as factors that impose severe constraints on mineral compositions; these oxides are no exception to the general rule. Because bulk compositions do play an important role in mineral chemistry and because the crust is a highly fractionated component of the mantle, elemental behavior among coexisting mineral species should be reflected in the compositions of individual minerals and in the compositions of coexisting phases which are derived from these contrasting source regions. The Buddington and Lindsley (1964) oxygen geobarometer-geothermometer, and the Boyd (1975) pyroxene geotherm are adequate testament to the fact that the environment of equilibration is locked into mineral chemistry. In these studies the application of mineral chemistry to natural systems hinges on experimental data, but in the absence of such data we are forced to the position of examining the system, the geological setting, and of deducing patterns of mineral equilibration indirectly. A key issue in such an approach is the question of how well the mineral reflects its environment of formation and what major effects minor constituents may have on properties such as solubility and stability. These uncertainties, and perhaps many others that have not been considered, are crucial to the conclusions of this study.

The theme of this study, and the empirical approach employed, is an outgrowth of a review on the compositional variations of spinels in low and high pressure rock types (Haggerty, 1976). Data reported here are based primarily on this review and relate specifically to high pressure regimes and to mafic and ultramafic suites. Although the study has not advanced to the point of cluster analysis techniques, it has identified a number of relationships that have not previously been observed. The approach throughout has been simple and it is governed by the crystallochemical characteristics of spinel cationic site occupancy. An extensive data base (840 spinel analyses) has been employed in this study and the major objectives

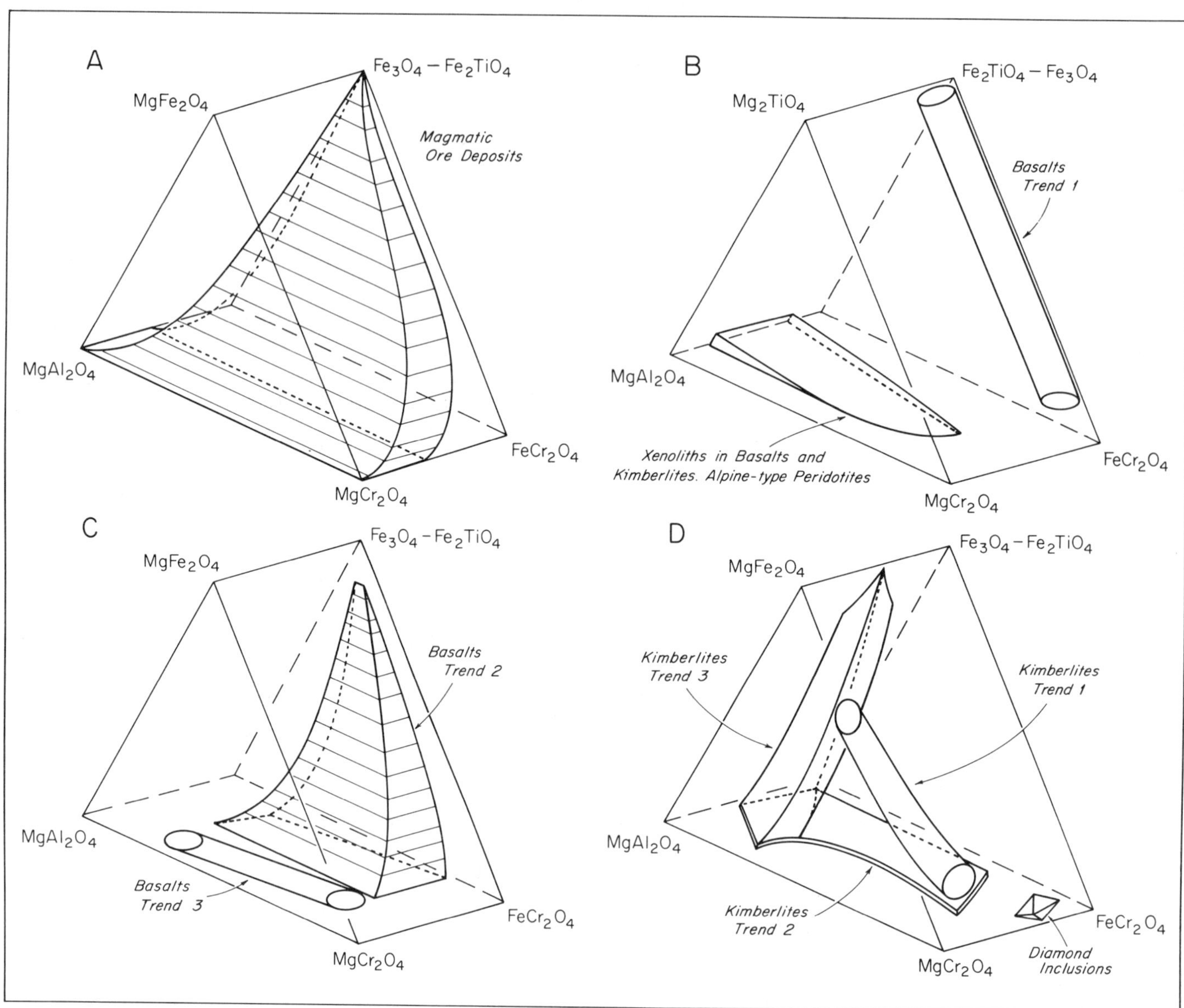

Fig. 1. Spinel distributions for a variety of rock types and geological settings. End members for each prism base are constant, but the apices differ and are dependent on Fe^{2+}, Fe^{3+}, and Ti. Where Fe^{3+} is dominant, the series is given as Fe_3O_4-Fe_2TiO_4; where Fe^{2+} and Ti are dominant, the end members are reversed. Basalt trends 1, 2 and 3 are respectively for subaerially extruded basalts, island-arc volcanics, and deep sea basalts. Kimberlite trends 1, 2, and 3 are for groundmass kimberlitic spinels, although some components along the base of the prism may have had a high pressure derivation.

are: (a) to define the distribution of [illegible]l compositions in lherzolite suites; (b) to distinguish between spinels from deep upper mantle and high level upper mantle source regions; and (c) to evaluate the apparent discrepancy of chromite inclusions in diamonds and the absence of spinels in garnet lherzolites.

It is emphasized that the conclusions reached in this study are tentative. However, the systematics of the trends established do appear to separate suites of spinels from different P-T environments, and it is hoped that these differences will provide a stimulus for future experimental research.

Spinel Distributions

The generalized distributions of spinels for a variety of rock types and geological settings are summarized in Fig. 1 within a series of multicomponent spinel prisms of the type employed by Stevens (1944), Irvine (1967) and Jackson (1969).

Several characteristic trends emerge from these distributions and these are related to progressively early and late stage crystallization paths. These trends are: (a) an early enrichment of Cr and a later enrichment of Al--the typical kimberlite trend; (b) an enrichment of Mg and a later enrichment of Fe^{2+}--the stratiform magmatic ore deposit trend; and (c) a range of compositions which are enriched in Al and Cr--exemplified by xenolith suites.

It is significant that a preferred directional sweep towards the face of the Fe-ternary (Fe_2TiO_4-Fe_3O_4-$FeAl_2O_4$-$FeCr_2O_4$) is shown by all suites with the exception of kimberlitic spinels which continue to retain their relatively high MgO contents with increasingly higher concentrations of Fe^{3+} and Ti. These spinels are most typical of kimberlitic groundmass crystallization, but there is some evidence which suggests that nucleation was initiated at high pressures. For example, spinels included in diamonds are high in Cr and Mg and low in Al and Fe. Xenolith and peridotite suites are confined to the spinel prism base and these show little or no enrichment in either Fe^{3+} or Ti. Three basalt trends are also illustrated in Fig. 1: the first is typical of subaerially extruded basalts; the second is characteristic of island arc volcanism; and the third is for spinels in deep sea basalts. The common factor between basaltic trends 1 and 2 is the final enrichment of Fe_3O_4-Fe_2TiO_4, and the common factor between trends 2 and 3 is the wide variation in Cr/Al ratios. For deep sea basalts the most aluminous spinels are in picritic basalts, intermediate Al and higher Cr spinels are present in olivine basalts, and spinels with the lowest Al and highest Cr are observed in tholeiitic suites. In summary, these data show that the earliest crystallized spinel components lie on the base on the spinel prism ($FeCr_2O_4$-$FeAl_2O_4$-$MgCr_2O_4$-$MgAl_2O_4$), that late stage crystallization results in Fe^{3+} + Ti enrichment, and that two major contrasting trends are present in high pressure suites. Kimberlitic spinels represent the first of these with high concentrations of $MgCr_2O_4$-$FeCr_2O_4$ at an early stage of paragenesis and a subsequent late stage trend towards $MgAl_2O_4$-$FeAl_2O_4$. The second contrasting trend is exemplified by peridotite and xenolith suites and this trend is a reversal of the kimberlite trend (i.e., early $MgAl_2O_4$- $FeAl_2O_4$ and later $FeCr_2O_4$-$MgCr_2O_4$). The differences in trends are most obvious in their variable Cr/Al behavior, but Fe^{2+}/Mg is an equally important but subtle parameter.

Cationic Distributions

In the generalizations summarized above the most significant points that emerge are that the compositions of early crystallized spinels in all mafic and ultramafic rocks, and all spinels in high pressure suites fall within the quaternary system $FeAl_2O_4$-$FeCr_2O_4$-$MgAl_2O_4$-$MgCr_2O_4$. This system is characterized, therefore, by Fe^{2+}/Mg and Cr/Al variations. The most widely used variables that express the distributions of spinel compositions in high pressure regimes are those characteristic of the spinel prism base. To locate these compositions the relevant expressions are Cr/Cr+Al *vs.* Fe^{2+}/Fe^{2+}+Mg. Each of these ratio sets are characteristic of cations in octahedral and tetrahedral coordination respectively, and for spinels on the prism base, which are *normal* spinels, the general formula is $R_8^{2+}[R_{16}^{3+}]O_{32}$. *Inverse* spinels at the apices of the prism have the formula $R_8^{2+}R_8^{3+}[R_8^{3+}]O_{32}$; where the brackets indicate octahedral coordination; R^{2+} = Fe, Mg, and R^{3+} = Cr, Al. Titanium substitution is accomplished by $R^{2+} + Ti^{4+} \rightleftarrows 2R^{3+}$ and ulvöspinel (Fe_2TiO_4) is expressed as $R_8^{2+}[R_8^{2+}Ti_8^{4+}]O_{32}$, which is an *inverse* spinel. In the data sets which follow the three variables that are examined are Fe *vs.* Mg, Cr *vs.* Al and the ratio Fe/Fe+Mg+Mn *vs.* Cr/Cr+Al. Manganese is a minor constituent of these suites but is included as a tetrahedral component. Vanadium is an octahedral component and is excluded because it is usually not reported.

The spinel suites considered in this study are divided into the following categories: (a) layered series; (b) peridotites; (c) podiform chromites; (d) kimberlites; and (e) xenoliths in volcanic rocks. Peridotite suites include spinels in harzburgites and spinels in dunites. Kimberlitic spinels include spinels from xenoliths and spinels present as inclusions in diamonds; the data do not include spinels from kimberlite groundmass studies because of the uncertainties of precisely which proportions are high pressure and which proportions are due to high crustal precipitation. With two exceptions, all data presented are from the compilation by Haggerty (1976); additional values for spinels in kimberlites are from Sobolev (1976) and from Ferguson et al. (1977).

Spinel prism, base-projection plots of Cr/Cr+Al *vs.* Fe/Fe+Mg for the five suites of data are shown in Fig. 2. Layered series spinels exhibit a wide scatter in tetrahedral and octahedral cations, whereas those in peridotite suites show a relatively well-developed positive correlation for low values of Cr and Fe but with an increasingly larger divergence from this curve as these cations increase. Approximately 80% of the podiform chromite data fall within the divergent field of peridotitic spinels; this population is concentrated at R^{3+} = 0.75 cations/32 oxygen and R^{2+} = 0.38 cations/32 oxygen. The compositions of spinels in kimberlitic xenoliths and spinels in diamonds closely parallel those of spinels in peridotites; the highest Cr contents are present in spinel diamond inclusions. Xenoliths in volcanic suites are distinctive in their low Cr contents, high Al contents, and variable Fe/Mg contents; there is a greater dispersion of the data with increasing Cr and this factor is shared by all suites.

Octahedral Cr *vs.* Al concentrations for the same data sets are shown in Fig. 3. The octahedral spinel site contains a maximum of 16

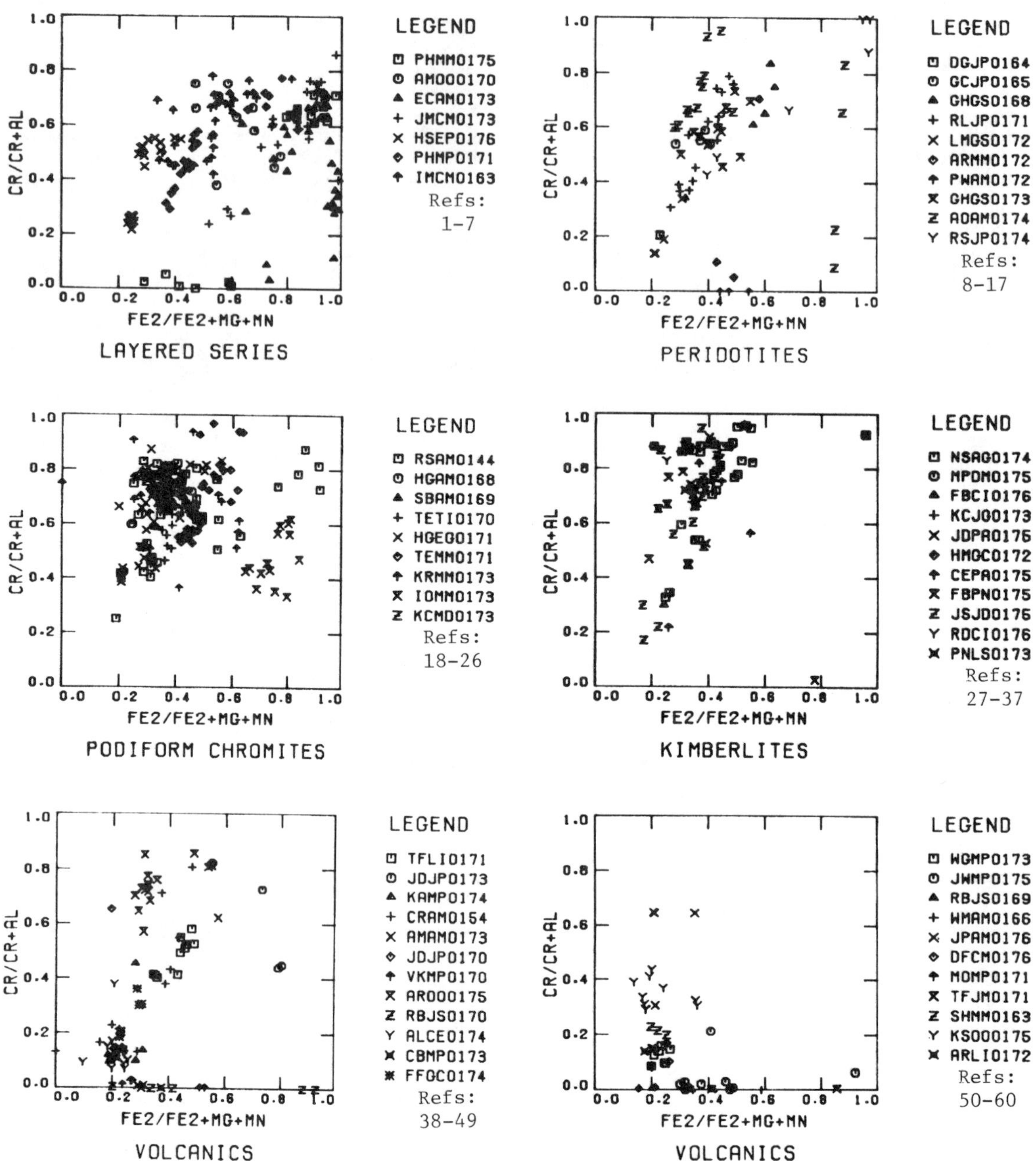

Fig. 2. Projections onto the base of the spinel prism expressed as R^{2+} *versus* R^{3+}. The data sets are divided into spinels from layered intrusives, peridotites, podiform chromite deposits, xenoliths in kimberlites, and the last two sets are for spinels in xenoliths from volcanic rocks. The references cited (1-60) are listed in the appendix.

cations/32 oxygen; thus the negative slope defined by Cr = 16 cations and Al = 16 cations is a B-site spinel control line. Deviation from this control line is indicative of one, or a combination of the following: (a) non-stoichiometry; (b) the presence of cations other than Cr or Al (e.g., Fe^{3+}, Ti^{4+} and V^{3+}); and (c) analytical error. Factor (c) can be dismissed because all data were subjected to internal consistency checks at the 99% confidence level. Taken as a group, the data sets conform rigorously to the B-site spinel control line for high values of Al and low values of Cr. The layered series is distinctly parabolic and isolated analyses from other suites lie on the crest of the parabola and on the positive limb of the parabola. Spinels within the ranges of low to intermediate Al contents contain substantial amounts of either Fe^{3+} or Ti, indicative of magnetite-ulvöspinel solid solutions or alternatively, *ferritchromit*. The former is characteristic of spinels in stratiform intrusions and the latter of spinels in alpine-type peridotites.

Spinels in the peridotite suites span virtually the entire range of Cr/Al although the data are sparse at the two extremes. Podiform chromites show a strong preferred population density which is high in Cr and low in Al. Kimberlitic spinels are also high in Cr and the highest Cr concentra-

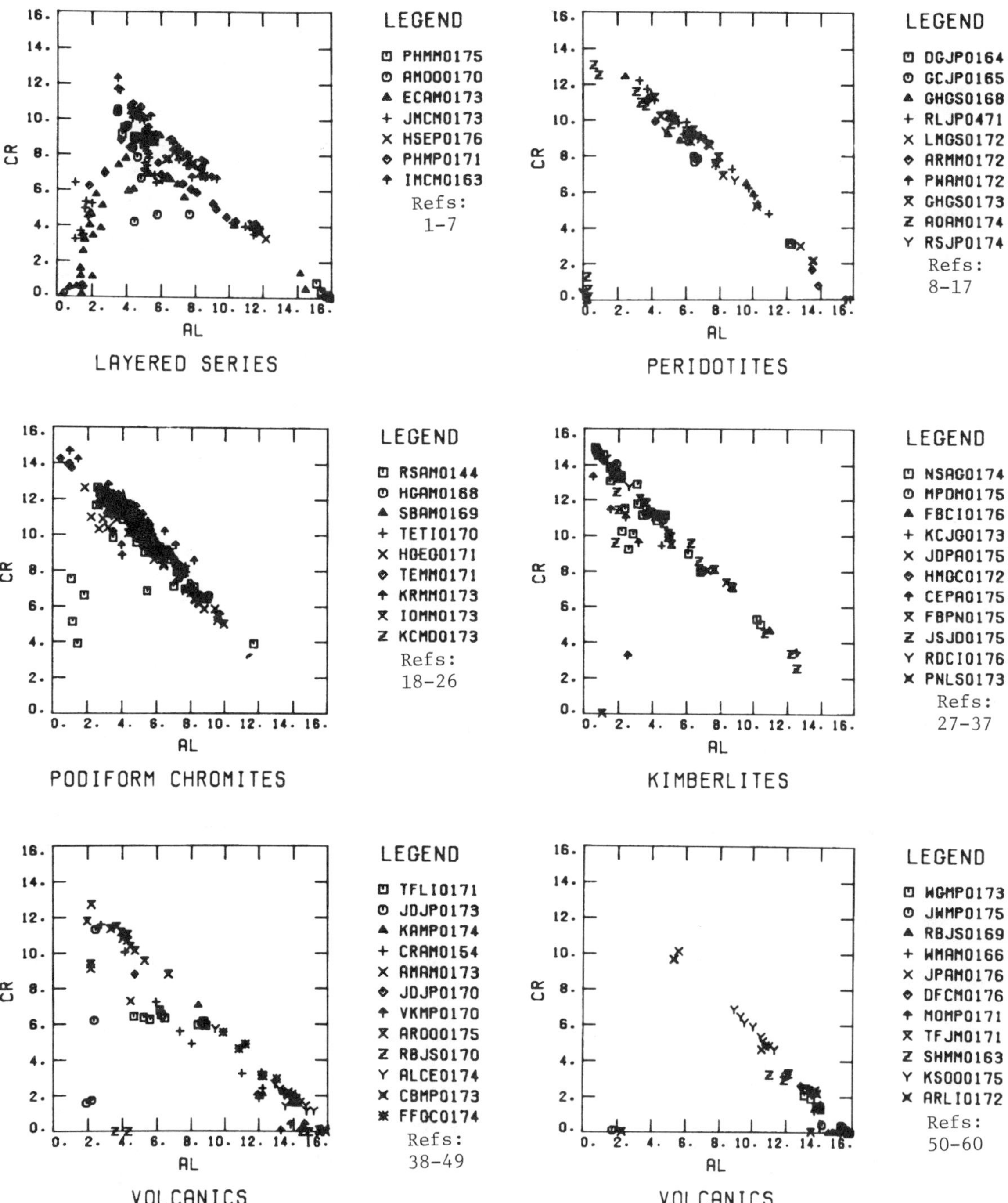

Fig. 3. Trivalent cations plotted for the same data sets as shown in Fig. 2. Comments to Fig. 2 apply here also.

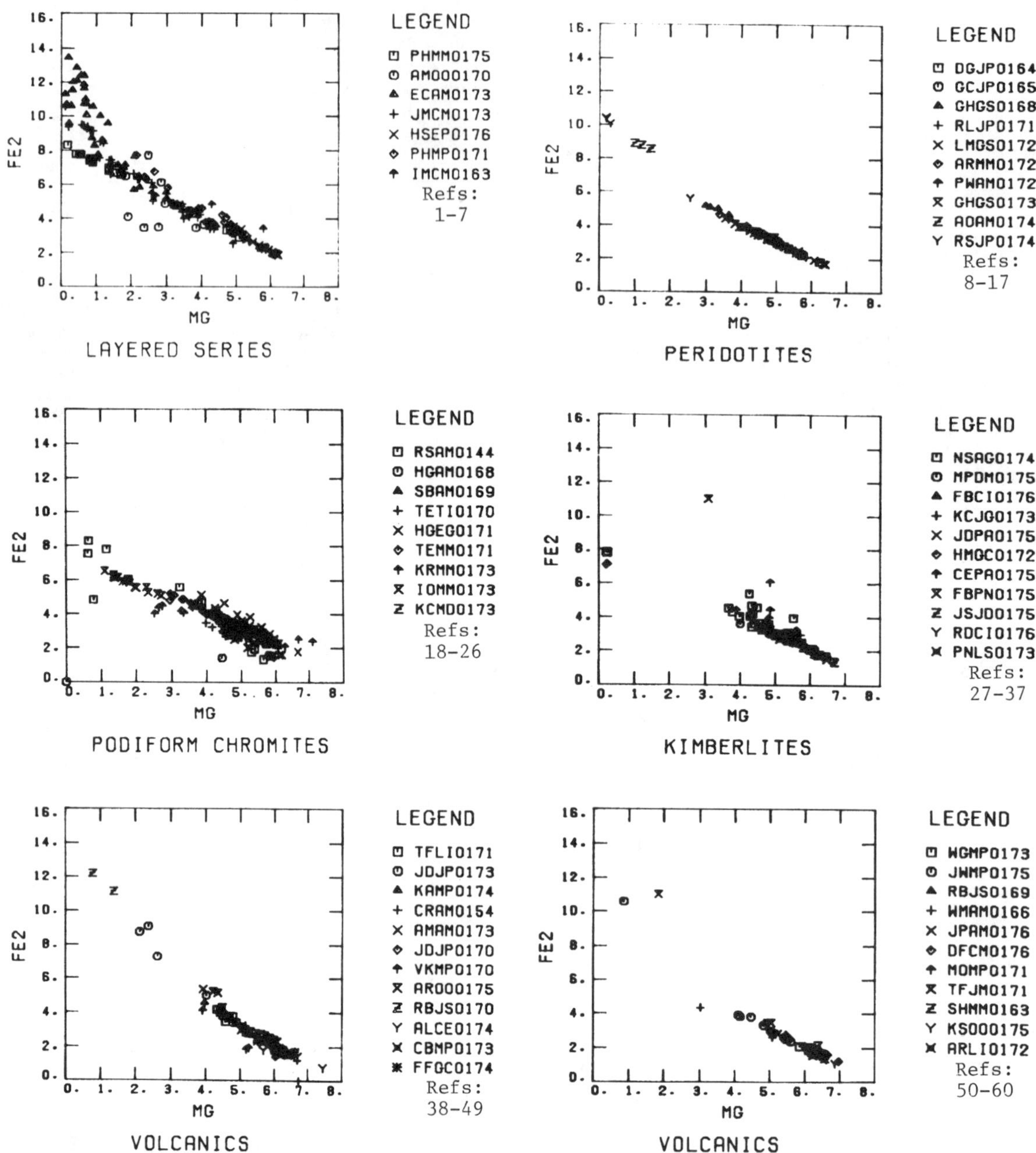

Fig. 4. Divalent cations plotted for the same data sets as shown in Fig. 2. Comments to Fig. 2 apply here also.

tions are present in diamond inclusions; other xenolith data trend towards increased Al concentrations. For xenoliths in volcanic rocks the reverse of the kimberlite trend is apparent and in these the concentration is at high Al contents with fewer data towards Cr. Note in particular that spinels in volcanic xenoliths occupy a specific region of Al = 12-16 cations/32 oxygen, a zone which is kimberlite absent; and that diamond inclusion spinels (Cr = 14-16 cations/32 oxygen) occupy a region which is free of spinels in volcanic xenoliths. Note also that diamond inclusion spinels deviate from stoichiometry.

Tetrahedral Fe *vs.* Mg concentrations, again for the same data sets, are illustrated in Fig. 4. For *normal* spinels the R^{2+} site contains 8 cations/32 oxygen, and for *inverse* spinels this site contains 16 cations/32 oxygen. The normal spinel control line is therefore between Fe = 8 cations/32 oxygen, and Mg = 8 cations/32 oxygen; the upper limit for Mg is fixed because of tetrahedral site preference energies and because the maximum R^{2+} occupancy in the A-site is 8 cations/32 oxygen. Thus Fe^{2+} is also limited to 8 cations/32 oxygen and values greater than this constant signify a departure from *normal* spinel cationic distributions, to a site distribution which is between that of *normal* and *inverse*. Spinels in layered series' suites lie largely on the *normal* spinel control line but these trend continuously towards higher values of Fe^{2+} because Ti^{4+} is present. These reach a maximum of 13.8 Fe cations/32 oxygen, indicative of a large ulvö-spinel (Fe_2TiO_4) component which has Fe cations = 16.0/32 oxygen. All of the remaining suites contain a paucity of data which lie beyond the A-site control line. Close inspection of the limits attained by Mg concentrations shows that there is a progressive increase in average magnesium contents, and that these are in the following order: layered series = 6.3; peridotites = 6.5; podiform chromites = 6.8; kimberlites = 6.9; and volcanic xenoliths = 7.0 Mg cations/32 oxygen. However, in all cases there are dense populations between 4 and 6, or between 5 and 6 Mg cations/32 oxygen.

Tetrahedral and octahedral (A- and B-site respectively) distributions for the five data sets may be summarized as follows: spinels in diamond inclusions contain high Cr contents and deviate from stoichiometry; spinels in kimberlitic xenoliths have a trend from high Cr to intermediate Al; spinels in volcanic xenoliths have exceptionally high Al contents and the trend is towards intermediate concentrations of Al+Cr; spinels in alpine-type peridotites, harzburgites and dunites have a wide range in Cr and Al but these do not reach the limits of either Cr in diamond inclusions or of Al in volcanic xenoliths; layered series spinels exhibit a pronounced parabolic relationship, with a negative limb (high Al and low Cr) which corresponds to the high pressure suites, a crest at intermediate values of Cr and Al, and a positive limb which terminates at low Cr+Al indicative of high Ti contents. For Fe/Mg relationships there is a close conformity to the *normal* spinel A-site control line, and the only suite which deviates markedly from this line are spinels in layered suites which have high Ti contents and a large *inverse* component.

Discussion

An initial premise in this study is that mineral compositions ought to reflect their P-T environment of formation and equilibration. Spinels from high crustal layered intrusive suites can be

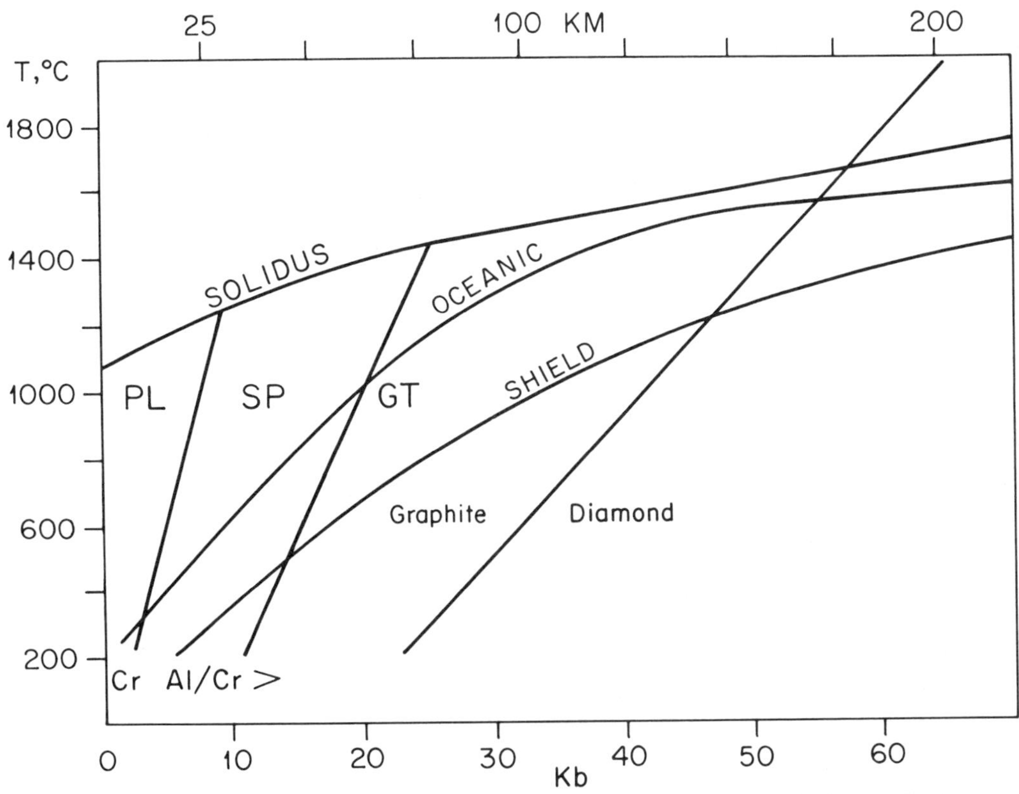

Fig. 5. P-T relationships for plagioclase (PL), spinel (SP) and garnet (GT) lherzolites, with respect to the peridotite dry solidus, oceanic and shield geotherms, and the diamond-graphite stability curve. Cr refers to chromite in plagioclase lherzolite, with Al largely in feldspar; Al/Cr > implies that this ratio should increase with increasing pressure. Spinels are not stable in the garnet lherzolite stability field but are stable as inclusions in diamonds.

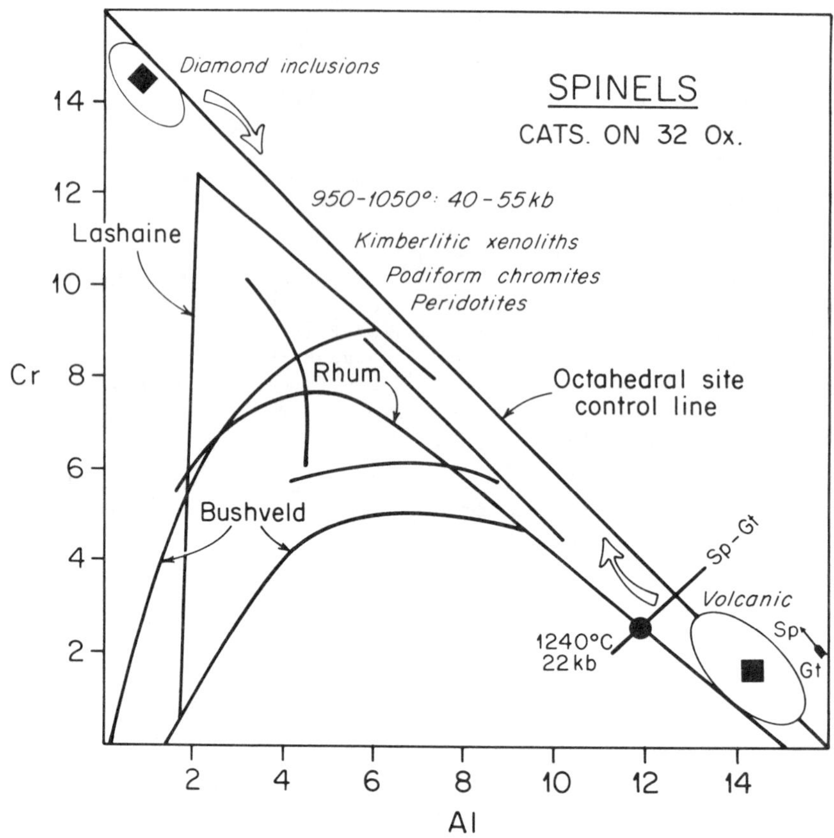

Fig. 6. A generalized compilation of Cr *versus* Al variations for all of the suites illustrated in Fig. 3. Two alternative propositions arise for Cr-Al systematics as an implied function of pressure: (1) an increase in Al, from spinels in diamond inclusions towards those in volcanic suites; and (2) an increase in Cr from spinels in xenolith suites in volcanic rocks towards those in diamond inclusions. The Gt-Sp trend represents the equilibration of garnet lherzolites into the spinel lherzolite field. The value at 22kb and 1240^{0}C is from Ferguson et al. (1977), and is an estimate for coexisting garnet and spinel. Other layered intrusive trends are centered on Rhum and the Bushveld.

distinguished from those of upper mantle origin based on high Ti and Fe^{3+} contents. However, from the data presented above a simple pattern does not emerge for spinels in high pressure regimes. This is so particularly for Fe/Mg relationships where a considerable overlap of compositional data for spinels from a wide variety of settings and rock types is present. For Cr/Al variations the relationships are even more enigmatic, notwithstanding the fact that there is now abundant evidence to show that the distributions of spinels in lherzolite suites conform to systematic patterns which are P-T dependent. For example, in low pressure plagioclase lherzolites Al is locked up largely in feldspar, and spinels in these rocks have high Cr/Al ratios. At intermediate pressures the spinels in spinel lherzolites have variable Cr/Al contents, whereas in high pressure garnet lherzolites both Cr and Al are preferentially partitioned into diopside and garnet; the resulting consequence is that spinel is absent. Given this overall and generalized distribution, it would appear that spinel compositions ought to vary systematically within the spinel lherzolite field if bulk compositions remain constant. The expectation is that the Al/Cr ratio should increase with increasing pressure as shown in Fig. 5. Spinel data for xenoliths in volcanic rocks reflect this distribution in Al/Cr in the sense that the more aluminous spinels are those which have formed in symplectic associations around garnet. The interpretation commonly held is that this reaction results from a lower pressure re-equilibration of a garnet lherzolite into the spinel lherzolite field. Thus, the garnet to spinel transition produces spinels which are high in Al and low in Cr (Fig. 6). These compositions are close to, and overlap, the compositions of spinel lherzolites in volcanic suites and the trend is towards increasing Cr contents. One analysis is available of a lherzolite in kimberlite which has primary garnet and spinel (Ferguson et al., 1977). This assemblage lies on the quasi-univariant spinel-garnet transition curve and the conditions estimated for its formation are 1240°C and 22 kb (Fig. 6). In terms of Cr/Al contents this value is extremely close to that of equilibrated garnet lherzolites (i.e., those yielding secondary symplectic spinels), and to spinel lherzolites in volcanic xenolith suites (Fig. 6). There are, however, a number of volcanic xenoliths with spinel compositions that are intermediate in Cr and Al (Fig. 4), and a plausible explanation for these is that they have equilibrated with the magma on their passage to the surface. Estimates do in fact show pressures as low as 5-10 kb and temperatures between 800-900°C. In summary, spinels in volcanic xenoliths are apparently consistent with the systematics of spinel distributions implied by an isochemical expression of the P-T lherzolite grid (Fig. 5).

At the opposite end of the Cr/Al scale and at pressures substantially higher than that of the lherzolite grid are spinel inclusions in diamond. These contain among the highest chrome contents of any spinels known; Al is extremely low and Fe/Mg is variable. Spinels in kimberlite xenoliths are commonly also high in Cr and estimates for their formation are 40-55 kb and 950-1050°C (Fig. 6), according to Boyd and Nixon (1973). Many xenoliths in kimberlites do not, of course, contain spinels because Cr and Al are partitioned into coexisting silicates. The Cr/Al trend in kimberlitic xenoliths is exactly the reverse of those in volcanic suites, with aluminum increasing systematically from spinels in diamond inclusions to spinels in xenoliths. It can be reasonably argued, therefore, that the spinel-pressure trend is from Cr to Al, based on diamond inclusions, if it were not for the fact that podiform chromites occupy a field which is relatively high in Cr and intermediate in Al. In addition, the transition from garnet to spinel appears to be well founded. Because this reaction produces high Al spinels, and because plagioclase lherzolites also contain high Cr spinels, two entirely contrasting partitioning processes must be in effect.

A similar set of conclusions for the relative distributions of Cr and Al in spinels from kimberlites and for spinels from volcanic xenoliths were reached by Basu and MacGregor (1975). Their estimates of pressure for the former is 40-50 kb and for the latter is 10-15 kb. However, although these estimates are consistent with the inferred distribution that high Cr spinels are indicative of high pressure regimes they note that the Al_2O_3 contents of coexisting orthopyroxene and spinel correlate positively, with kimberlitic xenoliths having low Al_2O_3 concentrations and volcanic xenoliths proportionately higher concentrations. These data suggest, therefore, that the elemental systematics may merely be a reflection of bulk composition and that magmas derived from the lower regions of the upper mantle are Cr-rich, whereas those from higher levels are Al-rich. This implied distribution may be correct but has not been unequivocally demonstrated.

A prominent feature of all spinels considered in the context of high pressure regimes is the low to virtual absence of Fe^{3+}. Experimental data by Ulmer (1969), for phases in the multicomponent spinel prism (Fig. 1), show that neither chromite ($FeCr_2O_4$) nor hercynite ($FeAl_2O_4$) are stable at 1300°C and an fO_2 value of 10^{-9} atms; in fact

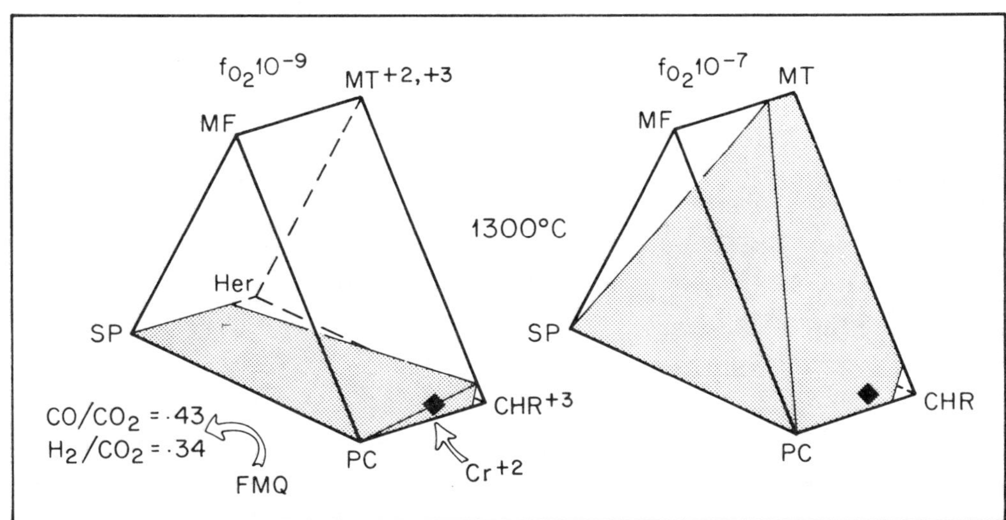

Fig. 7. The shaded volumes within the spinel prisms represent the stable and mutual solid solubilities that are possible at 1300°C and at values of $fO_2 = 10^{-7}$ and 10^{-9} atms. respectively (Ulmer, 1969). At the latter value, Cr^{2+} is known to be present in these spinels (Ulmer, 1969). Gas mixtures equivalent to the FMQ buffer are shown to the left. The compositions of diamond inclusion spinels are illustrated by solid diamond symbols. MF = magnesioferrite; Mt = magnetite; Sp = spinel; Her = hercynite; PC = picrochromite; and Chr = chromite.

chromite was shown to be unstable for the entire range of oxygen fugacities between $10^{-0.21}$ to 10^{-9} atms. Spinel compositions *equivalent* to those of inclusions in diamonds plot within a volume of the prism which is stable at 10^{-9} atms based on the assumption that chromium is present as Cr^{3+} as indicated in Fig. 7. The experimental conditions are equivalent to a $CO:CO_2$ ratio of 0.43 and a $H_2:CO_2$ ratio of 0.34 (Deines et al., 1974). This T°C and fO_2 environment is close to that of the FMQ buffer curve, a condition which is clearly incompatible with the oxidation states that are necessary for the nucleation and growth of diamond. If chromium is not in the Cr^{3+} state, as is commonly assumed, but rather in the divalent state, its partitioning behavior will be very different between A- and B-sites in the spinel structure and between coexisting oxides and silicates. Tetrahedral coordination of Cr^{2+} has been demonstrated by Mao and Bell (1974) for a kimberlitic spinel which was heat treated at 1400°C and $10^{-11.5}$ atms. A test of the available data on spinels in diamonds shows that there is a marked departure from the octahedral B-site control line (Figs. 3 and 6). High Al spinels in the volcanic suites, however, lie precisely on the B-site control line. The departure from B-site conformity strongly suggests, therefore, that some proportion of the chromium in diamond inclusion spinels is in the divalent state and is tetrahedrally (A-site) coordinated. If this proposal is correct, then the demise of spinel in the garnet lherzolite field, and its resurrection again in the diamond stability field, can be adequately explained on the basis of a dramatically different partitioning of Cr^{2+} and Cr^{3+} between oxides and silicates. In addition, if Cr^{2+} is the stable chromium species in the lower portion of the upper mantle, and Cr^{3+} is characteristic of the more highly oxidized states of the crust and higher levels of the upper mantle, then the contrasts in spinel pressure trends (Fig. 6) from Cr to Al (diamond inclusions to xenoliths) and Al to Cr (garnet lherzolites to spinel lherzolites) are reconciled. It should be noted, however, that the transition of Cr^{3+} to Cr^{2+} need not necessarily result from a decrease in fO_2 conditions but may result simply from the pressure stabilization of Cr^{2+} as has been demonstrated in Fe^{2+} and Fe^{3+} bearing minerals in which Fe is stabilized as metallic iron (Bell and Mao, 1975). Additional support for the likelihood of Cr^{2+} in diamond inclusion spinels is provided by the data on olivines in diamonds which show that Cr^{2+} ions are stabilized and that this results from pressure-induced reduction (Burns, 1975).

Conclusions

The widely employed projection onto the spinel prism base, which is expressed as the relationship of Cr/Cr+Al *vs.* Fe/Fe+Mg+Mn, is less informative than the individual A and B ratios (Fe *vs.* Mg, and Cr *vs.* Al, respectively) in determinations of crystallization trends, and in deducing the fields occupied by spinels in high and low pressure regimes. Tetrahedral and octahedral relationships for spinels in stratiform intrusives, peridotites, podiform chromites, kimberlites, and for spinels in volcanic xenoliths show: (1) that the highest pressure spinels in diamond inclusions have exceptionally high chromium contents, that these may contain Cr^{2+}, and that based on exclusive Cr^{3+} contents that these spinels deviate from B-site stoichiometry; (2) a continuous trend from high Cr to intermediate Al is present in spinels from diamond inclusions to spinels in kimberlite xenoliths, and this trend is inferred to result from decreases in P and T, below that of diamond stability; (3) the kimberlite spinel trend contrasts with the volcanic xenolith spinel trend in that the latter spinels are high in Al contents and decrease systematically towards Cr+Al concentrations; (4) garnet lherzolites on re-equilibration into the spinel lherzolite field precipitate highly aluminous spinels, suggesting that decreasing P and T yields a spinel trend from high Al to high Cr; (5) spinels in alpine-type peridotites, harzburgites and dunites display wide ranges in Cr and Al but in neither case do these elemental abundances attain the proportions of Cr in spinel diamond inclusions, or of Al in spinels from volcanic xenoliths; (6) spinels from stratiform intrusions display a distinctly parabolic relationship in Cr *vs.* Al, and a hyperbolic relationship in terms of Fe *vs.* Mg; (7) the parabolic trend may be subdivided into the negative limb (high Al and low Cr) which is equivalent to spinels from high pressure suites, a crest at values of intermediate Cr and Al, which is displayed by some spinels in volcanic xenoliths, of presumed intermediate P-T equilibration, and a positive limb which is characteristic only of low pressure, high crustal level spinel precipitation; (8) spinels from most high pressure suites are dominated by *normal* spinel structures and are free of Fe^{3+}, whereas those from low pressure rocks are characterized by large *inverse* spinel components, and high concentrations Ti and Fe^{3+}.

The major conclusion of this study is that Cr^{2+} may be a significant species of chromium in the lower reaches of the upper mantle. This model of Cr^{2+} and Cr^{3+} distribution may account for the spinel-absent interval of the garnet lherzolite field and the restabilization of chromite under the conditions of diamond stability. It may also account for the distribution of Cr-bearing spinels in general, in the sense that Al-poor chromites can also be present in low pressure plagioclase lherzolites given the condition that all chromium is Cr^{3+}. Diamond inclusion studies have always appeared to be anomalous with the generally held tenet that high Al spinels are characteristic of high P-T regimes. That assumption is generally correct in an isochemical system and spinels precipitated on equilibration through the garnet to spinel transition appear to provide unequivocal evidence for their highly aluminous nature. However, the observed trend from high Cr to high Al that has emerged from the kimberlite data, and the reverse trend for xenoliths in volcanic rocks is clearly indicative of two radically contrasting pressure regimes. Many other factors, and bulk composition in particular, may also influence these differences and the proposed model, therefore, requires experimental confirmation of this empirical and tentative conclusion. Because the volcanic xenoliths, and some xenoliths in kimberlites are in accord with the expectations of Cr/Al variations, within the framework of the lherzolite petrogenetic grid, and because current experimental and thermodynamic data are in conflict concerning the sensitivity of Al in enstatite as a pressure indicator, it would appear that spinels may offer an alternative to P-T derivations for this class of mantle derived rocks.

Perhaps the most compelling evidence for Cr^{2+} in the upper mantle is to be found in the high pressure experimental data summarized by Burns (1975) for a variety of transition elements in which the pressure induced stabilization of $Fe^{3+} \rightarrow Fe^{2+}$, $Mn^{3+} \rightarrow Mn^{2+}$ and $Cu^{2+} \rightarrow Cu^{+}$ has been demonstrated. Spectral measurements (Burns, 1975) show at the very minimum that Cr^{2+} is present in olivine inclusions in diamond.

Acknowledgments. This study was supported by NSF under grant EAR 76-23787, and by the University of Massachusetts Computer Center. Programming was undertaken by V. Congdon. Enthusiastic and dedicated support was provided by B.M. McMahon and R.B. Hardie III. Prolonged but deserved badgering from D. Rumble III aided in completing the review which has led to this study. Although the major conclusions of this study received a 2:1 vote of confidence from the reviewers, I am nevertheless grateful for the thought-provoking and detailed comments made by R. Brett, I.D. MacGregor, and B.R. Lipin. Credit for skillful drafting is due to Marie Litterer. To all I express my appreciation.

Appendix I

Key to references listed in Figs. 1-3.

Layered Series

1. Hamlyn (1975)
2. Mall and Rao (1970)
3. Cameron and Glover (1973)
4. Muir and Naldrett (1973)
5. Sigurdsson and Schilling (1976)
6. Henderson and Suddaby (1971)
7. MacGregor and Smith (1963)

Peridotites

8. Green (1964)
9. Challis (1965)
10. Himmelberg and Coleman (1968)
11. Loney, Himmelberg and Coleman (1971)
12. Medaris, Jr. (1972)
13. Rothstein (1972)
14. Whitney (1972)
15. Himmelberg and Loney (1973)
16. Onyeagocha (1974)
17. Springer (1974)

Podiform Chromites

18. Stevens (1944)
19. Golding and Bayliss (1968)
20. Bilgrami (1969)
21. Engin and Hirst (1970)
22. Golding and Johnson (1971)
23. Engin and Aucott (1971)
24. Rodgers (1973)
25. Oen, Kieft and Westerhof (1973)
26. Chakraborty (1973)

Kimberlites

27. Sobelov (1968)
28. Prinz, Manson, Hlava and Keil (1975)
29. Boyd, Fujii and Danchin (1975)
30. Cox, Gurney and Harte (1973)
31. Dawson and Smith (1975)
32. Meyer and Boyd (1972)
33. Emeleus and Andrews (1975)
34. Boyd and Nixon (1975)
35. Smith and Dawson (1975)
36. Danchin and Boyd (1976)
37. Nixon and Boyd (1973)

Volcanics

38. Frisch (1971)
39. Dawson and Smith (1973)
40. Aoki and Prinz (1974)
41. Ross, Foster and Myers (1954)
42. McBirney and Aoki (1973)
43. Dawson, Powell and Reid (1970)
44. Kutolin and Frolova (1970)
45. Reid, Donaldson, Brown, Ridley and Dawson (1975)
46. Binns, Duggan and Wilkinson (1970)
47. Littlejohn and Greenwood (1973)
48. Bacon and Carmichael (1973)
49. Frey and Green (1974)
50. Griffin (1973)
51. Wilkinson (1975)
52. Binns (1969)
53. Melson and Switzer (1966)
54. Pike (1976)
55. Francis (1976)
56. O'Hara, Richardson and Wilson (1971)
57. Frisch and Wright (1971)
58. Hamad (1963)
59. Suwa, Yusa and Kishida (1975)
60. Reid and Dawson (1972)

References

Aoki, K., and M. Prinz, Chromian spinels in lherzolite inclusions from Itinome-gata, Japan, *Contr. Min. Pet.*, *46*, 249-256, 1974.

Bacon, C.R., and I.S.E. Carmichael, Stages in the P-T path of ascending basalt magma: An example from San Quintin, Baja California, *Contr. Min. Pet.*, *41*, 1-22, 1973.

Basu, A.R., and I.D. MacGregor, Chromite spinels

from ultramafic xenoliths, Geochim. Cosmochim. Acta, 39, 937-945, 1975.

Bell, P.M., and H.K. Mao, Preliminary evidence of disproportionation of ferrous iron in silicates at high pressures and temperatures, Carnegie Inst. Washington Yr. Bk., 74, 557-559, 1975.

Bilgrami, S.A., Geology and chemical mineralogy of the Zhob Valley chromite deposits, West Pakistan, Am. Min., 54, 134-148, 1969.

Binns, R.A., High-pressure megacrysts in basanatic lavas near Armidale, New South Wales, Am. Jour. Sci., 267-A, 33-49, 1969.

Binns, R.A., M.B. Duggan, J.F.G. Wilkinson, High pressure megacrysts in alkaline lavas from northeastern New South Wales, Am. Jour. Sci., 269, 132-168, 1970.

Boyd, F.R., A pyroxene geotherm, Geochim. Cosmochim. Acta, 37, 2533-2546, 1973.

Boyd, F.R., T. Fujii, and R.V. Danchin, A non-inflected geotherm for the Udachnoya kimberlite pipe, U.S.S.R., Carnegie Inst. Wash. Yr. Bk., 75, 523-529, 1975.

Boyd, F.R., and P.H. Nixon, Origins of the ultramafic nodules from some kimberlites of northern Lesotho and the Monastery Mine, South Africa, Phy. Chem. Earth, 9, 431-454, 1975.

Buddington, A.F., and D.H. Lindsley, Iron-titanium oxide minerals and synthetic equivalents, J. Pet., 5, part 2, 310-357, 1964.

Burns, R.G., On the occurrence and stability of divalent chromium in olivines included in diamonds, Contrib. Min. and Pet., 51, 213-221, 1975.

Cameron, E.N. and E.D. Glover, Unusual titanian-chromian spinels from the eastern Bushveld complex, Am. Min., 58, 172-188, 1973.

Chakraborty, K.K., Some characters of the bedded chromite deposits at Kalrangi, Cuttack district, Orissa, India, Min. Depos. (Berl.), 8, 73-80, 1973.

Challis, G.A., The origin of New Zealand ultramafic intrusions, J. Pet., 6, 322-364, 1965.

Cox, K.G., J.J. Gurney, and B. Harte, Xenoliths from the Matsoku Pipe, Lesotho National Development Corporation, Maseru, in Lesotho Kimberlites, edited by P.H. Nixon, pp. 76-100, 1973.

Danchin, R.V., and F.R. Boyd, Ultramafic nodules from the Premier kimberlite pipe, South Africa, Carnegie Inst. Washington, Yr. Bk., 75, 531-535, 1976.

Dawson, J.B., D.G. Powell, and A.M. Reid, Ultrabasic xenoliths and lava from the Lashaine Volcano, Northern Tanzania, J. Pet., 11, 519-548, 1970.

Dawson, J.B., and J.V. Smith, Alkalic pyroxenite xenoliths from the Lashaine Volcano, Northern Tanzania, J. Pet., 14, 113-131, 1973.

Dawson, J.B., and J.V. Smith, Chromite-silicate intergrowths in upper-mantle peridotites, Phy. Chem. Earth, 9, 339-350, 1975.

Deines, P., R.H. Nafziger, G.C. Ulmer, and E. Woermann, Temperature-oxygen fugacity tables for selected gas mixtures in the system C-H-O at one atmosphere total pressure, Bull. Earth and Min. Sci., Experimental Station, Penn. State Univer., 88, 1-129, 1974.

Emeleus, C.H., and J.R. Andrews, Mineralogy and petrology of kimberlite dyke and sheet intrusions and included peridotite xenoliths from south-west Greenland, Phy. Chem. Earth, 9, 179-197, 1975.

Engin, T., and J.W. Aucott, A microprŏbe study of chromites from the Andizlik-Zimparalik area, south-west Turkey, Min. Mag., 38, 76-82, 1971.

Engin, T., and D.M. Hirst, The Alpine chrome ores of the Andizlik-Zimparalik area, Fethiye, southwest Turkey, Trans. Inst. Min. Metal., Sec. B, 79, 16-29, 1970.

Ferguson, J., D.J. Ellis, and R.N. England, Unique spinel-garnet lherzolite inclusion in kimberlite from Australia, Geology, 5, 278-280, 1977.

Francis, D.M., Corona-bearing pyroxene granulite xenoliths and the lower crust beneath Nunivak Island, Alaska, Can. Min., 14, 291-298, 1976.

Frey, F.A., and D.H. Green, The mineralogy, geochemistry and origin of lherzolite inclusions in Victorian basanites, Geochim. Cosmochim. Acta., 38, 1023-1059, 1974.

Frisch, T., Alteration of chrome spinel in a dunite nodule from Lanzarote, Canary Islands, Lithos, 4, 83-91, 1971.

Frisch, T., and J.B. Wright, Chemical composition of high-pressure megacrysts from Nigerian Cenozoic lavas, Jahr. f. Min. Mon., 19, 289-304, 1971.

Golding, H.G., and P. Bayliss, Altered chrome ores from the Coolac serpentine belt, New South Wales, Australia, Am. Min., 53, 162-183, 1968.

Golding, H.G., and K.R. Johnson, Variation in gross chemical composition and related physical properties of podiform chromite in the Coolac district, N.S.W., Australia, Econ. Geol., 66, 1017-1027, 1971.

Green, D.H., The petrogenesis of the high-temperature peridotite intrusion in the Lizard Area, Cornwall, J. Pet., 5-1, 134-188, 1964.

Griffin, W.L., Lherzolite nodules from the Fen Alkaline Complex, Norway, Contr. Min. Pet., 38, 135-146, 1973.

Haggerty, S.E., Opaque mineral oxides in terrestrial rocks. Oxide Minerals (short course notes), edited by D. Rumble III, Min. Soc. Amer., 3, 101-300, 1976.

Hamad, El. D., The chemistry and mineralogy of the olivine nodules of Calton Hill, Derbyshire, Min. Mag., 33, 483-497, 1963.

Hamlyn, P.R., Chromite alteration in the Panton sill, East Kimberley region, Western Australia, Min. Mag., 40, 181-192, 1975.

Henderson, P., and P. Suddaby, The nature and origin of the chrome-spinel of the Rhum layered intrusion, Contr. Min. Pet., 33, 21-31, 1971.

Himmelberg, G.R., and R.G. Coleman, Chemistry of primary minerals and rocks from the Red Mountain-Del Puerto ultramafic mass, California, U.S.G.S. Prof. Paper, 600, C18-C26, 1968.

Himmelberg, G.R., and R.A. Loney, Petrology of the Vulcan Peak alpine-type peridotite, southwestern Oregon, G.S.A. Bull., 84, 1585-1600, 1973.

Irvine, T.N., Chromian spinel as a petrogenetic indicator. Part 2. Petrologic applications, Can. J. Earth Sci., 4, 71-103, 1967.

Jackson, E.D., Chemical variation in coexisting chromite and olivine in chromite zones of the Stillwater complex, in Magmatic Ore Deposits, edited by H.D.B. Wilson, Econ. Geol. Mono., 4, 41-71, 1969.

Kutolin, V.A., and V.M. Frolova, Petrology of ultrabasic inclusions from basalts of Minusa and Transbaikalian Regions (Siberia, U.S.S.R.), Contr. Min. Pet., 29, 163-179, 1970.

Littlejohn, A.L., and H.J. Greenwood, Lherzolite nodules in basalts from British Columbia, Canada, Can. J. Earth Sci., 11, 1288-1308, 1973.

Loney, R.A., G.R. Himmelberg, and R.G. Coleman, Structure and petrology of the Alpine-type peridotite at Burro Mountain, California, U.S.A., J. Pet., 12, 245-309, 1971.

MacGregor, I.D., and C.H. Smith, The use of chrome spinels in petrographic studies of ultramafic intrusions, Can. Min., 7, 403-412, 1963.

Mall, A.P., and M.K. Rao, Distribution of iron and magnesium between chromites and orthopyroxenes in ultrabasics from Ganginemi, India, Lithos, 3, 113-121, 1970.

Mao, H.K., and P.M. Bell, Crystal field effects in spinel: oxidation states of iron and chromium, Carnegie Inst. Washington Yr. Bk., 73, 332-341, 1974.

McBirney, A.R., and K. Aoki, Factors governing the stability of plagioclase at high pressures as shown by spinel-gabbro xenoliths from the Kerguelen Archipelago, Am. Min., 58, 271-276, 1973.

Medaris, L.G., Jr., High-pressure peridotites in southwestern Oregon, G.S.A. Bull., 83, 41-58, 1972.

Melson, W.G., and G. Switzer, 1966, Plagioclase-spinel-graphite xenoliths in metallic iron-bearing basalts, Am. Min., 51, 664-676, 1966.

Meyer, H.O.A., and F.R. Boyd, Composition and origin of crystalline inclusions in natural diamonds, Geochim. Cosmochim. Acta., 36, No. 11, 1255-1273, 1972.

Muir, J.E., and A.J. Naldrett, A natural occurrence of two-phase chromium-bearing spinels, Can. Min., 11, 930-939, 1973.

Nixon, P.H., and F.R. Boyd, Petrogenesis of the granular and sheared ultrabasic nodule suite in kimberlites, Lesotho National Development Corporation, Maseru, in Lesotho Kimberlites, edited by P.H. Nixon, pp. 48-56, 1973.

Oen, I.S., C. Kieft, and A.B. Westerhof, Composition of chromites in cordierite- and mica-bearing Cr-Ni ores from Malaga Province, Spain, Min. Mag., 39, 193-203, 1973.

O'Hara, M.J., S.W. Richardson, and G. Wilson, Garnet peridotite stability and occurrence in crust and mantle, Contrib. Min. and Pet., 32, 48-68, 1971.

Onyeagocha, A.C., Alteration of chromite from the Twin Sisters dunite, Washington, Am. Min., 59, 608-612, 1974.

Pike, J.E.N., Pressures and temperatures calculated from chromium-rich pyroxene compositions of megacrysts and peridotite xenoliths, Black Rock Summit, Nevada, Am. Min., 61, 725-731, 1976.

Prinz, M., D.V. Manson, P.F. Hlava, and K. Keil, Inclusions in diamonds: garnet lherzolite and eclogite assemblages, Phys. Chem. Earth, 9, 797-815, 1975.

Reid, A.M., and J.B. Dawson, Olivine-garnet reaction in peridotites from Tanzania, Lithos, 5, 115-124, 1972.

Reid, A.M., C.H. Donaldson, R.W. Brown, W.I. Ridley, and J.B. Dawson, Mineral chemistry of peridotite xenoliths from the Lashaine Volcano, Tanzania, Phys. Chem. Earth, 9, 525-543, 1975.

Rodgers, K.A., Chrome-spinels from the Massif du Sud, southern New Caledonia, Min. Mag., 39, 326-339, 1973.

Ross, C.S., M.D. Foster, and A.T. Myers, Origin of dunites and of olivine-rich inclusions in basaltic rocks, Am. Min., 39, 693-736, 1954.

Rothstein, A.T.V., Spinels from the Dawros peridotite, Connemara, Ireland, Min. Mag., 38, 957-960, 1972.

Sigurdsson, H., and J.-G. Schilling, Spinels in mid-Atlantic ridge basalts: chemistry and occurrence, Earth Planet. Sci. Letters, 29, 7-20, 1976.

Sobolev, N.V., Deep seated inclusions in kimberlites and the problem of the composition of the upper mantle, Pub. Amer. Geophys. Union, 279 p., 1977.

Smith, J.V., and J.B. Dawson, Chemistry of Ti-poor spinels, ilmenites and rutiles from peridotite and eclogite xenoliths, Phy. Chem. Earth, 9, 309-322, 1975.

Springer, R.K., Contact metamorphosed ultramafic rocks in the Western Sierra Nevada foothills, California, J. Pet., 15-1, 160-195, 1974.

Stevens, R.E., Composition of some chromites of the western hemisphere, Am. Min., 29, 1-34, 1944.

Suwa, K., Y. Yusa, and N. Kishida, Petrology of peridotite nodules from Ndonyuo Olnchoro,

Samburu District, Central Kenya, Phy. Chem. Earth, 9, 273-286, 1975.
Ulmer, G.C., Experimental investigations of chromite spinels, in Magmatic Ore Deposits, edited by H.D.B. Wilson, Econ. Geol. Mono, 4, 114-131, 1969.
Wilkinson, J.F.G., An Al-spinel ultramafic-mafic inclusion suite and high pressure megacrysts in an analcimite and their bearing on basaltic magma fractionation at elevated pressures, Contr. Min. Pet., 53, 71-104, 1975.
Whitney, P.R., Spinel inclusions in plagioclase of metagabbros from the Adirondack Highlands, Am. Min., 57, 1429-1436, 1972.

PERIDOTITE XENOLITHS AND THE DYNAMICS OF KIMBERLITE INTRUSION

Jean-Claude C. Mercier

Department of Earth and Space Sciences, SUNY - Stony Brook
Stony Brook, New York 11794

Abstract. The time for garnet-peridotite xenoliths in kimberlite to reach the surface (4-6 h) and the kimberlite average intrusion velocity (40-70 km/h) are estimated from annealed olivine-tablet sizes in porphyroclastic peridotites and from the growth rate for annealing calculated on the basis of experimental data: during primary annealing recrystallization, tablet-shaped olivine neoblasts grow at a constant rate proportional to the annealing temperature and to the strain energy stored in the paleoblasts. This energy is proportional to the differential stress immediately prior to annealing, this stress being known through calibration of its effects on mean olivine-subgrain sizes and grain sizes resulting respectively from syntectonic recovery and recrystallization. In addition, experimentally determined flow laws for dry olivine are used to derive strain-rates from the stress and temperature data, hence the duration of the late deformation whenever strain can be estimated from internal deformation of olivine or recrystallized enstatite laminae. As these estimates range from a few hours to a few tens of years, the deformation cannot be related to major tectonic phenomena such as the ascent of large diapirs or convection-related flow, a conclusion in accord with the unrealistic plate-velocity estimates (3 to 30 km/yr) implied by such models. These deformations are therefore ascribed to kimberlite-conduit formation, the longest times being regarded as artifacts possibly due to heterogeneous deformation. Stresses experienced just prior to sampling (.4 - 1.2 kbar) are combined with estimates of the diameter (10 - 50m) of the zone mechanically perturbed by the kimberlite ascent, to estimate the energy released through conduit formation ($\sim 10^{14}$J, or 0.05 megaton).

Introduction

Wagner suggested as early as 1914 that kimberlites intrude the continental lithosphere very rapidly through an explosive-boring process. Although the emplacement is now ascribed to fluidization and/or gas-solid streaming (Dawson, 1962; McGetchin *et al*., 1973), it is still regarded as occurring at very high speeds. On the basis of the stability field for carbonate melts resulting from fractional crystallization of liquids with SiO_2/CaO less than 1/2, McGetchin and Ullrich (1973) calculated models for the eruption of kimberlite. The thermal history of the latter would then correspond to an adiabatic rise from the magma chamber up to about 3 km depth, above which point, the kimberlite transforms into a mixture of supercooled gas in expansion and hot silicate inclusions (xenoliths or xenocrysts). In their various dynamic models, McGetchin and Ullrich (in McGetchin *et al*., 1973) find velocities on the order of 20 m/s (72 km/h) increasing near the surface to about 380 m/s in a few seconds. Whereas the former estimate is generally regarded as reasonable (metastability of diamond) there is much skepticism about the effect of the subsurface supercooling on the velocity. In any instance, the late dynamics of the kimberlite eruptions will not be discussed here as it has virtually no effect on the textures, and the term velocity will be restricted to the pre-supercooling velocity.

In any presently available model, but McCallister's (this volume), the velocities are average values for kimberlite intrusions taken as a whole and do not yield comparative estimates for specific diatremes as only general chemical or physical constraints are considered. With the method presented here, the speed of intrusion at depth may be determined independently for any single diatreme which contains porphyroclastic (Harte, 1977) xenoliths with tablet-shaped neoblasts resulting from static annealing recrystallization. Temperature and pressure (or depth) of equilibration can be determined from the Ca and Al solubility in pyroxenes (Mercier, 1976; 1978) and the differential stress which produced the texture can be derived from the average subgrain spacing in the paleoblasts or from the size of the equant neoblasts ascribed to syntectonic recrystallization. These data (temperature, depth and stress) are then used to compute growth rates for the tablet-shaped neoblasts, the duration of the ascent and ultimately, the pre-supercooling velocity of eruption. Limitations of this technique include the effects of the thermal history and of the water content for some xenoliths. Geopiezometry also yields new cons-

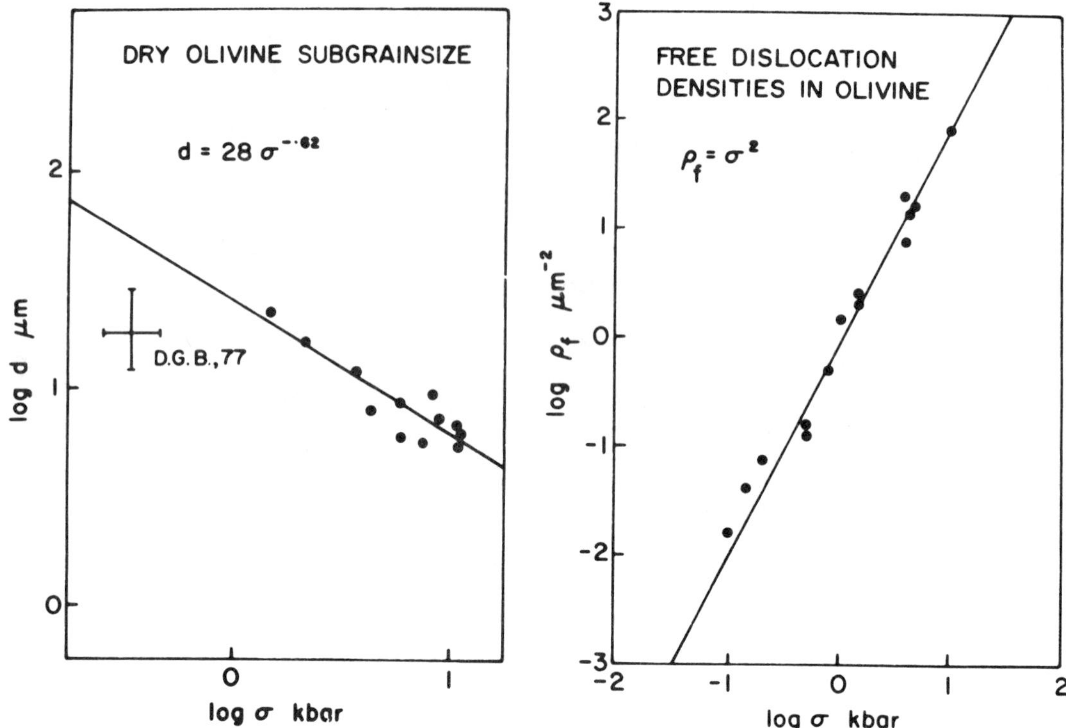

Fig. 1. Stress dependence of the dry-olivine substructure during deformation (dynamic recovery). A. Subgrain-size (average wall-spacing normal to [010]) for olivine decorated by oxidation. Data points and equation from Ross et al. (1978). The range of subgrain sizes and stresses reported by Durham et al. (1977) is shown for comparison.
B. Free-dislocation densities. Data from Goetze (1975).

traints on the nature of the late deformation which produced the sheared lherzolites and contributed to the eruption-channel formation, as recognized earlier by Goetze (1975).

Unless specified otherwise, the following units are used throughout this paper: pressure and differential stress in kilobars (1 kbar = 0.1 GPa), energies in calories (1 cal = 4.18 J) or megatons ($4.18\ 10^{15}$ J), temperatures in Kelvin degrees (K) subgrain and grain size in micrometers (μm), time in kiloseconds (ks) depths in kilometers (km), velocities in meters per second (m/s) and forces in newtons (N).

Recovery and recrystallization: geopiezometers

Recovery comprises all thermally activated processes through which the density and distribution of crystal defects can change to lower the total strain energy of crystals. The best understood and most obvious recovery process is polygonization (e.g. Carter, 1975) in which free dislocations climb and/or cross-slip to form low-angle intragranular walls bounding subgrains. Recovery is either 1) a softening process which is active during high-temperature deformation and allows steady-state creep to be achieved by compensating for work-hardening or 2) a static annealing process which relieves any stored strain energy (e.g. after cold working).

The width d of the subgrains produced by recovery generally depends on the flow stress σ ($=\sigma_1-\sigma_3$) applied during either cold-working or dynamic recovery, and is given by the empirical relation

$$d = A_d\ \sigma^m \qquad (1)$$

where A_d is a parameter proportional to the shear modulus and to the Burgers' vector. The exponent m is an empirical constant generally nearly equal to unity (Mercier et al., 1977), though significantly lower for "dry" olivine (i.e. with AlSiMag assembly and starting material with ∿0.3% H_2O bound in serpentine) as shown in Figure 1A (Ross et al., 1978):

$$d = 28\ \sigma^{-0.62} \qquad (2)$$

where d is the average width of the subgrains decorated by oxidation (Kohlstedt et al., 1976; Fig. 2A). This relation is well documented for increases in flow stress (Raleigh and Kirby, 1970; Mercier, 1977; Mercier et al., 1977; Durham et al., 1977; Ross et al.,1978), but if the applied stress is reduced after subgrains form, the subgrain size remains

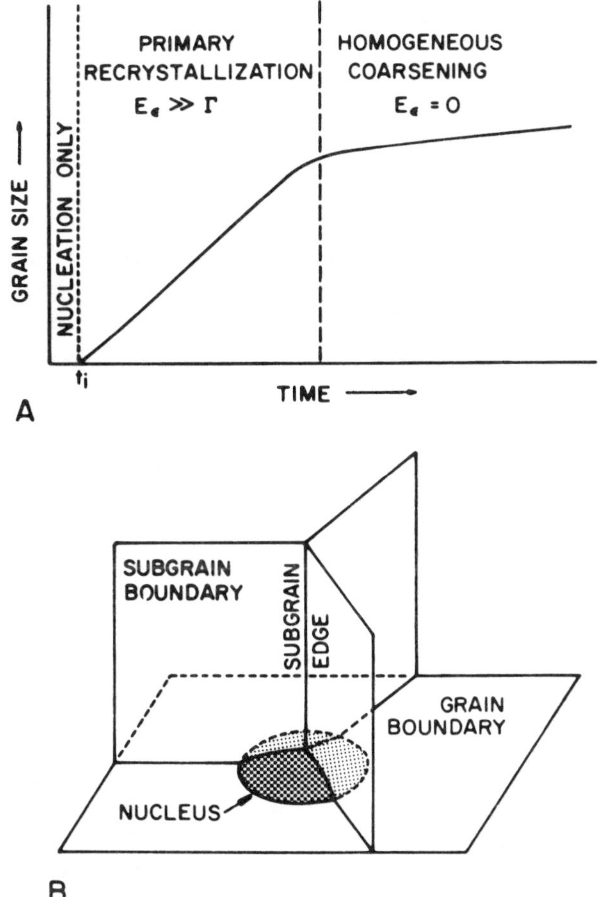

Fig. 3. Dynamics of annealing recrystallization.
A. Average neoblast size as a function of time. t_i is the induction period. Textures typical of primary recrystallization and of coarsening are illustrated in Fig. 2C and 2D.
B. Tridimensional model of nucleation at a multiple junction of grain and subgrain boundaries. To reduce the increase in energy during nucleation, the embryo would originally grow flattened along the grain boundary.

apparently unaffected, even after several days (Ross et al., 1978)

Free dislocation densities (ρ_f; total length of dislocations within subgrains per unit volume; in μm^{-2}) are also proportional to the square of the applied deviatoric stress during dynamic recovery

$$\rho_f = A_f \cdot \sigma^2 \qquad (3)$$

where the parameter A_f, which also depends on the shear modulus and on the Burgers vector, is equal to unity for olivine (data from Goetze, 1975; Fig. 1B).

Recrystallization produces a generation of strain free crystals (neoblasts) which progressively replaces the original strained crystalline material through grain-boundary migration. The process entails formation or individualization of new grains (nucleation s.l.) and their growth. Although the same terminology is commonly applied to static (annealing) and dynamic (syntectonic) recrystallization, there are major thermodynamic, kinetic and textural differences between the two. Annealing recrystallization results in complete relief of the strain energy through fast and complete recrystallization (within minutes in most experimental conditions; Mercier, 1977) of the strained material ("primary" recrystallization); following this cycle some crystals may still grow, thereby reducing the total surface energy in a process termed "secondary" recrystallization or coarsening. On the other hand, in dynamic conditions, continual nucleation will limit grain growth, thereby yielding a steady-state grain size with a log-normal distribution.

Three processes have now been observed (op.cit.) for the formation of neoblasts: nucleation s. str., subgrain rotation and bulging. Nucleation s. str. results from slow accretion of molecules (or atoms) which become disconnected simultaneously from the crystals across the nuclei boundaries. Because it first creates a surface energy much higher than the strain-energy released, nucleation is generally catalysed at multiple subgrain/grain boundary junctions as is observed in experiments (Ross et al., 1978) and in natural rocks; growth of flattened nuclei further minimizes the energy increase (Fine, 1965; Fig. 3B). Subgrain rotation (Poirier and Nicolas, 1975) consists of progressive polygonization: as more and more free dislocations are generated and move to subgrain boundaries, the tilt angle between subgrains progressively increases until the latter become independent grains. This process may be active crystal-wide (Nicolas et al.,1971)or may be limited to the most strained zones (Cahn, 1970). Bulging (Mercier, 1972; Etheridge and Kirby, 1977) is the growth of strain-free crystal appendices at the expense of neighboring more strained material. Bulges may ultimately separate from the crystals from which they grow so as to reduce the surface energy. Although these formation processes may apply to either static or dynamic recrystallization, further evolution of the neoblasts depends on the stress conditions.

Primary annealing recrystallization is characterized by the growth rate ($Y = \partial D/\partial t$, where D is the grain size) of neoblasts which form at the expense of strained paleoblasts. Growth occurs by grain-boundary diffusion (experimental observation) and would result from the difference between the activation energies on each side of the paleoblast/neoblast boundary for an atom (or molecule) to cross this boundary. To a first approximation, the energy barrier preventing an atom from jumping freely from a paleoblast to the neighboring neoblast is the activation energy (ΔG_a) for self-diffusion across the grain-boundary. The probability for such an atom (or molecule) to jump over the boundary may thus be written as

$$p = \nu \exp(-\Delta G_a/RT) \qquad (4)$$

where the characteristic frequency ν, i.e. the number of times per seconds (ca 10^{13}) that a given particle tries to cross the energy barrier is close to the Debye frequency set by the Eyring theory as {RT/hN} (h = Plank's constant; N = Avogadro's number; Christian, 1970). The increase in diameter related to such jumps is equal to this probability times the distance ℓ between two neighboring sites across the boundary, or $\{\ell\nu \exp(-\Delta G_a/RT)\}$. The energy barrier for atoms to return to the paleoblast

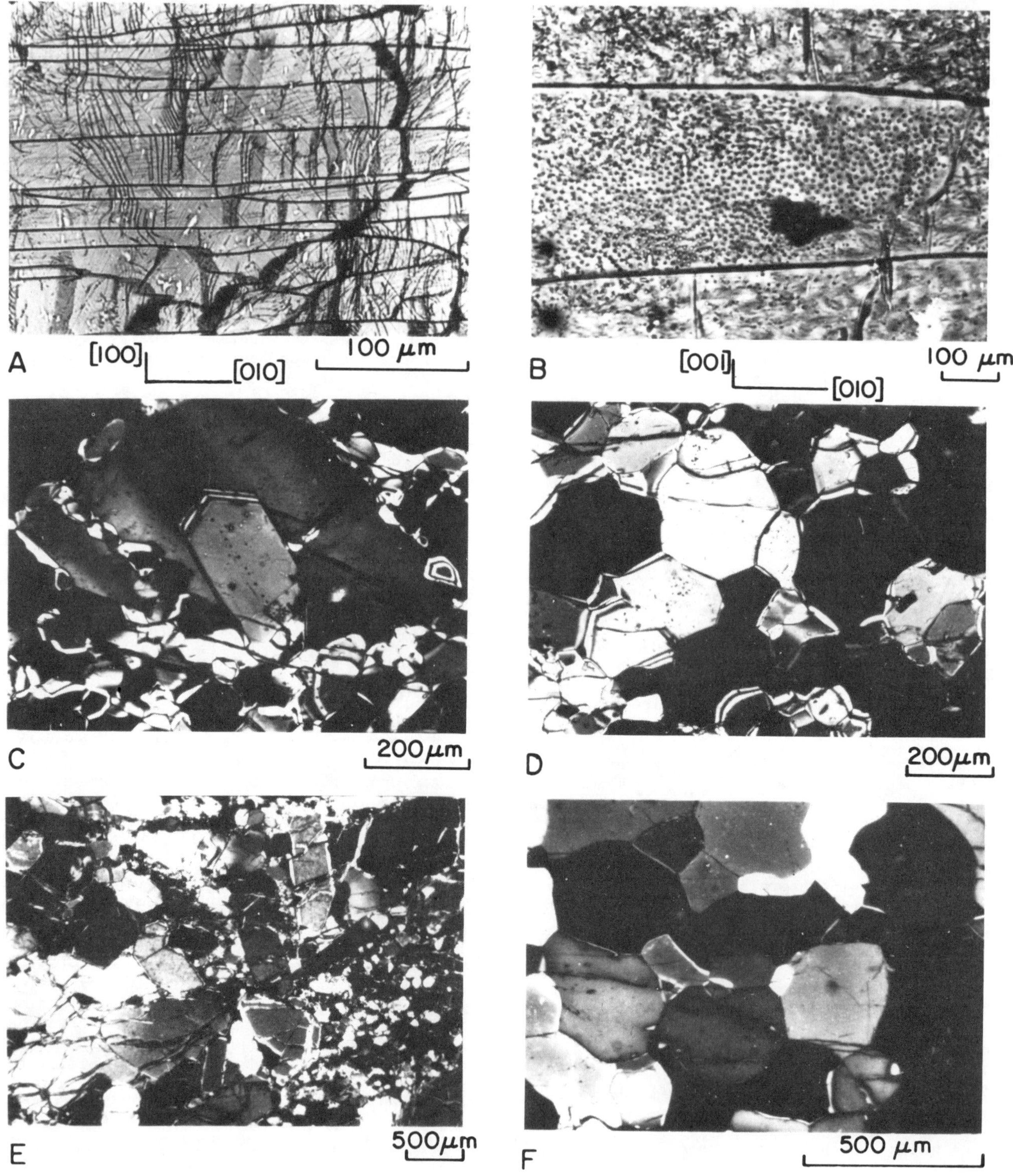

Fig. 2. Olivine substructure and recrystallization.
A. Subgrains (east-west bands) viewed parallel to the [001] direction with screw dislocations (north-south lines) normal to subgrain (100)-walls. Naturally deformed lherzolite from San Quintin, Baja California.
B. Subgrain (east-west bands) viewed parallel to the [100] direction. [100] screw-dislocations now appear as dots. Naturally deformed lherzolite from Kilbourne Hole, New Mexico.
C. Primary annealing recrystallization: typical tablet-shaped neoblast growing at the expense of a highly strained host crystal. Experimental sample.
D. Coarsening (or secondary annealing recrystallization) follows primary recrystallization. Grain boundaries are moving toward their center of curvature. Experimental sample.
E. Primary annealing recrystallization (tablets) in a xenolith from kimberlite (sample 69-KI-14; Kimberley, South Africa). The matrix comprises remnants of porphyroclasts and small syntectonic neoblasts (lower right).
F. Syntectonic recrystallization (σ = 0.7 kbar). Olivine grains have a substructure, the smallest crystals do not have the typically convex boundaries shown in D, and the grain size is stable even for long-duration experiments.

---

is increased by an amount E_ε, the lattice strain-energy represented by dislocations. By analogy with the previous relation, the decrease in diameter related to such jumps will be equal to $\{\ell\nu \exp[-(\Delta G_a+E_\varepsilon)/RT]\}$. In other words, a larger number of atoms do jump to the neoblast, the net rate of growth being given by the difference

$$Y = \ell\nu\exp[-\Delta G_a/RT] - \ell\nu\exp[-(\Delta G_a+E_\varepsilon)/RT]$$

$$\text{or} \quad Y = \frac{\ell RT}{hN}[\exp(-\Delta G_a/RT)][1-\exp(-E_\varepsilon/RT)] \tag{5}$$

The surface energy, so far neglected, can be introduced in a similar way, yielding the general relation (Mercier, 1977)

$$Y = [\frac{\ell RT}{hN} \exp(-\Delta G_a/RT)][1-\exp(-(E_\varepsilon \pm \Gamma)/RT)] \tag{6}$$

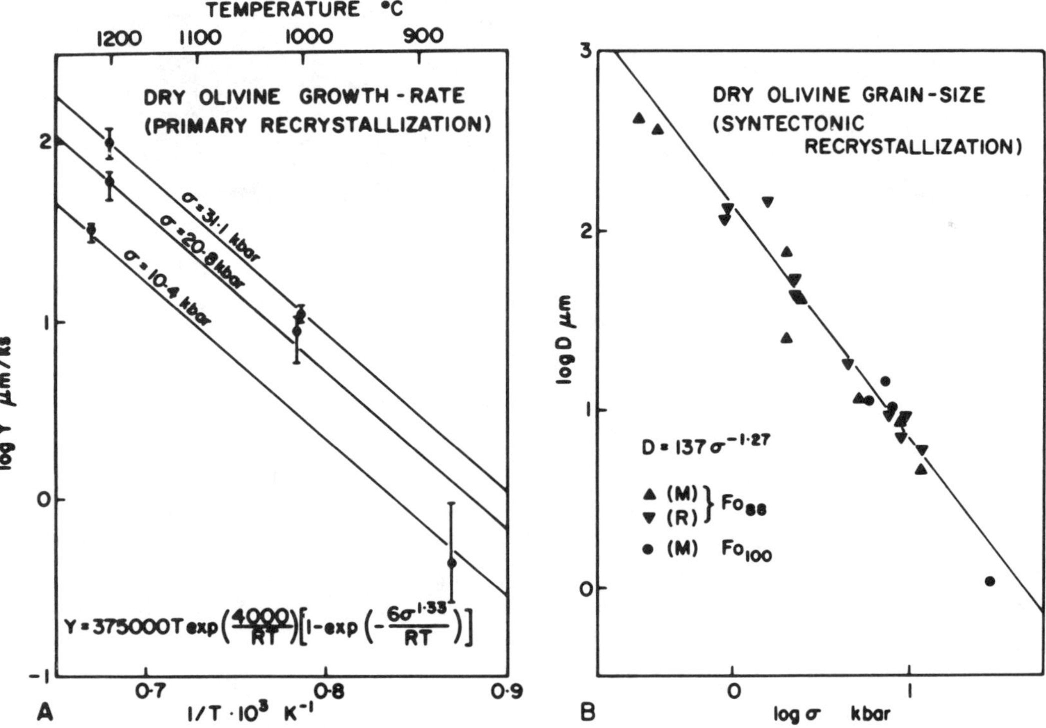

Fig. 4. Experimental recrystallization of dry olivine.
A. Growth rate ($Y=\partial D/\partial T$) during primary annealing recrystallization as a function of the deviatoric tress prior to annealing and of the anneal temperature. Error bars have been extended (where necessary) to include all possible values (see text).
B. Syntectonically recrystallized grain-size as a function of the applied deviatoric stress for Fo_{88} and Fo_{100}. Data from Mercier et al. (1977; "M") revised in part by Ross et al. (1978; "R"). The equation is based on the revised data.

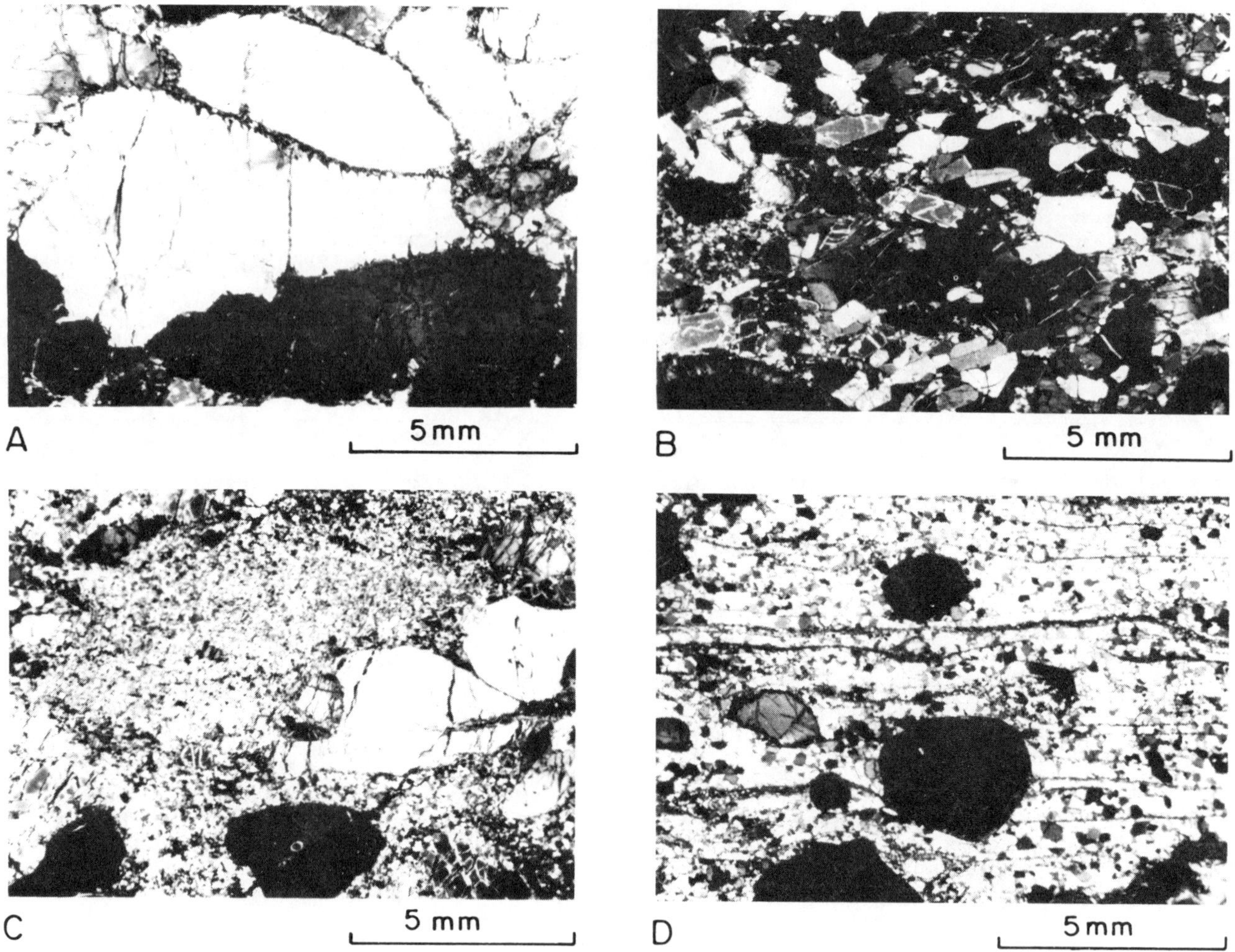

Fig. 5. Textures of peridotite xenoliths from kimberlites (in order of increasing depth).
A. Coarse-granular texture (KNC-1; Kimberley, South Africa). Such textures would be typical of the upper mantle (low deviatoric stress) as only the microstructure has been modified during xenolith sampling.
B. Annealed porphyroclastic texture (69-KI-14; Kimberley, South Africa). Tablet-shaped neoblasts in highly strained paleoblasts are ascribed to annealing and the fine-grained matrix of equant crystals to syntectonic recrystallization. The deformation was mostly accomodated by crystal glide (high stress). Fig. 2E shows details of this texture.
C. Porphyroclastic texture (J-71; Bultfontein, South Africa). This peridotite mostly recrystallized during deformation (lower stress than for B) and the strain energy in the paleoblasts was insufficient to enhance widespread annealing recrystallization.
D. Lesotho-type texture (PHN-1611; Thaba Putsoa, Lesotho). Olivine has entirely recrystallized and the enstatite-neoblast microlayering produces a strong foliation. Strain is much greater than 1000%.

where Γ is the surface energy per unit volume of the polycrystalline material. This energy acts against growth of neoblasts (hence the variable sign) during early recrystallization as a result of the increase in total grain-boundary area. During primary recrystallization, the strain energy available (if sufficient to enhance recrystallization) is more than two orders of magnitude higher than the surface energy (Taylor and Quinney, 1937) and relation (5) is therefore sufficient for characterization of the growth rate.

Annealing experiments were performed in a small solid-pressure-medium apparatus (Griggs, 1967) to define the growth rate of olivine during primary recrystallization as a function of the annealing temperature T and of the deviatoric stress σ applied prior to annealing. The assemblies used were similar in geometry to those described by Carter and Ave Lallemant (1970) but with a confining pressure medium of AlSiMag powder produced by room temperature compaction (in the pressure vessel and after sample insertion) of oversized AlSiMag-222 assemblies (Mercier, 1977). The material used was "dry" Mount-Burnet dunite, a medium-grained partially-recrystallized olivine cumulate (Fo_{88}) with about 0.3% H_2O bound in serpentine (Carter and Ave Lallemant, 1970). For all experiments, the strain required to enhance recrystallization was produced at constant temperature (900°C) and constant strain-rate (10^{-6}/s). When the differential stress was reached, this stress was immediately lowered and the temperature increased to the desired annealing temperature (1100-1300°C) to achieve relaxation (σ=0) in less than 4% of the total annealing time. Samples quenched after primary recrystallization had already ended (Fig.2D) were discarded. No correction was made in the course of data reduction for the induction period of Figure 3A: nucleation had started before the annealing conditions were reached, as was checked in prematurely aborted experiments.

Growth rates ($\partial D/\partial t$) should be computed from the size of the largest tablet-shaped neoblasts (measured at 45° to their long axis) enclosed in paleoblasts (Fig. 2C). However, error bars in Fig. 4A are extended to include still larger neoblasts (if any) in the recrystallized matrix since such neoblasts (whatever their shape) may have been early-nucleated olivine-tablets now in a state of coarsening (a slower growth process than primary recrystallization). Applying equation 6 to two experiments performed at a same temperature but different stress levels yields

$$Y_1/Y_2 = (1-\exp(E_{\varepsilon 1}/RT))/(1-\exp(E_{\varepsilon 2}/RT)) \quad (7)$$

In the absence of strong theoretical evidence for the general form of the dislocation strain-energy density (E_ε), the relation $E_\varepsilon=K\sigma^m$ is suggested as it satisfies the available experimental data (Mercier, 1977). From the numerical expression of (7), one can then define an empirical bijective relation $K\leftrightarrow m$, i.e. a curve in K-m space. The common intersection of curves based on various sets of experiments then yields the actual values of K and m (Mercier, 1977),

$$E_\varepsilon = 6\ \sigma^{1.33} \quad (8)$$

Using this relation, the growth-rate equation is then derived by plotting the stress-compensated growth rate $\{Y/[1-\exp(E_\varepsilon/RT)]\}$ against 1/T. From the data in Figure 4A,

$$Y = 3.75\ 10^5\ T\ \exp(\frac{40000}{RT})\ 1-\exp(\frac{6\sigma^{1.33}}{RT}) \quad (9)$$

indicating an activation energy for grain-boundary diffusion of 40 kcal.

After the primary recrystallization cycle ends, the average grain size still increases with time (coarsening), but at a much lower rate: the driving force is now a reduction in surface energy Γ (as E_ε = 0). Because Γ is small (Γ < 100 cal; op. cit.), the general equation (6) may be simplified as follows

$$Y_c = \frac{\ell\Gamma}{hN}\exp(-\Delta G_a/RT) \quad (10)$$

For soap-bubble experiments (Cahn, 1970), the surface energy per unit volume is a simple function of time: $\Gamma \propto t^{-\frac{1}{2}}$. During coarsening, each boundary migrates toward its center of curvature: the largest crystals grow in all directions, the smallest are consumed by surrounding grains and those of intermediate size are consumed on some sides and grow on others (Figure 2D). The grain size distribution remains log-normal at all times during this process (Feltham, 1957). However, coarsening is negligible in most natural rocks as its effect on grain size (e.g. the time for a 50% increase of D) is inversely proportional to the original grain size ($t_{50\%} \propto D_o^{-2}$; i.e. hours for fine-grained experimental samples mean years for natural peridotites at ~ 1200°C).

By analogy with annealing recrystallization of cold-worked materials, nucleation and recrystallization under dynamic conditions are generally regarded as being directly related to the substructure developed duting deformation (e.g. Lutton and Sellars, 1969). Since syntectonic recrystallization occurs essentially at grain boundaries for both experimentally and naturally deformed specimens, a model has been developed (Mercier, 1977) in which the only potential sites for nucleation are the multiple junctions of subgrains with grain boundaries (Fig. 3B), where the nucleation energy is lowest. The new grains are then pinned at their periphery by a network of subboundaries. If all sites were successfully generating new neoblasts one would expect formation of a number of neoblasts equal to the number of original subgrains (of spacing *d*). If *p* is the probability for a site to be activated and to yield a neoblast, the new average and steady-state grain-size D is given by

$$D = N_D^{-1/3} = (N_d \cdot p)^{-1/3} = d/p^{1/3} \quad (11)$$

where N_D is the number of neoblasts formed as steady-state textural equilibrium is achieved and N_d the original number of subgrains, both given per unit volume. It should be noted that this expression is actually quite model-independent: Lutton and Sellars (1969) and Twiss (1977) obtain similar relations for bulge-nucleation ($3/D_o d^2$ potential sites per unit volume) and subgrain rotation (energetic equilibrium between free and boundary dislocations), respectively. All three models predict a stress exponent ranging from 1.2 to 1.5. Furthermore, depending on the assumptions inherent in the details of each model (Mercier, 1977), a slight temperature dependence may be found for the steady-state grain size *D*, due to an apparent activation energy ranging from $+\Delta G_a/3$ to $-\Delta G_a/3$ (i.e. +14 to -14kcal, since ΔG_a = 40kcal; equation 9) and depending on the relative sensitivity of nucleation and crystal-growth processes to temperature,

$$D = A_D\sigma^{-1.35\pm.15}\exp[(0\pm14000)/RT] \quad (12)$$

Experimental studies on olivine show some evidence of temperature effects (+12.3 kcal, "dry", Mercier 1977; -14.2 kcal, wet, Ross et al., 1978) though the latest data for "dry" olivine (Ross et al., 1978 Figure 4B) suggest no significant dependence as proposed earlier by Kohlstedt and others (1976), and D is given simply by

$$D = 137\ \sigma^{-1.27} \qquad (13)$$

In conclusion, experimental data are now providing several geopiezometers for natural peridotites, including free-dislocation densities, subgrain-boundary spacing and syntectonically recrystallized grain sizes. However, dislocation densities are not a reliable geopiezometer as static recovery may occur in xenoliths while in hot magmas, thereby reducing free-dislocation densities and sharpening kink bands (as observed for xenoliths in basalts), and because high densities require the use of electron microscope and are therefore of questionable statistical significance due to the sample size. On the other hand, subgrains and neoblasts (if any) provide reliable information on the late high-stress deformations often ascribed to emplacement (obduction through low-temperature shear of basal peridotites; magma-conduit formation and/or sampling of xenoliths) while the original grain size reflects the steady-state mantle conditions (Figure 9). However, because some facies may have extensively recrystallized under late high stresses, detailed textural studies are critical to the interpretation of the data.

Annealed porphyroclastic textures as geochronometers.

Porphyroclastic peridotite xenoliths in kimberlites (Figure 5B & 2E) have a complex texture with three generations of olivine, including large strained paleoblasts, an intergranular matrix of equant grains, and tablet-shaped crystals usually cross-cutting the paleoblasts, the misorientation angle between the latter two being in the range 20-30° (Boullier, 1975). These tablets are typically strain and dislocation-free and contrast with mosaics of smaller equant crystals which have relatively high dislocation densities. These two neoblast generations are interpreted as having been respectively produced by primary annealing recrystallization (i.e. static grain growth) and syntectonic recrystallization. Accordingly, in annealing experiments, growth of olivine tablets is a much faster recovery process for high strain-energies and high temperatures, and such tablets can replace most of the paleoblasts before appreciable internal recovery of the grains can occur (Mercier, 1977). In some instances, a mosaic made of large tablets and nearly-equant crystals is observed. Both of these crystal types are then dislocation-free and the large equant crystals are also ascribed to annealing. Because they nucleated after the tablet-shaped crystals, their growth was inhibited; the annealed aggregate was in a state of coarsening when it was quenched. Because of the total absence of deformation of the tablet-shaped crystals, this static annealing is thought to have occurred while the xenoliths were transported to the surface by the kimberlite, in agreement with the calculations presented below. The absence of tablets growing within the equant-neoblast mosaic is ascribed to the lower strain-energy of these neoblasts.

Following McGetchin and Ullrich's (1973) model, the host kimberlite would rise adiabatically until it reaches a depth of a few kilometers (Fig. 6) where sudden expansion of the volatile-rich magma produces supercooling, a phenomenon also evidenced by the presence of diamonds xenocrysts: indeed, even large crystals might be completely transformed into graphite in a few hours upon decompression at high temperature. This supercooling applies also to the xenoliths despite their low thermal conductivity as 1) highly strained olivine paleoblasts show virtually no recovery in contrast with observations for xenoliths in basalts, and 2) diamond has also been found in xenoliths. This fast cooling of the xenoliths may be explained by the supercooled fluids circulating through the xenoliths as cracks open along grain boundaries as a result of contraction of elastically anisotropic crystals (behavior similar to that destroying the cohesion in xenoliths from basalts). The circulation of these fluids is evidenced by 1) the strong post-tectonic and pre-serpentinization metasomatism (e.g. K; Mercier and Carter, 1975; Boyd, 1975) and 2) the intensive serpentinization which occurred even though the host kimberlite remains only a few minutes in the temperature range where serpentinization is possible. However, growth of the neoblasts produced by annealing is virtually over before a significant temperature contrast between melt and inclusions is created through cooling and the xenoliths may be regarded as being "dry" during annealing recrystallization; equation 8 should apply (Figure 6).

The thermal history of the xenoliths is actually more complex than a thermally passive (i.e. without heat exchange) upward transport followed by quenching near the surface: peridotite xenoliths are slowly heated to the kimberlite temperature, but this temperature drops with time possibly due to the cooler inclusions. Appreciable cooling should occur at depths less than 120 km, since an adiabatic path would intersect the dry peridotite solidus (Kushiro, 1973), but no evidence of possible late partial melting is found in most xenoliths the only exception being volatile-bearing xenoliths from Kimberley (F.R. Boyd, writt. comm., 1978). Thermal models are therefore limited by the high-temperature composite path formed by the adiabat and the dry solidus (Fig. 6). Heating of xenoliths by the host kimberlite can be modelled by analogy with the cold sphere case treated by Carslaw and Jaeger (1959), using their non-dimensional graphic solution, i.e. $\{\kappa t/\ell^2\}$ isopleths in $\{(T-T_s)/(T_h-T_s)\}$ vs $\{x/\ell\}$ space with κ the thermal diffusivity ($1.2\ mm^2/s$ for peridotites), ℓ the sphere radius, T_s the initial sphere temperature, T_h the initial host-medium temperature and T the temperature in the sphere at a time t and a distance x from the center. Thermal models for $T_s = 1375^oC$ and $T_h = 1725^oC$ (Fig. 7) illustrate the variability of possible thermal profiles. In any instance, for small xenoliths (1-10 cm) and near the surface of large ones, the heating period is short compared to the total time for the peridotite to be brought up. The temperature to be used in the growth-rate equation should in this case be about the kimberlite temperature T_k and not that of the xenolith at the time of sampling. Quantitative study of tablet-size distribution across large porphyroclastic xenoliths should yield critical constraints on the initial temperature of the kimberlite and on its thermal history. In absence of such data, a series of thermal models (pattern in Fig.6) are used in subsequent velocity calculations. As the kimberlite must have been originally concentrated in zones of the upper mantle below the deepest xenoliths sampled by the kimberlite under study, initial temperatures up to 100°C higher (1625-1725°C; 1900-2000°K) than those of these xenoliths are tentatively assumed.

Defining D_s as the average size of the equant (syntectonic) neoblasts, the strain energy stored in the paleoblasts is given (from relations 7 and 12) by

$$E_\varepsilon = 6\sigma^{1.33} = 6\left[48D_s^{-.787}\right]^{1.33} = 1033D_s^{-1.05} \quad (14)$$

Since natural stresses for late deformations prior to sampling are at most of a few kilobars (Fig.9C) equation 5 may be simplified

$$Y = \frac{\ell}{hN} E_\varepsilon \exp(-\Delta G_a/RT) \quad (15)$$

that is, from relations 8 and 13,

$$Y = 1.95\ 10^8\ D_s^{-1.05}\exp(-20131/T) \quad (16)$$

This equation defines an instantaneous growth rate Y (for the olivine tablets) which is function of the temperature at any time, and therefore a function of time (t) and depth (Fig. 6). An equivalent growth-rate $\bar{Y}$ is defined as the constant rate which, applied for the total time of the kimberlite ascent would yield the observed grain size. At any given depth Z, this equivalent growth-rate is equal to the surface area below the curve for Y, divided by the depth,

$$\bar{Y} = t^{-1} \int \partial Y.\partial t = Z^{-1} \int_{300}^{T_k} \partial Y.\partial T \quad (17)$$

where T_k is the initial temperature of the kimberlite, 300°K an approximation for the final temperature, and Z the depth calculated through pyroxene geobarometry (Mercier, 1976; 1977) as

$$Z = 4.25+(534.64\ell nK_w^{\prime}-2139.39\ell nK_a^{\prime}-802.73)/(\ell nK_w^{\prime}.\ell nK_a^{\prime}-6.208\ell nK_w^{\prime}+2.26\ell nK_a^{\prime}+31.037) \quad (18)$$

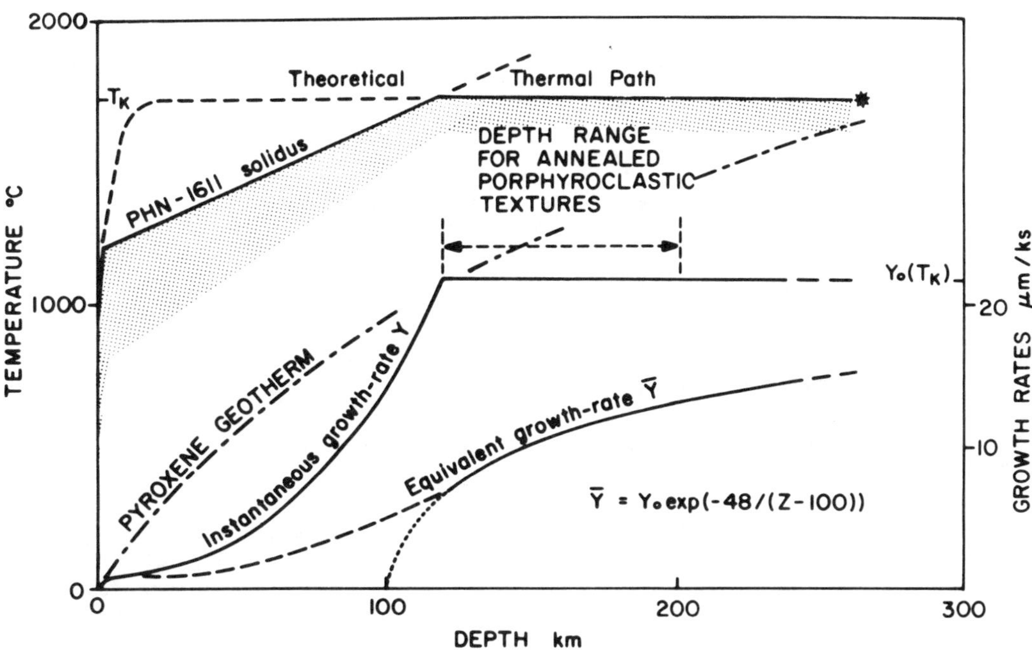

Fig. 6. Annealing-neoblast growth-rates as a function of depth. The thermal path is adapted from McGetchin and Ullrich (1973) by rescaling to an original temperature (asterisk) 100°C higher than the equilibration conditions of the deepest peridotite xenoliths in Southern-Africa kimberlites. Representative dry-lherzolite (PHN-1611) solidus from Kushiro (1973). The pattern covers the range of tested thermal-path models. Instantaneous and equivalent growth-rates shown apply to the high-temperature thermal path.

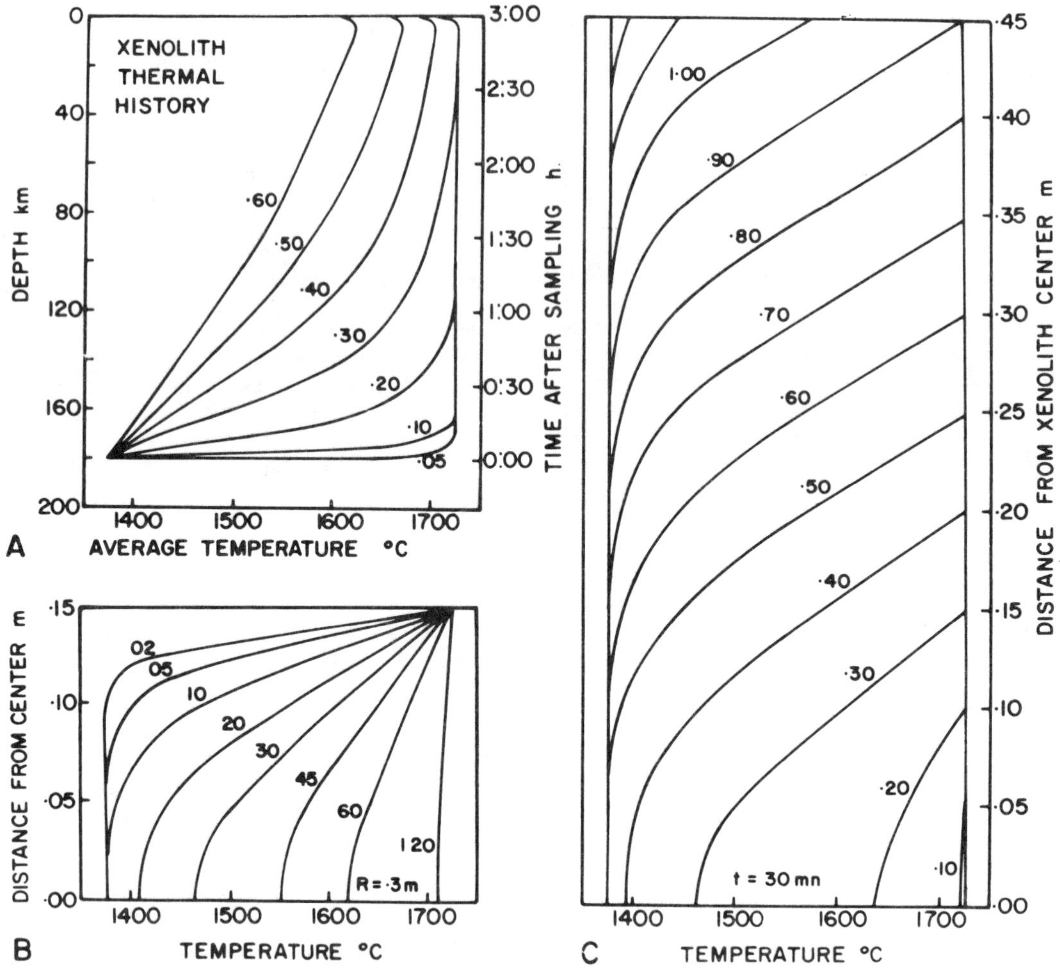

Fig. 7. Thermal history of the xenoliths.
A. Average-temperature variations for xenoliths of various sizes (numbers on lines; in meters) sampled at the average depth (180 km) at which annealed porphyroclastic textures originate. An average kimberlite velocity of 60 km/h has been used for scaling the model originally calculated as a function of time.
B. Temperature profiles across a 0.3 m sample. Curves are for various times (in mn) after sampling by the kimberlite. Near the xenolith surface, the curve slopes are independent of the xenolith size (if greater than 0.25 m in diameter).
C. Temperature profiles across samples of various sizes (numbers on curves, in meters), 30 mn after sampling by the kimberlite.

where, the element symbols representing atomic fractions per 6-oxygen formula unit,
- for enstatite,

$$K_w' = 14.493W/(1-2W)$$

$$K_a' = A.(1-A).(6.004-3.025Mg+0.702\ln K_w')$$

- for diopside,

$$K_w' = (1-2W)/(0.862+0.276W)$$

$$K_a' = A.(1-A).(3.298-1.781Mg+0.128\ln K_w')$$

with

$$W = Ca/(Ca+Mg+Fe^{2+}+Mn)$$

$$A = (Al+Cr-0.82Na)/2$$

For samples with a depth of origin greater than 120 km (case of most porphyroclastic samples), the equivalent growth rate obtained through incremental construction is well fit by the empirical relation (Fig. 6)

$$\overline{Y} = Y \exp[-48/(Z-100)] \qquad (19)$$

From the definition of the equivalent growth rate, the total time for kimberlite eruption from the magma reservoir up to the surface is given by

$t_k = D_A/\bar{Y}$ where D_A is the maximum diameter (measured at 45° to their long axis) for the tablet-shaped neoblasts formed by annealing recrystallization. The average speed of intrusion (V_k, m/s) in the pre-supercooling regime is therefore given by

$$V_k = Z\ Y/D_A = 1.95\ 10^8\ Z\ D_s^{-1.05} D_A^{-1} \exp\{-20131/T\}\exp\{-48/(Z-100)\} \quad (20)$$

or, considering the uncertainty on the exponent of D_s,

$$V_k = 2\ 10^8\ Z\ \exp(-\frac{20131}{T} - \frac{48}{Z-100})/(D_s D_A) \quad (21)$$

Velocities of eruption estimated on the basis of the above equation applied to small xenoliths (or samples near the xenolith surface) range from 40 (11 m/s) to 70 km/h (20 m/s) for Premier Mine and Thaba Putsoa (Table 1), depending on the model chosen for the initial temperature of the kimberlite (1625 to 1725°C, respectively). These values are in gross agreement with independent estimates derived from exsolution rates in pyroxenes (McCallister et al., 1977) and from the critical conditions inferred for kimberlite intrusions (Artyushkov and Sobolev, 1977). Although velocities for Precambrian Premier-Mine and Cretaceous Thaba-Putsoa kimberlites are virtually identical, the late stresses prior to annealing were several times higher for Premier-Mine peridotites. The paleogeotherms being the same for both localities,

TABLE 1. Kimberlite velocities and related parameters inferred from Thaba Putsoa (PHN), Lesotho, and Premier Mine (RVD), South Africa, xenoliths.

Sample #		PHN-1596	RVD-157	RVD-169
Z	km	222	239	186
D_s	μm	55	141	94
D_A	μm	900	376	353
@ T_k=1625°C				
Y	μm/h	167	65	81
t	h:mn	5:23	5:46	4:22
V	km/h	41.3	41.5	42.6
@ T_k=1675°C				
Y	μm/h	220	86	106
t	h:mn	4.06	4.23	3.20
V	km/h	54.2	54.5	55.9
@ T_k=1725°C				
Y	μm/h	285	111	137
t	h:mn	3:10	3:23	2:34
V	km/h	70.2	70.6	72.4

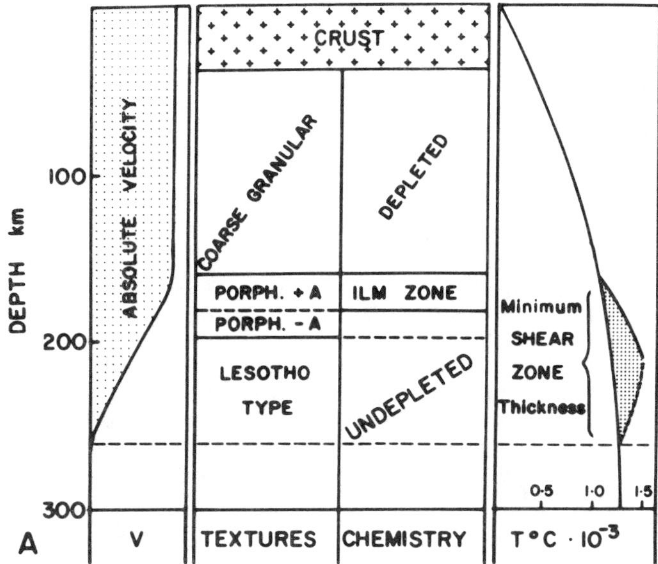

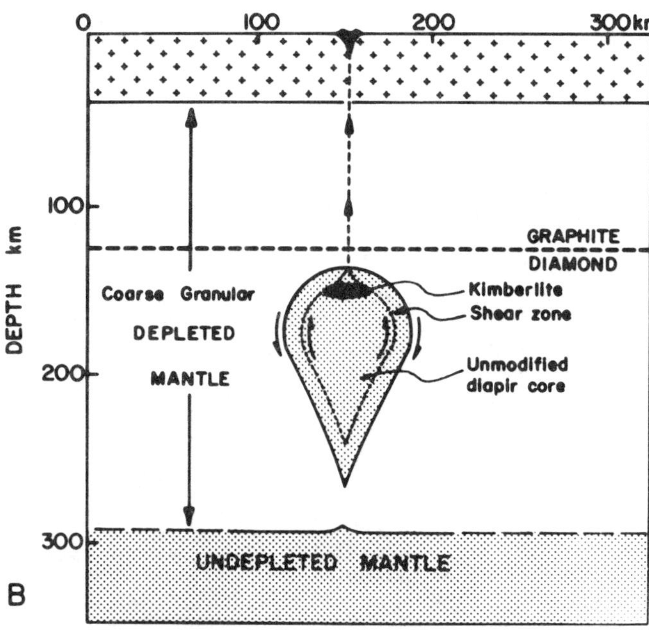

Fig. 8. Previous models for the formation of the sheared lherzolites.
A. Boyd's (1973) model. If the sheared lherzolites were the structural expression of the low velocity zone >100 km thick, minimum velocities would be 0.3 or 30 km/yr to account for Boyd's shear heating or late-stress estimates, respectively.
B. Gueguen's model (Green and Gueguen, 1974). If the sheared lherzolites were the outer layer of a rising diapir, stress estimates would yield unrealistic upward velocities of 0.01 and 3 km/yr for a minimum shear-zone thickness (based on size and homogeneity of xenoliths) and Gueguen's postulated thickness (10 km), respectively.

the velocities inferred are truly sensitive to the dynamics of eruptions and can be compared for different pipes despite the assumption of the initial kimberlite temperature. Therefore, quantitative studies of porphyroclastic textures may yield useful hints on conditions which have a direct bearing on the stability and presence of diamond.

Sheared textures and eruption-channel formation.

Following Boyd's (1970) early observations of contrasting pyroxene compositions, Boyd and Nixon (1972; 1973) recognized two varieties of peridotite xenoliths in kimberlites, differing in both chemistry and textures. The depleted "granular" type, with purple-red (5RP4/6) Cr-pyrope and bright emerald green (10GY5/6) Cr-diopside, is generally depleted both chemically ($Mg/\{Mg+Fe\}>0.933$) and paragenetically (Di<5%), though minor phlogopite is common; texturally, such peridotites range in olivine grain-size from 5 to 10 mm. The undepleted "sheared" lherzolites with reddish-brown (5R4/4) pyrope and dark-green (10GY5/2) subcalcic diopside, have a bulk composition relatively close to pyrolite (esp. PHN-1611; Nixon and Boyd, 1973) and their diopside content is generally more than the critical 5% (true lherzolites: Streikeisen, 1973); texturally, these facies have typical bimodal grain-size distributions, with olivines about 0.002 mm in diameter, forming a matrix with pyroxene and garnet crystals which are even larger than for the granular peridotites. As depleted granular peridotites generally appear to represent shallower upper mantle material (Nixon _et al_., 1973; Mercier, 1977) and because the textures are obviously "upper-mantle textures" in the sense that they were produced in the solid state before sampling by the kimberlite, the models proposed so far interpret these textures as being produced by relative displacement of various zones of the upper mantle.

Boyd's (1973) original model (Fig. 8A) is based on the kinked geotherm derived from geothermometry and geobarometry techniques which assume a pressure independent Di(En) solvus and use raw Al_2O_3 contents of enstatite, as corrections attempted then were regarded as not changing the geotherm shape. This geotherm closely follows the geotherms previously inferred from geophysical considerations (Clark and Ringwood, 1964), but steepens suddently around 1000°C (_i.e._ at a depth of about 170 km), thereby forming a kink interpreted as evidence for a thermal perturbation at depth, not long before the kimberlite eruption. Boyd and Nixon (1973) ascribed this perturbation to intense shearing at depth resulting "from movements at the base of the African plate which occurred during the break-up and dispersal of Gondwanaland". Mercier and Carter (1975) showed that appreciable shear heating was unlikely under the conditions imposed by this model: a strain rate of about 10^{-10}/s is necessary to perturb appreciably the geotherm assuming shear heating during most of the Mesozoic era. The lowest plate-velocity estimate under such conditions, considering the minimum possible thickness for the shear zone on the basis of Boyd's geotherm and assuming homogeneous shear, is about 30 m/yr, _i.e._ two orders of magnitude above any reasonable figure.

Green and Gueguen (1974) presented an alternative model (Fig. 8B), that of an adiabatically rising diapir, to explain both the peridoti formation of the kimberlites. Although deep diapirs are not _a priori_ discarded for kimberlite formation, the proposed model is incompatible with stress estimates derived from the sheared lherzolites. On the basis of these (800 to 1400 bars; Fig. 9C) and of pressures and temperatures derived from pyroxene geothermometry and geobarometry, most strain-rate estimates for the sheared lherzolites fall in a narrow range of 1 to 2 10^{-8}/s (Fig. 9D). Considering Green and Gueguen's original diapir (Fig. 8B), a 10-km representative thickness for the sheared zone would correspond to an adiabatic rise of 10 meters a day (3 km/yr). Varying the size of the shear zone by a factor of ten changes the speed estimate by the same factor and the model is therefore unacceptable. When applied to Boyd's model, already rejected on a different basis, strain-rate calculations based on sample textures yield plate velocities in the order of 30 km/yr or more, which is equally unacceptable. Other models are similar to these (_e.g_. MacGregor, 1974; Parmentier and Turcotte, 1974) and need not be discussed here. The arguments given above also apply to MacGregor's (1975) model in which a sharp change in thermal conductivity resulting from increasing deformation is suggested. Xenolith textures were therefore reexamined for their rheological implications.

The two types of coarse-grained textures (protogranular and tabular) share a relative absence of deformation, and represent equilibrated textures in which all minerals have similar grain sizes. The protogranular texture (Fig. 5) is ascribed to partial melting (Mercier and Nicolas, 1975), as supported by the curved boundaries, the chemical and modal depletion and the weak preferred orientation (Boullier and Nicolas, 1973; 1975). Due to the relatively low temperatures of equilibration, the coarse grain-size implies extremely low strain-rates ($>10^{-18}$/s; Fig. 9B) and once formed, the texture may stay unmodified for hundreds of millions of years during which the total strain will hardly reach a few percent. Coarse-tabular (Harte, 1977) textures are characterized by a similar grain size but dominantly straight boundaries and cannot result from simple polygonization with large rotations of the subgrains. In this case, a low-stress nucleation-recrystallization process must be involved to yield the relatively strong preferred orientations observed for strain rates which are still very low (10^{-16} to 10^{-17}/s; Fig. 9B).

The porphyroclastic texture described in the previous section may be considered as a transitional type produced by high-stress high-temperature deformation of coarse-grained types. Indeed, the paleoblast sizes are comparable, though slightly coarser (Fig. 9C), to the grain size of the former whereas stress estimates derived from equant neo-

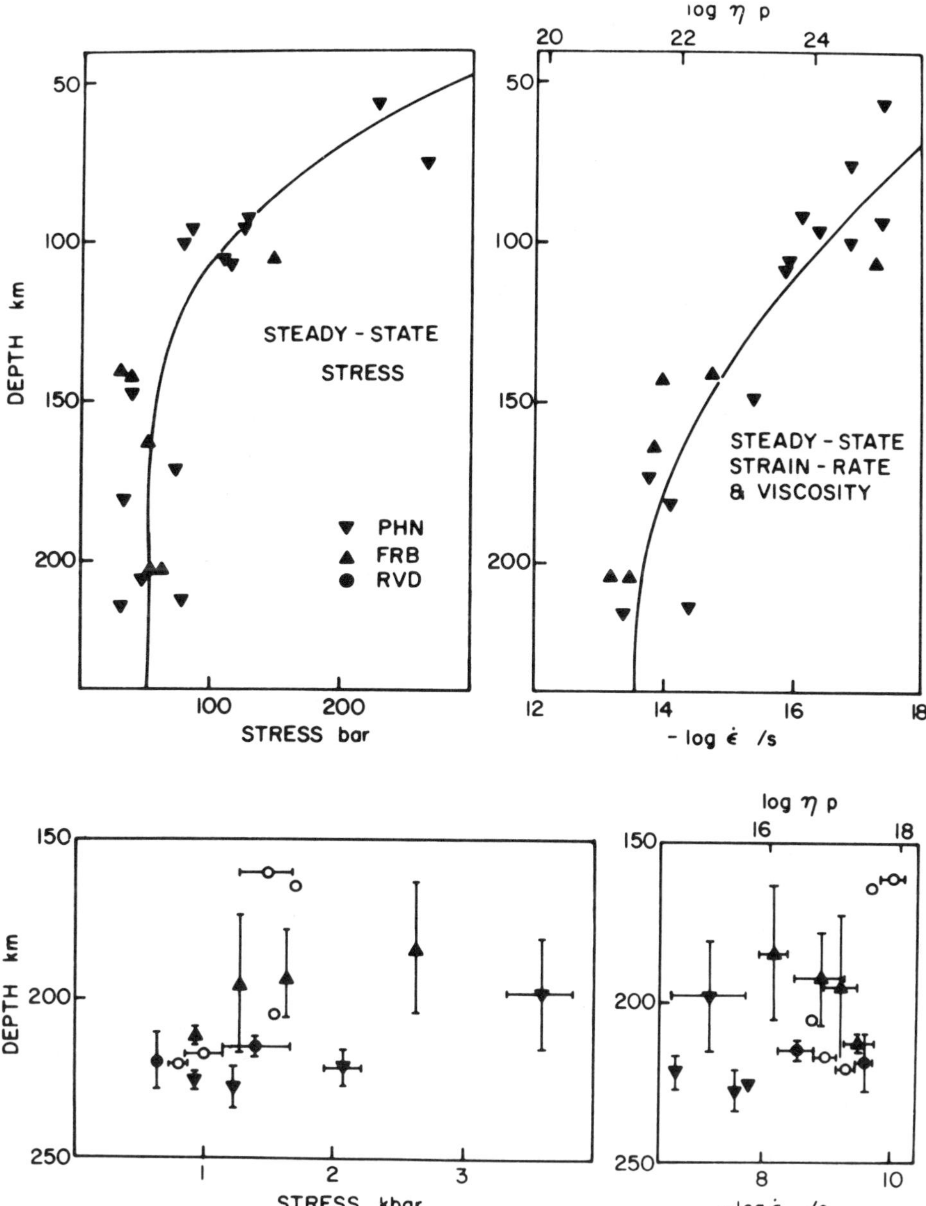

Fig. 9. Stress and strain-rate profiles based on textures of peridotite xenoliths in Premier-Mine, Thaba-Putsoa and Frank-Smith kimberlites (Southern Africa). Pressures and temperatures are based on Mercier's (1976a) solubility model for pyroxenes, stresses on equation 12, and strain-rates on Post and Griggs' (1973) dry-olivine flow-law, as yet the only relation to yield reasonable grain sizes for an asthenospheric strain rate in the range 10^{-15} to 10^{-13}/s (Mercier, 1977). The curves are optimum profiles calculated by iteration for a best fit of both stresses and strain rates simultaneously (Mercier, 1977). The viscosity scale is based on these two profiles. PHN (for P.H. Nixon's) samples from Thaba Putsoa, Lesotho, FRB (for F.R. Boyd's) samples from Frank Smith, South Africa, and RVD (for R.V. Danchin's) samples from Premier Mine, South Africa.

A. Steady-state stresses.

B. Steady-state strain-rates and equivalent-viscosities.

C. Late-deformation stresses. Depth error-bars represent combined analytical and phase equilibrium uncertainty (Mercier, 1976a; error unknown for open symbols) and stress error-bars the range of possible representative grain sizes.

D. Late-deformation strain-rates and equivalent-viscosities. Error bars as in C.

blast sizes are comparable, though slightly lower to those computed from subgrain sizes of coarse-grained types. It follows that both types of peridotites have seen comparable stress changes prior to sampling by the kimberlite, but the peridotite responded differently as a function of temperature. As strain increases with depth*, the porphyroclastic texture evolves into a sheared one, in which most of the olivine recrystallized dynamically. Enstatites start gliding parallel to {(100), [001]} and the most deformed ones commonly attain ratios of 1:6 to 1:8. The preferred orientation is typical of such high-strain textures, with strong maxima for both types of olivine and for enstatite, as a result of plastic flow and recrystallization.

Strained enstatite may ultimately recrystallize at both extremities and transform into extremely elongated enstatite aggregates, the texture and fabric remaining unchanged up to strains in the order of 1000%. The fabric is readily recognized in thin section by the maximum extinction for the whole rock, subparallel to the trace of the foliation and parallel to the flow direction. For the deepest samples and strains over 1000%, enstatite recrystallizes without appreciable flow of the paleoblasts. The preferred orientation for olivine (Mercier and Carter, 1975) is now completely different with no intermediate known, illustrating an entirely new deformation process. Under a polarizing microscope, an almost complete extinction of the olivine matrix is obtained with the foliation at 45° from the nicols. This unusual fabric has not yet been duplicated experimentally, but again corresponds to syntectonic recrystallization on the basis of the dislocation density and of detailed fabric analysis (Mercier and Carter, 1975).

The sheared-lherzolite facies being incompatible with the large-scale deformation involving relative displacement of domains of the upper mantle, are tentatively ascribed to late events occurring immediately prior to the sampling by the lava and related to its intrusion mechanism through the uppermost mantle. Considering a strain rate of 10^{-8}/s and temperatures on the order of 1400 to 1600°C, the olivine is likely to be entirely recrystallized in a matter of hours or days due to the fast kinetics of grain growth under such conditions. However, some of the sheared peridotites preserve a few olivine paleoblasts indicating a very short-lived deformation event (hours) in agreement with the maximum value of a few tens of years, estimated by combining strains derived from enstatites and calculated strain-rates for olivine, assuming an unlikely ($D_{en} << D_{ol}$) homogeneous deformation. The sheared textures were therefore produced immediately prior to sampling by the kimberlite, a conclusion also reached by Goetze (1975): the extremely high temperature combined with high stresses allowed deformation yielding strains above 100 percent in a few hours (at most). Such a penetrative deformation may account for the channel opening at depths, *i.e.* at high temperature. However, for rocks equilibrated at shallower levels the strength of the peridotite increases as the temperature decreases and ultimately, for the coarse-granular types, deformation was limited to narrow shear zones (observed in a few samples) or ultimately produced fractures (Mercier, 1977; Basu, 1978). As samples were broken off the walls along these fracture/shear zones, abrasion of the xenolith destroyed most of the evidence of such a process. Therefore, because of the deformation processes related to eruption-channel formation or sampling by the lava, the textures are a function of depth and if any large scale chemical or paragenetic zoning exists in the uppermost mantle, an apparent close relationship between chemistry and texture is expected, as actually observed in most kimberlites.

Attempts to model the eruption channel formation by using strain-rate estimates or eruption velocities yield unrealistically large widths for the perturbed zone, possibly as a result of transient conditions and heterogeneous deformations. However, applying stress estimates to scaled models provides useful constraints on the energy required for the late deformations which contributed to the conduit opening. Due to the rapidity of the intrusion process, the increase in stress with decreasing depth can hardly be explained by continuous total-force or energy increase and is more likely to indicate stress concentration and upward narrowing of the disturbed zone. Considering the case of Premier Mine, the stresses at the greatest recorded depths (*ca* 250 km) are about 350 bars and increase to 1250 bars near 170 km, which corresponds to a diameter reduction by a factor of about 2 for 80 km. Because of the meter-sized peridotite xenoliths recovered at some localities, the conduit should nowhere get smaller than 1 or 2 m, implying a minimum width of about 10 to 15m at 250 km depth for a deviatoric stress of 350 bars, *i.e.* a minimum force of 3 to 7 10^9N and an energy release in the range 2 10^{12} (homogeneous deformation) to 2 10^{15}J (boring process).

From the volume of material brought up, a value of about 5 10^{13} J may be reasonable for the minimum perturbed zone width, *i.e.* the actual energy release would be in the range 5 10^{13} to 5 10^{14} J. A kimberlite-eruption precursor might therefore be a deep-seated earthquake of magnitude 6 to 6.7 on the basis of Stacey's (1977) magnitude equation.

Conclusions

Experimental geopiezometers and preliminary growth-rate data for primary annealing-recrystal-

* with the exception of the low-temperature (900-1000°C) sheared xenoliths from Bultfontein; these rocks are excluded from the present discussion as their textures and the spatial relations between textural types recall those of two unusual spinel-facies xenolith series in basalts (Le Pouget, France; Mercier, 1972: San Quintin, Baja California; Basu, 1975; Mercier, 1977) and are not typical of kimberlites.

lization of olivine yield a few constraints on the dynamics and mechanisms of kimberlite eruption when applied to peridotite xenoliths. High-stress sheared textures characteristic of the high-temperature deepest facies would form within a few hours (at most) prior to sampling by the rising magma, this deformation contributing to the conduit formation. The energy released during this process would be in the range 0.1 to 0.01 megaton ($5\ 10^{13}$ to $5\ 10^{14}$ J), and corresponds to eathquake magnitudes 6 to 6.7. Late stresses which triggered the eruption are about 20 times larger than steady-state stresses for convection at any depth. Thus, the perturbed zone would increase in width with depth, doubling every 80 km, to attain an estimated 15 to 50 m width near 250 km. Porphyroclastic xenoliths offer a wealth of information including original temperature and thermal evolution of the kimberlite (under study) and its pre-supercooling velocity (40 to 70 km/h, depending on the thermal path assumed). This last type of information may be the most important as it is a potential prospecting-criterion that might be linked to the relative abundance of diamonds.

Acknowledgements. The author is especially indebted to F.R. Boyd and R.V. Danchin for free access to their xenolith collections and communication of unpublished data, to N.L. Carter, F.R. Boyd, D.L. Kohlstedt and D.A. Anderson for their critical reviews, and to C. Ronnat-Mercier for the illustration and typing. This research was supported by National Science Foundation Grant # EAR 76-04129 to N.L. Carter.

References

Artyushkov E.V. and S.V. Sobolev, Mechanism of the uplift of kimberlite magma from the depth, Extended Abstracts, Second International Kimberlite Conference, Santa Fe, New Mexico, 1977.

Basu A.R., Petrogenesis of xenoliths from the San-Quintin volcanic field, Baja California, Ph.D. Thesis, Univ. California at Davis, California, 1975.

Basu A.R., Jointed blocks of peridotite xenoliths in basalt: implications for mantle dynamics, Geology, in press, 1978.

Boullier A-M., Structure des peridotites en enclaves dans les kimberlites d'Afrique du Sud, These 3e Cycle, Nantes, France, 1-147, 1975.

Boullier A-M. and A. Nicolas, Texture and fabric of peridotite nodules from kimberlite at Mothae, Thaba Putsoa and Kimberley. In "Lesotho Kimberlites", P.H. Nixon ed., L.N.D.C., Maseru, Lesotho, 57-66, 1973.

Boullier A-M. and A. Nicolas, Classification of textures and fabrics of peridotite xenoliths from South African kimberlites. Phys. Chem. Earth 9, 467-476, 1975.

Boyd F.R., Garnet peridotites and the system $CaSiO_3$-$MgSiO_3$-Al_2O_3, Mineral. Soc. Amer. Sp. Paper 3, 65-75, 1970.

Boyd F.R., The pyroxene geotherm. Geochem. Cosmoch. Acta 37, 2533-2546.

Boyd F.R., Stress heating and compositional variations in enstatites from sheared lherzolites. Carnegie Inst. Washington Yearb. 74, 525-528, 1975.

Boyd F.R. and P.H. Nixon, Ultramafic nodules from the Thaba Putsoa kimberlite pipe. Carnegie Inst. Washington Yearb. 71, 363-373, 1972.

Boyd F.R. and P.H. Nixon, Structure of the upper mantle beneath Lesotho. Carnegie Inst. Washington Yearb. 72, 431-445, 1973.

Cahn R.W., Recovery and recrystallization. In "Physical Metallurgy", R.W. Cahn ed., Elsevier, New York, 1129-1198, 1970.

Carslaw H.S. and J.C. Jaeger, Conduction of heat in solids, 2nd ed., Oxford Univ. Press, London, 1-510, 1959.

Carter N.L., High temperature flow of rocks, Rev. Geophys. Space Phys. 13, 344-349, 1975.

Carter N.L. and H.G. Ave Lallemant, High temperature flow of dunite and peridotite. Geol. Soc. Amer. Bull. 81, 2181-2202, 1970.

Christian J.W., Phase transformations. In "Physical Metallurgy", R.W. Cahn ed., Elsevier, New York, 471-531, 1970.

Clark S.P.Jr and A.E. Ringwood, Density distribution and constitution of the mantle. Rev. Geophys. Space Phys. 2, 35-88, 1964.

Dawson J.B., Basutoland kimberlites. Geol. Soc. Amer. Bull. 73, 545-560, 1962.

Durham W.B., C. Goetze and B.Blake, Plastic flow of oriented single crystals of olivine, Part II: observations and interpretations of the dislocation structures. Tectonophys. 23, 1977.

Etheridge M.H. and S.H. Kirby, Experimental deformation of rock-forming pyroxenes: recrystallization mechanisms and preferred-orientation development (abstract). Eos 58, 513, 1977.

Feltham P., Grain growth in metals, Acta Met. 5, 97-105, 1957.

Fine M.E., Introduction to phase transformations in condensed systems. MacMillan Co., New York, 1-133, 1965.

Goetze C., Sheared lherzolites: from the point of view of rock mechanics. Geology 3, 172-173, 1975.

Green H.W. and Y. Gueguen, Origin of kimberlite pipes by upwelling in the upper mantle. Nature 249, 617-619, 1974.

Griggs D.T., Hydrolytic weakening of quartz and other silicates. Geophys. J. Roy. Astron. Soc. 14, 19-31, 1967.

Harte B., Rock nomenclature with particular relation to deformation and recrystallization textures in olivine-bearing xenoliths, J. Geol. 85, 279-288, 1977.

Kohlstedt D.L., C. Goetze and W.B. Durham, Experimental deformation of single-crystal olivine with application to flow in the mantle. In "The Physics and Chemistry of Minerals and Rocks", R.G.J. Strens ed., Wiley, New York, 35-49, 1976.

Kohlstedt D.L., C. Goetze, W.B. Durham and J.B. Van der Sande, A new technique for decorating dislocations in olivine, Science 191, 1045-1046, 1976.

Kushiro I., Partial melting of garnet lherzolites

from kimberlite at high pressures, In "Lesotho Kimberlites", P.H. Nixon ed., L.N.D.C., Maseru, Lesotho, 294-299, 1973.

Lutton M.J. and C.M. Sellars, Dynamic recrystallization in nickel and nickel-iron alloys during high-temperature deformation, Acta Met. 17, 1033-1043, 1969.

McCallister R.H., H.O.A. Meyer and R. Aragon, Thermal history and exsolution microstructures of two clinopyroxenes from the Thaba Putsoa kimberlite pipe, Lesotho, Extended Abstracts, Second International Kimberlite Conference, Santa Fe, New Mexico, 1977.

McGetchin T.R. and G.W. Ullrich, Xenoliths in maars and diatremes with inferences for the Moon, Mars and Venus, J. Geophys. Res. 78, 1833-1853, 1973.

McGetchin T.R., Y.S. Nikhanj and A.A. Chodos, Carbonatite-kimberlite relations in the Cane-Valley diatreme, San Juan County, Utah, J. Geophys. Res. 78, 1854-1869, 1973.

MacGregor I.D., The sytem $MgO-Al_2O_3-SiO_2$: Solubility of Al_2O_3 in enstatite for spinel and garnet peridotite compositions, Amer. Mineral. 59, 110-119, 1974.

MacGregor I.D., Petrologic and thermal structure of the upper mantle beneath South Africa in the Cretaceous. Phys. Chem. Earth 9, 455-466, 1975.

Mercier J-C.C., Structure des peridotites en enclaves dans quelques basaltes d'Europe et d'Hawaii. Regards sur la constitution du manteau superieur. These Doct. 3^{e} Cycle, Nantes, France, 1-229, 1972.

Mercier J-C.C., Single-pyroxene geothermometry and geobarometry, Amer. Mineral. 61, 603-615, 1976a.

Mercier J-C.C., Average velocities for kimberlite eruptions (abstract), Eos 57, 762, 1976b.

Mercier J-C.C., Natural peridotites: chemical and rheological heterogeneity of the upper mantle. Ph.D. Thesis, S.U.N.Y. Stony Brook, New York, 1-669, 1977.

Mercier J-C.C., Single-pyroxene geothermometry and geobarometry: a reappraisal. J. Geophys. Res., in press, 1978.

Mercier J-C.C. and N.L. Carter, Pyroxene geotherms, J. Geophys. Res. 80, 3349-3362, 1975.

Mercier J-C.C. and A. Nicolas, Textures and fabrics of upper-mantle peridotites as illustrated by xenoliths from basalts. J. Petrol. 16, 454-487, 1975.

Mercier J-C.C., D.A. Anderson and N.L. Carter, Stress in the lithosphere: inference from steady-state flow of rocks, Pure Appl. Geophys. (PAGEOPH) 115, 199-226, 1977.

Nicolas A., J-L. Bouchez, F. Boudier and J-C.C. Mercier, Textures, structures and fabrics due to solid flow in some European lherzolites. Tectonophys. 12,55-86, 1971.

Nixon P.H., F.R. Boyd and A-M. Boullier, The evidence of kimberlite and its inclusions on the constitution of the Earth. In "Lesotho Kimberlites", P.H. Nixon ed., L.N.D.C., Maseru, Lesotho, 312-318, 1973.

Parmentier E.M. and D.C. Turcotte, An explaination of the pyroxene geotherm based on plume convection in the upper mantle. Earth Planet. Sc. Lett. 24, 209-212, 1974.

Poirier J-P. and A. Nicolas, Deformation-induced recrystallization due to progressive misorientation of subgrains with special reference to mantle peridotites, J. Geol. 83, 707-720, 1975.

Post R.L. Jr and D.T. Griggs, The Earth's mantle: evidence of non-Newtonian flow. Science 118, 1242-1244, 1973.

Raleigh C.B. and S.H. Kirby, Creep in the upper mantle. Mineral. Soc. Amer. Spec. Paper 3, 113-121, 1970.

Ross J.V., H.G. Ave Lallemant and N.L. Carter, Stress dependence of recrystallized grain and subgrain size in olivine. J. Geophys. Res., in press, 1978.

Stacey F.D., Physics of the Earth (2nd ed.), Wiley, New York, 1-414, 1977.

Streikeisen A.L., Plutonic rocks: classification and nomenclature recommended by the IUGS subcommission on the systematics of igneous rocks, Geotimes 18, 26-30, 1973.

Taylor G.I. and H. Quinney, Proc. Roy. Soc. 163A, 157, 1937.

Twiss R.J., Theory and applicability of a recrystallized grain-size paleopiezometer. Pure Appl. Geophys. (PAGEOPH) 115, 227-244, 1977.

Wagner P.A., The diamond fields of Southern Africa (reed. 1971), Struik Ldt, Cape Town, 1-355, 1914.

III. MEGACRYSTS

MEGACRYST ASSEMBLAGES IN KIMBERLITE FROM NORTHERN COLORADO AND SOUTHERN WYOMING: PETROLOGY, GEOTHERMOMETRY-BAROMETRY, AND AREAL DISTRIBUTION

David H. Eggler

Geophysical Laboratory, Carnegie Institution of Washington, Washington, D. C. 20008 and The Pennsylvania State University, University Park, Pennsylvania 16802

M. E. McCallum

Department of Earth Resources, Colorado State University, Fort Collins, Colorado 80523 and U.S. Geological Survey, Denver, Colorado 80225

C. B. Smith

Department of Earth Resources, Colorado State University, Fort Collins, Colorado 80523

Abstract. Two compositionally distinct suites of megacrysts (crystals > 1 cm diameter) occur in kimberlites of the Colorado-Wyoming State Line and Iron Mountain, Wyoming districts, U.S.A. One suite is Cr-rich and one Cr-poor. The megacrysts are garnet, clinopyroxene, orthopyroxene, olivine, ilmenite, and clinopyroxene-ilmenite intergrowths. The Cr-poor megacrysts are characterized by lower, more variable Mg/(Mg + Fe) than their Cr-rich counterparts and by somewhat higher TiO_2. The Cr-rich garnet, olivine, and orthopyroxene megacrysts, with relatively high and constant Mg/(Mg + Fe), are similar chemically to comparable minerals from lherzolites, both granular and sheared. Clinopyroxenes from lherzolites, however, contain more jadeite component and somewhat more Cr_2O_3 than Cr-rich clinopyroxene megacrysts.

Temperatures and pressures of equilibration of orthopyroxene megacrysts from the Cr-rich suite occur in an array that is interpreted as a steady state geothermal gradient, a gradient also defined by peridotite nodules. Equilibration temperatures of some Cr-poor orthopyroxenes also fall in that array, but temperatures of other Cr-poor orthopyroxenes occur in a subarray of higher, perturbed temperatures. The Cr-poor megacrysts are interpreted to have crystallized from partial melts (or fractionated derivatives of partial melts) of relatively undepleted, peridotitic diapirs. The Cr-rich megacrysts, as well as an anomalous group of lower-temperature, relatively magnesian Cr-poor megacrysts, are interpreted to have crystallized from liquids derived by melting of a mixture of undepleted peridotite and depleted, lithospheric wallrocks.

Two clusters of pipes in the State Line district and the third cluster of pipes in the Iron Mountain district each contain suites of Cr-poor megacrysts that have distinctive chemical compositions. These distinctions are easiest to understand if distinct conduit systems for these clusters of pipes extended to 140-200 km, the depths of origin of the megacrysts.

Introduction

Kimberlite pipes and dikes in the State Line district, Colorado-Wyoming, and the Iron Mountain district, Wyoming (Fig. 1) contain a variety of nodular xenoliths derived from the upper mantle and the lower crust. These nodules have been described previously (Eggler and McCallum, 1974, 1976; McCallum *et al.*, 1975; McCallum and Eggler, 1976) and have been assembled into an approximate stratigraphic column (Fig. 2). On the basis of abundance of nodules, the upper mantle section is thought to have consisted primarily of garnet peridotite (lherzolite and harzburgite), which underwent a phase transition, just below the lower crust, to spinel peridotite. One garnet peridotite nodule contains diamonds. These peridotites have been depleted of less refractory elements and enriched in more refractory elements, such as Mg and Cr, relative to a more primitive upper

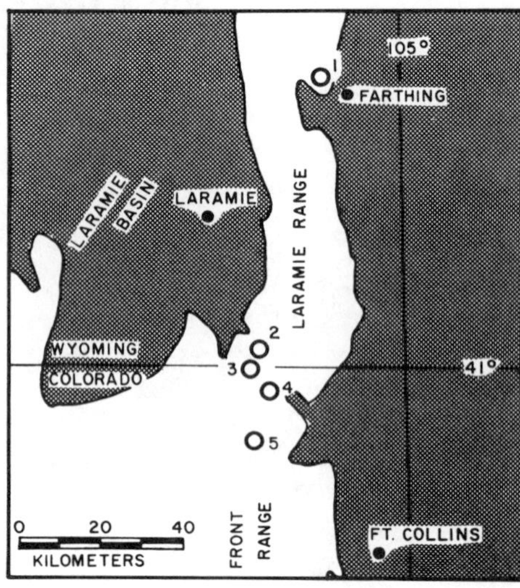

Fig. 1. Locations of clusters of Colorado-Wyoming kimberlite pipes and dikes: 1. Iron Mountain district; 2. Ferris (1-2) group and the Aultman pipe; 3. Schaffer (1-16) group; 4. Nix (1-4) and Moen (1-4) groups. 5. Sloan 1 and 2 pipes. The stippled areas are Paleozoic and younger rocks, and the clear areas are predominantly Precambrian rocks.

mantle material (ref. Boyd and Nixon, 1973). Depletion presumably occurred during previous episodes of partial melting in Precambrian time. These lherzolites and harzburgites have been recovered from nearly all of the Colorado-Wyoming pipes, although only xenoliths from the Sloan and Nix groups of pipes contain unaltered orthopyroxenes and olivines, in addition to the garnets and clinopyroxenes that have survived alteration in nearly all xenoliths. The textures of the depleted peridotites range from granular to sheared, although in all the highly-sheared (fluidal mosaic) nodules, orthopyroxene and olivine have not survived near-surface alteration. Other mantle-derived xenoliths include eclogites and a group of garnet clinopyroxenites, garnet websterites, spinel websterites, and olivine garnet websterites, called the websterite group (McCallum and Eggler, 1976).

Another group of nodular xenoliths abundant at many of the pipes consists of single crystal megacrysts (> 1 cm and up to 13 cm). Many megacrysts are rounded to subangular; surfaces of such megacrysts are polished and may also have grooves, striations, and pits that were developed during emplacement of kimberlite. Some megacrysts have shattered, however, along internal fractures, parting planes, or exsolution lamellae, producing angular fragments. Each megacryst is invariably homogeneous chemically, insofar as electron microprobe analysis can detect, although occasionally small mineral inclusions are present. From the compositions of megacrysts and, where available, inclusions, two suites have been distinguished, one relatively Cr-rich and one relatively Cr-poor (Eggler and McCallum, 1974, 1976). The Cr-poor suite consists of olivine, orthopyroxene, clinopyroxene, garnet, and ilmenite. Associated with the Cr-poor suite are clinopyroxene-ilmenite intergrowths. The Cr-rich suite consists of olivine, orthopyroxene, clinopyroxene, and garnet.

Concentration of megacrysts is different in different pipes, both in terms of relative abundance and of type. These differences in concentration are in part attributable to differential weathering (e.g., orthopyroxene megacrysts have survived near-surface alteration only at the Sloan and Nix pipes). The order of relative abundance of megacryst minerals (Cr-rich minerals indicated by Cr subscript) is, in the State Line district: garnet $\simeq$ ilmenite > garnet$_{Cr}$ > cpx$_{Cr}$ > cpx > opx$_{Cr}$ > opx > olivine > olivine$_{Cr}$ > cpx-ilmenite intergrowths; in the Iron Mountain district: ilmenite > garnet > cpx > cpx-ilmenite intergrowths > garnet$_{Cr}$.

Quite apart from their scientific interest, megacrysts are important because they have been frequently used as tracer minerals in locating kimberlite pipes. Cr-poor megacrysts have been described from Lesotho (Boyd and Nixon, 1973, 1975; Nixon and Boyd, 1973; Bloomer and Nixon, 1973; Rolfe, 1973), South Africa (e.g., Boyd and Dawson, 1972; Boyd and Nixon, 1975; Danchin and Boyd, 1976), and the USSR (Frantsesson, 1970;

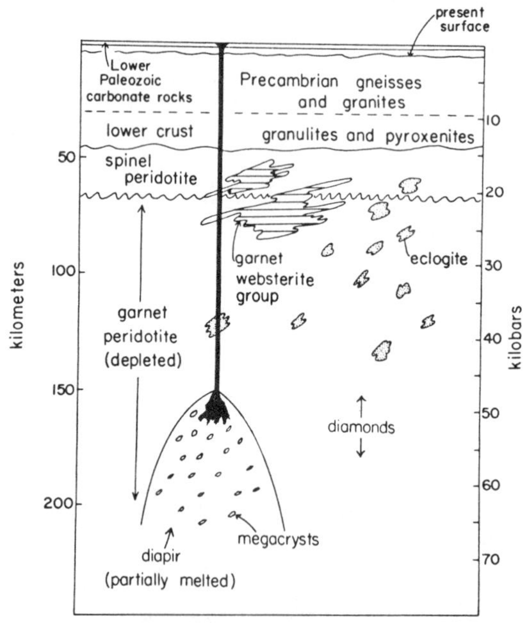

Fig. 2. A schematic cross section of the crust and upper mantle beneath Colorado-Wyoming at the time of kimberlite eruptions (Devonian), modified from McCallum and Eggler (1976). The diapir may have been upwelling or may have been associated with convective overturn.

Boyd et al., 1976). Relative to Lesotho and South African megacrysts, Colorado-Wyoming megacrysts have higher Cr_2O_3 contents and slightly higher Mg/(Mg + Fe). Colorado-Wyoming pipes also do not contain the very subcalcic clinopyroxenes (presumed to coexist with very calcic orthopyroxenes) that equilibrated at temperatures > 1300°C and that are characteristic of Lesotho (Nixon and Boyd, 1973) and the Monastery Mine (Gurney et al., 1977). Cr-rich megacrysts are virtually unknown outside of Colorado-Wyoming, although Cr diopsides have now been reported from Orapa, Botswana (Shee and Gurney, 1977).

Petrography and Chemistry of Megacrysts

Analytical Methods

In this paper, the chemical analyses, with the exception of analyses of rare earths, were performed on the MAC 400 automated microprobe at the Geophysical Laboratory, using the methods of Finger and Hadidiacos (1972). Analyses of major elements are believed to be accurate to within 2%, relative. Although total iron is analyzed on a microprobe, proportions of ferrous and ferric iron can be calculated by assuming cation and anion stoichiometry (i.e., 2 cations and 3 oxygens in ilmenite) (Finger, 1972). Although the amount of ferric iron in mantle-derived silicates is typically so small that it is within the uncertainty limits of analysis, the amount of ferric iron in ilmenites is statistically significant (Finger, 1972).

Garnet

The typical size of the Cr-poor garnet megacrysts is 1-4 cm, although the group includes the largest (13 cm) megacrysts found. They are typically reddish-brown, although some smaller megacrysts are orange. Cr-rich megacrysts, which rarely exceed 3 cm, range in color from lavender to deep purple or black. Both Cr-poor and Cr-rich varieties are typically fractured. The Cr-poor garnets commonly have weakly-developed shear planes. Incipient alteration is present along fractures or parting planes and along rims of some crystals.

The ranges of chemical parameters of garnets are displayed in Table 1 and Figures 3 and 4. The Cr-rich megacrysts are more calcic than the Cr-poor (Fig. 3), reflecting higher Cr_2O_3 contents, have higher and much more restricted Mg/(Mg + Fe) (Figs. 3 and 4), and are somewhat more titaniferous.

Clinopyroxene

The Cr-rich clinopyroxene megacrysts are emerald green, whereas Cr-poor clinopyroxenes are bluish-gray to gray-green. Both types range up to about 3 cm, but the Cr-poor varieties are generally larger, probably because they are less susceptible to shattering: Cr-rich megacrysts are typically angular fragments of originally larger nodules, whereas Cr-poor megacrysts are well-rounded and ellipsoidal to ovoid (although 1 or 2 cleavage faces may be present). Shattering of Cr-rich megacrysts was accompanied, moreover, by partial alteration, along cleavage or parting planes, to mixtures of serpentine, chlorite, and calcite. Alteration in the Cr-poor megacrysts was confined to films that coat fractures, cleavage planes, and rounded portions of nodules.

Chemical analyses of the clinopyroxenes are shown in Table 1 and Figures 3 and 5. Like the garnets, Cr-rich clinopyroxenes have higher, more restricted Mg/(Mg + Fe) (Figs. 3 and 5) and lower TiO_2 (Table 1) than Cr-poor clinopyroxenes.

TABLE 1. Ranges of Chemical Parameters of Megacrysts and of Minerals in Peridotites.

	no.	Ca/(Ca+Mg)	Mg/(Mg+Fe)	TiO_2	Cr_2O_3	Na_2O
		mol ratios		wt percent		
Megacrysts						
Cr-rich garnets	34	0.14 -0.27	0.818-0.841	0.22-0.94	6.3 -13.0	0.00-0.09
Cr-poor garnets	50	0.13 -0.22	0.683-0.836	0.23-1.3	0.03- 4.8	0.00-0.12
Cr-rich clinopyroxenes	25	0.41 -0.48	0.920-0.931	0.09-0.22	0.83- 2.4	0.9 -1.6
Cr-poor clinopyroxenes	91	0.36 -0.47	0.826-0.908	0.18-0.48	0.08- 1.0	1.0 -1.7
Cr-rich orthopyroxenes	73	0.016-0.026	0.923-0.934	0.01-0.16	0.39- 0.69	0.07-0.20
Cr-poor orthopyroxenes	19	0.013-0.029	0.886-0.921	0.08-0.28	0.06- 0.31	0.10-0.25
Cr-rich olivines	3		0.906-0.919	0.00-0.00	0.06- 0.11	
Cr-poor olivines	6		0.878-0.898	0.02-0.05	0.01- 0.04	
ilmenites (Cr-poor)	85		0.079-0.651		0.01-10.6	
Lherzolite-harzburgite						
garnets	29	0.14 -0.25	0.817-0.859	0.07-1.2	3.5 - 13.3	0.00-0.15
clinopyroxenes	25	0.40 -0.47	0.909-0.932	0.07-0.43	1.0 - 2.7	1.0 -2.3
orthopyroxenes	8	0.010-0.024	0.918-0.930	0.04-0.19	0.25- 0.66	0.07-0.24
olivines	8		0.903-0.919	0.00-0.05	0.04- 0.11	

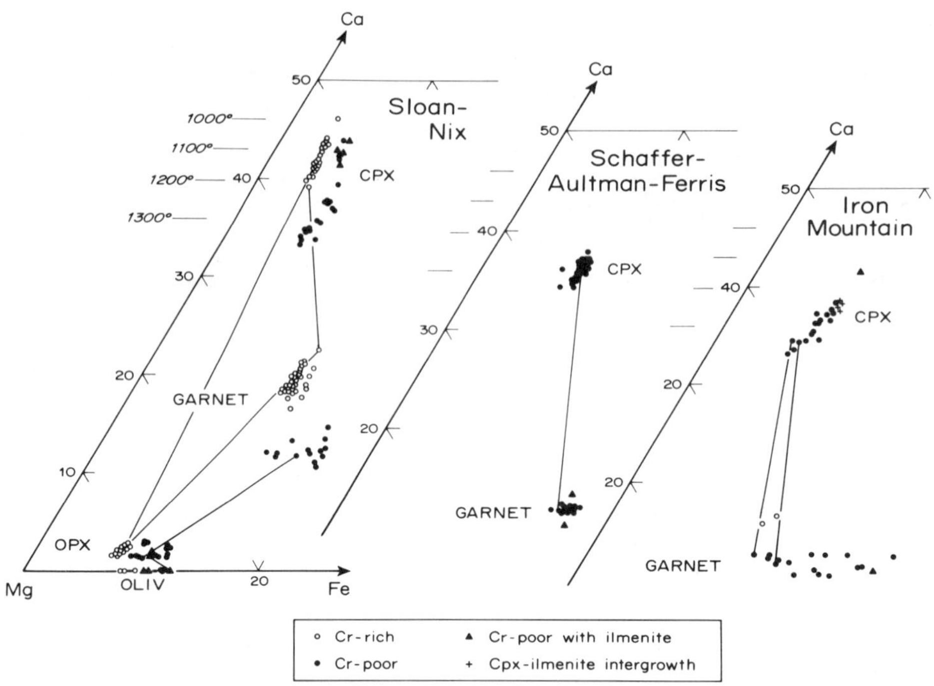

Fig. 3. Compositions of clinopyroxene, garnet, orthopyroxene, and olivine megacrysts from groups of Colorado-Wyoming kimberlite occurrences, plotted in a portion of the Ca-Mg-Fe (mol) projection. The Davis-Boyd (1966) pyroxene solvus is shown on the Ca-Mg tieline. Lines connect compositions of megacryst hosts and inclusions.

The Cr-poor clinopyroxenes extend to more subcalcic compositions (Fig. 3), reflecting higher temperatures of equilibration.

Orthopyroxene

Orthopyroxene megacrysts have been recovered virtually exclusively from the Sloan 2 pipe in Colorado (Fig. 1), where the Cr-rich type is about four times as abundant as the Cr-poor. Cr-rich orthopyroxenes are typically bright to pale green, in contrast to gray-green to gray-brown Cr-poor varieties. Megacrysts of both types range up to about 3 cm in maximum dimension. Orthopyroxenes are less rounded than other types of megacrysts. This angularity is apparently a function of cleavage-controlled fragmentation during emplacement, producing "cockscomb" terminations and ridges on the surfaces of megacrysts, as illustrated by Frantsesson (1970) and described by Bloomer and Nixon (1973, p. 29) and Nixon and Boyd (1973, p. 68). Most of the orthopyroxene nodules are free of alteration, although a few samples exhibit minor serpentine and calcite along cleavage planes.

Chemical analyses of orthopyroxenes are shown in Table 1 and Figure 6. The trends in Mg/(Mg + Fe) and TiO_2 reflect trends in the garnets and orthopyroxenes.

Olivine

Olivine megacrysts that are single crystals are relatively rare. The megacrysts, which generally do not exceed 2 cm in size, are typically pale greenish-brown to yellow-brown. They are well rounded where not broken along planes of conchoidal fracture. Considerably more common are dunite nodules, with aggregate textures, that are interpreted to be recrystallized olivine megacrysts. These nodules, ranging up to 3 cm in size, are typically pale yellow-brown and intensely fractured. Polyhedral grains of olivine comprising the aggregates are generally less than 1 mm in diameter. In many nodules the aggregates form a granulated, mosaic matrix containing olivine porphyroclasts as much as 1.5 cm in size; this relationship is similar to that of granulated olivine in the sheared lherzolites of southern Africa (Nixon and Boyd, 1973, p. 74; Boyd and Nixon, 1975, p. 443), and aggregate mosaic textures present in many ilmenite nodules from Colorado-Wyoming and African kimberlites. Megacrysts and dunites have been recovered only from the Sloan 2 pipe.

Analyzed olivines (Table 2) include both single crystals, aggregate dunites (2 Cr-poor, 1 Cr-rich), and inclusions in other megacrysts. Chemical compositions of the dunites are comparable to those of single crystals and inclu-

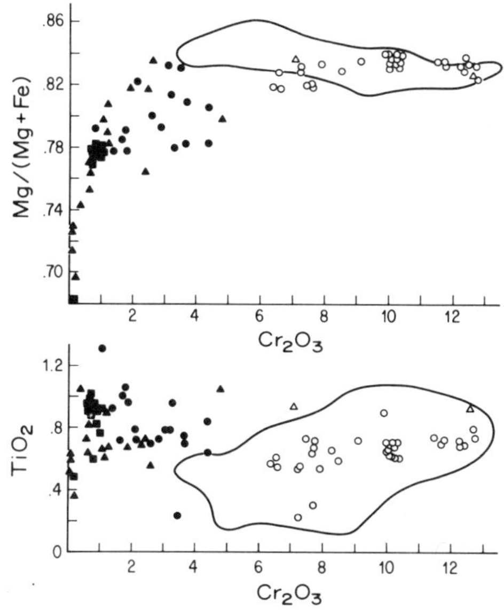

Fig. 4. Compositions of garnet megacrysts. Mg/(Mg + Fe) is a mol ratio; TiO_2 and Cr_2O_3 are wt%. The circled fields delimit compositions of garnets from Colorado-Wyoming lherzolites and harzburgites, with the exception of a diamondiferous peridotite discussed by Eggler and McCallum (1976). Circles denote megacrysts from the Sloan-Nix pipes, squares from the Schaffer-Aultman-Ferris pipes, and triangles from the Iron Mountain occurrences. Filled symbols denote Cr-poor megacrysts and open circles Cr-rich megacrysts.

sions. Note that Cr-poor olivines, while they are less magnesian, are also actually poorer in chrome.

Ilmenite

Ilmenite megacrysts are typically 1-3 cm in maximum dimensions, although one sample exceeds 13 cm. The megacryst suite includes both single crystals and polycrystalline aggregates, the latter ranging from equigranular to inequigranular with polygonal, interlobate or amoeboid granoblastic textures. These textures are similar to those observed by Frick (1973) and Mitchell (1973) in South African ilmenite nodules.

Ilmenite megacrysts are considerably less fractured than associated garnet megacrysts; in some samples grain boundaries and fracture planes are filled with quartz, calcite and/or hematite, and more rarely with small grains of covellite, chalcopyrite, or pyrite.

Although Cr_2O_3 contents of ilmenites reach 10.6 wt% (Fig. 7), all the ilmenites appear to be associated with Cr-poor silicates. In particular, the two ilmenites with highest Cr_2O_3 (Fig. 7) contain inclusions of olivine whose Mg/(Mg + Fe) (0.896-0.898) fall in the Cr-poor group: consider that an olivine with Mg/(Mg + Fe) = 0.898 would be in equilibrium with an orthopyroxene with Mg/(Mg + Fe) = 0.912, based upon a partition coefficient calibrated from peridotite nodules and from an orthopyroxene-olivine inclusion pair from in an ilmenite host (Fig. 3); an orthopyroxene with Mg/(Mg + Fe) = 0.912 is clearly Cr-poor (Table 1).

The parabolic relationship between Cr_2O_3 and MgO (Fig. 7) is remarkably similar to the relationship observed by Haggerty (1975, p. 304) in ilmenites from African kimberlites.

Clinopyroxene-Ilmenite Intergrowths

Megacrysts of intergrown clinopyroxene and ilmenite have been found at four localities in the Wyoming Iron Mountain district (IM4, 7, 8 and 26 pipes) and at one pipe in Colorado (Sloan 2). These megacrysts represent the second authenticated occurrence of such intergrowths in North America (Smith et al., 1976), similar nodules having been reported from Riley County, Kansas, by Brookins (1971), Gurney et al. (1973), and McCallister et al. (1975). The megacrysts are similar to those described from a number of African kimberlite localities (e.g., Wagner, 1914; Williams, 1932; MacGregor et al., 1937; Dawson and Reid, 1970; Boyd, 1971; Boyd and Dawson, 1972; Boyd and Nixon, 1975; Gurney

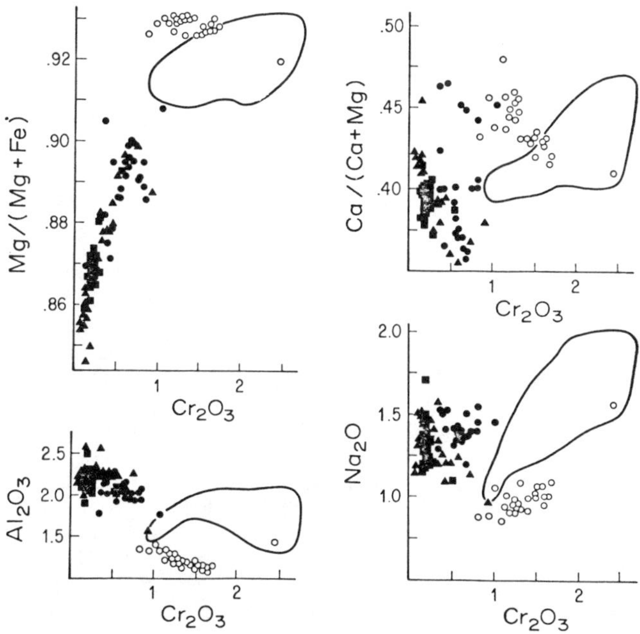

Fig. 5. Compositions of clinopyroxene megacrysts. Mg/(Mg + Fe) and Ca/(Ca + Mg) are mol rations; all other parameters are wt%. The circled fields delimit compositions of clinopyroxenes from Colorado-Wyoming lherzolites and harzburgites. Symbols are the same as on Figure 4.

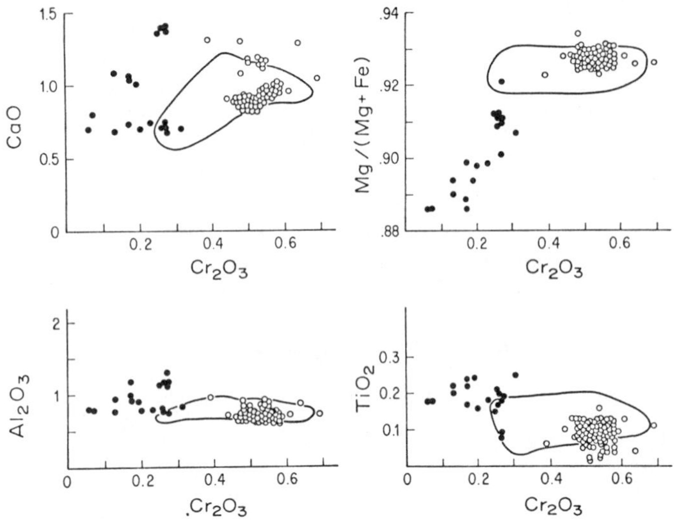

Fig. 6. Compositions of orthopyroxene megacrysts. Mg/(Mg + Fe) is a mol ratio; other parameters are wt. %. Circled fields delimit compositions of orthopyroxenes from Colorado-Wyoming lherxolites and harzburgites. Symbols are the same as on Figure 4.

et al., 1973) and from the USSR (e.g., Frantsesson, 1970; Ilupin et al., 1974).

One graphic, one granular, and 26 lamellar intergrowths have been collected. Pyroxene is completely altered to a mixture of serpentine, chlorite, and calcite in most of the intergrowths, but five intergrowths contain unaltered diopside. Ilmenite is slightly altered to hematite along fractures and pyroxene-ilmenite interfaces. The nodules are as large as 4.5 cm in maximum dimension, but most are approximately 1 cm in size.

The lamellar intergrowths are composed of ilmenite platelets (0.5-1.0 mm thick) and rods regularly distributed in a single-crystal diopside host (Smith, 1977). The ranges of compositions among minerals in these intergrowths are remarkably restricted (clinopyroxenes - Fig. 3 and ilmenites - Fig. 7). The mineral compositions (e.g., nodule IM7-IC1, Table 2) are similar to those in lamellar intergrowths from Riley County, Kansas and Monastery Mine, South Africa (Gurney et al., 1973, p. 244). Compositions of clinopyroxenes fall in a group of very Fe-rich, Cr-poor megacrysts from the Iron Mountain district (Figs. 3 and 5). Ilmenites are more magnesian and contain less Fe_2O_3 than other ilmenite megacrysts from the Iron Mountain district, although considerably more magnesian ilmenites are abundant in the pipes of the State Line District (Fig. 7).

The graphic intergrowth (IM4-IC1, Table 2) consists of a single (?) altered diopside (?) crystal with ilmenite lamellae that intersect one another at angles of 120 degrees, producing an open V-shaped pattern that is similar to that described by Dawson and Reid (1970) in a graphic intergrowth from the Monastery Mine, South Africa. Ilmenite in the nodule has a distinctive composition compared to ilmenite in lamellar intergrowths (Table 2).

The granular intergrowth, from the Sloan 2 pipe, is the only pyroxene-ilmenite nodule recovered from the State Line district. This nodule is characterized by somewhat rounded, irregular-shaped grains of ilmenite in an altered diopside (?) crystal. No chemical data are available.

Mineral Inclusions

A number of megacrysts of both the Cr-rich and Cr-poor mineral suites contain small rounded to irregular-shaped mineral inclusions (Table 3). It is evident from the configuration of host-inclusion tielines in Fig. 3 that the compositions of these hosts and inclusions are completely consistent -- i.e., a Cr-poor, Fe-rich host contains a Cr-poor, Fe-rich inclusion. This consistency strongly suggests that the minerals in each suite coexisted in chemical equilibrium.

The host-inclusion relations listed in Table 3 are not necessarily representative of the entire population of megacrysts, because the group of megacrysts analyzed was selected from the population collected with a bias towards inclusions. The frequencies of occurrence within specific minerals are probably reasonably accurate -- that is, Cr-poor garnet and ilmenite appear to have more inclusions than other types of minerals, whereas no inclusions have been observed in clinopyroxenes or olivines.

Exsolved Megacrysts

Several pyroxene megacrysts with coarse lamellae (0.1 - 1.0 mm thick) of clinopyroxene or orthopyroxene and garnet have been collected from the Sloan, Nix, and Moen groups of pipes. In addition to garnet, some contain spinel. These nodules are grey and as much as 3 cm in size. They exhibit the same "cockscomb" surface texture present on many orthopyroxene megacrysts. Clinopyroxenes and garnets in the nodules are low in Cr_2O_3 (0.07-0.29 and 0.07-0.39 wt %, respectively); clinopyroxenes are aluminous (3.1 - 5.9 wt % Al_2O_3), and the Ca/(Ca + Mg) (0.505 - 0.513) reflect very low temperatures of exsolution. The nodules are relatively magnesian (Mg/(Mg + Fe) in orthopyroxene = 0.890 - 0.910). The bulk composition of these nodules, as best as it can be reconstructed, does not correspond to any of the single-crystal megacrystic pyroxenes.

A clinopyroxene megacryst, 4 cm in diameter and containing exsolution lamellae of orthopyroxene and garnet, has been described by Boyd et al. (1976, p. 528) from the Obnazhënnaya pipe, USSR. The texture of the nodule and composi-

TABLE 2. Compositions of Megacrysts from Colorado-Wyoming Kimberlite Pipes.

Type of Megacrysts[2]	Analyses in weight percent[1]												
	SiO_2	TiO_2	Al_2O_3	Cr_2O_3	Fe_2O_3	FeO	MnO	NiO	MgO	CaO	Na_2O	K_2O	total
Cr-rich garnets													
SD2-G20	40.7	0.70	14.9	10.2	--	6.7	0.39	0.01	19.0	7.6	0.01	0.00	100.2
SD2-G15(2 pxs)	39.5	0.74	12.6	12.7	--	6.7	0.38	0.10	17.6	8.9	0.03	0.00	99.3
Cr-poor garnets													
SD2-G151	42.2	0.72	20.4	2.13	--	8.3	0.32	0.00	21.4	4.9	0.05	0.00	100.4
SD2-G2 (opx)	41.3	0.64	19.1	4.4	--	8.8	0.38	0.02	20.5	4.6	0.09	0.00	99.8
SH3-9	41.6	0.98	21.2	0.65	--	10.6	0.33	0.00	19.8	5.0	0.06	0.00	100.2
IM20-G1 (cpx)	42.9	0.56	20.3	2.61	--	7.5	0.28	0.00	21.3	5.2	0.03	0.00	100.7
IM26-G17 (ilm)	41.7	0.63	21.9	0.07	--	13.0	0.44	0.01	18.2	4.4	0.06	0.00	100.4
Cr-rich clinopyroxenes													
SD-20A	54.4	0.13	1.20	1.50	--	2.6	0.11	0.06	18.8	19.8	0.99	0.04	99.6
SD2-G15 [gar]	55.4	0.19	1.43	2.43	--	2.9	0.12	0.06	18.4	17.8	1.55	0.04	100.3
Cr-poor clinopyroxenes													
SD2-04	55.4	0.22	2.07	0.63	--	4.1	0.14	0.08	20.0	16.3	1.34	0.02	100.3
SD2-I17 [ilm]	55.0	0.32	1.93	0.83	--	3.7	0.09	0.01	17.6	19.3	1.44	0.02	100.2
SH16-C7	54.9	0.38	2.06	0.18	--	5.1	0.14	0.07	18.5	17.4	1.25	0.02	100.0
IM20-G1 [gar]	55.1	0.18	2.23	0.73	--	4.0	0.16	0.07	20.1	16.3	1.13	0.02	100.0
Cr-rich orthopyroxenes													
SD2-G15 [gar]	57.2	0.13	0.71	0.61	--	4.9	0.14	0.11	35.3	0.95	0.20	0.00	100.3
Cr-poor orthopyroxenes													
SD2-02	56.5	0.15	1.13	0.25	--	5.9	0.16	0.13	34.2	1.36	0.22	n.a.	100.0
SD2-G2 [gar]	56.4	0.25	0.81	0.31	--	6.3	0.15	0.13	34.2	0.70	0.12	0.00	99.4
Cr-rich olivines													
SD2-D1	41.2	0.00	0.04	0.11	--	8.1	0.10	0.30	50.8	0.06	0.01	n.a.	100.7
Cr-poor olivines													
SD2-I16 [ilm]	40.0	0.04	0.02	0.01	--	12.0	0.16	0.20	48.2	0.04	0.00	n.a.	100.7
SD2-I18 [ilm]	40.3	0.03	0.02	0.04	--	10.0	0.16	0.27	49.5	0.03	0.02	0.00	100.3
Ilmenites (Cr-poor)													
SD2-I18 (ol)	0.00	48.6	1.0	10.6	9.1	14.8	0.23	0.20	15.5	0.02	0.21	0.00	100.3
SD2-I16 (ol,opx)	0.17	52.9	0.44	2.8	7.5	21.5	0.25	0.05	14.2	0.02	0.12	n.a.	100.0
SH3-28	0.29	48.3	0.41	0.12	15.2	25.8	0.34	0.04	9.8	0.01	n.a.	n.a.	100.3
IM26-G17 [gar]	0.13	49.4	0.75	0.01	12.5	26.5	0.27	0.08	9.8	0.02	0.05	0.00	99.5
IM7-I15	0.11	39.1	0.32	1.08	28.1	26.1	0.20	0.02	5.0	0.00	0.00	n.a.	100.0
Lamellar intergrowths													
IM7-IC1 Cpx	54.9	0.37	2.16	0.10	--	5.3	0.16	0.04	18.0	17.8	1.32	0.02	100.2
IM7-IC1 ilm	0.16	50.6	0.88	0.43	10.4	26.2	0.23	0.00	11.0	0.02	0.03	0.00	100.0
Graphic intergrowths													
IM4-IC1 ilm	0.16	52.1	0.88	0.56	4.0	24.9	3.25	0.17	11.8	0.03	0.00	n.a.	97.9

[1]Analyses by electron microprobe; total iron reported as FeO, except for ilmenites, for which Fe_2O_3 is calculated from stoichiometry. n.a. = not analyzed. [2]y (x) = y contains an inclusion of x. y[x] = y is an inclusion in a host of x.

tions of the phases are similar to those in the nodules described above.

Rare Earth Elements

Rare Earth Element (REE) analyses were obtained, by mass spectrometry isotope dilution, of selected samples of clinopyroxene and garnet megacrysts. One Cr-rich diopside, two Cr-poor diopsides, one Cr-rich garnet, and one Cr-poor garnet were analyzed. Small but distinct differences in abundance and chondrite-normalized fractionation patterns between the two nodule groups are evident (Fig. 8). Cr-rich diopside is slightly enriched in the light and intermediate REE (La to Gd), and both Cr-rich diopside and garnet are depleted in the heavy REE (Dy to Yb), relative to their Cr-poor counterparts (Fig. 8). The patterns are similar to those of garnet and clinopyroxene mineral separates from Lesotho granular and sheared garnet lherzolite (Shimizu, 1974). REE's of a Cr-poor clinopyroxene intergrown with ilmenite in a nodule from the Monastery Mine, South Africa (Mitchell et al., 1973) are appreciably higher than those from both Cr-poor and Cr-rich diopside megacrysts from Colorado-Wyoming kimberlite.

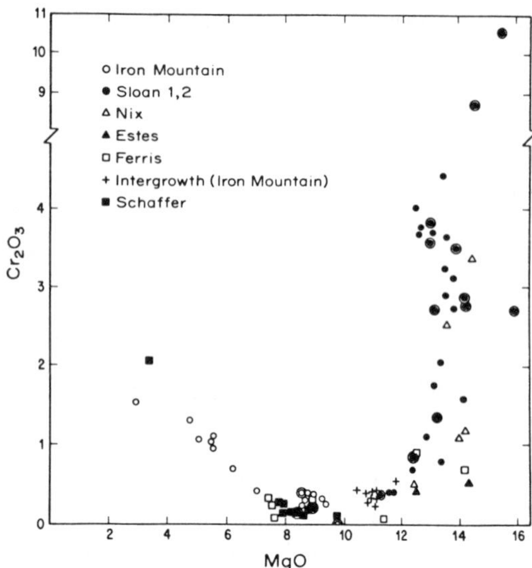

Fig. 7. Compositions of ilmenite megacrysts, in wt %. Note the different scales above and below the break in the ordinate. Circled points are hosts with silicate inclusions or inclusions in silicate hosts.

Chemical Variations between Groups of Kimberlite Pipes

The Cr-rich megacryst suites in the Colorado-Wyoming kimberlite districts appear to belong to one population, but distinctive groups of Cr-poor megacrysts are associated with three groups of pipes: pipes of the Iron Mountain district; a group of pipes including the Schaffer 1-16, Ferris 1-2, and Aultman; and a group of pipes including the Sloan 1-2 and Nix 1-4.

The distinctions between the three groups of pipes are best seen in compositions of ilmenites. Ilmenites from the Iron Mountain district are restricted to the low-MgO limb of the MgO-Cr_2O_3 parabola (Fig. 7) (except for intergrowth ilmenites), although there is overlap with other groups in the range of 7-12 wt % MgO. Sample IM7-I15 (Table 2) is an example of a low - MgO ilmenite. Ilmenites from the Sloan-Nix group are also distinctive, in that they are the most magnesian, occurring alone on the high-MgO limb of the parabola (Fig. 7). Ilmenites from the Schaffer-Aultman-Ferris group are restricted, with one exception, to the trough of the parabola.

The other two Cr-poor minerals that occur in unaltered form in all the groups of pipes are garnets and clinopyroxenes. Distinctions in compositions of these minerals exist also. For instance, extremely Fe-rich, Cr-poor garnets and clinopyroxenes occur only in the Iron Mountain pipes (Figs. 4 and 5) (an example is garnet IM26-G17, Table 2); the compositions of megacrysts from the Schaffer-Aultman-Ferris pipes are very restricted (Fig. 3) (examples are garnet SH3-9 and clinopyroxene SH16-C7, Table 2); and calcic clinopyroxenes from the Iron Mountain district are distinctively less magnesian and chromian than calcic clinopyroxenes from the Sloan-Nix pipes (Figs. 3 and 5). It is particularly interesting to note the distinctions between the Sloan-Nix pipes and the geographically-proximate Schaffer-Aultman-Ferris pipes. For instance, there is no overlap in Cr_2O_3 content between garnets from the two groups of pipes (Fig. 4).

A Megacryst Geotherm

Pressures and temperatures of equilibration of peridotites and of orthopyroxene megacrysts are shown in Fig. 9. All of the nodules are from the Sloan 2 and Nix pipes, because they are the only pipes containing unaltered orthopyroxene. Paleotemperatures of peridotites were established from the Ca/(Ca + Mg) of clinopyroxenes, based on the solvi of Davis and Boyd (1966) and Lindsley and Dixon (1976). Paleopressures were calculated by the method of Wood (1974), as modified by Wood (1977), from the alumina content of orthopyroxene and the Ca/(Ca + Mg + Fe) and Cr/(Cr + Al) of garnet. In calculating equilibration conditions of orthopyroxene megacrysts, it was assumed that each orthopyroxene crystallized in equilibrium with a fictive clinopyroxene and a fictive garnet, based on the occasional observation of inclusions in megacrysts (Table 3) and on the chemical consistency of those inclusions (see the section above). The composition of the fictive clinopyroxene coexisting with each orthopyroxene does not have to be estimated, because the Ca/(Ca + Mg) of the orthopyroxene is also sensitive to temperature. Temperatures of equilibration of Cr-poor orthopyroxene megacrysts were calculated by the Ca/(Ca + Mg) thermometer of Boyd and Nixon (1973, p. 265), an empirical thermometer

TABLE 3. Frequency of Inclusions within Megacrysts

		Inclusions.								
Host Megacrysts	No. of Samples	Cr-rich gar.	Cr-poor gar.	Cr-rich opx.	Cr-poor opx.	Cr-rich cpx.	Cr-poor cpx.	Cr-poor ilm.	Cr-poor olv.	Carbonate
Cr-rich gar.	35	0	0	1	0	1	0	0	0	1
Cr-poor gar.	50	0	0	0	1	0	3	2	0	0
Cr-rich opx.	73	0	0	0	0	1	0	0	0	0
Cr-poor opx.	19	0	0	0	0	0	0	1	0	0
Cr-poor cpx.	91	0	0	0	0	0	0	0	0	1
Cr-poor ilm.	85	0	3	0	2	0	5	0	5	0

Only chemically-analyzed megacrysts and only the classes of megacrysts that have hosts or inclusions are listed.

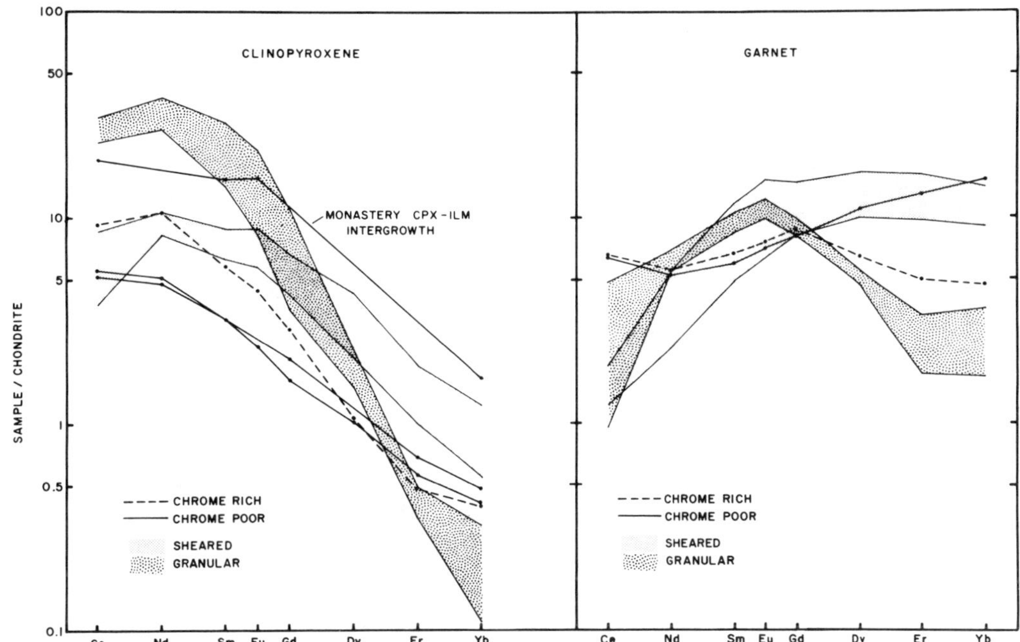

Fig. 8. Chondrite-normalized rare earth element distribution patterns of Colorado-Wyoming clinopyroxene and garnet megacrysts. Also shown are ranges of clinopyroxene and garnet REE from Lesotho sheared and granular lherzolites (Shimizu, 1974, 1975) and clinopyroxene from a Monastery Mine, South Africa, clinopyroxene-ilmenite intergrowth (Mitchell et al., 1973). Analyses by isotope dilution mass spectrometry at the U.S. Geological Survey, Denver, under the direction of C.E. Hedge.

calibrated from coexisting Cr-poor orthopyroxenes and clinopyroxenes in Lesotho lherzolites. Because those pyroxenes are essentially the same compositions as Colorado-Wyoming Cr-poor pyroxene megacrysts, the Boyd-Nixon thermometer was used for Cr-poor megacrysts. A thermometer for Cr-rich orthopyroxenes from the Sloan-Nix pipes was developed by plotting compositions of five pyroxene pairs from lherzolites (Cr-rich) from the Sloan 2 pipe and two pyroxene pairs from Cr-rich megacrysts with inclusions. A line drawn parallel to that of Boyd and Nixon (1973), but displaced 50°C down-temperature, fits the 7 pyroxene pairs within 25°C. Clinopyroxene temperatures for that thermometer were based on the solvi mentioned above. The composition [Ca/(Ca + Mg + Fe) and Cr/(Cr + Al)] of the fictive garnet coexisting with each orthopyroxene is also needed for Wood's (1974) procedure. It was found that the composition can be estimated with sufficient accuracy, however, from the composition of the orthopyroxene itself, because both Ca/(Ca + Mg + Fe) and Cr/(Cr + Al) of garnets (Cr-rich and poor) from lherzolites and megacrysts were found to correlate with Cr/(Cr + Al) in coexisting orthopyroxenes.

Temperatures and pressures of equilibration of peridotites shown in Fig. 9 are the same as those presented earlier (McCallum and Eggler, 1976), with two differences: the points are somewhat shifted because the new correction of Wood (1977) was used; and four harzburgites have been included. Inclusion of the harzburgites is justified only if their orthopyroxenes are saturated in Ca. That assumption is valid for at least one of the rocks, which contains rare clinopyroxene that was not encountered during analysis. Garnets in the other harzburgites fall on the normal Ca-Cr trend for garnets in lherzolites (Danchin and Boyd, 1976, fig. 15), and so the coexisting orthopyroxenes are also assumed to be Ca-saturated (ref. Danchin and Boyd, 1976). The megacryst data base (Fig. 9) is somewhat expanded from that shown by Eggler and McCallum (1976), and the array of points for Cr-rich megacrysts is shifted from the array shown by Eggler and McCallum (1976) because of the improved thermometer described above.

In both halves of Fig. 9 a steady state, noninflected geotherm has been drawn with a slope subparallel to the well-known 'shield geotherm' of Clark and Ringwood (1964), much as Boyd (1973) drew a steady state geotherm through part of his array of points. The geotherm shown is about 30°C cooler than the 'shield geotherm' but about 60°C hotter than the steady state geotherm of Boyd (1973). The peridotite points fall near that geotherm; one of the porphyroclastic lherzolites may be significantly hotter than the steady state gradient. The bulk of megacryst points also fall near the steady state geotherm, but a number, in particular the Cr-

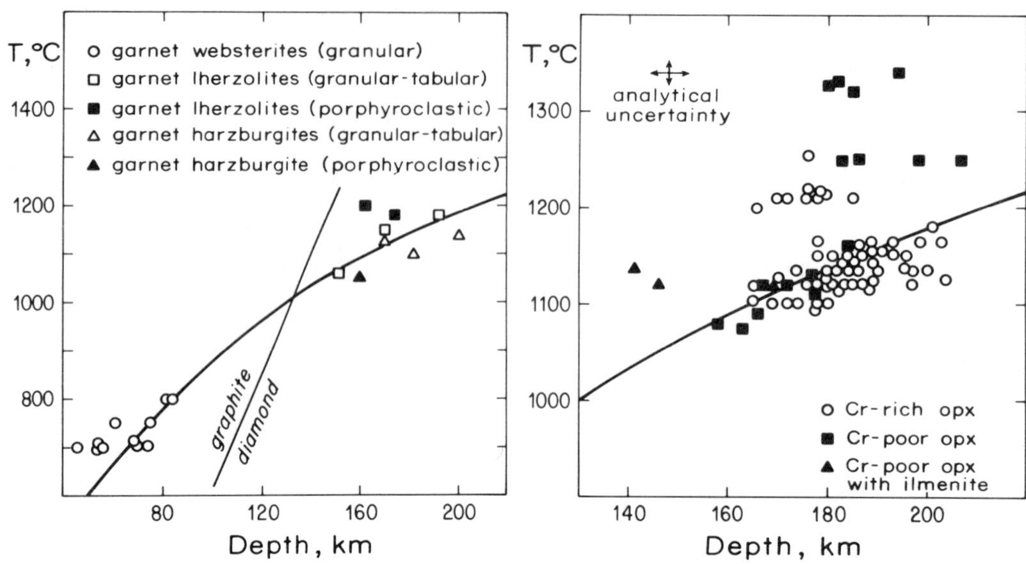

Fig. 9. Pressures and temperatures of equilibration of Colorado-Wyoming peridotite nodules (left) and orthopyroxene megacrysts (right). Note the different temperature scales. All the peridotite nodules and all but one of the megacrysts are from the Sloan 2 pipe. The ranges of uncertainty for megacrysts due to analytical error were computed by assuming 5% relative error in analyses of CaO, MgO, and Al_2O_3 in orthopyroxenes.

poor megacrysts, lie well above the geotherm in a region of disturbed geothermal gradient. This region is reminiscent of inflected geotherms established for Lesotho (Boyd, 1973), Montana (Hearn and Boyd, 1975), and the Premier mine, South Africa (Danchin and Boyd, 1976). Unlike those P-T arrays, however, points on the megacryst array at depths in excess of 170-180 km fall along a steady state geotherm as well as at temperatures above that geotherm.

All the orthopyroxene points shown in Figure 9, it should be remembered, are from the Sloan-Nix pipes. It is assumed that Cr-poor orthopyroxene megacrysts from other pipes, too altered for analysis , would plot in the same array, because equilibration temperatures of Cr-poor clinopyroxenes from all pipes (Fig. 3) fall in the same range as temperatures of Sloan-Nix orthopyroxenes (1025°-1310°C). It will be recalled, however, that lower-temperature (1025°-1125°C) Cr-poor pyroxenes from the Sloan-Nix pipes, that plot along the steady state geotherm, are anomalously chromian and magnesian. It is conceivable that Cr-poor orthopyroxenes from other pipes, should they be available for analysis, would plot only as an array of higher, perturbed temperatures.

Discussion

The Geometry of Kimberlite Conduits

As discussed above, there are three groups of pipes that contain suites of Cr-poor megacrysts with distinctive chemistries, the Iron Mountain group, the Schaffer-Aultman-Ferris group, and the Sloan-Nix group. Because the Cr-poor megacrysts are derived from depths of 150-200 km (Fig. 9), each group of pipes must have had a separate conduit that extended to at least that depth. The Sloan-Nix and Schaffer-Aultman-Ferris groups of pipes are particularly interesting in that their megacryst compositions are distinctly different, but they are located only a few kilometers apart (Fig. 1). These relationships suggest that while their conduits must also have been only a few kilometers apart, the conduits tapped distinctly different source materials. The separate main conduits presumably branched at upper levels, feeding the pipes within each group.

Megacrysts--Disaggregated Peridotites?

Megacrysts of both the Cr-rich and Cr-poor suites are believed to have crystallized as assemblages of coexisting olivine, orthopyroxene, clinopyroxene, garnet, and ilmenite, which are lherzolite assumblages. The megacrysts are not, however, believed to be products of disaggregation of such polymineralic lherzolites. The following factors support the contention that the megacrysts are distinct suites: (1) As indicated by Nixon and Boyd (1973) and Eggler and McCallum (1974, 1976), megacrysts are of large size (1-13 cm for Colorado-Wyoming megacrysts, and in excess of 20 cm for some African megacrysts), whereas minerals in lherzolites have a much smaller grain size (< 3 mm). (2) Neither the Cr-rich nor Cr-poor suites of megacrysts shows complete or near-complete compositional overlap with minerals of lherzolites. There is no question that there is no overlap in the case of Cr-poor megacrysts (Figs. 4, 5, 6). These relatively iron-rich megacrysts have no analogues among minerals of the depleted lherzolites. (In Lesotho, Cr-poor megacrysts are similar chemically to sheared, relatively undepleted lherzolites [Nixon and Boyd, 1973]. As noted in the Introduction, in Colorado-Wyoming all lherzolites have mineral compositions that reflect depletion.) The Cr-rich megacrysts, on the other hand, are somewhat similar to minerals of lherzolites. No significant differences exist, in fact, between compositions of Cr-rich orthopyroxene, olivine, and garnet megacrysts and orthopyroxenes, olivines, and garnets from lherzolites (Figs. 3 and 6, Table 1). The other principal mineral of the Cr-rich suite, clinopyroxene, however, is not as similar to clinopyroxenes in lherzolites: clinopyroxenes from lherzolites have distinctly more jadeite component, reflected in higher Na_2O and an accompanying amount of Al_2O_3 (Fig. 5), than Cr-rich megacrysts (the Tschermaks content of both sets of pyroxenes is about the same), and the clinopyroxenes from lherzolites are more chromian at particular levels of Mg/(Mg + Fe) and Ca/(Ca + Mg) (Fig. 5).

Fractionation Trends in Megacrysts

Both Cr-rich and Cr-poor megacrysts have many of the characteristics that led Nixon and Boyd (1973) to infer that Cr-poor megacrysts from Lesotho grew in the presence of liquid -- the grain size is ultracoarse, and inclusions and intergrowths among minerals are uncommon. Evidence for the Cr-poor group is especially strong, because that assemblage includes clinopyroxene-ilmenite intergrowths. Such intergrowths from Africa have been interpreted to have crystallized from liquid (e.g., Williams, 1932; Boyd and Dawson, 1972; Gurney et al., 1973), a process verified by experiments at 38 kbar pressure (Wyatt et al., 1975; Wyatt, 1977).

Many workers have interpreted chemical trends among Cr-poor megacrysts -- principally the pronounced variations in Mg/(Mg + Fe) -- to indicate that the megacrysts grew in a fractionating liquid (Boyd and Nixon, 1973; Mitchell, 1973; Frick, 1973; Gurney et al., 1973). The trend in Cr-poor clinopyroxene megacrysts -- subcalcic, higher-temperature clinopyroxenes are more magnesian than calcic, lower-temperature clinopyroxenes -- is consistent with fractionation. The Cr-poor clinopyroxenes from the Iron Mountain district and the higher-temperature (1150°-1310°C) clinopyroxenes from the Sloan-Nix pipes follow that trend (Fig. 3), which may be called the "normal" trend. Garnets show sympathetic variation in Mg/(Mg + Fe) (Fig. 3). It is apparent, from Fig. 3, that ilmenites crystallized with garnets and clinopyroxenes from at least the middle of the clinopyroxene trend to the magnesia-poor end. On the MgO-Cr_2O_3 parabola (Fig. 7), such ilmenites fall in the range of MgO = 14-7 and Cr_2O_3 = 1-0. (This interpretation, that ilmenites crystallized with garnets and clinopyroxenes and, presumably, with orthopyroxenes and olivines, is inconsistent with the view of Mitchell [1973, p. 308-9] that ilmenites from African kimberlites crystallized before silicate minerals. Mitchell may have been misled by the fact that quite magnesian ilmenites, such as the ilmenites in intergrowths, can crystallize with relatively iron-rich pyroxenes.) It is not known what silicates, if any, crystallized with the less-magnesian ($MgO < 7$) ilmenites.

Lower-temperature (1025°-1125°C) Cr-poor clinopyroxenes from the Sloan-Nix pipes do not follow the "normal" trend defined above. They are anomalously magnesian (Fig. 3) and, to a more variable extent, chromian (Fig. 5). An analogous anomaly can be seen in compositions of Cr-poor orthopyroxenes from the Sloan-Nix pipes, inasmuch as a group of subcalcic (lower-temperature) Cr-poor orthopyroxenes are as magnesian (Fig. 3) and chromian (Fig. 6) as the group of calcic (higher-temperature) Cr-poor orthopyroxenes. Many of the very high-chrome ilmenites (>2.5% Cr_2O_3) are hosts for or inclusions in the anomalous pyroxenes. The ilmenites, with olivine inclusions, discussed in a preceding section are among those ilmenites. It would appear, from Figures 3, 5, and 6, that these anomalous Cr-poor pyroxenes (and 2 olivines) from the Sloan-Nix pipes are somewhat transitional to Cr-rich pyroxenes and olivines. Examples of anomalous megacrysts (Table 2) are garnet SD2-G2, with an included orthopyroxene, clinopyroxene SD2-I17, and ilmenite SD2-I18, with an included olivine.

The Cr-rich megacrysts have pronounced variations in Cr_2O_3 (Figs. 4, 5, 6) but have restricted ranges of Mg/(Mg + Fe) (Table 1, Figs. 4, 5, 6), unlike Cr-poor megacrysts. Their range of temperatures of equilibration does not extend to temperatures as high as the Cr-poor megacrysts, either (Figs. 3 and 9). If these megacrysts are interpreted to have grown in the presence of liquids, like the Cr-poor megacrysts, the liquids could not have undergone much fractionation. It is possible that the liquids were produced by partial melting, the restricted Mg/(Mg + Fe) reflecting "buffering" by minerals of the peridotite source. In that case, however, temperatures of equilibration should be higher than temperatures of Cr-poor megacrysts, because the Cr-poor megacrysts presumably crystallized from liquids that fractionated from primary melts. The temperatures are not higher (Fig. 3). Another possible explanation of their origin is that the peridotite from which their mother liquids were extracted had been contaminated by contact with the refractory lithosphere. Such a process could also explain the anomalously magnesian nature of the calcic Cr-poor clinopyroxene and associated Cr-poor orthopyroxene, olivine, and ilmenite megacrysts from the Sloan-Nix pipes.

Megacrysts, Geotherms, and Diapirs

The array of pressures and temperatures of equilibration of orthopyroxene megacrysts (Fig. 9) has been interpreted, in a previous section, as a steady state geotherm with a superimposed subarray of perturbed temperatures. In contrast, the megacryst geotherms presented by Nixon and Boyd (1973) and by Boyd (1974) are distinctly inflected at depths below about 150 km, whereas a geotherm from the Udachanaya pipe, USSR, is non-inflected (Boyd et al., 1976). Nixon and Boyd (1973) suggested that the inflected limb of the Lesotho geotherm was caused by stress-heating in the low-velocity zone. This type of heating has been discounted by Green and Gueguen (1974) and by Goetze (1975). Alternative models have been presented for heating by an upwelling diapir (Green and Gueguen, 1974), by a mantle plume (Parmentier and Turcotte, 1974), and by a diapir associated with convective overturn (Boyd, 1976). It is conceivable that suites of nodules could be recovered from a diapir and from its wallrocks that would yield an inflected geotherm, a non-inflected geotherm, or both.

Diapirs can be related to megacrysts if it is assumed that diapirs melt at some stage in their ascent (Green and Gueguen, 1974). The product of the partial melting of a peridotitic diapir, at 150-200 km depth, will be a kimberlitic liquid (Eggler and Wendlandt, 1977). It has been repeatedly proposed that megacrysts have crystallized from some protokimberlitic liquid (e.g., Dawson and Stephens, 1975; Mitchell and Brunfelt, 1975; Mitchell and Clarke, 1976; Gurney et al., 1977; Eggler and Wendlandt, 1977). Probably many of the megacrysts, in particular most of the Cr-poor megacrysts, did not actually grow in the partial melt itself, but rather in a fractionating magma separated from its peridotite source. Nevertheless, all the liquids were kimberlitic, and so the megacrysts are, in a sense, cognate to kimberlite.

The lherzolites and harzburgites in Colorado-Wyoming represent the depleted, thermally undisturbed lithosphere. Their depths of equilibration, 150-200 km, indicate that they probably were not picked up above the diapir, which can be assumed to have risen to about 150 km, the depth of the shallowest megacrysts. They either were wallrocks to the diapir or were entrained in the diapir. If they were entrained, producing a hybrid diapir, melting of hybridized regions could produce locally homogeneous and magnesian liquids from which the Cr-rich megacrysts and the anomalous Cr-poor megacrysts from the Sloan-Nix pipes could have crystallized. Because entrained pieces of lithosphere would have been cool, contamination may also explain the relatively cool temperatures of the Cr-rich megacrysts and of those Cr-poor megacrysts from the Sloan-Nix pipes that fall along the steady state geotherm (Fig. 9) and that are anomalously magnesian.

The key to the formation of kimberlite may be the coincidence of an asthenospheric diapir or convective system and a fracture system in the lithosphere. Pressure can be released along the fracture system, initiating rise of the melted portion of a diapir, a kimberlitic liquid, up the conduits. That liquid composition is modified, to varying extents, by inclusions of wallrocks, both from the asthenospheric diapir and the lithosphere, and by low-pressure processes such as fractionation, liquid immiscibility, and interaction with near-surface waters. (The kimberlite matrices of Colorado-Wyoming pipes, whose mineral assemblages and textures have been shaped by such low-pressure processes, are described in a companion paper in this volume by McCallum et al.) The only unmodified evidence of the original kimberlitic liquids is the megacrysts and, perhaps, the diamonds that grew in them.

Acknowledgments. We are indebted to the ranchers in the study area who allowed us access to the kimberlites, and without whose generous cooperation no nodules would have been available. Appreciation is extended to Colorado State University students, research colleagues and others, especially C. D. Mabarak, F. Yaussi, and P. Brownell, who kindly provided us with some of our sample material. The aid provided by the Geophysical Laboratory in Washington, D. C., in obtaining microprobe data and by C. E. Hedge of the U.S. Geological Survey in Denver, Colorado, in analyzing REE is gratefully acknowledged. J. D. Pasteris, B. C. Hearn, Jr., M. A. Kuntz, G. L. Snyder, and H. S. Yoder, Jr., kindly reviewed the manuscript. Special thanks are due to R. B. Thompson, W. K. Camp, and M. Olafsson for their efforts in collating data and drafting figures.

This study was supported by the Earth Sciences Section of the National Science Foundation (contracts DES 74-13098 and EAR 74-13098A01), the Carnegie Institution of Washington, the U.S. Geological Survey, and the Colorado State University Development Fund. The paper was approved for publication by the Directors of the U.S. Geological Survey and the Geophysical Laboratory.

References

Bloomer, A. G. and P. H. Nixon, The geology of the Letseng-la-terae kimberlite pipes, in Lesotho Kimberlites, edited by P. H. Nixon, pp. 20-32, Lesotho National Development Corporation, Maseru, Lesotho, 1973.

Boyd, F.R., Enstatite-ilmenite and diopside-ilmenite intergrowths from the Monastery mine, Carnegie Inst. Wash. Year Book 70, 134-138, 1971.

---------- A pyroxene geotherm, Geochim. Cosmo. Acta, 37, 2533-2546, 1973.

---------- Ultramafic nodules from the Frank Smith kimberlite pipe, South Africa, Carnegie Inst. Wash. Year Book 73, 285-294, 1974.

---------- Inflected and noninflected geotherms, Carnegie Inst. Wash. Year Book 75, 521-523, 1976.

Boyd, F. R. and J. B. Dawson, Kimberlite garnets and pyroxene-ilmenite intergrowths, Carnegie Inst. Wash. Year Book 71, 373-378, 1972.

Boyd, F. R., T. Fujii, and R. V. Danchin, A non-inflected geotherm for the Udachanaya Kimberlite Pipe, USSR, Carnegie Inst. Wash. Year Book 75, 523-30, 1976.

Boyd, F. R. and P. H. Nixon, Origin of the ilmenite-silicate nodules in kimberlites from Lesotho and South Africa, in Lesotho Kimberlites, edited by P. H. Nixon, pp. 254-268, Lesotho National Development Corporation, Maseru, Lesotho, 1973.

Boyd, F. R. and P. H. Nixon, Origins of the ultramafic nodules from some kimberlites of northern Lesotho and the Monastery Mine, South Africa, Phys. Chem. Earth, 9, 431-54, 1975.

Brookins, D. G., Ilmenite-(serpentinized) pyroxene nodules from the Stockdale kimberlite pipe, Riley County, Kansas (abstract), Abstr. with Program, Geol. Soc. Amer., 3, 233, 1971.

Clark, S. P., and A. E. Ringwood, Density distribution and constitution of the mantle, Rev. Geophys., 2, 35-88, 1964.

Danchin, R. V. and F. R. Boyd, Ultramafic nodules from the Premier kimberlite pipe, South Africa, Carnegie Inst. Wash. Year Book 75, 531-37, 1976.

Davis, B. T. C. and F. R. Boyd, The join $Mg_2Si_2O_6$-$CaMgSi_2O_6$ at 30 Kb pressure and its application to pyroxenes from kimberlite, J. Geophys. Res., 71, 3567-3576, 1966.

Dawson, J. B. and A. M. Reid, A pyroxene-ilmenite intergrowth from the Monastery Mine, South Africa, Contr. Mineral. Petrol., 26, 296-301, 1970.

Dawson, J. B., and W. E. Stephens, Statistical classification of garnets from kimberlite and associated xenoliths, J. Geology, 83, 589-607, 1975.

Eggler, D. H. and M. E. McCallum, Preliminary upper mantle-lower crust model of the Colorado-Wyoming Front Range, Carnegie Inst. Wash. Year Book 73, 295-300, 1974.

---------- A geotherm from megacrysts in the Sloan kimberlite pipes, Colorado, Carnegie Inst. Wash. Year Book 75, 538-41, 1976.

Eggler, D. H. and R. F. Wendlandt, Experimental studies on the relationship between kimberlite magmas and partial melting of peridotite, Extended Abstracts, Second International Kimberlite Conference, Santa Fe, N. M., 1977.

Finger, L. W., The uncertainty in the calculated ferric iron-content of a microprobe analysis, Carnegie Inst. Wash. Year Book 71, 600-603, 1972.

Finger, L. W., and C. G. Hadidiacos, Electron microprobe automation, Carnegie Inst. Wash. Year Book 71, 598-600, 1972.

Frantsesson, E. V., The petrology of the kimberlites, translated by D. A. Brown, Camberra, Department of Geology, Australian National University, A.C.T., Publication No. 150, p. 194, 1970.

Frick, C., Kimberlite ilmenites, Geol. Soc. South Africa Trans., 76, 85-94, 1973.

Goetze, C., Sheared lherzolites: from the point of view of rock mechanics, Geology, 3, 172-173, 1975.

Green, H. W. II, and Y. Gueguen, Origin of kimberlite pipes by diapiric upwelling in the upper mantle, Nature (London), 249, 617-620, 1974.

Gurney, J. J., H. W. Fesq, and E. J. D. Kable, Clinopyroxene-ilmenite intergrowths from kimberlite: a re-appraisal, in Lesotho Kimberlites, edited by P. H. Nixon, pp. 238-253, Lesotho National Development Corporation, Maseru, Lesotho, 1973.

Gurney, J. J., W. R. O. Jakob, and J. B. Dawson, Megacrysts from the Monastery Mine, Extended Abstracts, Second International Kimberlite Conference, Santa Fe, N. M., 1977.

Haggerty, S. E., The chemistry and genesis of opaque minerals in kimberlite, Phys. Chem. Earth, 9, 295-307, 1975.

Hearn, B. C., and F. R. Boyd, Garnet peridotite xenoliths in a Montana, U. S. A., kimberlite, Phys. Chem. Earth, 9, 247-256, 1975.

Ilupin, I. P., S. F. Sobolev, B. P. Zolotarev, and A. A. Lebedev-Zinovyev, Geochemical specialization of kimberlites from various parts of Yukutia, Geochem. Int., 11, no. 2, 357-70, 1974.

Lindsley, D. H., and S. A. Dixon, Diopside-enstatite equilibria at 850° to 1400°C, 5 to 35 kb, Amer. J. Sci., 276, 1285-1301, 1976.

MacGregor, A. M., J. C. Ferguson, and F. L. Amm, The geology of the country around the Queen's Mine, Bulawayo district, Bull. Geol. Surv. S. Rhodesia, 30, 1-175, 1937.

McCallister, R. H., H. O. A. Meyer, and D. G. Brookins, "Pyroxene"-Ilmenite xenoliths from the Stockdale pipe, Kansas: chemistry, crystallography, and origin, Phys. Chem. Earth, 9, 287-293, 1975.

McCallum, M.E. and D. H. Eggler, Diamonds in an upper mantle peridotite nodule from kimberlite in southern Wyoming, Science, 192, 253-56, 1976.

McCallum, M. E., D. H. Eggler, and L. K. Burns, Kimberlitic diatremes in northern Colorado and southern Wyoming, Phys. Chem. Earth, 9, 149-161, 1975.

Mitchell, R. H., Magnesian ilmenite and its role in kimberlite petrogenesis, J. Geol., 81, 301-11, 1973.

Mitchell, R. H., and A. O. Brunfelt, Rare earth element geochemistry of kimberlite, Phys. Chem. Earth, 9, 671-686, 1975.

Mitchell, R. H., D. A. Carswell, and A. O. Brunfelt, The Monastery Mine kimberlite pipe. I. Mineralogy and rare earth geochemistry of an ilmenite-clinopyroxene xenolith from the

Monastery Mine, in Lesotho Kimberlites, edited by P. H. Nixon, pp. 224-229, Lesotho National Development Corporation, Maseru, Lesotho, 1973.

Nixon, P. H. and F. R. Boyd, The discrete nodule association in kimberlites from northern Lesotho, in Lesotho Kimberlites, edited by P. H. Nixon, pp. 67-75, Lesotho National Development Corporation, Maseru, Lesotho, 1973.

Parmentier, E. M., and D. L. Turcotte, An explanation of the pyroxene geotherm based on plume convection in the upper mantle, Earth Planet. Sci. Lett., 24, 209-212, 1974.

Rolfe, D. G., The geology of the Kao kimberlite pipes, in Lesotho Kimberlites, edited by P. H. Nixon, pp. 101-106, Lesotho National Development Corporation, Maseru, 1973.

Shee, S. R., and J. J. Gurney, The mineralogy of xenoliths from Orapa, Botswana, Extended Abstracts, Second International Kimberlite Conference, Santa Fe, N. M., 1977.

Shimizu, N., Rare earth elements (REE) in garnets and clinopyroxenes from garnet lherzolite nodules in kimberlite, Carnegie Inst. Wash. Year Book 73, 954-61, 1974.

Smith, C. B., Kimberlite and mantle derived xenoliths at Iron Mountain, Wyoming, unpubl. M.S. thesis, Colorado State University, 1977.

Smith, C. B., M. E. McCallum, and D. H. Eggler, Clinopyroxene-ilmenite intergrowths from the Iron Mountain kimberlite district, Wyoming, Carnegie Inst. Wash. Year Book 75, 542-544, 1976.

Wagner, P. A., The Diamond Fields of Southern Africa, 355 pp., The Transvaal Leader, Johannesburg, Second impression 1971, Cape Town, Struik (Pty) Ltd., 1914.

Williams, A. F., The Genesis of the Diamond, 2 vols., London, E. Benn Ltd., 1932.

Wood, B. J., The solubility of alumina in orthopyroxene coexisting with garnet, Contr. Mineral. Petrol., 46, 1-15, 1974.

---------- The influence of Cr_2O_3 on the relationships between spinel- and garnet-peridotites, Extended Abstracts, Second International Kimberlite Conference, Santa Fe, N. M., 1977.

Wyatt, B. A., The melting and crystallization behavior of a natural clinopyroxene-ilmenite intergrowth, Contrib. Mineral. Petrol., 61, 1-9, 1977.

Wyatt, B. A., R. H. McCallister, F. R. Boyd, and V. Ohashi, An experimentally produced clinopyroxene-ilmenite intergrowth, Carnegie Inst. Wash. Year Book 74, 536-39, 1975.

MEGACRYSTS FROM THE MONASTERY KIMBERLITE PIPE, SOUTH AFRICA

J.J. Gurney

Department of Geochemistry, University of Cape Town,
Private Bag, Rondebosch, 7700, South Africa

W.R.O. Jakob

Atomic Energy Board, Pelindaba, Pretoria,
South Africa

J.B. Dawson

Department of Geology, St. Andrews University,
Scotland

Abstract. Megacrysts of garnet, clinopyroxene, orthopyroxene, olivine and ilmenite from the Monastery Mine show systematic variations in chemistry over a wide range of compositions. These are interpreted to be due to igneous fractionation processes, taking place over a wide range of declining temperature at essentially constant pressure in a magma body of limited extent.

Introduction

Kimberlite has long been known to contain rock fragments and minerals of Upper Mantle origin (e.g. Wagner, 1914). Harte (1978) has grouped the ultramafic inclusions into five categories:- (i) peridotites, (ii) garnet-pyroxenites, (iii) eclogites, (iv) megacrysts and (v) metasomatised rocks. He defines the megacrysts, with which this study is concerned, as large monomineralic single crystals, much coarser (2 - 20 cm) than the minerals found in ultramafic nodules (<1 cm). These large crystals may consist of olivine, ortho- and clinopyroxene, garnet, ilmenite, phlogopite and rare zircon. The association of ilmenite and a silicate is very common and rarely one of the silicate minerals is a host to another silicate mineral.

The diopside ilmenite lamellar intergrowths have received special attention in recent years. Clinopyroxene-ilmenite intergrowths have been reported from Monastery Mine, the Frank Smith and Jagersfontein Pipes (Williams, 1932), Premier Mine (Gurney and Haggerty, 1975) and the Uintjesburg and Sonop Kimberlites in South Africa (Boyd and Dawson, 1972), Thaba Putsoa, Solane, Lemphane, Sekameng, Letseng-la-terae and Pipe 200 in N. Lesotho (Nixon and Boyd, 1973a), from Artur de Paiva, Angola (Boyd and Danchin, 1974), from Kansas (Brookins, 1971), Kentucky (Gurney et al., 1973) and Wyoming (Smith et al.,1976) in the U.S.A. and most recently from Barbaru'u, Malaita (Nixon and Boyd, 1977), where the host rock is alnoitic.

Other megacryst phases, especially garnet, clinopyroxene and orthopyroxene are always present wherever the lamellar intergrowths are found and since some or all are found at many other localities too, the megacryst suite is a very common component of kimberlite. Certain megacrysts are found, for instance, at Bellsbank, Barkly West, the Koffiefontein group of pipes, at Premier, in Kimberley and adjacent areas, and at many of the kimberlites in the Victoria West and Britstown Districts of South Africa, as well as in the Gibeon kimberlite province in South West Africa, Orapa in Botswana, and at Sloan, U.S.A. (Eggler and McCallum, 1976).

There are two distinct megacryst populations in some kimberlites, based on chemical compositions. One group is termed the chrome-poor and the other the chrome-rich grouping, as described by Smith, McCallum and Eggler (1976). Our study is only concerned with the first of these since the chrome-rich group of megacrysts were not found at Monastery Mine.

Two previous studies on the lamellar intergrowths have suggested a paragenesis involving re-equilibration at low pressures of a single high pressure phase - a titanium-rich garnet (Ringwood and Lovering, 1970),or an ilmenite structured pyroxene (Dawson and Reid, 1970).

TABLE 1. Mineral Abundances At Monastery.

Mineral	wt.%
Ilmenite nodules	47
Ultrabasic nodules	17
Ilmenite-diopside intergrowths	13
Garnet nodules	10
Dunite nodules	7
Garnet-diopside, diopside, bronzite, phlogopite etc.	4
Basement gneiss	2

Other investigators prefer an origin by eutectic or cotectic crystallisation (Williams, 1932; MacGregor and Wittkop, 1970; Gurney et al., 1973; Frick, 1973 and Wyatt, 1977), whilst Mitchell (1977) draws attention to the possibility of simultaneous crystallisation of cpx and ilmenite under vapour-rich conditions and analagous to the crystallisation of quartz and feldspar in graphic granites.

However, as mentioned previously, the lamellar cpx/ilmenite intergrowths are only a single unit in the megacryst suite and are not always present at every locality where megacrysts occur. In general (Boyd and Nixon, 1975) report that the relative abundances of the silicates are garnet > diopside > enstatite > olivine, and that all four minerals occur with and without ilmenite, which may be the most common of all megacrysts.

Considering the megacryst suite as a whole, Nixon and Boyd (1973) suggested that these minerals crystallised as phenocrysts in crystal mush magmas in the low velocity zone, that they co-existed with deformed lherzolites over a depth range of ∿50 km, that the zones of sheared lherzolite and of magma were believed to be of large dimension so that chemical equilibrium was

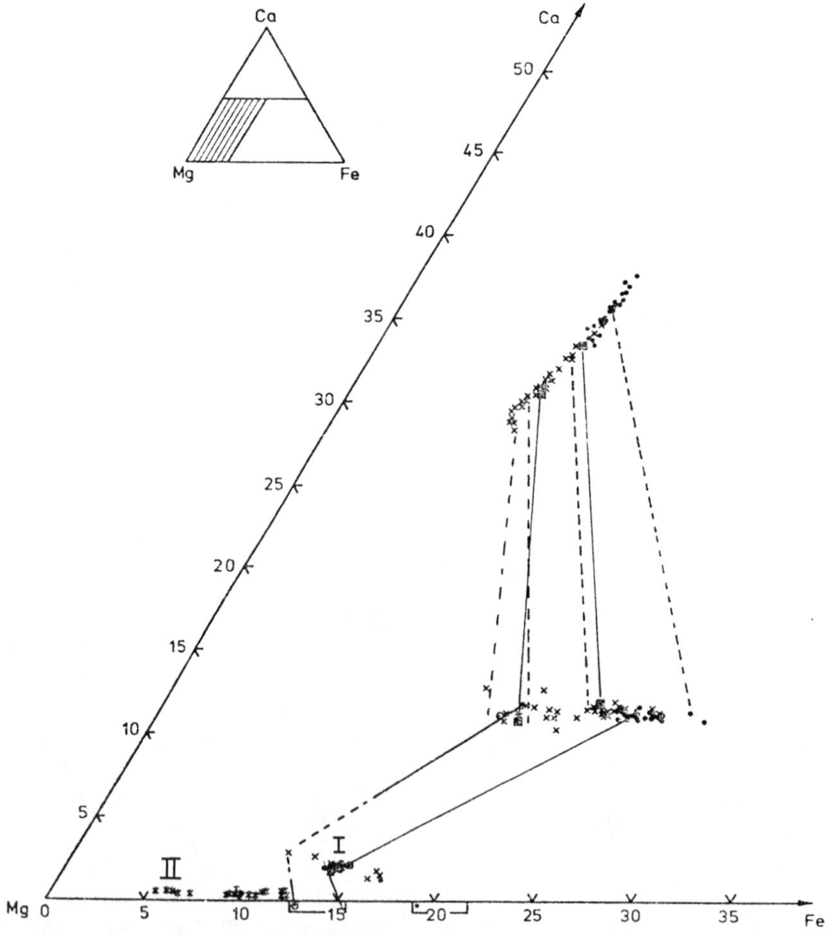

Figure 1: A Ca:Mg:Fe diagram for clinopyroxene, garnet, orthopyroxene and olivine megacrysts from the Monastery Mine.

Symbols: × : Discrete Megacrysts ○ : Olivine/garnet

● : Ilmenite Association ▲ : Garnet in orthopyroxene/ilmenite

■ : Diopside/garnet

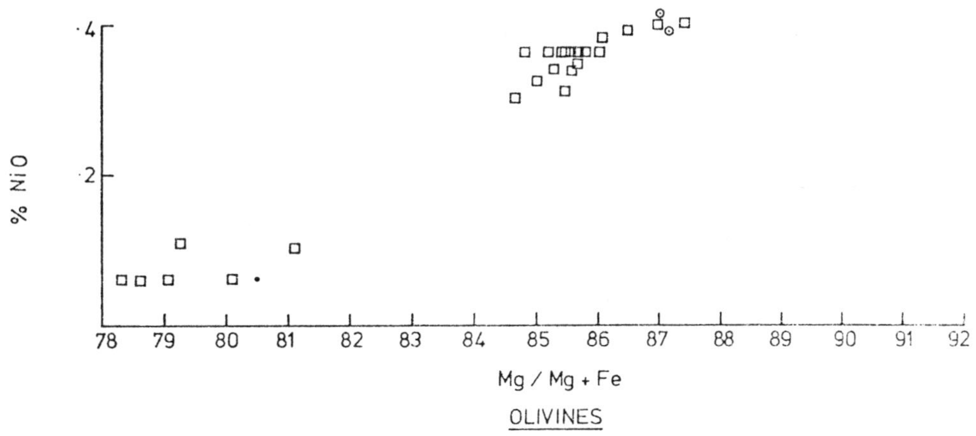

Figure 2: The NiO contents and Mg/Mg+Fe ratio of the two groups of olivine megacrysts at the Monastery Mine.

Symbols: □ : Olivine Megacrysts

○ : Olivine Garnet Intergrowths

● : Olivine Ilmenite Intergrowths

not established between them, and that liquids seeped through the crystal mush following a P-T path down the geotherm. Whilst these liquids would be represented as a component of the kimberlite ground mass, the large range in equilibration temperatures noted was thought to make it most unlikely that the megacrysts were phenocrysts from a single magma chamber (Nixon and Boyd, 1973c).

This model was developed from the evidence of megacrysts from a number of localities.

The current study is concerned entirely with

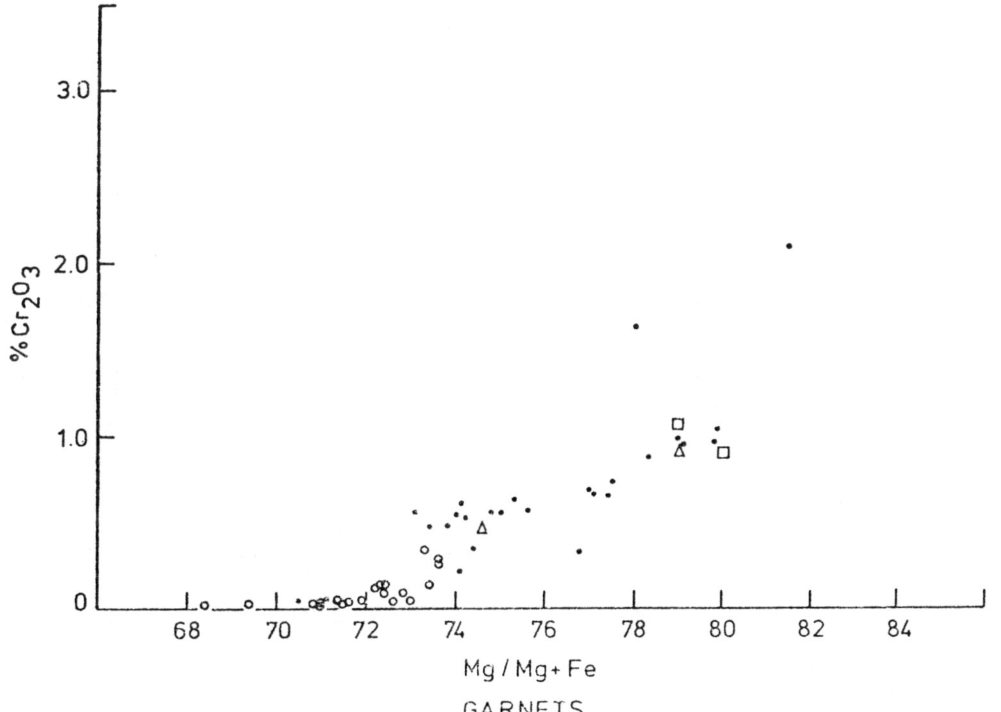

Figure 3: The variation of Cr_2O_3 with Mg/Mg+Fe ratio for megacryst garnets from the Monastery Mine.

Symbols: ● : Garnet Megacrysts △ : Garnet-diopside Association

○ : Garnet-ilmenite Association □ : Garnet-olivine Association

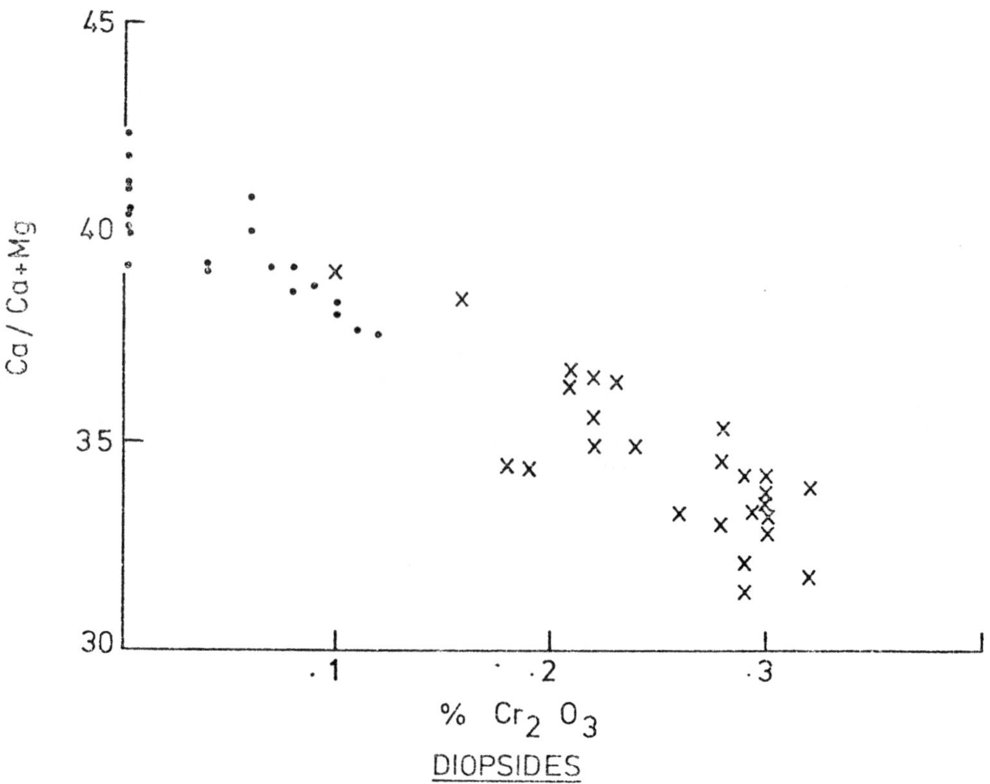

Figure 4: The variation of Cr_2O_3 with Ca/Ca+Mg ratio in the clinopyroxene megacrysts from the Monastery Mine.
Symbols: × : Clinopyroxene Megacrysts
● : Clinopyroxene/ilmenite Intergrowths

the megacryst assemblage found at the Monastery Mine, O.F.S., R.S.A. The study was initiated to conduct a systematic and comprehensive sampling of the megacrysts from one single kimberlite pipe. The Monastery Kimberlite Pipe was selected, for it was well known that megacrysts were present in greater abundance than is found at any other locality in South Africa or possibly in the world.

The Monastery Kimberlite is situated between Marquard and Clocolan in the south eastern portion of the Winburg District in the Orange Free State. It was described by Wagner (1914), and more recently in some detail by Whitelock (1973).

Sampling and Locality Description

The Monastery Kimberlite intrusion is oval in shape with maximum dimensions of 180 x 70 m. The country rocks are near horizontal shales and mudstones alternating with feldspathic sandstones all belonging to the Red Beds of the Stormberg Group in the Karroo Supergroup. Fresh mica megacrysts in the Monastery Kimberlite gave a rubidium strontium age of 90 ± 4 million years (Allsopp and Barrett, 1975) and Davis et al. (1976) obtained a zircon age of 90.4 million years. Davis et al. (op. cit.) have obtained similar ages for the kimberlites of the Kimberley District, Northern Lesotho and Botswana.

Whitelock (1973) gives a detailed description of the mining history and geology of the Pipe and recognises four types of kimberlite of which two predominate:

(a) The Quarry Type Kimberlite which is the most abundant variety and from which most of the megacrysts are derived. Diamond grades of 50 carats per 100 tons were established.

(b) The Breccia Type Kimberlite is the second major component and it consists of up to 80% of medium sized sedimentary xenoliths in a soft serpentinous micaceous matrix in which megacrysts of garnet are common.

(c) The East End Kimberlite is a kimberlite rich in carbonate xenoliths and characterised by altered dunite nodules and sporadic ilmenite. Garnets and gem quality zircon might also occur.

(d) The Fine Grained Kimberlite. This was only exposed by mining in the northern area and was found to lack xenoliths.

The heavy mineral concentrate from the mine dumps of the mine has been investigated by Nixon and Boyd (1973b) with the observed abundances reported in Table 1.

During a recent visit to the Monastery Pipe, the following abundances of megacrysts were observed in decreasing order:-
1. ilmenite 2. diopside-ilmenite 3. garnet 4. diopside 5. olivine 6. mica 7. orthopyroxene 8. garnet-ilmenite 9. orthopyroxene-ilmenite 10. orthopyroxene-diopside 11. rare finds of garnet-diopside and olivine-ilmenite.

Megacrysts (>2cm) were sorted into mineral groups and up to 25 minerals were selected at random from the different populations. In rare instances inclusions of another silicate phase in a megacryst were found. All such examples were selected for analysis on the assumption that the minerals represent equilibrium assemblages. Megacrysts of phlogopite were not sampled because they were vermiculitised and zircons were not found by us, although known to be present, and to be associated with ilmenite.

Analytical Methods

The bulk of the analyses in this study were performed with the Microscan 5 electron microprobe microanalyser manufactured by Cambridge Scientific Instruments Limited. This instrument, together with instrumental conditions, standard and data reduction, as used in this study have been described in detail by Lawless (1974).

Analytical Results

The detailed analytical results have been recorded by Jakob (1977) and will not be given here except where particularly relevant to the interpretation of the origin of the Monastery megacryst suite. In general the megacrysts were found to be homogeneous and unzoned. Zoning was occasionally noted at ilmenite-silicate interfaces but was restricted to narrow zones of local chemical disequilibrium and is not considered further here. Exsolution phenomena were only noted in some of the ilmenites and in the orthopyroxenes designated Group II orthopyroxenes.

The major variations in the chemistry of the pyroxenes, the garnets and the olivines in the megacryst suite can be well represented in a Ca:Mg:Fe ternary plot since these three elements are the major components showing wide variations. Manganese, titanium and chromium are always minor components as is aluminium (except for the garnets) and sodium (except for the clinopyroxenes). In both these exceptions the observed ranges in content are small. The plot in Fig. 1 shows the following important features:-
1. The clinopyroxene megacrysts from a continuous linear trend towards higher calcium, slightly higher iron and lower magnesium contents with the clinopyroxenes in the clinopyroxene-ilmenite megacrysts at the high calcium, higher iron end of the trend.

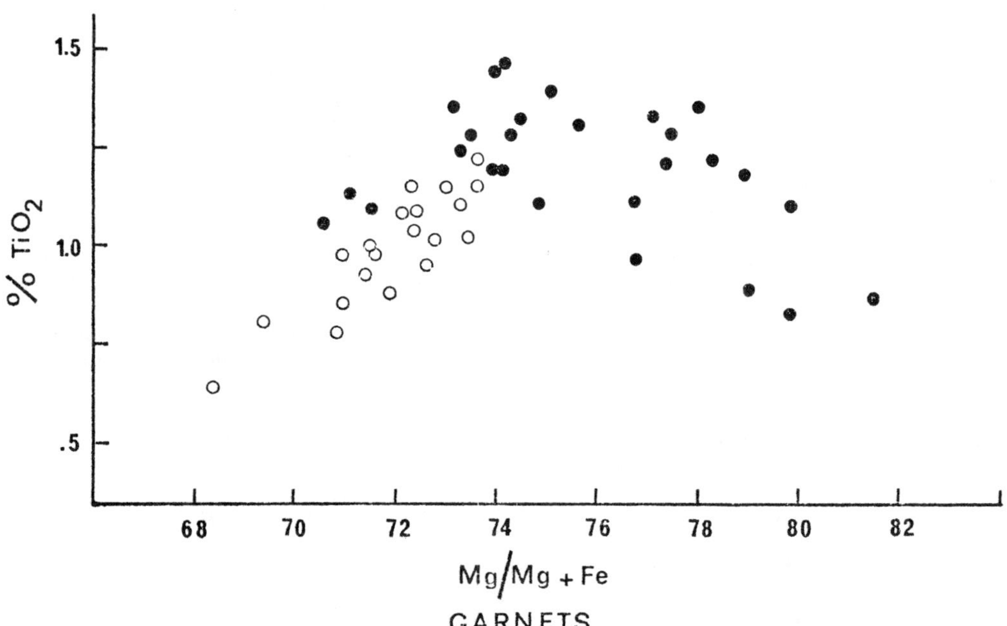

Figure 5: The variation of TiO_2 content with Mg/Mg+Fe ratio for garnet megacrysts from the Monastery Mine.
Symbols: ●: Garnet Megacrysts
○: Ilmenite Association

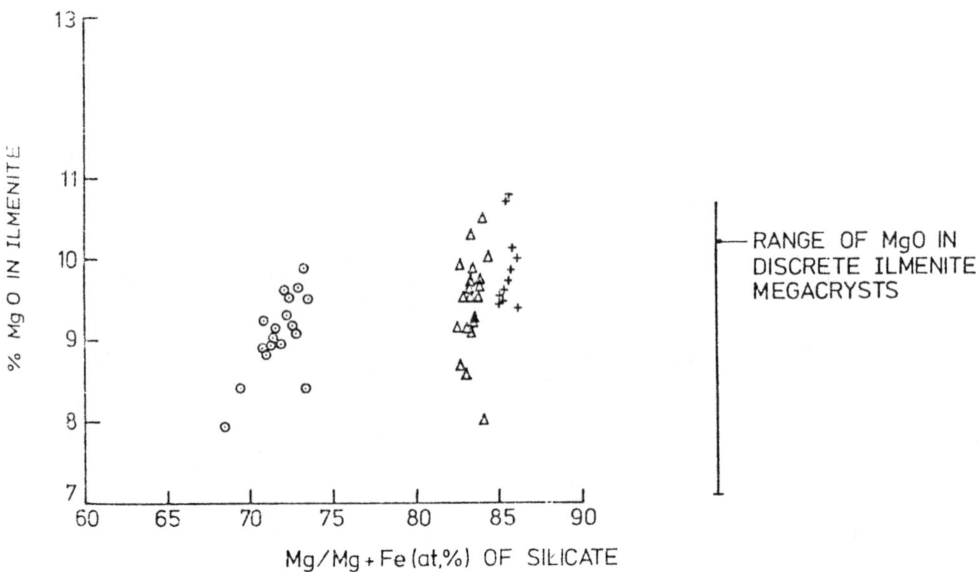

Figure 6: The variation of MgO in ilmenite with the Mg/Mg+Fe ratio of the co-existing silicate phase.
Symbols: +: Enstatite
△: Diopside
○: Garnet

2. The garnet megacrysts form a similar linear trend of increasing iron and decreasing magnesium and relatively constant calcium in association with the garnets found as small inclusions in ilmenite.
3. The orthopyroxenes define two population groups:-

Group I orthopyroxenes are clear glassy homogeneous orthopyroxenes with and without ilmenite which define a trend of decreasing calcium with increasing iron content.

Group II orthopyroxenes show a considerable spread of Mg/Fe ratio at constant calcium content. This second group is characterised by a platy appearance and by the appearance of chrome diopside exsolution lamellae and occasionally garnet exsolution lamellae along the cleavage planes.
4. The olivines fall into two population groups which can be discriminated on the basis of magnesium, iron or nickel contents as shown in Fig. 2.

Solid tie lines in Fig. 1 represent tie lines drawn between a host megacryst and a small inclusion of another mineral within that megacryst. The dotted lines are tie lines which we have inferred to be approximately correct on the basis of the natural assemblages and theoretical calculations which will be discussed in a later section.

With the exception of olivine, in which chromium was not detected, the highest chrome concentrations are found in minerals at the magnesian end of the compositional trends noted in Fig. 1, as shown for the garnets in Fig. 3. This corresponds to a trend for chromium to decrease with increasing Ca/Ca+Mg ratio in the clinopyroxenes as shown in Fig. 4.

Titanium shows similar trends in the garnets, clinopyroxenes and orthopyroxenes. Titanium increases in the silicate phase with increasing iron content up until the point where ilmenite appears in association with that mineral and from this point titanium decreases with increasing iron content. Since titanium is relatively concentrated in the garnets, the trend is best displayed by this mineral as shown in Fig. 5.

When a silicate phase and ilmenite co-exist, the Mg/Mg+Fe ratio of the silicate is covariant with the MgO content of the ilmenite as displayed in Fig. 6. Silicate phases which could be inferred to be in equilibrium with each other on the basis of the tie-lines drawn in Fig. 1 co-exist with ilmenites of similar composition, which, as noted by Mitchell (1977b) are within the compositional range of the discrete ilmenite megacrysts.

The orthopyroxenes fall into two groupings as shown in Fig. 1, with different Mg/Mg+Fe ratios. The Group I orthopyroxenes are glassy and homogeneous, even where they occur in association with ilmenite. The Group II orthopyroxenes are platy and always contain minor amounts of either cpx or cpx + gt. along cleavage partings. These latter are interpreted as exsolved phases. The two groupings are also readily distinguished with respect to their CaO contents. The variation of CaO with respect to Mg/Mg+Fe is shown in Fig. 7 and with respect to Al_2O_3 in Fig. 8.

The most magnesian and iron-rich mineral compositions for the five megacryst phases studied are given in Table 2, whilst the compositions of co-existing phases other than the common cpx/ilm., gt/ilm., and opx/ilm. are given in Table 3 and are for cpx/gt(2), ol/gt(1), ol/opx(1), ol/ilm.(1) and opx/ilm/gt(1).

The various figures and tables of data were compiled from electron microprobe analyses of 30 olivines, 27 cpx, 24 cpx/ilm, 27 garnet, 25 opx, 18 opx/ilm, and 46 ilmenite megacrysts. 19 ilmenites contained garnet inclusions which were also analysed, whilst the compositions of exsolved cpx and/or gt. were determined in 18 opx megacrysts.

Discussion

Previous studies, such as those acknowledged earlier have established the general compositional characteristics of the garnets, pyroxenes, olivines and ilmenites in the megacryst suite. The more detailed sampling of a single diatreme carried out in this study directs attention to the strong geochemical trends shown by most of the megacrysts which are consistent with igneous differentiation processes under the influence of declining temperature. All the garnet, clinopyroxene and ilmenite megacrysts, plus their inclusions and the cpx/ilmenite intergrowths fall into this category, together with the Group I orthopyroxenes and the high nickel, magnesian olivines (See Fig. 1). The low nickel olivines and the Group II orthopyroxenes must be considered separately.

The first grouping which includes most of the megacrysts displays many features which suggest that the minerals have formed in a single process. Amongst these are:-

(a) The strong linear trends shown by the clinopyroxenes, garnets and Group I orthopyroxenes in Fig. 1; in each case the trends are comprised of minerals with and without ilmenite.

(b) The similar behaviour of chromium and titanium in these three silicate phases, as shown in Figs. 3, 4 and 5, and discussed previously.

(c) The correlation in Mg/Mg+Fe ratio between coexisting ilmenites and silicates shown in Fig. 6.

(d) The similarity in the range of Mg/Mg+Fe ratio shown by ilmenites associated with cpx, garnet and opx as can also be seen in Fig. 6.

(e) Occasional finds of inclusions of one silicate in another as listed earlier.

(f) The similarity in observed maximum dimensions of the four silicate phases (15-25cm), their an-

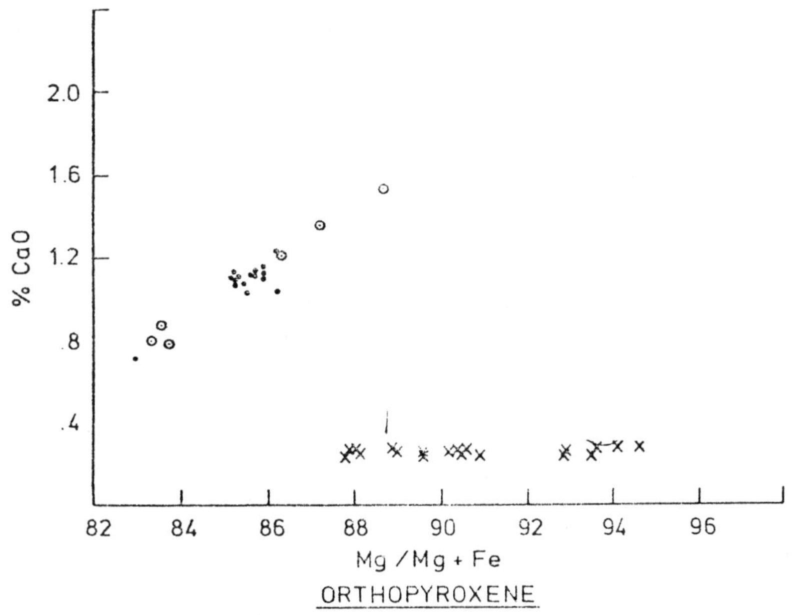

Figure 7: The variation of CaO with Mg/Mg+Fe ratio for Group I and Group II orthopyroxene megacrysts from the Monastery Mine.

Symbols: Group I { o: Opx Megacryst; ●: Opx - Ilmenite Intergrowth }

Group II ×: Opx Megacryst

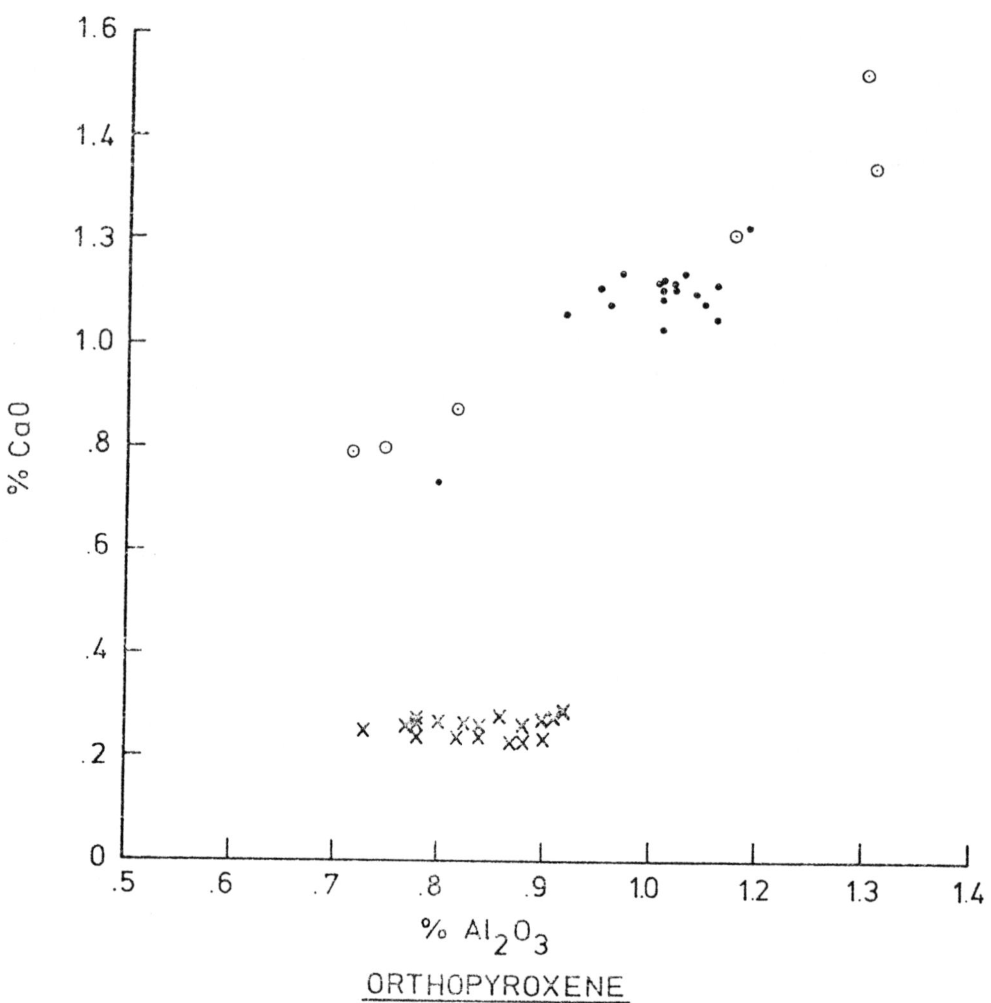

Figure 8: Plot of CaO vs Al_2O_3 variations in Group I and Group II megacrysts from Monastery Mine.

Symbols: Group I { ○ : Opx Megacryst; ● : Opx - Ilmenite Intergrowth

Group II × : Opx Megacryst.

hedral shapes and their observed presence in the single intrusive phase of the Quarry Kimberlite.

(g) Garnets which equilibrated in the presence of clino- and ortho-pyroxene have a Ca/Ca+Mg ratio close to 0.13 (O'Hara and Yoder, 1967) which can be represented by a buffer reaction of the form:-

$$\underset{cpx}{CaMgSi_2O_6} + \underset{pyrope}{Mg_3Al_2Si_3O_{12}} \leftrightarrow \underset{opx}{Mg_2Si_2O_6} + \underset{gross/pyrope}{CaMg_2Al_2Si_3O_{12}}.$$

Both the Monastery megacryst garnets and the garnet inclusions in ilmenite have calcium contents within the range 4.17-5.13 wt.% CaO, giving Ca/Ca+Mg ratios of 0.13-0.15. This is consistent with crystallisation in equilibrium with cpx and opx.

(h) The clinopyroxene, garnet and orthopyroxene compositions are entirely compatible in terms of probable temperatures of equilibration as discussed later.

Since both the olivine/ilmenite and olivine/silicate associations are rare, the direct evidence for the co-genetic origin of olivine is more tenuous than it is for the four other minerals discussed here. Nevertheless these rare associations do exist and the Mg/Mg+Fe ratio of the magnesian group of olivines is entirely within the range which it can be inferred would co-exist with the other megacryst silicates, by comparison with the Matsoku peridotite/pyroxenite suite (Cox et al., 1973). As mentioned by Boyd

TABLE 2: Megacrysts with the Highest (H) and Lowest (L) Mg/Mg + Fe Ratios in Each Group.

	1	2	3	4	5	6	7
SiO_2	40.30	39.70	38.92	38.16	55.80	55.31	41.66
TiO_2	.03	--	.02	.02	.35	.39	.88
Al_2O_3	.08	.06	.17	.04	2.46	2.53	20.66
Cr_2O_3	.03	--	--	--	.34	.02	2.09
FeO	11.70	13.20	17.78	20.40	5.34	6.36	8.11
MnO	.13	.10	.16	.20	.14	.13	.31
MgO	48.80	45.47	42.81	41.30	21.10	16.91	20.02
CaO	.09	.10	.04	.03	14.00	17.29	5.06
Na_2O	--	--	--	--	1.58	1.84	--
K_2O	--	--	--	--	ND*	.02	--
NiO	.37	.36	.10	.06	.00	.00	.00
Total	101.53	98.99	100.00	100.21	101.12	100.80	98.79
			Atomic Proportions Based on Selected No. of Oxygens				
Oxygen	4	4	4	4	6	6	12
Si	.984	1.000	.991	.983	1.976	1.991	3.012
Ti	.001	--	.000	.000	.009	.011	.048
Al	.002	.002	.005	.001	.103	.107	1.761
Cr	.001	--	--	--	.010	.001	.119
Fe	.239	.278	.379	.439	.158	.191	.490
Mn	.003	.002	.003	.004	.004	.004	.019
Mg	1.776	1.707	1.624	1.586	1.113	.907	2.157
Ca	.002	.003	.001	.001	.531	.667	.392
Na	--	--	--	--	.108	.128	--
K	--	--	--	--	--	.001	--
Ni	.007	.007	.002	.001	.000	.000	.000
Sum	3.014	2.999	3.006	3.016	4.013	4.009	8.000
	Fo 88.14 Fa 11.86	Fo 85.99 Fa 14.01	Fo 81.10 Fa 18.90	Fo 78.30 Fa 21.70	Wo 29.46 En 61.76 Fs 8.77 M: 87.56 C: 32.30	Wo 37.77 En 51.38 Fs 10.84 M: 82.57 C: 42.37	Ca 12.90 Mg 70.97 Fe 16.13 M: 81.48

Analysis No.		Description
1	H	High Ni olivine, FRB 220
2	L	High Ni olivine, RJ 44
3	H	Low Ni olivine, RJ 25
4	L	Low Ni olivine, RJ63
5	H	Clinopyroxene, FRB 5
6	L	Clinopyroxene, RJ 449
7	H	Garnet, RJ 112

*ND = not detected.

	8	9	10	11	12	13	14
SiO_2	41.13	57.31	55.49	58.85	57.05	--	.04
TiO_2	.64	.20	.16	ND*	.08	50.32	47.31
Al_2O_3	21.99	1.30	.75	.92	.78	1.17	.51
Cr_2O_3	--	.12	.02	.27	.17	.36	.06
FeO	14.19	7.30	10.94	3.63	8.28	37.23	45.22
MnO	.32	.10	.13	.09	.13	.14	.17

TABLE 2, (continued).

	8	9	10	11	12	13	14
MgO	17.22	32.12	30.89	35.63	33.44	10.03	7.06
CaO	4.34	1.53	.80	.29	.24	.04	.02
Na_2O	--	.27	.26	.07	--	--	--
K_2O	--	--	.07	ND*	--	--	--
NiO	.00	.00	.00	.00	.00	.00	.00
Total	99.83	100.25	99.51	99.77	100.17	99.29	100.39
			Atomic Proportions Based on Selected No. of Oxygens				
Oxygen	12	6	6	6	6	3	3
Si	2.999	1.984	1.970	2.003	1.979	--	.001
Ti	.035	.005	.004	--	.002	.912	.884
Al	1.890	.053	.031	.037	.032	.033	.015
Cr	--	.003	.001	.007	.005	.007	.001
Fe	.865	.211	.325	.103	.240	.751	.940
Mn	.020	.003	.004	.003	.004	.003	.004
Mg	1.871	1.657	1.634	1.808	1.729	.360	.261
Ca	.339	.057	.030	.011	.009	.001	.001
Na	--	.018	.018	.005	--	--	--
K	--	--	.003	--	--	--	--
Ni.	.000	.000	.000	.000	.000	.000	.000
Sum	8.020	3.992	4.020	3.977	4.000	2.068	2.107
	Ca 11.02	Wo 2.95	Wo 1.53	Wo .55	Wo .45	Il 60.50	Il 70.43
	Mg 60.84	En 86.07	En 82.14	En 94.07	En 87.40	Gk 39.50	Gk 29.57
	Fs 28.13	Fs 10.98	Fs 16.33	Fs 5.38	Fs 12.14	Hm .00	Hm .00
	M: 68.38	M: 88.69	M: 83.42	M: 94.59	M: 87.80		
		C: 3.31	C: 1.83	C: .58	C: .51		

Analysis No.	Description
8	L Garnet, RJ 262
9	H Group I orthopyroxene, RJ 130
10	L Group I orthopyroxene, RJ 122
11	H Group II orthopyroxene, RJ 322
12	L Group II orthopyroxene, RJ 128
13	H Ilmenite, RJ 409
14	L Ilmenite, RJ 187

ND = not detected.

and Nixon (1975) olivine is often not abundant in the megacryst suite, but there are a number of recorded instances of an olivine/silicate megacryst or olivine/ilmenite association from other localities (e.g. Boyd, 1974; Meyer et al., 1977). Furthermore the only alternative mantle derived source of olivines of the observed range of chemistry (Fo_{88-84}) would be the iron rich cumulates from Matsoku (Gurney et al., 1975) which have not been found at Monastery Mine and have much smaller grain size (<1cm). Consequently we interpret the evidence as favouring the grouping of the magnesian olivines with the clearly associated megacryst phases, that is, the clinopyroxenes, the garnets, the Group I orthopyroxenes and the ilmenites.

The second group of olivines have compositions in the range Fo_{78-81} and NiO<0.12 wt.% (See Fig. 2). They cannot be linked to the main group of megacrysts either by direct association or by inference from their chemistry. They are too iron rich to crystallise from the Quarry Kimberlite and we have no explanation for them except the somewhat evasive suggestion that they may be derived from one of the other kimberlite types found at this locality.

The Group II orthopyroxenes are the only other megacrysts not so far discussed. There appear to be at least two potential origins for these. They could be derived from the disaggregation of ultra-coarse xenoliths such as that described by Smith and Dawson (1973) or they could be orthopyroxenes which at formation had compositions which extended the Group I opx trend, and which have subsequently exsolved cpx, and sometimes garnet on cooling. The Group II orthopyroxenes exhibit a range in Mg/Mg+Fe ratios which at the magnesian end (En_{94}) corresponds to that found in coarse grained depleted garnet lherzolite or harzburgite, and which terminates at the magnesian end of the Group I opx trend (En_{88}). We favour an association of the Group II opx megacrysts with the Group I opx megacrysts at Monastery, because of the continuity of their Mg/Mg+Fe ratios which are thought to represent a differentiation index.

In this study therefore it has been inferred that all the megacrysts except the high iron low nickel olivines formed in a single process. Other workers (e.g. Boyd, 1974) have made similar interpretations, which justify the application of certain geothermometers to particular minerals. The results of such calculations are instructive and will now be discussed.

The clinopyroxene megacrysts display a range in Ca/Ca+Mg values which corresponds to a range in equilibration temperatures of 1385°-1220°C for the discrete megacrysts and of 1270°-1130°C for lamellar diopside/ilmenite intergrowths using an equation fitted to the data of Davis and Boyd (1966) for the diopside solvus.

Two clinopyroxene megacrysts contain garnet (RJ1, RJ466) and give diopside solvus temperatures of 1362°C and 1315°C respectively. Another temperature estimate can be made for these two assemblages based on the iron/magnesium distribution between cpx/gt. (Akella and Boyd, 1974). Using the equation

$$T = \frac{4620}{2.31+\ln k_D} \quad \text{which was}$$

fitted to their data, temperatures of 1345°C and 1280°C are obtained, in good agreement with the diopside solvus temperatures.

In Fig. 1 the tie lines between gt/cpx of RJ1 and RJ466 are drawn as solid lines. The close agreement between the two temperatures calculated by independent methods has encouraged us to generate inferred tie lines which are dotted lines in Fig. 1. These dotted lines are generated by calculating a diopside solvus temperature for a particular cpx megacryst and using the Akella and Boyd equation to predict the FeO/MgO ratio of the co-existing garnet. The tie lines generated in this way have very similar slopes to those for RJ1 and RJ466. Therefore the ranges in equilibrium temperature calculated for gt/cpx by the Akella and Boyd equation and by the diopside solvus method for all the cpx megacrysts are virtually identical (252°C and 255°C).

The Group I and Group II orthopyroxenes retain quite different thermal records. The former are glassy and homogeneous, and occur with and without ilmenite. Applying the empirical geothermometer of Boyd and Nixon (1973) to the Group I orthopyroxenes without ilmenite gives a range of equilibration temperatures of 1410°C-1190°C, which closely matches the range of 1385°C-1220°C observed for the discrete clinopyroxenes reported previously.

The Group I orthopyroxenes with ilmenite (17) appear in contrast, to have equilibrated at relatively constant temperature as can be inferred from the plot of CaO against Mg/Mg+Fe shown in Fig. 7. This restricted range would not match the range for ilmenite bearing clinopyroxene nor garnet in ilmenite. However it is suggested that the orthopyroxene composition is buffered by a reaction of the type:-

$$FeTiO_3 + Mg_2Si_2O_6 \leftrightarrow MgTiO_3 + MgFeSi_2O_6$$

and that the geothermometer therefore does not apply in the presence of ilmenite.

The Group II orthopyroxenes all contain apparently exsolved clinopyroxene and sometimes garnet. Except for one sample the mineral compositions suggest that these enstatites have re-equilibrated with sub-solidus exsolution to temperatures of between 920°-960°C.

On the basis of experiments in the systems MAS (MacGregor, 1974) and CMAS (Akella, 1976) the amount of aluminium in orthopyroxene has been used to estimate equilibration pressures. We have attempted to predict which orthopyroxene compositions would co-exist with particular garnet and diopside compositions on the basis of observed mineral associations, inferred tie lines and matched temperatures. We have then calculated equilibrium pressures for the full range of Group I orthopyroxenes using a calculation procedure devised by Fraser (per. comm.), used in Fraser and Lawless (1978) and similar to that described by Wood (1974). The calculations yield equilibration pressures of 45.3±1 kbars for all the Group I enstatites.

There are considerable uncertainties in the geothermometry and particularly in the geobarometry involved in these calculations which have been discussed elsewhere (e.g. O'Hara and Yarwood, 1978). These uncertainties will not greatly affect our interpretation of the chemistry of the Monastery megacrysts providing two major aspects can be sustained; namely that the mineral compositions reflect

1) Equilibrium temperatures with a range of several hundred degrees centigrade.

2) Equilibrium pressures which are essentially isobaric.

It is reasonable to expect that calculations of the equilibration conditions of the megacryst suite of minerals will be more accurate than for peridotite minerals because of the minor role of chromium in the megacrysts and because the higher concentrations of sodium in megacryst orthopyroxenes reduce the analytical uncertainty in the determination of this critical constituent.

TABLE 3. Coexisting Phases: Megacrysts (M) and Inclusions (I).

	1	2	3	4	5	6	7	8	9	10	11	12
SiO_2	55.39	41.87	54.18	41.52	39.60	41.27	39.60	56.28	39.46	--	55.54	41.10
TiO_2	.36	1.06	.46	1.01	.03	.99	.05	.31	.02	51.73	.21	.84
Al_2O_3	2.67	21.73	2.49	21.47	.03	21.31	.04	1.17	.04	.55	.92	21.80
Cr_2O_3	.26	.92	.13	.47	.04	1.06	ND*	.06	ND	.62	--	.07
FeO	5.61	9.64	6.21	11.28	12.38	9.48	14.40	8.77	18.37	33.79	9.47	12.20
MnO	.13	.23	.10	.23	.08	.20	.12	.10	.19	.33	.11	.25
MgO	19.82	20.35	18.71	18.74	46.82	20.02	45.20	32.21	43.09	12.47	32.42	18.00
CaO	14.32	4.39	15.53	4.78	.09	4.74	.07	1.36	.04	.06	1.06	4.46
Na_2O	1.87	--	1.81	--	--	--	--	.28	--	--	.22	--
K_2O	.04	--	.03	--	--	--	--	--	--	--	--	--
NiO	.00	.00	.00	.00	.41	.00	.36	.00	.00	.00	.00	.00
Total	100.47	100.19	99.65	99.50	99.48	99.07	99.85	100.54	101.22	99.55	99.95	98.72
Atomic Proportions Based on Selected No. of Oxygens												
Oxygen	6	12	6	12	4	12	4	6	4	3	6	12
St	1.980	2.988	1.968	3.006	.990	2.984	.995	1.960	.994	--	1.952	3.005
Ti	.010	.057	.013	.055	.001	.054	.001	.008	.000	.921	.006	.046
Al	.112	1.828	.107	1.832	.001	1.816	.001	.048	.001	.015	.038	1.879
Cr	.007	.052	.004	.027	.001	.061	--	.002	--	.012	--	.004
Fe	.168	.575	.189	.683	.259	.573	.303	.255	.387	.669	.278	.746
Mn	.004	.014	.003	.014	.002	.012	.003	.003	.004	.007	.003	.015
Mg	1.056	2.165	1.013	2.022	1.745	2.157	1.692	1.671	1.617	.440	1.698	1.962
Ca	.548	.336	.604	.371	.002	.367	.002	.051	.001	.002	.040	.349
Na	.130	--	.127	--	--	--	--	.019	--	--	.015	--
K	.002	--	.001	--	--	--	--	--	--	--	--	--
Ni	.000	.000	.000	.000	.008	.000	.007	.000	.000	.000	.000	.000
Sum	4.016	8.015	4.029	8.010	3.008	8.024	3.004	4.017	3.005	2.065	4.031	8.007
	Wo 30.95	Ca 10.92	Wo 33.47	Ca 12.06	Fo 87.08	Ca 11.85	Fo 84.83	Wo 2.57	Fo 80.69	Il 52.23	Wo 1.98	Ca 11.43
	En 59.58	Mg 70.38	En 56.08	Mg 65.74	Fa 12.92	Mg 69.64	Fa 15.17	En 84.52	Fa 19.31	Gk 47.77	En 84.22	Mg 64.16
	Fs 9.46	Fe 18.71	Fs 10.45	Fe 22.21		Fe 18.51		Fs 12.91		Hm .00	Fs 13.80	Fe 24.40
	M: 86.29	M: 79.00	M: 84.30	M: 74.75		M: 79.01		M: 86.75			M: 85.92	M: 72.45
	C: 34.19		C: 37.37					C: 2.95			C: 2.30	

Analysis No.	Description	
1	M	CPX RJ1
2	I	Garnet RJ1
3	M	CPX RJ466
4	I	Garnet RJ466
5	M	Olivine RJ474
6	I	Garnet RJ474
7	M	Olivine RJ473
8	I	OPX RJ473
9	M	Olivine RJ422
10	I	Ilmenite RJ422
11	M	OPX RJ460
12	I	Garnet RJ460

*ND = not determined.

TABLE 4. Major Element Analysis of Inclusion in Olivine RJ469 and of Monastery Quarry Kimberlite, PHN 1870

Sample No.	RJ 469 wt.%	PHN 1870[+] wt.%
SiO_2	31.4	27.98
TiO_2	5.5	4.22
Al_2O_3	2.63	2.64
Cr_2O_3	0.15	0.14
FeO*	13.9	11.58
MnO	0.15	0.18
MgO	25.9	26.17
CaO	8.97	9.16
Na_2O	0.38	0.64
K_2O	1.45	1.78
H_2O^-	N.D.	0.40
H_2O^+	N.D.	7.33
P_2O_5	N.D.	0.94
CO_2	present	5.83
S	N.D.	0.09
TOTAL	91.8	99.86

* Total iron as FeO

N.D. Not determined

+ From Gurney and Ebrahim (1973).

The Monastery olivine megacrysts are unstrained crystals which show no sign of recrystallisation. They form the two populations shown in Fig. 1. Boyd (1974) analysed 12 megacrysts from Monastery covering the range 0.80 - 0.88 forsterite but published only three analyses including the extreme values. These fall within the two populations established in this study. The magnesium rich olivine megacryst population has a similar range in composition to the groundmass olivines in the matrix of the Monastery Kimberlite as reported by Boyd (1974). This suggests a genetic link. During the present study, all sizes of olivines with Mg/Mg+Fe ratios near 0.86 have been found in the kimberlite enclosing some of the megacrysts.

The composition of the olivine in the two lherzolites from Monastery which we have analysed is more magnesian than any megacrysts or groundmass olivines. If the two lherzolites are representative of the ultrabasic nodules at Monastery then the megacrysts could not have formed by fragmentation of coarse grained lherzolites.

Use has been made of the iron magnesium distribution coefficient in an attempt to predict the olivine compositions which would be in equilibrium with certain liquids (Roeder and Emslie, 1970; Roeder, 1974). The distribution coefficient has been shown to be ~0.3 (Roeder and Emslie, 1970; O'Hara et al., 1975) and to be relatively insensitive to pressure, temperature and oxygen fugacity (D.G. Fraser, pers. comm.). For a constant ${}^{ol}_{liq}K_D$ Fe,Mg of 0.30 the calculated liquids in equilibrium with the Monastery olivine megacrysts are much more iron rich than those produced from two garnet lherzolites from kimberlite (Kushiro, 1973).

The distribution of nickel between olivine and liquid has a distribution coefficient which is generally taken to assume a value between 10 and 18 for melt of basaltic compositions (e.g. Hakli and Wright, 1967; Leeman, 1973 and Banno and Matsui, 1973). However, O'Hara et al. (1975) argue that the ${}^{ol}_{liq}K_D$ nickel is probably close to unity for melt of peridotitic composition and increases as the Fe/Mg ratio of the melt increases. Calculations suggest that for the magnesian olivine megacrysts to have crystallised from a basaltic melt of 250-800 ppm NiO ${}^{ol}_{liq}K_D$ Ni will assume a value between 16 and 5 which falls into the range expected. The iron-rich olivine megacrysts would require a ${}^{ol}_{liq}K_D$ Ni of 3.2-1.0 to have crystallised from such a melt. This range is lower than that expected from the iron-rich nature of the olivines. These considerations suggest that whilst the magnesian olivine megacrysts could have co-existed with a melt with an FeO/MgO ratio of 0.80-0.88 and a nickel concentration of 250-800 ppm NiO, the iron-rich olivines could not. The experiments of Kushiro (1973) indicate that such a magma could be produced by 10-25% partial melting of mantle peridotites with Fe/Mg+Fe ratios near 0.13.

Extrapolations of tie lines for the Monastery megacryst suite suggest that the compositional gap of the olivines corresponds to the crystallisation range of the other silicate-ilmenite intergrowths and only one olivine ilmenite has been found. The absence of olivine in the range in which ilmenite crystallised together with other silicates could be due to one of several reasons. Olivine could have stopped cyrstallising, it could have entered into a reaction relationship with the liquid or its composition

TABLE 5. Chemical Parameters for Minerals in a Kimberlite Inclusion in an Olivine, for Magnesian Olivine Megacrysts and for Lherzolites. (Inclusion Data from Haggerty and Boyd, 1975; Lherzolite Data from Boyd and Nixon, 1973; Boyd and Nixon, 1975).

	Inclusion	Megacrysts	Lherzolites[a]
Olivine			
Mg/Mg+Fe at.%	86-88	84.6-87.4	91.3 ; 93.8
Garnet			
TiO_2 wt.%	0.89	0.6- 1.5	0.48; 0.09
Cr_2O_3 wt.%	1.28	0-2	3.07; 5.00
Mg/Mg+Fe at.%	81.6	71-81	85; 87.9

[a] FRB 1 and 1859 N respectively.

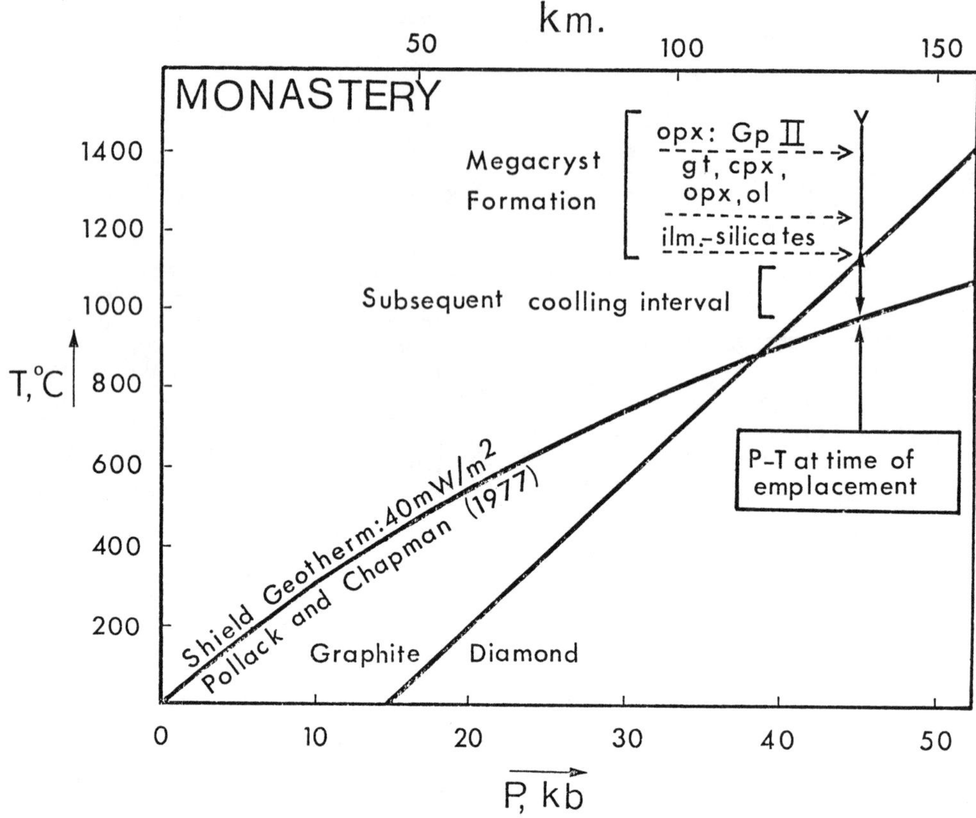

Fig. 9. Schematic diagram to illustrate the proposed conditions of formation of the megacrysts and the Quarry kimberlite at the Monastery Mine.

might be buffered by a reaction such as $FeSiO_4+MgTiO_3 \leftrightarrow MgSiO_4+FeTiO_3$. Attention is also drawn to the hematite content of the ilmenite (5%-27% He). As the proposed magma, from which the Monastery megacrysts crystallised, differentiates towards more iron rich compositions it is possible that the oxygen fugacity also increases with the resultant formation of successively greater amounts of ferric iron. Although the total iron content of the magma is therefore greater, it does not necessarily follow that the Fe^{2+}/Mg^{2+} ratio, which dictates the olivine composition, has also increased. Certainly in such a situation this ratio will increase more slowly than at constant oxygen fugacity and one can envisage that under certain circumstances it may not increase at all or may even decrease.

Such a situation does not explain the presence of olivine with lower iron contents although compositional gaps in olivine compositions are very common in differentiated rock series. One possibility is that the two olivine populations are derived from two kimberlite intrusions at Monastery.

A number of olivine megacrysts have been found to contain inclusions of kimberlite. One of these particular inclusions, approximately 4 mm in diameter and of tubular form, has been analysed. Its bulk chemistry is very similar to that of the Quarry Type Kimberlite from Monastery although the inclusion has slightly higher iron and lower sodium and volatile contents (Table 4). Several minerals crystallised from within the inclusion but no detailed work has been done on them. Some minerals in another similar inclusion have been analysed by Haggerty and Boyd (1975) and are compared to the Monastery megacrysts and lherzolites in Table 5. The similarity of the minerals in the inclusion to the megacrysts is noteworthy. Haggerty and Boyd (1975) favour an entrapped liquid hypothesis to that of infilling by later kimberlite. The entrapped liquid hypothesis assumes that liquids of the composition of the inclusion (i.e. kimberlitic) were present during crystallisation of the olivine megacrysts.

Interpretation

The iron rich, nickel poor olivines are interpreted as being of exotic origin.

The rest of the megacrysts are thought to have crystallised during a single igneous differentiation process of limited extent under essentially isobaric conditions.

It may be possible to relate the melting to

the events that gave rise to the Karoo volcanic event. A large partial melt produced by the increase in heat flow during the breakup of Gondwanaland is regarded by Cox (1970) to have produced the Karoo volcanism. Large volumes of this melt differentiated on the way up to shallower depths and gave rise to the Karoo tholeiites. Small pockets of this melt might have been trapped at depth. Mitchell and Brunfeld (1975) suggest that the decrease in heat flow after the breakup of Gondwanaland results in small degrees of melting of partially depleted Upper Mantle material after the basaltic melts that gave rise to the Karoo volcanism were removed. These small degrees of partial melts will give more undersaturated basaltic and alkaline melts. Partial melts produced from lherzolites when diopside is the major phase to melt (together with any hydrous phases present) is also compatible with the rare earth pattern found for the diopside and diopside ilmenite intergrowths at Monastery (Gurney et al., 1973; Mitchell et al., 1973).

The kimberlite inclusions in the olivine megacryst described earlier suggest that the melt present during crystallisation of the megacrysts reached CO_2 and water rich kimberlitic composition during the crystallisation sequence. The presence of CO_2 causes liquidus boundaries in the system diopside forsterite silica to move to more magnesium rich compositions (Eggler, 1974). Consequently, orthopyroxene and not olivine was the first mineral to crystallise. The now lamellar Group II enstatites (Mg/Mg+Fe 0.88 to 0.94) are therefore regarded to be the first megacryst minerals to have formed. The subsequent crystallisation sequence is summarised in Fig. 9. It is relevant to point out that at the temperatures and pressures which we have suggested for the formation of the megacrysts, diamond could not form and that only in the subsequent cooling interval is the system within the diamond stability field, thus providing a rational explanation for the absence of minerals with megacryst chemistry as inclusions in diamonds. The evidence for subsequent cooling after the crystallisation of the ilmenite/silicate association is provided by the Group II orthopyroxenes.

The time scale for the crystallisation sequence and the subsequent cooling of the system from which the megacrysts formed must have been large since the total decline in temperature is of the order of 450°C. The period between the cessation of Karoo volcanism and the pipe intrusion event is of the order of 60 m.y. and it is suggested that this is the period over which the cooling occurred.

The principal differences between this model and that of Nixon and Boyd (1973c) or Pasteris et al. (this volume) are in the vertical extent of the megacryst body from which the mineral suite is derived and the composition of the interstitial residual liquid. Irving (1974) discusses mechanisms of megacryst formation and states that "the implications are compelling that pyroxene megacrysts are fragments of loosely cemented, pegmatitic, polycrystalline aggregates precipitated from a basaltic magma". A pegmatitic intergrowth can also be suggested for the Monastery megacrysts, where only a small amount of liquid with kimberlitic affinities is dispersed among the megacrysts. Whether or not the interstitial liquid has sufficient volume to give rise to the Monastery kimberlite is difficult to assess but the presence of trapped inclusions in megacryst olivine with essentially the composition of the Quarry Type Kimberlite is strong evidence in favour of such a contention. Accordingly we favour a model in which the Quarry Type Kimberlite is the final evolved liquid from a magmatic differentiation event in the Mantle during which the megacrysts formed as phenocrysts, although this is difficult to reconcile with observed kimberlite and olivine megacryst MgO/FeO ratios.

The model has been criticised (Pasteris et al: this volume) on the grounds of the large range in equilibration temperatures found for the pyroxene discrete nodules in this and previous studies. Our contention is that such wide ranges are inevitable for any magma body which cools from high temperatures to low temperatures in situ in a peridotitic mantle whilst crystallising no hydrous phases. The melt in such a closed system will have a progressively increased water content and the large effects of water on the peridotite solidus are very well known (e.g. Kushiro et al., 1968).

In this study, we have not determined the composition of any of the phlogopite megacrysts which are also found in the Quarry Type Kimberlite at Monastery. This was due to the fact that where they are exposed in the kimberlite they are vermiculitised. The experimental results of Bravo and O'Hara (1975), on the melting of water saturated phlogopite-peridotites at 30 kilobars, suggests that the phlogopite megacrysts will only have crystallised at the lower end of the temperature scale of differentiation which has been outlined here, and therefore will not negate the above argument.

Acknowledgments. The authors wish to acknowledge the whole hearted support of the Anglo American Corporation both in the use of facilities and for financial assistance.

This project has benefited from discussions with colleagues in Cape Town and with other scientists, in particular Dr. B. Harte, E. Wyatt, Dr. S.E. Haggerty and Dr. D.G. Fraser.

References

Akella, J., Garnet-pyroxene equilibria in the system $CaSiO_3$ - $MgSiO_3$ - Al_2O_3 and in a natural mineral mixture, Amer. Mineralogist, 61, 589-598, 1976.

Akella, J., and F.R. Boyd, Petrogenic grid for garnet peridotites, Carnegie Inst. Year Book, 73, 269-273, 1974.

Allsop, H.L., and D.R. Barrett, Rb-Sr age determinations on South African kimberlite pipes, Phys. Chem. Earth, 9, 605-618, 1975.

Banno, S., and Y. Matsui, On the formulations of partition coefficients for trace element distribution between minerals and magma, Chem. Geol., 11, 1-15, 1973.

Boyd, F.R., Olivine megacrysts from the kimberlites of the Monastery and Frank Smith Mines, South Africa, Carnegie Inst. Year Book, 73, 282-285, 1974.

Boyd, F.R., and R.V. Danchin, Discrete nodules from the Artur De Paiva Kimberlite, Angola, Carnegie Inst. Year Book, 73, 278-282, 1974.

Boyd, F.R., and J.B. Dawson, Kimberlite garnets and pyroxene-ilmenite intergrowths, Carnegie Inst. Year Book, 71, 376-378, 1972.

Boyd, F.R., and P.H. Nixon, Origin of the ilmenite-silicate nodules in kimberlites from Lesotho and South Africa, in Lesotho Kimberlites, pp. 254-268, 1973.

Boyd, F.R., and P.H. Nixon, Origins of the ultramafic nodules from some kimberlites of N. Lesotho and the Monastery Mine, South Africa, Phys. Chem. Earth, 9, 431-455, 1975.

Bravo, M.S., and M.J. O'Hara, Partial Melting of Phlogopite-bearing synthetic spinel and garnet lherzolites, Phys. Chem. Earth, 9, 845-854, 1975.

Brookins, D.G., Ilmenite - (serpentinized) pyroxene nodules from the Stockdale kimberlite pipe, Riley county, Kansas (abstract), Abstr. with programs from 1971, Geol. Soc. Amer., 3, 233 p., 1971.

Cox, K.G., Tectonics and vulcanism of the Karroo period and their bearing on the postulated fragmentation of Gondwanaland, in African magmatism and tectonics, edited by T.N. Clifford and I.G. Gass, pp. 211-235, Oliver and Boyd, Edinburgh, 1970.

Cox, K.G., J.J. Gurney, and B. Harte, Xenoliths from the Matsoku pipe, in Lesotho Kimberlites, edited by P.H. Nixon, pp. 76-100, Cape and Transvaal, 1973.

Davis, B.T.C., and F.R. Boyd, The join $Mg_2Si_2O_6$-$CaMgSi_2O_6$ at 30 kb pressure and its application to pyroxenes from kimberlite, J. Geophys. Res., 71, 3567-3576, 1966.

Davis, G.L., T.E. Krogh, and A.J. Erlank, The Ages of zircons from kimberlites from South Africa, Carnegie Inst. Year Book, 75, 821-824, 1976.

Dawson, J.B., and A.M. Reid, A pyroxene-ilmenite intergrowth from the Monastery Mine, South Africa, Contrib. Mineral. Petrol., 26, 296-301, 1970.

Eggler, D.H., quoted in Thermodynamics in Geology, by B.J. Wood and D.G. Fraser, D. Reidel Publ. Co., Holland, 1974.

Eggler, D.H., and M.E. McCallum, A Geotherm from megacrysts in the Sloan kimberlite pipes, Colorado, Carnegie Inst. Year Book, 75, 538-541, 1976.

Fraser, D.G., and P.J. Lawless, Paleogeotherms: implications of disequilibrium in garnet lherzolite xenoliths, Nature, 273, 220-221, 1978.

Frick, C., Intergrowth of orthopyroxene and ilmenite from Frank Smith Mine, near Barkly West, South Africa, Trans. Geol. Soc. S. Africa, 76, 195-200, 1973.

Gurney, J.J., and S. Ebrahim, Chemical composition of Lesotho kimberlites, in Lesotho Kimberlites, edited by P.H. Nixon, pp. 280-284, Cape and Transvaal, 1973.

Gurney, J.J., H.W. Fesq, and E.J.D. Kable, Clinopyroxene-ilmenite intergrowths from kimberlite: A re-appraisal, in Lesotho Kimberlites, edited by P.H. Nixon, pp. 238-253, Cape and Transvaal, 1973.

Gurney, J.J., and S.E. Haggerty, An ilmenite-pyroxene intergrowth from the Premier Mine, long abstracts, Kimberlite Symposium, Cambridge, July, 1975.

Gurney, J.J., B. Harte, and C.G. Cox, Mantle xenoliths in the Matsoku kimberlite pipe, Phys. Chem. Earth, 9, 507-524, 1975.

Haggerty, S.E., and F.R. Boyd, Kimberlite inclusions in an olivine megacryst from Monastery, long abstracts, Kimberlite Symposium, Cambridge, July, 1975.

Hakli, T., and T.L. Wright, The fractionation of nickel between olivine and augite as a geothermometer, Geochim. Cosmochim. Acta, 31, 877-884, 1967.

Harte, B., Kimberlite nodules, upper mantle petrology and geotherms, Phil. Trans. R. Soc. Lond. A., 288, 487-500, 1978.

Irving, A.J., Megacrysts from the newer basalts and other basaltic rocks of South Eastern Australia, Geol. Soc. Amer. Bull., 85, 1503-1514, 1974.

Jakob, W.R., Geochemical aspects of the megacryst suite from the Monastery kimberlite pipe, unpublished M.Sc. thesis, University of Cape Town, 81 p., 1977.

Kushiro, I., Partial melting of garnet lherzolites from kimberlite at high pressure, in Lesotho Kimberlites, edited by P.H. Nixon, pp. 294-299, Cape and Transvaal, 1973c.

Kushiro, I., Y. Syono, and S. Akimoto, Melting of a peridotite nodule at high pressures and high water pressures, J. Geophys. Res., 73, 6023-6029, 1968.

Lawless, P.J., Some aspects of the geochemistry of kimberlite xenocrysts, unpublished M.Sc. thesis, University of Cape Town, 121 p., 1974.

Leeman, W.P., Partitioning of Ni and Co between olivine and basaltic liquid: an experimental study, EOS, 54, 1222-1223, 1973.

MacGregor, I.D., The system MgO-Al_2O_3-SiO_2: solubility of Al_2O_3 in enstatite for spinel and garnet peridotite composition, Amer. Mineralogist, 59, 110-119, 1974.

MacGregor, I.D., and R.W. Wittkop, Diopside-ilmenite xenoliths from the Monastery Mine, Orange Free State, South Africa (abstract), Abstr. with program for 1970, Geol. Soc. Amer., 2, 113 p., 1970.

Meyer, H.O.A., H.M. Tsai, and J.J. Gurney, Enstatite xenocryst containing co-existing Cr-poor and Cr-rich garnet, Weltevreden Floors, South Africa - Extended Abstracts, Second International Kimberlite Conference, Santa Fe, New Mexico, 1977.

Mitchell, R.H., Ultramafic xenoliths from the Elwin Bay kimberlite: the first Canadian paleogeotherm, Can. J. Earth. Sci., 14, 1202-1210, 1977a.

Mitchell, R.H., Geochemistry of magnesian ilmenites from kimberlites in South Africa and Lesotho, Lithos, 10, 29-37, 1977b.

Mitchell, R.H., A.O. Brunfelt, and P.H. Nixon, Part II: Trace elements in magnesian ilmenites from Lesotho kimberlites, in Lesotho Kimberlites, edited by P.H. Nixon, pp. 230-235, Cape and Transvaal, 1973.

Mitchell, R.H., and A.O. Brunfelt, Rare earth element geochemistry of kimberlite, Phys. Chem. Earth, 9, 671-686, 1975.

Mitchell, R.H., D.A. Carswell, and O.A. Brunfelt, Part I: Ilmenite association trace element studies, in Lesotho Kimberlites, edited by P.H. Nixon, pp. 224-229, Cape and Transvaal, 1973.

Nixon, P.H., and F.R. Boyd, Petrogenesis of the granular and sheared ultrabasic nodule suite in kimberlites, in Lesotho Kimberlites, edited by P.H. Nixon, pp. 48-56, Cape and Transvaal, 1973a.

Nixon, P.H., and F.R. Boyd, Notes on the heavy mineral concentrates (from Monastery) in Lesotho Kimberlites, edited by P.H. Nixon, pp. 218-220, Cape and Transvaal, 1973b.

Nixon, P.H., and F.R. Boyd, The discrete nodule (megacryst) association in kimberlites from northern Lesotho, in Lesotho Kimberlites, edited by P.H. Nixon, pp. 67-75, Cape and Transvaal, 1973c.

Nixon, P.H., and F.R. Boyd, garnet-bearing ultrabasic and discrete nodule suites from Malaita, Solomon Islands, S.W. Pacific, and their bearing on an oceanic geotherm, Extended Abstracts, Second International Kimberlite Conference, Santa Fe, New Mexico 1977.

O'Hara, M.J., M.J. Saunders, and E.L.P. Mercy, Garnet-peridotite, primary ultrabasic magma and eclogite; interpretation of upper mantle processes in kimberlite, Phys. Chem. Earth, 9, 571-604, 1975.

O'Hara, M.J., and G. Yarwood, High pressure-temperature point on an archaean geotherm, magma genesis by implied anatexis and consequences for garnet-pyroxene thermometry and barometry, Phil. Trans. R. Soc. Lond. A., 288, 441-453, 1978.

O'Hara, M.J., and H.S. Yoder, Formation and fractionation of basic magmas at high pressures, Scot. J. Geol., 3, 67-117, 1967.

Pasteris, J.D., F.R. Boyd, and P.H. Nixon, The Frank Smith Ilmenite association, this volume, 1978.

Ringwood, A.E., and J.F. Lovering, Significance of pyroxene-ilmenite intergrowths among kimberlite xenoliths, Earth Planet. Sci. Letts., 7, 371-375, 1970.

Roeder, P.L., and R.F. Emslie, Olivine-liquid equilibrium, Contr. Mineral. Petrol., 29, 275-289, 1970.

Smith, J.V., J.B. Dawson, and F.I. Bishop, Sulphide-oxide and silicate-oxide intergrowths in xenoliths of upper mantle peridotite, Extended Abstracts, First International Kimberlite Conference, Cape Town, South Africa, 1973.

Smith, C.B., M.E. McCallum, and D.H. Eggler, Clinopyroxene-ilmenite intergrowths from the Iron Mountain kimberlite district, Wyoming, Carnegie, Inst. Year Book, 75, 542-546, 1976.

Wagner, P.A., The diamond fields of southern Africa, The Transvaal Leader, Johannesburg, second impression, 1971, 355 p., Struik, Cape Town, South Africa, 1914.

Whitelock, T.K., The Monastery Mine kimberlite pipe, in Lesotho Kimberlites, edited by P.H. Nixon, pp. 214-218, Cape and Transvaal, 1973.

Williams, A.F., The genesis of the diamond, E. Benn Ltd., London, 2 vols., 1932.

Wyatt, B.A., The melting and crystallisation behaviour of a natural clinopyroxene-ilmenite intergrowth, Contr. Mineral. Petrol., 61, 1-9, 1977.

Wood, B.J., The solubility of alumina in orthopyroxene co-existing with garnet, Contr. Mineral. Petrol., 46, 1-15, 1974.

PARTIAL THERMAL HISTORY OF TWO EXSOLVED CLINOPYROXENES FROM THE THABA PUTSOA KIMBERLITE PIPE, LESOTHO

Robert H. McCallister, Henry O.A. Meyer and Ricardo Aragon

Department of Geosciences, Purdue University, West Lafayette, Indiana 47907

Abstract. Diopside solid solutions (Di_{ss}) from a sheared lherzolite and a discrete megacryst from the Thaba Putsoa Kimberlite Pipe, Lesotho, have been examined by means of transmission electron microscopy. The clinopyroxene in the sheared lherzolite (PHN 1611) has a Ca/(Ca+Mg) of 0.315, whereas the ratio is equal to 0.320 in the discrete megacryst (PHN 1600E4). Crushed grains of each sample were observed with a Siemens Elmiskop I, operating at 100kV. In both specimens exsolution lamellae of pigeonite are present and appear to be parallel to (001) of the diopside host. The average of at least 4 measurements yields values of lamellae thicknesses of 187±8 Å for PHN 1600E4 and 192±12 Å for PHN 1611. The scale of the exsolution microstructures or wavelength (λ) indicates that the process occurred rapidly.

From the recent experimental results of McCallister and Yund (1977) on exsolution mechanisms and microstructures in iron-free pyroxenes, it is possible to delineate the (001) coherent spinodal for diopside solid solutions with Ca/Ca+Mg $\geq$ 0.27. For a Ca/Ca+Mg = 0.317 the (001) coherent spinodal is located between 1275°-1300°C. At 1300°C, although well within the solvus, no exsolution occurred in the experimental runs; however, at and below 1275°C exsolution occurred by spinodal decomposition with a resulting exsolution microstructure remarkably similar to that observed in the natural clinopyroxenes.

In an attempt to arrive at a thermal history of the latter specimens, a series of continuous cooling experiments was made on a synthetic Di_{ss} (Ca/Ca+Mg = .317). For a cooling rate equivalent to 20°C/hr, a λ of 178±10 Å was obtained. This result compares favorably with that from the natural materials in which the average λ is 190±10 Å, and suggests that their cooling time from 1300°-1000°C was on the order of 15 hours.

Introduction

In the past few years a tremendous scientific effort has been directed toward understanding the various aspects associated with the intrusion of kimberlites and the xenoliths contained within them. Boyd (1973) initiated the first major study to determine pressures and temperatures of equilibration of lherzolite xenoliths by comparing the chemistry of component phases with the available phase equilibria in synthetic systems. As a consequence of this he was able to establish a fossil geotherm for northern Lesotho and suggest models to describe it. One interesting result of this study was that a number of clinopyroxenes in what were described as sheared lherzolites had Ca/(Ca+Mg) as low as 0.28. Also, discrete clinopyroxene megacrysts from the same locality had a range in Ca/(Ca+Mg) of 0.305 to 0.339 (Nixon and Boyd, 1973). In none of the descriptions of these specimens was any mention made of the appearance of intergrowths. However, an initial examination by single crystal X-ray techniques confirmed the presence of two clinopyroxenes in a Di_{ss} from a sheared lherzolite (PHN 1611) and in a discrete Di_{ss} megacryst (PHN 1600 E4) from Thaba Putsoa, Lesotho (Fig. 1). In both cases the exsolved pigeonite solid solution (Pig_{ss}) is parallel to (001) of the host Di_{ss}, and the bulk Ca/(Ca+Mg) ratios are 0.315 and 0.320, respectively. In order to observe the microstructures directly, crushed grains of the samples were examined by means of transmission electron microscopy. Average lamellae thicknesses are 187±12 Å for PHN 1600 E4 and 192±12 Å for PHN 1611. These low values indicate that the cooling rate was sufficiently rapid so as to prevent little coarsening of the microstructure.

Spinodal decomposition is an exsolution mechanism which occurs rapidly relative to nucleation and growth when the pyroxene is quenched into the area bounded by the coherent spinodal curve (Yund and McCallister, 1970). This mechanism appears to have been favored during the exsolution of the Lesotho pyroxenes, as will be shown later. As a result of the recent experimental study by McCallister and Yund (1977), it has been possible to determine a portion of the (001) coherent spinodal for Di_{ss}'s on the $CaMgSi_2O_6$-$Mg_2Si_2O_6$ join. Also, from the work of McCallister (1978) in which he studied the coarsening kinetics of a Di_{ss} with a composition

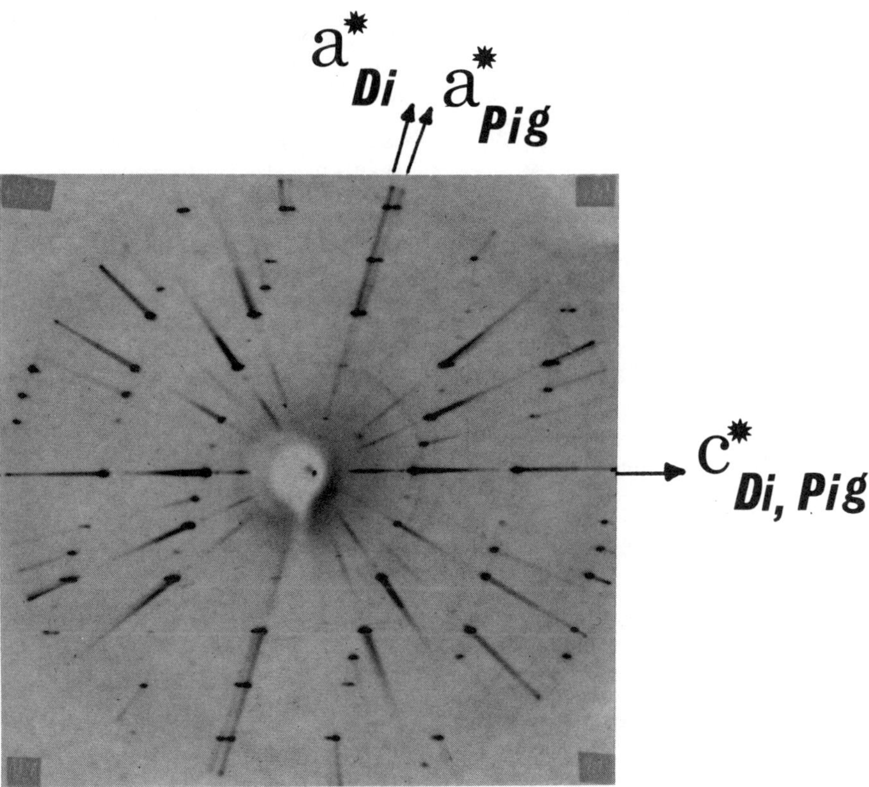

Fig. 1. $h0\ell$ precession photograph (MoKα; unfiltered X-radiation) of a grain of clinopyroxene from PHN 1611. Pig and Di are abbreviations for pigeonite solid solution and diopside solid solution, respectively.

0.541 $CaMgSi_2O_6$-0.459 $Mg_2Si_2O_6$ ($Di_{54.1}$) it was possible to obtain a quantitative relationship between exsolution wavelength or the interlamellar spacing and time at temperatures of 1300°, 1200° and 1100°C. These experiments coupled with continuous cooling runs using a synthetic Di_{ss} with Ca/Ca+Mg = 0.317 have made it possible to delineate a portion of the thermal history of the natural clinopyroxenes previously described. The results of this study suggest that the nodules in the kimberlite cooled rapidly through the temperature interval of approximately 1300°C, at which exsolution was initiated, to 1000°C where the process effectively ceased. This conclusion is consistent with the model calculations of kimberlite intrusion by McGetchin and Ullrich (1973), McGetchin, et al. (1973) and Mercier (this volume) which suggest a rapid rate of ascent.

Methods

The synthesis of all starting materials was carried out by crystallizing a glass of the appropriate composition at 20 Kbar and 1500°C for 1 hr in a 3/4: piston-cylinder apparatus. Platinum capsules were used to hold approximately 200 mg of charge which had been initially prepared by melting a gel, and W3Re-W25Re thermocouples were used to monitor temperature. Samples were annealed in Pt capsules either isothermally or by dropping the temperature every 15 minutes to obtain the desired cooling rate.

Both natural and synthetic materials were examined with a 100 kV Siemens Elmiskop I electron microscope without a tilting stage. Crushed grains were supported by a carbon substrate on a 200 mesh copper grid. For each grained photographed a corresponding electron diffraction pattern was taken; so that, apparent lamellar thicknesses could be corrected to true thicknesses. Average uncorrected thicknesses were determined by measuring distances approximately normal to the lamellae on the photographs from five grains and dividing by the number of lamellae observed. Absolute magnifications were obtained by photographing a carbon replica of a waffle grating with spacing of 0.50 μ at the beginning and end of each 12 plate cassette.

The natural samples were also photographed by an X-ray precession camera (Fig. 1). For both samples, pigeonite and the host diposide share (001) and are coherent; that is, _a_ and _b_ are identical in both phases while the respective c*'s are coincident but different in length. This is consistent with the electron microscope observation in which only (001) lamellae were found.

Results

The results of the recently completed coarsening study of McCallister (1978) using $Di_{54.1}$ indicates that λ or wavelength obeys the following relationship with time:

$$\lambda = \lambda_o + kt^{1/3}$$

where λ_o is the critical wavelength needed for growth and k is the kinetic constant at a particular temperature. The isothermal data were collected at 1300°, 1200° and 1100°C, and a plot $3\ell n(k)$ vs $1/T(°K)$ is linear with the slope corresponding to an activation energy for the process of 99 ± 2 Kcal/mole. A consequence of this is that k at 1000°C is sufficiently small, so as to contribute little to the final lamellar thickness of samples continuously cooled to that temperature. For comparison k at 1000°C = 17.8 Å/day$^{1/3}$, while at 1300°C k + 216.8 Å/day$^{1/3}$. A final point regarding this study is that $\lambda_o = \infty$ at the coherent spinodal which is approximately 1335°C for $Di_{54.1}$ (McCallister and Yund, 1977). Based on theoretical calculations λ_o is on the order of a few hundred Å's within a degree of the coherent spinodal and changes only by one order of magnitude through the next 100° (Cahn, 1968). This is, in general, what was observed in the experimental study: λ_o = 109 ± 39 Å at 1300°C; 89 ± 4 Å at 1200°C and 53 ± 10 Å at 1100°C.

In order to calculate the cooling rates of the two natural samples a number of assumptions had to be made. First, it was assumed that the dependence of k with temperature would be the same for the natural clinopyroxenes with Ca/(Ca+Mg) approximatley equal to 0.317 as it is for the synthetic material with Ca/(Ca+Mg) = 0.271. This assumption is probably not a source of considerable error in that the k's are reflecting diffusion rates which should vary only slightly over this composition range. A more serious problem arises when one tries to fix the value of λ_o for the natural pyroxenes. The only data which we have is that for an undercooling of 35°C below the coherent spinodal, λ_o = 109Å, for $Di_{54.1}$. From the experimental work of McCallister and Yund (1977) using a synthetic diopside with a Ca/(Ca+Mg) of 0.317 it was possible to infer that the coherent spinodal for this composition could be as high as 1300°C. Therefore assuming exsolution was initiated at an undercooling of 35° (1265°C) where λ_o = 109°A, calculations were made to determine the linear cooling rate needed to obtain a λ_{final} of 190 Å. This was done for a number of quench temperatures as low as 600°C. It became obvious that the calculated cooling rates are quite rapid (on the order of several hundred degrees per day), and that there is no change in the rate after a temperature of 1100°C is reached. However, because the final wavelength is only a factor of two different from the assumed λ_o a slight change in λ_o will make a marked difference in the cooling rate. For example, for the conditions of λ_o = 110 Å, T_o(°C) = 1300°, T_{quench} (°C) = 1000° and λ = 192 Å, the calculated cooling rate is 394°/day. If λ_o is changed to 90 Å, which is within 2σ while the other values are kept the same the rate is 207°/day. In both cases the inference is rapid cooling, however, the difference of 200°/day for a change in λ_o of 20 Å suggests that another approach is needed if more accurate values were to be obtained.

Using the calculated rates previously described as a guide, cooling experiments were conducted with a synthetic diopside solid solution having a Ca/(Ca+Mg) = 0.317 ($Di_{63.3}$). Furnace temperatures were lowered every 15 minutes through the range of 1300° - 1000°C. The rate of cooling for the initial experiment was 40°C/hr, and the average wavelength in several grains is 181 ± 15 Å. While the similarity in λ with the average for the two natural clinopyroxenes (190 ± 10 Å) is obvious, the amplitude of the fluctuations or compositional difference as reflected in the respective electron diffraction patterns appears greater for the natural samples. This result suggested that possibly a slower rate of cooling was needed to increase the compositional difference between the host and guest phases. Accordingly, an experiment was conducted in which the rate of cooling was 20°/hr. In this case the average wavelength was 178 ± 10 Å, which is quite similar to the 190 Å value for the naturally exsolved pairs. Also, the electron diffraction pattern showed separation of reflection parallel to c* instead of the streaking that was evident in the 40°/hr run. Although we were unable to directly compare the compositional differences in the natural and synthetic microstructures due to orientation limitations, it is clear that the 20°/hr run is a closer approximation to the true cooling rate in the natural materials. Interestingly, this value is only 86°/day different from the calculated rate of 394°/day with λ_o = 110 Å. Considering all of the assumptions and uncertainties associated with the calculated rates, the similarity with an actual cooling experiment is encouraging. In Figure 2 photomicrographs of the two natural exsolution microstructures are compared with that resulting from the 20°/hr run with the synthetic material. In all three photomicrographs c* is normal to the electron beam and the lamellae are parallel to (001), within the error of the measurement.

Conclusions

This study represents the first attempt at determining a quantitative, partial cooling history for naturally exsolved pyroxenes. Although there are some assumptions involved in applying the data on synthetic materials to the naturals, the result of 20°C/hr is not inconsistent with other studies directed at determining intrusion velocities of kimberlites. For example, McGetchin and Ullrich (1973) calculate velocities at 90km of 25m/sec with a gas temperature of

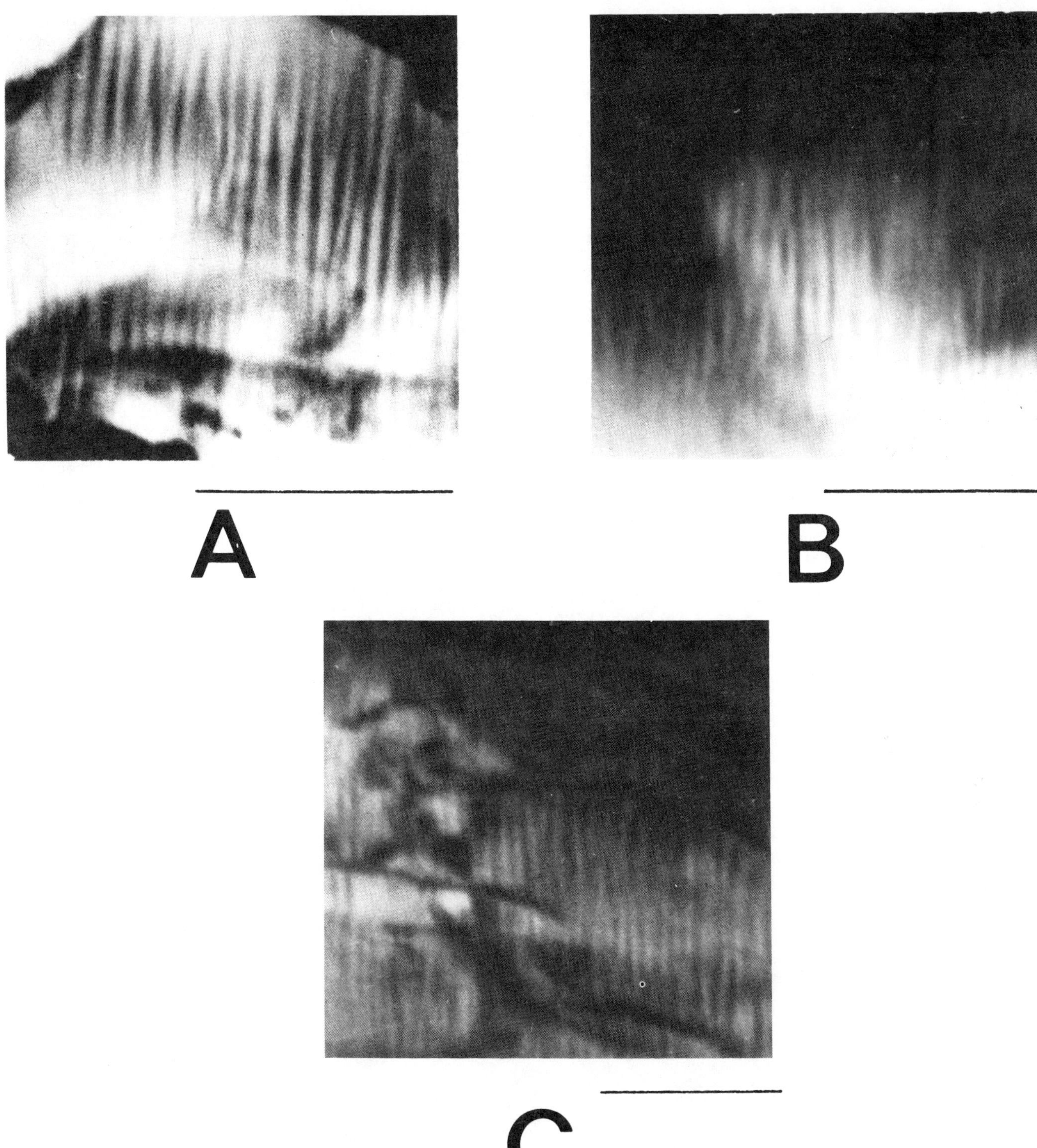

Fig. 2. A comparison of the exsolution microstructures in the natural and synthetic clinopyroxenes. In all electron photomicrographs the orientation is the same with c* common to both the pigeonite and diopside phases. The scale corresponds to 0.25μ. A. PHN 1611 - a clinopyroxene in a sheared lherzolite from Lesotho. Ca/(Ca+Mg) = 0.315. B. PHN 1600E4 - a clinopyroxene megacryst from Lesotho. Ca/(Ca+Mg) = 0.320. C. DE 163 - a synthetic clinopyroxene cooled from 1300 - 1000°C at 20°/hr in 15 minute. Ca/(Ca+Mg) = 0.317.

1000°C. This rate exponentially increases to 334m/sec at the surface. Also, Mercier (this volume) on the basis of olivine recrystallization studies has estimated intrusion velocities of 5m/sec in the region of pre-supercooling. Because we believe that the exsolution reaction effectively ceases at 1000°C a temperature which can be attained in nodule and megacryst within a few kilometers of the surface, the total time for development of the observed microstructure is on the order of 15 hrs.

While it is impossible to determine the thermal path of the kimberlite from our data, we are able to say that the cooling rate was extremely fast. Also, it is worth noting that from the Lesotho geotherm of Boyd (1973) the extimated depth at which both of the natural samples equilibrated is on the order of 210 km. Admittedly, exsolution was not initiated at this point as the temperature was 1400°C and consequently well above the coherent spinodal. However, if we assume that the rate of cooling was linear and exsolution began at 1300° at approximately 190 km (a value taken from Boyd's Lesotho geotherm), a comparison with calculated travel times normalized to this depth can be made. Having done this the estimated travel time for the Mercier model (this volume) from 190 km is 14.3 hrs, while the results of McGetchin et al. (1973) give 9.5 hrs, if their upper limit of 10 hrs through 200 km is used. Our range agrees to within a factor of two of those previously discussed; however, it must be remembered that we assumed a linear cooling rate from initial conditions of 1300°C and 190 kms depth. Although this assumption may be inappropriate and the agreement with other workers fortuitous, we can say with certainty that the total time for cooling from 1300° - 1000°C could not be much in excess of 15 hours.

Acknowledgements

The authors wish to thank Dr. F. R. Boyd for supplying us with the two natural samples used in this investigation and for reviewing this manuscript. Also we would like to thank Dr. Gordon Nord for a thoughtful review which contributed significantly to the final presentation. R.H.M. acknowledges the National Science Foundation for providing support for this project through Grant EAR76-03804.

References

Boyd, F.R., A pyroxene geotherm, Geochim. Cosmochim. Acta, 37, 2533-2546, 1973.

Cahn, J.W., Spinodal decomposition, Trans. AIME, 242, 166-180, 1968.

Kushiro, I., Determination of liquids relations in synthetic silicate systems with electron probe analysis: The system forsterite-diopside-silica at 1 atmosphere, Amer. Mineral., 57, 1260-1271, 1972.

McCallister, R.H., the coarsening kinetics associated with exsolution in an iron-free clinopyroxene, Contrib. Mineral. Petrol., 65, 327-331, 1978.

McCallister, R.H., and R.A. Yund, Coherent exsolution in Fe-free pyroxenes, Amer. Mineral., 62, 721-726, 1977.

McGetchin, T.R., and G. Wayne Ullrich, Xenoliths in maars and diatremes with inferences for the Moon, Mars and Venus, J. Geophys. Res., 78, 1832-1852, 1973.

McGetchin, T.R., Y.S. Niklanj, and A.A. Chodos, Carbonate-kimberlite relations in the Cane Valley Diatreme, San Juan County, Utah, J. Geophys. Res., 78, 1852-1868, 1973.

Nixon, P.H., and F.R., Boyd, the discrete nodule association in kimberlites from northern Lesotho, In Lesotho Kimberlites, edited by P.H. Nixon, pp. 67-75. Lesotho National Development Corporation, Maseru, Lesotho, 1973.

Yund, R.A. and R.H. McCallister, Kinetics and mechanisms of exsolution, Chem. Geol., 6, 5-30, 1970.

THE MINERAL CHEMISTRY OF ILMENITE NODULE ASSOCIATIONS FROM THE MONASTERY DIATREME

Stephen E. Haggerty, Richard B. Hardie III and Brendan M. McMahon

Department of Geology, University of Massachusetts, Amherst, Massachusetts 01003

Abstract. A comprehensive petrographic and electron microprobe study of an ilmenite nodule suite from the Monastery diatreme has shown that although two major compositional populations can be identified for discrete nodules and for nodules in pyroxene symplectites, that these diverse associations are derived from a common source region in the mantle which was saturated in ilmenite. The discrete nodule suite is initially less magnesian than ilmenite in symplectic associations but equilibrates to higher magnesian concentrations as a consequence of decreases in P_{Total} and in fO_2. Symplectic ilmenite is considered to form at a lower pressure and temperature than discrete ilmenite. These intergrowths are protected from a continuous interaction with the magma because of the pyroxene encasement and not because these assemblages are derived from a distinct source or because they were removed from the system. A marker horizon for nodule incorporation into the kimberlite is initiated by the onset of carbonatitic immiscibility which is manifested by high MnO contents in ilmenite. Exsolution of oxides in ilmenite is pervasive in the entire Monastery suite, and subsolidus reduction is evident in the crystallization trends of discrete nodules and of symplectic ilmenite. Experimental data on nodule ilmenite under controlled T°C and fO_2 atms, show that textures and compositions of subsolidus reduction assemblages lie within the stability fields demarcated by the buffers FMQ at the upper limit and graphite -CO-CO_2 at the lower limit. Among the four currently held proposals for symplectic origin (garnet decomposition, replacement, exsolution, and eutectoid crystallization), the following have not been factored into models of formation: (a) wide modal variations in silicate-oxide ratios; (b) the presence of clinopyroxene and orthopyroxene; (c) oxide exsolution and subsolidus reduction of ilmenite; (d) the presence of sulfide immiscible liquids during symplectite formation; or (e) the role of phlogopite and other phases at ilmenite-pyroxene interfaces.

Introduction

Ilmenite is widely recognized as a characteristic constituent of many kimberlites. The mineral is present as inclusions in diamond, as large discrete nodules, and also as a primary phase of kimberlitic groundmass assemblages. Among the most unusual and least understood associations are lamellar intergrowths of ilmenite and pyroxene which are relatively abundant in some kimberlites but are unknown in any other rock type.

We have undertaken a systematic petrographic and electron microprobe study of a large suite of ilmenite nodules from the Monastery pipe, South Africa. The objectives of the study were directed towards a chemical characterization of the suite, an analysis of reaction trends, exsolution intergrowths, contact and interface assemblages, and a comparison of discrete nodules with nodules displaying ilmenite-pyroxene intergrowths. The suite consists of 163 nodules which we have classified into the following categories: (a) symplectic ilmenite-pyroxene intergrowths (57 nodules); (b) discrete ilmenite nodules (66 samples); and (c) ilmenite nodules encased in a closely adhering kimberlitic matrix (40 nodules). The chemical data base is 1040 mineral analyses which employed an 8 element matrix for ilmenite and a 10 element matrix for pyroxene.

Discrete Nodule Suite

The size range of ilmenites in the discrete nodule suite varies between approximately 1 cm and 5 cm. These nodules are typically rounded and have smooth surfaces and this is attributed to conduit abrasion and jet streaming during fluidization. Of the 106 discrete nodules examined, 67% are polycrystalline and the remainder are optical single crystals. The polycrystalline nodules display pronounced polygonal textures with 120° dihedral angles characteristic of recrystallization under conditions of intense stress annealing. Approximately 80% of the nodules contain oriented lamellar intergrowths and these are of three types: (a) aluminous spinels; (b) magnesian-titanohematites; and (c) magnesian-titanomagnetites. All three types are oriented along basal {0001} ilmenite parting planes but in many cases the lamellae anastomose into braided domains which are interpreted to be decorated crystal dislocations in ilmenite. The aluminous spinels are fine grained (~1-5 μm) and are evenly distributed throughout the ilmenite. Mg-titanohematite is typically coarser with 10-20 μm wide lamellae and these also show an even distribution. By contrast, the Mg-titanomagnetites are concentrated at the edges of the nodules and the lamellae (5-20 μm) are frequently distorted. The aluminous spinels and the Mg-titanohematites are considered to result by exsolution *sensu stricto*, whereas the Mg-titanomagnetites are reaction products induced by subsolidus reduction. We have demonstrated this reaction experimentally and the results are discussed in a subsequent section. Five of the discrete nodules contain trapped ovoid inclusions of calcite + pyrrhotite + pentlandite. Reactions between the inclusions and the ilmenite host result in a consistent sequence which is: (a) magnetite at the inclusion contact; (b) an intervening rind which is a strongly zoned titaniferous magnetite; and (c) a magnesium-depleted ilmenite adjacent to the host ilmenite. Reaction contacts between the nodules and the kimberlite matrix bear a striking similarity to this sequence but with the exception that perovskite is a commonly associated constituent.

Representative electron microprobe analyses of the discrete nodules, of core-mantle zonation trends in these nodules, and of the exsolution and reaction products described above are listed in Table 1. Data for ilmenite and for Mg-titanohematite are summarized in Fig. 1. The cores of the discrete nodules form a tight cluster centered on $Geik_{30}$ Ilm_{45} Hem_{25}. Nodules which are polycrystalline are distinctly removed from this cluster and these have lower Fe_2O_3 contents (15 mole % hematite) than those which are single crystals. MgO contents are higher (45 mole % geikielite) and $FeTiO_3$ contents are lower (40 mole % ilmenite). Both suites of nodules exhibit pronounced zoning and these trend towards $MgTiO_3$ enrichment. We refer to this enrichment trend as the *magmatic reaction trend*. This magmatic trend differs from a second trend, towards $FeTiO_3$, which we have named the *kimberlite reaction trend*. An inference to be drawn from these trends is that the $MgTiO_3$ enrichment factor is probably indicative of a reaction induced by a decrease in pressure (Haggerty, 1976), whereas late stage matrix interaction is associated with the carbonatitic stage of the Monastery kimberlite. Carbonatitic ilmenites are enriched in MnO (1-5 wt%) and we believe that there is now strong evidence that these high manganese contents are the hallmark to carbonatites and to carbonate reactions in kimberlites (McMahon and Haggerty, 1976). The exsolved lamellae in the nodules have compositions which peak at 60 mole % hematite and the range extends to a minimum of Hem_{40}. Mole percent concentrations of ilmenite and geikielite vary within small limits and for each the range is Ilm_{20} $Geik_{20}$ to Ilm_{40} $Geik_{40}$.

Ilmenite-Pyroxene Intergrowths

The ilmenite-pyroxene nodules examined in this study are in the size range 1-5 cms and one nodule measures 15 x 5 cm. Proportions of ilmenite to pyroxene vary enormously and estimates range from a ratio of 1:1 to a ratio of at least 10:1 pyroxene:ilmenite. Three of the 57 nodules contain orthopyroxene and the remainder are clinopyroxene. The degree of serpentinization is variable but there are some cases in which the entire pyroxene host has been transformed. Interface assemblages at orthopyroxene-ilmenite contacts yield olivine + rutile + Ti-rich phlogopite, whereas the assemblage between clinopyroxene and ilmenite is Fe-rich phlogopite + orthopyroxene + perovskite. Phlogopite in both assemblages is prominently developed and growth of the mica is clearly at the expense of both ilmenite and pyroxene but tends to be concentrated in the ilmenite and at the ilmenite-pyroxene interface. There is no optical evidence for pyroxene exsolution although we have noted that the clinopyroxenes fall into two groups which we have designated *clean* and *sieve*. The latter does not appear to be related to serpentinization and we do not have an explanation for its spongy appearance or for the process responsible for the sieve texture. These two pyroxene types are compositionally distinct as discussed below. Exsolution within the ilmenite is similar to that of the discrete nodule suite although Mg-titanomagnetite has not been positively identified; exsolution lamellae are texturally identical and these bodies are Al-spinels and Mg-titanohematites. Of the 57 symplectic nodules 47% contain polycrystalline ilmenite but no structural distortions are observed in the associated pyroxenes. Our interpretation of these nodules is similar to that of the discrete nodule suite in which stress annealing and recrystallization has taken place. Two of the pyroxene-ilmenite nodules contain abundant sulfide-bearing tubes which are preferentially located at the terminal contacts of elongated ilmenite laths. In some instances the tubes pierce both the ilmenite and the pyroxene and we consider that the only reasonable explanation to account for their presence is to invoke an immiscible sulfide liquid which was present during the formation of the intergrowth. The sulfides are pyrrhotite + pentlandite but these have now been largely replaced by goethite + magnetite.

Analytical data for these nodules are listed in Table 2 and the compositions of ilmenites and Mg-titanohematites are plotted in the ternary Fe_2O_3-$FeTiO_3$-$MgTiO_3$ (Fig. 1). The symplectic ilmenites overlap the compositions of the discrete nodule suite but the preferred population of analyses is poorer in Fe_2O_3 and centers on $Geik_{35}$ Ilm_{45} Hem_{20}. At the extreme edges of a large proportion of the ilmenites is a zone which is depleted in Fe_2O_3 and we attribute this to partial subsolidus reduction during symplectite formation. The prominently developed magnesium-enrichment trend observed in the discrete nodules is absent and we suggest that the sheathing effect of the surrounding pyroxene provides an effective armoring which inhibits extensive and continuous interaction with the liquid. Only two of the nodules in the collection have an adhering kim-

TABLE 1. Representative Electron Microprobe Analyses of the Discrete Ilmenite Nodule Suite.

	1(C)	2(I)	3(M)	4(C)	5(I)	6(M)	7	8	9(H)	10(L)
SiO_2	0.00	0.00	0.00	0.00	0.00	0.00	0.00	0.00	0.00	0.00
TiO_2	45.95	52.57	53.51	51.03	51.96	53.38	51.77	50.75	53.26	40.25
Cr_2O_3	0.72	0.35	1.04	0.88	0.90	0.75	0.99	2.68	0.00	0.00
Al_2O_3	0.37	0.12	0.16	0.24	0.32	0.29	0.29	0.36	0.13	1.66
Fe_2O_3	18.34	8.63	10.10	10.29	9.71	10.52	9.59	10.54	6.47	28.22
FeO	27.38	24.61	19.81	25.40	24.71	17.64	23.88	24.26	29.42	20.25
MgO	7.12	12.50	15.63	11.32	12.18	16.76	12.53	11.83	10.21	8.79
MnO	0.23	0.31	0.40	0.27	0.26	0.43	0.32	0.27	0.25	0.22
CaO	0.01	0.04	0.03	0.02	0.03	0.03	0.01	0.01	0.00	0.03
Total	100.11	99.13	100.67	99.44	100.06	99.80	99.37	100.70	99.75	99.43
Si	0.000	0.000	0.000	0.000	0.000	0.000	0.000	0.000	0.000	0.000
Ti	0.824	0.928	0.903	0.898	0.903	0.901	0.903	0.879	0.941	0.723
Cr	0.014	0.007	0.019	0.016	0.016	0.013	0.018	0.049	0.000	0.000
Al	0.010	0.003	0.004	0.007	0.009	0.008	0.008	0.010	0.004	0.047
Fe^{3+}	0.329	0.135	0.171	0.181	0.169	0.178	0.167	0.183	0.114	0.507
Fe^{2+}	0.566	0.483	0.372	0.497	0.478	0.331	0.463	0.468	0.578	0.405
Mg	0.253	0.437	0.523	0.395	0.420	0.561	0.433	0.406	0.358	0.313
Mn	0.005	0.006	0.008	0.005	0.005	0.008	0.006	0.005	0.005	0.005
Ca	0.000	0.001	0.001	0.001	0.001	0.001	0.000	0.000	0.000	0.001
Total	2.000	2.000	2.000	2.000	2.000	2.000	2.000	2.000	2.000	2.000
Hem.	28.21	12.63	15.63	16.55	15.53	16.29	15.38	16.46	10.85	41.27
Ilm.	48.52	45.24	34.07	45.40	43.91	30.34	42.57	42.11	54.79	32.91
Geik.	21.71	40.94	47.91	36.07	38.58	51.40	39.80	36.57	33.88	25.45
Other	1.56	1.19	2.39	1.98	1.98	1.97	2.35	4.86	0.48	0.37

Core, intermediate, and margins are indicated by C, I and M, respectively.
Analyses 1-3, sample MIK-2; analyses 4-6, sample MIK-4. Analyses 7-8 are average analyses for discrete polycrystalline nodules, sample ML-36 and ML-51, respectively. Analyses 9-10 are host (H) ilmenite and exsolved magnesium-titanohematite lamellae (L), respectively, sample MIK-14. Hem., Ilm., and Geik. are mole %; Other refers to Cr + Al + Mn + Ca mole %.

berlite; one of these nodules has the reaction assemblage perovskite + Mg-titanomagnetite (discrete crystals) + Mn-ilmenite, and the other has rutile + sphene + Fe-phlogopite. Oriented lamellae of Mg-titanohematite in the ilmenite symplectites are present in 76% of the nodules. This mineral has lower MgO (~7.5 *vs.* 9.0 25%), lower FeO (~25.0 *vs.* 28 wt%) and lower TiO_2 (~44.0 *vs.* 49.0 wt%) contents, but higher Fe_2O_3 (~21.0 *vs.* 12.0 wt%) and Al_2O_3 (0.5-1.0 wt%) concentrations than those of the ilmenite host which forms the main body of the assemblage with pyroxene. There exists a region of compositional overlap between these lamellae and with those in the discrete nodule suite, but a large proportion of the lamellae are distinctly lower in Fe_2O_3, which is compatible with the slightly lower fO_2 values of formation as indicated by the zonal trend of the ilmenite bodies at pyroxene interfaces (Fig. 1). An additional 20% of the nodules contain well developed but fine (~1-5 μm wide) Al-spinel lamellae but these are beyond the resolution of the electron microprobe beam and of edge excitation effects.

Pyroxene compositions are illustrated for the quadrilateral components in Fig. 2. These occupy three regions: orthopyroxenes (Wo_1 En_{84} Fs_{15}) and two sets of clinopyroxenes. The *clean* pyroxenes (Wo_{35} En_{53} Fs_{12}) are less calcic than the *sieve* pyroxenes (Wo_{40} En_{52} Fs_8) and based on the Davis and Boyd (1966) diopside-enstatite solvus we estimate that the respective temperatures of formation using Ca/(Ca + Mg) ratios are 1200°C and 980°C respectively. Because of the uncertainty in the location of the solvus and because more recent determinations of the solvus yield conflicting results we emphasize that these temperatures are simply estimates.

Exsolution and Subsolidus Reduction

High resolution oil-immersion petrography of the nodule suites shows a pervasive development of exsolution-like bodies in ilmenite. We define this process as exsolution-like based on the following factors: (a) the bodies are sharply defined; (b) the lamellae terminate in tapered projections; (c) the lenses are crystallographically controlled in the host; and (d) synneusis textures, although rare, are observed. The criteria which are incompatible with exsolution *sensu stricto* are: (a) that the solute and the solvent (i.e., for spinel and for ilmenite) are crystallographically dissimilar; and (b) that there are

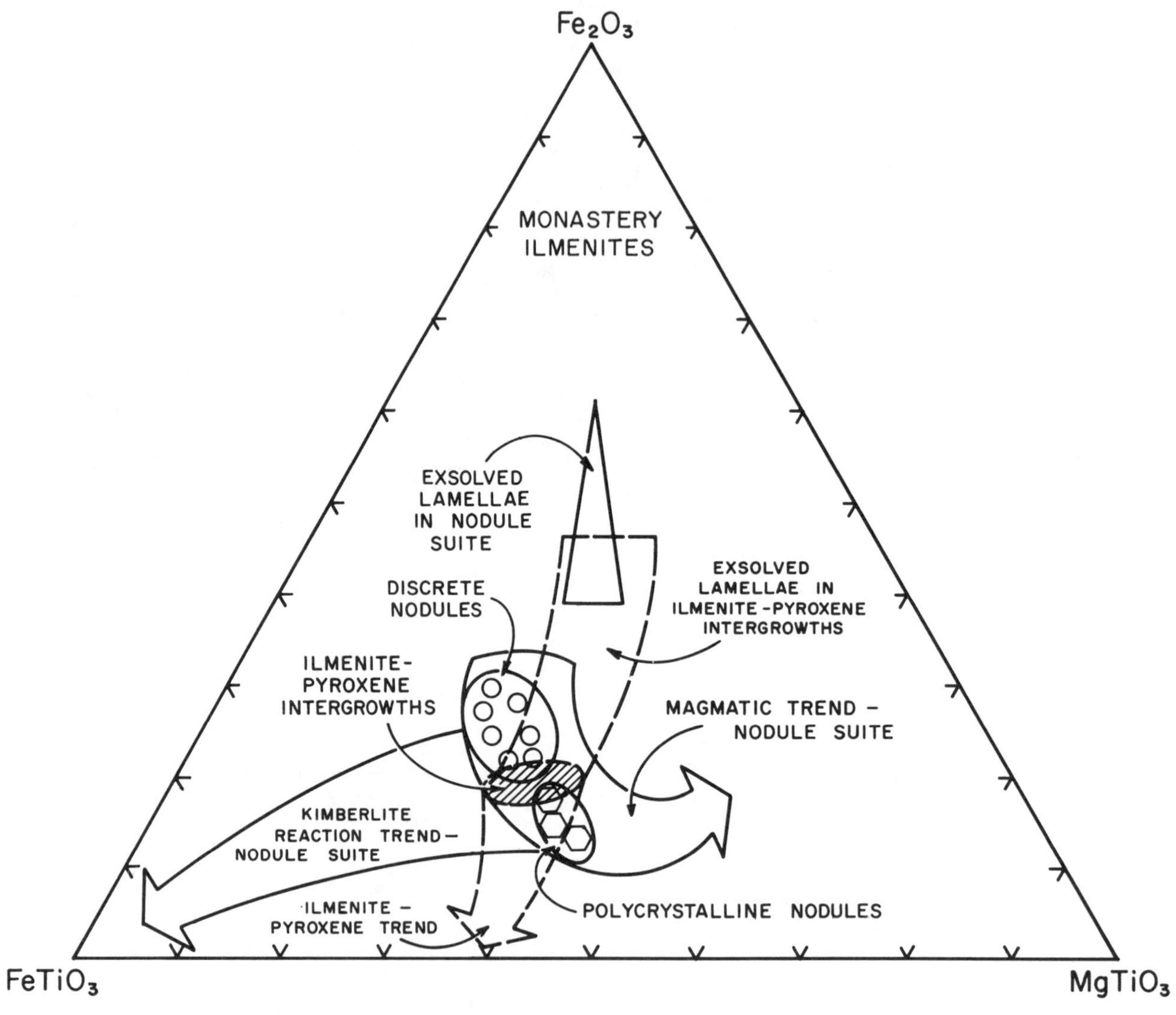

Fig. 1. An oxide ternary variation diagram that illustrates the compositional and reaction trends of ilmenites that are present in the Monastery kimberlite.

no data or evidence for multiple solvi in the system Fe_2O_3-$FeTiO_3$-$MgTiO_3$. Given these constraints we tentatively suggest that the Mg-titanohematites are most likely the result of a ternary solvi intersection under slow cooling and that the process is truly exsolution. For the Al-spinels we propose an unmixing process which is comparable to exsolution in the sense of diffusion. Because these spinels tend to occur in relatively low abundances a small degree of spinel-structured and ilmenite-structured solid solubility is not an unreasonable proposition at high temperatures and pressures. We have concentrated most of our effort in attempting to unravel the role of Mg-titanomagnetite which is particularly abundant in the discrete nodule suite and, as noted earlier, has a tendency to concentrate at the edges of nodules. Because of this preferred distribution their formation is most likely the result of alteration and in an experimental study to test this hypothesis we have successfully demonstrated that these lamellae can be induced under conditions of subsolidus reduction as suggested by Buddington and Lindsley (1964) and Haggerty (1976).

The experimental data, between 900° and 1200°C and for a range of fO_2 values between 10^{-14} and 10^{-8} atms, are shown in Fig. 3. Temperatures were monitored continuously using a Pt-Rh thermocouple, and fO_2 was measured using a calcia-stabilized zirconium electrolytic cell according to the Sato (1971) technique by gas mixing procedures with variable proportions of CO_2 and hydrogen. Run products were electrically quenched into cold water or into chilled mercury. The starting material consisted of fragments of a lamellar-free Monastery discrete nodule which does contain a barely perceptible peripheral veneer of alteration. This rim is sufficiently thin that it does not influence the result of the equilibrium experiment. The range of experimental runs we report here are governed in part by the electrolytic cell, in part by the range of reasonable conditions anticipated for Mg-ilmenite decomposition, and by data reported by Ulmer et al. (1976) for the Premier Mine kimberlite. These latter data are shown in Fig. 3 and the accompanying experimental buffer curves are Ni-NiO, FMQ, and WM. At the highest levels of fO_2 the characteristic phase produced is a member of the pseudobrookite (Fe_2TiO_5)-karrooite (Mg_2TiO_5) series. At lower values of fO_2 and for a range of temperatures, Mg-titanomagnetites in fine and in coarse lamellae were synthesized in all run products. Although the time was varied among the runs from 5 hours to 3 days, a characteristic group of runs, which parallel the buffer curves, resulted in exceptionally coarse lamellae and a least-squares fit to the experimental data points is illustrated. For these we assume that their production is not kinetically controlled but is related also to T°C and fO_2. For fO_2 values that approach the graphite-CO-CO_2 buffer, metallic iron was precipitated. The region of Mg-titanomagnetite production exactly duplicates those observed in the Monastery discrete ilmenite suite from a textural and compositional standpoint. Although P_{Total} has not been controlled we note that the relevancy of these experiments is with respect to the buffer curves. For an increase in P_{Total} the curves will shift towards higher values of fO_2 and the experimental data points will shift also, so that with respect to FMQ and G-CO-CO_2 the range of Mg-titanomagnetite production in kimberlitic ilmenite is below FMQ and above G-CO-CO_2. A more detailed discussion of these experimental data will appear elsewhere.

Discussion

Among the overall objectives of this study are attempts to characterize and to evaluate the setting of ilmenite nodule formation for ilmenites which are present as discrete nodules and for ilmenites in pyroxene intergrowths. Our analytical data base is extensive and our interpretation of the results suggest that there are at least four distinct events which can be identified. The first is the *magmatic trend* of the discrete nodule suite towards $MgTiO_3$; the second is the *ilmenite-pyroxene trend* towards the join $FeTiO_3$-$MgTiO_3$; the

TABLE 2. Representative Electron Microprobe Analyses of Ilmenites in Ilmenite-Pyroxene Intergrowths.

	1(C)	2(I)	3(M)	4(C)	5(I)	6(M)	7(H)	8(L)	9(H)	10(L)
SiO_2	0.00	0.00	0.00	0.00	0.00	0.00	0.00	0.00	0.00	0.00
TiO_2	47.58	51.05	53.61	44.54	46.10	54.56	53.51	47.39	48.77	40.86
Cr_2O_3	0.07	0.12	0.21	0.04	0.03	0.00	0.07	0.19	0.32	0.88
Al_2O_3	0.78	0.79	0.12	0.94	0.87	0.00	0.04	1.07	0.32	2.91
Fe_2O_3	17.05	11.11	5.51	23.15	20.04	3.68	7.03	17.80	14.42	26.18
FeO	23.63	26.39	28.74	21.98	23.55	30.76	28.08	24.30	25.15	19.08
MgO	10.62	10.84	10.77	9.99	9.92	10.10	11.12	10.17	10.39	9.77
MnO	0.19	0.14	0.16	0.21	0.21	0.19	0.17	0.17	0.16	0.23
CaO	0.02	0.03	0.08	0.03	0.01	0.07	0.02	0.00	0.01	0.01
Total	99.94	100.47	99.20	100.88	100.78	99.36	100.04	101.09	99.54	99.92
					Normalized Cations					
Si	0.000	0.000	0.000	0.000	0.000	0.000	0.000	0.000	0.000	0.000
Ti	0.838	0.891	0.948	0.783	0.811	0.967	0.937	0.828	0.865	0.721
Cr	0.001	0.002	0.004	0.001	0.001	0.000	0.001	0.004	0.006	0.016
Al	0.002	0.022	0.003	0.026	0.024	0.000	0.001	0.029	0.009	0.080
Fe^{3+}	0.301	0.194	0.097	0.407	0.353	0.065	0.123	0.311	0.256	0.462
Fe^{2+}	0.463	0.512	0.565	0.430	0.461	0.607	0.547	0.472	0.496	0.374
Mg	0.371	0.375	0.377	0.348	0.346	0.355	0.386	0.352	0.365	0.341
Mn	0.004	0.003	0.003	0.004	0.004	0.004	0.003	0.003	0.003	0.005
Ca	0.000	0.001	0.002	0.001	0.000	0.002	0.001	0.000	0.000	0.000
Total	2.000	2.000	2.000	2.000	2.000	2.000	2.000	2.000	2.000	2.000
Hem.	26.38	17.87	9.31	34.22	30.31	6.34	11.61	27.24	22.72	38.55
Ilm.	40.63	47.15	53.99	36.11	39.58	58.85	51.56	41.33	44.04	31.22
Geik.	32.55	34.52	36.03	29.25	29.71	34.43	36.38	30.84	32.43	28.48
Other	0.44	0.46	0.67	0.42	0.40	0.38	0.45	0.59	0.81	1.75

C, I, and M are core, intermediate and margins of ilmenite crystals. Analyses 1-3, sample MIP-33; 4-6, sample MIP-18. Analyses 7-10 are ilmenite host (H) and exsolved magnesian titanohematite (L), respectively; 7-8 sample MIP-55; 9-10 sample MIP-46. Hem., Ilm., Geik. are in mole %; Other refers to Cr + Al + Mn + Ca mole %.

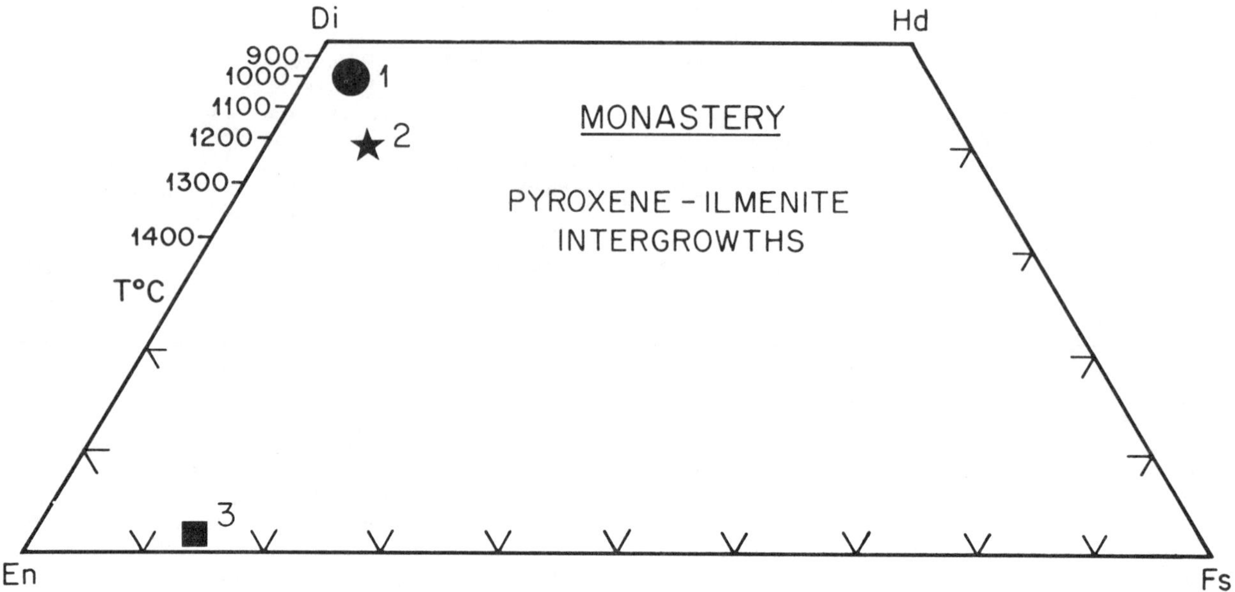

Fig. 2. Pyroxene quadrilateral showing the relative compositions of *sieve* (1), and *clean* (2) clinopyroxenes, and orthopyroxenes (3) in association with ilmenite. Temperatures along the enstatite (En) - diopside (Di) join are from Davis and Boyd (1966).

third is the *kimberlite reaction trend* towards $FeTiO_3$ which includes MnO enrichment; and the fourth is the episode of exsolution and subsolidus reduction. The magmatic nodule trend is a liquid-interaction trend which is assumed to result from a partial decrease in P_{Total}. A consequence of this trend is that fO_2 decreases, and this is supported by the observed decreases in Fe_2O_3 contents towards the nodule margins. Zonation trends in ilmenite-pyroxene symplectites are adequately explained in terms of a decrease in fO_2 but in these the enrichment in $MgTiO_3$ is not observed and our explanation is that the ilmenite is armored by the symplectic pyroxene from the surrounding magma. The relative displacements of preferred compositional populations for the discrete nodules and for the symplectites are in accord with those determined by Mitchell (1977) which show that the discrete nodule suite from Monastery are slightly more enriched in Fe_2O_3 than those ilmenites in symplectic associations. The data overlap, however, and within the constraints of the analytical uncertainty we conclude that both nodule types are almost certainly derived from a similar or the same ilmenite-saturated source region of the mantle. We regard the observation of the kimberlite reaction trend as a significant contribution to the role of carbonatitic reactions and our data now show an unequivocal relationship with MnO and late stage CO-CO_2 interactions. The enrichment of manganese is abrupt and we consider that this may be indicative of nodule incorporation into kimberlite and of carbonatite liquid immiscibility. These Mn-rich ilmenites may therefore be interpreted as marker horizons which identify the fluidization event of kimberlite eruptions.

The interface reactions between ilmenite and orthopyroxene, and ilmenite and clinopyroxene, contain phlogopite as a major constituent in coarse euhedral laths. Not all symplectic assemblages from Monastery exhibit this reaction. This suggests that although potassium metasomatism was active in some zones within the pre-eruptive kimberlite that it was not an all-encompassing process. It is of interest also to note that the assemblage depends on the composition of the pyroxene: clinopyroxene yields orthopyroxene + perovskite; and orthopyroxene yields olivine + rutile.

Within the context of the proposals suggested for the origin of ilmenite-pyroxene symplectitic intergrowths by: (a) decomposition of a high pressure titanium garnet (Ringwood and Lovering, 1970); (b) exsolution of ilmenite from an ilmenite structured pyroxene (Dawson and Reid, 1970); (c) ilmenite replacement of pyroxene; or (d) eutectic crystallization of ilmenite and pyroxene (MacGregor, 1970; MacGregor and Wittkop, 1970; Boyd, 1971; Wyatt et al., 1975; Wyatt 1977), we concur with the objections raised by Mitchell (1977) that none of these mechanisms are totally acceptable. We are attracted to the analogy between these symplectites and graphic intergrowths of quartz and feldspar (Mitchell, 1977) but here too it is our contention that the following points have not been adequately addressed in any of these models: (a) the fact that both clinopyroxene and orthopyroxene are present; (b) that there are wide variations in modal abundances between ilmenite and pyroxene; (c) that immiscible sulfide liquids are present during ilmenite-pyroxene formation; (d) that although we regard the presence of phlogopite as due to K-metasomatism that it may indeed be a primary interface mineral at ilmenite-pyroxene contacts; and (e) that subsolidus reduction and exsolution in ilmenite are key issues in cooling rates and in the environment of ilmenite-pyroxene formation. Because of the uncertainties in the experimental data for the location of the diopside-enstatite solvus, and because of the effects of minor element distributions on the solid solutions, our estimates of 1200°C for a large proportion of the Monastery ilmenite symplectites is tentative. Nevertheless, this temperature is in agreement with those estimated by Pasteris et al. (1977) and with those determined by Mitchell (1977), which are 1175-1275°C and 1200°C respectively. Our estimate for clinopyroxenes with sieve textures (~980°C) is most likely the result of either a lower temperature equilibration or the result of partial alteration resulting from a carbonate reaction to account for the observed increase in Ca/(Ca + Mg).

Summary and Conclusions

We have identified two major ilmenite compositional populations, one for the discrete nodule suite and a second for ilmenite-pyroxene intergrowths from the Monastery diatreme. There is sufficient overlap in our analytical data to suggest that both ilmenite types are derived from a common source region in the mantle which was ilmenite-saturated. On average, the symplectic assemblages have ilmenites which are more magnesian than the discrete nodules but the paths of equilibration result in terminal compositions which are substantially more magnesian (e.g., 15 wt% MgO *vs.* 8 wt% for discrete nodule edges and symplectic ilmenite edges respectively) in the discrete nodule suite. Growth of these nodules, therefore, took place over a relatively long period of crystallization and based on initial (i.e., core compositions) MgO contents the discrete nodule suite may predate the symplectites by a small mar-

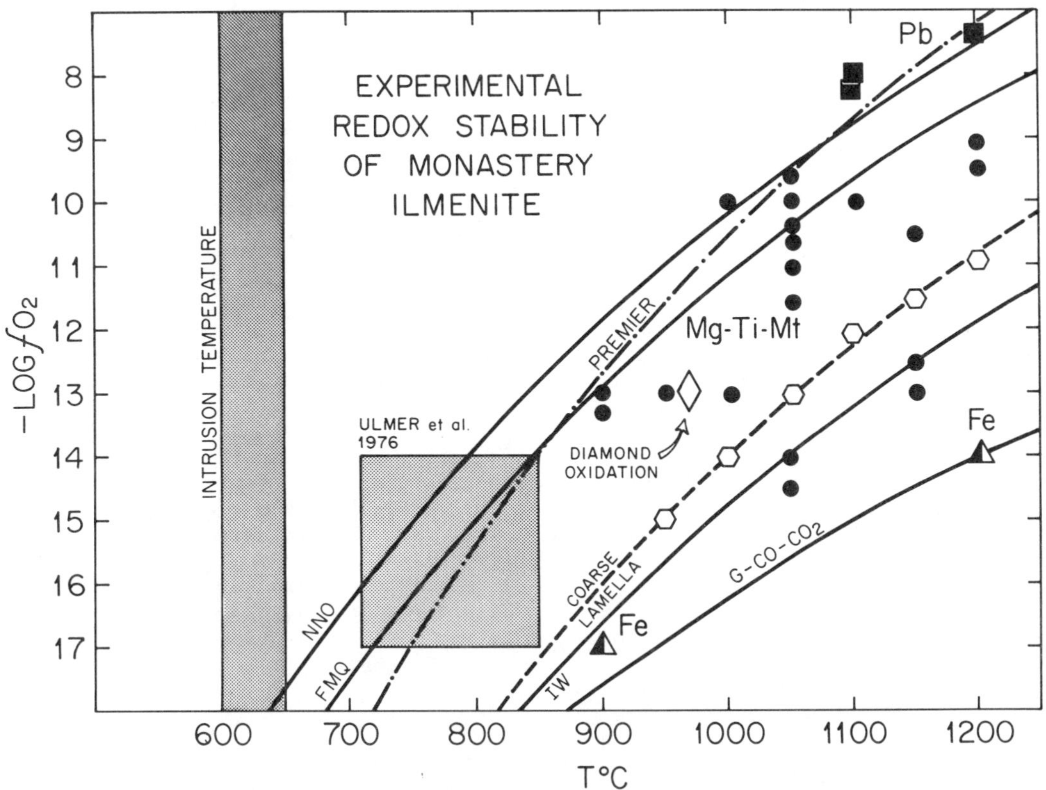

Fig. 3. Temperature-oxygen fugacity relationships for an ilmenite discrete nodule from the Monastery kimberlite, in which magnesium-titanomagnetite (Mg-Ti-Mt) is produced (full circles). The dashed curve is a least squares fit to experimental run products that resulted in very coarse Mg-Ti-Mt lamellae in decomposed ilmenite (hexagons). At low values of f_{O_2} metallic-iron (Fe) is formed (triangles), and at high values of f_{O_2}, members of the pseudobrookite mineral group (Pb) develop (full squares). The shaded fields labeled intrusion temperature, and Ulmer et al. (1976) are estimates of the ranges of T and f_{O_2} for kimberlites. Solid curves are experimental buffer curves, and the dashed-dot curve is an intrinsic T-f_{O_2} determination for the Premier kimberlite (Ulmer et al., 1976). The diamond oxidation estimate is also from Ulmer et al. (1976).

gin. In general this would suggest that specific Fe/Mg ratios may be necessary for symplectic formation, but the position of pyroxene on the liquidus is also clearly of related significance. We note, however, that ilmenite-free discrete pyroxene nodules have equilibration temperatures of approximately 1400°C so that we favor a partial controlling factor which may be dominated by ilmenite compositions. Pasteris et al. (1977) suggest that the chemical and possibly the physical environment under which symplectic intergrowths nucleated and crystallized were more constrained than those for ilmenite discrete nodules. Our data support this conclusion and we suggest that the symplectites formed at both lower P and T than those of the discrete nodules at Monastery. We consider that the magmatic trend for the discrete nodule suite towards higher concentrations of $MgTiO_3$ results from a decrease in P_{Total} and this is based on the results of autolith studies, on the compositions of ilmenites in diamond inclusion studies, on the groundmass compositions of ilmenite, and on molar volume changes for the reaction:

$$\text{Ilmenite} + \text{Forsterite} \rightleftarrows \text{Geikielite} + \text{Fayalite}$$

as summarized by Haggerty (1976).

We have drawn attention to a number of new observations on ilmenite-pyroxene intergrowths which we deem essential to a uniform interpretation of formation. Among these items the three most significant considerations are: (a) to account for the wide modal variations in ilmenite and pyroxene; (b) to account for the fact that both orthopyroxene and clinopyroxene are present; and (c) to provide for a resident time in the mantle which is sufficiently long to induce oxide exsolution in ilmenite and to initiate subsolidus reduction. We offer no explanation for factors (a) and (b). Item (c) may not be important and in the absence of ternary solvi data for the system Fe_2O_3-$FeTiO_3$-$MgTiO_3$ it cannot be assessed. Our subsolidus experimental data show that a uniform and widespread process of reduction is operative under mantle conditions during fluidization or at the intrusive event. The range of fO_2 stability for Mg-titanomagnetite formation is above the G-CO-CO_2 buffer and below that of FMQ.

Acknowledgments. Support to attend the 1st International Kimberlite Conference during which the sample collection was made by S.E.H. was provided by NSF and by the International Association of Geochemistry. Laboratory study was supported by NSF under grant EAR 76-23787 to S.E.H. The microprobe facility is supported by NSF (GA-74206) and by the University of Massachusetts. To all we express our appreciation. Fruitful discussions have resulted with R. Danchin, F. Boyd, B. Wyatt and the Anglo American Oxide Short Course Class of 1977. The manuscript was reviewed and improved upon by D. Rumble, J. Pasteris and N. Boctor. Credit for skillful drafting is due to Marie Litterer.

References

Boyd, F.R., Estatite-ilmenite and diopside-ilmenite intergrowths from the Monastery Mine, Carnegie Inst. Washington Yr. Bk., 70, 134-138, 1971.

Dawson, J.B., and A.M. Reid, A pyroxene-ilmenite intergrowth from the Monastery Mine, South Africa, Contr. Min. Pet., 26, 296-301, 1970.

Davis, B.T.C., and F.R. Boyd, The join $Mg_2Si_2O_6$-$CaMgSi_2O_6$ at 30 Kilobars pressure and its application to pyroxenes from Kimberlites. J. Geophys. Res., 71, 3567-3576, 1966.

Haggerty, S.E., Opaque mineral oxides in terrestrial rocks, in Oxide Minerals (short course notes), edited by D. Rumble III, Min. Soc. Amer., 3, 101-300, 1976.

MacGregor, I.D., An hypothesis for the origin of Kimberlite, in Mineralogy and petrology of the upper mantle, sulfides, mineralogy and geochemistry of non-marine evaporites, edited by B.A. Morgan, Min. Soc. Amer. Spec. Papers, 3, 51-62, 1970.

MacGregor, I.D., and R.W. Wittkop, Diopside-ilmenite xenoliths from the Monastery Mine, Orange Free State, South Africa, Geol. Soc. Amer. Abs. with Programs, 2, 113, 1970.

McMahon, B.M., and S.E. Haggerty, Oka carbonatite complex: oxide mineral zoning in mantle petrogenesis. Geol. Soc. Amer. Abs. with Programs, 8, No. 6, 1006, 1976.

Mitchell, R.H., Geochemistry of magnesian ilmenites from kimberlites in South Africa and Lesotho, Lithos, 10, 29-37, 1977.

Pasteris, J.D., F.R. Boyd, and P.H. Nixon, The ilmenite association from the Frank Smith Kimberlite, South Africa,(this volume).

Ringwood, A.E., and J.F. Lovering, Significance of pyroxene-ilmenite intergrowths among kimberlite xenoliths, Earth Planet Sci. Letters, 7, 371-375, 1970.

Sato, M., Electrochemical measurements and control of oxygen fugacity and other gaseous fugacities with solid electrolyte sensors, in Research techniques for high pressure and high temperature, edited by G.C. Ulmer, Springer-Verlag, New York, Chap. 3, pp. 43-99, 1971.

Ulmer, G.C., M. Rosenhauer, E. Woermann, J. Ginder, A. Drory-Wolff, and P. Wasilewski, Applicability of electrochemical oxygen fugacity measurements to geothermometry, Am. Min., 61, 653-660, 1976.

Wyatt, B.A., The melting and crystallization behaviour of a natural clinopyroxene-ilmenite intergrowth, Contrib. Min. Pet., 61, 1-9,1977.

Wyatt, B.A., R.H. McAllister, F.R. Boyd, and Y. Ohashi, An experimentally produced clinopyroxene-ilmenite intergrowth, Carnegie Inst. Washington Yr. Bk., 74, 536-539, 1975.

KIMBERLITIC CHROMIAN PICROILMENITES WITH INTERGROWTHS OF TITANIAN CHROMITE AND RUTILE

B. Wyatt

Anglo American Research Laboratories, P.O. Box 106, Crown Mines, 2025, South Africa.

Abstract. A suite of chromian picroilmenite macrocrysts (up to 13 wt% Cr_2O_3) from kimberlite which contain lamellar intergrowths of titanian chromites (23-42 wt% TiO_2, 7-16 wt% Cr_2O_3) and rutile are described. Using the mineral chemistry of coexisting phases in the natural assemblages as well as the preliminary results of an ongoing study on the system, ilmenite-spinel, it has been possible to construct a phase compatibility diagram for systems containing ilmenite-spinel-Cr_2O_3-rutile relevant to upper mantle conditions. A model is developed whereby the spinel and rutile lamellae intergrowths may be formed by an initial step involving reduction "exsolution" followed by isochemical breakdown during cooling.

Introduction

Picroilmenite, is one of the most diagnostic minerals found in kimberlite and is often characterized by high Cr_2O_3 contents (Haggerty, 1975; Mitchell, 1973, 1977). Other opaque oxide minerals which also occur in kimberlite include spinels of various compositions (Haggerty, 1973, 1975) and rutile. The chemistry of these minerals, in most cases, may be used to distinguish kimberlites from other terrestrial rock types, although they have similarities to many lunar opaque oxides. A notable exception is armalcolite $|(Fe,Mg)Ti_2O_5|$ which is common in lunar rocks (Anderson et al., 1970), but has only been recorded in kimberlite in one case (Haggerty, 1975).

A suite of chromian picroilmenite macrocrysts (the term "macrocrysts" is used with respect to kimberlites as suggested by Clement et al., 1977, as follows: macrocryst - crystal visible to the unaided eye and significantly larger than the surrounding matrix) containing lamellae of chromian spinel, some of which are also associated with rutile, are described. Using these data and the preliminary results of an ongoing experimental study of the spinel-ilmenite system, as well as previously published experimental work, phase relationships are described which may partly account for the petrogenesis of ilmenite, spinel, rutile and armalcolite in kimberlite.

Mineralogy

The mineral analyses of a suite of chromian ilmenite macrocrysts containing Cr-Ti-spinel "exsolution" lamellae (the term "exsolution" is used in the descriptive sense, and does not necessarily refer to exsolution senso stricto, see Haggerty, 1976, p.Hg 108) similar to those previously recorded by Danchin and D'Orey (1972), Boyd and Nixon (1975) and Haggerty (1975) are presented in Table 1. Several of the ilmenite macrocrysts also have rutile inclusions and lamellae intimately associated with the spinel lamellae (Fig. 1a,b). Some of the macrocrysts are suspected to contain armalcolite but this has yet to be conclusively confirmed. All the ilmenite macrocrysts which have lamellae of spinel and rutile are polycrystalline with a decussate to granoblastic polygonal texture (Spry, 1969). Spinel and rutile also form segregations along grain boundaries (Fig. 1a). Those ilmenite grains of similar composition which are not polycrystalline, however, do not have "exsolved" spinel and rutile.

A common feature of the picroilmenite grains having "exsolution" lamellae of spinel and rutile, is the exceptionally high Cr_2O_3 content of the grains, up to 13 wt% at times. The Cr_2O_3 content of the ilmenite however, decreases markedly immediately adjacent to the spinel lamellae (Table 1, L1/10/33). In addition, the Al_2O_3 content of the grains are also abnormally high when compared to most ilmenite macrocrysts found in kimberlite. The spinels are rich in both chromium and titanium, and correspond to titanian chromites. Similar Cr-Ti-spinels are commonly found on the moon (Agrell et al., 1970) but rarely found terrestrially, and then they are mostly associated with kimberlite intrusives (Haggerty, 1973, 1975; Dawson and Hawthorne, 1973; Donaldson and Reid, in press). The rutile exsolutions are also chromium rich, but their low totals suggest that other elements are present.

TABLE 1. Microprobe Analyses of Ilmenite Macrocrysts Containing Lamellae of Spinel and Rutile.

Sample No.	LI/10/33				LI/10/14			LI/10/39			LI/10/35		
Mineral[1]	ILM	SP	RUT	ILM[2]	ILM	SP	RUT	ILM	SP	RUT	ILM	SP	RUT
No. Grains	5	3	8	1	3	2	4	4	3	2	3	1	1
SiO_2	0.01	0.16	0.05	0.00	0.00	1.34	0.15	0.00	0.06	0.04	0.00	0.08	0.00
TiO_2	48.46	8.11	87.86	54.70	56.31	13.34	93.64	51.09	7.37	79.23	48.99	6.78	89.81
Al_2O_3	1.02	11.40	0.21	0.27	0.35	10.40	0.07	0.90	12.35	0.40	1.12	12.34	0.16
Cr_2O_3	11.18	41.56	4.18	4.14	3.31	25.22	2.16	8.62	40.80	5.41	10.64	42.17	4.07
Fe_2O_3 x	7.35	4.93	1.78	5.40	2.42	10.03	0.58	4.53	60.07	4.30	6.23	7.30	0.90
FeO	18.48	19.25		19.84	20.07	22.83		21.07	18.23		19.01	16.56	
MnO	0.18	0.26	0.03	0.21	0.39	0.49	0.00	0.22	0.35	0.06	0.19	0.29	0.02
MgO	13.96	14.22	0.48	16.30	16.86	15.43	0.10	13.80	14.32	1.58	13.92	15.40	0.11
CaO	0.03	0.00	0.11	0.06	0.09	0.10	0.02	0.04	0.01	0.03	0.03	0.00	0.01
Total[4]	100.67	99.89	94.70	100.92	99.80	99.18	96.72	100.27	99.56	91.05	100.13	100.92	95.08
FeO(Total)	25.10	23.69	1.60	24.70	22.25	31.86	0.52	25.14	23.69	3.87	24.61	23.13	0.81
Oxygens	3	4	2	3	3	4	2	3	4	2	3	4	2
Si^{4+}	0.000	0.005	0.001	0.000	0.000	0.043	0.002	0.000	0.002	0.001	0.000	0.003	0.000
Ti^{4+}	0.824	0.195	0.919	0.915	0.946	0.321	0.965	0.872	0.177	0.851	0.836	0.160	0.940
Al^{3+}	0.027	0.430	0.003	0.007	0.009	0.392	0.001	0.024	0.465	0.007	0.030	0.456	0.003
Cr^{3+}	0.200	1.051	0.046	0.073	0.058	0.638	0.023	0.155	1.031	0.061	0.191	1.046	0.045
Fe^{3+} x	0.125	0.119	0.019	0.090	0.041	0.242	0.006	0.077	0.146	0.046	0.106	0.172	0.009
Fe^{2+}	0.349	0.515	0.000	0.369	0.375	0.611	0.000	0.400	0.487	0.000	0.361	0.434	0.000
Mn^{2+}	0.003	0.007	0.000	0.004	0.007	0.013	0.000	0.004	0.009	0.001	0.004	0.008	0.000
Mg^{2+}	0.470	0.678	0.010	0.540	0.561	0.736	0.002	0.467	0.682	0.034	0.471	0.720	0.002
Ca^{2+}	0.001	0.000	0.002	0.001	0.002	0.003	0.000	0.001	0.000	0.000	0.001	0.000	0.000
Total xx	2.000	3.000	1.000	2.000	2.000	3.000	1.000	2.000	3.000	1.000	2.000	3.000	1.000

Sample No.	LI/10/2			LI/10/12		LI/10/19		LI/10/1[3]		M77/192	
Mineral[1]	ILM	SP	RUT	ILM	SP	ILM	SP	RUT	ILM	ILM	SP
No. Grains	6	1	4	3	1	4	1	4	3	4	4
SiO_2	0.02	0.38	0.07	0.00	0.02	0.00	0.22	0.12	0.12	0.02	0.01
TiO_2	49.42	15.29	86.24	47.24	10.89	49.30	8.42	96.73	56.86	52.49	16.31
Al_2O_3	1.05	10.29	0.20	1.02	9.03	1.16	10.41	0.41	0.87	0.04	0.67
Cr_2O_3	10.22	36.86	4.44	13.06	36.45	11.23	43.38	0.64	0.66	1.64	23.21
Fe_2O_3 x	5.96	0.00	1.80	6.64	6.95	5.43	3.57	0.73	0.00	6.25	16.31
FeO	19.19	20.08	0.00	18.70	20.46	19.48	18.71	0.00	37.35	29.71	36.82
MnO	0.20	0.25	0.02	0.17	0.26	0.19	0.28	0.01	0.03	0.34	0.37
MgO	14.04	17.18	0.44	13.23	14.38	13.81	14.64	0.00	4.17	9.62	6.13
CaO	0.03	0.03	0.01	0.02	0.00	0.03	0.02	0.00	0.00	0.02	0.01
Total[4]	100.13	100.36	93.22	100.08	98.44	100.63	99.65	98.64	100.06	100.13	99.84
FeO(Total)	24.56	17.18	1.62	24.67	26.71	24.73	21.93	0.66	37.35	35.33	51.48
Oxygens	3	4	2	3	4	3	4	2	3	3	4
Si^{4+}	0.000	0.012	0.001	0.000	0.001	0.000	0.007	0.002	0.003	0.000	0.000
Ti^{4+}	0.843	0.358	0.917	0.811	0.268	0.838	0.203	0.978	1.022	0.928	0.438
Al^{3+}	0.028	0.378	0.003	0.027	0.348	0.031	0.393	0.006	0.024	0.001	0.028
Cr^{3+}	0.183	0.909	0.050	0.236	0.943	0.201	1.100	0.007	0.012	0.031	0.656
Fe^{3+} x	0.102	0.000	0.019	0.114	0.171	0.092	0.086	0.007	0.000	0.111	0.438
Fe^{2+}	0.364	0.523	0.000	0.357	0.560	0.368	0.502	0.000	0.746	0.584	1.101
Mn^{2+}	0.004	0.007	0.000	0.003	0.007	0.004	0.008	0.000	0.001	0.007	0.011
Mg^{2+}	0.475	0.798	0.009	0.450	0.702	0.465	0.700	0.000	0.149	0.337	0.327
Ca^{2+}	0.001	0.001	0.000	0.000	0.000	0.001	0.001	0.000	0.000	0.001	0.000
Total xx	2.000	2.986	1.000	2.000	3.000	2.000	3.000	1.000	1.957	2.000	3.000

ILM = ilmenite; SP = spinel; RUT = rutile; x = calculated from mineral formula (Finger, 1972); xx = Normalised during calculation (Finger, 1972); (1) Ilmenite analyses between spinel lamellae except where indicated. Spinel analyses selected as those having highest Cr_2O_3, see text; (2) Ilmenite adjacent to spinel lamellae; (3) Rutile host to ilmenite lamellae; (4) Low totals for rutiles due to incomplete analyses.

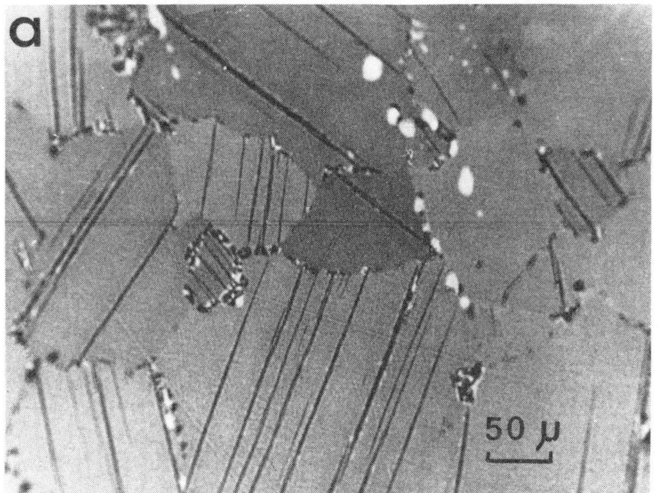

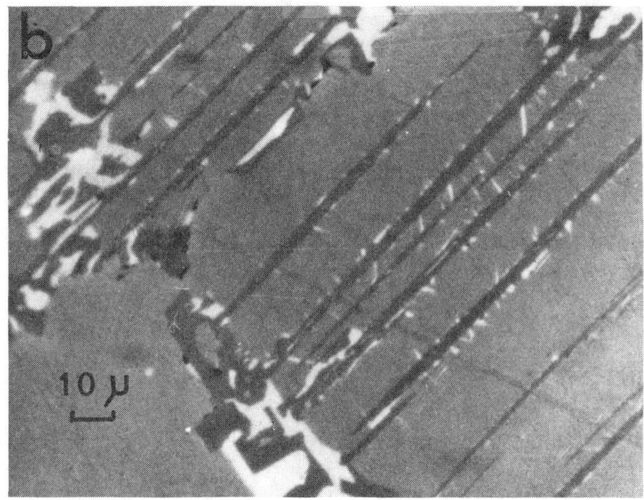

Figs. 1a,b. Photomicrographs of lamellae and segregations of titanian-chromite (dark grey) and rutile (white) in polycrystalline ilmenite (light grey) with a decussate texture (Spry, 1969).

Tie lines between naturally coexisting ilmenite, spinel and rutile in the macrocrysts are shown as solid lines in Figure 2. Also included are other natural assemblages taken from the literature (broken lines). Tie-lines between spinel$_{ss}$-ilmenite$_{ss}$, ilmenite$_{ss}$-rutile$_{ss}$ and spinel$_{ss}$-rutile$_{ss}$ are defined in addition to a three phase field, ilmenite$_{ss}$-rutile$_{ss}$-spinel$_{ss}$.

Discussion

Using data from the natural assemblages it is possible to construct a phase compatibility diagram (Fig. 3) for the system, ilmenite-spinel-Cr_2O_3-rutile. The preliminary results of high pressure (15-35kb) and temperature (1000-1550°C) experiments relevant to the system ilmenite-spinel, indicate that this diagram applies to pressures of probably more than 20 kb, and is therefore applicable to upper mantle conditions. It is emphasised that this diagram is schematic and that the effect of temperature, pressure and oxygen fugacity on the position of the three phase assemblage, spinel-ilmenite-rutile is not yet precisely known. Based on the composition of co-existing phases (Fig. 2) the ilmenite$_{ss}$ composition in the three phase assemblage ilmenite$_{ss}$-spinel$_{ss}$-rutile$_{ss}$ would, however, be expected to become Cr-depleted with decreasing temperature. Knecht et al. (1977) have shown that in experiments conducted at 1 atmosphere, the three phase assemblage, ferropseudobrookite$_{ss}$-ilmenite$_{ss}$-spinel$_{ss}$ rotates according to the reaction:

Fe_2TiO_4(spinel)+Cr_2O_3(ilmenite) ⟶ $FeCr_2O_4$(spinel) + $FeTiO_3$(ilmenite).

Additional experimentation is required to determine whether this relationship holds for the assemblage rutile$_{ss}$-ilmenite$_{ss}$-spinel$_{ss}$ at elevated pressures. Further complications may arise along the spinel join as Al_2O_3 becomes an important constituent in the R_2O_3 component. For example, experimental work at atmospheric pressures (Muan et al., 1972; Schreifels and Muan, 1975) as well as crystallographic predictions (Takeda et al., 1974) indicate that immiscibility, in both the $FeAl_2O_4$-Fe_2TiO_4-$FeCr_2O_4$ and $MgAl_2O_4$-Mg_2TiO_4-$MgCr_2O_4$ systems for the temperature range 1000-1300°C, originates along the titanate-aluminate join and extends part-way toward the chromite end of the join, whereas miscibility is complete between Mg_2TiO_4 - $MgCr_2O_4$ and Fe_2TiO_4 - $FeCr_2O_4$. Although immiscibility is expected to diminish with increasing pressure, phase compatibility may be complex at low temperatures (less than approximately 1000°C(?)). Also, the pseudo-brookite-type minerals, such as armalcolite, appear to be restricted to relatively low pressures. Pure armalcolite ($Fe_{0.5}Mg_{0.5}Ti_2O_5$) is unstable below 1010 ±20°C at atmospheric pressure (Lindsley et al., 1975), but with increasing pressure breaks down (dT/dP = 20°C/kb; Friel et al., 1977) to Mg-armalcolite, rutile and ilmenite. The effect of adding Al^{3+}, Cr^{3+} and Ti^{3+} ions is to stabilize armalcolite to lower temperatures, and to raise the decomposition by approximately 35°C/kb (Kesson and Lindsley, 1975). Thus, the projected curves for armalcolite stability would intersect reasonable mantle solidii at pressures less than 20 kb depending on the composition of armalcolite. The presence of armalcolite in any suite of mantle minerals would therefore indicate pressures of formation of less than 20 kb. Studies in the synthetic systems MgO-FeO-TiO_2-Al_2O_3-Cr_2O_3-Fe_2O_3 conducted at atmospheric pressures (Woerman et al., 1969; Muan et al., 1971; Johnson et al., 1971; Muan et al., 1972; Lipin and Muan, 1974, 1975; Schreifels and Muan, 1975) all show pseudobrookite$_{ss}$ minerals co-existing with spinel$_{ss}$ and/or ilmenite$_{ss}$, whereas in the high pressure system rutile appears to replace pseudobrookite-type minerals.

The observation, at least for the suite of grains examined, that the coarsely crystalline ilmenites are homogeneous, while only those which are polycrystalline (Fig. 1a) contain lamellae and inclusions of spinel and/or rutile,

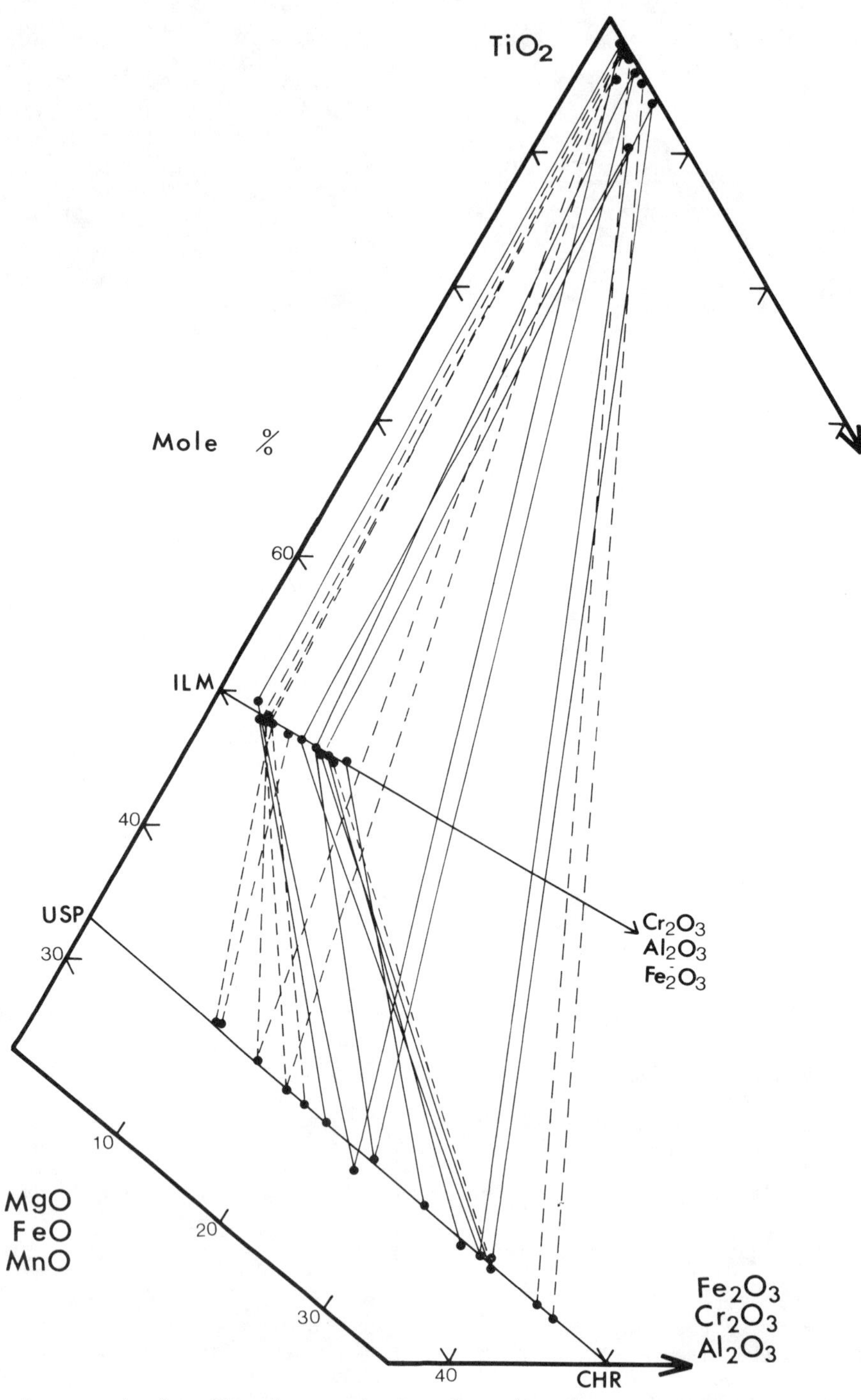

Fig. 2. Tie-lines for coexisting ilmenite, spinel and rutile from natural assemblages. Solid lines - this study, broken lines - data from Smith and Dawson (1975); Dawson, and Smith,(1977); Boyd and Nixon (1975); Haggerty (1973); Donaldson and Reid (in press). ILM = ilmenite; CHR = chromite; USP = ulvöspinel.

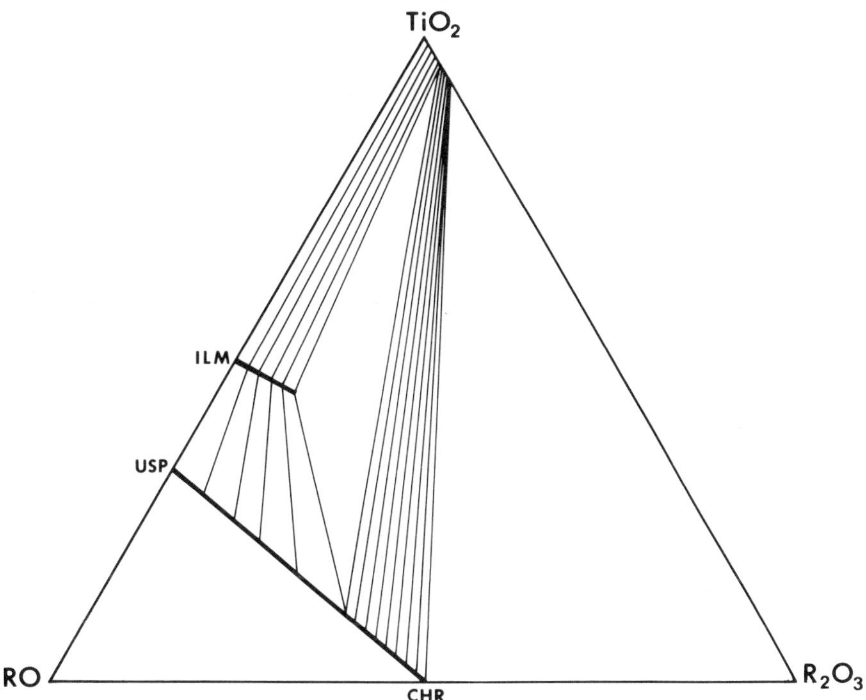

Fig. 3. Schematic phase compatibility diagram (mole %) constructed from Figure 3 and relevant to upper mantle conditions (> 20 kb). End-members in the three-phase triangle, usp_{ss}-TiO_2-ilm_{ss} will vary with temperature. Abbreviations as for Figure 2.

suggests that the "exsolution-like" process may be associated with the annealing and recrystallization of earlier, deformed ilmenites. Mitchell (1972) infers that polycrystalline, kimberlitic ilmenites are derived from annealed grains which have a prehistory of deformation. Mitchell (1972) also points out that "exsolution" in such ilmenites may result from less rapid quenching than the majority of homogeneous, kimberlitic ilmenites.

It is emphasised, however, that for the model outlined below, deformation and recrystallization are not prerequisites for the occurrence of spinel and rutile intergrowths in ilmenite although deformation and subsequent recrystallization may enhance their formation. In fact, Boctor (personal communication) has described such intergrowths in ilmenite macrocrysts which do not show any signs of recrystallization. Nevertheless, the lamellae of spinel and/or rutile in the natural assemblages, suggest that chromian ilmenite appears to be unstable at relatively low temperatures and under subsolidus conditions.

Where spinel lamellae occur in the host ilmenite without rutile, it is possible that reduction "exsolution" (Haggerty, 1971a,b 1972; El Goresy et al., 1972; El Goresy and Ramdohr, 1975) precedes the more advanced dissociation products also including rutile. Following Haggerty (1972), it is unlikely that the spinel lamellae in ilmenite are true exsolution products as there is no evidence of extensive spinel solubility in the rhombohedral ilmenite structure, and that the possibility of an initially exsolved rhombohedral phase inverting to cubic spinel, seems unlikely.

A model, which obviously requires experimental verification, is proposed to account for the spinel and rutile lamellae in the ilmenite macrocrysts whereby a possible initial reduction stage is followed by isochemical decomposition (Fig. 4). With decreasing temperature, a chromian ilmenite (point X) under a mildly reducing environment would move to point Y thereby precipitating spinel according to the idealised reaction:

$$\underbrace{FeTiO_3+Cr_2O_3+Fe_2O_3}_{\text{chromian ilmenite}_{ss}} \longrightarrow \underbrace{Fe_2TiO_4+FeCr_2O_4+\tfrac{1}{2}O_2}_{\text{chromian ulvöspinel}_{ss}}$$

With a further decrease in temperature, the boundaries of the three phase field would move towards the RO-TiO_2 join such that the bulk composition (point Z) lies in the field, spinel-ilmenite-rutile. If the three phase boundaries move further towards the RO-TiO_2 join with decreasing temperature, then it is possible that spinel and rutile may be the final breakdown products. Thus the complete breakdown would

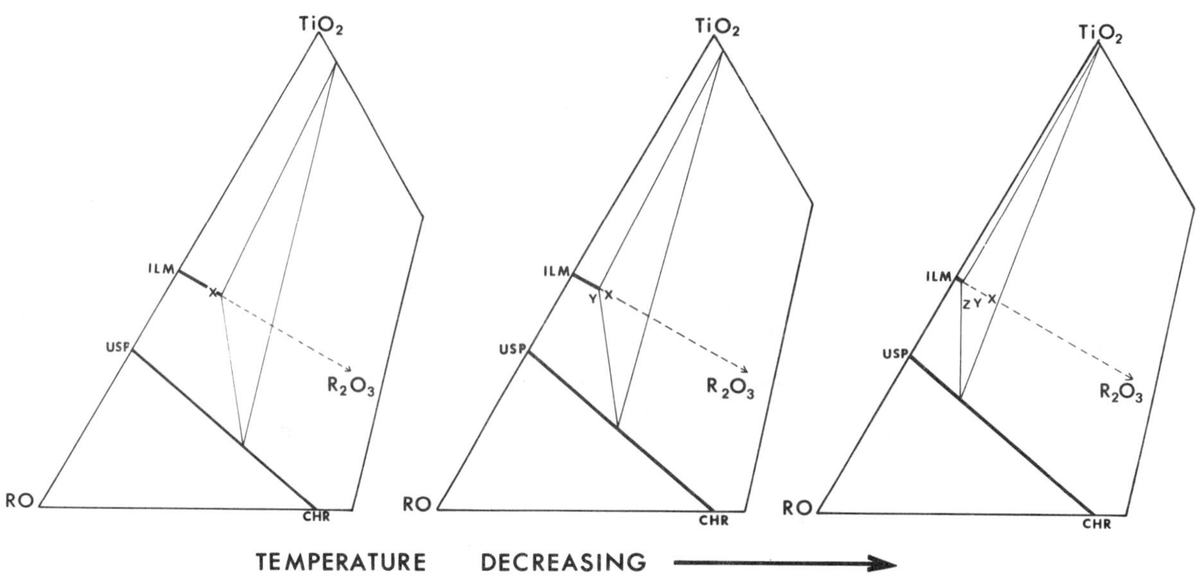

Fig. 4. Schematic diagram illustrating the isobaric decomposition of chromium-bearing ilmenite (X) with decreasing temperature and oxygen fugacity. Positions Y and Z are successive bulk compositions during reduction and cooling such that Ti/R(total) = constant.

take place in the following stages:

(i) Cr-ilmenite —> Cr-ilmenite + spinel (possible initial reducing step on drop in temperature);
(ii) Cr-ilmenite —> Cr-depleted ilmenite + spinel + rutile;
(iii) Cr-depleted ilmenite —> Cr-spinel + rutile.

While an ilmenite-spinel assemblage may be the result of an initial reduction process on cooling, the final assemblages associated with rutile can be accounted for simply by isochemical breakdown. Indeed, Kesson and Lindsley (1975) and Haselton and Nash (1975) regard the more advanced breakdown of many lunar spinels to ilmenite-spinel-FeO assemblages as being due to isochemical breakdown and not necessarily as products of reduction as advocated by Haggerty (1971a,b, 1972), El Goresy *et al.*, (1972) and El Goresy and Ramdohr (1975). The isochemical model proposed here for mantle ilmenites is somewhat similar to that of Haselton and Nash (1975) but takes place at higher oxygen fugacities than the lunar material. If the decrease in temperature is also accompanied by a decrease in pressure, it is possible that pseudobrookite minerals may occur as breakdown products. A number of the macrocrysts appear to have armalcolite associated with the spinel and rutile although verification is still required.

Ulvöspinel apparently has a reaction relationship to ilmenite of the type:

$$Fe_2TiO_4 + TiO_2 \text{ (from melt)} \longrightarrow 2FeTiO_3.$$

Woerman *et al.* (1969), studying the binary system $MgO-TiO_2$, determined that the spinel, Mg_2TiO_4, has a reaction relationship to ilmenite at 1630°C. Also, several experimental and mineralogical studies (Haggerty *et al.*, 1970; O'Hara *et al.*, 1970; Green *et al.*, 1971; Longhi *et al.*, 1974; Kesson, 1975) on Ti-rich lunar basalts endorse that result. Experiments currently being undertaken suggest that the above reaction relationship persists at high pressures in the Cr_2O_3-rich system. The absence, or apparent rarity, of chromian ulvöspinel as discrete macrocrysts in kimberlite, but common occurrence of chromian picroilmenite may be explained by such a reaction relationship.

In addition, the rarity of ilmenite as inclusions in diamond (Meyer and Tsai, 1976; Harris and Gurney, in press), may be purely a function of phase equilibria. For example, a characteristic of many inclusions in diamond is the high Cr_2O_3 content of the minerals (Meyer, 1975), and in particular, almost all the spinel inclusions have Cr_2O_3 contents of greater than 60 wt%. According to the phase compatibilities presented in Figure 3, if such spinels are associated with Ti-bearing minerals, then the latter may be expected to be rutile and not ilmenite.

Acknowledgements. My colleagues, Dr. R.V. Danchin, Dr. B.H. Scott and Mr. D.N. Robinson at the Anglo American Research Laboratories are thanked for critically reading the manuscript. Comments made by the reviewers on an early draft contributed to substantially improving the manuscript. The financial support and cooperation of the Anglo American Corporation provided for this study is greatly appreciated.

References

Agrell, S.O., A. Peckett, F.R. Boyd, S.E. Haggerty, T.E. Bunch, E.N. Cameron, M.R. Dence, J.A.V. Douglas, A.G. Plant, R.J. Traill, O.B. James, K. Keil, and M. Prinz, Titanian chromite, aluminian chromite and chromian ulvospinel from Apollo 11 rocks, Proc. Apollo 11 Lunar Sci. Conf. Suppl. 1 Geochim. Cosmochim. Acta 1, 81-86, 1970.

Anderson, A.T., F.R. Boyd, T.E. Bunch, E.N. Cameron, A. El Goresy, L.W. Finger, S.E. Haggerty, O.B. James, K. Keil, M. Prinz, and P. Ramdohr, Armalcolite: A new mineral from Apollo 11 samples, Proc. Apollo 11 Lunar Sci. Conf. Suppl. 1, Geochim. Cosmochim. Acta 1, 55-64, 1970.

Boyd, F.R. and P.H. Nixon, Origins of the ultramafic nodules from some kimberlites of northern Lesotho and the Monastery Mine, South Africa, Physics Chem. Earth 9, 431-454 1975.

Clement, C.R., E.M.W. Skinner, and B.H. Scott, Kimberlite redefined, Extended Abstracts, Second International Kimberlite Conference, Santa Fe, New Mexico, 1977.

Danchin, R.V. and F. D'Orey, Chromium spinel exsolution in ilmenite from the Premier Mine, Transvaal, South Africa, Contr. Mineral. Petrol. 35, 43-49, 1972.

Dawson, J.B. and J.B. Hawthorne, Magmatic sedimentation and carbonatitic differentiation in kimberlite sills at Benfontein, South Africa, J. Geol. Soc. Lond. 129, 61-85, 1973.

Dawson, J.B. and J.V. Smith, The MARID (mica-amphibole-rutile-ilmenite-diopside) suite of xenoliths in kimberlite, Geochim. Cosmochim. Acta 41, 309-323, 1977.

Donaldson, C.H. and A.M. Reid, Multiple intrusion of a kimberlite dyke, (in press).

El Goresy, A. and P. Ramdohr, Subsolidus reduction of lunar opaque oxides: Textures, assemblages, geochemistry and evidence for a late endogenic gas mixture, Proc. Sixth Lunar Sci. Conf. Suppl. 6, Geochim. Cosmochim. Acta 1, 729-755, 1975.

El Goresy, A., L.A. Taylor, and P. Ramdohr, Fra Meuro crystalline rocks: Mineralogy, geochemistry and subsolidus reduction of opaque oxides, Proc. Third Lunar Sci. Conf. Suppl. 3, Geochim. Cosmoshim. Acta 1, 333-349, 1972.

Finger, L.W., The uncertainty in the calculated ferric iron-content of microprobe analysis, Carnegie Inst. Wash. Yearb. 71, 600-603, 1972.

Friel, J.J. R.I. Harker, and G.C. Ulmer, Armalcolite stability as a function of pressure and oxygen fugacity, Geochim. Cosmochim. Acta 41, 403-410, 1977.

Green, D.H., A.E. Ringwood, N.G. Ware, W.O. Hibberson, A. Major, and E. Kiss, Experimental petrology and petrogenesis of Apollo 12 basalts, Proc. Second Lunar Sci. Conf. Suppl. 2, Geochim. Cosmochim. Acta 1, 601-615, 1971.

Haggerty, S.E. Compositional variations in lunar spinels, Nature Phys. Sci. 233, 156-160, 1971a.

Haggerty, S.E. Subsolidus reduction of lunar spinels, Nature Phys. Sci. 234, 113-117, 1971b.

Haggerty, S.E. Apollo 14: Subsolidus reduction and compositional variation of spinel,. Proc. Third Lunar Sci. Conf. Suppl. 3, Geochim. Cosmochim. Acta 1, 305-332, 1972.

Haggerty, S.E., Spinels of unique composition associated with ilmenite reactions in the Liqhobong kimberlite pipe, Lesotho, In: Lesotho Kimberlites (editor P.H. Nixon), 350 Lesotho National Development Corporation, Maseru, 1973.

Haggerty, S.E. The chemistry and genesis of opaque minerals in kimberlites, Physics. Chem. Earth 9, 195-307, 1975.

Haggerty, S.E. Opaque mineral oxides in terrestrial igneous rocks, In: Oxide Minerals (Editor, D. Rumble). Mineral Soc. Am. Short Course Notes 3, Hg 101-Hg 300, 1976.

Haggerty, S.E., F.R. Boyd, P.M. Bell, L.W. Finger, and W.B. Bryan, Opaque minerals and olivine in lavas and breccias from Mare Tranquillitatis, Proc. Apollo 11 Lunar Sci. Conf. Suppl. 1, Geochim. Cosmochim. Acta 1, 513-538, 1970.

Harris, J.W. and J.J. Gurney, Inclusions in diamonds. In: Properties of Diamonds (Editor J.E. Field), Academic Press, in press.

Haselton, J.D. and W.P. Nash, A model for the evolution of opaques in mare lavas. Proc. Sixth Lunar Sci. Conf. Suppl. 6, Geochim. Cosmochim. Acta 1, 747-755, 1975.

Johnson, R.E., E. Woermann, and A. Muan, Equilibrium studies in the system MgO-"FeO"-TiO_2, Am. J. Sci. 271, 278-292, 1971.

Kesson, S.E. Mare basalt: Melting experiments and petrogenetic interpretations, Proc. Sixth Lunar Sci. Conf. Suppl. 6, Geochim. Cosmochim. Acta 1, 921-944, 1975.

Kesson, S.E. and D.H. Lindsley, The effects of Al^{3+}, Cr^{3+} and Ti^{3+} on the stability of armalcolite, Proc. Sixth Lunar Sci. Conf. Suppl. 6, Geochim. Cosmochim. Acta 1, 911-920, 1975.

Knecht, B., B. Simons, E. Woermann, and A. El Goresy, Phase relations in the system Fe-Cr-TiO-O and their application in lunar thermometry, Proc. Eighth Lunar Sci. Conf. Suppl. 8, Geochim. Losmochim. Acta 1, 2125-2135, 1977.

Lindsley, D.H., S.E. Kesson, M.J. Hartzman, and M.K. Cusman, The stability of armalcolite: Experimental studies in the system MgO-Fe-Ti-O, Proc. Fifth Lunar Sci. Conf. Suppl. 5, Geochim. Cosmochim. Acta 1, 521-534, 1974.

Lipin, B.R. and A. Muan, Equilibria bearing on the behaviour of titanate phases during crystallization of iron silicate melts under strongly reducing conditions, Proc. Fifth Lunar Sci. Conf. Suppl. 5, Geochim. Cosmochim. Acta 1, 535-548, 1974.

Lipin, B.R. and A. Muan, Equilibrium relations among iron-titanium oxides in silicate melts: The system $CaMgSi_2O_6$-"FeO"-TiO_2 in equilibrium with metallic iron, Proc. Sixth Lunar Sci. Conf. Suppl. 6, Geochim. Cosmochim. Acta 1, 945-958, 1975.

Longhi, J., D. Walker, T.L. Grove, E.M. Stolper, and J.F. Hays, The petrology of Apollo 17 mare basalts, Proc. Fifth Lunar Sci. Conf. Suppl. 5, Geochim. Cosmochim. Acta 1, 447-469, 1974.

Meyer, H.O.A. Chromium and the genesis of diamond, Geochim. Cosmochim. Acta 39, 929-936, 1975.

Meyer, H.O.A. and H. Tsai, The nature and significance of mineral inclusions in natural diamond: a review, Minerals Sci. Engng. 8, 242-261, 1976.

Mitchell, R.H. Magnesium ilmenite and its role in kimberlite petrogenesis. J. Geol. 81, 301-311, 1973.

Mitchell, R.H. Geochemistry of magnesian ilmenites from kimberlites in South Africa and Lesotho, Lithos, 10, 29-37, 1977.

Muan, A., J. Hauk, and T. Lofall, Equilibrium studies with a bearing on lunar rocks, Proc. Third Lunar Sci. Conf. Suppl. 3 Geochim. Cosmochim. Acta. 1, 185-196, 1972.

Muan, A., J. Hauk, E.F. Osborn, and J.F. Schairer, Equilibrium relations among phases occurring in lunar rocks, Proc. Second Lunar Sci. Conf. Suppl. 2, Geochim. Cosmochim. Acta 1, 497-505, 1971.

O'Hara, M.J., G.M. Biggar, S.W. Richardson, C.E. Ford, and B.G. Jamieson, The nature of the seas, mascons and the lunar interior in the light of experimental studies, Proc. Geochim. Cosmochim. Acta 1, 695-710, 1970.

Schreifels, W.A. and A. Muan, Liquid-solid equilibria involving spinel ilmenite, and ferropseudobrookite in the system "FeO"-Al_2O_3-TiO_2 in contact with metallic iron, Proc. Sixth Lunar Sci. Conf. Suppl. 6, Geochim. Cosmochim. Acta 1, 973-985, 1975.

Smith, J.V. and J.B. Dawson, Chemistry of Ti-poor spinels, ilmenites and rutiles from peridotites and eclogite xenoliths, Physics Chem. Earth 9, 309-322, 1975.

Spry, A. Metamorphic Textures, Pergamon Press, New York, 350p, 1969.

Takeda, H., M. Muyamoto, and A.M. Reid, Crystal chemical control of ilmenite partitioning for co-existing chromite-ulvöspinel and pigeonite-augite in lunar rocks, Proc. Fifth Lunar Sci. Conf. Suppl. 5, Geochim. Cosmochim. Acta 1, 727-743, 1974.

Woermann, E., B. Brezny, and A. Muan, Phase equilibria in the system MgO-iron oxide-TiO_2 in air, Am. J. Sci. 267-A, 463-479, 1969.

THE ILMENITE ASSOCIATION AT THE FRANK SMITH MINE, R.S.A.

Jill Dill Pasteris

Department of Geology and Geophysics, Yale University, New Haven, Connecticut 06520

F. R. Boyd

Geophysical Laboratory, 2801 Upton Street, N.W., Washington, D.C. 20008

P. H. Nixon

Department of Geology, The University, P.O. Box 4820, Port Moresby, Papua, New Guinea

Abstract. Discrete nodules of ilmenite from the Frank Smith kimberlite pipe, South Africa, are polygranular with a range in grain size of 0.2-25 mm. They are characterized in polished section by surface pitting that delineates grain boundaries and that is variably developed within grains, depending on crystallographic orientation. The ilmenites usually contain spinel lamellae (~1μ wide) of two types: short, black rods of pleonaste and longer, tan lamellae of titanomagnetite. Pyroxene-ilmenite lamellar intergrowths differ from those found at other pipes in that they are commonly sheared and in that enstatite-ilmenite intergrowths are as abundant as diopsidic varieties.

Compositions of ilmenites and intergrown silicates are similar to those from other South African kimberlites except that the Frank Smith ilmenites are distinctively rich in Mg. Discrete ilmenites have a range in Mg/(Mg + Fe^{2+}) of 0.376-0.525. Ilmenites from lamellar intergrowths are more restricted in Mg/(Mg + Fe^{2+}) with a range of 0.463-0.499.

The fact that both the ilmenite discrete nodules and lamellar intergrowths at Frank Smith are distinctively rich in Mg is strong evidence that they are consanguineous in origin. It is suggested that at the time of eruption the discrete and lamellar ilmenites were equilibrated with diopside, enstatite, garnet, possibly olivine and small amounts of liquid with variable Mg/(Mg + Fe^{2+}) over a depth range of the order of several tens of kilometers. The interstitial liquids may have had an aggregate composition like kimberlite and they and the nodules are believed to have been mixed together during eruption.

Introduction

The relationship between kimberlite and the abundant ilmenite nodules that many kimberlites contain needs clarification in order to better understand the origin of kimberlite magma. Ilmenite nodules are most commonly monomineralic, but a variety of ilmenite intergrowths with silicates, including host-inclusion combinations, lamellar intergrowths with pyroxenes, and granular intergrowths of various types have been discovered. A problem that has been especially difficult to resolve is whether the ilmenite discrete nodules formed as high-pressure phenocrysts in kimberlite or protokimberlite magma (Gurney, et al., 1977; Mitchell, 1977; Frick, 1973a) or whether they are accidental mantle inclusions. It is a further problem to understand whether or not there is a consanguinity between the ilmenite discrete nodules, the lamellar intergrowths, and perhaps some of the granular intergrowths.

Ilmenite discrete nodules and ilmenite-silicate intergrowths are especially abundant in the Frank Smith Mine and associated Weltevreden diatremes, R.S.A. These two blows, connected by a dike, contain a variety of basaltic and micaceous kimberlites (Wagner, 1914; Field Guide, Kimberlite Conference, 1973). Previous studies have shown that ilmenites from Frank Smith tend to be rich in MgO in comparison with ilmenites from many other kimberlites (Boyd and Nixon, 1973; Mitchell, 1977). The Frank Smith ilmenite association is also distinctive in the relative abundance of enstatite-ilmenite lamellar intergrowths and in the presence of exotic intergrowths (Meyer, et al., 1977; Rawlinson and Dawson,

Table 1: Compositions of Ilmenite Discrete Nodules from the Frank Smith Mine, Showing High and Low Values for Mg/(Mg + Fe), Cr_2O_3, Al_2O_3, and Fe_2O_3, weight percent.

	FRB 50/1	FRB 50/2	FRB 50/11	FRB 50/18	FRB 50/19	PHN 2353C	PHN 2353E
SiO_2	0.12	0.11	0.17	0.25	0.22	0.10	0.10
TiO_2	52.3	47.0	51.0	47.1	50.5	49.4	43.7
Al_2O_3	0.64	0.37	0.81	0.34	0.57	1.75	0.35
Cr_2O_3	0.21	0.80	1.03	0.78	<0.03	0.72	0.93
Fe_2O_3*	9.35	16.0	11.2	15.2	11.2	13.9	20.2
FeO	23.9	26.2	22.4	25.9	25.1	21.5	23.8
MnO	0.25	0.27	0.21	0.25	0.25	0.24	0.27
MgO	12.8	8.87	13.0	9.20	11.3	12.7	8.49
CaO	0.04	<0.03	<0.03	<0.03	<0.03	0.04	<0.03
NiO	0.12	0.06	0.23	0.08	0.05	0.15	0.15
Total	99.7	99.7	100.1	99.1	99.2	100.5	98.0
				O = 3000			
Si	3	3	4	6	5	2	2
Ti	906	842	879	845	888	848	799
Al	17	10	22	10	16	47	10
Cr	4	15	19	15	0	13	18
Fe^{3+}*	162	286	193	274	197	239	369
Fe^{2+}	460	522	429	517	492	411	484
Mn	5	5	4	5	5	5	6
Mg	440	315	445	327	395	431	308
Ca	1	0	0	0	0	1	0
Ni	2	1	4	2	1	3	3
Total†	2000	2000	2000	2000	2000	2000	2000
$Mg/(Mg+Fe^{2+})$	0.489	0.376	0.509	0.388	0.455	0.512	0.389

*Calculated (Finger and Hadidiacos, 1972).

†Cation totals normalized in course of ferric iron calculation.

1977; Clarke, et al., 1977). Discrete nodules from Frank Smith, including lamellar intergrowths, differ in texture from those found elsewhere in that they are commonly deformed (Frick, 1973b).

A petrographic and electron probe study of the Frank Smith ilmenite association has been carried out to extend understanding of these perplexing mantle samples. All of the specimens studied in this investigation were collected from the Frank Smith dumps, mainly in the form of chips of coarse concentrate with maximum dimensions of 4-5 cm. Representative analyses of these ilmenites and associated silicates are given in Tables 1-4; additional analyses can be found in Nixon and Boyd (1973a), Frick (1973b), Boyd (1974a), and Mitchell (1977).

Petrography

Virtually all the Frank Smith ilmenite discrete nodules are polygranular (Fig. 1A), with a grain size ranging from 0.2 to 25 mm. The most distinctive optical feature of the ilmenite nodules is surface pitting--visible macroscopically as well as microscopically (Fig. 1B). This characteristic can sometimes be seen before polishing, inasmuch as Dawson (1962, personal communication) noted small, open tubules as well as finer pits on freshly fractured surfaces of Lesotho ilmenites. The pitting appears to have a distinct crystallographic orientation (Fig. 1C) the shape of the pits as well as their density vary with grain orientation. Individual nodules usually contain both pitted and non-pitted grains. The coarsest pits generally delineate grain boundaries; also the presence of internal pitting commonly defines one optically continuous grain. More rarely, however, the pit boundary encloses two grains whose interface is recognized only by differences in pleochroism, or anisotropism, or both (Fig. 1D). Electron microprobe analyses and Vickers' hardness tests on adjacent pitted and pit-free

TABLE 2: Compositions of Ilmenite Discrete Nodules and Included Silicates from the Frank Smith Mine, weight per cent.

	FRB 17		PHN 2349		*PHN 2351 b	
	Ilmenite	Garnet	Ilmenite	Diopside	Ilmenite	Enstatite
SiO_2	0.13	41.5	0.10	54.9	0.09	56.4
TiO_2	51.4	0.74	49.4	0.27	54.1	0.22
Al_2O_3	0.68	21.9	0.59	2.21	0.70	1.24
Cr_2O_3	<0.03	0.06	0.77	0.36	1.18	0.10
Fe_2O_3	10.6†	...	12.5†	...	7.21†	...
FeO	25.3	11.0‡	25.9	4.45‡	22.9	8.28‡
MnO	0.27	0.40	0.24	0.13	0.25	0.14
MgO	11.7	19.7	10.2	16.8	14.2	32.8
CaO	<0.03	4.39	<0.03	20.0	<0.03	0.79
Na_2O	n.d.§	0.05	n.d.	1.79	n.d.	0.29
NiO	n.d.	n.d.	0.13	n.d.	0.20	n.d.
Total	100.1	99.7	99.8	100.9	100.8	100.3
	O = 3000	O = 12000	O = 3000	O = 6000	O = 3000	O = 6000
Si	3	2990	2	1978	2	1962
Ti	895	40	872	7	917	6
Al	19	1861	16	94	19	51
Cr	0	3	14	10	21	3
Fe^{3+}	185†	...	220†	...	122†	...
Fe^{2+}	489	665‡	508	134‡	432	241‡
Mn	5	24	5	4	5	4
Mg	403	2111	359	900	478	1700
Ca	0	339	0	772	0	29
Na	n.d.	7	n.d.	125	n.d.	20
Ni	n.d.	n.d.	2	n.d.	4	n.d.
Total	2000‖	8040	2000‖	4024	2000‖	4016
$Mg/(Mg + Fe^{2+})$	0.452	0.761	0.414	0.870	0.525	0.876

*Listed by Nixon and Boyd (1973a) as a granular intergrowth.

†Calculated (Finger and Hadidiacos).

‡Total Fe as Fe^{2+}.

§Not determined.

‖Cation totals normalized in course of ferric iron calculation.

grains failed to show any consistent physical or chemical inhomogeneities. This pitting may be an alteration feature because it is crystallographically controlled and enhanced at grain boundaries. Although the pitting is not wholly produced by polishing it may be augmented or enhanced in the section-making process.

Most of the discrete ilmenites contain fine (~1μ wide) spinel lamellae. These lamellae are too fine-grained for quantitative electron microprobe analysis. A combination of broad beam analysis and fine beam traverses across the exsolved phases yielded qualitative data. The spinels are of two types: (1) short, black rods, the most abundant phase, are very low-Cr pleonaste; (2) longer, tan lamellae are Cr-bearing (probably only a few weight %) titanomagnetite. The spinels always have a single common orientation within the basal plane of an ilmenite grain (Fig. 1E). Where both spinels are present the black rods are often enclosed within the tan lamellae (Fig. 1F). Exsolution of the spinels probably resulted from: (1) subsolidus reduction of ilmenite to produce the tan lamellae of titanomagnetite, and (2) true exsolution from ilmenite under lowered temperature, or pressure, or both, to produce the black spinel. There appears to be no relationship between these spinels and the pitting feature previously described (Fig. 2A).

A distinctive feature of the spinels is that they often are concentrated along curved trends, i.e., subgrain boundaries within the ilmenite grains (Figs. 2B, 2C). In spite of the curved nature of these boundaries, the spinels remain parallel within a given optically continuous grain of ilmenite. This configuration appears to result from concentration of exsolution

Table 3: Compositions of Pyroxene and Ilmenite in Lamellar Intergrowths from the Frank Smith Mine, weight percent.

	FRB 402/1		FRB 406		FRB 476	
	Ilmenite	Diopside	Ilmenite	Enstatite	Ilmenite	Enstatite
SiO_2	0.14	54.3	0.17	56.7	0.12	56.7
TiO_2	51.9	0.49	51.6	0.29	52.3	0.22
Al_2O_3	0.76	2.25	0.94	1.00	0.59	0.89
Cr_2O_3	0.38	0.12	1.58	0.16	0.28	0.05
Fe_2O_3	9.96*	...	9.77*	...	9.25*	...
FeO	24.5	4.92†	23.1	7.23†	25.1	7.19†
MnO	0.21	0.15	0.20	0.18	0.21	0.16
MgO	12.3	18.6	12.9	33.8	12.2	33.2
CaO	0.04	16.7	<0.03	1.17	<0.03	0.91
Na_2O	n.d.‡	1.56	n.d.	0.20	n.d.	0.21
NiO	0.11	n.d.	0.18	n.d.	0.14	n.d.
Total	100.3	99.1	100.4	100.7	100.2	99.5
	O = 3000	O = 6000	O = 3000	O = 6000	O = 3000	O = 6000
Si	3	1977	4	1957	3	1975
Ti	897	13	885	8	906	6
Al	21	96	25	41	16	37
Cr	7	3	29	4	5	1
Fe^{3+}	172*	...	168*	...	160*	...
Fe^{2+}	471	150†	442	209†	484	210†
Mn	4	5	4	5	4	5
Mg	422	1008	440	1740	418	1725
Ca	1	652	0	43	0	34
Na	n.d.	110	n.d.	13	n.d.	14
Ni	2	n.d.	3	n.d.	3	n.d.
Total	2000§	4014	2000§	4020	2000§	4006
$Mg/(Mg + Fe^{2+})$	0.473	0.871	0.499	0.893	0.463	0.892

*Calculated (Finger and Hadidiacos, 1972).
†Total Fe as Fe^{2+}.
‡Not determined.
§Cation totals normalized in course of ferric iron calculation.

along dislocations in the ilmenite (S. E. Haggerty, pers. comm.). Detailed optical study reveals small differences in reflection pleochroism between some of these ilmenite subgrains, suggesting slight rotation between them.

Grain boundary textures, i.e., the degree of equilibrium, degree of boundary definition, and grain size and orientation, vary widely even within a single ilmenite nodule (Figs. 2D, 2E). Neither the degree of pitting nor the presence or absence of spinels (whether or not they decorate subgrain boundaries) appears related to grain boundary equilibrium.

The following textural features are interpreted as indications of partial or total recrystallization of ilmenite in the discrete nodules: (1) twin lamellae are occasionally present near the edges of a nodule; (2) equilibrium grain boundary intersections are common; (3) bands of fine mosaic ilmenite sometimes traverse coarse grains; (4) there are occasionally bands or long, tapering wedges within grains that have a slightly different extinction position from the host; (5) spinel lamellae appear to lie along dislocations; and (6) a few specimens show a foliation due to alignment of elongated, highly sutured grains.

Lamellar intergrowths are abundant at the Frank Smith mine. Haggerty, et al. (1977), have estimated that about 10% of such intergrowths in the Monastery kimberlite are enstatite-ilmenite and the balance are diopside-ilmenite, but at Frank Smith the proportions of the two types are more nearly equal. The lamellar intergrowths at Frank Smith are also distinctive in that they commonly show deformation textures with variable degrees of strain (Fig. 2F and Frick, 1973b).

Textures of the pyroxene-ilmenite inter-

Table 4: Compositions of Minerals in Ilmenite-Silicate Intergrowths of Problematic Origin from the Frank Smith Mine, weight percent.

(Ranges for inhomogeneous oxides are given in parentheses. See text for petrographic descriptions.)

	PHN 3244 D/1				PHN 3244 D/2			PHN 3248		
	Ilmenite	Enstatite	Garnet	Olivine	Ilmenite	Enstatite	Olivine	Ilmenite	Enstatite	Diopside
SiO_2	0.14	56.8	42.5	40.5	0.12	56.2	40.5	0.23	56.7	54.2
TiO_2	52.7	0.16	1.05	0.04	51.8	0.34	0.04	55.0	0.25	0.29
Al_2O_3	0.39	1.16(0.83-1.53)	21.2	0.05	0.38	1.40	0.05	0.12	1.60(1.16-1.96)	2.08
Cr_2O_3	3.31	0.12	1.24	<0.03	2.89	0.22	<0.03	0.98	0.07	0.33
Fe_2O_3	5.86*	...	...	...	7.83*	...	...	2.59*	...	...
FeO	24.9	6.78†	10.1†	11.8†	25.1	6.67†	11.8†	28.2	7.21†	4.37†
MnO	0.28	0.13	0.40	0.17	0.26	0.15	0.17	0.27	0.13	0.12
MgO	12.5	33.5	19.9	46.5	11.8	32.8	46.5	11.9	32.9	16.7
CaO	0.03	0.68(0.50-0.87)	4.72	0.08	<0.03	1.25	0.08	<0.03	0.64(0.46-0.85)	19.9
Na_2O	n.d.‡	0.14	0.07	n.d.	n.d.	0.41	n.d.	n.d.	0.32	1.63
NiO	0.05	n.d.	n.d.	0.25	0.19	n.d.	0.25	0.11	n.d.	n.d.
Total	100.2	99.5	101.2	99.4	100.4	99.4	99.4	99.4	99.8	99.6
	O = 3000	O = 6000	O = 12000	O = 4000	O = 3000	O = 6000	O = 4000	O = 3000	O = 6000	O = 6000
Si	3	1974	3017	1007	3	1960	1007	5	1966	1977
Ti	911	4	56	1	898	9	1	961	6	8
Al	11	48	1773	1	10	58	1	3	65	90
Cr	60	3	70	0	53	6	0	18	1	10
Fe^{3+}	101*	...	...	...	136*	...	...	45*	...	...
Fe^{2+}	477	197†	601†	246†	485	195†	246†	548	208†	133†
Mn	5	4	24	4	5	4	4	5	3	4
Mg	429	1736	2102	1725	407	1707	1725	411	1702	908
Ca	1	25	359	2	0	47	2	0	23	778
Na	n.d.	9	10	n.d.	n.d.	28	n.d.	n.d.	20	115
Ni	1	n.d.	n.d.	5	4	n.d.	5	2	n.d.	n.d.
Total	2000§	4001	8012	2991	2000§	4014	2991	2000§	3994	4023
$Mg/(Mg+Fe^{2+})$	0.474	0.898	0.778	0.859	0.457	0.898	0.875	0.429	0.891	0.872

*Calculated (Finger and Hadidiacos, 1972).

†Total Fe as Fe^{2+}.

‡Not determined.

§Cation totals normalized in course of ferric iron calculation.

Fig. 1

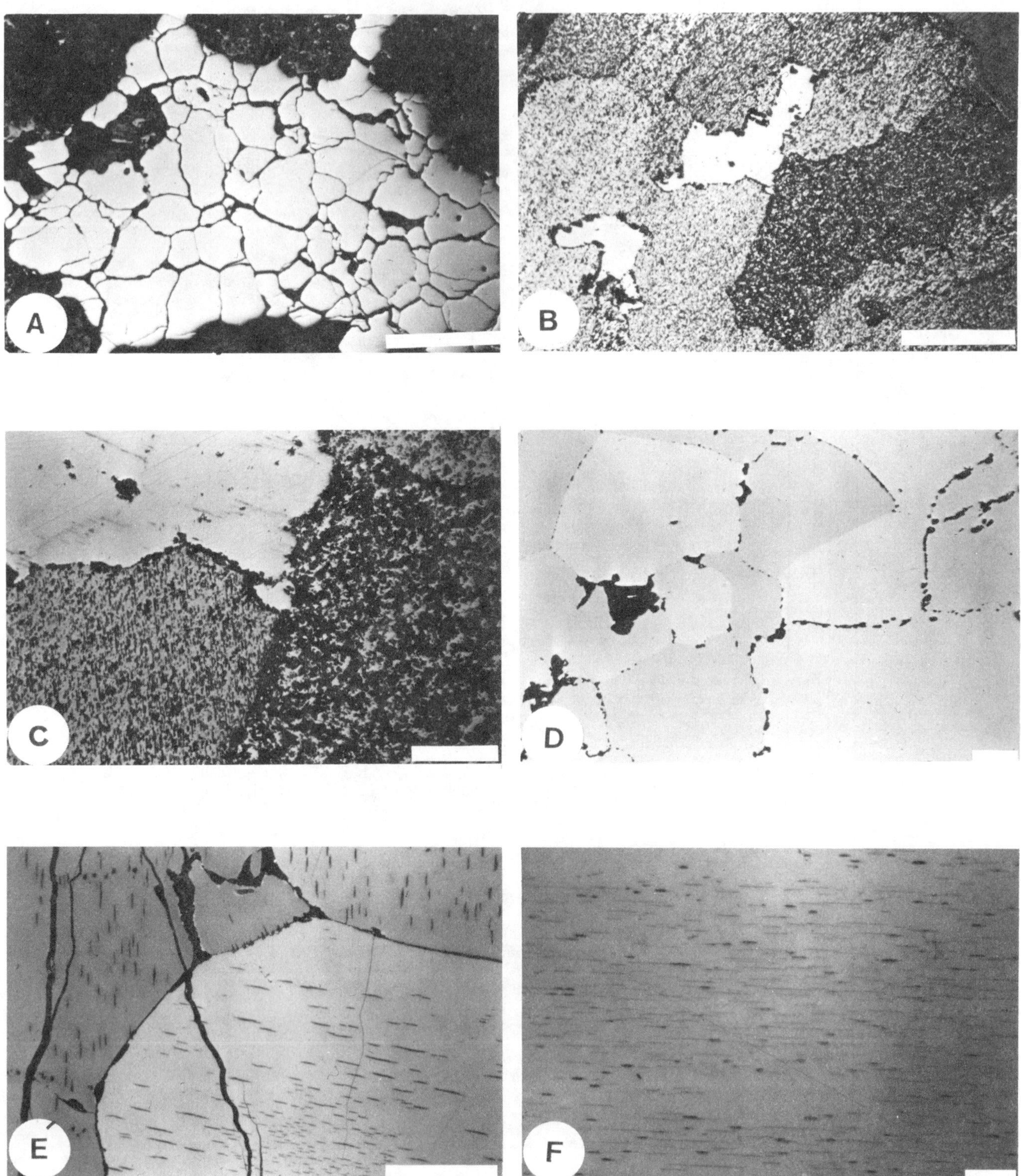

growths vary widely. At one extreme the intergrowths contain straight, parallel lamellae of pyroxene and ilmenite, in which each phase has a single optical orientation. Exsolved spinels define subgrains (c.f. Clarke, et al., 1977). The spinels are aligned parallel to the host ilmenite lamella, i.e., along the (0001) plane. Even in the strictly lamellar intergrowths, however, ilmenite blades are often discontinuous. Where they intersect kimberlite, the ilmenite lamellae usually become misaligned, polygranular, and altered to perovskite. In some specimens the lamellar texture is not well developed and unstrained, single crystals of pyroxene may contain blebs as well as lamellae of ilmenite (Meyer, et al., 1977).

Deformation and subsequent recrystallization tend to destroy the primary lamellar texture. In deformed nodules the ilmenite is recrystallized to polygranular aggregates and in some cases the lamellae are transformed into irregularly shaped blebs. It appears that exsolution is enhanced by annealing inasmuch as spinels preferentially exsolve along dislocations; i.e., along high-energy, subgrain boundaries. The original, single crystals of pyroxene are broken into subgrains with slightly differing optical orientation and with undulate extinction. More extreme deformation causes recrystallization of enstatite as neoblasts. In one specimen (FRB 402/1, Table 3) a single crystal of diopside with sharp extinction containing well-formed ilmenite lamellae grades abruptly into strained and disoriented diopside subgrains with irregular blebs of ilmenite.

There are a number of intergrowths of problematic origin that have varied textural relations. Specimen PHN 3248 (Table 4) was a nodule 40 cm in diameter in hardebank when discovered at the Frank Smith mine. This nodule is composed primarily (~90%) of enstatite that is partly granular and partly in sheaf-like aggregates, resembling quench crystals. The acicular crystals of enstatite in the sheaf-like aggregates range up to 1 mm in diameter and are broken into subgrains with undulate extinction. It is unclear whether this texture is entirely the result of rapid crystallization or whether there has been subsequent deformation. The ilmenite mainly forms polygranular, irregular, interstitial blebs but in one portion of the nodule it is aggregated in a patch about 3 cm in diameter. The enstatite in this nodule is markedly inhomogeneous in Ca and Al (Table 4) but the ilmenite appears homogeneous within a single thin section cut from the nodule. This intergrowth may be akin to nodule BD2027 (Rawlinson and Dawson, 1977).

Two specimens have textural relations that suggest the possibility of metasomatic introduction of ilmenite (e.g., Harte and Gurney, 1975). Both are chips picked from coarse concentrate, 3-4 cm in their maximum dimension. PHN 3244D/1 (Table 4) consists primarily of enstatite with a variable grain size ranging up to 1 cm in a mesotasis of polygranular ilmenite. The ilmenite forms approximately 25% of the rock. Olivine and garnet are present as small grains in subordinate amount. The more coarse grains of enstatite appear deformed, being broken into subgrains with undulate extinction. Ilmenite permeates these coarse enstatite crystals, forming both irregular blebs and crude lamellae. In contrast, the less coarse crystals of enstatite are unstrained and free of included ilmenite. The enstatite in this specimen is markedly inhomogeneous in Al and Ca (Table 4).

Specimen PHN 3244D/2 is a sheared harzburgite that is permeated by irregular patches of intergranular, polycrystalline ilmenite. Enstatite porphyroclasts range up to 7 mm in diameter. Olivine is present both in strain-free neoblasts and 3-4 mm porphyroclasts. The ilmenite forms about 10% of the rock and is associated with small flakes of strongly pleochroic, red-brown phlogopite.

Compositional Relations

The range in Mg/(Mg + Fe^{2+}) for the Frank Smith ilmenite discrete nodules is 0.376-0.525 (Tables 1, 2). Most of the specimens from Frank Smith analyzed by Mitchell (1977) fall

Fig. 1. Microphotographs of ilmenite-bearing nodules from the Frank Smith mine in reflected light. (A): Polygranular ilmenite (light gray) discrete nodule with silicate inclusion (black). Specimen FRB 50/5. Scale bar = 500μ.
(B): Ilmenite discrete nodule with both pitted and non-pitted grains. Scale bar = 500μ.
(C): Enlargement of part of Figure 1b. Depending upon grain orientation, the pits appear coarse and irregular or oriented along parallel lineations. Specimen FRB 50/12. Scale bar = 100μ.
(D): A long, continuous pit boundary encloses two optically distinct grains toward the middle of the photograph (distinguished by reflection pleochroism). Specimen FRB 50/1. Scale bar = 100μ.
(E): Polygranular aggregate of ilmenite within a pyroxene-ilmenite intergrowth. The black spinels have a single orientation within any given grain. Specimen 3008. Scale bar = 50μ.
(F): Long, tan titanomagnetite lamellae enclose short rods of black spinel. A single ilmenite grain is the host. Specimen 3011. Scale bar = 10μ.

Fig. 2

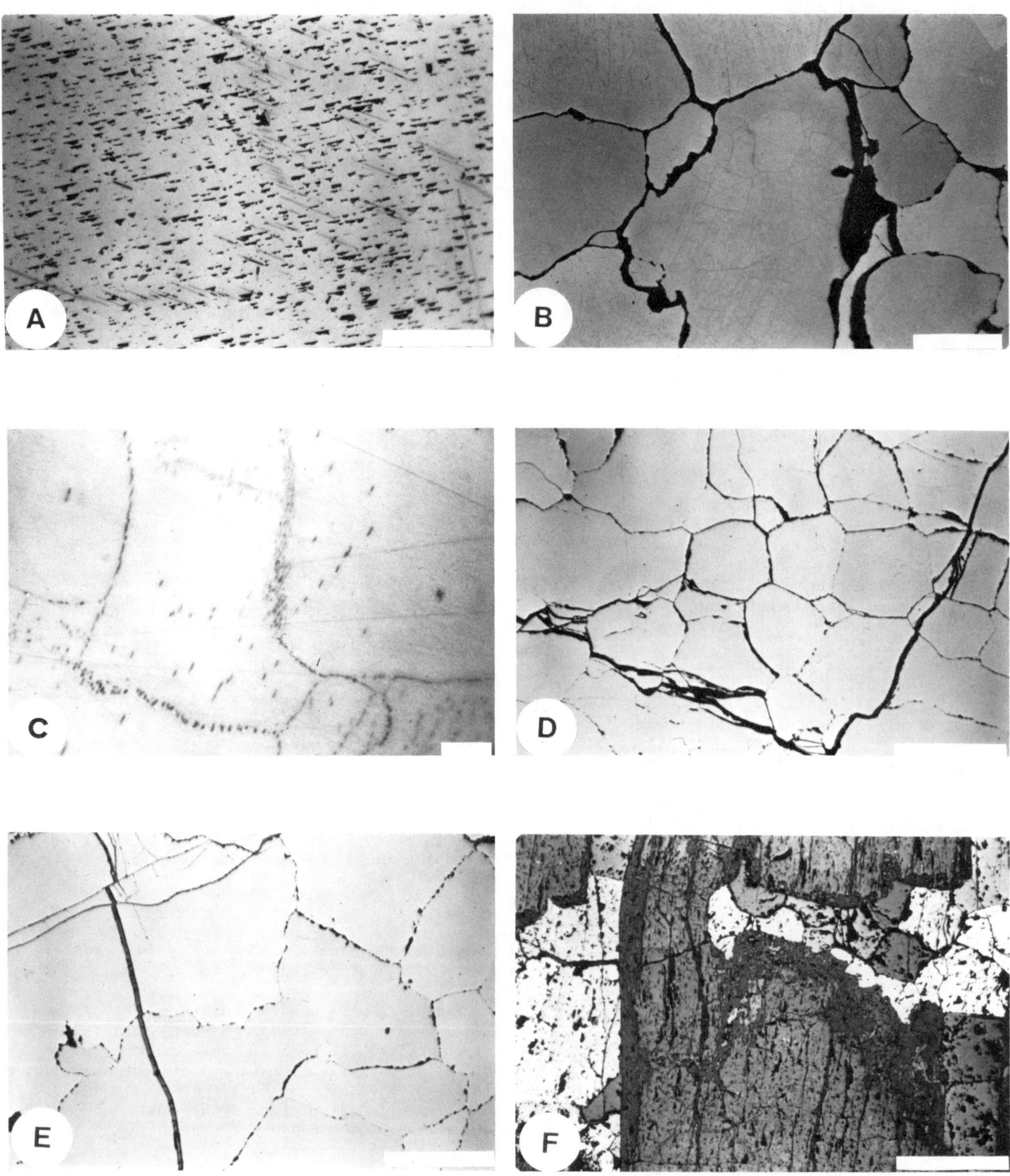

in or close to this range with the exception of one (F1, Mitchell, op. cit., Table 1) for which Mg/(Mg + Fe^{2+}) is 0.744. There is an apparent positive correlation between calculated Fe_2O_3 (Finger, 1972) and $FeTiO_3$ (Fig. 3). Only one nodule was checked in detail for homogeneity, but marginal zoning like that found by Haggerty, et al. (1977), in Monastery ilmenites was not noted.

The Cr_2O_3 in the Frank Smith discrete ilmenites ranges from <0.03 to 1.18 wt % (Fig. 4A) and there is no correlation with Mg/(Mg + Fe^{2+}). The Al_2O_3 contents have the range 0.34-1.75 wt % with a crude positive correlation with Mg/(Mg + Fe^{2+}) (Fig. 4B). A similar correlation may exist for NiO (Fig. 4C) but there is much scatter. The MnO (Fig. 4D) (0.2-0.3 wt %) does not vary systematically with Mg/(Mg + Fe^{2+}).

Significant SiO_2 was found in all the ilmenites (Fig. 4E), ranging from 0.09-0.34 wt.%. The SiO_2 varied from 0.27 to 0.47 wt.% in the most Si-rich ilmenite, but inasmuch as all values are higher than for most other ilmenites this variation is more likely due to inhomogeneity than to the presence of submicroscopic inclusions. Ilmenites in nodule suites from Malaita, Solomon Islands, and East Griqualand, South Africa, do not contain significant SiO_2 (unpublished data). These two nodule suites are believed to have originiated at shallower depths than the Frank Smith suite (Nixon and Boyd, 1977). Such relations could mean that solution of Si in ilmenite is enhanced by pressure, but the problem may be complex because recent analyses of lunar ilmenites commonly contain values ranging up to 0.1 wt.% SiO_2 (e. g., Warner, et al., 1977).

Six specimens of ilmenite intergrown with garnet or pyroxene in a host-inclusion relationship have mineral compositions that are mainly concordant with those of other Frank Smith discrete nodules and with those found for analogous specimens from northern Lesotho. Two garnet inclusion in ilmenite discrete nodules have TiO_2 in the range 0.5-1.0 wt. % and Cr_2O_3 <0.1 wt. % (Table 2, FRB 17). These two garnets

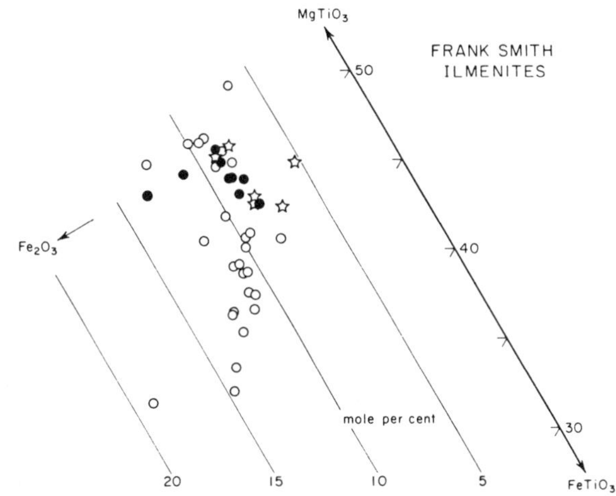

Fig. 3. Ilmenites from the Frank Smith mine plotted in a portion of the ternary $MgTiO_3$-$FeTiO_3$-Fe_2O_3. Open circles = ilmenite discrete nodules; solid circles = ilmenite/pyroxene lamellar intergrowths; stars = granular intergrowths, some apparently metasomatic and some of problematic origin.

are slightly more magnesian than comparable specimens from Lesotho, correlating with the fact that Frank Smith ilmenites are more magnesian than ilmenites from many other pipes. Ilmenite inclusions have been found in two bronzite discrete nodules and one specimen of bronzite included in an ilmenite nodule has been studied (PHN 2351b, Table 2). These bronzites have Mg/(Mg + Fe^{2+}) in the range 0.87-0.89; TiO_2 is 0.2-0.3 wt. % and Cr_2O_3 is <0.1 wt. %. One sample of diopside included in an ilmenite nodule (PHN 2349, Table 2) has a composition very similar to a diopside included in ilmenite from the Solane pipe in northern Lesotho (PHN 1620A, Nixon and Boyd, 1973b, Table 23).

The ilmenites of the lamellar intergrowths have a much more restricted Mg/(Mg + Fe^{2+}), in the range 0.463-0.499, than do the ilmenite

Fig. 2. Microphotographs of ilmenite-bearing nodules from the Frank Smith mine in reflected light.
(A): Tan titanomagnetite lamellae showing high-angle intersections with lineations of pits in an ilmenite grain. This shows that pitting is not the result of plucked spinel lamellae. Specimen FRB 50/12. Scale bar = 50μ.
(B) and (C): These photographs show the same discrete nodule ilmenite grain at two different magnifications. Exsolved spinels delineate ilmenite subgrains. Specimen FRB 50/5. (B) scale bar = 100μ; (C) scale bar = 8μ.
(D): Well-equilibrated, polygranular ilmenite from discrete nodule FRB 50/4. Grain boundaries are regular and slightly curved; 120° triple junctions are common. Scale bar = 500μ.
(E): Disequilibrium grain boundaries are abundant. Within the same nodule as (D), only a few millimeters from the location of Figure 2D. Scale bar = 500μ.
(F): The ilmenite lamellae in this ilmenite-orthopyroxene intergrowth are polygranular. Crossed nicols. Specimen FRB 476. Scale bar = 500μ.

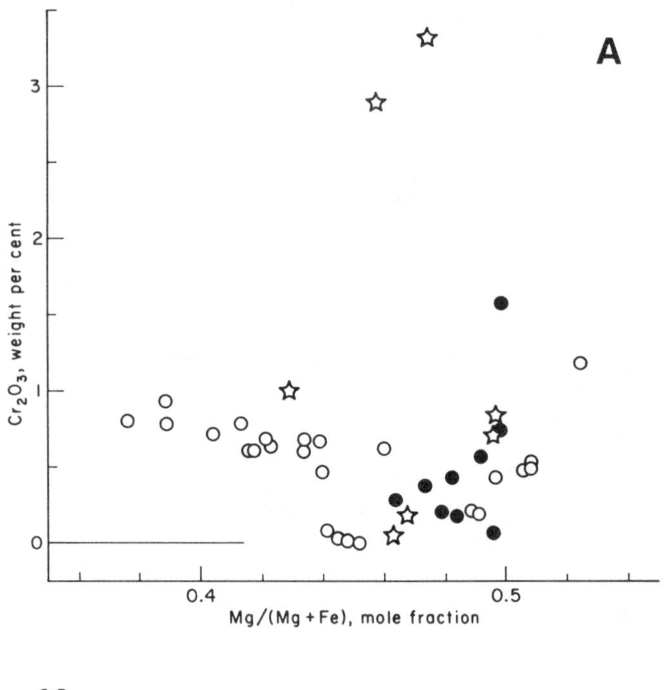

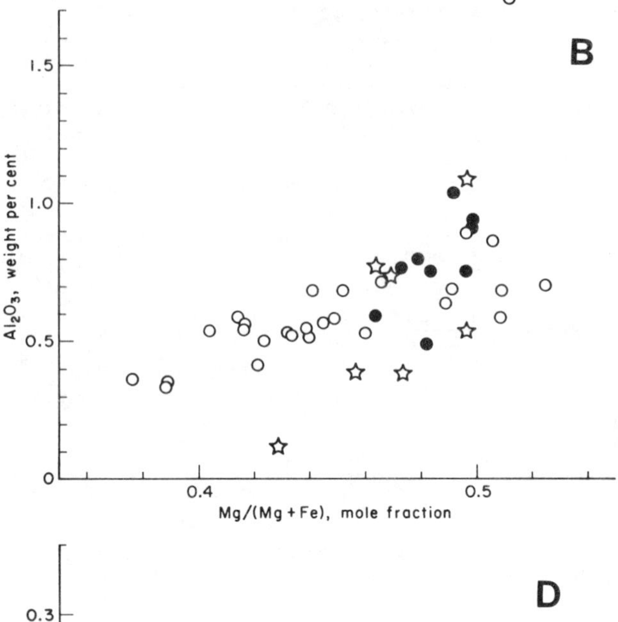

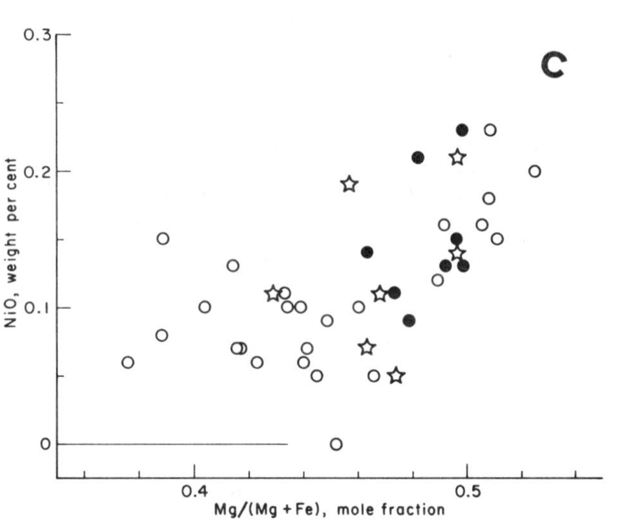

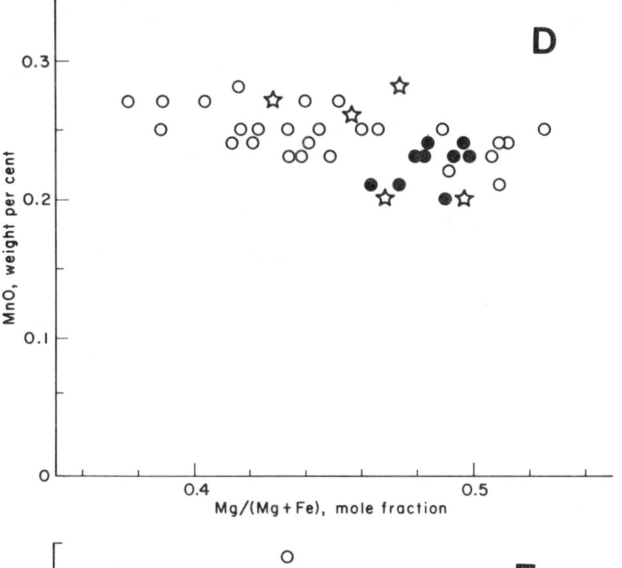

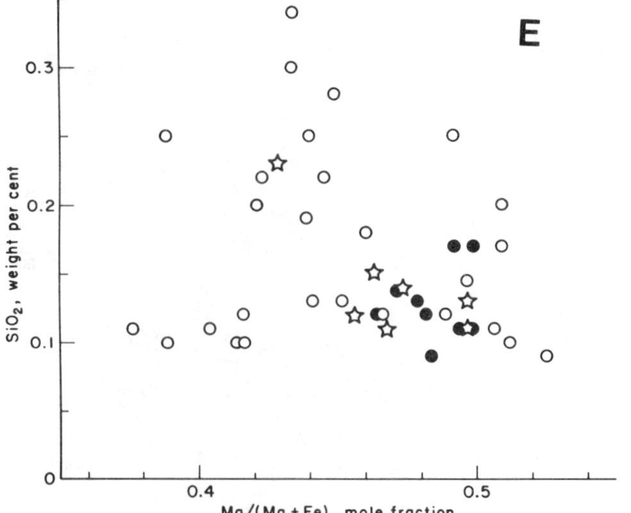

Fig. 4A-E. Variations of Cr_2O_3, Al_2O_3, NiO, MnO and SiO_2 with Mg/(Mg + Fe) for Frank Smith ilmenites. See legend of Fig. 3 for explanation of points.

discrete nodules, although this range is overlapped by the more magnesian discrete nodules (Fig. 10; also Mitchell, 1977). Analytical data for Fe_2O_3, MnO, SiO_2, Cr_2O_3, and Al_2O_3 in the lamellar ilmenites have no anomalies that would serve to distinguish them from the discrete nodule ilmenites (Fig. 11A-E). One of the ilmenites from a lamellar intergrowth (FRB 406, Table 3) contains a uniform 1.58 wt. % Cr_2O_3, a value that is substantially higher than previously observed for ilmenite in similar nodules.

Diopside and enstatite in the Frank Smith lamellar intergrowths are also magnesian. For example, the range in $Mg/(Mg + Fe^{2+})$ for five diopsides is 0.868-0.873, compared with 0.824-0.850 for similar specimens from Pipe 200, Lesotho (Boyd and Nixon, 1973), Kao, Lesotho (Nixon and Boyd, 1973c), Thaba Putsoa, Lesotho (Boyd and Nixon, 1975), Monastery, South Africa (Boyd, 1971; Boyd and Nixon, 1975), Uintjes Berg, South Africa (Boyd and Dawson, 1972), and Artur De Paiva, Angola (Boyd and Danchin, 1974). Cr_2O_3 in the Frank Smith lamellar pyroxenes is uniformly <0.2 wt. % while TiO_2 is 0.2-0.3 wt. % in the enstatites and 0.4-0.5 wt. % in the diopsides. The jadeite component in the diopsides is 10-15 mole % (Table 3).

Compositions of ilmenites from intergrowths of problematic origin have no anomalies that would serve to differentiate them, other than high Cr in two that are believed to be metasomatic (PHN 3244D/1, 2; Table 4). Compositions of silicates in these nodules are also similar to those found in other ilmenite intergrowths. The $Mg/(Mg + Fe^{2+})$ in the olivine (0.875) and enstatite (0.897) from the harzburgite nodule PHN 3244D/2 are more Fe-rich than would be expected in a residual peridotite. Presumably these minerals were metasomatized during introduction of ilmenite.

Discussion

The consanguinity of the lamellar intergrowths and the ilmenite discrete nodules is strongly supported by the fact that both are distinctively magnesian at the Frank Smith pipe. Compositions of pyroxenes and garnets intergrown with ilmenite as discrete nodules from the Frank Smith kimberlite are also distinctively magnesian.

The hypothesis that all ilmenite, pyroxene, garnet, and possibly olivine (Boyd, 1974b) discrete nodules from a given pipe are related in origin is difficult to establish in an unequivocal manner. The hypothesis is supported, however, by the fact that the discrete nodules have ranges of composition, including $Mg/(Mg + Fe^{2+})$, that relate them to each other and distinguish them from similar minerals in peridotite and pyroxenite nodules, and by host-inclusion relations of many specimens from a number of kimberlite pipes. This hypothesis is at variance with the proposal of Mitchell (1977) that " . . . discrete nodule magnesian ilmenite is a single early liquidus phase (phenocryst) . . ."

Boyd and Nixon (1973) and Nixon and Boyd (1973b) have proposed that these discrete nodules were crystals in liquid at the time of eruption. The fact that the pyroxene discrete nodules have equilibration temperatures that are generally above 1100°C and that characteristically they are not intergrown with coarse phlogopite are consistent with an origin in liquid. Discrete nodules from many kimberlites in southern Africa are characteristically unstrained, but it is clear from the present investigation and the earlier work of Frick (1973) that the pyroxene-ilmenite lamellar intergrowths and enstatite discrete nodules from Frank Smith are commonly sheared. Recrystallization of the ilmenite discrete nodules may also have been induced by stress. The presence of deformation textures in the Frank Smith discrete nodule suite certainly weakens, but perhaps does not invalidate the hypothesis that these crystals were in liquid at the time of eruption.

Grain size and microstructure studies have established that the differential stresses that deformed peridotite and associated nodules from kimberlites were as high a 1 kb (Goetze, 1975; Nicolas and Ricoult, 1977). The deformation is believed to have occurred immediately prior to eruption, perhaps in the course of conduit formation (Mercier, 1977). In the presence of such high stresses it seems possible that crystals in liquid might be deformed if the proportion of liquid was small. Moreover, it may be useful to speculate that impact stresses during violent eruption might have produced some of the observed deformation features.

The wide range in $Mg/(Mg + Fe^{2+})$ of 0.376-0.744 (Tables 1, 2, and Mitchell, 1977) for the Frank Smith ilmenite discrete nodules could be due to crystallization from liquids of variable composition. Garnet and pyroxene discrete nodules from a large number of pipes from southern Africa have analogous variations (e.g., Nixon and Boyd, 1973B; Boyd and Nixon, in press). Twelve garnet discrete nodules from Frank Smith have a range in $Mg/(Mg + Fe^{2+})$ of 0.750-0.838 and for twenty bronzite nodules the range is 0.876-0.921 (Fig. 12.).

The $Mg/(Mg + Fe^{2+})$ for Frank Smith bronzite and bronzite-ilmenite discrete nodules tends to increase with increasing $Ca/(Ca + Mg)$, i.e., with increasing equilibration temperature (Fig. 5). There is considerable scatter and one point is a gross misfit, but the crude trend in Fig. 5 appears analogous to the relations found for diopside discrete nodules from Letseng-la-terae, Lesotho (Nixon and Boyd,

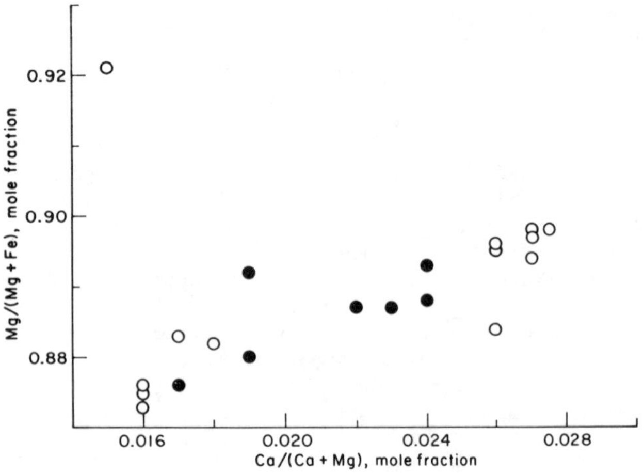

Fig. 5. Variations of Mg/(Mg + Fe) with Ca/(Ca + Mg) in bronzite nodules from the Frank Smith mine. Open circles are for bronzite discrete nodules; solid circles are bronzite-ilmenite intergrowths, some of which are lamellar and some host-inclusion pairs.

1973b, Fig. 20). Moreover, ilmenite-bearing pyroxene discrete nodules (both lamellar and host-inclusion varieties) seem restricted to the lower range of equilibration temperature exhibited by all pyroxene discrete nodules found at Frank Smith (Fig. 5) and in northern Lesotho (Boyd and Nixon, 1973).

These relations were originally interpreted to indicate that the upper part of the Low Velocity Zone beneath northern Lesotho was enriched in Ti and probably Fe and that crystallization of ilmenite in the discrete nodule associations was restricted to these shallower levels (Boyd and Nixon, op cit.). This interpretation could reasonably be extended to a rising, partially-molten diapir as well as a Low Velocity Zone model.

The Frank Smith lamellar intergrowths have notably restricted ranges in equilibration temperature as well as in Mg/(Mg + Fe^{2+}). Diopsides in five lamellar intergrowths have Ca/(Ca + Mg) of 0.391-0.408, corresponding to a temperature of 1200°C based on the solvus of Davis and Boyd (1966) or 1250°C based on the solvus of Lindsley and Dixon (1976). The four enstatite intergrowths have Ca/(Ca + Mg) of 0.019-0.024 corresponding to temperatures of 1205°-1280°C based on the empirical thermometer of Boyd and Nixon (1973). This thermometer should properly be used only for nodules from northern Lesotho, because of the pressure effects on the pyroxene solvi, but the depths of origin of the Frank Smith discrete nodules are about the same as those for Lesotho discrete nodules (Boyd, 1974a).

Equilibration temperatures for the enstatite intergrowths can be independently calculated with the enstatite-ilmenite thermometer of Bishop (in press), with pressure estimated from the Al_2O_3 content of the enstatites (MacGregor, 1974). The temperatures so obtained are 1185°C-1255°C. The range is identical to that estimated above and the values from the two thermometers are in agreement.

The circumstances that promote the crystallization of lamellar intergrowths rather than host-inclusion pairs or separate crystals are not understood, although Wyatt (1977) has demonstrated that these circumstances are probably igneous. Mitchell (1977) and Frick (1973b) may well be correct in suggesting the possible importance of kinetic factors in the formation of lamellar intergrowths.

The hypothesis that discrete nodules similar

Table 5. Ranges for equilibration temperatures of pyroxene discrete nodules from individual kimberlite pipes in southern Africa.

Pipe	Pyroxene	Range °C	Reference
Frank Smith	En + Di	200	this paper
Monastery	Di	255	Gurney, et al., 1977
Letseng-la-terae	En + Di	215	Bloomer and Nixon, 1973
Bultfontein	En + Di	265	Boyd and Nixon, in press
De Beers	En + Di	275	Boyd and Nixon, in press
Kimberley	En	300	Boyd and Nixon, in press
Camatue, Angola	Di	200	Boyd, unpublished

to those found in northern Lesotho originated as high-pressure phenocrysts in kimberlite magma has received support from the experimental studies of Eggler and Wendlandt (1977) and from the observation of Gurney, et al. (1977), that tubular inclusions of kimberlite are common in olivine megacrysts from the Monastery mine. A model involving a magma chamber that is circumscribed in volume, however, does not fit well with the large ranges in equilibration temperature found for pyroxene discrete nodules from individual kimberlite pipes. For kimberlite pipes in southern Africa these ranges in equilibration temperature of pyroxene discrete nodules including ilmenite-bearing types are commonly 200°-300°C (Table 5). Although the absolute values estimated for equilibration temperatures vary with different thermometers and have substantial uncertainties, the ranges for particular suites of pyroxenes are less subject to uncertainty.

Lesotho-type pyroxene discrete nodules do not exhibit exsolution textures that are visible under the microscope. McCallister, et al. (in press), have shown that submicroscopic exsolution lamellae in a diopside discrete nodule from Thaba Putsoa formed during a cooling period of about 15 hours. This diopside has Ca/(Ca + Mg) of 0.320, corresponding to an equilibration temperature of approximately 1375°C (Davis and Boyd, 1966). Clearly such a pyroxene could not have cooled 200°-300°C below its equilibration temperature for a period of time sufficient for the crystallization of other, coarse, homogeneous, discrete nodules.

A model analogous to that proposed by Boyd and Nixon (1973) and Nixon and Boyd (1973b) for northern Lesotho provides for a wide range in equilibration temperature of discrete nodules. It is suggested that discrete nodules coexisted with small amounts of interstitial liquids (kimberlite?) of variable composition, including $Mg/(Mg + Fe^{2+})$, over a depth interval of the order of several tens of kilometers, and that these nodules and interstitial liquids were then mixed during eruption.

The compositional relations of the ilmenite and ilmenite-bearing discrete nodules from the Frank Smith mine appear consistent with such a model.

Acknowledgements. The authors would like to thank Mr. Roelf Versluis, owner of the Frank Smith mine, for granting permission to collect samples at the mine on a number of occasions. The assistance of J. Barry Hawthorne in arranging the collecting trips is gratefully acknowledged. Asish Basu, Henry O. A. Meyer, Roger Mitchell, and Hatten S. Yoder, Jr., have provided critical commentary that has been most helpful in preparing this manuscript.

References

Bishop, F. C., The distribution of Fe^{2+} and Mg between coexisting ilmenite and pyroxene with applications to geothermometry, Amer. Jour. Sci., in press.

Bloomer, A. G., and P. H. Nixon, The geology of Letseng-la-terae kimberlite pipes, in Lesotho Kimberlites, edited by P. H. Nixon, Lesotho National Development Corporation, Maseru, Lesotho, 20-36, 1973.

Boyd, F. R., Ultramafic nodules from the Frank Smith kimberlite pipe, South Africa, Carnegie Institution of Washington Year Book 73, 285-293, 1974a.

Boyd, F. R., Olivine megacrysts from the kimberlites of the Monastery and Frank Smith mines, South Africa, Carnegie Institution of Washington Year Book 73, 282-285, 1974b.

Boyd, F. R., Enstatite-ilmenite and diopside-ilmenite intergrowths from the Monastery mine, Carnegie Institution of Washington Year Book 70, 134-138, 1971.

Boyd, F. R., and R. V. Danchin, Discrete nodules from the Artur De Paiva kimberlite, Angola, Carnegie Institution of Washington Year Book 73, 278-282, 1974.

Boyd, F. R., and J. B. Dawson, Kimberlite garnets and pyroxene-ilmenite intergrowths, Carnegie Institution of Washington Year Book 71, 373-378, 1972.

Boyd, F. R., and P. H. Nixon, Ultramafic nodules from the Kimberley pipes, South Africa, Geochimica et Cosmochimica Acta, in press.

Boyd, F. R., and P. H. Nixon, Origins of the ultramafic nodules from some kimberlites of northern Lesotho and the Monastery mine, South Africa, Physics and Chemistry of the Earth, 9, 431-454, 1975.

Boyd, F. R., and P. H. Nixon, Origin of the ilmenite-silicate nodules in kimberlites from Lesotho and South Africa, in Lesotho Kimberlites, edited by P. H. Nixon, Lesotho National Development Corp., Maseru, Lesotho, 254-268, 1973.

Clarke, D. B., G. G. Pe, R. M. Mackay, K. R. Gill, M. J. O'Hara, and J. A. Gard, A new potassium-iron-nickle sulphide from a nodule in kimberlite, Earth and Planetary Sci. Letters, 35, 421-428, 1977.

Davis, B. T. C., and F. R. Boyd, The join $Mg_2Si_2O_6$-$CaMgSi_2O_6$ at kb pressure and its application to pyroxenes from kimberlite, J. Geophys. Res., 71, 3567-3576, 1966.

Dawson, J. B., Basutoland kimberlites, Bull. Geol. Soc. Amer., 73, 545-560, 1962.

Eggler, D. H., and R. F. Wendlandt, Experimental studies on the relationship between kimberlite magmas and partial melting of peridotite, Extended Abstracts, Second International Kimberlite Conference, Santa Fe, New Mexico, 1977.

Finger, L. W., The uncertainty in the calculated ferric iron content of a microprobe analysis, Carnegie Institution of Washington Year Book 71, 600-603, 1972.

Frick, C., Kimberlitic ilmenites, Trans. Geol. Soc. South Africa, 76, 195-200, 1973b.

Frick, C., Intergrowth of orthopyroxene and ilmenite from Frank Smith Mine, near Barkley West, South Africa, Trans. Geol. Soc. South Africa, 76, 195-200, 1973b.

Goetze, C., Sheared lherzolites: from the point of view of rock mechanics, Geology, 3, 172-173, 1975.

Gurney, J. J., W. R. O. Jakob, and J. B. Dawson, Megacrysts from the Monastery mine, Extended Abstracts, Second International Kimberlite Conference, Santa Fe, New Mexico, 1977.

Haggerty, S. E., R. B. Hardie, III, and B. M. McMahon, The mineral chemistry of ilmenite nodule associations from the Monastery diatreme, Extended Abstracts, Second International Kimberlite Conference, Santa Fe, New Mexico, 1977.

Harte, B., and J. J. Gurney, Ore mineral and phlogopite mineralization within ultramafic nodules from the Matsoku kimberlite pipe, Lesotho, Carnegie Institution of Washington Year Book 74, 528-536, 1975.

Lindsley, D. H., and S. A. Dixon, Diopside-enstatite equilibria at 850° to 1400°C, 5 to 35 kb, Amer. Jour. Sci., 276, 1285-1301, 1976.

MacGregor, I. D., The system $MgO-Al_2O_3-SiO_2$: solubility of Al_2O_3 in enstatite for spinel and garnet peridotite compositions, Amer. Mineral., 59, 110-119, 1974.

McCallister, R. H., H. O. A. Meyer, and R. Aragon, Partial thermal history of two exsolved clinopyroxenes from the Thaba Putsoa kimberlite pipe, Lesotho, Proceedings of the Second International Kimberlite Conference, in press.

Mercier, J. C., Peridotite xenoliths and the dynamics of kimberlite intrusion, Extended Abstracts, Second International Kimberlite Conference, Santa Fe, New Mexico.

Meyer, H. O. A., H-m. Tsai, and J. J. Gurney, Enstatite xenocryst containing coexisting Cr-poor and Cr-rich garnet, Weltevreden Floors, South Africa, Extended Abstracts, Second International Kimberlite Conference, Santa Fe, New Mexico, 1977.

Mitchell, R. H., Magnesian ilmenite and its role in kimberlite petrogenesis, J. Geol., 81, 301-311, 1973.

Mitchell, R. H., Geochemistry of magnesian ilmenites from kimberlites in South Africa and Lesotho, Lithos, 10, 29-37, 1977.

Nicholas, A., and D. Ricoult, Estimates of stress and recovery conditions in various types of mantle peridotites, Extended Abstracts, Second International Kimberlite Conference, Santa Fe, New Mexico.

Nixon, P. H., and F. R. Boyd, Garnet bearing ultrabasic and discrete nodule suites from Malaita, Solomon Islands, S. W. Pacific, and their bearing on an oceanic geotherm, Extended Abstracts, Second International Kimberlite Conference, Santa Fe, New Mexico, 1977.

Nixon, P. H., and F. R. Boyd, Discrete nodules (megacrysts) and lamellar intergrowths in Frank Smith kimberlite pipe, Extended Abstracts, International Conference on Kimberlites, Rondebosch, South Africa, 1973a.

Nixon, P. H., and F. R. Boyd, The discrete nodule association in kimberlites from Northern Lesotho, in Lesotho Kimberlites, edited by P. H. Nixon, Lesotho National Development Corp., Maseru, Lesotho, 67-75, 1973b.

Nixon, P. H., and F. R. Boyd, Deep-seated nodules, in Lesotho Kimberlites, edited by P. H. Nixon, Lesotho National Development Corp., Maseru, Lesotho, 106-109, 1973c.

Rawlinson, P. J., and J. B. Dawson, A quench orthopyroxene-ilmenite xenolith from kimberlite--evidence for Ti-rich liquid in the upper mantle, Extended Abstracts, Second International Kimberlite Conference, Santa Fe, New Mexico, 1977.

Wagner, P. A., The Diamond Fields of Southern Africa, The Transvaal Leader, Johannesburg, 1914 /2nd impression C. Struick (PTY) Ltd., Cape Town, 1971/.

Warner, R. D., K. Keil, and G. J. Tayler, Coarse-grained basalt 71597: a product of partial olibine accumulation, Proc. 8th Lunar Sci. Conf., 1429-1442, 1977.

Wyatt, B. A., The melting and crystallization behavior of a natural clinopyroxene-ilmenite intergrowth, Contrib. Mineral. Petrol., 61, 1-9, 1977.

A UNIQUE ENSTATITE MEGACRYST WITH COEXISTING CR-POOR AND CR-RICH GARNET, WELTEVREDEN FLOORS, SOUTH AFRICA

Henry O. A. Meyer and Hsiao-ming Tsai

Department of Geosciences, Purdue University, West Lafayette, Indiana 47907

John J. Gurney

Department of Geochemistry, University of Cape Town, Rondebosch, C. P. South Africa

Abstract. A large (17 cm) single crystal of enstatite (En_{88}) contains Mg-ilmenite (MgO $\sim$ 12 wt%) and inclusions of polyphase garnets. The major garnet is orange in color and is predominantly pyrope-almandine ($Py_{71}Al_{21}Gr_{8}$) whereas inside this garnet is a second one that is pink in color, and Cr-rich ($Cr\text{-}Py_{33}Py_{28}Al_{19}Gr_{20}$). This Cr-rich garnet itself contians inclusions of Cr-diopside, chromite and ilmenite. The enclosing orange almandine garnet at the boundary with the pink Cr-rich garnet contain rounded Mg-ilmenites together with a string of elongated olivines (Fo_{86}) that are in optical continuity. Also present in the same zone is diopside. Within the enstatite host and in close association with the polyphase garnet assemblage are calcite, Ti-phlogopite and serpentine. Although it is conceivable the Cr-rich garnet and related phases represent the remnant of an earlier preexisting rock that underwent considerable partial melting it must also be considered that the second garnet may be metamorphic in origin. However, irrespective of the origin, based on orthopyroxene geothermometry and Al_2O_3 content possible P and T of equilibration are in the region of 60kb and 1200°C.

Introduction

A common feature in kimberlite is the presence of xenoliths of upper mantle origin. Detailed mineralogical and chemical investigations of such xenoliths provide significant information on the constitution and processes of the upper mantle (Nixon and Boyd, 1973a,b,c; MacGregor and Basu, 1974; Wilshire and Jackson, 1975). The majority of these xenoliths are either ultramafic (garnet-lherzolite, pyroxenite, harzburgite) or eclogitic (griquaite) in nature, and both rock types are generally coarse-grained and may contain minerals several cm in size. Also found in kimberlite are large monominerallic fragments (xenocrysts, megacrysts or discrete nodules) that are in general comparable to the mineral constituents of the xenoliths (Nixon and Boyd, 1973a). Typically the minerals garnet, diopside, enstatite and ilmenite occur as xenocrysts or megacrysts (the term megacryst will be used throughout to remove any genetic connotation). The mechanism by which these large monominerallic grains form in the mantle is not readily understood, although Nixon and Boyd (1973a) and Gurney et al. (1977) have suggested they represent phenocrysts in crystal-mush magmas in the low velocity zone. The relation of these megacrysts to the ultramafic xenoliths and to kimberlite itself is not fully established. To answer these, and other questions concerning kimberlite, xenoliths and the genesis of diamond it is necessary to characterize in detail the mineralogy and chemistry of significant samples of these exotic fragments of the mantle.

Several of the kimberlites in Southern Africa produce a wide variety of xenoliths and megacrysts(e.g. Matsoku, Lesotho-Gurney et al., 1975; Thaba Putsoa, Lesotho-Nixon and Boyd, 1973b; Artur de Paiva, Angola-Boyd and Danchin, 1974; and Frank Smith, South Africa-Boyd, 1974). Frank Smith and Weltevreden kimberlites are two closely related blows connected by a short kimberlite dike . In overall outcrop the two blows have a dumb-bell shape. Weltevreden Floors is the area over which many years ago the kimberlite from Weltevreden pipe was spread in order to decompose. These floors are usually exceptionally good collecting grounds for xenoliths and megacrysts. At Weltevreden Floors a very large (17cms) megacryst of enstatite was found. This megacryst is a single crystal of enstatite (En_{88}) and contains inclusions of partially oriented Mg-ilmenite. However, by far the most unique and important feature of this megacryst is the inclusions of garnet. These inclusions are polyphase inasmuch as the major garnet is an orange-colored pyrope-almandine which surrounds almost perfectly circular pink colored Cr-rich pyrope garnet. This latter garnet itself contains Cr-diopside and chromite. Within the

Fig. 1a. Enstatite xenocryst enclosed in kimberlite. Note the irregular scattered grains of ilmenite.

orange garnet, at the boundary with the pink garnet, are small, round Mg-ilmenites, and diopside, as well as a string of several elongated olivine grains that are all in optical continuity. The Mg-ilmenite, diopside and olivine completely surround the pink Cr-rich garnet. This enstatite megacryst thus contains two suites of chemically distinct inclusions -- a Cr-rich suite and a Cr-poor suite.

Mineralogy

The enstatite megacryst was ovoid in shape with the longest dimension being 17cms (Fig. 1a). When collected it was surrounded by kimberlite but there is no indication of kimberlitic material having intruded into the megacryst, apart from some thin veinlets of serpentine. Easily visible on the surface of the specimen, and especially on cut-surfaces, is ilmenite which is partially oriented within the enstatite (Fig. 1b). One area of ilmenite is very similar to the ilmenite-pyroxene intergrowths that occur as discrete assemblages in several kimberlites (Gurney et al., 1973; McCallister et al., 1975; Boyd and Nixon, 1973).

Randomly scattered throughout the megacryst are irregular inclusions of predominantly orange garnet up to 2 cms in size. These garnets themselves contain round inclusions of pink garnet as well as small round grains of ilmenite that together with small elongate crystals of olivine and diopside abut the pink garnet. Within the pink garnet, inclusions of emerald green diopside, chromite and chrome-rich ilmenite are present (Fig. 2). Exterior to the orange garnet and usually associated with this garnet and ilmenite are small regions of Ti-phlogopite and calcite. Small veins of serpentine are also present within the megacryst.

All phases observed in the megacryst have been analyzed using an automated M.A.C. microprobe with on-line computer reduction of data via a modified Bence-Albee (1968) and Albee and Ray (1970) matrix refinement which includes background, deadtime, drift, fluorescence and atomic number effects. Standards were predominantly glasses except for K, Cr, Fe and Mn for which minerals were used. The analyses are believed correct to within ±2% of the concentration for major elements. For minor elements the accuracy is much less but reproducibility is usually to within 0.02 wt. percent. Cell dimensions of the enstatite and garnet were obtained using back-reflection X-ray diffraction techniques, and for ilmenite using precession methods. All X-ray data were refined using least squares refinement procedures of Burnham (1965) modified by Finger (pers. comm.).

Enstatite Megacryst

The megacryst is essentially a single crystal of enstatite with a composition of En_{88} (Table 1). The enstatite is remarkably uniform in composition throughout the specimen with aluminum and calcium being the most noteworthy minor elements, although solid solution towards garnet and diopside is very limited. Compositionally this orthopyroxene is more Fe-rich than most that are

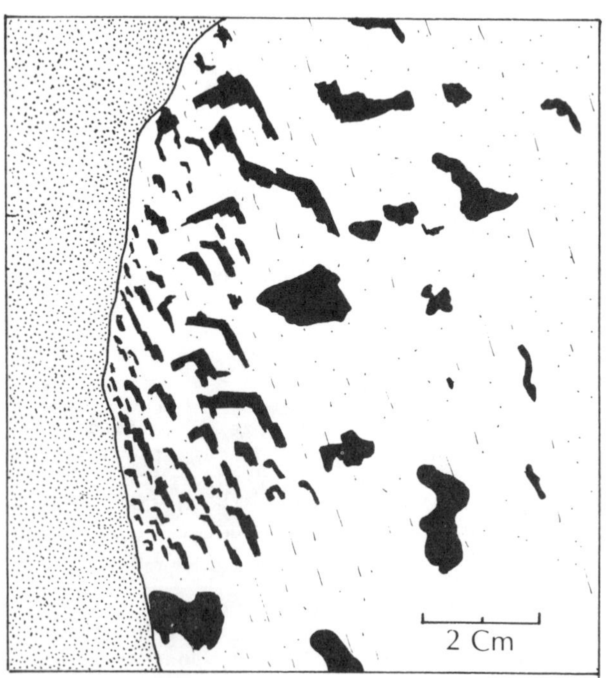

Fig. 1b. Ilmenite occurring as discrete, apparently random crystals in enstatite as well as a regular lamella intergrowth.

constituents of the ultramafic xenoliths (Boyd and Nixon, 1975). However, it does resemble other enstatite occurring as discrete grains in the Frank Smith kimberlite (Boyd, 1974), as well as enstatite inclusions in an olivine-ilmenite intergrowth from Thaba Putsoa, Lesotho (Boyd and Nixon, 1973).

Ilmenite Inclusions in Enstatite

Ilmenite is the most common inclusion in the megacryst and varies from irregular, rounded blebs to angular, parallel to sub-parallel lamellae. The inclusion of ilmenite in silicate crysts is not uncommon and Boyd and Nixon (1973) have listed several types of occurrences from widely dispersed kimberlite pipes in Southern Africa. However, this specimen from Weltevreden Floors appears to be the first that combines both irregular and lamellar intergrowths.

The ilmenite is typical of those from kimberlite and associated xenoliths in that it has a

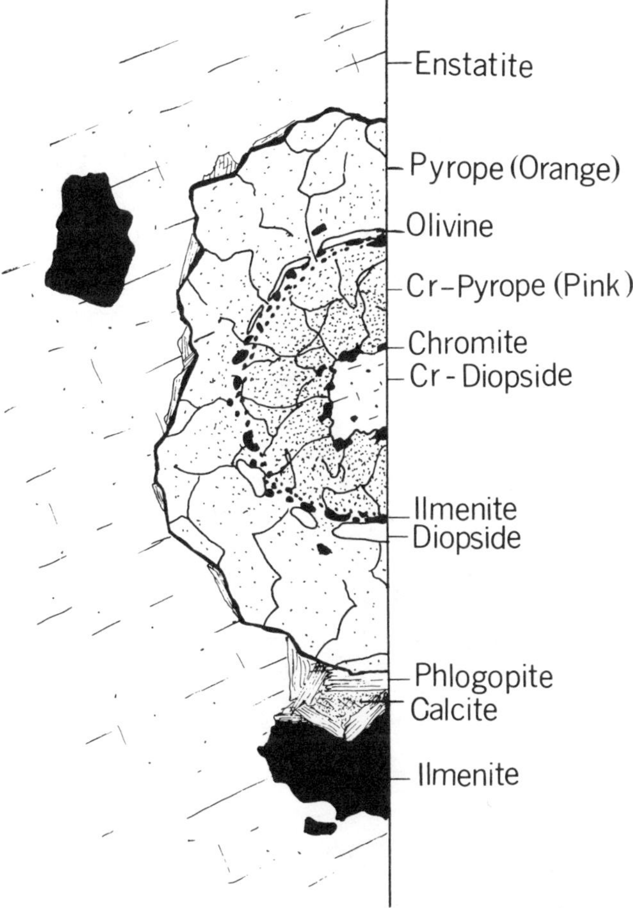

Fig. 2. Diagram showing complete assemblage of all phases and their mutual relationships in the enstatite host.

TABLE 1. Analysis of Enstatite Megacryst (73-21006) from Weltevreden kimberlite, South Africa, plus comparable analyses of enstatite in kimberlites.

Oxides	73-21006	S*	A	B
SiO_2	56.7	(0.98)	56.9	55.7
TiO_2	0.23	(0.03)	0.17	0.22
Al_2O_3	0.82	(0.05)	0.74	0.88
Cr_2O_3	0.02	(0.03)	<0.03	0.18
FeO	8.09	(0.11)	8.51	8.54
MnO	0.16	(0.02)	0.20	0.18
MgO	33.2	(0.66)	33.0	32.3
CaO	0.93	(0.02)	0.76	0.85
NiO	0.02	(0.03)	-	-
Na_2O	0.16	(0.09)	0.12	0.19
K_2O	0.01	(0.01)	-	-
Totals	100.3		100.4	99.1
Cations on the basis of 6 Oxygens (x1000)				
Si	1969		1976	1965
Ti	6		4	6
Al	34		30	37
Cr	0		0	5
Fe	235		247	252
Mn	5		6	5
Mg	1718		1708	1700
Ca	35		28	32
Ni	0		-	-
Na	10		8	13
K	0		-	-
Totals	4012		4007	4010
Mg/Mg+Fe	0.88		0.87	0.87
	a 18.28Å			
	b 8.83			
	c 5.20			

A: Discrete enstatite xenocryst (PHN2786E1), Frank Smith kimberlite. (Boyd, 1974)
B: Enstatite inclusion in dunite (1582), Thaba Putsoa. (Boyd and Nixon, 1973)
*: Standard deviation of 8 analyses.

relatively high content of MgO (Mitchell, 1973; Gurney et al., 1973; Haggerty, 1975), Table 2, corresponding to about 42% geikielite ($MgTiO_3$) molecule. Boyd and Nixon (1973), Mitchell (1977), and Pasteris et al. (this volume) have suggested that ilmenite from Frank Smith kimberlite is substantially more Mg-rich than ilmenites from other kimberlites. The geikielite content for this specimen is in agreement with the range 36.7 - 49.2 for Frank Smith ilmenite determined by Boyd and Nixon. Chromium and aluminum are relatively minor. Numerous thin, minute oriented lamellae in ilmenite are believed to be spinels

TABLE 2. Analyses of Ilmenite associated with Enstatite Megacryst.

Oxide	A	B	C	D
SiO_2	0.15	0.29	0.13	0.26
TiO_2	53.5	49.8	38.8	47.0
Al_2O_3	0.32	1.07	1.46	1.13
Cr_2O_3	0.19	1.89	13.6	5.00
FeO	33.5	35.1	35.7	36.1
MnO	0.35	0.28	0.16	0.27
MgO	12.4	10.8	9.42	10.9
CaO	0.06	0.09	0.03	0.12
NiO	-	<0.01	0.01	0.01
Na_2O	-	<0.01	<0.01	0.05
K_2O	-	<0.01	<0.01	<0.01
Totals	100.4	99.3	99.3	100.8
Fe_2O_3*	7.88	9.88		
FeO	26.4	26.2		

Cations on the basis of 3 Oxygens (x1000)

	A	B
Si	3	7
Ti	923	876
Al	9	29
Cr	3	34
Fe'''*	134	169
Fe''	500	507
Mn	1	6
Mg	419	371
Ca	7	2
Totals	1999	2001
Ilm.	50.5	51.3
Geik.	42.3	37.3
Hem.	7.2	11.5

Ilmenite A Cell Dimensions
a = 5.05
c = 13.89

* Fe''' calculated from total Fe, and based on formula ABO_3 assuming $B = R^{+4} + R^{+3}$ and assigning equal amount R^{+3} to A.

A: Major ilmenite inclusions in enstatite.
B: Round ilmenite in garnet.
C: Ilmenite with continuous chromite lamellae
D: Ilmenite with discontinuous chromite lamellae
Analyses C and D were obtained using broad beam techniques.

since marked increase in Cr occurs when the electron beam is scanned across the lamellae.

A. Garnets

A.i. Orange Garnet (Host). Inclusions of orange garnet occur randomly throughout the enstatite megacryst and appear to be crystallographically unrelated to the host enstatite. Generally the garnet is irregular to rounded in habit although in some instances the presence of angular faces indicates the possibility of a euhedral habit. The largest garnet inclusion observed was almost 2 cms in length. In all cases observed the orange garnet enclosed either a single or several circular fragments of pink garnet.

Several garnets were analyzed and apart from calcium little variation in element content was observed (Table 3). Calcium showed a small variation (3.14 - 4.93 wt. % CaO) among grains, and an increase in FeO was noted in one garnet near a sulfide inclusion, although this may have been due to secondary flouresence. Compositionally the garnet is equivalent to $pyrope_{72}$ $almandine_{17}$ with minor andradite, grossular and insignificant uvarovite and spessartite.

Compared with other garnets from ultramafic xenoliths and megacrysts this garnet resembles the discrete garnet nodules (xenocrysts)

TABLE 3. Analyses of Garnet Inclusions in Enstatite Megacryst.

Oxides	Orange	S**	Pink	S**
SiO_2	41.8	(0.36)	40.7	(0.39)
TiO_2	0.96	(0.11)	0.71	(0.21)
Al_2O_3	21.5	(0.49)	15.5	(0.48)
Cr_2O_3	0.33	(0.44)	9.96	(0.56)
FeO	10.6	(0.25)	7.72	(1.27)
MnO	0.32	(0.02)	0.38	(0.03)
MgO	20.2	(0.43)	17.3	(0.49)
CaO	4.05	(0.80)	8.36	(0.61)
NiO	<0.01		<0.01	
Na_2O	0.12	(0.08)	0.10	(0.06)
K_2O	<0.01		<0.01	
Totals	99.9		100.6	
*Fe_2O_3	2.38		1.00	
FeO	8.43		6.82	

Cations on the basis of 12 Oxygens (x1000)

	Orange	Pink
Si	2985	2986
Ti	52	40
Al	1812	1336
Cr	19	578
*Fe'''	127	55
Fe''	504	419
Mn	20	24
Mg	2150	1887
Ca	309	656
Na	17	14
Totals	7995	7995

a: 11.533 ± 0.003Å   a: 11.604 ± 0.003Å

*Fe''' Calculated from total Fe to satisfy garnet stoichiometry.
** Standard deviation: orange garnet - 7 anals; pink garnet - 5 anals.

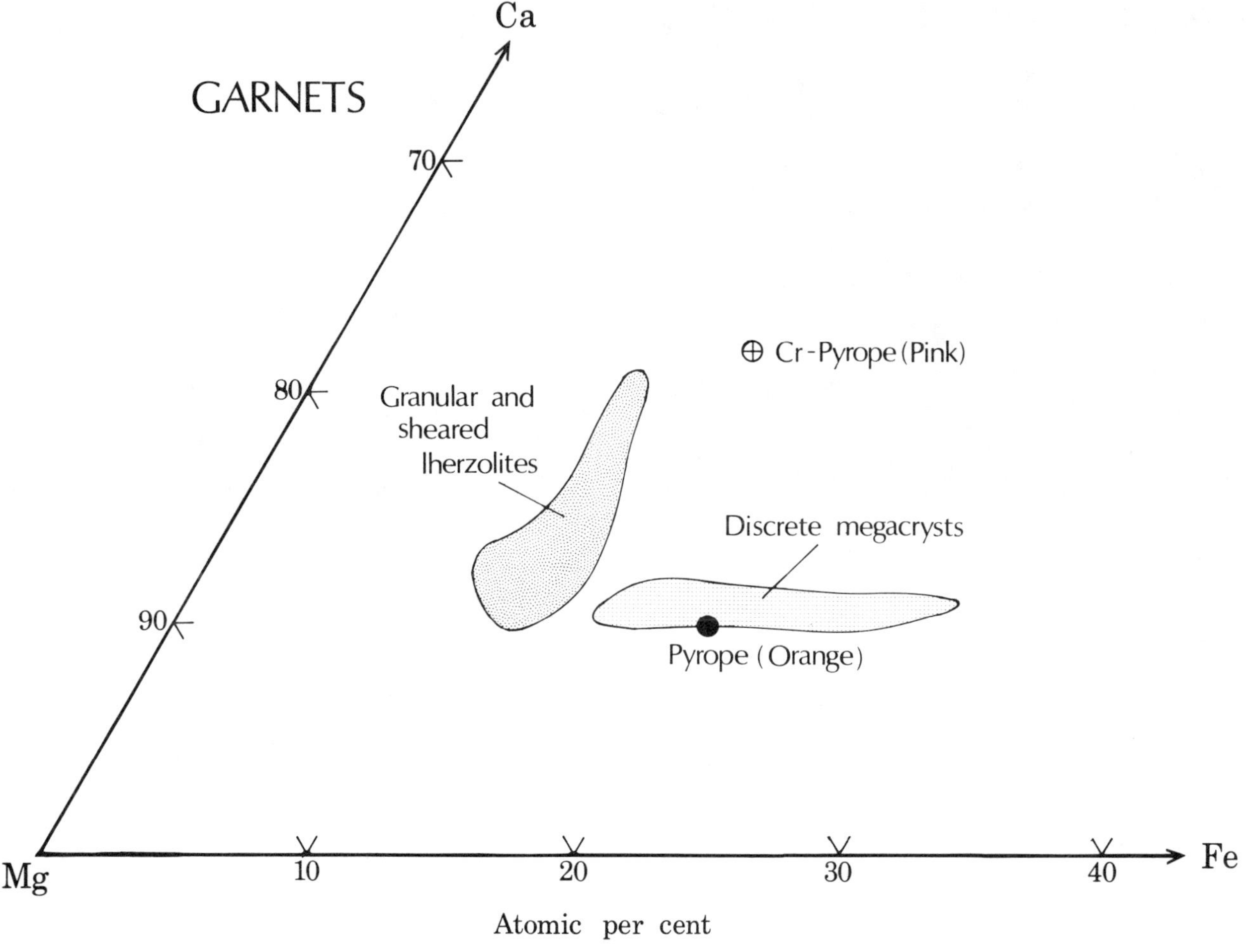

Fig. 3. Compositions of Cr-rich and Cr-poor garnets in terms of Ca-Fe-Mg. Fields for granular, sheared and discrete xenoliths after Nixon and Boyd (1973a,b).

discussed by Nixon and Boyd (1973a). They showed that garnet megacrysts were generally more Fe-rich than their counterparts in sheared and granular lherzolites (Fig. 3). However, although this garnet does coexist with ilmenite it is not so Fe-rich as those observed by Nixon and Boyd to that coexist with ilmenite. A garnet megacryst of comparable composition to that discussed here is presented by Boyd (1974) from Frank Smith kimberlite.

A.ii. Pink Garnet (Inclusion in Orange Garnet). The pink to purple colored garnets often occur as almost perfectly round inclusions within the orange garnet, and are up to 0.5 cm in diameter.

These garnets differ in composition from the orange garnet in being considerably more chromium-rich as well as containing about twice as much calcium (Table 3). The pink garnet appears to be slightly more inhomogeneous than the orange garnet in that FeO showed variation between 6 and 9 wt. %. Chromium and calcium also showed variation but over a much smaller range than iron.

This garnet is different from those occurring in garnet-lherzolite xenoliths in kimberlite (Fig. 3) (c.f. Nixon and Boyd, 1973b) and can be matched only by rare garnets in the heavy mineral concentrate from Kao, Lesotho (Hornung and Nixon, 1973) and from Finsch kimberlite, South Africa (Gurney and Switzer, 1973).

Although it is possible to distinguish optically and chemically between the two garnets it appears that there is a narrow zone between them in which chemical gradation from one to the other occurs. This region or zone is also characterized by containing most of the small rounded Mg-ilmenites that surround the pink garnet (Fig. 4). Compositionally the zone is one in which a marked decrease in Cr occurs from the pink garnet towards the orange counterpart. Corresponding to this change in chromium, Al and

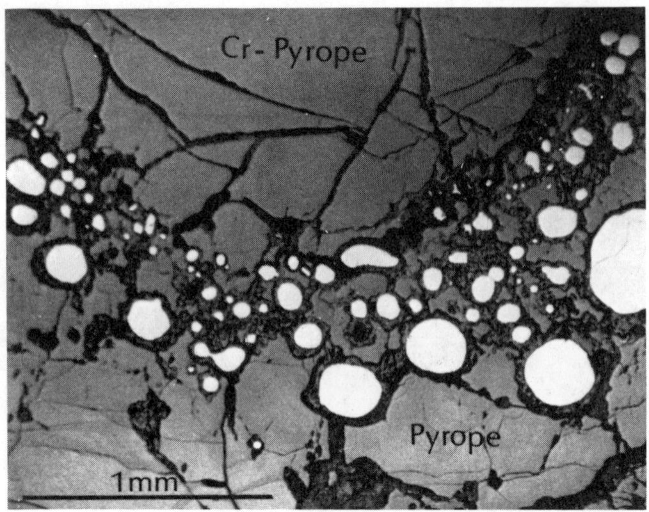

Fig. 4. Mg-ilmenites in the zone between the Cr-rich garnet and Fe-rich garnet. Note the rounded and globular nature of the ilmenites and the overall increase in size away from the Cr-rich garnet.

Mg increase and Ca decreases in the same direction. These changes are shown for a traverse from pink to orange garnet in terms of Cr_2O_3/(Cr_2O_3 + Al_2O_3) and CaO/(CaO + MgO) in Figure 5.

It is noteworthy that the small, round ilmenites generally occur where the Cr_2O_3/(Cr_2O_3 + Al_2O_3) value changes abruptly and these ilmenites are not found where the value has reached a minimum in the orange garnet. The transition zone in which the ilmenite occur corresponds to a change in Cr_2O_3 content of the garnets of about 6 to 2 wt %.

B. Ilmenite

On textural criteria there are two groups of ilmenite. One group contains numerous, closely spaced continuous exsolution lamellae whereas the second, and more common group, have thin, short and discontinuous lamellae. The lamellae in both groups are believed from optical and chemical data to be predominantly chromite. Lamellae similar to the discontinuous type have been noted by Danchin and d'Orey (1972) and Haggerty (1975).

All the ilmenite occurs in the transition zone between the pink and orange garnet but those with dense chromite lamellae are confined to the more chrome-rich parts of the zone (Cr_2O_3 in garnet > 6 wt. %) and are thus closer to the pink garnet than the other group. A characteristic of the ilmenites is that they are present as small rounded bodies that vary in size from <20μm to about 500μm. In some specimens the grains are smallest closer to the pink garnet and become

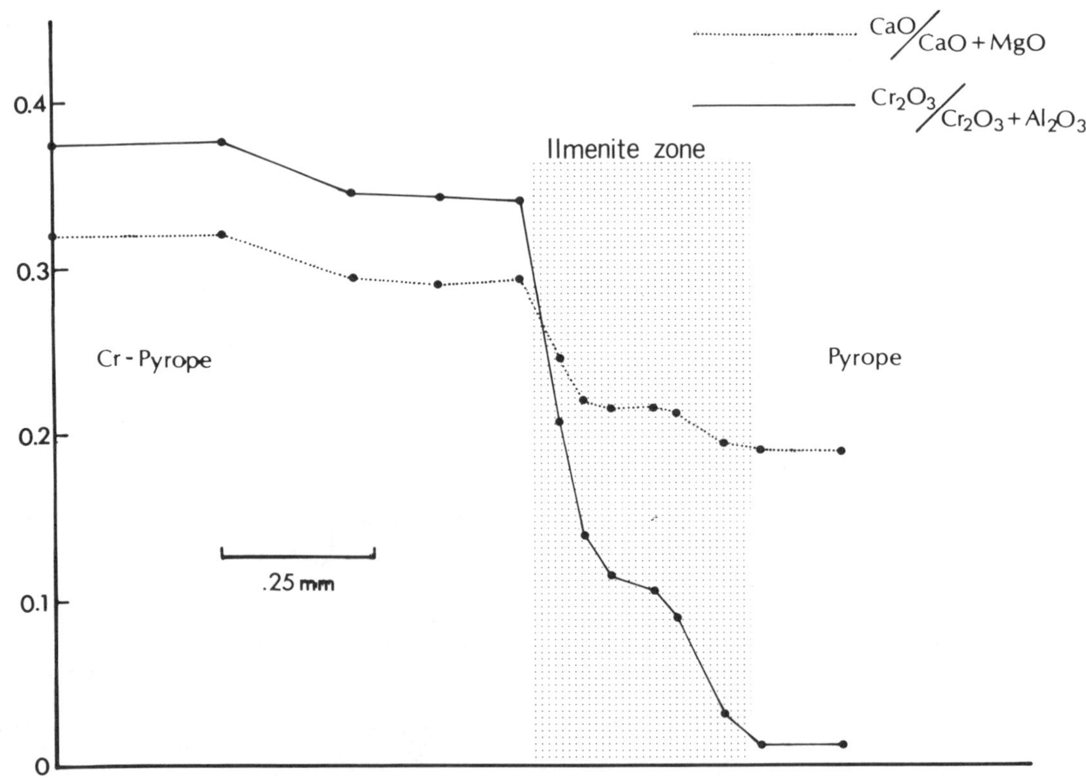

Fig. 5. Chemical profiles across zone between the Cr-rich garnet (Cr-pyrope) and the Fe-rich garnet (Pyrope). Note the marked break at the edge of the Cr-pyrope where ilmenite occurs.

progressively larger towards the orange garnet (Fig. 4).

The spacing of the continuous lamellae in the ilmenite in the Cr-rich region is sufficiently close that the electron beam is unable to resolve the separate phases. Accordingly, the beam size was increased to cover a larger area and approximate bulk analysis obtained. A similar technique was applied to the ilmenite with discontinuous lamellae, and the results are shown in Table 2. It is readily seen that the ilmenite with the continuous lamellae is much more chrome-rich than the other. Magnesium and aluminum contents are similar and are comparable with the "pure" ilmenite that lies between the discontinuous lamellae (Table 2). Since chromium and titanium are the major elements which show variation, and beam scans across lamellae show corresponding variations in these two elements, it is concluded the lamellae are predominantly chromite. Prior to exsolution of these lamellae it is thus possible that some ilmenite contained about 13 wt. % Cr_2O_3, which is in excess of the known solubility of Cr in ilmenite at 1 atmosphere (Muan et al., 1973).

The small, round ilmenites with the discontinuous chromite lamellae could be analyzed free of the chromite and this composition is given in Table 2. The areas analyzed were examined at x1500 magnification and no exsolved phase was visible and it thus appears that chromium and aluminum are actually present in solution in the ilmenite. As mentioned previously, the magnesium and aluminum contents are comparable with the bulk broad-beam analyses of the ilmenites. MgO displayed some variation between different grains but on average was 11 wt. %, which is slightly less than MgO in the large discrete ilmenite inclusions in the enstatite. Aluminum and chromium are also lower in the discrete ilmenite than in these rounded grains in the garnet inclusion (Table 2). However, compositionally these ilmenites are all comparable with those occurring in kimberlite (Mitchell, 1973; Boyd and Nixon, 1973; Haggerty, 1975).

C. Diopsides

Two compositionally distinct diopsides are present in the garnets. The largest diopsides are emerald green in color, Cr-bearing and occur within the Cr-rich pink garnet (Fig. 2). Associated with these chrome-diopsides are small irregular titanium-chromites.

The other diopside are associated with the interface between the orange and pink garnets. These diopsides are almost colorless; form somewhat elongated oval grains, often containing rounded ilmenite, and occur at the Cr-poor side of the garnet transition zone (Fig. 2). Characteristically, they have very low contents of Cr_2O_3(<0.6 wt. %).

Compositions of the two diopsides are presented in Table 4, Figure 6. Notable is the higher Cr content of the diopside in the pink garnet, whereas the other diopside contains a greater percentage of FeO. Although the Cr-diopside is relatively homogeneous, the diopside in the orange garnet showed a variation in composition both within and between grains. Expressed in terms of Ca/(Ca + Mg) one grain showed a range of 0.406 to 0.420; whereas between grain variation was from 0.359 to 0.429 (Table 4). This range in Ca/(Ca + Mg), if expressed in terms of temperatures of equilibration using the Davis and Boyd (1966) diopside-enstatite solvus, corresponds to a variation of 200°C. This is unrealistic and it is most probable that the diopside was not in equilibrium with enstatite. Furthermore, the position of these diopside grains on the edge of the garnet transition zone, in a region of relatively steep chemical gradients, is the most likely cause of the variation in element contents. Generally, the larger grains showed the most inhomogeneity, and this is probably a reflection of the fact that these grains would span greater ranges of element variation in the garnets abutting the diopside. An alternative explanation is that the inhomogeneity is due to the later alteration that can be seen to have invaded some of the diopsides; optically the extent of such alteration is difficult to assess.

Cr-diopsides comparable in mineral chemistry to these observed in the pink garnet (Table 4) have been described by Cox et al. (1973) in coarse grained granular lherzolites from the Matsoku kimberlite, Lesotho, and interestingly is also comparable to a diopside inclusion in diamond from Jagersfontein Mine, South Africa. Boyd (1974) has described a somewhat compositionally similar diopside from a sheared lherzolite from the Frank Smith Mine.

Analyses comparable to the diopside inclusions in the orange garnet are relatively common in diopside megacrysts, and diopside/garnet or diopside/ilmenite intergrowths from Frank Smith kimberlite (Nixon and Boyd, 1973c; Boyd, 1974).

D. Olivine

Olivine occurs in the same region of the orange garnet as the diopside. However, it differs inasmuch as the grains are more elongated and are often in optical continuity (Fig. 2). Of interest in one specimen is that the olivine is confined to one side of the pink garnet whereas the diopside occurs on the other side (Fig. 2). Compositionally the olivine is Fo_{86} (Table 5), and has a fairly large content of chromium. In this feature the olivine resembles olivine from sheared lherzolite xenoliths (Nixon and Boyd, 1973b) and olivine that occurs as inclusions in diamond (Meyer and Boyd, 1972; Meyer, 1975). However, the high iron content of this olivine is typical of olivine inclusions from discrete megacrysts (Nixon and Boyd, 1973a) and is in excess of those occurring as inclusions in diamond.

TABLE 4. Diopside Inclusions in Garnets from Enstatite Megacryst (73-21006), Weltevreden Kimberlite.

	Cr-diopside		Diopside			
Oxides	(3 anals)	S	1	2	Av.(15 anals)	S
SiO_2	54.5	(1.02)	55.7	55.4	55.8	(0.67)
TiO_2	0.33	(0.02)	0.69	0.40	0.52	(0.15)
Al_2O_3	2.13	(0.13)	2.92	1.84	2.50	(0.59)
Cr_2O_3	2.50	(0.24)	0.54	0.36	0.44	(0.12)
FeO	3.20	(0.10)	5.49	4.39	4.86	(0.36)
MnO	0.08	(0.03)	0.20	0.11	0.13	(0.05)
MgO	17.0	(0.26)	18.7	17.4	17.6	(0.51)
CaO	17.8	(0.17)	14.6	18.2	16.7	(1.24)
NiO	0.06	(0.05)	0.05	0.02	0.02	(0.03)
Na_2O	2.06	(0.21)	2.47	1.82	2.07	(0.45)
K_2O	0.05	(0.02)	0.05	0.06	0.03	(0.01)
Totals	99.7		101.4	100.0	100.7	
Cations on the basis of 6 Oxygens (x1000)						
Si	1959		1977	2000		
Ti	9		18	10		
Al	90		122	77		
Cr	109		14	9		
Fe	96		162	132		
Mn	2		5	3		
Mg	910		990	935		
Ca	682		554	702		
Ni	2		1	0		
Na	143		169	127		
K	2		2	2		
Totals	4004		4014	3999		
Ca/Ca+Mg	0.428		0.359	0.429		
Fs%	5.7		9.5	7.5		
En	53.9		58.0	52.9		
Wo	40.4		32.5	39.7		

E. Chromite

Chromite is present at the junction between the emerald green Cr-diopside and pink garnet (Fig. 2). The grains are relatively small (<0.1mm), and are very irregular in shape. Slight subtle color differences between core and margins can be detected at x1500 magnification. These color differences are due to compositional variation and the grains are particularly inhomogeneous. However, although this inhomogeneity is believed to be original and not due to later alteration, the fact that the cores are more Fe and Ti-rich, and Al-poor than the margins appear to be the reverse of that observed by Haggerty (1975) for spinels from kimberlites in Lesotho. Typical analyses for one grain are presented in Table 5 and these variations are fairly representative of all grains analyzed (Fig. 7). Aluminum and titanium show the greatest variation, followed by iron and chromium. Magnesium in all spinels is fairly constant averaging about 14 wt. % MgO.

Spinels approaching those described here both in composition and diversity have been reported previously by Haggerty (1975) and Clarke and Mitchell (1975) in groundmass samples from kimberlites in Southern Africa, and Northern Canada respectively. In both instances the spinels are either xenocrystic or from the groundmass of the kimberlite.

F. Sulfides

Small inclusions of sulfide occur within and marginal to the orange garnet. The grains vary in size from about 0.2 to 1.5mm, and are generally elongated and rounded. In reflected light the grains can be observed to consist of

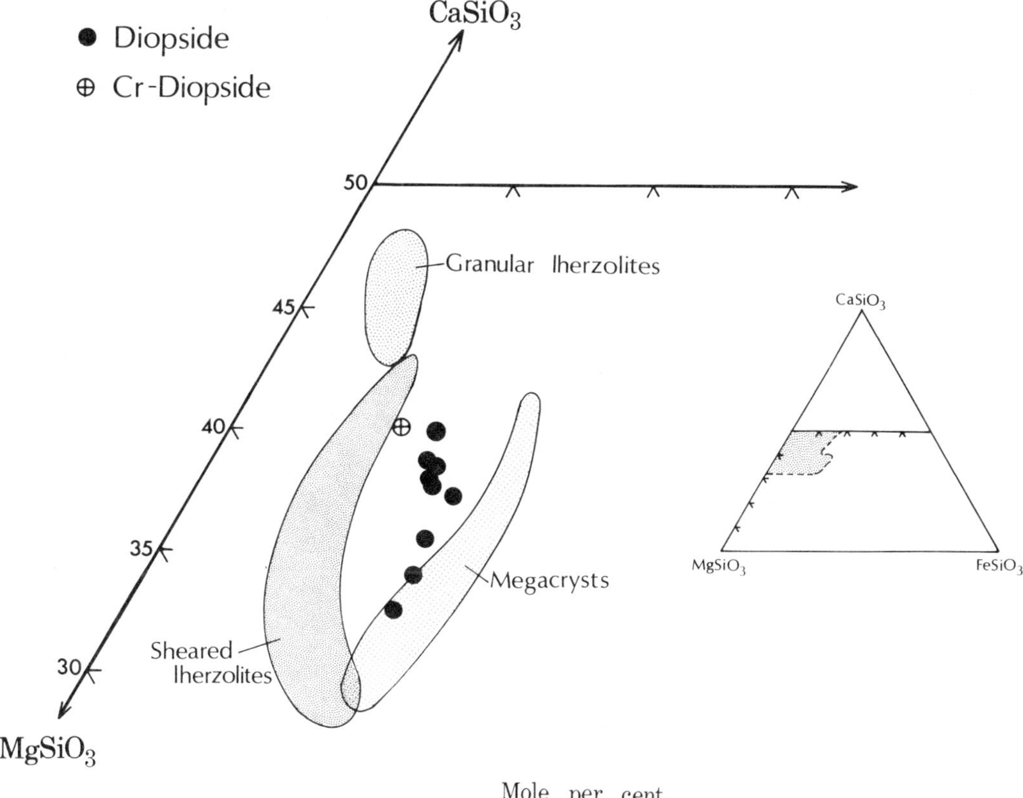

Fig. 6. Compositions of diopsides associated with the two garnets shown in terms of Ca-Mg-Fe. Fields for diopsides from sheared and granular lherzolites and discrete megacrysts after Nixon and Boyd (1973a,b).

an intergrowth of pentlandite, pyrrhotite, chalcopyrite and a border phase which is a K-rich sulfide phase, containing approximately 6 wt. % K. This phase is similar to that reported by Clarke et al. (1977) in pyroxene-ilmenite intergrowths from Frank Smith mine. A similar phase has also been observed by Tsai et al. (1977) in xenoliths from Obnazhennaya, Siberia.

Secondary Minerals

Small veinlets of serpentine exist throughout the enstatite xenocryst as well as occurring within the garnet inclusions. Generally the serpentine is more Fe-rich and Cr-poor when it occurs within the orange garnet in contrast to its occurrence in the pink garnet (Table 6).

Secondary phlogopite is plentiful in association with the garnet inclusions, and is particularly common at the margin of the orange garnet with the surrounding enstatite. Large "nests" of phlogopite plates occur between the orange garnet and the large ilmenite inclusions in the enstatite (Fig. 2). Often in the center of these nests are small regions of calcite (Fig. 2). Phlogopite also is present as thin stringers or veinlets in the pink garnet. Interestingly, the phlogopite in the pink garnet is much more Cr-rich than that which is associated with the orange garnet (Table 6).

Discussion

The pertinent features of the mineralogy of the enstatite xenocryst from Weltevreden Floors are illustrated diagrammatically in Fig. 2. The importance of the specimen is the coexistence of Cr-rich phases (Cr-diopside, chromite, Cr-rich ilmenite and Cr-garnet) as inclusions in the more Fe-rich garnet and all enclosed in the enstatite host.

Although the formation of the two garnet assemblage is intrinsically part of the genesis of the whole enstatite xenocryst, it is possible to suggest two modes of origin for the orange Fe-rich garnet:

i. The Fe-rich garnet formed during an igneous event and crystallized on the Cr-rich garnet which acted as a nucleus.

ii. The Fe-rich garnet formed metasomatically and is the result of diffusion plus reaction between enstatite and the Cr-rich garnet.

To differentiate between these two possibil-

TABLE 5. Analyses of Olivine and Chromite Inclusions in Garnets from Enstatite Megacryst (73-21006).

	Olivine		Chromites			
Oxides	(7 anals)	S	Core	Margin	A1	C1
SiO_2	39.7	(0.89)	0.34	0.29	0.25	0.28
TiO_2	0.05	(0.02)	7.70	4.96	4.16	4.20
Al_2O_3	0.04	(0.04)	10.7	19.9	19.7	22.9
Cr_2O_3	0.06	(0.03)	35.5	34.0	35.4	34.5
FeO	13.1	(0.25)	30.5	26.8	24.5	23.9
MnO	0.12	(0.01)	0.38	0.42	0.36	0.26
MgO	47.5	(0.78)	13.4	14.5	13.5	14.8
CaO	0.05	(0.02)	0.09	0.07	0.02	0.07
NiO	0.07	(0.03)	<0.01	0.11	<0.01	<0.01
Na_2O	0.03	(0.03)	<0.01	<0.01	0.02	0.02
K_2O	<0.01		<0.01	<0.01	0.01	<0.01
Totals	100.7		98.6	101.0	98.0	100.9
	Cations on the Basis of 4 Oxygens (x1000)					
Si	983		11	8	7	8
Ti	1		194	116	100	96
Al	1		422	732	747	823
Cr	1		940	838	899	833
Fe	272		854	699	659	610
Mn	3		10	11	9	6
Mg	1751		668	676	645	673
Ca	1		3	1	0	1
Ni	1		0	1	0	0
Na	1		0	0	0	0
	3015		3103	3083	3066	3051
Fo%	87					

ities on the basis of major element chemistry is not possible. Certainly there is no gradient for Al, Fe or Ca in the enstatite adjacent to the garnet such as one may expect if the garnet had formed as a result of enstatite reequilibration. However, in favor of this metasomatic origin may be cited the observations that the orange garnet has euhedral relationship to the enclosing enstatite, and that sulfide grains have no apparent compositional difference whether they are inside the orange garnet or enstatite.

In contrast to this evidence for a metamorphic origin, observations supporting genesis in an igneous process seem more abundant. Especially pertinent are the occurrences of elongated and optically-continuous bead-like strings of olivine and diopside between the two garnets (Fig. 2) as well as the presence of small rounded ilmenites in this same zone (Figs. 4 and 2). These ilmenites are slightly different in composition from the major ilmenite which occurs as discrete or lamellar intergrowth with the enstatite (Table 2). Furthermore, the rounded shape of the Cr-rich garnet, in one instance with a crescent-shape embayment, strongly suggests the effects of assimilation.

Considering the Cr-rich assemblage alone, it is possible that this represents the remnant of an incompletely digested (or molten) preexisting rock, or phase. Whether this preexisting material is consanguinous with the magmatic material from which the enstatite megacryst formed, or whether it was derived from the mantle "chamber" walls is unknown. Conceivably, the rareness of Cr-pyrope in kimberlite may be due to it not being in equilibrium with later magmatic environments. Apart from unique events such as the present instance, or being armored in diamond, the Cr-pyrope is reassimilated and thus evidence of its presence is lost. Gurney and Switzer (1973) and Nixon and Hornung (1968) have shown that Cr-pyrope is very rare in kimberlite concentrate.

Texturally it is apparent that the Cr-rich assemblage predates the host enstatite and was probably in existance prior to the formation of the enclosing Fe-rich garnet. If the Cr-garnet owes its present shape to partial assimilation (or melting) and was thus enveloped by melt it

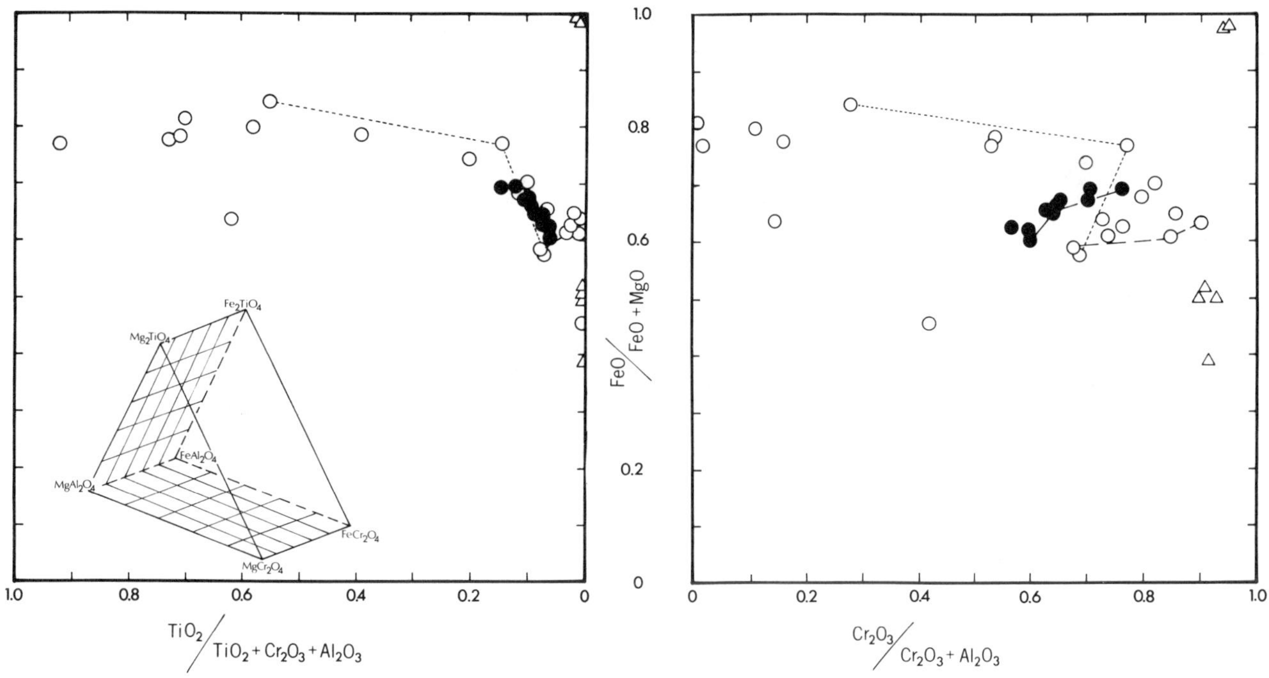

Fig. 7. Compositional variations in chromites associated with the Cr-diopside and Cr-pyrope. Open circles are spinel data from Haggerty (1975), Clarke and Mitchell (1975). Triangles are chromites occurring as diamond inclusions (Meyer and Boyd, 1972).

is possible it acted as a nucleus for subsequent crystallization. Just how the mechanism operated to produce the bead-like strings of olivine and diopside around the nucleus as well as the almost perfectly round ilmenites can only be guessed.

Attempts to determine pressures and temperatures of equilibration for the megacryst using various geothermometers have produced fairly consistent results, within the limits of applicability for each technique. For example, using the enstatite-solvus curve (Boyd and Nixon, 1973) and alumina content, a temperature of 1220°C at 64kb is obtained. Interestingly, this is also the temperature obtained for the Cr-diopside -- Cr-pyrope pair using the calibration of Akella and Boyd (1974) for Mg-Fe partitioning. The orthopyroxene-ilmenite geothermometer (Bishop, 1977) also gives a similar temperature of equilibration (∿1150°C) provided a pressure of approximately 60kb (equivalent to that obtained from the Al_2O_3 content of the enstatite) is assumed.

TABLE 6. Analyses of Secondary (Alteration) Minerals in Enstatite Megacryst (73-21006) from Weltevreden Kimberlite

	Phlogopite		Serpentine	
Oxide	In Pyrope	In Cr-Pyrope	In Pyrope	In Cr-pyrope
SiO_2	39.1	39.4	45.4	45.3
TiO_2	7.38	5.89	<0.01	<0.01
Al_2O_3	14.5	14.8	0.09	0.05
Cr_2O_3	0.07	2.14	<0.01	0.05
FeO	8.45	7.12	3.10	1.64
MnO	0.07	0.06	0.16	0.18
MgO	17.3	18.3	38.4	40.2
CaO	<0.01	<0.01	0.16	0.15
NiO	<0.01	<0.01	0.01	<0.01
Na_2O	0.43	0.22	0.12	0.14
K_2O	9.41	9.12	0.04	0.04
Totals	96.7	97.1	87.5	87.8

Acknowledgements

Support for this work was provided by National Science Foundation grants Earth Sciences Section EAR76-22698. HMT gratefully acknowledges a David Ross Fellowship, Purdue Research Foundation. The manuscript was greatly improved by comments from Drs. Boyd, MacGregor, Prinz and Mitchell, to whom we give our thanks.

References

Akella, J. and F.R. Boyd, Petrogenetic grid for garnet peridotites, Carnegie Inst. Wash. Yrbk., 73, 269-273, 1974.

Albee, A.L. and L. Ray, Correction factors for electron probe microanalysis of silicates, oxides, carbonates, phosphates and sulphates, Anal. Chem., 42, 1408-1414, 1970.

Bence, A.E. and A.L. Albee, Empirical correction

factors for the electron microanalysis of silicates and oxides, Jour. Geol., 26, 382-403, 1968.
Boyd, F.R., Ultramafic nodules from the Frank Smith kimberlite pipe, South Africa, Carnegie Inst. Wash. Yrbk., 73, 285-294, 1974.
Bishop, F.C., $Mg-Fe^{2+}$ partitioning between orthopyroxene and ilmnite: A high-temperature geothermometer, (Abstr.) EOS, 58, 521, 1977.
Boyd, F.R. and R.V. Danchin, Discrete nodules from the Artur de Paiva kimberlite, Angola, Carnegie Inst. Wash. Yrbk., 73, 278-282, 1974.
Boyd, F.R. and R.H. McCallister, Densities of fertile and sterile garnet peridotites, Geophys. Res. Letts., 3, 509-512, 1976.
Boyd, F.R. and P.H. Nixon, Origin of the ilminite-silicate nodules in kimberlites from Lesotho and South Africa, in Lesotho Kimberlites, Ed. P.H. Nixon, Lesotho Kimberlites, Ed. P.H. Nixon, Lesotho Nat. Dev. Corp., Maseru, 254-288, 1973.
Boyd, F.R. and P. H. Nixon, Origins of the ultramafic nodules from some kimberlites of Northern Lesotho and the Monastery Mine, South Africa Phys. Chem. Earth, 9, 431-454, 1975.
Burnham, C.W., Refinement of lattice parameters using systematic correction terms, Carnegie Inst. Wash. Yrbk., 64, 200-202, 1965.
Clarke, D.B. and R.H. Mitchell, Mineralogy and Petrology of the kimberlite from Somerset Island, N.W.T., Canada, Phys. Chem. Earth, 9, 123-135, 1975.
Clarke, D. B., G.G. Pe, R.M. Mackay, K.R. Gill, M.J. O"Hara, and J.A. Gard, A new potassium-iron-nickel sulphide from a nodule in kimberlite, Earth Planet. Sci. Letts., 35, 421-428, 1977.
Craig, J.R. and G. Kullerud, Phase relationships in the Cu-Fe-Ni-S system and their application to magmatic ore deposits, Econ. Geol. Mono., 4, 343-358, 1969.
Danchin, R.V. and F. D'Orey, Chromian spinel exsolution in ilmenite from the Premier Mine, Transaal, South Africa, Contr. Mineral. Petrol., 35, 43-49, 1972.
Davis, B.T.C. and F.R. Boyd, The join $Mg_2Si_2O_6$-$CaMgSi_2O_6$ at 30 kilobars pressure and its application to pyroxenes from kimberlites, J. Geophys. Res., 14, 3567-3576, 1966.
Dawson, J.B., and J.B. Hawthorne, Magmatic sedimentation and carbonatitic differentiation in kimberlite sills, Jour. Geol. Soc. London, 129, 61-85, 1973.
Gurney, J.J. and G. S. Switzer, The discovery of garnets closely related to diamonds in the Finsch Pipe, South Africa, Contr. Mineral. Petrol., 39, 103-116, 1973.
Gurney, J.J., H.W. Fesq, and E.H.D. Kable, Clinopyroxene-ilmenite intergrowths from kimberlite: A reappraisal, In Lesotho Kimberlites, Ed. P.H. Nixon, 238-253, Lesotho Nat. Dev. Corp., Maseru, 1973.
Gurney, J.J., B. Harte, and K.G. Cox, Mantle xenoliths in the Matsoku kimberlite pipe, Phys. Chem. Earth, 9, 507-524, 1975.
Gurney, J.J., W.R.D. Jakob, and J.B. Dawson, Megacrysts from the Monastery Mine, (Abstr.) 2nd Int. Kimberlite Conf., Santa Fe,, 1977.
Haggerty, S.E., Spinels of unique composition associated with ilmenite reactions in the Liqhobong Kimberlite pipe, Lesotho, in Lesotho Kimberlites, Ed. P.H. Nixon, 149-158, Lesotho Nat. Dev. Corp., Maseru, 1973.
Haggerty, S.E., The chemistry and genesis of opaque minerals in kimberlites, Phys. Chem. Earth, 9, 295-307, 1975.
Hornung, G. and P.H. Nixon, Chemical variations in the knorringite-rich garnets, in Lesotho Kimberlites, Ed. P.H. Nixon, 122-127, Lesotho Nat. Dev. Corp., Maseru, 1973.
Kullerud, G., R.A. Yund, and G.H. Moh, Phase relations in the Cu-Fe-S, Cu-Ni-S and Fe-Ni-S systems, Econ. Geol. Mono, 4, 323-343, 1969.
MacGregor, I.D. and A.R. Basu, Thermal structure of the lithosphere: A petrologic contribution, Science, 185, 1007-1011, 1974.
McCallister, R.H., H.O.A. Meyer, and D.G. Brookins, Pyroxene-ilmenite xenoliths from the Stockdale pipe, Kansas: Chemistry, crystallography and origin, Phys. Chem. Earth, 9, 287-294, 1975.
Meyer, H.O.A., Chromium and the genesis of diamond, Geochim. Cosmochim. Acta., 39, 929-936, 1975.
Meyer, H.O.A. and F.R. Boyd, Composition and origin of crystalline inclusions in natural diamond, Geochim. Cosmochim. Acta, 36, 1255-12733, 1972.
Meyer, H.O.A. and D.P. Svisero, Mineral inclusions in Brazilian diamonds, Phys. Chem. Earth, 9, 785-795, 1975.
Meyer, H.O.A. and H.M. Tsai, Mineral inclusions in natural diamond -- their nature and significance: A review, Mineral Sci. Eng., 8, 242-261, 1976.
Meyer, H.O.A. and H.M. Tsai, Inclusions in diamond and the mineral chemistry of the upper mantle, Proc. 2nd Symp. Origin and Distribution of the Elements, UNESCO, Paris (In Press), 1977.
Mitchell, R.H., Magnesian ilmenite and its role in kimberlite petrogenesis, Jour. Geol., 81, 301-311, 1973.
Mitchell, R.H., Geochemistry of magnesian ilmenites from kimberlites in South Africa and Lesotho, Lithos, 10, 29-37, 1977.
Mitchell, R.H. and D.B. Clarke, Oxide and sulphide mineralogy of the Penyuk Kimberlite, Somerset Island, N.W.T., Canada, Contr. Mineral. Petrol., 56, 157-172, 1976.
Muan, A., J. Hauch and T. Lofall, Equilibrium studies with a bearing on lunar rocks, Proc. Third Lunar Sci. Cong., Supp. 3. Geochim. Cosmochim. Acta., 1, 185-196, 1972.
Nixon, P.H. and F.R. Boyd, The discrete nodule (megacryst) association in kimberlites from Northern Lesotho, in Lesotho Kimberlites, Ed. P.H. Nixon, 67-75, Lesotho, Nat. Dev. Corp., Maseru, 1973a.

Nixon, P.H. and F.R. Boyd, Petrogenesis of the granular and sheared ultrabasic nodule suite in kimberlite, in Lesotho Kimberlites, Ed. P.H. Nixon; 48-56, Lesotho Nat. Dev. Corp., Maseru, 1973b.

Nixon, P.H. and F.R. Boyd, Discrete nodules (megacrysts) and lamellar intergrowths in Frank Smith kimberlite pipe, Int. Conf. in kimberlites, Extended Abstracts, Univ. Cape Town, Rondebosch, South Africa, 243-246, 1973c.

Nixon, P.H. and G. Hornung, A new chromium garnet end member, knorringite, from kimberlite, Am. Mineral, 53, 1833-1840, 1968.

Nixon, P.H., O. Von Knorring and J.M. Rooke, Kimberlites and associated inclusions of Basutoland: A mineralogical and geochemical study, Am. Mineral. 48, 1090-1132, 1963.

Pasteris, J.D., F.R. Boyd, and P.H. Nixon, Frank Smith ilmenite association, (this volume), 1978.

Tsai, H.M., Y.N. Shieh and H.O.A. Meyer, Mineralogy and S^{34}/S^{32} ratios of sulfides associated with kimberlite, xenoliths and diamonds, (Abstr.) 2nd Int. Kimberlite Conf., Santa Fe, 1977.

Wilshire, H.G. and E.D. Jackson, Problems in determining mantle geotherms from pyroxene compositions of ultramafic rocks, Jour. Geol., 83, 313-329, 1975.

A QUENCH PYROXENE-ILMENITE XENOLITH FROM KIMBERLITE: IMPLICATIONS FOR PYROXENE-ILMENITE INTERGROWTHS.

Penelope J. Rawlinson* and J.B. Dawson

Department of Geology, University of St.Andrews, Scotland.

*Present address: Department of Geology, Parks Road, Oxford.

Abstract. A xenolith from the Weltvreden Mine, South Africa, consists mainly of acicular bronzite, intergrown irregularly with magnesium ilmenite, and rarer large bronzites, pyrope and diopside-ilmenite intergrowths. Three stages of cooling can be recognised in this non-equilibrium assemblage which contains (though on a finer scale) phases typical of the discrete nodule and intergrowth suites found in many kimberlites. Comb-layering and spherulitic texture of bronzite-ilmenite intergrowths indicate quenching of a high-titanium liquid. After crystallisation of garnet, pyroxene and ilmenite, the residual fluids crystallized serpentine, phlogopite, diopside, ilmenite, perovikite and calcite - an assemblage similar to many kimberlite groundmasses.

Introduction

In recent years there has been considerable debate on the origin of the pyroxene-ilmenite intergrowths found as nodules in kimberlite pipes. Previously regarded as a rare phenomenon, they have now been found in numerous kimberlite intrusions in southern Africa (Gurney et al. 1973), more rarely in Yakutian and American kimberlites (Ilupin et al. 1973; Smith et al. 1976) and also in an alnoitic breccia on the island of Malaita, Solomon Islands (Allen and Deans, 1965). The ilmenite in the more common intergrowths with clinopyroxene is of regular, lamellar habit, but in the rarer orthopyroxene-ilmenite intergrowths the ilmenite tends to be irregular. Clinopyroxene/ilmenite intergrowths from the Monastery Mine were originally believed to be of eutectic origin having crystallized from a liquid (Williams, 1932), an origin subsequently supported on chemical grounds by Gurney et al. (1973) and by Wyatt(1977) who has successfully reproduced the textures experimentally. Other hypotheses, proposing exsolution of the ilmenite from a high pressure precursor such as ilmenite-structured pyroxene or garnet, have also been proposed (Dawson and Reid, 1970; Ringwood and Lovering, 1970), but recent experimental work by Akella and Boyd (1972) and Green and Sobolev (1975) indicates that only limited amounts of TiO_2 (up to 1.8 wt.% - representing not more than 4 wt.% ilmenite) is found in clinopyroxenes (and garnets) synthesized in runs up to 30 kb. Here we describe a nodule from a South African kimberlite that provides firm textural evidence that both orthopyroxene/ilmenite and clinopyroxene/ilmenite intergrowths can result by cotectic crystallization from a titanium-rich liquid.

Sample Description

The nodule, oval in shape, and measuring 14 x 10 x 10 cm before sectioning, was found in the kimberlite of the Weltvreden Mine, the "sister" pipe of the more famous Frank Smith Mine in the Harts River valley, 65 km N.N.W. of Kimberley. Fine-grained radiating aggregates of 2 mm acicular crystals could be seen on the surface, and the nodule was cut by a 10-15mm-wide vein of ilmenite. Fig. 1 is a large thin-section cut from the centre of the nodule, and illustrates the variable textural domains found within the nodule. On either side of the ilmenite vein are relatively coarse anhedral, equidimensional crystals of abundant orthopyroxene (up to 3 mm) and very rare pale-yellow garnet (2mm) (only 5 grains were seen in the large thin section). Further from the vein, on one side of the specimen, isolated coarse crystals of orthopyroxene form the cores of macrospherulites consisting of acicular orthopyroxene (up to 3mm long; length:breadth ratios ~ 8:1) intergrowth with irregular blebs of ilmenite (Fig. 2 and 3); bright-orange mica occurs in the interstices between pyroxenes in the macrospherulites and also in the spaces between mutually-interfering macrospherulites where it may contain rounded blebs of ilmenite. Further from the vein is comb-layered, finer-grained acicular orthopyroxene intergrown with irregular ilmenite; the growth direction in the comb-layered units is from the present edge of the specimen in towards the area of macro-

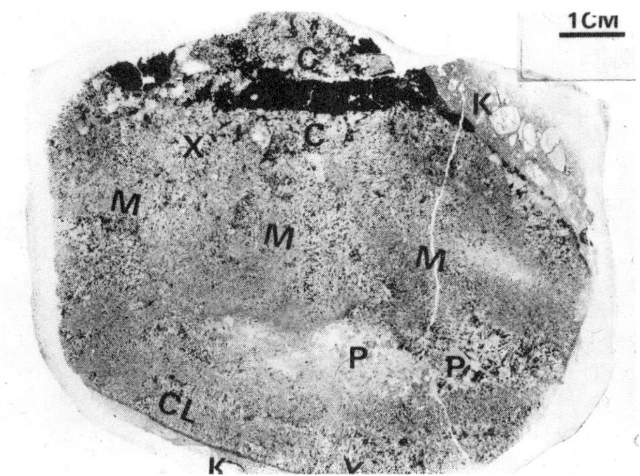

Fig. 1. Thin section of BD2027 illuminated from rear. The various textural domains are CL - comb-layered zone, M - macrospherulite zone, C - zone of coarser crystals, P - coarse patches. Areas centred on points X and Y are enlarged in Figs. 2 and 4.
K - area of enclosing kimberlite.

spherulites (Fig. 4). Both the macrospherulites and the comb-layered units could only have arisen by rapid crystallization of the phases from a liquid (c.f. Lofgren and Donaldson, 1975). Within the zone of macrospherulites are coarser areas containing serpentine (pseudomorphs ?after olivine), phlogopite (sometimes with bright-red rims), calcite and rounded blebs of ilmenite of considerably larger size than those intergrown with acicular orthopyroxene (see Fig. 1). In addition, around these coarse patches the orthopyroxenes become larger-grained and there are relatively large clinopyroxenes in

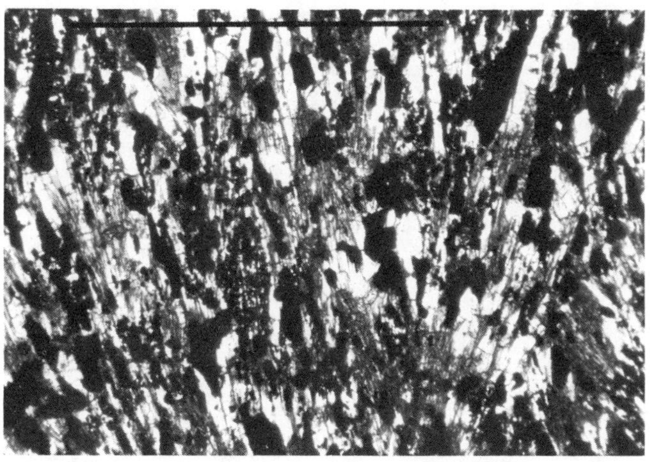

Fig. 3. Irregular intergrowths of acicular orthopyroxene and ilmenite in macrospherulite zone. Scale-bar = 0.5mm, p.p.l.

regular intergrowths with ilmenite; the morphology of some of the ilmenite in the regular intergrowths with clinopyroxene strongly resembles that of ilmenite in clinopyroxene/ ilmenite nodules (Fig. 4). Where they protrude into the coarse patches, both types of pyroxene are highly altered. Finally, again within the zone of macrospherulites, are tiny patches of fine-grained (<0.5mm) serpentine + mica + ilmenite + perovskite + calcite + diopside that, together, comprise an assemblage very similar to kimberlite groundmass (Dawson et al. 1977). These, together with the coarser patches, are interpreted as the result of crystallization within an essentially closed system of the volatile -rich late-stage residuum from a

Fig. 2. Enlargement of area around X (Fig. 1). Acicular orthopyroxenes radiate out from a coarser earlier crystal of orthopyroxene (O). Scale bar = 0.5mm. Crossed polarizers.

Fig. 4. Enlargement of area around Y (Fig. 1). Comb-layered and branching orthopyroxene crystals. Scale bar = 0.5mm. Crossed polarizers.

liquid that had earlier crystallized the garnet, pyroxenes, ilmenite and interstitial micas. The specimen is not symmetrical as the zones of macrospherulites, comb-layered units and coarse patches are found only on one side of the specimen, though there is a distinct impression that the rock represents a former essentially symmetrical system of units on either side of the zone of coarser orthopyroxene and garnet, with most of one series of zones having been broken off.

Bulk Chemistry

An analysis of a 1cm-thick slab of essentially the same modal composition as the section illustrated in Fig. 1 (excluding the attached kimberlite) is given in Table 1 where it is compared with the bulk analysis of an orthopyroxene-ilmenite intergrowth from the neighbouring Frank Smith Mine. The principal significant differences in the two analyses are in Al_2O_3, CaO, K_2O and H_2O that can be directly attributed to modal phlogopite and minor diopside, calcite and serpentine in the Weltvreden rock. A significant feature is the low oxidation state of the iron in the Weltvreden nodule.

Phase chemistry

Mineral analyses were made by electron microprobe at the Universities of Edinburgh, Cape Town (Geoscan V), Oxford (Geoscan IX) and Chicago (ARL-ENX), using the Reed-Ware correction programme.

Garnet. The garnets (Table 2), enclosed within the coarser orthopyroxene areas, was one of the first phases to crystallize. It is pyropic, and its high TiO_2 but low Cr_2O_3 content categorize it as high-titanium pyrope (classification of Dawson and Stephens, 1975), a type

Fig. 5. Regular ilmenite intergrown with clinopyroxene in vicinity of a pegmatite patch. Scale bar = 0.5mm. p.p.l.

TABLE 1. Bulk Analysis of BD 2027.

	1	2
SiO_2	37.98	36.96
TiO_2	12.77	15.36
Al_2O_3	2.01	1.19
Cr_2O_3	0.24	n.d.
Fe_2O_3	1.41	n.d.
FeO	13.32	15.48*
MnO	0.16	0.21
MgO	25.19	26.23
CaO	2.27	1.13
Na_2O	0.43	0.67
K_2O	1.32	0.18
H_2O^+	1.90	n.d.
H_2O^-	0.29	n.d.
P_2O_5	0.05	0.03
Total	99.34	97.41

1. BD 2027 Analyst: Geochemical Laboratories, University of Cape Town.
2. Orthopyroxene-ilmenite intergrowth, Frank Smith Mine (Frick, 1973).

* Converted from original value of total iron as Fe_2O_3 17.19

found previously mainly as large megacrysts in the kimberlite environment. Although containing relatively low CaO, the values are within the range for high-titanium pyropes.

Orthopyroxene. The orthopyroxenes are high-titanium bronzites (terminology of Stephens and Dawson, 1977) with TiO_2 > 0.2 wt.% and mg (Mg/(Mg+Fe)) < 0.90. The coarser equidimensional crystals (Table 3) have only small compositional differences from grain to grain; in most grains all Al is in four-fold coordination and mg values are very constant. The acicular pyroxenes, intergrown with blebs and ovoid patches of ilmenite, are sometimes nucleated upon the coarser, equidimensional grains, and there is a compositional overlap between the two morphologically-different types of bronzite, though the acicular types show the greater variation as a group and have significant Al in six-fold coordination. Most coarse crystals are homogeneous compositionally - the typical range of composition within acicular crystals is shown in Analyses 9a-9b, Table 3. The range shown is inter-grain, and most acicular grains are remarkably uniform along their length. Compared with orthopyroxene in orthopyroxene/ilmenite intergrowths from the Monastery and Frank Smith

TABLE 2 Garnets in 2027.

	BD2027 Range	Mean (n 6)	High Ti pyrope*	Cr pyrope*
SiO_2	41.5-42.3	41.8		
TiO_2	1.20-2.17	1.48	1.09	0.17
Al_2O_3	20.8-22.4	21.5		
Cr_2O_3	0.10-0.25	0.20	0.91	3.47
FeO^T	10.7-11.7	11.1	9.84	8.01
MnO	0.27-0.36	0.32		
MgO	20.4-21.0	20.7	20.3	20.0
CaO	2.63-3.22	3.01	4.52	5.17
Na_2O	0.08-0.13	0.10		

* From Dawson and Stephens (1975).

Mines, the Weltvreden orthopyroxenes tend to be more magnesian and, in the case of the acicular pyroxenes, slightly more calcic (Fig. 6).

Clinopyroxenes in regular intergrowths with ilmenite are found in the vicinity of the coarse patches and also as discrete grains in small fine-grained areas (Dawson et al. 1977). The intergrowth pyroxenes (Table 4), vary in composition from point to point, especially in Al_2O_3 and TiO_2 which may possibly be due to localized competition for various elements, and are unlike those in the discrete clinopyroxene-ilmenite intergrowths from localities, such as Monastery, Uintjes Berg, Sonop (see summary by Gurney et al. 1973) which are low-Cr diopsides (Stephens and Dawson, op. cit.), in having higher Ca/(Ca+Mg) ratios and TiO_2, but lower total FeO and lower Ma_2O plus Al_2O_3 (i.e. lower jadeite molecule). On a Ca-Mg-Fe plot (Fig. 7), the main differences are apparent; the Weltvreden clinopyroxenes show strong similarities to MARID-suite clinopyroxenes and, together with the clinopyroxene in the fine-grained patches, also resemble kimberlite groundmass clinopyroxenes. These similarities are further emphasized by an insufficiency of Al to fill the Al^{IV} sites, in itself suggesting that the Na (in the absence of Al^{IV}) is probably linked with Fe^{3+} in acmite molecule. This suggestion of increased oxidation correlates with increasing Fe^{3+} in the associated ilmenite (below).

Ilmenites. All ilmenites in 2027 are highly magnesian (Table 5), thereby resembling most other kimberlite ilmenites. Throughout the established crystallization sequence, there is a distinct trend with a change in Mg/Fe^{2+} and Fe^{2+}/Fe^{3+} from ilmenite in opx-il intergrowths to ilmenite in cpx-il intergrowths in coarse areas, suggesting (a) falling temperature and

TABLE 3. Analyses of Equidimensional and Acicular Orthopyroxenes in BD2027.

	BD2027 Equidimensional grains					BD2027 Acicular grains						11	
	1	2	3	4	5	6	7	8	9a	9b	10	Range	Mean
SiO_2	53.3	55.7	52.7	55.5	55.6*	55.8	56.2	56.4	56.9	56.8	56.4		
TiO_2	0.25	0.23	0.24	0.21	0.23	0.49	0.37	0.35	0.31	0.28	0.36	0.04-0.35	0.19
Al_2O_3	2.09	2.20	1.98	1.32	1.90	1.89	1.57	1.88	1.59	1.35	1.66	0.71-2.26	1.06
Cr_2O_3	0.09	0.06	0.10	0.09	0.08	0.14	0.12	0.12	0.14	0.11	0.12	0-0.27	0.10
FeO^T	8.97	8.83	8.80	8.52	8.78	7.97	7.96	7.95	7.88	8.01	7.95	6.70-10.89	8.64
MnO	0.17	0.12	0.15	0.15	0.15	0.14	0.19	0.18	0.14	0.12	0.15		
MgO	32.6	32.5	32.8	33.1	32.8	31.3	32.0	31.5	32.1	32.5	31.9	30.4-33.9	32.2
NiO	0.03	0.04	0.09	0.05	0.05	0.09	0.20	n.d.	n.d.	n.d.	-		
CaO	1.07	0.98	1.15	1.00	1.05	1.74	1.07	0.81	0.92	0.71	1.03	0.63-1.60	1.07
Na_2O	0.24	0.23	0.29	0.21	0.24	0.38	0.23	0.28	0.29	0.25	0.27	0.03-0.34	0.22
K_2O	0.03	0.03	0.02	0.03	0.03	-	-	-					
Total	98.8	100.9	98.3	100.2		100.0	99.9	99.6	100.2	100.1			

1-4. Individual grains of coarse orthopyroxene grains. 5. Mean of analyses 1-4. * Mean of analyses 2 and 4. Analyst R. Hervig (Chicago). 6. In intergrowth A with ilmenite; 7. In intergrowth B with ilmenite; 8. In intergrowth C with ilmenite; 9a, 9b. Widest variations in opx crystals in intergrowth D; 10. Average of analyses 6-9; 11. Range and mean of high-Ti bronzite (Stephens and Dawson, 1977).

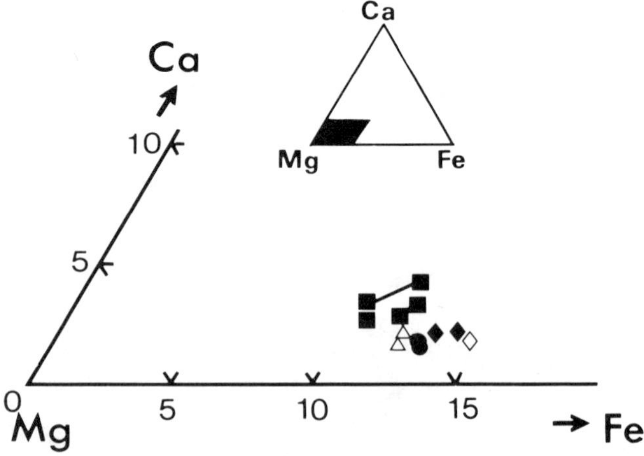

Fig. 6. Ca-Mg-Fe diagram of orthopyroxenes intergrown with magnesian ilmenite. Solid circles - coarse equideimensional grains 2027; solid squares - acicular grains 2027 (tie line joins two most extreme compositions found within a single grain); solid diamonds - Monastery (new data); open diamond - Monastery (Boyd, 1971); open triangles - Frank Smith Mine (Frick, 1975).

(b) increased oxidation. The ilmenite in the massive vein extends the Mg/Fe depletion trend but is less oxidised. Whereas most of the ilmenites broadly resemble those in discrete nodules in kimberlite, the low Al_2O_3 and Cr_2O_3 in the ilmenite intergrown with clinopyroxene is most similar to that in ilmenites in MARID-suite nodules (Dawson and Smith, 1976). The micas in the macrospherulitic and comb-layered areas are titanian, low-Cr phlogopites (TiO_2 up to 4.2%; Cr_2O_3 < 0.2%). By contrast the micas in the coarse areas are higher in TiO_2, and changes in FeO^T/Al_2O_3 contents in foxy-red rims on these micas suggest much of the iron in the rims is probably Fe^{3+} (Table 6).

Experimental Work

Wyatt (1977) has reproduced the textures of clinopyroxene/ilmenite intergrowths experimentally from melted crushed natural intergrowths. High P-T experiments were carried out on BD2027 at the Grant Institute of Geology, Edinburgh, mainly to attempt to reproduce the spherulitic and comb-layered texture. Single-stage piston cylinder apparatus was used for runs at 10, 20 and 30 kbrs; powders were loaded into platinum tubes and runs lasted 10-15 minutes as a compromise between iron loss and equilibrium time. Quenched charges were examined both optically and by X-ray. No attempt was made to control fO_2 and H_2O and this, together with the fact that the bulk charges contain small amounts of water and CO_2 (from crushed mica, serpentine and calcite) must be borne in mind when assessing the results. It should be appreciated

TABLE 4. Analyses of BD2027 Clinopyroxenes, intergrown with Ilmenite, compared with other kimberlite Pyroxenes.

	1	2	3	4	5
SiO_2	53.9	51.9	-	54.2	53.5
TiO_2	1.16	2.66	0.50	0.34	0.83
Al_2O_3	0.37	1.32	3.19	0.50	0.57
Cr_2O_3	0.11	0	0.09	0.44	0.10
FeO^T	3.65	3.70	5.86	4.62	3.81
MnO	0.06	0.07	-	0.13	-
MgO	16.6	15.9	16.9	17.0	16.3
NiO	0.06	0	-	-	-
CaO	23.5	23.5	17.5	20.6	23.3
Na_2O	0.48	0.62	1.85	1.27	0.53
Total	99.7	99.7			

1.2. BD2027 clinopyroxenes, intergrown with ilmenite.

3. Mean of 28 low-chrome diopsides, most are intergrown with ilmenite (Dawson and Stephens, 1977)

4. Mean of 7 MARID-suite diopsides (Dawson and Smith, 1977)

5. Mean of 8 kimberlite groundmass diopsides Dawson et al., 1977)

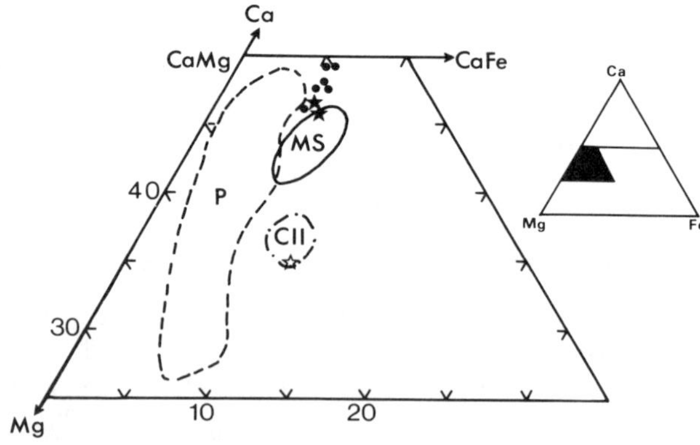

Fig. 7. Ca-Mg-Fe diagram for clinopyroxenes. Solid stars - intergrown with ilmenite 2027; open star intergrown with ilmenite, Monastery (new data). CII in clinopyroxene-ilmenite intergrowths from Monastery, Sonop and Uintjes Berg (Gurney et al., 1975). P - in peridotite and pyroxenite xenoliths from kimberlite (various sources). MS - in MARID-suite nodules from kimberlite (Dawson and Smith, 1977). Solid circles - groundmass diopsides in kimberlite (Dawson et al. 1977).

TABLE 5. Analyses of Ilmenites in BD2027.

	1	2	3	4	5	6	7	8
SiO_2	0.06	0	0.2	0.05	0.02	0.07	0	0
TiO_2	53.9	54.3	53.2	52.1	53.3	54.0	48.5	51.5
Al_2O_3	0.08	1.05	0.05	0.02	0.96	0.05	0.10	1.00
Cr_2O_3	0.96	0.56	0.67	0.97	0.60	0.36	0.89	0.29
FeOT	32.8	32.2	34.5	34.6	32.0	34.7	41.2	36.1
MnO	0.25	0.31	0.25	0.24	0.29	0.59	0.20	0.22
MgO	11.6	12.2	11.9	11.0	12.5	9.19	10.9	11.8
NiO	0.14	0.10	n.d.	n.d.	n.d.	0.06	n.d.	n.d.
Total	99.9	100.8	100.8	99.0	99.7	99.3	101.5	101.7

1. With opx, intergrowth A; 2. with opx intergrowth B; 3. with opx, intergrowth C; 4. with opx, intergrowth D; 5. large isolated ilmenite in macrospherulite zone; 6. intergrowth with cpx; 7,8. in coarse vein.

	1	2	3	4	5	6	7	8
$FeTiO_3$	52.9	49.8	50.7	52.9	48.7	62.9	45.5	47.5
$MgTiO_3$	40.0	41.3	40.7	38.3	42.6	32.5	36.9	40.0
$MnTiO_3$	0.5	0.6	0.4	0.1	-	1.0	0.4	0.4
Fe_2O_3	5.2	5.3	7.5	7.4	6.3	3.0	16.3	10.5
Al_2O_3	-	1.4	-	-	1.3	-	0.1	1.4
Cr_2O_3	0.8	-	0.6	0.9	0.5	0.3	0.8	0.2

that the experiments were not intended as a rigorous phase-relationship study; the main object was to determine the solidus of the melted rock, prior to the textural quenching experiments. The main results, summarized in Fig. 8, are (a) no garnet or clinopyroxene were detected in the charges, probably because (if present) the amounts were below the detection limits; (b) olivine is the liquidus phase at all pressures, but near the solidus reacts with liquid to produce orthopyroxene; (c) enstatite and ilmenite crystallize together over a temperature range of >200°C and enstatite replaces olivine just above the solidus; (d) pseudobrookite and rutile appeared in the charges as a function of water content and fO_2 in the experiments; (e) the enstatite field increases with increasing P (at the expense of olivine) and suggests that enstatite and ilmenite may be liquidus phases at pressures in excess of ∿32 kb. Hence, the absence of olivine in the natural rock could be ascribed to P > 30 kbr for solidification of the rock. In his analagous experiments, Wyatt (1977) also found olivine to be the liquidus phase at 20 kbrs but it reacted out at lower temperatures to give clinopyroxene and liquid.

Experiments to simulate the texture were carried out at 20 kbrs. The powders were melted by heating to above the liquidus, holding this

TABLE 6. Representative Analyses of Core and Red Rims of Micas in Coarse Areas.

	Core	Rim
SiO_2	40.1	39.5
TiO_2	5.14	6.83
Al_2O_3	11.6	7.90
Cr_2O_3	0.15	0
FeO^T	6.58	11.4
MnO	0.05	0.08
MgO	20.4	19.8
NiO	0.04	n.d.
CaO	-	0.03
Na_2O	0.26	0.34
K_2O	9.73	9.50
Total	94.1	95.4

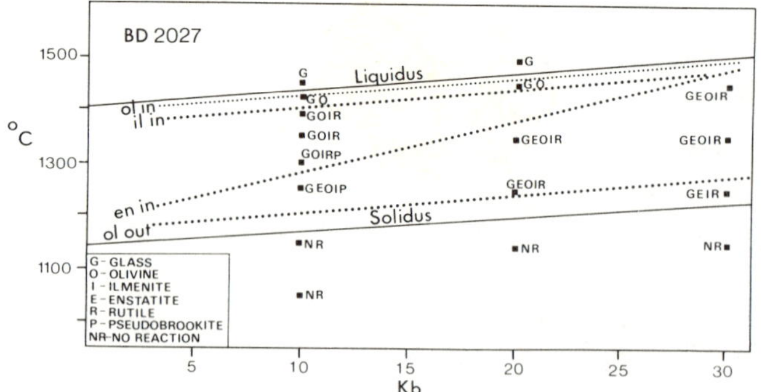

Fig. 8. Phase relations in BD2027.

temperature to allow complete melting, then cooled rapidly to the solidus temperature. The charges were held at the solidus for either 0.5 minutes or 10 minutes, then quenched to room temperature to preserve the assemblage and texture obtained at the solidus. In runs when charges were held at the solidus for 0.5 minutes, an olivine cumulate was found in the base of the charge together with acicular orthopyroxene intergrown with ilmenite, and minor glass. In runs held at on the solidus for 10 minutes, olivine was absent and the charge consisted of intergrown lamellar orthopyroxene and ilmenite, the texture of which strongly resembles the acicular orthopyroxene with intergrown ilmenite in BD 2027 (Fig. 9).

Cooling History of BD2027

Combining the petrographic and experimental evidence we suggest the following cooling history for the rock BD2027. A high-titanium silicate liquid started to precipitate high-titanium pyrope and high-Ti bronzite probably at pressure in excess of 30 kb. Green and Sobolev (1975) synthesized high-Ti pyropes between 21 and 40 kb and at 900-110°C from pyrolite (less 40% olivine) and olivine basanite and noted that these high-Ti garnets coexisted with liquid (and, in some cases clinopyroxene) at the higher pressures and temperatures, but that, for a given pressure, ilmenite will only join the assemblage at lower temperatures. Bearing in mind the bulk compositional differences between BD2027 and the rocks investigated by Green and Sobolev (op. cit.) the initial crystallization of garnet + orthopyroxene (rather than garnet + clinopyroxene) before the appearance of ilmenite is not unexpected. The liquidus temperature may not be as high as shown in Fig. 8 if more volatiles had been initially present. Following the initial crystallization of garnet and orthopyroxene, the liquid was then supercooled fairly rapidly to solidus (or subsolidus) temperatures, at which point the bulk of the liquid crystallized acicular orthopyroxene and ilmenite, in the macrospherulites and comb-layered areas; in some macrospherulites, the acicular bronzites nucleated on the earlier orthopyroxene, whereas the comb-layered units suggests a planar nucleation surface. Finally residual liquids, in which CaO, K_2O and volatiles were enhanced, formed the coarser patches and small areas of serpentine (?after late-precipitating olivine), phlogopite, ilmenite, calcite and diopside; some of the diopsides were small and discrete (Dawson et al. 1977) but larger ones were intergrown with regularly oriented magnesian ilmenite.

The observable distinct growth stages for the opx-il and cpx-il intergrowths show we are not dealing with an equilibrium assemblage, and current pyroxene thermometry and barometry are therefore inapplicable. However, the textural evidence strongly suggests that the orthopyroxene-ilmenite and clinopyroxene-ilmenite intergrowths equilibrate at different temperatures. In the past some workers (e.g. Boyd and Nixon, 1975) have suggested that opx-il and cpx-il intergrowths formed essentially in equilibrium and hence PT estimates derived from pyroxenes in individual intergrowths could, in the absence of xenoliths containing two pyroxenes, be used to provide supplementary points in the construction of paleogeothermal gradients. The three stages of growth of the pyroxene in BD2027 (two accompanied by ilmenite) indicates that these suggestions cannot be accepted unreservedly.

Conclusions

Petrographic study and experimental investigation of specimen BD2027 indicate that irregular orthopyroxene-ilmenite and regular clinopyroxene intergrowths can be produced during

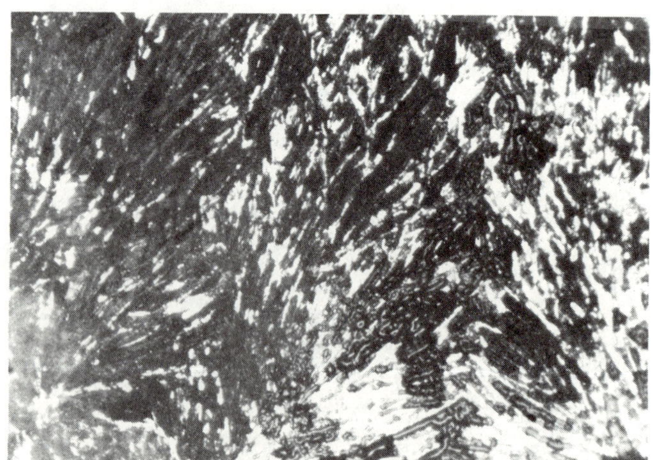

Fig. 9. Pyroxene-ilmenite intergrowth resulting from quenching of BD2027 after holding 10 minutes at the solidus. Compare with Fig. 3 for textural similarity. Reflected light; light areas are ilmenite. Scale bar = 0.5mm.

cooling of a titanium-rich liquid. High-titanium pyrope and high-Ti bronzite, often found as megacrysts in many kimberlites also occur in BD2027, and its final residual liquids precipitate phases commonly found in the matrix of kimberlites. In short, BD2027 appears to be a small-scale version of a crystal fractionation process that, on a larger scale, would have produced a set of phases (discrete, intergrown and interstitial), that appear typical of a kimberlite event. Due to the planar cooling surface for the comb-layered nucleation, we infer that the cooling of this liquid took place within a small upper-mantle dyke, and suggest that the larger, discrete high-titanium pyropes and the coarser pyroxene-ilmenite intergrowths found in other kimberlites may be derived from coarser (possibly pegmatitic) upper-mantle bodies in which the larger grain-size could be attributed to slower cooling, higher volatile concentrations, or both.

Acknowledgements. The specimen was collected and preliminary probe analyses were made by JBD during a visit to South Africa as a C.S.I.R. Senior Visiting Research Fellow in the Department of Geochemistry, University of Cape Town. Further probe-work and the high-pressure experimental work were carried out at the Grant Institute of Geology; P.J.R. wishes to especially acknowledge the assistance received from C. Ford, C. Begg and P. Hill. R. Hervig (University of Chicago) provided some orthopyroxene analyses. Finally, we wish to thank S. Fell, and J. Allan for secretarial and technical help.

References

Akella, J., and F.R. Boyd, Partitioning of Ti and Al between pyroxenes, garnets and oxides, Carnegie Inst. Wash. Year Book 71, 378-384 (1972).

Allen, J.B., and T. Deans, Ultrabasic eruptives with alnöitic-kimberlite affinities from Malaita, Solomon Islands, Min. Mag., 34, 16-34, 1965.

Boyd, F.R., Enstatite-ilmenite and diopside-ilmenite intergrowths from the Monastery Mine, Carnegie Inst. Wash. Year Book 70, 134-143, 1971.

Boyd, F.R., and P.H. Nixon, Origins of the ultramafic nodules from some kimberlites of northern Lesotho and the Monastery Mine, South Africa, Phys. Chem. Earth 9, 431-454, 1975.

Dawson, J.B., and A.M. Reid, A pyroxene-ilmenite intergrowth from the Monastery Mine, Contr. Min. Petrol., 26, 276-301, 1970.

Dawson, J.B., and W.E. Stephens, Statistical classification of garnets from kimberlite and associated xeniliths, J. Geol., 83, 589-607, 1975.

Dawson, J.B., and J.V. Smith, The MARID (mica-amphicole-rutile-ilmenite-diopside) suite of xenoliths in kimberlite, Geochim. Cosmochim. Acta, 41, 309-323, 1977.

Dawson, J.B., Smith, J.V., and R. Hervig, Late-stage diopsides in kimberlite, Neues Jahrb. Mineral. Monatsheft, 1977 (12), 529-543, 1977.

Frick, C., Intergrowths of orthopyroxene and ilmenite from the Frank Smith Mine, nr. Barkly West, S. Africa, Trans. Geol. Soc. S. Africa 76, 195-200, 1973.

Green, D.H., and N.V. Sobolev, Coexisting garnets and ilmenites synthesized from pyrolite and olivine basanite and their significance for kimberlitic assemblages, Contr. Min. Petrol. 50, 217-229, 1975.

Gurney, J.J., Fesq, H.W. and E.J.D. Kable, Clinopyroxene-ilmenite intergrowths from kimberlite: a reappraisal, in Lesotho kimberlites edited by P.H. Nixon, pp. 238-253, Lesotho Dev. Corpn, Maseru, 1973.

Ilupin, I.P., Kaminskiy, F.V. and N.V. Troneva, Pyroxene-ilmenite graphic inclusions from the Mir pipe (Yakutic) and their origin, Int. Geol. Rev., 16, 1298-1305, 1973.

Lofgren, G.E., and C.H. Donaldson, Curved branching crystals and differentiation in comb-layered rocks. Centr. Min. Petr., 50, 1-11, 1975.

Ringwood, A.E., and J.F. Lovering, Significance of pyroxene-ilmenite xeniliths amongst kimberlite xenoliths, Earth Planet. Sci. Letters 7 371-375, 1970.

Smith, C.B., McCallum, M.E., and D.H. Eggler, Clinopyroxene-ilmenite intergrowths from the Iron Mountain kimberlite district, Wyoming, Carnegie Inst. Wash. Year Book 75, 542-544, 1976.

Stephens, W.E., J.B. Dawson, Statistical comparison between pyroxenes in kimberlites and associated xenoliths, J. Geol., 85, 433-449, 1977.

Williams, A.F.: The genesis of the diamond, E. Benn, London, 1932.

Wyatt, B.A., The melting and crystallization behaviour of a natural clinopyroxene-ilmenite intergrowth, Contr. Min. Petr., 61, 1-9, 1977.

SYNTHESIS OF NICKELOAN DJERFISHERITES AND THE ORIGIN OF POTASSIC SULPHIDES AT THE FRANK SMITH MINE

D. B. Clarke

Department of Geology, Dalhousie University, Halifax, N. S., Canada B3H 3J5

Abstract. Many alkali metal-transition element chalcogenides have been synthesized, but only three potassic sulphides are known to occur as minerals. One of these, a nickeliferous djerfisherite, occurs in a clinopyroxene-ilmenite nodule from the Frank Smith mine. To investigate its conditions of formation, a range of djerfisherites with 100 Fe/(Fe+Ni) from 100-41 has been synthesized at temperatures of 649°, 452° and 352°C at one-atmosphere pressure. The experimental work shows the ability of nickeliferous djerfisherites to form in the subsolidus by a reaction between pre-existing sulphides and a potassium-rich vapour phase. Textural evidence from the Frank Smith occurrence points to the formation of nickeliferous djerfisherite by this mechanism at temperatures less than 610°C where pentlandite is stable. Another, chlorine-free potassium-iron-nickel-copper sulphide has been discovered in the Frank Smith nodule, and has probably formed by the same process as the nickeliferous djerfisherite. It is proposed that the term 'djerfisherite' be redefined as the nickel- and copper-free iron end-member with a formula approximating $K_6Fe_{26-x}S_{26}Cl_1$, and that essentially Fe-Ni solid solutions be called nickeloan djerfisherite (Ni < Fe), and Fe-Cu solid solutions be termed cuprian djerfisherite (Cu < Fe).

Introduction

To date, only three naturally-occurring, potassium-transition element sulphides have been described, namely djerfisherite (Fuchs, 1966), rasvumite (Sokolova et al., 1970) and bartonite (G. K. Czamanske, personal communication), although Dobrovol'skaya et al. (1975a) have referred to two other potassium sulphides "of complex composition". However, in spite of their scarcity in nature, there exists a large number of synthetic alkali metal-transition element chalcogenides such as $KFeS_2$ (Schneider, 1869), KCu_3S_2 (Burschka and Bronger, 1977), $Na_2Cr_2S_4$ (Groger, 1880), $NaSbS_2$, Na_3SbS_3, $Na_6Sb_4S_9$ (Berul' et al., 1971), $K_2Ag_4S_3$, $Rb_2Ag_4S_3$ (Bronger and Burschka, 1976), $CsAg_3S_2$, $RbAg_4S_3$ (Eyck et al., 1973), Rb_2PdS_4 (Huster and Bronger, 1974), K_6HgS_4 (Sommer et al., 1976), (Li,Na,K,Rb,Cs) $Sb(S,Se)_2$ (Bazakutsa et al., 1973) and many others. Most of these compounds have been synthesized in inert atmospheres and at high temperatures (700-900°C) from fusion mixtures of alkali carbonates, powdered transition-element metals and sulphur. Goettel (1976) has reviewed the evidence for the chalcophilic behaviour of potassium and has discussed the implications this has for the existence of potassium in the Earth's core.

The occurrence of potassium-iron-nickel sulphide in a clinopyroxene-ilmenite nodule from the Frank Smith mine has already been described (Clarke et al., 1977). In that paper two varieties of potassic sulphide were described, a homogeneous tan-coloured variety and a spongy, brown variety. Both were thought to be the same mineral (FSKS, or Frank Smith potassic sulphide) but further chemical work has shown these to be two different phases. The spongy, brown mineral appears to be a nickeliferous variety of djerfisherite and, in fact, unknown to Clarke et al. (1977), had already been described in diatremes in Yakutia by several Russian workers (Tsepin et al., 1972; Dobrovol'skaya et al., 1975a, b). The optically homogeneous, tan variety is chemically different from the djerfisherite in apparently having variable potassium contents and no chlorine (Fig. 1, Table 1). X-ray data are not yet available for this unidentified phase.

This paper describes the synthesis of a series of djerfisherites of variable Fe/(Fe+Ni) ratio. Those synthesis experiments are then discussed in relation to the origin of K-sulphides which occur in the Frank Smith kimberlite.

Experimental Procedures

Starting materials consisted of: potassium thioferrite ($KFeS_2$) prepared according to standard techniques (Brauer, 1965; see also Boon and MacGillavry, 1942); Fe sponge and powdered Ni metal, both reduced in a stream of dry hydrogen before use; synthetic FeS; synthetic NiS; and KCl. Bulk compositions, lying in the 'plane'

Fig. 1A. Polysulphide bleb 190/186 showing a large core of pyrrhotite (Po) with exsolved flames of pentlandite (Pn). Some Pn flames cross the boundary between Po and the two potassic sulphides, djerfisherite (Df) and the unknown sulphide (U). The box defines the region of X-ray scanning photographs shown in Fig. 1B.

K-(Fe,Ni)S-Cl and roughly equal to the natural compositions plus excess potassium and chlorine, were prepared from the starting materials (Fig. 2). Bulk composition F* had 100 Fe/(Fe+Ni) (atomic) = 100; F had 100 Fe/(Fe+Ni) = 95; and $\overline{F}$ had 100 Fe/(Fe+Ni) = 50. Approximately 50 mg of the starting materials were loaded and welded in Au capsules in air, and then three capsules were sealed in an evacuated silica glass tube which was placed in a horizontal resistance furnace at temperature. The temperature control was ±2°C, and all runs were done at one-atmosphere pressure. At the conclusion of each run the glass tube was removed from the furnace with tongs and dropped into a large container of cold water; measured quenching times were less than five seconds.

Experimental Results

Preliminary identification of the run products (Table 2) was made by X-ray powder diffraction techniques. Since the yield of the runs was never pure djerfisherite, only the most intense djerfisherite peaks could be reliably measured. Run products were also analyzed by electron microprobe and some representative data appear in Table 3. The remainder are plotted on Figure 3, which also includes analyzed djerfisherites from earlier runs (nos. 1-21, not reported here) of shorter duration which were not quenched, but nevertheless generated a wide range of djerfisherite compositions.

The important features of the synthesis runs are:

1. Djerfisherites have formed at one-atmosphere pressure over a wide range of temperatures (649-352°C), i.e. at temperatures certainly below (and possibly above) the solidus in this sulphide system.

2. The reaction in F bulk compositions approximates to:

$$10KFeS_2 + 10Ni + 5FeS + 5NiS + 1.6KCl \rightarrow \underset{Mss}{2(Fe,Ni)S} + \underset{Hz}{(Ni,Fe)_3S_2} + 5K + 0.6KCl$$

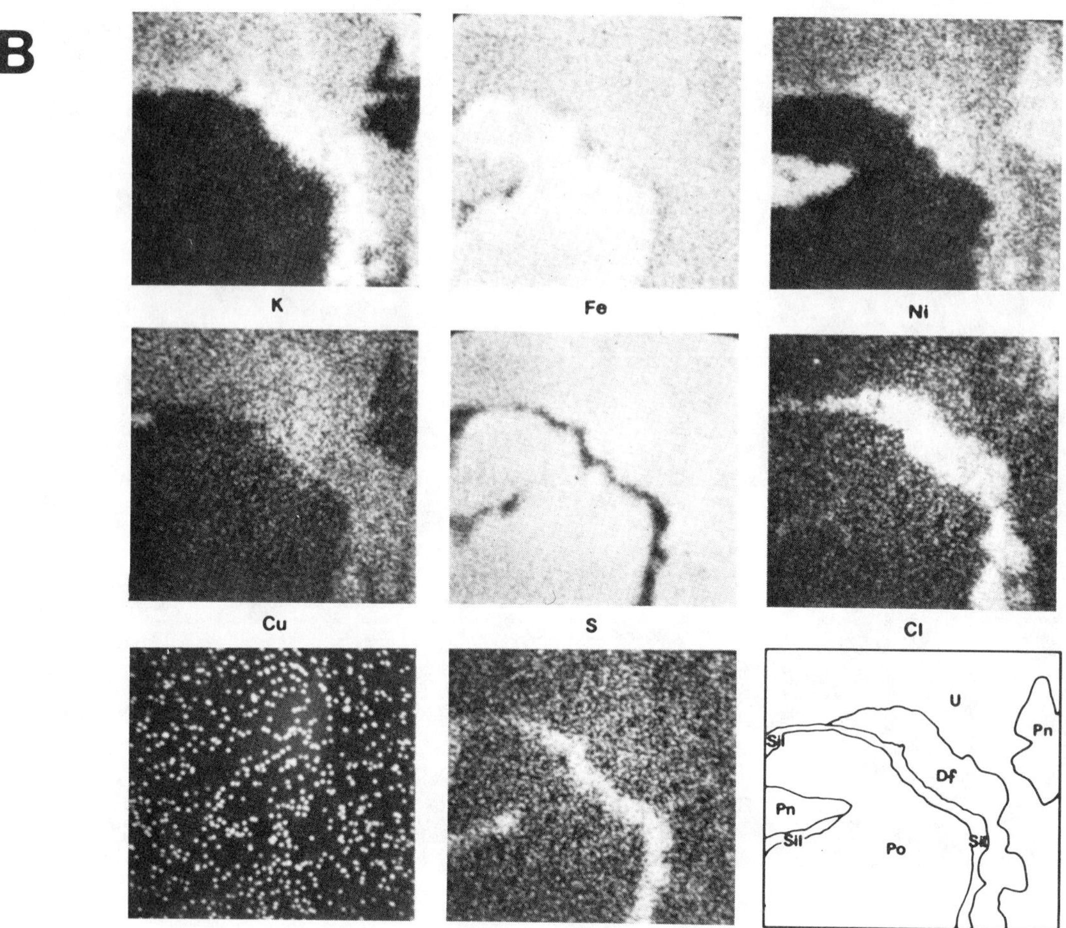

Fig. 1B. X-ray scanning photographs of the area marked in Fig. 1A. (The exposure times for Cl and F were six and fifteen times longer, respectively, than for the other photographs, and therefore show greater background effects--the intensely white area on the Cl photograph represents only 1.40 wt.% Cl.) Note that a small fracture filled with silicate material separates the Po core from the rim of potassic sulphides.

$$+\underset{Df}{K_6(Fe,Ni)_{26-x}S_{26}Cl_1} \qquad (1)$$

where some of the potassium is in solid solution in the Mss (KMS of Table 2; measured K contents are 0.1 - 5.0 wt.% in Mss, but analytical totals are low for this phase (Table 3) and are not considered very reliable at present) and the rest may have exited as discrete K blebs (Ganguly and Kennedy, 1977) which were lost during sample preparation.

3. The absence of djerfisherite at 296° C may be a result of the failure of the $KFeS_2$ and KCl to decompose, and may not therefore be a limit to the formation of djerfisherite under natural conditions.

4. X-ray diffraction data show that the djerfisherite compositions produced appear to belong to an isomorphous solid solution series (Table 4).

Origin of the Frank Smith Sulphides

Textural Evidence

As discussed by Clarke et al. (1977), the ellipsoidal polysulphide blebs appear to have been an original feature of the host clinopyroxene-ilmenite intergrowth. They exhibit a close spatial relationship to the ilmenite lamellae and show curved boundaries with the oxide grains. In their original solid form it is likely that they consisted of a homogeneous Fe-Ni monosulphide solid solution. Only at a temperature of less than 610°C could the pentlandite flames have exsolved, and this implies that the exsolution took place at a late stage in the intrusion and cooling of the diatreme. Misra and Fleet (1973) and others have shown that pentlandite can form from Mss in relatively short times, i.e. ca. 20 days. The extremely low level of Ni in the Po coexisting with the Pn, suggests that exsolution of Pn

TABLE 1. Analyses (wt %) of Sulphides from Frank Smith Nodule.

	1	2	3	4	5
K	0.00	0.00	9.06	8.03	4.83
Na	0.00	0.00	0.07	0.00	0.00
Fe	62.89	37.36	39.64	41.05	42.21
Ni	0.10	28.59	11.54	14.51	11.09
Cu	0.09	0.14	2.77	1.82	3.45
Mg	nd	0.00	0.00	0.00	0.00
S	36.13	33.15	32.39	34.79	34.22
Cl	0.00	0.00	1.35	0.00	0.00
	99.21	99.24	96.82	100.20	95.80

1. Po $(Fe_{>0.99}, Ni_{<0.01}, Cu_{<0.01})_{1.00}\ S_{1.00}$
2. Pn $(Fe_{0.58}, Ni_{0.42}, Cu_{<0.01})_{8.96}\ S_{8.00}$
3. Df $(K_{0.99}, Na_{0.01})_{6.04}\ (Fe_{0.74}, Ni_{0.21}, Cu_{0.05})_{24.46}\ S_{26.02}\ Cl_{0.98}$
4. Unknown $K_{5.11}\ (Fe_{0.73}, Ni_{0.24}, Cu_{0.03})_{24.22}\ S_{27.00}$
5. Unknown (avg. of 8) $K_{3.13}\ (Fe_{0.76}, Ni_{0.19}, Cu_{0.05})_{25.27}\ S_{27.00}$

Analyses 1-4 by wave-length spectrometry (15kV, 0.05μA, 80-second counting time, standards for K, Na, Mg (Kakanui kaersutite), Fe (synthetic magnetite), Ni (nickel metal), S (synthetic chalcopyrite), Cl (sodium chloride).

Analyses represented in 5 by energy dispersion (15kV, 15nA, 100-second counting time, same standards as above).

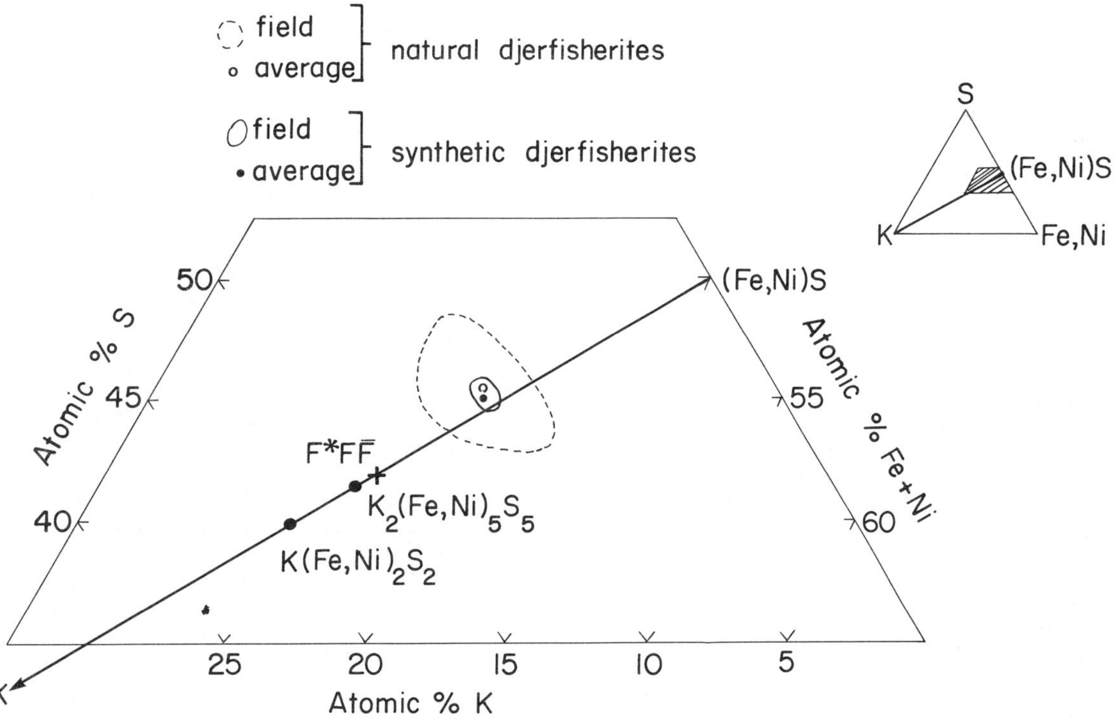

Fig. 2. Projection from chlorine into the 'plane' K-FeNi-S showing the location of the bulk compositions F*, F and F̄, the range of all known, naturally-occurring djerfisherites, the range of synthetic djerfisherites reported here, and some hypothetical compounds $K(Fe,Ni)_2S_2$ ($\equiv KFeS_2 + Ni$) and $K_2\ (Fe,Ni)_5\ S_5$ ($\equiv 2KFeS_2 + 2Ni + FeS$).

TABLE 2. Summary of Products Identified in Experimental Runs

Temp (°C)	Time (h)	Run No.	Products
851	20	23F*	KMS(?), Po
		23F	MS
		$23\bar{F}$	MS, KMS(?), Hz
649	20	24F*	Df, Po, KCl, KMS(?)
		24F	Df(?), Po, KCl
		$24\bar{F}$	Df, Hz, KCl
452	70	22F*	Df, KCl
		22F	Df, KCl
		$22\bar{F}$	Df, KCl
352	192	26F*	Df, KCl, FeS
		26F	Df, KCl, FeS, Ni
		$26\bar{F}$	Df, KCl, Hz, $KFeS_2$(?)
296	192	25F*	no reaction
		$25\bar{F}$ (leak?)	Hz, KCl

Df - djerfisherite
Hz - heazlewoodite
Po - pyrrhotite (usually nickeliferous)
MS - monosulphide solid solution
KMS - potassic monosulphide solid solution
KCl - potassium chloride ⎫
Ni - nickel ⎬ starting materials
$KFeS_2$ - potassium thioferrite ⎪
FeS - troilite ⎭

continued to very low temperatures, probably less than 300°C (Misra and Fleet, 1973). The spatial relationship of the potassic sulphides as rims replacing the Po-Pn intergrowth suggests their formation after the Po-Pn exsolution texture had started developing.

Experimental Evidence

The experimental work bears mainly on the formation of nickel-bearing djerfisherites, and not necessarily on the origin of the unknown K-sulphide. The synthesis experiments show that djerfisherite can form over a wide range of temperature, i.e. 649-352°C, but most significantly that it can form at one-atmosphere pressure (i.e. in a high level diatreme) and at temperatures below the sulphide solidus (i.e. metasomatically through the reaction of normal sulphides and a potassium-bearing vapour phase).

Origin of the Polysulphide Blebs

The following steps are believed to have taken place in the origin and development of the polysulphide blebs in the Frank Smith nodule.

1. Immiscible sulphide droplets were enclosed in an intergrowth of ilmenite and clinopyroxene at high temperatures and at mantle depths.
2. At lower temperatures, possibly in the mantle, but perhaps not until eruption in the diatreme, the sulphide droplets crystallized to a homogeneous Fe-Ni monosulphide solid solution.
3. When the temperature of the emplaced diatreme reached some value below 610°C, exsolution of the Pn flames began.
4. Probably during this process of exsolution the entire nodule was subject to alteration by metasomatic fluids, the presence of which is indicated by the incipient breakdown of the clinopyroxene-ilmenite intergrowth to biotite and uralite (?). These fluids were undoubtedly rich in H_2O and CO_2 and, in addition, they must have contained significant amounts of K, Cl and possibly Cu. Naughton et al. (1974, 1975) have measured significant concentrations of K, HCl, and Cu (as CuCl gas) in the fumes emanating from Hawaiian volcanoes, and it is reasonable to expect that such vapour species might be very highly concentrated in the post-magmatic gaseous

TABLE 3. Representative Analyses (wt%) of Potassic Sulphides from the Experimental Runs.

	1	2	3	4	5
K	9.08	8.98	8.70	9.22	1.97
Fe	55.60	51.45	29.66	55.87	35.05
Ni	0.00	3.83	27.58	0.00	22.12
S	32.86	32.38	31.65	32.94	35.33
Cl	1.40	1.39	1.43	1.39	0.06
Total	98.94	98.03	99.02	99.42	94.53

1. Df, Run 24F* 649°C $K_{5.89}Fe_{25.25}\ S_{26.00}\ Cl_{1.00}$
2. Df, Run 24F 649°C $K_{5.91}\ (Fe_{0.93}Ni_{0.07})_{25.39}\ S_{25.99}\ Cl_{1.01}$
3. Df, Run 24$\overline{F}$ 649°C $K_{5.85}\ (Fe_{0.53}Ni_{0.47})_{26.30}\ S_{25.94}\ Cl_{1.06}$
4. Df, Run 22F* 452°C $K_{5.97}\ Fe_{25.32}\ S_{26.01}\ Cl_{0.99}$
5. KMS, Run 23$\overline{F}$ 851°C $K_{1.23}\ (Fe_{0.62}Ni_{0.38})_{24.57}\ S_{26.96}\ Cl_{0.04}$

Analytical techniques as per Table 1, analyses 1-4.

phase associated with a kimberlite. (The Cu may have been an original constituent of the Mss which, during the processes of exsolution and metasomatism, has been partitioned almost exclusively into the potassic sulphides. Or, because it exists virtually entirely in the metasomatic phases, the copper could be regarded as a constituent of the late-stage metasomatic fluid. Interestingly, if CuCl is responsible for providing both the Cu and Cl for the djerfisherite, then the weight ratio of Cu/Cl should be ∿ 1.79, and in fact this is close to analyzed values in djerfisherite, but this argument does not apply to the Cu-bearing, Cl-free unknown potassic sulphide).

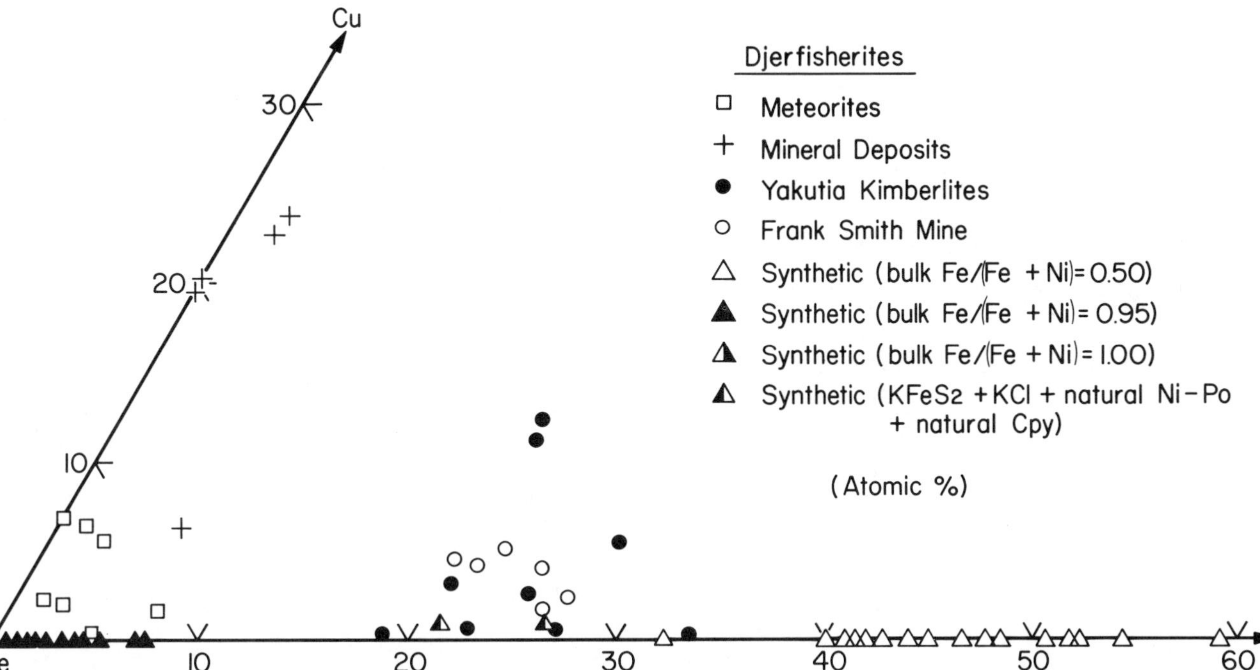

Fig. 3. A compilation of all known djerfisherite compositions, both natural and synthetic. As suggested by X-ray diffraction data in Table 4, there is probably an isomorphous solid solution series of djerfisherites from 100 Fe/(Fe+Ni) = 100 to at least 100 Fe/(Fe+Ni) = 41. The apparent discontinuities in the series appear to be a function of the limited number of bulk compositions in the synthesis runs.

TABLE 4. A Comparison of the Chemistry and X-Ray Diffraction Data for Two Djerfisherites from the Experimental Run at 450°C.

	Df Run 22F	Df Run $22\bar{F}$
K	8.91	9.18
Fe	53.05	30.41
Ni	2.33	26.51
S	32.52	33.02
Cl	1.20	1.38
Total	98.01	100.50
Miller	dA° (I)	dÅ (I)
(100)	10.297 (35)	10.285 (60)
(111)	5.976 (35)	5.909 (30)
(310)	3.264 (35)	3.239 (40)
(311)	3.121 (55)	3.089 (65)
(222)	2.984 (70)	2.957 (80)
(331)	2.373 (65)	2.380 (100)
(440)	1.830 (100)	1.833 (95)

Run 22F $K_{5.87}(Fe_{0.96}Ni_{0.04})_{25.49} S_{26.13} Cl_{0.87}$

Run $22\bar{F}$ $K_{5.93}(Fe_{0.55}Ni_{0.45})_{25.16} S_{26.02} Cl_{0.98}$

5. The reactions to produce potassic sulphides began before exsolution of pentlandite had strongly depleted the original Mss (?) in nickel, and terminated before reaching completion, as evidenced by the cores of pyrrhotite in some large grains (Fig. 1A).

On Djerfisherite

The name 'djerfisherite' was first given to a K-Fe-Cu sulphide occurring in meteorites (Fuchs, 1966). Since that time, the name has been applied to more copper-rich varieties occurring in copper mineral deposits (Genkin et al., 1970) and recently to Ni-rich, Cu-poor varieties occurring in kimberlites (Tsepin et al., 1972).

The average of 24 natural compositions (excluding chlorine, since it is frequently not reported) is $K_{5.96}(Fe,Ni,Cu)_{25.29} S_{26.00}$, and for 25 synthetic compositions from these experiments is $K_{5.99} (Fe,Ni)_{25.42} S_{26.00} Cl_{1.00}$. These structural formulae tend to support the site-occupancy studies of Dmitrieva and Ilyukhin (1975), and Dmitrieva (1976), in that K/S/Cl = 6/26/1. However, it is less clear from these analyses how many transition metal ions are present per formula unit, i.e. a fixed number of 25 or 26, or possibly a variable number 26-x, where x depends on the ratio of Fe/Ni/Cu and thus possibly on the number of "d" shell electrons required to stabilize the structure.

In view of the experiments reported here which show that neither Cu nor Ni is essential to the formation of djerfisherite it would seem reasonable to redefine 'djerfisherite' as the pure iron end-member in the system K-Fe-Ni-Cu-S-Cl (see analyses of Runs 22 F* and 24 F* in Table 3) with a formula approximating $K_6Fe_{26-x}S_{26}Cl_1$. The essentially binary solid solutions should be called cuprian and nickeloan djerfisherites until such time as the discovery of sufficiently Cu-rich (Cu > Fe) or nickel-rich (Ni > Fe) samples in nature warrants the definition of new end-member terms.

The Unknown Sulphide

The unknown sulphide has not been studied in much detail. In simple terms, its chemistry is like a nickeliferous pyrrhotite with variable amounts of potassium. Within the limitation of the generally low analytical totals, a general formula for this phase might be $K_{6-y}(Fe,Ni,Cu)_{26-x}S_{27}$. Written in this way with twenty-seven anions, it suggests a relationship to djerfisherite in that the transition element/sulphur ratio is 0.90-0.94 (Df = 0.98), and indeed this phase may represent a transition from a 'normal' monosulphide solid solution to djerfisherite. In this respect, it may bear the same relationship to djerfisherite as do the chlorine-free and chlorine-bearing (1.36 wt.%) varieties of bartonite.

Various plots of the chemical relationships of this phase to the other sulphides in the Frank Smith nodule are shown in Fig. 4. The variation in the concentration of transition elements (Fig. 4D) suggests that a substitution of the type 2Ni $\rightleftarrows$ Fe + Cu may be taking place in order to maintain a constant number of 'd' shell electrons in the structure of the mineral, as Dmitrieva (1976) has suggested as the reason for non-stoichiometry in djerfisherite.

Finally, the unknown phase may have an analogue in the form of KMS (Tables 2, 3) in the experimental runs, but investigation of this possible correlation remains to be done.

Conclusions

1. A series of djerfisherite compositions with 100 Fe/(Fe+Ni) from 100 to 41 has been synthesized. Neither copper nor nickel is essential to the formation of djerfisherite and thus djerfisherite should be redefined as the pure iron end-member $K_6Fe_{26-x}S_{26}Cl_1$.
2. Experimental and textural evidence show that the Frank Smith potassium-iron-nickel-copper sulphides could have formed from pre-existing sulphides in the subsolidus by interaction with a metasomatic fluid rich in K, Cl and possibly Cu.
3. Another potassium-iron-nickel-copper sulphide, without Cl, and with a chemical composition intermediate between monosulphide solid solution and djerfisherite, has been found in the Frank Smith assemblage. It may represent a transitional stage in the formation of djerfisherite from more normal sulphides.

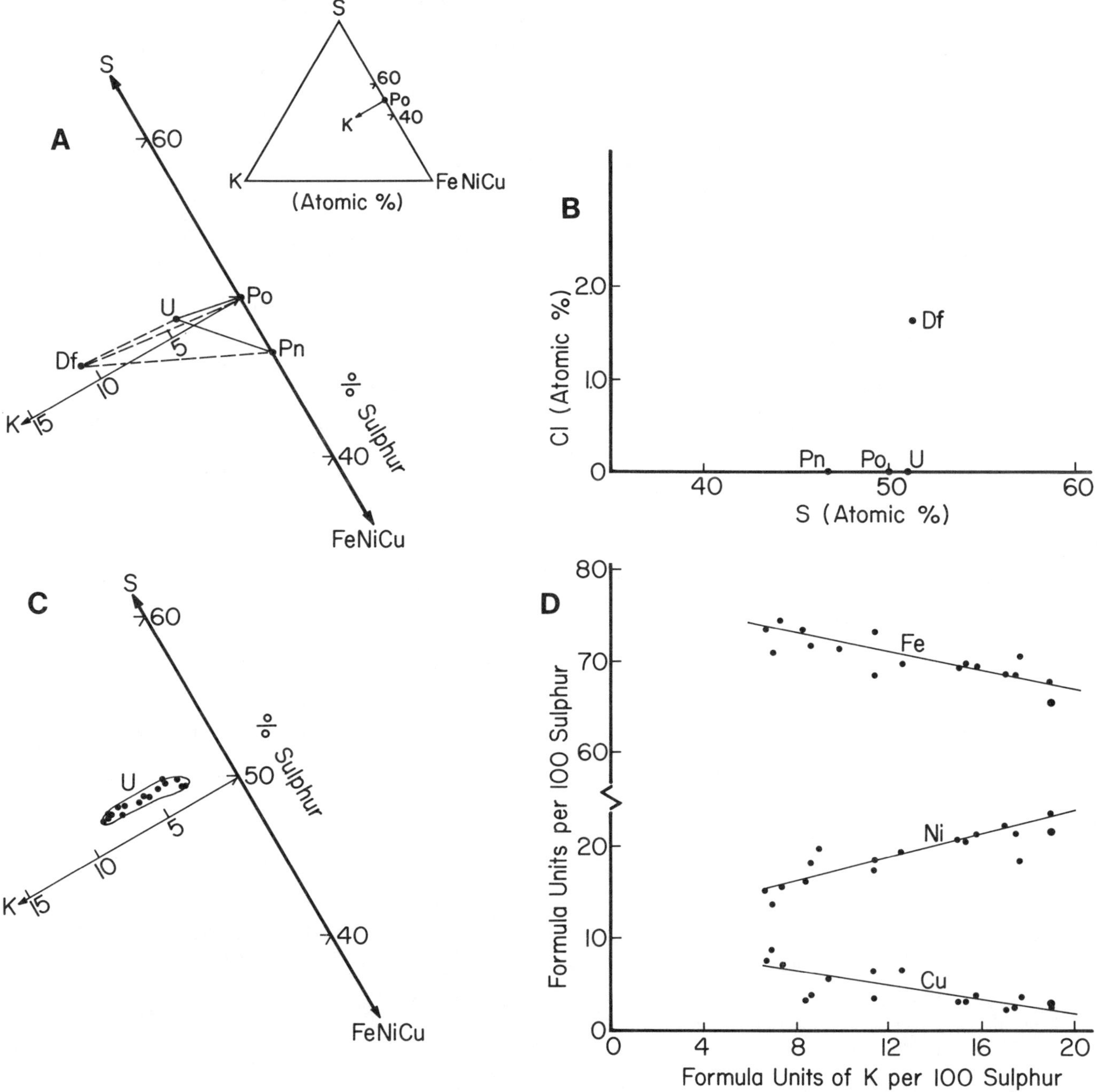

Fig. 4A. Coexisting sulphides in the polysulphide bleb 207/144 (Table 1) projected from Cl. Po, Pn and U lie in the plane of the diagram whereas Df is directed towards the Cl apex.
Fig. 4B. Projection from K onto the plane (Fe,Ni,Cu)-S-Cl.
Fig. 4C. A compilation of analyses of the unknown sulphide U. The analyses display a trend of K values from 2.76-8.03 wt.%, but not exceeding the K value of ∿ 9.0 wt.% found in djerfisherite.
Fig. 4D. Chemical variation in the analyses of the unknown sulphide. Potassium content appears to be positively correlated with nickel and negatively correlated with copper and iron.

Acknowledgements. For permission to use their laboratories for the experimental work, I wish to thank Dr. M. J. O'Hara (Edinburgh), Dr. S. Whiteway (NRC, Atlantic Regional Laboratory, Halifax) and Dr. J. J. Fawcett (Toronto). Mr. C. Begg provided valuable assistance in the Edinburgh laboratory. Dr. P. G. Hill and Mr. R. M. MacKay were responsible for the analytical facilities in Edinburgh and Halifax. Ms. P. Scrutton and Ms. H. Corrigan provided the much-needed translations of Russian papers. Part of this work was conducted while on sabbatical leave in Edinburgh and was supported by NRC travel and operating grants.

References

Bazakutsa, V. A., N. I. Gnidash, V. B. Lazarev, E. I. Rogacheva, A. V. Salov, L. N. Sukhorikova, M. P. Vasil'eva, and S. I. Berul', Lithium, sodium, potassium, rubidium or cesium antimony sulphide or selenide ($MSbX_2$) ternary chalcogenide compounds. Zhurnal Neorganicheskoi Khimii, 18, 3234-3239, 1973.

Berul', S. I., V. B. Lazarev, and A. B. Sivlov, The sodium sulphide-antimony sulphide system. Russian J. Inorg. Chem., 16, 1072-1073, 1971.

Boon, J. W. and C. H. MacGillavry, The crystal structure of potassium thioferrite $KFeS_2$ and sodium thiochromite $NaCrS_2$. Recueil des Travaux Chimiques de Pays-Bas, 61, 910-920, 1942.

Bronger, W. and C. Burschka, Potassium silver sulphide ($K_2Ag_4S_3$) and rubidium silver sulphide ($Rb_2Ag_4S_3$), synthesis and crystal structure, Zeits. fuer Anorg. Allge. Chemie, 425, 109-116, 1976.

Brauer, G., Handbook of Preparative Inorganic Chemistry Vol. 2, 1507, Academic Press, New York, 1965.

Burschka, C. and W. Bronger, Potassium tricopper disulphide, a new thiocuprate, Zeits. fuer Naturforschung, 32B 11-14, 1977.

Clarke, D. B., G. G. Pe, R. M. MacKay, K. R. Gill, M. J. O'Hara and J. A. Gard, A New Potassium-Iron-Nickel Sulphide from a Nodule in Kimberlite, Earth Planet. Sci. Lett. 35, 421-428, 1977.

Dmitrieva, M., Crystal chemical formulas for djerfisherites from structural positions. Azvest. Akad. Nauk SSSR, Ser. Geol., 4, 97-101, 1976.

Dmitrieva, M. T. and V. V. Ilyukhin, The crystal structure of djerfisherite, Doklady Akad. Nauk SSSR, 223, 343-346, 1975.

Dobrovol'skaya, M. G., A. I. Tsepin, L. N. Vyalsov, I. P. Lupin, I. V. Muravyeva, G. V. Basova and G. P. Belyayevskaya, Iron, nickel and copper isomorphism in djerfisherite in Isomorphism in Minerals, edited by F. V. Chukrov, F. V. Borutzky and N. N. Mozgova, pp. 162-169, Nauka, Moscow, 1975a.

Dobrovol'skaya, M. G., A. I. Tsepin, I. V. Ilupin, and A. I. Ponomarenko, Djerfisherite from the kimberlites of Yakutia. Academy of Science of USSR, "Nauka", Leningrad, 1975b.

Eyck, J., C. Burschka and W. Bronger, New ternary silver sulphides, Naturwissenschaften, 60, 518, 1973.

Fuchs, L. H., Djerfisherite, alkali copper-iron sulphide: a new mineral from enstatite chondrites, Science, 153, 166-167, 1966.

Ganguly, J. and G. C. Kennedy, Solubility of K in Fe-S liquid, silicate-K-$(Fe\text{-}S)^{liq}$ equilibria, and their planetary implications, Earth Planet. Sci. Lett., 35, 411-420, 1977.

Genkin, A. D., N. V. Troneva and Zhuravlev, The first occurrence in ores of the sulphide of potassium, iron and copper-djerfisherite. Geochem. Int., 7, 693-701, 1970.

Goettel, K. A., Models for the origin and composition of the Earth, and the hypothesis of potassium in the Earth's core, Geophys. Surveys, 2, 369-397, 1976.

Groger, M., Beitrag Zur Kenntnis der Schwefelverbindunger des Chroms. Monats. für Chemie., 1, 242-249, 1880.

Huster, J. and W. Bronger, The structure of potassium palladium sulphide ($K_2Pd_3S_4$) and rubidium palladium sulphide ($Rb_2Pd_3S_4$). J. Solid State Chem., 11, 254-260, 1974.

Misra, K. C. and M. E. Fleet, The chemical compositions of synthetic and natural pentlandite assemblages, Econ. Geol., 68, 518-539, 1973.

Naughton, J. J., V. A. Lewis, D. Hammond and D. Nishimoto, The chemistry of sublimates collected directly from lava fountains at Kilauea volcano, Hawaii, Geochim. Cosmochim. Acta, 38, 1679-1690, 1974.

Naughton, J. J., V. Lewis and D. Thomas, Fume compositions found at various stages of activity at Kilauea volcano, Hawaii, J. Geophys. Res., 80, 2963-2966, 1975.

Schneider, T., Ann. Physik, 136, 460 p., 1869.

Sokolova, M. N., M. G. Dobrovol'skaya, N. I. Organova, M. Ye Kazakova and A. L. Dmitrik, A Potassium-Iron Sulfide, the New Mineral Rasvumite, Vses. Mineral. Obshchest, Zap., 99, 712-720, 1869.

Sommer, H., R. Hoppe and M. Jansen, 1976, The first thiomercurate (II) with an island structure: potassium thiomercurate (II) $K_6(HgS_4)$. Naturwissenschaften, 63, 194-195, 1976.

Tsepin, A. I., M. G. Dobrovol'skaya, A. I. Ponomarenko and I. P. Ilupin, The first discovery of djerfisherite in the kimberlites of Yakutia, Abs. of Works of Employees of IGEM AN USSR for 1971 (Moscow), 1972.

IV. XENOLITHS FROM THE COLORADO PLATEAU

ECLOGITE, PYROXENITE AND AMPHIBOLITE INCLUSIONS IN THE SULLIVAN BUTTES LATITE, CHINO VALLEY, YAVAPAI COUNTY, ARIZONA

Richard J. Arculus

Research School of Earth Sciences, Australian National University, Canberra, A.C.T. 2600

Douglas Smith

Department of Geological Sciences, University of Texas at Austin, Austin, Texas 78712

Abstract. Mafic and ultramafic inclusions are abundant in potassic latite about 25 my in age that crops out on the southwest structural border of the Colorado Plateau. About 60% are eclogite composed of garnet ($Py_{40}Alm_{37}Gross_{23}$) and clinopyroxene (5-30 mol.% jadeite) with minor amphibole, apatite, rutile, Fe-Ti oxides and rare altered clinozoisite. Phase layering is common and eclogites with contrasting gnt/cpx ratios occur in sharp contact. Layering in general appears to reflect a cumulus origin rather than metamorphic segregation or tectonic or intrusive juxtaposition. Thirty percent of the inclusions are dominated by pargasitic amphibole (up to 90 modal %) with minor diopside, garnet, phlogopite, apatite and oxides.

Websterites and pyroxenites form 2-4% of the inclusions, and some contain strained pyroxene porphyroclasts with exsolved garnet and pyroxene in a mosaic groundmass of orthopyroxene and garnet. Crustal schists, quartzites and granulites constitute the remainder of the inclusion population. All rock types have metamorphic fabrics, and only rare phlogopite-apatite-clinopyroxenite inclusions could represent cumulates from the host latite. Amphibole and phlogopite appear primary in many samples, but secondary growth also occurred. Late stage partial melting in some rocks led to distinctive textures and to formation of plagioclase, corundum and possibly amphibole and rare kyanite.

Distribution coefficients of Fe and Mg between gnt-cpx pairs ((FeO/MgO) gnt/(FeO/MgO)cpx) range from 3.3 to 7.0 with generally lower values for websterites than for eclogites and amphibolites. Typical orthopyroxene coexisting with garnet contains 1.4 to 2.5 wt.% Al_2O_3. These data suggest temperatures and pressures of equilibration in the range 700-1000°C, 10-20 kbar. The inclusions may be re-equilibrated samples of an eclogite-rich cumulate sequence formed during Precambrian (1720-1820 my) igneous activity. Highly fractionated rare earth element abundances of the latite suggest that garnet remained in the source material; the latites might have formed by partial melting of part of the cumulate sequence.

Introduction

The rounded dense inclusions in the Sullivan Buttes Latite of Chino Valley, Yavapai County, Arizona were first reported by Krieger (1965). We present here a summary of inclusion types, detailed descriptions of selected samples, and preliminary petrogenetic interpretations: mineralogic, trace element, and isotopic investigations are continuing.

The host rock crops out in the southwestern structural transition zone of the Colorado Plateau (Fig. 1), and it has been dated by K-Ar methods at approximately 25 my (Krieger et al., 1971). The host rock forms shallow intrusions prominent as hills in the valley, and it also occurs in pyroclastic breccias, flows, and derived gravels. The country rocks are of Precambrian and Paleozoic age. The eruption of latite preceded the effusion of an extensive series of basaltic and differentiated rocks and also the development of Basin and Range-type faulting in the period 15-5 my.

Phenocrysts of biotite (Mg/Mg + Fe = 0.65), clinopyroxene ($Ca_{44}Mg_{46}Fe_{10}$), feldspar ($An_{55}Ab_{43}Or_2$) and apatite are present in the latite (Table 1); amphibole is a prominent phenocryst in other varieties of the host rock. The inclusions (1-50 cm maximum dimensions) are distributed on the order of 1 per 0.25 sq. meters on host rock surfaces. Freshest samples are near country rock contacts, presumably due to the chilling that must have occurred there.

Estimates of rock type abudance are based on grid counts at the chief locality (1950 USGS Paulden 15' quadrangle, NE¼ section 11, T16N, R1W) and are given in the abstract. The dominant eclogites are composed of variable proportions of garnet and clinopyroxene; some samples

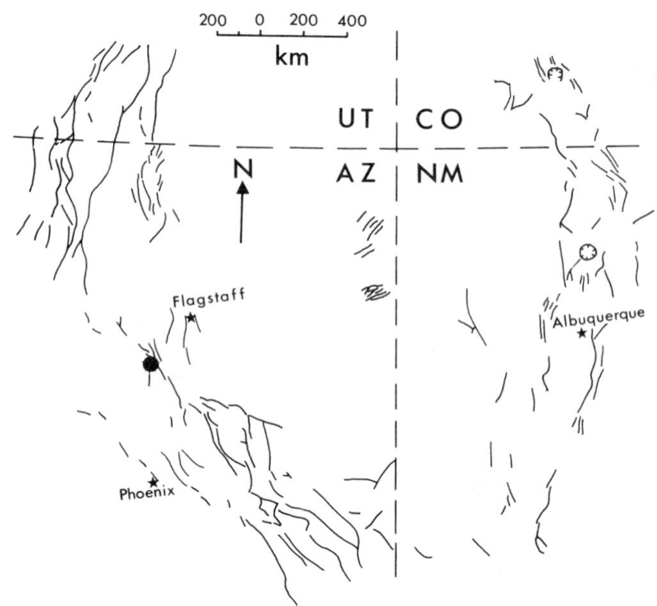

Figure 1. Location of the Chino Valley latite (Black dot) and fault systems bounding the Colorado Plateau, adapted from King (1969).

contain up to 95 modal % garnet (i.e. garnetites). Assemblages grade from eclogites with trace amphibole to those with 90 modal % amphibole and minor clinopyroxene and garnet. Amphibolites are more prominent at other localities. Websterites, orthopyroxenites and clinopyroxenites are a minor inclusion type and include rare phlogopite orthopyroxenite, biotite clinopyroxenite, and quartz-sphene clinopyroxenite.

The recognizable crustal xenoliths are schists, quartzites (local Precambrian Mazatzal) and 2 pyroxene-feldspathic granulites. No olivine-bearing inclusions have been found but some highly altered, tan-coloured inclusions may be altered lherzolites. All inclusion types except the crustal xenoliths are discussed in more detail below. Representative mineral analyses given in Table 2 to 4 were obtained by wavelength-dispersive electron microprobe techniques at UT-Austin, described by Smith and Levy (1976), and by energy-dispersive techniques at ANU, as described by Reed and Ware (1975): since energy-dispersive and wavelength-dispersive analyses have different sensitivities for trace amounts of light elements, the analyses have been differentiated by superscripts "e" and "w", respectively.

Eclogites

These inclusions are typically granular mosaic-textured rocks with crystal sizes averaging 2 mm but ranging up to 5 mm. Inequant grains in some rocks define a moderate foliation. The proportion of garnet ranges from greater than 90% to less than 10%; it commonly exceeds 50%. Phase-layering on scales of a few mm to at least a few cm is present, and assemblages of strikingly different garnet/clinopyroxene proportions are in sharp contact. Analyzed garnet-poor and garnet-rich layers in Table 2 from one rock (149A, C, Table 1) have these measured modes: cpx

TABLE 1. Host latite and selected inclusion bulk analyses.

	155	149A	149C	M7615	175
SiO_2	61.61	51.62	45.24	51.29	43.89
TiO_2	0.76	0.44	0.41	0.13	1.23
Al_2O_3	13.89	6.59	15.62	3.27	12.79
Fe_2O_3	4.14	-	-	2.58	-
FeO	0.72	6.30*	11.59*	3.83	11.71*
MnO	0.08	0.06	0.29	0.12	0.13
MgO	3.69	12.45	12.85	14.41	13.31
CaO	4.63	19.88	12.77	21.50	10.72
Na_2O	2.79	1.91	0.85	1.34	2.41
K_2O	5.49	0.02	0.04	0.16	0.50
P_2O_5	0.36	0.09	0.11	-	0.13
H_2O^+	0.73	0.30	0.36	-	1.43
H_2O^-	0.31	0.06	0.06	-	0.20
Total	99.20	99.72	100.19	99.13	98.54

Rock analyses by modified rapid method by G.K. Hoops, University of Texas, Austin, except for M7615, by a combination of XRF and wet chemistry by Elsheimer, Espos and Tilman, U.S.G.S. (courtesy of R.B. Moore).

*Total Fe as FeO

155 = host latite; 149A = cpx-rich eclogite layer; 149C = gnt-rich eclogite layer; M7615 = green clinopyroxenite; 175 = gnt-cpx-amphibolite.

TABLE 2. Microprobe analyses of primary and secondary phases in the inclusions.

	9A-CP[w]	9C-CP[w]	9A-GT[w]	9C-GT[w]	74-GT[e]	74-CPX[e]	7[w]	8[w]
SiO_2	53.7	54.4	39.9	40.3	40.00	54.17	40.0	46.2
TiO_2	0.18	0.14	0.07	0.08	0.00	0.00	0.33	0.40
Al_2O_3	4.31	3.88	22.4	22.6	22.70	2.87	16.2	12.7
FeO*	5.04	3.50	18.6	16.8	15.30	2.97	15.3	6.14
MnO	0.03	0.03	0.45	0.50	0.30	0.00	0.00	0.00
MgO	13.7	14.8	8.48	11.9	12.80	15.64	10.8	17.1
CaO	21.7	22.1	11.1	8.29	8.51	23.19	11.2	11.4
Na_2O	2.01	1.71	<0.02	<0.02	0.00	0.95	1.90	3.35
K_2O	<0.01	<0.01	<0.01	<0.01	0.00	0.00	1.93	0.29
Total	100.6	100.5	101.0	100.5	100.70	100.25	97.7	97.6

*Total Fe as FeO

9A-CP = cpx in 149A; 9C-CP = cpx in 149C; 9A-GT = gnt in 149A; 9C-GT = gnt in 149C; 74-GT = gnt in PR 74, total includes CR203 0.15 wt.%; 74-CPX = cpx in PR 74, total includes Cr203 0.21 wt.%; 7 = amph in eclogite PR 193 near host-latite contact; 8 = primary amph in eclogite PR 149.

(90%)-gar(9%)-rutile, amphibole(tr) compared to cpx(30%)-gar(52%)-amph(18%). The sharp contact is transected by a faint foliation. In other rocks, contacts are diffuse and there is a complete gradation from well-defined layers to irregular lenses. In a few samples, clinopyroxene-rich veins typically 1-2 mm wide, cut garnet-rich layers. A few thin layers are almost solid garnet, and inclusions 5 cm in diameter with up to 95% modal garnet may be disrupted fragments of thicker garnet-rich lenses. Individually homogeneous garnet and clinopyroxene crystals in adjacent layers are not identical in composition (Fig. 2, Table 2), though differences are small in comparison to the total compositional ranges in the suite.

Garnet compositions range from $Pyr_{32}Alm_{40}Gr_{28}$ to $Pyr_{47}Alm_{33}Gr_{20}$; all plot in field B (Fig. 3)

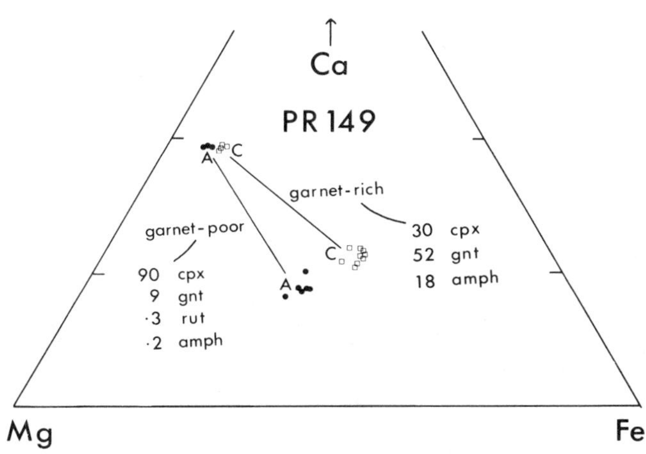

Figure 2. Ca:Fe:Mg ratios in pyroxenes and garnets from a layered eclogite. The garnet-poor and garnet-rich layers are in sharp contact. The dispersion of points for each layer considerably exceeds analytical reproducibility: the clusters include core and rim analyses from a number of crystals. Bulk layer and mineral analyses are in Table 2. The compositional differences might be "primary" or a result of post-layering re-equilibration.

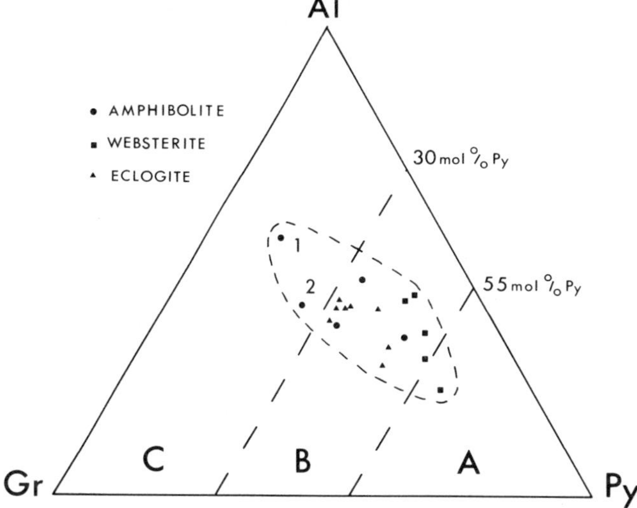

Figure 3. Representative garnet analyses. Each point represents a different rock. Fields A, B, and C were defined for eclogite garnets by Coleman et al. (1965) for eclogites from inclusions in kimberlites, basalts, and layers in ultramafic rocks (A), from bands or lenses within migmatitic gneissic terrains (B), and from bands or lenses within alpine-type metamorphic rocks (C). Points 1 and 2 are amphibolites PR 63 and PR 105, respectively.

TABLE 3. Microprobe analyses of garnet-biotite-amphibolite PR 63.

	1	2	3	4	5	6	7
SiO_2	40.30[e]	40.35[e]	37.06[e]	36.67[e]	0.00[e]	45.13[e]	50.68[e]
TiO_2	0.67	0.85	2.17	0.00	0.55	0.00	0.00
Al_2O_3	15.12	15.05	15.40	22.17	48.39	34.75	5.28
Cr_2O_3	0.00	0.15	0.00	0.00	0.11	0.00	0.00
FeO*	15.84	14.99	17.25	24.20	43.96	0.29	26.07
MnO	0.14	0.00	0.00	2.43	0.46	0.00	0.24
MgO	9.87	11.37	13.63	3.32	5.46	0.00	20.32
CaO	10.94	10.85	0.13	10.75	0.00	17.93	0.66
Na_2O	2.76	2.11	0.36	0.39	0.00	1.19	0.25
K_2O	0.84	1.11	8.66	0.00	0.00	0.12	0.00
Total	96.48	96.82	94.66	100.99	98.93	99.41	101.74

*Total Fe as FeO
1 = amph. core; 2 = interstitial amph; 3 = biotite core; 4 = garnet core;
5 = hercynite in amph; 6 = feldspar in gnt; 7 = opx in mica.

of Coleman et al. (1965), defined for garnets of eclogites which occur in bands or lenses within migmatite gneiss terrains. The garnets show little zoning, and they vary little in composition within a given layer. Typical analyses with the wavelength dispersive probe show less than 0.11% TiO_2 and less than 0.02% Na_2O, in contrast to the larger amounts typical of garnets from diamondiferous eclogites (Reid et al., 1976). Clinopyroxenes contain from 5 to 30 mole percent jadeite, a range similar to that of "B-type" eclogites. The presence of rare, partly altered clinozoisite is also consistent with "B-type" classification.

Pargasitic amphibole is present in most but not all eclogites, and its abudance equals or exceeds that of garnet and clinopyroxene in rocks which grade into the amphibolites considered below. In some eclogites amphibole grains appear to be primary, as the grains are anhedral and share a common foliation with garnet and pyroxene of similar grain size. In other eclogites, however, pargasite occurs as flame-like inclusions within clinopyroxene and as small crystals in fracture-zones which transect the foliation defined by garnet and clinopyroxene. In many such rocks amphibole seems to partly replace clinopyroxene. In one specimen (PR193) with a selvage of host latite, amphibole and biotite are developed at garnet-pyroxene grain boundaries near the contact with latite, and amphibole without biotite occurs along fracture zones within the nodule. Interior parts of the inclusion contain no hydrous phases. Primary amphibole may be less potassic than later amphibole, but evidence is incomplete (Table 2).

Rutile and apatite are common accessory phases, comprising as much as two percent of some rocks; magnesian ilmenite is less common. The rutile typically contains well-developed lamellae and intergrowths of magnesian pseudobrookite$_{ss}$.

Textural evidence of late partial melting is present in a few samples. Some nodules, particularly those with abundant amphibole, contain pockets (50µm) with interstitial brown glass and quench crystals of plagioclase, amphibole, and clinopyroxene. Other samples contain rare pockets of plagioclase associated with resorbed garnet; textures suggest that this plagioclase is also a late melting product. One altered, amphibole-free eclogite contains abundant kyanite associated with altered brown glass, zoned plagioclase, skeletal pinacoidal crystals of corundum, relict garnet, and mostly-altered clinopyroxene: clinopyroxene alteration is like that described by Switzer and Melson (1969) in partly-melted kyanite eclogites from Roberts Victor mine. The presence of euhedral, unaltered kyanite crystals in sharp contact and apparent equilibrium with the brown glass is evidence for melting at depth.

Late-stage oxidation effects are evident in most samples. Garnets commonly have opaque rims, and in some samples only opaque pseudomorphs of garnet remain: the opaque areas are fine-grained symplectite intergrowths of oxides and silicates, with bulk compositions very close to original garnet, differing only in increased Na_2O (0.05-0.85 wt%) and in slightly low analysis sums suggestive of partial hydration. This garnet oxidation is most prominent in inclusions from the interiors of latite intrusions, and it likely occurred after host rock emplacement. The magnesian pseudobrookite associated with rutile may have formed at the same time.

Original cumulus layering is a plausible explanation for the compositional layering described above, though no cumulus textures were recognized. Moreover, in only rare cases was evidence observed for subsolidus compositional re-equilibration of high temperature phases, like that described below for the websterites. Such re-equilibration would lead to slightly different pyroxene and garnet compositions in adjacent layers with different original phase proportions; such differences are documented in Table 2 and Fig. 2. Metamorphic textures such as granulated fabrics with ~120° grain boundaries now dominate the rocks, however, and metamorphic segregation or intrusive or tectonic juxtaposition remain as explanations for some of the observed modal contrasts. The thin clinopyroxene-rich veins may have formed in the solid state during deformation of more brittle garnet-rich layers. The secondary hydrous phases may have formed during this later deformation.

Amphibolites

The amphibolites are also phase-layered; some grade over distances of mm from zones of eclogitic character to those with large (up to 15 mm) amphibole-dominated zones. There is an overlap in the proportion of Pyr:Alm: Gross contents of garnets present in amphibole-rich inclusions and those present in eclogitic inclusions (Fig. 3). However, the garnets in amphibolites range up to considerably more almandine-rich compositions (e.g. $Pyr_{14}Alm_{55}Gr_{31}$, Table 3) than in eclogites. The amphibolites with garnets of almandine-rich nature have not been found to grade in hand specimen into eclogitic zones.

Phlogopite, apatite and Fe-Ti oxides are present in many of the amphibole-rich inclusions. As in eclogites, rutile is an important accessory, and it forms up to 20 modal % of some rocks, again consistent with cumulus enrichment. In one relatively Fe-rich amphibolite (PR105 - Fig. 3) the following Fe-Ti oxide assemblages have been analyzed: in amphibole blebs of rutile and rutile with intergrown $pseudobrookite_{ss}$, and $ulvospinel_{ss}$ plus $ilmenite_{ss}$; in garnet, $pseudobrookite_{ss}$ plus $ilmenite_{ss}$ and $ulvospinel_{ss}$ plus $pseudobrookite_{ss}$. All of the Fe-Ti oxides are highly magnesian with up to 10-13 wt% MgO in ilmenite and spinel. The magnesian nature of the ilmenite is similar to kimberlitic occurrences (e.g. Mitchell, 1977) but it is not certain that the oxide assemblage of the Chino Valley inclusions represents mantle conditions.

In the most Fe-rich amphibolite studied (PR63 - Fig. 3), with amphibole (55%), biotite (30%), garnet (12%) and trace clinopyroxene (Table 3), the Fe-Ti oxide assemblage consists of rutile with pseudobrookite rims and $ilmenite_{ss}$. The ilmenite is associated with feldspar in patches in the garnet and both may result from low-pressure garnet breakdown and reaction. Hercynite is also present as perfect euhedral grains in amphibole. Amphibole crystals in this rock have euhedral faces where in contact with the host latite, even though amphibole is not a phenocryst phase in this particular latite.

Websterites and Pyroxenites

Many websterites have large (1-5 cm) low-Ca pyroxene porphyroclasts containing exsolution lamellae of garnet, orthopyroxene and minor clinopyroxene. Broad beam microprobe analysis suggests that the assemblage results from the reaction of an original aluminous pyroxene to an opx-gnt-cpx assemblage (1-4, Table 4). Diopside porphyroclasts with garnet lamellae are present but less common. Some of the porphyroclasts display an undulatory schistosity defined by the garnet and opx-cpx lamellae, with superimposed kink deformation. A mosaic textured, granular groundmass of orthopyroxene, garnet and clinopyroxene is present between the porphyroclasts and in zones transecting what appear to have been continuous crystals. Mineral compositions are

TABLE 4. Microprobe analyses of phases in garnet websterite (PR 35), amphibole-phlogopite-orthopyroxenite (PN-1-47) and green clinopyroxenite (PR 31).

	1^w	2^e	3^e	4^w	5^w	6^w	7^w	8^w	9^e	10^e
SiO_2	51.4	55.02	39.80	54.4	48.8	40.5	56.7	52.5	51.42	51.82
TiO_2	na	0.00	0.00	0.11	0.21	0.49	0.03	0.30	0.00	0.19
Al_2O_3	6.68	0.97	22.98	3.50	9.86	14.8	1.48	1.80	1.97	6.69
Cr_2O_3	na	0.00	0.24	0.30	0.44	0.42	na	0.04	0.22	0.00
FeO*	15.2	13.44	19.66	4.50	4.31	4.53	8.37	6.91	9.18	8.34
MnO	0.28	0.10	0.62	na	0.06	0.01	na	0.23	0.00	0.00
MgO	24.7	30.0	12.80	14.5	19.4	23.6	33.8	16.1	12.95	10.36
CaO	1.66	0.07	4.88	20.8	10.8	<0.02	0.30	21.5	22.75	19.49
Na_2O	na	0.17	0.00	2.34	2.23	0.38	na	0.46	0.65	2.45
K_2O	na	0.00	0.00	<0.02	1.00	9.91	na	na	0.00	0.00
Total	99.9	99.77	100.40	100.4	97.1	94.7	100.7	99.9	99.64	99.34

*Total Fe as FeO. na - not analyzed.
1 = broad beam average of px porphyroclast in PR 35; 2 = opx lamella in clast;
3 = garnet lamella in clast; 4 = cpx lamella in clast; 5 = PN-1-47 amph;
6 = phlogopite in PN-1-47; 7 = opx in PN-1-47; 8 = cpx in PN-1-47;
9 = green cpx core in PR 31; 10 = brown cpx core in PR 31.

TABLE 5. Kd's, solution characteristics and P,T estimates for garnet websterites (rows 1-4), and Kd's for eclogites and amphibolite.

	Kd Gar-Cpx	Ca/(Ca Mg) cpx	Ca/4 cat. opx	% Al_2O_3 opx	T,P*
PN-1-20	3.4	.498	.016	1.4	950, 20
PN-1-48	3.7	.494	.012	1.5	900, 20
PR 35	5.5	.505	.006	1.4	700,<10
PN-1-24	5-6	.507	.006	2.2	<700 <10
PR 192	6.4-7.0		eclogite		
PR 161	6.1-6.8		eclogite		
PN-1-43	4.0-4.4		gnt. amphibolite		
PR 149A	6.1-7.1		cpx-rich eclogite		
PR 149C	5.6-6.5		gnt-rich eclogite		

*From an appoximation using gar-opx (Wood, 1974) and Råheim and Green (1974) gar-cpx.

the same in the groundmass as in the clasts.

Other websterites range from rocks with large undeformed, interlocking pyroxene crystals with garnet and pyroxene lamellae to rocks with all phases in mosaic intergrowths. In one phase-layered sample, amphibole and garnet appear to have an antipathetic relationship forming garnet-rich and amphibole-rich layers, although they also occur together in contact. The amphibole, a typical pargasite, occurs both as separate grains and as equant inclusions within pyroxene, and it is most probably primary. An amphibole-phlogopite orthopyroxenite (analyses 5-8, Table 4) contains equigranular, tabular grains of all phases in an apparent equilibrium metamorphic fabric. Rare garnet orthopyroxenites contain radiating sprays of prismatic orthopyroxene crystals with interstitial amphibole. In some websterites, however, amphibole also occurs in textures indicative of late introduction and growth such as anhedral interstitial grains and rims on other phases. Rutile, ilmenite$_{ss}$, pseudobrookite$_{ss}$ and rare aluminous spinel rimmed by garnet form the oxide assemblages, and zircon is present in one sample of orthopyroxenite. The Mg/(Mg + Fe) ratios of the garnets and clinopyroxenes of the websterites and orthopyroxenites show some overlap with the same phases in eclogites and amphibolites, but also range up to higher Mg/(Mg + Fe) ratios than these other rock types (Fig. 3).

Bright green clinopyroxenite, a distinctive minor inclusion type, is characterized by mosaic textured clinopyroxene with a range in Na and Al content (Analyses 9,10, Table 4). Apatite, quartz, feldspar and sphene form accessory phases, locally present as apparently primary crystals; feldspar and quartz also occur as veins and patches of secondary appearance. Hematite has been observed in one clinopyroxenite but in general, Fe-Ti oxides are lacking. The Mg/(Mg + Fe) ratios of the clinopyroxenes are typically lower than in the majority of the other inclusions.

Rare, undeformed phlogopite-apatite-clinopyroxenites represent the only possible cumulates from the host magma. These rocks contain apparent cumulus crystals of biotite, phlogopite (Mg/(Mg+Fe) = .75), apatite, amphibole, rare rutile, and fragments of altered eclogite. Clinopyroxene crystals have normal and sector zoning; they are relatively titaniferous (.5-1% TiO_2) and contain .8-1.3% Na_2O. Typical eclogite pyroxenes are less titaniferous and more sodic, but similar phases are phenocrysts in the host latite.

Discussion

The simplest hypothesis for the origin of the inclusion suite is that the distinctive assemblages are genetically related in some way. This possibility is examined in more detail below, but other possibilities such as the formation of metamorphic eclogites with segregation banding are being studied. The modal gradations of garnet, clinopyroxene and amphibole between the eclogites and amphibolites together with chemical gradations of the phases themselves is strong evidence in support of consanguinity. However, the relationship of the pyroxenites to the eclogites and amphibolites is less clear.

Some preliminary estimates of pressures and temperatures of equilibration may be made based on the experimental and empirical geobarometers/thermometers of Wood (1974) and Råheim and Green (1974). For the 2 pyroxene-garnet assemblages, the Kd (distribution coefficient for Fe and Mg between garnet and clinopyroxene) and solution characteristics of the pyroxenes give temperatures and pressures in the range 700-1000°C, 10-20 kbar (Table 5). This would place the equilibration of the websterites and orthopyroxenites in the upper mantle and lower crust. The Kd's for the eclogites and amphibolites are generally larger (Table 5) and would suggest lower temperatures and/or pressures of equilibration. In any case, the presence of pargasitic amphibole would restrict the pressures of equilibration to less than 25 kbar (Mysen and Boettcher, 1975).

Irving and Green (1970) and Irving (1974) were able to reproduce the garnet websterite and pyroxenite assemblages of the Delegate Pipes (N.S.W., Australia) in the region of 12-15 kbar and 1000-1100°C. They suggested that the garnet and orthopyroxene in the garnet websterites formed by exsolution from original higher temperature aluminous clinopyroxene. The latter phase is predicted to be in equilibrium with basalt melts in the 12-15 kbar range (e.g. Green and Ringwood, 1967). Similar exsolution textures are developed in some Chino Valley nodules, and similar aluminous clinopyroxenes appear to have been present before exsolution. The garnet websterites of New South Wales are generally distinctive in comparison with the Chino Valley samples however, on account of the low-Na (high-CaTs) content of the clinopyroxenes (White, 1964; Lovering and White, 1969). Additional constraints on equilibration conditions cannot be applied in the absence of experimental studies of coexisting garnet-clinopyroxene-amphibole-phlogopite assemblages.

The textural and compositional relations suggest the tentative history outlined below.

1. Formation of a cumulus sequence, primarily of eclogite, and probably including websterites and amphibole-rich rocks in the upper mantle-lower crust. Pargasite, phlogopite, and apatite were likely hydrous cumulus phases. Late-stage hydrous melts may have been injected into earlier cumulates.
2. Metamorphism, re-equilibration, and tectonic disruption of the sequence.
3. Introduction of secondary hydrous phases into some nodules.

TABLE 6. Major and trace element analysis of host latite PR 42.

SiO_2	58.58	Sc	14.2	La	64.7
TiO_2	0.78	Cr	143	Ce	135
Al_2O_3	14.47	Co	12.8	Sm	9.4
FeO*	5.21	Ni	108	Eu	2.12
MnO	0.09	Hf	8.1	Tb	0.72
MgO	3.59	Ta	0.90	Yb	1.4
CaO	4.62	Th	50		
Na_2O**	2.57	Ba	1340		
K_2O	5.15				
P_2O_5	0.36				
H_2O^+	1.46				
H_2O^-	1.66				
CO_2	0.01				
	98.55				

Major elements by XRF, except as noted, and trace elements (in ppm) by INAA, all courtesy of A.J. Irving, Lunar and Planetary Institute. Water and carbon dioxide by G.K. Hoops, University of Texas, Austin.

*Total Fe as FeO **by INAA

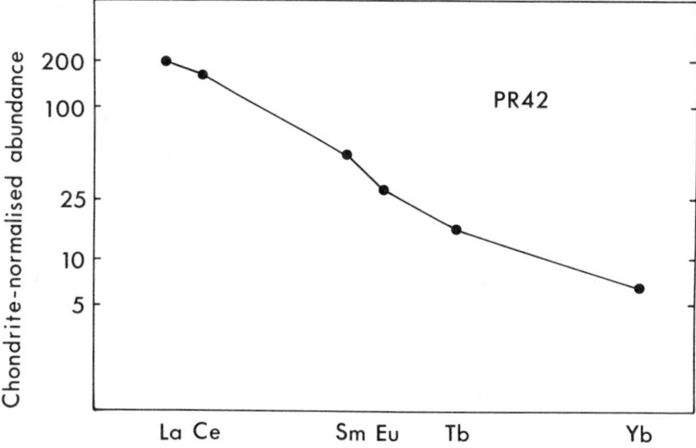

Figure 4. Rare-earth profile for host latite sample PR 42. Analytical data for the rock are in Table 6.

4. Development of partial-melting textures in a few nodules at depth.
5. Transport to the surface and subsequent local oxidation.

Studies of simple systems ($CaO-MgO-Al_2O_3-SiO_2$) have demonstrated olivine and orthopyroxene reaction relationships with haplobasaltic melts (Kushio, 1968; Yoder and Kushiro, 1974; O'Hara and Yoder, 1967). The occurrence of garnet websterite and eclogite assemblages may be explained as a sequence of cumulates from a basaltic melt, equilibrated at successively lower temperatures. The Mg/(Mg + Fe) ratios of the pyroxenite and eclogitic assemblages are compatible with a sequence of declining temperatures.

The importance of the hydrous phases however, together with the ubiquitous appearance of Fe-Ti oxides in the inclusions precludes further direct comparison of the natural with synthetic parageneses.

The magma from which the bulk of the cumulates precipitated must have preceded formation of the 25 my latite. The youngest earlier igneous activity known in the immediate region was that which accompanied and closely followed formation of the Yavapai Series, a Precambrian greenstone belt. The activity occurred from approximately 1720 to 1820 m.y. ago (Anderson and Silver, 1976), and associated magmas may have been parental to the inclusions studied here. Intriguingly, a fragment of gold (~100 micron size) has been discovered during mineral separations preparatory to trace analysis of phases in a garnet amphibolite. Massive sulfide deposits in the Yavapai Series at nearby Jerome have yielded significant amounts of gold.

It is possible that the host latite may represent a partial melt of an amphibole-phlogopite-garnet-clinopyroxene source rock like some of the inclusions. The evidence for partial melting at depth in the kyanite-bearing eclogite lends credence to the hypothesis, but isotopic and experimental evidence is not yet available to test it. Rare-earth element (REE) abundances in the latite (Table 6, Fig. 4) do suggest equilibration of the melt with garnet. The pattern of strong relative enrichment in the light versus heavy REE is compatible with garnet fractionation. In any event the latite must have been generated at depths as great if not greater than the equilibration pressures indicated by the inclusions. Consequently, Sullivan Buttes latite-inclusion relationships are important, in view of the current debate over the possibility of generation of quartz-normative melts in the upper mantle (e.g. Green, 1976).

If the eclogites, amphibolites, and pyroxenites are related to Precambrian igneous activity from 1720 to 1820 my ago, they could be of local importance elsewhere in the upper mantle below the Plateau, complicating the models of McGetchin and Silver (1972) and Smith and Levy (1976). The eclogites are unlike the unusual lower-temperature eclogites included in some of the Colorado Plateau kimberlitic diatremes in the Navajo field. The diatreme eclogites, classed in Group C of Coleman et al. (1965), contain highly-zoned almandine-rich garnet and jadeite-rich clinopyroxene and lawsonite (O'Hara and Mercy, 1966; Watson and Morton, 1969; Essene and Ware, 1970; Smith and Zientek, 1977): they are also quite different in bulk composition. A somewhat similar eclogite has been found, however, in a felsic minette in the Navajo field (McDowell et al., 1978). Such felsic minettes, unusual variants of the more mafic minettes on the central Plateau, are similar in many respects to the Sullivan Buttes latite, and the two types are very similar in age (McDowell et al., 1978). The thermal event responsible for genesis of the latite probably also produced the other potassic volcanic rocks of the Plateau.

Acknowledgements

Howard G. Wilshire brought the locality to our attention and generously contributed some of the inclusions studied here. The assistance of the Organizing Committee of the Second International Kimberlite Conference is greatly appreciated. Smith acknowledges support from the National Science Foundation, Earth Sciences Section, Grant EAR 76-12368 and from the Geology Foundation of the University of Texas at Austin.

References

Anderson, C.A., and L.T. Silver, Yavapai Series - a greenstone belt, Arizona Geol. Digest, 10, 13-26, 1976.

Coleman, R.G., D.E. Lee, L.B. Beatty, and W.W. Brannock, Eclogites and eclogites: their differences and similarities, Geol. Soc. Am. Bull., 76, 483-508, 1965.

Essene, E.J., and N.G. Ware, The low temperature xenolithic origin of eclogites in diatremes, N.E. Arizona (abstr.), Abstr. Prog. Geol. Soc. Am., 2, 547-548, 1970.

Green, D.H., Experimental testing of "equilibrium" partial melting of peridotite under water-saturated, high-pressure conditions, Can. Mineral., 14, 255-268, 1976.

Green, D.H., and A.E. Ringwood, The genesis of basaltic magmas, Contrib. Mineral. Petrol., 15, 103-190, 1967.

Irving, A.J., Geochemical and high pressure experimental studies of garnet pyroxenite and pyroxene granulite xenoliths from the Delegate basaltic pipes, Australia, J. Petrol., 15, 1-40, 1974.

Irving, A.J., and D.H. Green, Experimental duplication of mineral assemblages in basic inclusions of the Delegate breccia pipes, Phys. Earth Planet Interiors, 3, 385-389, 1970.

King, P.B., Tectonic map of North America, U.S. Geol. Surv. 1:2,500,000, 1969.

Krieger, M.H., Geology of the Prescott and Paulden quadrangles, Arizona, U.S. Geol. Surv. Prof. Paper, 467, 1965.

Krieger, M.H., S.C. Creasey, and R.F. Marvin, Ages of some Tertiary andesitic and latitic volcanic rocks in the Prescott-Jerome area, north-central Arizona, U.S. Geol. Surv. Prof. Paper, 750-B, 157-160, 1971.

Kushiro, I., Compositions of magmas formed by partial zone melting of the earth's upper mantle, J. Geophys. Res., 73, 619-634, 1968.

Kushiro, I., and H.S. Yoder, Jr., Formation of eclogite from garnet lherzolite-liquidus relations in a portion of the system $MgSiO_3$-$CaSiO_3$-Al_2O_3 at high pressures, Carnegie Inst. Wash. Yb., 73, 266-269, 1976.

Lovering, J.F., and A.J.R. White, Granulitic and eclogitic inclusions from basic pipes at Delegate, Australia, Contrib. Mineral. Petrol., 21, 9-52, 1969.

McDowell, F.W., M.F. Roden, R.J. Arculus and D. Smith, Potassic volcanism and associated inclusions on the Colorado Plateau, Geol. Soc. Amer. Abstr. with Programs, 10, 116, 1978.

McGetchin, T.R., and L.T. Silver, A crustal-upper mantle model for the Colorado Plateau based on observations of crystalline rock fragments in the Moses Dike rock, J. Geophys. Res., 77, 7022-7037, 1972.

Mitchell, R.H., Geochemistry of magnesian ilmenites from kimberlites in South Africa and Lesotho, Lithos, 10, 29-37, 1977.

Mysen, B.O., and A.L. Boettcher, Melting of a hydrous mantle: I, Phase relations of natural peridotite at high pressures and temperatures with controlled activities of water, carbon dioxide and hydrogen, J. Petrol., 16, 520-548, 1975.

O'Hara, M.J., and E.L.R. Mercy, Peridotite and pyrope from the Navajo Country, Arizona and New Mexico, Am. Mineral., 51, 336-352, 1966.

O'Hara, M.J., and H.S. Yoder, Jr., Formation and fractionation of basic magmas at high pressures, Scott. J. Geol., 3, 67-117, 1967.

Råheim, A., and D.H. Green, Experimental determination of the temperature and pressure dependence of the Fe-Mg partition coefficient for coexisting garnet and clinopyroxene, Contrib. Mineral. Petrol., 48, 179-203.

Reed, S.J.B., and N.G. Ware, Quantitative electron microprobe analysis of silicates using energy-dispersive X-ray spectrometry, J. Petrol., 16, 499-519, 1975.

Reid, A.M., R.W. Brown, J.B. Dawson, G.G. Whitfield, and J.C. Siebert, Garnet and pyroxene compositions in some diamondiferous eclogites, Contr. Mineral. Petrol., 58, 203-220, 1976.

Smith, D., and S. Levy, Petrology of the Green Knobs diatreme and implications for the upper mantle below the Colorado Plateau, Earth Planet. Sci. Lett., 29, 107-125, 1976.

Smith, D., and M. Zientek, Garnet-pyroxene growth in eclogite inclusions from the Garnet Ridge kimberlitic diatreme, Arizona, Proceedings of the Second International Kimberlite Conference, in press.

Switzer, G., and W.G. Melson, Partially melted kyanite eclogite from the Roberts Victor mine, South Africa, Smithson, Contr. Earth Sci., 1, 1-9, 1969.

Watson, K.D., and D.M. Morton, Eclogite inclusions in kimberlite pipes at Garnet Ridge, northeastern Arizona, Am. Mineral., 54,267-285, 1969.

White, A.J.R., Clinopyroxenes from eclogites and basic granulites, Am. Mineral., 49, 883-888, 1966.

Wood, B.J., The solubility of alumina in orthopyroxene coexisting with garnet, Contrib. Mineral. Petrol., 46, 1-15, 1974.

GARNET PYROXENITE AND ECLOGITE XENOLITHS FROM THE SULLIVAN BUTTES LATITE, CHINO VALLEY, ARIZONA

Daniel J. Schulze[1] and Herwart Helmstaedt

Department of Geological Sciences, Queen's University, Kingston, Ontario, Canada

Abstract. An unusual suite of ultramafic xenoliths from Tertiary latites in Chino Valley near Prescott, Arizona, is comprised of garnet pyroxenites and eclogites and represents the first occurrence of eclogite xenoliths in non-kimberlitic rocks on the Colorado plateau. The mineral assemblage of the garnet pyroxenites (calcic clinopyroxene, orthopyroxene, garnet, and Fe-Ti oxide) is the product of a complex subsolidus reequilibration of three aluminous pyroxenes involving exsolution of garnet and an Fe-Ti oxide, deformation, and recrystallization. Unmixing of subcalcic clinopyroxene, originally coexisting with calcic clinopyroxene and orthopyroxene, resulted in a lamellar intergrowth of orthopyroxene and calcic clinopyroxene. Eclogites (omphacite, garnet) have reacted with the host magma and may show evidence of partial melting. The absence of metamorphic eclogite xenoliths in basaltic rocks in the southwestern United States does not prove that eclogites are absent at depth. Basaltic magmas were probably too hot to preserve the relatively low-temperature eclogite assemblages.

Introduction

The Tertiary Sullivan Buttes latite in Chino Valley on the southwestern margin of the Colorado plateau contains numerous garnet-bearing ultramafic xenoliths. Eclogites, many of which are highly altered, are most abundant, but locally garnet pyroxenites and amphibole rocks (with or without garnet) may dominate. No xenoliths with primary olivine have been found.

The xenoliths were first noted by Krieger (1965). Schulze (1977) distinguished eclogites (garnet, omphacite) from garnet pyroxenites (garnet, calcic clinopyroxene ± orthopyroxene). Schulze et al. (1978) described exsolution of Fe-Ti oxides and garnet from the pyroxenes of the garnet pyroxenites. In the present paper we have concentrated on the petrology of the garnet pyroxenites and eclogites. Arculus and Smith (in press) have described a similar xenolith suite collected at a different locality in Chino Valley.

With the exception of a single garnet clinopyroxenite in basaltic ejecta from Dish Hill, California (Shervais et al., 1973) and garnet websterites from the State Line kimberlites in Colorado and Wyoming (M. McCallum, personal communication, 1977), no rocks similar to the Chino Valley garnet pyroxenites have been found as xenoliths in the western United States. They have been found elsewhere, as xenoliths (e.g., Beeson and Jackson, 1970; Lovering and White, 1969) and as layers and lenses in alpine ultramafic bodies (e.g., Dickey, 1970; Lasnier, 1971). The garnet pyroxenites from Chino Valley merit attention because they differ from the other occurrences in the following respects:

1. No primary olivine has been found in any Chino Valley xenoliths, though other garnet pyroxenites are generally associated with peridotites.
2. Three aluminous pyroxenes apparently coexisted prior to subsolidus reequilibration.
3. Some pyroxenes in the Chino Valley samples exhibit lamellae of an Fe-Ti oxide which exsolved (along with garnet) prior to deformation and recrystallization. The only other garnet pyroxenite with such oxide lamellae, known to us, is a xenolith from a kimberlite in New York (Schulze et al., 1978).

The presence of eclogites as xenoliths in the latites is unusual. Elsewhere on the Colorado plateau xenoliths of such rocks are known only from the kimberlite diatremes of the Navajo field (e.g., Watson and Morton, 1969).

Host Rocks

The Sullivan Buttes Latite has been mapped and described by Krieger (1965) and Krieger et al. (1971). It outcrops primarily in the Simmons and Paulden quadrangles, with minor exposures in the Prescott and Clarkdale quadrangles. Two latite

[1] present address: Geosciences Department, University of Texas at Dallas, Richardson, Texas, 75080, U.S.A.

TABLE 1. Chemical Compositions of Latites.

Sample	43-A[1]	54-A[1]
SiO_2	60.60	62.80
TiO_2	0.91	0.78
Al_2O_3	13.00	13.70
Fe_2O_3[1]	2.28	2.24
FeO[1]	2.54	2.12
MnO	0.06	0.06
MgO	4.80	3.25
CaO	4.45	4.90
Na_2O	1.60	5.12
K_2O	5.15	3.10
P_2O_5	0.29	0.28
H_2O^-	-	-
H_2O^+	-	-
CO_2	-	-
Total	95.68	98.35
estimated LOI[2]	3.50	1.00
$\frac{FeO}{FeO+Fe_2O_3}$ (wt.%)[1]	.527	.485

1 X-ray fluorescence determinations, F. Dunphy (Queen's), analyst, total Fe reported as Fe_2O_3. FeO, Fe_2O_3, FeO/FeO+Fe_2O_3 calculated by Le Maitre's (1976) method for volcanic rocks.

2 Loss on ignition estimated from fused glass bead, N.A. = not analyzed.

units have yielded K-Ar ages of 23.4 and 26.7 million years (Krieger et al., 1971). The latites, which occur mostly as plugs and flows, have intruded and overlie Mazatzal Quartzite, Prescott Granodiorite, and other rocks of Precambrian age, Lower Paleozoic sedimentary rocks, as well as basalt and gravel of unknown age. They are, in turn, overlain by younger basalts, sedimentary rocks, and gravels.

Krieger (1965) recognized two main varieties of latite (then called andesite). The hornblende latite contains hornblende phenocrysts up to 1 cm long and smaller biotite plates and pyroxene phenocrysts in a matrix of cryptocrystalline to glassy material containing plagioclase microlites. The biotite-pyroxene latite consists of phenocrysts of biotite and clinopyroxene in cryptocrystalline to glassy material with microphenocrysts and microlites of plagioclase (some skeletal in habit). The biotite-pyroxene latite grades into "basaltic andesite" with the disappearance of biotite and pyroxene and the appearance of olivine.

Most of the xenoliths in this study come from two occurrences (both biotite-pyroxene latite) which Krieger (1965) identified as vents (see her Plate 2, cross section G-G'). They are petrographically similar, though the latite from the western vent contains more biotite and is more glassy. The latite in the eastern vent (USGS Paulden, Arizona 15' quadrangle, Township 16 North, Range 1 West, centre of Section 11) is more highly oxidized; the biotite plates are commonly replaced by an aggregate of magnetite particles. It is red in hand specimen vs. the buff colour of the western latite (USGS Paulden, Arizona 15' quadrangle, Township 17 North, Range 1 West, Section 31 NW1/4,SE1/4).

Chemical analyses of the two latites (43-A, eastern vent; 54-A, western vent) are presented in Table 1.

Xenoliths

The xenoliths in our collection range in length from less than 1 mm to about 15 cm. Eclogites are largest, averaging approximately 5 cm in diameter. Garnet pyroxenites rarely exceed 5 cm and average about 1-2 cm in diameter. Xenoliths in which fresh garnet is preserved, have been observed in only four of many latite units examined. Most xenoliths of our collection come from the two vents on cross-section G-G' of Krieger (1965). Those from the eastern vent (sample series 43 and 44) are mostly altered eclogites. The western vent (sample series 1 and 54) contains the nodules with best preserved garnets, both eclogites and garnet pyroxenites.

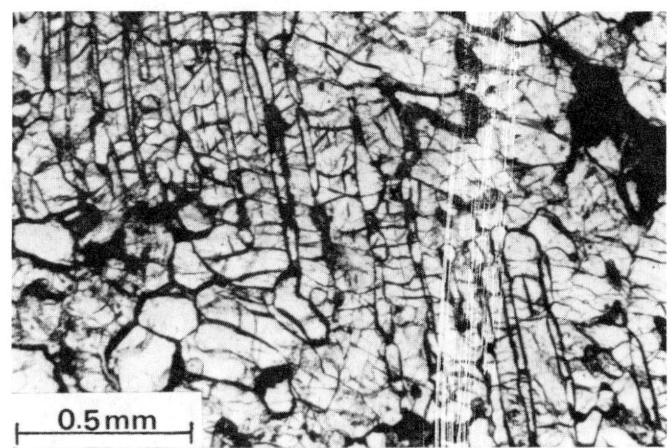

Fig. 1. Remnant of calcic clinopyroxene porphyroclast with high relief garnet lamellae and thin, opaque Fe-Ti oxide lamellae in a matrix (upper right and lower left) of granular calcic clinopyroxene, garnet and opaque. Plane polarized transmitted light.

Arculus and Smith (in press) described larger eclogite xenoliths and observed that well preserved eclogites are confined to the chilled margins of the latite intrusions.

Garnet pyroxenites

All xenoliths in this group consist of garnet and clinopyroxene and may or may not contain orthopyroxene. They have been classified as garnet clinopyroxenites and garnet websterites.

Xenoliths from the western vent show little reaction with the host magma. Some have narrow (2-4 mm) garnet-free margins without garnet alteration products and some garnets, especially those in contact with the latite, have an opaque kelyphyte rim. In the eastern vent, only pyroxene is preserved in the few xenoliths recognized as belonging to this group. Garnet is altered completely. Some secondary amphibole and biotite is common in most of the xenoliths, replacing the pyroxenes or rimming the garnets. A two-phase Fe-Ti oxide or rutile may also be present.

Textural evidence indicates that the garnet pyroxenites originated through subsolidus unmixing of three aluminous pyroxenes; calcic clinopyroxene (cpx), subcalcic clinopyroxene (scpx), and orthopyroxene (opx). Cpx and scpx are much more common than opx. The scpx is represented by grains of lamellar intergrowth of cpx and opx. In most cases these intergrowths are not well preserved, but electron microprobe analyses of better preserved examples (Table 5) suggest that, prior to unmixing into cpx and opx lamellae this phase consisted of ~50 mole % cpx and 50% mole % opx ($Ca/Ca+Mg+Fe^{+2} \approx 0.24$).

Garnet clinopyroxenites. These rocks consist of a fine-grained (<1 mm) matrix of

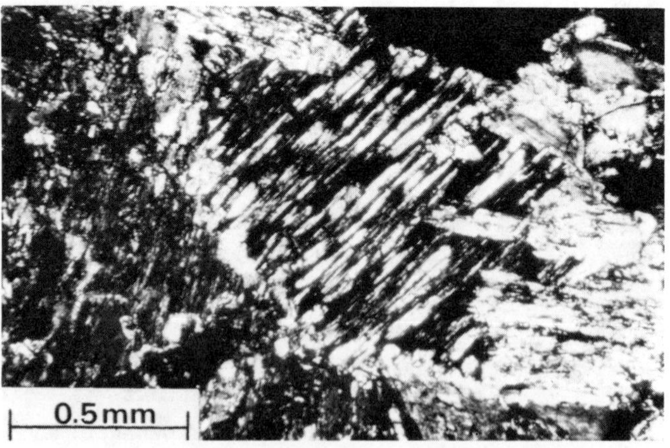

Fig. 2. Granular subcalcic clinopyroxene, now unmixed into a lamellar intergrowth of cpx (extinct) and opx. Transmitted light, crossed nicols.

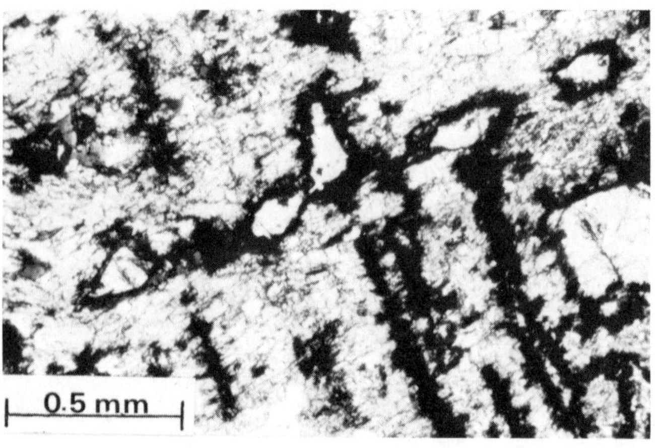

Fig. 3. Garnet lamellae (highly altered) in a matrix of recrystallized subcalcic clinopyroxene. Transmitted light, plane polarized.

granular cpx and garnet with uncommon coarse (up to 2 cm) porphyroclasts of cpx (Fig. 1). Lamellar intergrowths of these porphyroclasts with one or all of the minerals garnet, opx, and Fe-Ti oxide have been recognized. The opx lamellae are too narrow to be analyzed accurately, and analysis of the oxide lamellae is hindered by their narrow width and the fact that they consist of two phases, one nearly pure Fe-oxide, the other containing more than 40 wt.% TiO_2 and several wt.% MgO.

Three types of intergrowths exist in the cpx porphyroclasts (Schulze et al., 1978): 1. Clinopyroxene with planar or bead-like lamellae of garnet and fine Fe-Ti oxide lamellae parallel to (100). Host pyroxenes and lamellae have been deformed together, as evidenced by bent lamellae, and undulatory extinction and kink bands in the host grains. All stages between deformed and completely recrystallized cpx host grains can be recognized. In the latter case, remnants of garnet lamellae are the only indication that larger cpx grains existed previously. All minerals are equigranular in completely recrystallized xenoliths. 2. Clinopyroxenes with garnet lamellae only. Except for the absence of the Fe-Ti oxide lamellae, host grains and garnet lamellae resemble Group 1 in all respects. 3. Clinopyroxenes with oxide lamellae only. These grains are smaller than the porphyroclasts, but may also be strained and partly recrystallized. Lamellae are generally confined to the centres of grains and resemble the oxide lamellae in Group 1 in all respects.

Garnet websterites. These xenoliths may consist of large grains of cpx and opx with exsolved garnet, of a lamellar intergrowth of cpx and opx resulting from unmixing of scpx, or of any combination of cpx, unmixed scpx and opx.

TABLE 2. Chemical Composition of Xenoliths[1]

	Pyroxenites		Eclogites			
Sample	54-1	54-45	43-33	44-6	54-97	54-127
SiO_2	48.10	49.60	45.60	45.50	44.20	43.80
TiO_2	0.90	0.63	1.98	1.37	0.74	1.90
Al_2O_3	8.90	8.20	12.50	14.30	16.80	16.05
Fe_2O_3	3.20	2.70	3.15	2.81	3.30	3.36
FeO[2]	10.42	9.72	9.81	9.71	11.70	11.74
MnO	0.25	0.13	0.26	0.27	0.28	0.25
MgO	14.85	17.50	9.75	9.40	8.50	7.40
CaO	12.00	8.42	13.40	14.60	13.20	14.05
Na_2O	<0.50[4]	<0.50[4]	1.63	1.00	0.50	0.95
K_2O	0.73	0.34	0.22	0.15	0.57	0.22
P_2O_5	0.16	0.06	0.20	0.06	0.17	0.13
Total	99.33	97.30	98.50	99.17	99.96	99.85
estimated LOI[3]	<0.50	2.00	1.00	1.00	<0.50	N.A.
$\frac{100\ Mg}{Mg+Fe^{+2}}$ (mol)	72	76	64	63	56	53

1. X-ray fluorescence determinations, F. Dunphy (Queen's), analyst; total Fe reported as Fe_2O_3.
2. Calculated from total Fe by LeMaitre's (1976) method for plutonic rocks.
3. Loss on ignition estimated from fused glass bead, N.A. = not analyzed.
4. Numbers reported as <"X" not reported in total, though Na_2O estimated as 0.30 for calculation of oxidation ratio.

Websterites derived from unmixing of scpx consist of granular garnet and grains of lamellar intergrowths of opx and cpx (Fig. 2). Lamellae average 0.1 - 0.5 mm in width. The lamellar intergrowth is recrystallized locally into a very fine-grained cpx and opx mass. Prior to recrystallization and unmixing, the parent scpx exsolved garnet lamellae, now partially altered to an opaque mass. The example illustrated in Fig. 3 gives the impression that 2 sets of garnet lamellae were exsolved. Granular Fe-Ti oxide within the lamellar intergrowths may represent recrystallized oxide lamellae. Websterites derived from scpx are commonly more amphibolitized than other varieties of garnet pyroxenites.

Whole rock chemistry. Table 2 includes bulk chemical analyses of two websterites derived from cpx and scpx. A higher CaO content in 54-1 than in 54-45 is due to the higher proportion of cpx in 54-1. The relatively high K_2O contents (0.73 and 0.34 wt.%) can be attributed to secondary amphibole and biotite, as the primary silicate phases are virtually potassium-free. The rocks were not analyzed for Cr_2O_3 but they are probably chromium-poor because the average Cr_2O_3 of the mineral phases does not exceed ~0.2 wt.% (Tables 3, 4, and 5).

Aside from their high Mg/Fe values (100 Mg/Mg+Fe = 72.76) and somewhat low Al_2O_3 and alkali content, these rocks are roughly basaltic in composition.

Mineral Chemistry. All minerals were analyzed on an ARL-EMX electron microprobe at Queen's University. Raw data were reduced according to Bence-Albee correction procedures as modified by P. L. Roeder.

Chemical compositions of the silicate minerals from three garnet pyroxenites are presented in

TABLE 3. Electron Microprobe Analyses of Minerals from Garnet Pyroxenite 1-2 (Clinopyroxenite).

	1	2	3	4	5	6
average of	8	10	16	16	17	14
SiO_2	53.3	53.7	52.7	54.1	39.2	39.1
TiO_2	0.2	0.1	0.2	0.1	0.1	0.1
Al_2O_3	2.7	1.9	3.0	1.7	21.8	21.7
FeO*	6.3	5.5	7.2	15.3	23.6	22.9
MnO	0.1	0.1	0.1	0.2	0.8	0.8
MgO	14.6	15.1	14.8	27.4	9.3	9.2
CaO	21.3	22.1	20.3	0.8	5.6	5.7
Na_2O^+	1.7	1.4	1.3	0.5	0.2	0.2
K_2O	0.0	0.0	0.0	0.0	0.0	0.0
Cr_2O_3	0.2	0.2	0.2	0.0	0.2	0.2
Total	100.4	100.1	99.8	100.1	100.8	99.9
mole % Mg	45	46	46	75	36	36
Ca	48	48	45	2	16	16
Fe^{+2}	7	6	9	23	49	48

\+ effective detectability limit of Na_2O is approximately 0.5 wt.%.
* total Fe expressed as FeO.
1. lamellae-free clinopyroxene
2. lamellae-bearing clinopyroxene
3. clinopyroxene lamellae in unmixed scpx
4. orthopyroxene lamellae in unmixed scpx
5. granular garnet
6. lamellar garnet

TABLE 4. Electron Microprobe Analyses of Minerals From Garnet Pyroxenite 54-17 (Websterite).

	1	2	3	4	5	6	7	8	9	10
average of	9	6	9	6	10	9	10	6	6	14
SiO_2	53.9	54.3	52.3	54.4	54.3	55.9	39.4	39.5	39.5	39.4
TiO_2	0.1	0.1	0.3	0.0	0.1	0.1	0.1	0.1	0.1	0.1
Al_2O_3	2.0	1.6	4.0	1.1	1.2	1.8	21.8	21.8	21.9	21.8
FeO*	5.4	5.2	6.5	15.7	16.2	11.4	22.1	22.1	22.1	22.2
MnO	0.0	0.1	0.1	0.2	0.2	0.1	0.7	0.9	0.7	0.8
MgO	15.1	15.5	15.0	27.8	27.8	30.3	10.4	10.3	10.8	10.3
CaO	22.1	22.3	20.4	0.2	0.2	1.1	5.3	5.3	5.1	5.2
Na_2O^+	1.3	1.3	1.1	0.5	0.5	0.5	0.2	0.2	0.3	0.2
K_2O	0.0	0.0	0.0	0.0	0.0	0.0	0.0	0.0	0.0	0.0
Cr_2O_3	0.3	0.3	0.2	0.1	0.1	0.1	0.3	0.3	0.4	0.4
Total	100.2	100.7	99.9	100.0	100.6	101.3	100.3	100.5	100.9	100.4
mole % Mg	46	46	46	76	76	81	40	40	41	40
Ca	48	48	45	0	0	2	15	15	14	14
Fe^{+2}	7	6	8	24	24	17	46	46	45	46

+ effective detectability limit of Na_2O is approximately 0.5 wt.%.
* total Fe expressed as FeO

1. lamellae-free clinopyroxene
2. lamellae-bearing clinopyroxene
3. clinopyroxene lamellae in unmixed scpx
4. lamellae-free orthopyroxene
5. lamellae-bearing orthopyroxene
6. orthopyroxene lamellae in unmixed scpx
7. granular garnet associated with clinopyroxene
8. lamellar garnet in clinopyroxene
9. granular garnet associated with orthopyroxene
10. lamellar garnet in orthopyroxene

Tables 3, 4, and 5. Sample 1-2 is a garnet clinopyroxenite with minor unmixed scpx, 54-17 is a garnet websterite (opx + cpx + unmixed scpx) and 54-3 is a garnet websterite (unmixed scpx). Sample 1-2 consists of two calcic clinopyroxenes and an orthopyroxene (+ garnet and oxide), 54-17 consists of two clinopyroxenes and two orthopyroxenes (+ garnet and oxide), and 54-3 has one clinopyroxene and one orthopyroxene (+ garnet and oxide). The positions of these phases on a ternary Ca-Mg-Fe plot are shown in Fig. 4. The clinopyroxenes fall into three textural categories (lamellae-bearing porphyroclasts, recrystallized grains, lamellae in unmixed scpx) and two chemical categories. The porphyroclasts which contain garnet and/or oxide lamellae differ little in chemical composition from the smaller, granular cpx grains which contain no lamellae.

The third cpx is found as lamellar intergrowths with opx (the unmixed scpx). Where analyzed (samples 1-2 and 54-17) the cpx of this intergrowth is richer in Al_2O_3 and FeO, and poorer in CaO than the cpx of the other textural categories. Chemical equilibrium has thus not been achieved since the scpx unmixed into cpx and opx.

In sample 54-45, where three corresponding textural varieties of opx exist, the opx from the unmixed scpx is similarly higher in Al_2O_3, MgO and CaO and lower in FeO than the primary opx.

The garnets in all textural modes are very similar within one xenolith. Between xenoliths the main variation is in the MgO and FeO content, the CaO component being essentially constant at approximately 15 mole %. Garnets of samples 1-2, 54-45 and 54-3 contain 36, 40 and 51 mole % pyrope and 49, 46 and 35 mole % almandine, respectively, and are representative of the range found in the pyroxenite group (Fig. 5).

Eclogites

Petrography. Both massive and layered eclogites have been found. Layering is defined by variations in garnet:pyrope:amphibole ratio and has been interpreted as planar tectonic fabric. Garnet-rich regions are discontinuous and lens out into garnet-poor regions.

Garnet has been preserved only in the eclogites from the western vent. It is texturally distinct from the recrystallized garnet granules of the pyroxenites and occurs as discrete grains (up to 3 mm in diameter) some of which display euhedral shapes (Fig. 6). Garnet never occurs as exsolution lamellae. Most garnets are pseudomorphed by a fine symplectic intergrowth made nearly opaque by finely dispersed magnetite. X-ray powder diffraction patterns suggest that the intergrown silicate phases are plagioclase, orthopyroxene, and probably olivine. Fig. 7 shows the typical "feathery" texture of the intergrowths that appears to have resulted from the dendritic growth of alteration minerals from expansion fractures towards remnants of fresh garnet. The fractures are thought to have formed by thermal expansion when the xenoliths were heated in the magma.

Clinopyroxenes commonly exhibit clear, unaltered cores, surrounded by "porous"-looking rims (Fig. 8). Cores and rims are in optical continuity, but slight changes in extinction angles and birefringence may exist. Cores are somewhat brownish compared to the colourless rims. Some clinopyroxenes show marked dispersion and are optically zoned, even where unaltered. Plagioclase is present interstitial to and within these altered rims and accounts for the

TABLE 5. Electron Microprobe Analyses of Minerals from Garnet Pyroxenite 54-3 (Websterite)

	1	2	3	4
average of	6	6	27	5
SiO_2	52.6	54.4	54.7	40.4
TiO_2	0.2	0.1	0.1	0.1
Al_2O_3	2.6	2.3	3.0	22.4
FeO*	6.5	12.0	9.0	17.7
MnO	0.1	0.1	0.2	0.5
MgO	15.4	28.8	20.6	13.7
CaO	20.7	1.1	10.8	5.6
Na_2O^+	1.2	0.7	1.1	0.3
K_2O	0.0	0.0	0.1	0.0
Cr_2O_3	0.3	0.1	0.2	0.3
Total	99.6	99.7	99.8	101.0
mole % Mg	47	79	62	51
Ca	45	2	24	15
Fe^{+2}	8	19	14	35

* total Fe reported as FeO

\+ effective detectability limit of Na_2O is approximately 0.5 wt.%.

1. Clinopyroxene lamellae in unmixed scpx.
2. Orthopyroxene lamellae in unmixed scpx.
3. Bulk analysis of scpx. Sample was moved under beam on several grains for 3240 seconds total counting time.
4. Garnet.

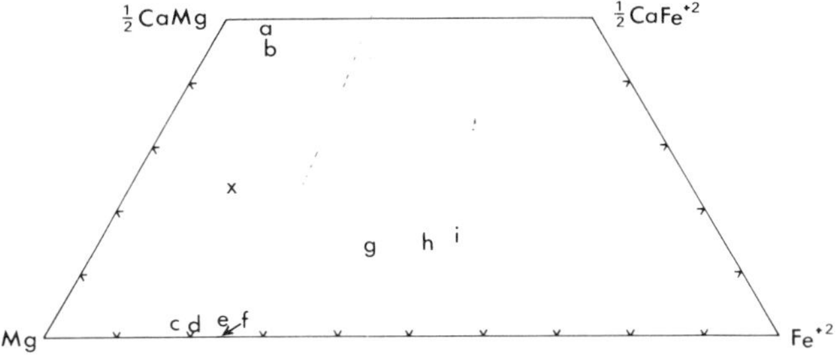

Fig. 4. Lower portion of a Ca-Mg-Fe^{+2} ternary diagram on which pyroxene and garnet analyses from garnet pyroxenites are plotted.
a) lamellae-free and lamellae-bearing cpx from 1-2 and 54-17; b) cpx lamellae in unmixed scpx from 1-2, 54-17, 54-3; c) opx lamellae in unmixed scpx from 54-17; d) opx lamellae in unmixed scpx from 54-3; e) opx lamellae in unmixed scpx from 1-2; f) lamellae-free and lamellae-bearing opx from 54-17; g) granular garnet from 54-3; h) all modes of garnet from 54-17; i) granular and lamellar garnets from 1-2; x) reconstructed scpx from 54-3.

"porous" appearance. Many pyroxenes are totally "porous" and have no clear cores.

Amphiboles and phlogopite are also present in some of these nodules. Some samples contain two generations of amphibole, one of which commonly appears as small, fresh euhedra. The other may make up a significant portion of the nodule but is usually highly altered. Either of these generations may be the only amphibole present in a xenolith. Rutile is a ubiquitous accessory mineral and is commonly rimmed or entirely replaced by ferropseudobrookite. A two-phase Fe-Ti oxide is present in a few of these samples. Apatite occurs in some specimens as discrete grains or as inclusions in garnet. In several samples it comprises up to 10% of a xenolith. A few grains of quartz have been found as inclusions in garnet or (rarely) clinopyroxene.

Whole Rock Chemistry

Table 2 includes bulk chemical analyses of four eclogite xenoliths, two with preserved (54-97, 54-127) and two with altered garnets (43-33, 44-6). All samples are olivine normative and relatively low in alkalis; three samples are ultrabasic in terms of their silica content. However, all samples have low $Mg/Mg+Fe^{+2}$ ratios (64 to 53) which indicates a clear affinity to a basic rather than ultrabasic protolith. The low silica content of these samples may be a function of the locally high garnet to pyroxene ratio, as this group also includes garnet-poor eclogites and pure pyroxenites. The low alkali content may reflect depletion in these highly mobile elements during heating in the latite magma.

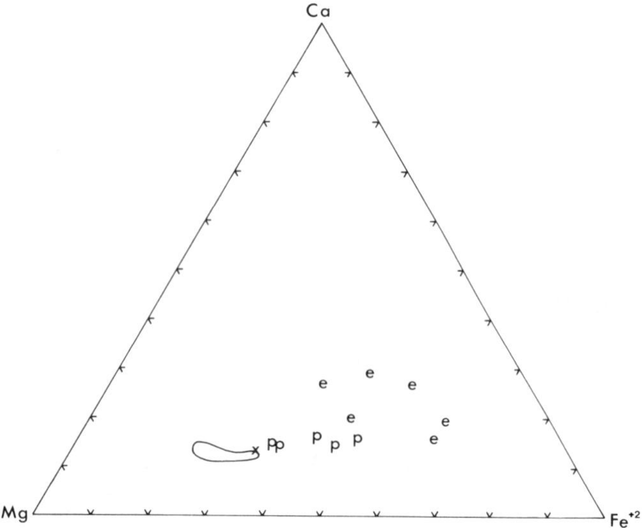

Fig. 5. Ca-Mg-Fe^{+2} plot of garnets from Prescott eclogites (e), Prescott garnet pyroxenites (p), and Roberts Victor eclogite SA 11-2 (x) (Helmstaedt and Schulze, unpublished data). Enclosed field represents garnets from Dish Hill (Shervais et al., 1973) and Salt Lake Crater (Beeson and Jackson, 1970) garnet pyroxenites.

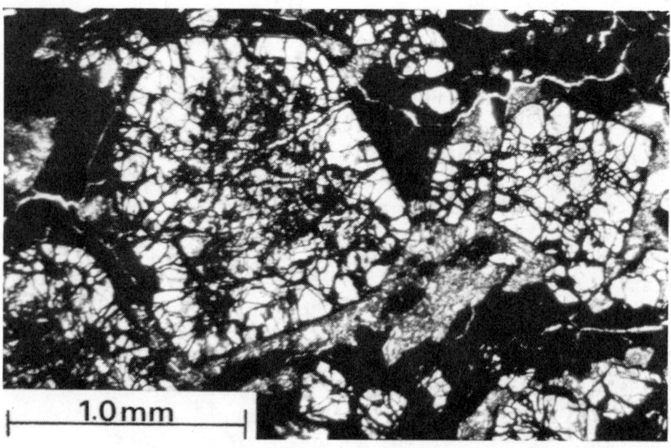

Fig. 7. Euhedral garnet porphyroblasts from eclogite 54-127. Note the opaque alteration along radial expansion fractures. Plane polarized transmitted light.

Mineral Chemistry

Chemical compositions of garnets, core pyroxenes, rim pyroxenes, and amphiboles are presented in Tables 6 and 7.

Clinopyroxene cores (omphacite) differ from the rims primarily in their jadeite content. The cores are relatively rich in jadeite (11-26 mole %) while the rims are depleted (0-7 mole %) and enriched in MgO, FeO and CaO (Fig. 9). Some rims contain still a substantial amount of Al_2O_3 (up to approximately 8 wt.%), now likely in the form of Tschermak's component.

The range of garnet compositions is presented in Table 6. Garnets are almandine-rich and contain a significant grossular component. On a ternary Ca-Mg-Fe diagram they do not overlap the garnet field from the garnet pyroxenites (Fig. 5). Analyses of the garnet alteration products plagioclase, orthopyroxene, and magnetite are not available due to their minute grain size and the fine nature of the intergrowths.

The ferropseudobrookite, the alteration product of the rutile, contains approximately 60 mole % $FeTi_2O_5$ (ferropseudobrookite) and 40 mole % $FeTi_2O_5$ (pseudobrookite).

Plagioclase intergrown with the altered clinopyroxene rims is very fine-grained. A prelim-

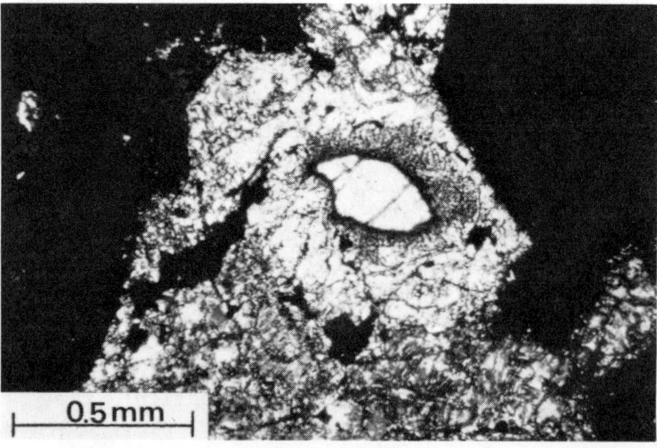

Fig. 6. Clear, unaltered omphacite core surrounded by "porous" partially melted rim of diopside and plagioclase in eclogite xenolith. The garnets are completely altered in this sample. Plane polarized transmitted light.

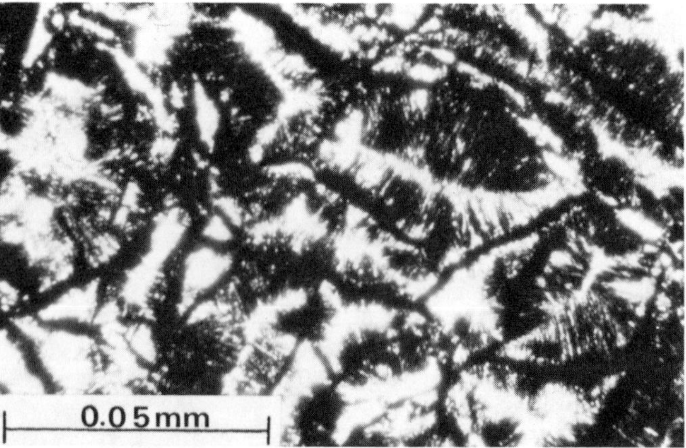

Fig. 8. Detail of dendritic growth pattern of the alteration assemblage replacing garnet in an altered eclogite, highlighted by opaque magnetite. Plane polarized transmitted light.

TABLE 6. Electron probe analyses of garnet, opx and cpx rim from eclogite 54-127, and ranges in other Chino Valley eclogites.

	Garnet	Range in other eclogites	Cpx core	Range in other eclogites	Cpx rim	Range in other eclogites
SiO_2	38.5	37.7 - 39.6	54.0	51.4 - 54.0	52.4	49.9 - 52.8
TiO_2	0.1	0.0 - 0.2	0.1	0.1 - 0.8	0.2	0.2 - 0.5
Al_2O_3	21.5	21.1 - 22.2	5.5	5.0 - 10.4	3.8	2.6 - 7.7
FeO^a	24.4	18.4 - 29.4	6.5	3.4 - 11.2	7.8	4.2 - 13.0
MnO	0.5	0.3 - 0.5	0.0	0.0 - 0.1	0.0	0.0 - 0.1
MgO	5.3	4.7 - 9.4	11.5	9.2 - 13.1	13.4	11.1 - 15.5
CaO	9.6	6.9 - 10.7	18.6	16.6 - 21.3	20.8	19.1 - 22.6
Na_2O^b	0.1	0.1 - 0.3	3.6	2.1 - 4.7	1.6	0.7 - 2.3
K_2O	0.0	0.0 - 0.0	0.0	0.0 - 0.1	0.0	0.0 - 0.0
Cr_2O_3	0.0	0.0 - 0.5	0.1	0.1 - 0.2	0.2	0.1 - 0.4
Total	100.0	99.9 - 100.8	99.9	99.6 - 100.9	100.2	99.6 - 100.9
mole % Ca	27.0	19.4 - 29.3				
Mg	20.6	18.4 - 35.6				
Fe^{+2}	52.4	37.2 - 62.2				
mole % Jd^c			20	11 - 26	4	0 - 7
Ac			6	4 - 16	7	6 - 14
Aug			74	67 - 85	89	84 - 95

a total Fe reported as FeO

b effective detectability limit of Na_2O is approximately 0.5 wt.%

c Jd calculated as total Na - $2Al^{IV}$, Ac calculated as Na - Jd, Aug calculated as 100% - (Jd + Ac)

inary analysis (54-97) yielded a composition of $Ab_{51}Or_3An_{46}$.

Preliminary analyses of amphiboles (Table 7) indicate that the altered (primary) amphiboles contain more Na_2O, MgO and SiO_2, but less Al_2O_3, K_2O, and FeO than the secondary (fresh) amphiboles. Analyses of the primary amphiboles correspond well with an analysis of barroisite from a Naustdal, Norway eclogite (Binns, 1967).

Discussion

Garnet pyroxenites

The subsolidus history of the pyroxenites outlined by Schulze et al. (1978) suggests that the present mineral assemblages (except late hydrous phases) originated by exsolution of garnet and an Fe-Ti oxide from three originally aluminous pyroxenes. Cpx exsolved garnet, partially recrystallized, exsolved Fe-Ti oxide ($\pm$ more garnet), deformed and recrystallized further. Opx was somewhat more resistant to recrystallization but has a similar exsolution and deformation history. The scpx also exsolved garnet, recrystallized, and unmixed to form the lamellar opx-cpx intergrowths. Because no porphyroclasts of scpx are preserved, it is not certain whether it exsolved an Fe-Ti oxide as well. Fe-Ti oxide grains within the opx-cpx intergrowths may represent recrystallized oxide lamellae.

Aluminum may be accommodated in the pyroxene in the hypothetical Tschermak's component $MAlAlSiO_6$, exsolving as garnet by the breakdown reaction:

$$\underbrace{MAlAlSiO_6 + 2MSiO_3}_{\text{pyroxene}} \rightarrow \underbrace{M_3Al_2Si_3O_{12}}_{\text{garnet}}$$

where M represents Ca, Mg, Fe^{+2}, etc.

Further garnet exsolution would accompany the expulsion of Ti from the pyroxene structure as ilmenite:

$$\underbrace{FeTiAlAlO_6 + 3MSiO_3}_{\text{pyroxene}} \rightarrow \underbrace{FeTiO_3}_{\text{ilmenite}} + \underbrace{M_3Al_2Si_3O_{12}}_{\text{garnet}}$$

The iron oxide accompanying the ilmenite may have been dissolved in the cpx as $Fe^{+2}Fe^{+3}AlSiO_6$

exsolving, along with $FeTiO_3$ (in solid solution) via the reaction:

$$\underbrace{2Fe^{+2}Fe^{+3}AlSiO_6 + FeSiO_3}_{\text{pyroxene}} \rightarrow \underbrace{Fe_3Al_2Si_3O_6}_{\text{garnet}} + \underbrace{Fe_2O_3}_{\text{oxide}}$$

In all of the pyroxenites, secondary amphibole (+ biotite) formed last. The websterites with unmixed scpx are more amphibolitized than the other pyroxenites, perhaps because they are more fine grained and therefore have a greater grain boundary surface area.

The pyroxenites have strong textural resemblances to garnet pyroxenite xenoliths described from basalts in other localities, e.g., Dish Hill, California (Shervais et al., 1973) and Salt Lake Crater, Hawaii (Beeson and Jackson, 1970), but differ from these in several respects. The Chino Valley samples are finer grained, contain Fe-Ti oxide lamellae, an inferred third pyroxene was present, and they have lower $Mg/Mg + Fe^{+2}$ values.

Campbell and Borley (1974) documented the partitioning of Ti into pyroxene with decreasing Cr content late in the crystallization history of a basaltic intrusion as the Cr content of the melt became depleted. This corresponds well with the correlation of "ilmenite" exsolution in pyroxenes with lower $Mg/Mg{+}Fe^{+2}$ values. The mean

TABLE 7. Electron Microprobe Analyses of Amphiboles from Eclogites.

	1	2	3	4	5
average of	16	3	1		5
SiO_2	49.8	48.6	37.3	49.68	40.9
TiO_2	0.3	0.3	0.2	0.43	1.1
Al_2O_3	9.1	11.5	17.5	9.92	15.4
Fe_2O_3				3.79	
FeO*	9.0	6.9	14.6	9.57	12.6
MnO	0.1	0.0	0.2	0.03	0.1
MgO	15.3	15.0	10.5	13.39	12.0
CaO	10.1	9.8	12.4	7.70	11.2
Na_2O	3.5	3.9	1.9	3.86	3.1
K_2O	0.7	0.4	1.5	0.15	1.1
Cr_2O_3	0.1	0.1	0.1	-	0.1
Total	98.0	96.5	96.2	98.52	97.6

* total Fe reported as FeO for 1, 2, 3, 5
1. primary amphibole, sample 54-127
2. primary amphibole, sample 54-63
3. secondary amphibole, sample 54-63
4. barroisite from eclogite, Naustdal, Norway, Binns (1967)
5. primary amphibole in garnet amphibole rock sample 54-82

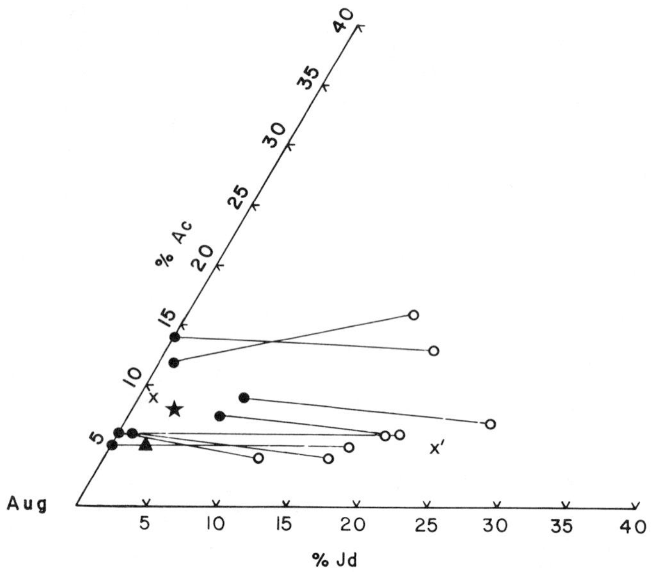

Fig. 9. Jadeite-Acmite-Augite (mole percent) plot of pyroxenes from Prescott. The open circles represent unaltered omphacite cores and are linked to their corresponding partially melted rim pyroxenes, denoted by filled circles. The x and x' represent rim and core of the pyroxenes in a partially melted Roberts Victor eclogite SA 11-2 (Helmstaedt and Schulze, unpublished data); the star represents the average clinopyroxene from three garnet pyroxenite xenoliths from Prescott; the triangle represents the average composition of the diopside phenocrysts in the latite host to the Prescott nodules.

TiO_2 content of the two Chino Valley pyroxenites is 0.77 wt.%, and the mean 100 $Mg/Mg{+}Fe^{+2}$ = 74, whereas in the Dish Hill sample (Shervais et al., 1973) the TiO_2 content is lower (0.31 wt.%) and the 100 $Mg/Mg{+}Fe^{+2}$ value is higher (83). The four Salt Lake pyroxenites which have exsolution textures (Beeson and Jackson, 1970) have a mean TiO_2 content of 0.61 wt.% and 100 $Mg/Mg{+}Fe^{+2}$ = 84. A sample without exsolution features has a low $Mg/Mg{+}Fe^{+2}$ value (74) and a higher TiO_2 content (1.76 wt.%).

Thus, the presence of Fe-Ti oxide lamellae in pyroxenes from garnet pyroxenites which have lower $Mg/Mg{+}Fe^{+2}$ values than most other garnet pyroxenites may indicate that these rocks represent a basic magma (or crystals accumulated from such a magma) which was more differentiated than that which produced the Dish Hill and Salt Lake pyroxenes.

Three pyroxenes have not been described previously from garnet pyroxenites, although clinopyroxenes, orthopyroxenes and subcalcic clinopyroxenes (individually or in pairs) are known in many other occurrences. Table 5 lists mineral compositions from a websterite derived from scpx (54-3). The constituent opx and cpx from the unmixed scpx are Al_2O_3-rich and the cpx is CaO-poor and the opx is CaO-rich relative to the porphyroclastic and granular cpx and opx of the other pyroxenites. The third pyroxene composition listed is a bulk electron probe analysis (the high Al_2O_3 and K_2O, which are higher than either the opx or cpx constituents, may be due to slight contamination by secondary amphibole). Plotted on a ternary Ca-Mg-Fe diagram (Fig. 4), the reconstructed scpx plots halfway between the opx and the cpx. Although three-pyroxene regions in other studies (e.g., Ross and Heubner, 1976; Lindsley et al., 1974) position the subcalcic pyroxene on the Fe-rich side of the opx-cpx join, our reconstructed scpx plots near the Mg-rich side of this join. Further reconstruction (to a pre-garnet exsolution stage) might change the position of the scpx relative to the cpx-opx join, but is not considered realistic on the basis of our present samples. We do not imply that the three pyroxenes described here must have been an equilibrium assemblage. The textural evidence shows only that the three aluminous pyroxenes were present together on a handspecimen scale. The pyroxenes may also have originated as the crystallization path of their source magma passed from one two-pyroxene field into another.

Subcalcic clinopyroxenes have been found in other garnet pyroxenites. Kornprobst (1969) described scpx with up to approximately 40% exsolved orthopyroxene from clinopyroxene in the Beni Bouchera ultramafic massif. Wilkinson's (1976) reconstructed parent subcalcic clinopyroxenes from Salt Lake Crater contain approximately 15 wt.% CaO that later unmixed to opx lamellae in a cpx which now contains 19 wt.% CaO. Green (1966) reported (subcalcic) clinopyroxene in equilibrium with orthopyroxene (both of which were aluminous) near the liquidus of an olivine basalt at 1335°C and 18 kb that contained 10.8 wt.% CaO. The Al_2O_3 content of this scpx (9.9 wt.%) is similar to that of the whole rock of our websterite 54-45 is (8.2 wt.%) which is thought to be representative of an aluminous subcalcic clinopyroxene prior to garnet exsolution. It is therefore concluded that the Chino Valley clinopyroxenites originated by partial or complete crystallization of a basic magma at high pressures where aluminous pyroxenes $\pm$ garnet $\pm$ an Fe-Ti oxide were the stable assemblage. Upon cooling it reverted to a pyroxene (Al-poor) + garnet + Fe-Ti oxide assemblage.

Eclogites. Although no major changes occurred in the garnet pyroxenites, the eclogites underwent drastic mineralogical changes when heated in the latite magma. Garnet altered in the solid state, but portions of the clinopyroxenes may have melted, yielding a Na-Al-rich liquid and leaving a jadeite-poor diopside residue. The plagioclase in the interstices of the diopside residue may be the result of crystallization of this liquid upon cooling. However, as no glass is preserved, it is not certain whether the heating caused actual partial melting or merely alteration in the solid state.

Switzer and Melson (1969) and Chinner and Cornell (1974) have described eclogites and a grospydite from the Roberts Victor kimberlite (South Africa) in which the pyroxenes have undergone similar textural and chemical changes. Switzer and Melson who identified glass attributed the textures to partial melting during adiabatic rise within the relatively cool kimberlite. Our example represents a breach of the omphacite breakdown curve by heating and decompression in the rising latite magma.

P-T conditions. Preliminary estimation of pressures and temperatures of equilibration of the pyroxenite and eclogite assemblages prior to incorporation in the magma is restricted to calculations of the distribution coefficients (K_D) of Fe^{+2} and Mg between clinopyroxenes and garnets. Using the average garnet and average granular and porphyroclastic clinopyroxene compositions of the pyroxenites, K_D values of 8.4 (54-17) and 10.1 (1-2) have been calculated for these assemblages. If the unaltered cores of omphacites represent the composition of the clinopyroxenes in equilibrium with garnet, K_D values for the eclogites are 7.3 (54-30), 8.2 (54-2), and 11.4 (54-127). The K_D values for pyroxenites and eclogites are very similar and correspond to temperatures of about 550-700°C in the pressure range of 10-20 kb (Råheim and Green, 1974). This is definitely below any reasonable liquidus temperature inferred for the host latite.

Eclogite-pyroxenite relationships

Our data are compatible with the interpretation that igneous pyroxenites at depth intruded metamorphic eclogites or their protoliths. Upon cooling the pyroxenites exsolved garnet and Fe-Ti oxide, were deformed, recrystallized, and later hydrated (secondary amphibole and biotite) along with the eclogites prior to incorporation as xenoliths in the latite. The pyroxenites changed relatively little during transport in the latite magma, while the eclogites were highly altered and may have been partially melted.

Whereas we interpret the layering in the eclogites as metamorphic feature, Arculus and Smith (in press) have proposed that it may represent a cumulate texture. The pyroxenites are considered as part of the cumulate sequence. A nodule collected by us during the Second International Kimberlite Conference at the locality studied by Arculus and Smith may help to solve the problem. The composite nodule shows a sharp contact between a garnet pyroxenite and an eclogite that is parallel with a foliation emphasized by amphibole-rich portions. Both cpx and scpx (unmixed) can be recognized in the pyroxenite portion less than 1 cm from the eclogite contact. The occurrence of cpx, scpx, and eclogitic cpx within less than 1 cm of each other, shows that the present layering can hardly be a cumulate feature. It does not rule out, however, that the present layering may represent a former igneous layering transposed by deformation and metamorphism.

Relationships to other eclogites

Although ultramafic xenoliths are common in basaltic rocks of the western United States, xenoliths of metamorphic eclogites have not been identified among them. However, highly sodic eclogites are known from the Four Corners kimberlitic diatremes (Watson, 1960; O'Hara and Mercy, 1966; Watson and Morton, 1969). They are of metamorphic origin and have been interpreted by Helmstaedt and Doig (1975) to represent fragments of oceanic lithosphere. According to such an hypothesis, eclogites should be common under the southwestern United States, and should occur as xenoliths elsewhere on the Colorado plateau. Do the Chino Valley eclogites represent such occurrence? In their present state they differ from the Four Corners suite in several respects. The Four Corners eclogitic clinopyroxenes are more sodic, containing up to approximately 70 mole percent jadeite vs. only 20-25 mole percent maximum in the Chino Valley rocks. The garnets from the Four Corners eclogites are somewhat more iron-rich (approximately 55-75 mole percent almandine vs. approximately 35-65 mole percent from Chino Valley). Amphibole is not common in the Four Corners eclogites, lawsonite is present in the Four Corners rocks and the accompanying members of the Four Corners ultramafic suite are similar to meta-ophiolites (Helmstaedt and Schulze, in press). They include garnet peridotites, spinel peridotites (deformed and undeformed), spinel websterites, garnet (pyrope-rich) clinopyroxenites, and various hydrated ultramafic rocks. The differences between the two types of eclogites are essentially those between Group C and Group B eclogites (Coleman *et al*., 1965).

If the two occurrences are related, the differences between the eclogites could be due to the higher temperature of metamorphism of the Chino Valley rocks, to sodium loss during heating in the latite, or perhaps to derivation from a different depth or a different protolith. On the other hand, the xenoliths may be samples of the Precambrian basement of the Colorado plateau or part of a continental upper mantle of Precambrian age. Compelling reasons for or against a common mode of origin of the two eclogites do not exist at present. Chances of finding low-temperature eclogites as xenoliths elsewhere on the plateau are slim. The eclogites from Chino Valley were highly altered in the latite magma. It is most likely that they would not have survived the transport to the surface in more basic magmas. The absence of other eclogite xenolith occurrences in the southwestern United States, therefore, must not mean that such rocks were not present elsewhere in the lower crust or upper mantle of this region, but may reflect the lack of sufficiently cool transport media needed to preserve them during their rise to the surface.

Acknowledgements. We are thankful to the 1975 field school staff of the University of Arizona, Phoenix, for bringing the Chino Valley nodules to our attention, and to Christine Coffey for her help in collecting the samples on a subsequent visit to Chino Valley in 1976.

This research was supported by National Research Council of Canada Grant No. A8375. Financial assistance to attend the Second International Kimberlite Conference from the Organizing Committee and the Graduate Faculty of Queen's University is gratefully acknowledged. Critical reviews by Steve Ehrenberg and Douglas Smith helped to improve the manuscript.

References

Arculus, R. J. and D. Smith, Dense inclusions in the Sullivan Buttes Latite, Chino Valley, Yavapai County, Arizona, Proceedings of the Second International Kimberlite Conference, in press.

Beeson, M.H., and E.D. Jackson, Origin of the garnet pyroxenite xenoliths at Salt Lake Crater, Oahu, Mineral. Soc. Amer. Spec. Paper 3, pp. 95-112, 1970.

Binns, R.A., Barroisite-bearing eclogite from Naustdal, Sogn og Rjordane, Norway, J. Petrol., v. 8, 349-371, 1967.

Campbell, I.H., and G.D. Borley, The geochemistry of pyroxenes from the lower layered series of the Jimberlana Intrusion, Western Australia, Contrib. Mineral. Petrol., 47, 281-297, 1974.

Chinner, G.A., and D.H. Cornell, Evidence of kimberlite-grospydite reaction, Contrib. Mineral. Petrol., 45, 153-160.

Coleman, R.G., D.E. Lee, L.B. Beatty, and W.W. Brannock, Eclogites and eclogites: Their differences and similarities, Geol. Soc. Amer. Bull., 76, 483-508, 1965.

Dickey, J.S., Partial fusion products in alpine-type peridotites: Serrania de la Ronda and other examples, Miner. Soc. Amer. Special Paper 3, 33-49, 1970.

Green, D.H., The origin of the "eclogites" from Salt Lake Crater, Hawaii, Earth Planet. Sci. Letters, 1, 414-420, 1966.

Helmstaedt, H., and R. Doig, Eclogite nodules from kimberlite pipes of the Colorado Plateau - Samples of subducted Franciscan-type oceanic lithosphere, Phys. Chem. Earth, 9, 95-111, 1975.

Helmstaedt, H., and D.J. Schulze, Garnet clinopyroxenite-chlorite eclogite transition in a xenolith from Moses Rock: further evidence for metamorphosed ophiolites under the Colorado plateau, Proceedings of the Second International Kimberlite Conference, in press.

Kornprobst, J., Le massif ultrabasique des Beni Bouchera (Rif Interne, Maroc): Etude des peridotites de haute temperature et de haute pression, et des pyroxénolites, à grenat ou sans grenat, qui leur sont associées, Contrib. Mineral. Petrol., 23, 283-322, 1969.

Krieger, M.H., Geology of the Prescott and Paulden Quadrangles, Arizona, U.S. Geological Survey Professional Paper 467, 127, 1965.

Krieger, M.H., S.C. Creasy, and A.F. Marvin, Ages of some Tertiary andesite and latitic rocks in the Prescott-Jerome area, north-central Arizona, in Geological Survey Research, 1971: U.S. Geol. Survey Prof. Paper 570-B, B157-B160, 1971.

Lasnier, B., Les peridotites et pyroxénolites à grenat du Bois des Feuilles (Monts du Lyonnais), France, Contrib. Mineral. Petrol., 34, 29-42, 1971

LeMaitre, R.W., Some problems of the projection of chemical data into mineralogical classifications, Contrib. Mineral. Petrol., 56, 181-189, 1976.

Lindsley, D.H., H.E. King, Jr., A.C. Turnock, and J.E. Grover, Phase relations in the pyroxene quadrilateral of 980°C and 15 kbar, Geol. Soc. Amer. Abstracts with programs, 6, 846, 1974.

Lovering, J.F., and A.J.R. White, Granulitic and eclogitic inclusions from basic breccia pipes of Delegate, Australia, Contrib. Mineral. Petrol., 21, 9-52, 1969.

O'Hara, M.J., and E.L.P. Mercy, Peridotite and pyrope from the Navajo Country, Arizona and New Mexico, Amer. Mineral., 51, 336-52, 1966.

Råheim, A., and D.H. Green, Experimental determineration of the temperature and pressure dependence of the Fe-Mg partition coefficient for coexisting garnet and clinopyroxene, Contrib. Mineral. Petrol., 48, 179-203, 1974.

Ross, M. and J. S. Heubner, A pyroxene geothermometer based on composition-temperature relationships of naturally occurring orthopyroxene, pigeonite, and augite, Extended Abstracts, Int. Conf. Geothermom. Geobarom. Penn State Univ., 1975.

Schulze, D.J., H. Helmstaedt, and R.M. Cassie, Pyroxene-ilmenite intergrowths in garnet pyroxenite xenoliths from a New York kimberlite and Arizona latite, Amer. Mineral., 63 258-265, 1978.

Shervais, J.W., H.G. Wilshire, and E.D. Schwartzman, Garnet clinopyroxenite xenoliths from Dish Hill, California, Earth Planet. Sci. Letters, 19, 120-130, 1973.

Switzer, G., and W.G. Melson, Partially melted kyanite eclogite from the Roberts Victor Mine, South Africa, Smithsonian Contrib. Earth Sci., 1, 1-9, 1969.

Watson, K.D., Eclogite inclusions in serpentine pipes at Garnet Ridge, northeastern Arizona, Geol. Soc. Amer. Bull., 71, 2082-2083, 1960.

Watson, K.D. and D.H. Morton, Eclogite inclusions in Kimberlite pipes at Garnet Ridge, northeastern Arizona, Amer. Mineral., 54, 267-285, 1969.

Wilkinson, J.F.G., Some subcalcic clinopyroxenites from Salt Lake Crater, Oahu, and their petrogenetic significance, Contrib. Mineral. Petrol., 58, 181-201, 1976.

GARNETIFEROUS ULTRAMAFIC INCLUSIONS IN MINETTE FROM THE NAVAJO VOLCANIC FIELD

Stephen N. Ehrenberg

Department of Earth and Space Sciences, University of California, Los Angeles, California 90024

Abstract. Ultramafic xenoliths at The Thumb, a small minette neck near Shiprock, New Mexico, are divisible into three overlapping textural/compositional groups: coarse garnet peridotites with minerals relatively low in Fe and Ti and high in Cr, sheared garnet peridotites with minerals relatively high in Fe, Ti, and Cr, and megacrystalline rocks (medium- to ultracoarse-grained intergrowths of clinopyroxene, orthopyroxene, olivine, garnet, phlogopite, and ilmenite) with minerals very rich in Fe and Ti and low in Cr. The megacrystalline rocks are compositionally similar to the "discrete nodule" suite of Lesotho kimberlites, but are fragments of polycrystalline aggregates rather than phenocryst-like single crystals. Pyroxene compositions indicate that the inclusions equilibrated at temperatures ranging from 930-1230°C within a narrow vertical interval at around 130 km depth.

The extreme textural and compositional heterogeneity of this restricted source terrane and the correlation between xenolith textures and mineral compositions are explained by the following model: The megacrystalline rocks are interpreted as precipitates from liquids introduced into an older, previously depleted terrane consisting of the coarse garnet peridotites. The Ti-Fe-rich sheared peridotites are thought to be the products of deformation and Ti-Fe-metasomatism of the coarse peridotite country rocks in tectonically disrupted aureoles surrounding the intrusive megacrystalline regions. The suggested metasomatizing medium is a residual liquid intermediate in composition between clinopyroxene and ilmenite.

The model for The Thumb's inclusion suite may also be extended to explain the texturally and compositionally analogous suite of xenoliths described by Boyd and Nixon (1974) from Lesotho kimberlites. The Lesotho sheared garnet peridotites are viewed as the products of extensive metasomatism of previously less enriched mantle in response to intrusion of magma from which the discrete nodule association crystallized.

Introduction

The Navajo volcanic field formed in Late Oligocene time (25-32 m.y.b.p.) as small volumes of potassic magma and abundant diatreme-forming volatiles erupted throughout an area of ~20,000 km near the center of the Colorado Plateau. This region was tectonically stable throughout Paleozoic and Mesozoic time and was broken into major upwarps and basins in response to Laramide deformation (~60 m.y. ago; Kelly, 1955). Smith and Levy (1976) suggest that the apparent localization of the volcanic centers along monoclines bounding the Laramide upwarps may reflect greater ease of conduit formation along these deep preexisting fault zones. The volcanic rocks occur as numerous dikes, necks and breccia pipes, as well as several maar-type craters with associated flows.

Two fundamentally distinct types of volcanic rocks are present: minette (potassic phlogopite-diopside-sanidine lamprophyre) and "kimberlite" (an unusual variety relatively low in incompatible elements). The minette is thought to have formed by small percentage of melting of garnet peridotite (Kay and Gast, 1973), whereas the kimberlite appears to be the product of subsolidus eruption of volatiles plus altered peridotite (McGetchin and Silver, 1970). The kimberlites and their high-pressure xenoliths have been extensively studied for information about the upper mantle underlying the Colorado Plateau (McGetchin and Silver, 1970; Smith and Levy, 1976).

Recently it was recognized that the minettes also contain xenoliths from the upper mantle (Baldridge, et al., 1975), and subsequent field studies located peridotite xenoliths at 16 of 22 minette centers investigated. The presence of garnet-bearing peridotites at several of these localities is especially interesting because garnet peridotite fragments are extremely rare in the Navajo kimberlites (although pyrope xenocrysts are abundant). Peridotite xenoliths at most of the minette localities are small

Table 1. Microprobe analyses of minerals in minette from the Thumb. Numbers of grains averaged are indicated in parentheses. Cation proportions for phlogopite, clinopyroxene, and feldspar are based on 22, 6, and 8 oxygens, respectively.

	PHLOG (11)	CPX (10)	FSP (5)
SiO_2	40.5	54.6	66.0
TiO_2	3.55	.46	.17
Al_2O_3	13.1	.84	18.7
Cr_2O_3	.51	.12	.02
FeO*	5.86	3.90	.92
MnO	.05	.09	.00
MgO	21.9	18.0	.01
CaO	.06	21.5	.19
Na_2O	.32	.37	4.33
K_2O	9.97		10.5
	95.8	99.8	100.8
Si	5.773	1.984	2.979
Ti	.380	.013	.006
Al	2.205	.036	.995
Cr	.057	.003	.001
Fe	.677	.119	.035
Mn	.006	.003	.000
Mg	4.656	.975	.001
Ca	.009	.837	.009
Na	.088	.026	.379
K	1.812		.605
	15.663	3.996	5.009

(<4 cm) and either highly altered or scarce, but a small neck called The Thumb was found to contain abundant little-altered garnet peridotite fragments ≤30 cm in diameter. This neck is located 2 km southeast of the Redrock Trading Post (accessible via the Redrock Highway, southwest from the town of Shiprock, New Mexico). Of 120 garnet peridotite and 42 megacrystalline-group inclusions cataloged from this locality, 46 and 22, respectively, were selected for analysis by electron probe and other techniques. Six garnet peridotites from other minette localities have also been analyzed.

Minette Host Rock

The Thumb is an oval neck of uniform non-vesicular minette 100 m in diameter. As noted by Williams (1936), xenoliths mostly <50 cm in diameter compose around 10% of the neck. Most of these are sedimentary rocks, acid plutonic rocks, and basic garnet granulites. Peridotites comprise less than a percent of the total.

The minette consists of ~23% phlogopite and 17% diopside phenocrysts in an aphanitic (<10μm) groundmass of biotite, chlorite, Ti-magnetite, clinopyroxene, and sanidine, along with minor calcite and glass (Table 1). On interelement plots of Ti, Cr, and Fe the phenocrysts fall within the range of variation of the phlogopites and clinopyroxenes in the ultramafic xenoliths from this locality (Fig. 2). However, the clinopyroxene phenocrysts are much lower in jadeite and enstatite content than the clinopyroxenes in the xenoliths, probably reflecting lower pressures and temperatures of crystallization of the phenocrysts.

Abundant xenocrysts of clinopyroxene and phlogopite derived from the megacrystalline-group inclusions (described below) may be recognized by their large size and jagged outlines compared to the phenocrysts. Most thin-sections contain 2 to 6 grains each of xenocrystic mica and clinopyroxene which are overgrown by thick mantles of the phenocryst phlogopite and diopside, respectively. Xenocrysts of magnesian ilmenite are also common.

Peridotite Xenoliths

The ultramafic xenoliths at The Thumb are almost exclusively of garnet lherzolite mineralogy. Only one spinel lherzolite and one websterite have been found. The garnet peridotites are quite variable in modal proportions (Fig. 1 and Table 2) and include lherzolites, harzburgites, and a few dunites. Phlogopite,

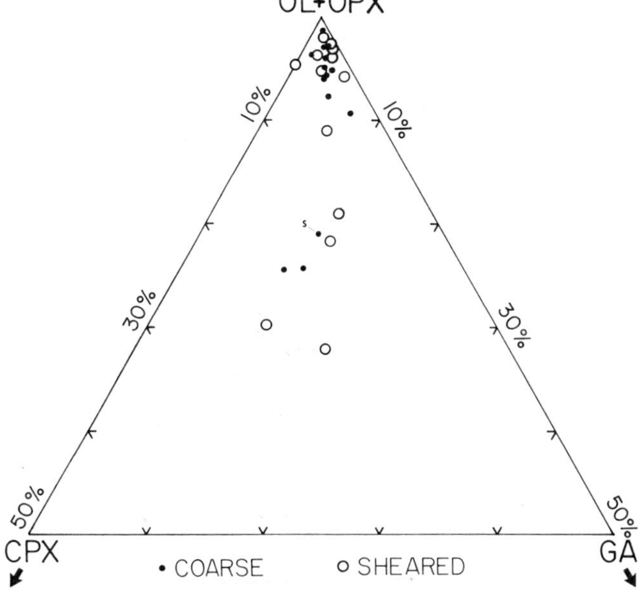

Fig. 1. Modal proportions of 26 garnet peridotite xenoliths from The Thumb and one from Ship Rock(s).

Table 2. Microprobe and modal analyses of minerals in garnet peridotite xenoliths. Cation proportions for pyroxenes, garnets, phlogopites, and ilmenites calculated on the basis of 6, 12, 22, and 3 oxygens, respectively. The S-ratio (defined in text) is given for each element following its abundance. Numbers of analytic points averaged for single and replicate analyses are given in parentheses. Elements not analyzed are left blank. Sample names ending in THM, SPR, and BRD are from The Thumb, Ship Rock, and Black Rock Dike (near Fort Defiance), respectively. Analytic techniques are given in Ehrenberg (1978).

	N077THM (COARSE)								I078THM (COARSE)					
	CPX	(6)	OPX	(4)	GA	(12)	Ph	(4)	CPX	(3 & 6)	OPX	(3 & 6)	GA	(8)
SiO_2	54.5	3	56.7	2	41.2	2	40.7	2	54.6	2	57.2	2	40.9	2
TiO_2	.42	2	.19	3	.12	5	2.85	2	.03	1	.02	2	.07	1
Al_2O_3	2.05	5	1.40	1	20.8	13	13.4	1	2.00	2	1.50	1	20.4	2
Cr_2O_3	1.37	6	.58	1	3.77	23	1.05	3	1.17	1	.50	2	4.56	3
FeO*	2.61	2	5.02	1	6.85	2	3.74	1	2.73	1	5.27	1	6.74	1
MnO	.09	1	.11	2	.41	1	.01	1	.08	1	.12	1	.33	1
MgO	18.0	3	35.1	4	20.0	3	23.4	1	18.8	2	34.8	1	20.0	2
CaO	20.2	3	.79	1	6.08	3	.12	2	20.2	4	.98	1	5.80	1
Na_2O	1.39	4	.11	1	.01	1	.22	6	.98	1	.10	3	.01	1
K_2O							9.86	4						
	100.6		100.0		99.3		95.3		100.6		100.5		98.7	
Si	1.959		1.951		2.975		5.768		1.961		1.959		2.970	
Ti	.011		.005		.007		.304		.0008		.0005		.004	
Al	.087		.057		1.769		2.233		.085		.061		1.743	
Cr	.039		.016		.215		.118		.033		.014		.262	
Fe	.078		.144		.413		.443		.082		.151		.409	
Mn	.003		.003		.025		.001		.002		.003		.020	
Mg	.965		1.800		2.152		4.945		1.003		1.776		2.163	
Ca	.776		.029		.470		.018		.778		.036		.451	
Na	.097		.007		.001		.060		.068		.007		.001	
K							1.784							
	4.015		4.012		8.027		15.675		4.014		4.007		8.024	
Modal %	1.7		19		3.5		1		3.3		19		4.4	
Olivine:														
Mg/(Mg+Fe)	90.9								91.0					
wt. % CaO	0.08								0.10					

	R077THM (COARSE)						A082THM (COARSE)						140THM (SHEARED)							
	CPX	(3 & 6)	OPX	(3 & 6)	GA	(6)	CPX	(3 & 6)	OPX	(3 & 4)	GA	(8)	CPX	(4)	OPX	(5)	GA	(7)	ILM	(6)
SiO_2	54.6	2	57.0	1	41.4	2	54.2	2	56.8	2	41.1	2	54.3	2	57.2	2	40.8	1	.02	1
TiO_2	.28	1	.15	1	.37	1	.61	1	.30	2	.27	5	.63	1	.28	1	.70	1	51.0	6
Al_2O_3	3.16	1	1.80	1	21.9	3	3.03	1	1.51	3	21.5	4	3.34	1	1.54	1	21.9	5	.96	43
Cr_2O_3	.87	2	.29	1	2.04	1	1.11	1	.37	2	2.80	4	.39	5	.09	1	.96	11	1.39	12
FeO*	3.46	1	5.99	2	7.49	1	3.50	2	6.39	2	8.22	3	5.34	1	8.71	1	11.5	1	27.49	3
MnO	.11	1	.12	1	.29	1	.10	1	.11	1	.39	1	.08	1	.10	1	.33	1	.21	2
MgO	18.4	2	34.0	2	21.2	1	17.4	1	33.8	2	20.7	2	16.8	4	31.2	1	19.1	3	10.24	4
CaO	18.3	3	1.07	3	4.79	1	19.1	2	.90	3	5.06	1	17.1	3	1.03	1	4.91	1	.02	1
Na_2O	1.67	1	.21	2	.03	1	1.83	1	.17	1	.03	1	2.06	1	.22	1	.07	1		
K_2O																			**9.27	
	100.8		100.6		99.6		100.9		100.4		100.0		100.1		100.4		100.3		100.6	
Si	1.951		1.956		2.960		1.945		1.958		2.945		1.965		1.986		2.943		.000	
Ti	.008		.004		.020		.016		.008		.016		.017		.007		.038		.892	
Al	.133		.073		1.848		.128		.061		1.816		.142		.063		1.867		.026	
Cr	.025		.008		.115		.031		.010		.159		.011		.002		.055		.026	
Fe	.103		.172		.448		.105		.184		.493		.162		.253		.693		.534	
Mn	.003		.003		.018		.003		.003		.024		.002		.003		.020		.004	
Mg	.981		1.738		2.261		.933		1.736		2.211		.907		1.614		2.058		.355	
Ca	.700		.039		.367		.733		.033		.389		.662		.038		.380		.000	
Na	.116		.014		.004		.127		.011		.004		.144		.015		.010			
K																			Fe^{3+} .162	
	4.020		4.007		8.040		4.022		4.005		8.055		4.013		3.982		8.063		1.999	
Modal %	14		18		11		15		16		9		5		13		6			
Olivine:																				
Mg/(Mg+Fe)	90.1						89.5						84.3							
wt. % CaO	0.10						0.10						0.14							

	B082THM (SHEARED)						U078THM (SHEARED)						D076THM (SHEARED)							
	CPX(3 & 6 & 6)		OPX	(3 & 6)	GA	(8)	CPX		OPX	(3 & 8)	GA	(6)	CPX	(6)	OPX	(6)	GA	(7)	PHLOG	(3 & 5)
SiO_2	53.9	2	55.9	2	40.7	2	54.8	1	56.6	1	40.9	2	54.9	1	57.1	1	40.5	2	40.9	2
TiO_2	.66	1	.35	1	.70	2	.77	1	.36	1	.78	2	.68	2	.35	2	.90	3	3.57	1
Al_2O_3	3.57	2	2.10	2	21.6	3	3.58	1	1.72	4	21.3	1	3.37	1	1.71	3	18.2	2	13.6	7
Cr_2O_3	.55	2	.22	2	1.78	7	.95	1	.30	1	2.40	1	2.37	2	.78	1	6.29	4	1.64	1
FeO	5.40	2	8.54	2	10.1	1	3.83	2	6.64	1	8.28	4	3.00	1	5.19	1	6.87	1	3.66	2
MnO	.14	1	.15	1	.31	1	.10	1	.12	1	.28	1	.10	1	.27	1	.32	1	.01	1
MgO	18.1	2	31.7	1	19.9	2	17.3	1	33.5	1	20.8	3	17.1	2	34.6	2	20.2	3	22.8	3
CaO	16.5	2	1.34	1	4.72	2	17.7	1	.98	2	5.19	2	17.5	1	.95	2	5.76	2	.04	1
Na_2O	1.80	1	.27	1	.05	1	2.13	2	.22	1	.08	1	2.29	2	.37	1	.08	1	.44	29
K_2O																			9.58	7
	100.5		100.6		99.8		101.2		100.5		99.9		101.4		101.4		99.1		96.2	
Si	1.939		1.945		2.937		1.953		1.952		2.933		1.953		1.944		2.959		5.736	
Ti	.018		.009		.038		.021		.009		.042		.018		.009		.049		.377	
Al	.152		.086		1.836		.150		.070		1.804		.141		.068		1.563		2.251	
Cr	.016		.006		.102		.027		.008		.136		.067		.021		.363		.182	
Fe	.163		.248		.612		.114		.192		.497		.089		.148		.419		.430	
Mn	.004		.004		.019		.003		.004		.017		.003		.008		.020		.001	
Mg	.969		1.642		2.143		.920		1.721		2.220		.907		1.758		2.200		4.772	
Ca	.637		.050		.365		.677		.036		.399		.667		.035		.450		.006	
Na	.126		.018		.007		.147		.015		.011		.158		.024		.011		.120	
K																			1.716	
	4.022		4.009		8.059		4.012		4.007		8.060		4.004		4.015		8.035		15.591	
Modal %	20		10		10		10		16		12		2.3		18		1.8		1	
Olivine:																				
Mg/(Mg+Fe)	85.4						88.9						91.3							
wt.% CaO	0.17						0.12						0.09							

	0011SPR (COARSE)		altered		H051BRD (COARSE)			
	CPX (5&4)*	OPX (6)* (CIT)	GA (7&3)* (ave.)	PHLOG (5&3)* (ave.)	CPX	(6)	GA	(6)
SiO_2	54.7	57.3	41.4	41.4	55.1	1	41.6	1
TiO_2	.18	.10	.20	1.40	.43	2	.49	1
Al_2O_3	1.69	.70	21.6	13.3	3.47	2	22.0	1
Cr_2O_3	1.00	.20	2.04	.63	.92	2	1.50	2
FeO	2.97	6.26	9.50	4.30	3.50	1	8.51	2
MnO	.09	.13	.40	.03	.07	1	.30	2
MgO	17.0	34.9	19.0	24.3	16.4	1	21.1	2
CaO	21.2	.44	4.82	.05	19.1	1	4.77	2
Na_2O	1.41	.07	.38	.72	2.44	1	.06	2
K_2O			.11	9.97				
	100.3	100.1	99.5	96.1	101.4		100.4	
Si	1.981	1.969	2.995	5.833	1.965		2.960	
Ti	.005	.003	.011	.148	.012		.026	
Al	.072	.028	1.842	2.211	.146		1.848	
Cr	.029	.005	.167	.070	.026		.084	
Fe	.090	.180	.575	.507	.104		.506	
Mn	.003	.004	.025	.036	.002		.018	
Mg	.917	1.788	2.047	5.104	.872		2.237	
Ca	.822	.016	.373	.008	.727		.364	
Na	.099	.005	.053	.197	.169		.008	
K			.010	1.792				
	4.018	3.998	8.098	15.906	4.022		8.051	
Modal %	10	13	10	3.5				
Olivine:								
Mg/(Mg+Fe)	90.1							
wt.% CaO	0.03							

*Average of analyses done at California Institute of Technology by W.S. Baldridge and at U.C.L.A., except for OPX (C.I.T. analysis).

**Fe_2O_3 (calculated by method of Carmichael, 1967)

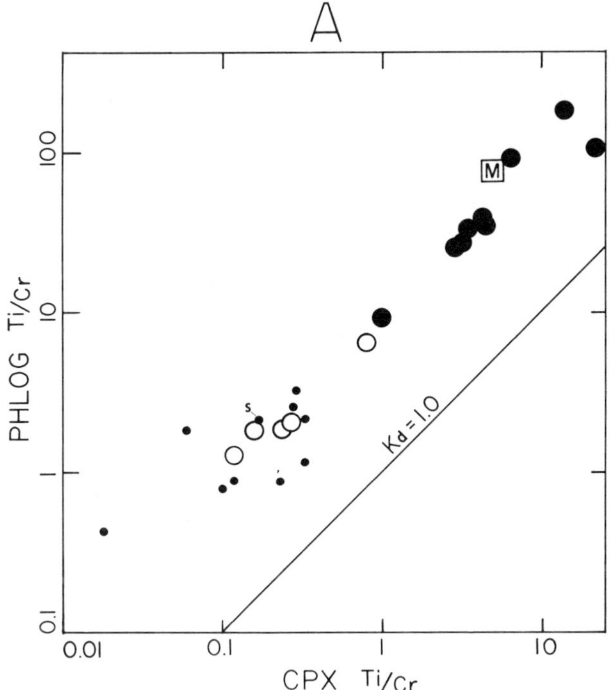

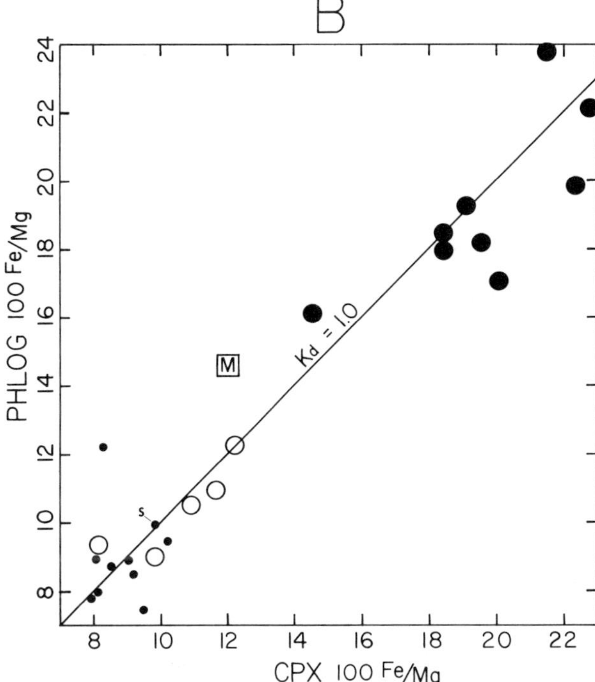

Fig. 2. Corvariation of phlogopite and clinopyroxene compositions in ultramafic xenoliths from The Thumb and Ship Rock (s). "M" is average for minette phenocyrsts at The Thumb (Table 1). Other symbols are explained in the caption to Fig. 3.

chromium-rich spinel, and ilmenite are common accessory minerals. Exsolution on the scale visible with the petrographic microscope is insignificant or absent in the pyroxenes of the peridotite xenoliths.

Garnets in all nodules are altered along their margins to kelyphytic intergrowths of aluminous spinel, pyroxenes, and in some cases feldspar. Edges of clinopyroxenes are commonly altered to turbid clinopyroxene of lower jadeite and higher diopside content. This effect is especially pronounced in the jadeite-rich clinopyroxenes of the megacrystalline nodules.

Textures of 71 peridotite xenoliths were classified using the terminology of Hart (1977). These include 45 coarse peridotites (9 of which contain minor regions of fine-grained mosaic texture olivine), 6 porphyroclastic rocks, and 20 mosaic-porphyroclastic rocks. Several peridotites display modal inhomogeneities, such as garnet- or pyroxene-rich lenses and variations in modal proportions from one side of a xenolith to the other, and a few nodules contain veins 1-2 millimeters thick or discontinuous vein-like concentrations composed of variable combinations of phlogopite, chromite, clinopyroxene, orthopyroxene, and rutile. Minerals within the veins are similar in composition to the minerals in the rest of the rock.

Phlogopite in the peridotites typically occurs as subhedral crystals a few millimeters in longest dimension resembling the "primary" micas illustrated by Carswell (1974, Fig. 1A). However, in two nodules the garnets are rimmed by fine-grained phlogopite resembling the "secondary" phlogopite illustrated by Carswell (1974, Fig. 1B). Commonly the phlogopite appears to have formed after crystallization of the anhydrous minerals because it occurs in veins or is localized along the margins of garnets or clinopyroxenes. Phlogopite compositions show rough linear correlations with the compositions of the coexisting clinopyroxenes (Fig. 2A and B), orthopyroxenes, garnets, and olivines.

Minerals in some of the peridotites appear to be homogeneous for all major elements, but inhomogeneous distribution of one or more elements is observed in many samples. The magnitudes of these inhomogeneities may be appreciated from the S-ratios given for each element in Table 2. The S-ratio ($s/\sqrt{x}$) (Boyd, 1967) is the ratio of the observed variation in the numbers of X-ray counts accumulated on different grains (the standard deviation) divided by the variation attributable to counting statistics (the square root of the mean number of counts). For elements having concentrations greater than 1% S-ratios of 1 to 3 are consistent with homogeneous distribution in the points analyzed, while S-ratios >3 suggest inhomogeneity.

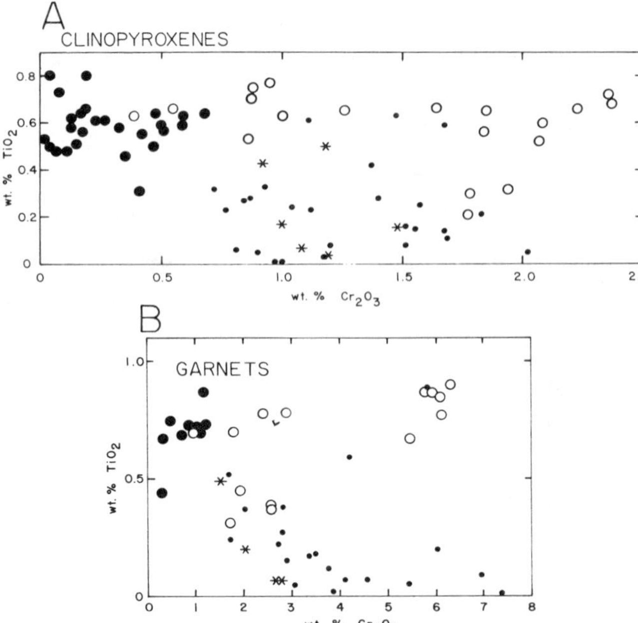

Fig. 3. Wt.% TiO_2 vs. wt.% Cr_2O_3 in clinopyroxenes and garnets from ultramafic xenoliths mostly from The Thumb. Dots represent coarse garnet peridotite minerals. Large open circles are sheared garnet peridotite minerals. Large filled circles are megacrystalline minerals. Asterisks are minerals of coarse garnet peridotites from localities other than The Thumb.

The peridotite xenoliths display a strong overall correlation between texture and mineral chemistry, although several samples provide exceptions to the general pattern. The samples in Table 2 span the range of compositional variation of the analyzed peridotites. Garnets and clinopyroxenes in the sheared peridotites (porphyroclastic and mosaic-porphyroclastic) tend to be richer in Ti than in the coarse peridotites, although Cr contents of the two groups overlap (Fig. 3). Pyroxenes and olivines in the sheared peridotites have generally lower Mg/(Mg+Fe) ratios, and the clinopyroxenes of the sheared rocks are more sodic (Figs. 4 and 5). The two groups are best distinguished by the TiO_2 contents of the clinopyroxenes (Fig. 3A); most sheared rocks have CPX TiO_2 >0.45 wt. % (16 samples), and most coarse rocks have CPX TiO_2 <0.45 wt.% (23 samples). Three sheared peridotites, however, have mineral compositions similar to the coarse group, and three coarse peridotites fall within the compositional fields of the sheared group.

No distinction between the coarse and sheared peridotite groups is apparent in terms of the modal proportions of the essential minerals (Fig. 1), but the common oxide phase in the coarse peridotites is chromite, whereas both ilmenite and chromite are typical of the sheared peridotites (Table 2).

All spinel and garnet peridotites found in minette localities other than The Thumb have coarse texture and for the most part fall within the range of compositional variation of the coarse peridotites from The Thumb (Table 2).

Megacrystalline Group

The megacrystalline group at The Thumb includes both large (≤9 cm) single-crystal nodules and medium- to ultracoarse-grained peridotite and pyroxenite intergrowths. Nomenclature based on modal proportions is meaningless or at least misleading for many of these intergrowths because grain size is typically comparable to sample size. The name "megacrystalline rocks" was chosen for the polycrystalline inclusions of this group in order to emphasize their close similarity in mineral chemistry and grain size to the large single-crystal inclusions ("megacrysts" or "discrete nodules") found in alkali basalts and kimberlites. Irving (1974) observed that although pyroxene megacrysts in alkali basalts typically occur as single crystals, these commonly appear to be disaggregated parts of coarse-grained polycrystalline aggregates. Similarly, many of the large single-crystal inclusions at The Thumb are probably fragments of polycrystalline intergrowths.

The megacrystalline rocks consist of variable assemblages of the minerals clinopyroxene, orthopyroxene, garnet, olivine, phlogopite, and ilmenite. Individual examples of nodules consisting predominantly of each of these minerals have been found, but clinopyroxene is by far the most abundant mineral in this

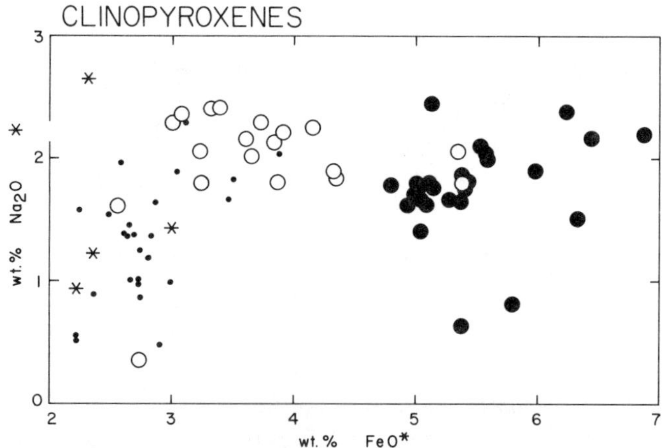

Fig. 4. Wt.% Na_2O vs. wt.% total Fe as FeO in clinopyroxenes of ultramafic xenoliths from The Thumb and other minette localities. Symbols as in Fig. 3.

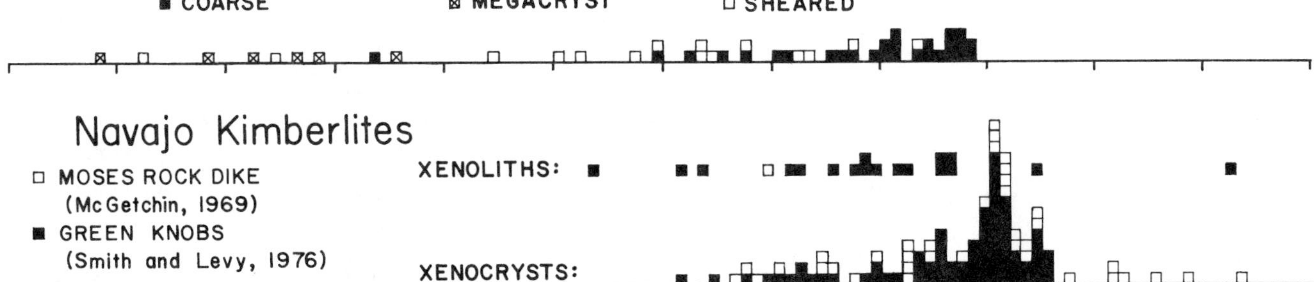

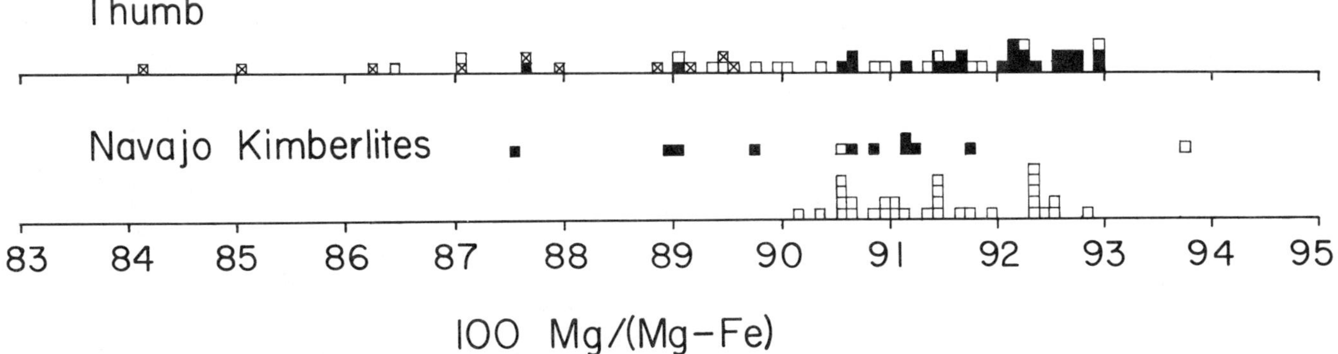

Fig. 5. 100Mg/(Mg+Fe) of olivines and orthopyroxenes in xenoliths from The Thumb and xenocrysts and spinel lherzolite xenoliths from Navajo kimberlites.

group. Small pieces of the megacrystalline clinopyroxene and phlogopite are strewn throughout the minette in remarkable abundance, most thin-sections containing several xenocrysts of each. Two nodules consisting of lamellar intergrowths of ilmenite in large (4-8 cm) clinopyroxene crystals are of particular interest because they underscore the close similarity between the megacrystalline group at The Thumb and the discrete nodule suite from kimberlites.

In contrast to the nearly exsolution-free clinopyroxenes of the peridotite xenoliths, the clinopyroxenes of the megacrystalline rocks commonly contain rounded blebs or thick lamellae of exsolved orthopyroxene. Interstitial garnet in several medium-grained clinopyroxenites may also have formed by exsolution. Phlogopite and ilmenite typically occur interstitially or as inclusions within the other minerals. In many nodules, originally ultracoarse-grained olivine and orthopyroxene are locally recrystallized to very fine-grained mosaic texture. Phlogopite-rich nodules ("glimmerites") also exhibit intense deformation and recrystallization.

The minerals of the megacrystalline group have Ti contents similar to the sheared peridotites, but are distinguishable from all of the peridotites except two sheared lherzolites (140THM and B082THM, Table 2) by their high Fe and low Cr abundances (Table 3; Figs. 2, 3, 4, and 5).

Temperatures and Pressures of Xenolith Equilibration

Temperatures of equilibration were estimated for 53 inclusions from The Thumb using two different methods. The first set of temperatures was obtained by using microprobe determinations of clinopyroxene Ca/(Ca+Mg) to read temperatures directly from the diopside-enstatite solvus determined experimentally for the pure Ca-Mg-Si-O system. The solvus position determined at 30 kbar by Mori and Green (1976) was used because of the relatively

Table 3. Microprobe analyses of megacrystalline-group inclusions from the Thumb.

	CF76THM CPX (4)	CPX (3) altered rim	OPX (4) exsolved incl.	106THM CPX (3)	I051BRD CPX (4)	250THM GA (5&5)	G073THM CPX (6)	OPX (6)	GA (6)	144THM CPX (1&3)	OPX (3)	PHLOG (3)	ILM (4&3)	175THM CPX (3&6) altered	GA (5)	PHLOG (3)	ILM (4)
SiO_2	54.4 2	54.2 1	56.2 1	54.5 3	54.7 2	42.4 4	54.4 1	56.3 1	41.0 2	54.9 3	55.9 1	39.7 1	.02 1	54.2 1	42.0 3	40.0 1	.02 1
TiO_2	.63 1	.34 1	.33 1	.55 1	.37 1	.85 2	.52 1	.29 1	.73 2	.50 1	.27 1	4.26 2	51.7 4	.56 1	.44 1	3.36 2	50.1 1
Al_2O_3	3.69 2	1.82 12	2.16 1	4.49 1	3.70 1	21.4 1	3.67 1	2.42 1	21.9 3	4.93 1	2.41 1	13.9 2	1.81 8	3.89 2	22.2 1	13.5 5	.40 18
Cr_2O_3	.59 1	.62 2	.23 1	.42 1	.46 1	1.15 5	.51 1	.26 1	1.22 4	.47 1	.25 1	.43 1	.39 7	.18 1	.30 2	.12 2	.48 1
FeO	4.80 1	4.42 1	7.66 1	5.12 1	4.63 1	9.39 1	5.09 1	7.55 1	8.83 1	3.60 1	6.62 1	6.04 2	25.5 5	6.23 1	12.2 1	8.00 2	26.1 6
MnO	.09 2	.07 1	.08 1	.09 1	.11 1	.26 1	.11 1	.26 1	.26 1	.10 1	.08 2	.02 1	.22 2	.14 1	.36 1	.03 1	.19 2
MgO	17.4 2	18.0 2	31.3 2	16.9 2	19.9 2	20.0 1	19.2 1	32.4 1	20.6 2	19.0 1	31.7 1	21.1 2	10.3 9	15.3 2	17.5 1	20.3 3	8.67 12
CaO	16.7 2	19.0 7	1.26 1	14.9 2	14.0 2	4.97 1	15.8 1	1.56 1	4.93 2	15.3 1	1.63 1	.10 3	.00 1	17.9 2	5.57 2	.79 1	.04 1
Na_2O	1.79 1	.94 9	.28 1	2.45	1.49 1	.08 2	1.63 2	.39 1	.05 1	1.62 1	.29 1	.30 1		2.38 2	.06 1	.15 3	
K_2O													**10.8			9.49 2	**14.7
	100.0	99.5	99.6	99.5	99.3	100.4	100.9	101.3	99.5	100.4	99.1	95.9	100.7	100.8	100.6	96.0	100.6
Si	1.958	1.975	1.964	1.968	1.965	3.020	1.942	1.936	2.946	1.961	1.953	5.660	.000	1.959	3.023	5.762	.000
Ti	.017	.009	.009	.015	.010	.046	.014	.008	.040	.013	.007	.458	.894	.015	.024	.362	.881
Al	.157	.078	.089	.191	.157	1.794	.154	.098	1.859	.152	.099	2.337	.049	.166	1.879	2.283	.011
Cr	.017	.018	.006	.012	.013	.065	.014	.007	.069	.013	.007	.049	.007	.005	.017	.014	.009
Fe	.145	.135	.224	.155	.139	.559	.152	.217	.531	.147	.194	.722	.491	.188	.733	.959	.510
Mn	.003	.002	.002	.003	.003	.016	.003	.008	.016	.003	.002	.002	.004	.004	.022	.004	.004
Mg	.935	.978	1.630	.908	1.064	2.122	1.021	1.659	2.205	1.010	1.653	4.478	.352	.826	1.873	4.334	.303
Ca	.644	.740	.047	.578	.537	.379	.603	.058	.380	.587	.061	.016	.000	.693	.431	.121	.001
Na	.125	.066	.019	.171	.104	.011	.113	.026	.007	.112	.020	.083		.167	.008	.042	
K												1.854	Fe^{3+} .187			1.735	Fe^{3+} .259
	4.000	4.000	3.990	4.001	3.992	8.012	4.016	4.017	8.053	3.998	3.996	15.658	1.984	4.023	8.010	15.616	1.978
Olivine:																	
Mg/(Mg+Fe)										86.5							
wt.% CaO										0.18							

CF76THM - 0.5 cm xenocryst
106THM and I051BRD - 1 cm megacrysts
250THM - 2 cm megacryst
G073THM - medium-grained clinopyroxenite with exsolved garnet and orthopyroxene
144THM - ultracoarse-grained olivine - clinopyroxene intergrowth
175THM - 5 x 7 x 3 cm garnet nodule with included minerals

**Fe_2O_3 (calculated by method of Carmichael, 1967)

high pressure of this work and the good agreement with the work of Lindsley and Dixon (1976) which was carried out mainly at 15 kbar. A second set of temperature estimates was calculated using equation 5 of Wells (1977), which provides a correction to the temperature estimates for the effect of Fe content on the solubility of Mg in the clinopyroxene.

Pressures of equilibration were estimated using the experimental data of Akella (1976). Approximately one third of the orthopyroxenes from The Thumb contain <1.4 wt% Al_2O_3, the lowest alumina content observed by Akella, and many of the xenoliths yield pressures >44 kbar, the highest pressure experimentally investigated. Therefore, in order to make the pressure estimates, Akella's alumina isopleths were extrapolated to higher pressures and lower alumina values using his Figure 3 (Akella, 1976, p. 593). Following the procedure used by Wood (1974), the alumina contents of the orthopyroxenes were corrected for the contribution of jadeite component (corrected molecular Al = total Al - Na + Cr). This correction applies only to the rocks with the more Fe-rich minerals because Na/Cr is typically <1 for the orthopyroxenes with lower Fe contents.

With the exception of one peridotite and two megacrystalline rocks which lack garnet, all samples contain coexisting clinopyroxene, orthopyroxene, and garnet, as required by the method of pressure estimation. Based on compositional and petrographic similarity with the garnetiferous samples, it may be assumed that these three rocks also equilibrated with garnet.

Mineral analyses of many of the xenoliths have S-ratios >3 for one or more elements (Table 2), suggesting that at least minor inhomogeneity and therefore departure from equilibrium is common in these rocks. However, S-ratios for Ca and Mg in clinopyroxene and Al in orthopyroxene, the quantities of which the P-T estimates are most strongly dependent, are <3 in 32 of the samples, and in only four rocks are the S-ratios for any of these three parameters ≥6. Because the P-T estimates for the samples with the higher S-ratios are not notably different from the P-T values obtained for the more homogeneous samples, it did not seem necessary to exclude the former values from the data set.

The very limited range of compositional variation observed for both mineral grains picked from crushed portions of large xenoliths and minerals in duplicate polished sections prepared from opposite ends of individual xenoliths suggests that the type of compositional variation described by Wilshire and Jackson (1975) is insignificant in the xenoliths from The Thumb. "Composite" xenoliths like those treated by Wilshire and Jackson have not been observed at The Thumb. Minerals in and adjacent to thin (≤2 mm) veins in several nodules are not notably different in composition from minerals further away from these veins.

The temperatures and pressures estimated based on the Fe-free solvus appear to define a steep geotherm (Fig. 6A), whereas correction of the temperatures for the effect of variable Fe contents using the Wells equation collapses the spread of both T and P and suggests a range of 300°C at a single depth. The latter estimates are preferred because of their lesser spread. Possibly the wider range of T and P in Fig. 6A is the result of the highly variable Fe contents of the samples. Random error in the temperature determination for a set of peridotites equilibrated at the same P and T will produce a positive P-T slope paralleling the isopleths of Al_2O_3 content in orthopyroxene (Ernst, in press). This relation may be the source of the apparent steep geotherm in Fig. 6A.

Both methods of calculation show an overall correlation between temperature, texture type, and abundance of easily fusible components (represented in Fig. 6 by the wt.% TiO_2 in clinopyroxene). The coarse peridotites show lower average temperatures of equilibration than the sheared peridotites. The megacrystalline rocks display the highest equilibration temperatures of any of the nodules.

The spread of points in Fig. 6B may be the result of a number of factors, including (1) effects of compositional variations not accounted for by the method of P-T estimation (e.g., variable Cr abundances, as discussed by Wood, 1977), (2) limited precision of microprobe analyses, (3) sample inhomogeneity, and (4) real variations in the physical conditions of the xenoliths' source terrane.

The effect of microprobe precision on the P-T estimates has been evaluated by repeated analysis over a one year period of a homogeneous diopside internal standard. Twenty-five analyses representing 87 analytic spots show standard deviations of 0.7% for the Ca/(Ca+Mg) ratio and 2.5 relative % for the Al_2O_3 content (2.8 wt%). At the two sigma level, these errors would produce maximum variations of ±25 C (Mori and Green solvus) and ±0.7 kbar, respectively, for a sample equilibrated at 1200°C and 50 kbar. The effects of mineral inhomogeneity and variable minor element abundances are difficult to evaluate. However, inasmuch as the pressure estimates are dependent on the temperature estimates, the narrow range in pressures compared to temperatures in Fig. 6B suggests that the xenoliths' source terrane indeed contained wide variations in temperature within a very narrow vertical interval.

Discussion

Petrogenesis of the Inclusions

Based on the correspondence between textures and mineral compositions (Figs. 2, 3, 4, and 5), the ultramafic xenoliths at The Thumb are divisible into three overlapping textural/compositional groups: coarse garnet peridotites, sheared garnet peridotites, and megacrystalline rocks of garnet lherzolite mineral assemblage. The narrow range of the pressure estimates for these nodules (Fig. 6B) indicates that all were derived from the same limited volume of the upper mantle. The rarity of other ultramafic lithologies at The Thumb, in particular spinel peridotite and websterite, implies that the exposed magma volume sampled only the one restricted vertical interval of the mantle at around 130 km depth and then incorporated almost no other xenoliths until it reached the base of the crust.

A petrogenetic model for the xenoliths must account for not only their correlation between mineral chemistry and texture type, but also the intimate coexistence of the three nodule types within the same terrane at depth. The model presented here explains the megacrystalline rocks as the products of crystallization of Fe- and Ti-rich magma, the coarse garnet peridotites as the preexisting country rocks into which this magma was intruded, and the sheared garnet peridotites as the products of Fe-Ti-alkali-metasomatism of the coarse peridotites by the intrusive magma or by residual liquids produced by its crystallization.

The coarse garnet peridotites are viewed as possible residua of partial melting of an originally more fertile upper mantle composition such as Ringwood's (1975) pyrolite. This suggestion is based on the low modal percentages of clinopyroxene and garnet and high Mg/(Mg+Fe) of the minerals in most of the coarse peridotites (Figs. 1 and 5). Alternatively, the coarse peridotites may be interpreted as relatively barren primordial mantle inasmuch as they do not appear highly depleted in comparison to pyrolite, and it is not clear that all regions of the upper mantle were initially similar to pyrolite in composition.

The association of the depleted coarse peridotites and the very enriched megacrystalline rocks within the same limited volume of the upper mantle (Fig. 6) is most plausibly explained by intrusion of the liquids from which the megacrystalline rocks crystallized into a preexisting terrane consisting of the coarse peridotites (Fig. 7A). The spatial relations of these rocks are unknown; the megacrystalline rocks may be envisaged as forming either a single large intrusive body or a stockwork of narrow dikes and veins as has been suggested for the Al-augite group inclusions characteristic of alkali basalt localities (Wilshire and Pike, 1975). Indeed, the megacrystalline group is highly analogous with the Al-augite group and probably reflects the operation of similar processes, but under conditions of much lower geothermal gradient and possibly greater depth.

Because the intrusive liquids would presumably be derived from some depth >130 km they would initially be at higher temperature than the coarse peridotites. This relation may explain the common presence of exsolved orthopyroxene and garnet in the clinopyroxene of the megacrystalline nodules (indicating substantial cooling at depth following initial crystallization) and the lack of appreciable exsolution in the pyroxenes of both the coarse and sheared peridotites (consistent with heating in response to the intrusive event). Thus the wide temperature range of the points in Fig. 6B may partly reflect local thermal perturbation of the xenolith source terrane from an earlier position along a normal continental geotherm.

The sheared peridotites do not appear to be either residua of partial melting or precipitates from Fe-Ti-rich liquids because they include rocks with very high abundances of both Ti and Cr. For example, the garnet in sample D076THM (Table 2) contains 6.3 wt.% Cr_2O_3 and 0.9 wt.% TiO_2. The minerals of the sheared peridotites have Ti contents comparable to the megacryatalline minerals, Cr contents which overlap with the coarse peridotites, and Fe contents intermediate between these groups (Figs. 3, 4, and 5). On the basis of these relations, it is suggested that the sheared peridotites formed by metasomatism and deformation of the coarse peridotites in the regions immediately adjacent to the intrusive megacrystalline masses (Fig. 7A). According to this model, the Cr contents of the sheared peridotites reflect the variable Cr contents of their coarse-texture peridotite protoliths, whereas Ti and Fe have been metasomatically introduced by liquids associated with the megacrystalline rocks. The lack of correlation between peridotite texture and modal proportions (Fig. 1) suggests that the only major elements added by the metasomatism in amounts which are significant compared to the original abundances are Fe, Ti, and alkalis.

The metasomatic model is supported by the experimental data of Mysen and Boettcher (1975), which show that Ti and Fe partition readily into the liquid phase during partial melting

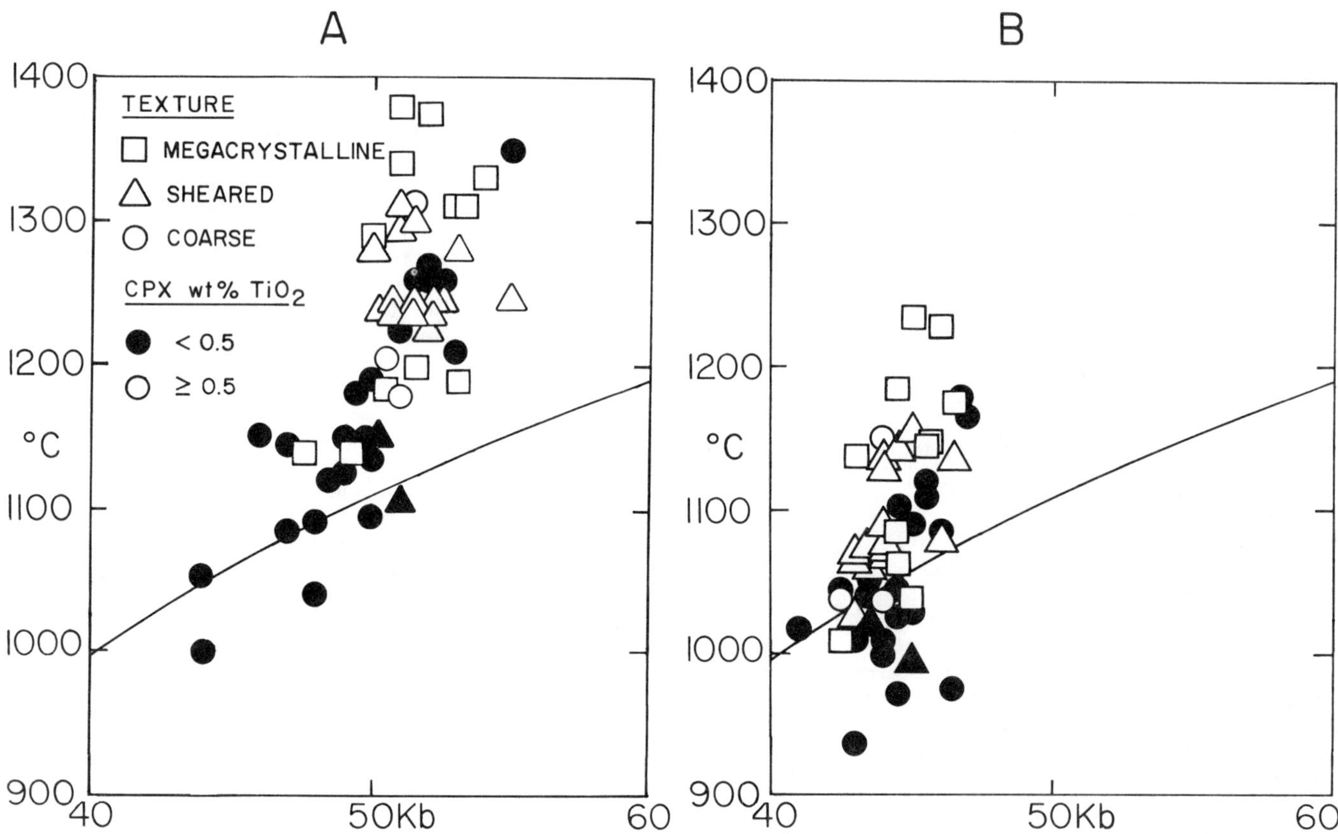

Fig. 6. Estimated pressures and temperatures of equilibration of ultramafic xenoliths from The Thumb. Curved line is segment of continental shield geotherm of Clark and Ringwood (1964). Shapes of symbols indicate textures; filled and open symbols indicate CPX TiO_2 content.

A. Temperatures from solvus of Mori and Green (1976) and pressures from Al_2O_3 isopleths of Akella (1976).

B. Temperatures from equation 5 of Wells (1977) and pressures from Al_2O_3 isopleths of Akella (1976).

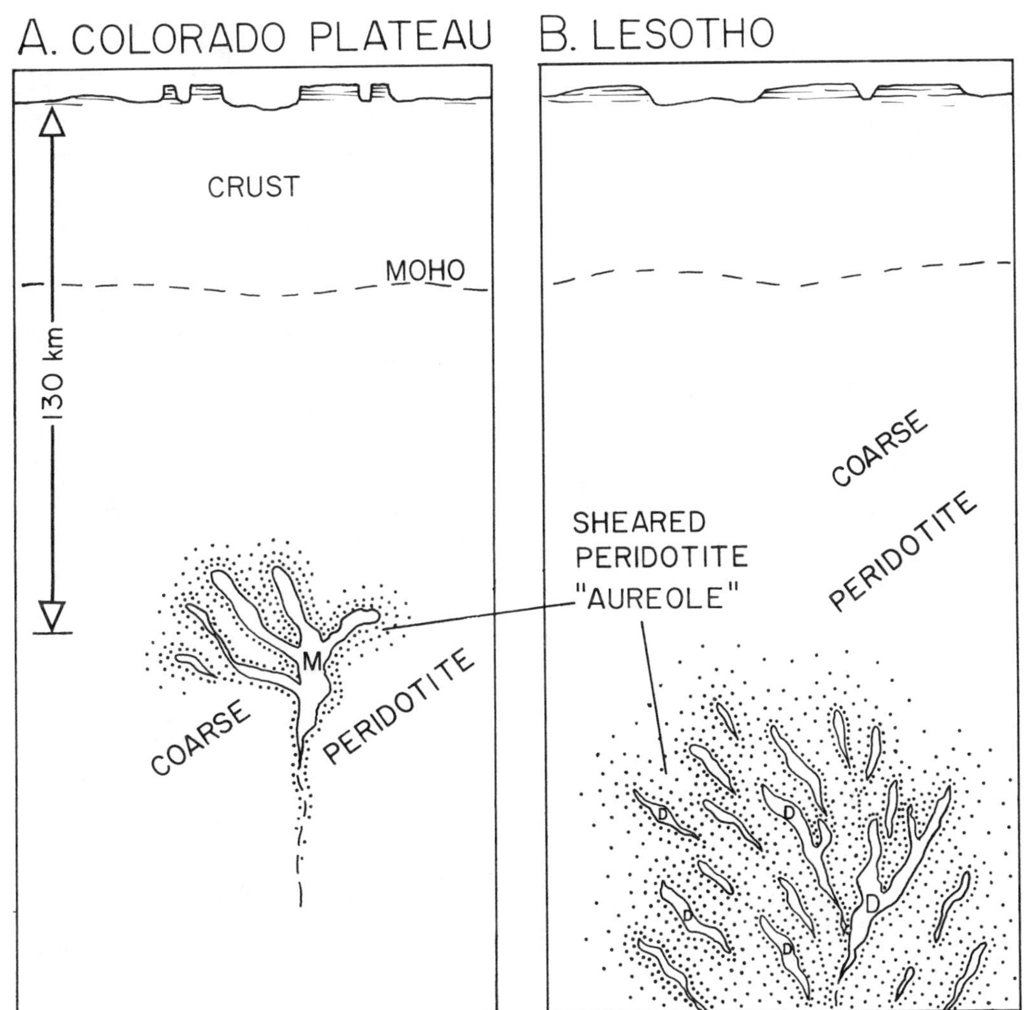

Fig. 7. Diagrammatic cross-sections showing inferred petrologic relations in source terranes of ultramafic xenoliths at time of eruption of (A) minette of the Colorado Plateau (Late Oligocene) and (B) kimberlite of Lesotho (Cretaceous). "M" represents megacrystalline rocks (solidified intrusive magma). "D" represents partly molten regions of magma and discrete nodule "phenocrysts". Stippled pattern represents metasomatic/tectonic aureoles of Ti-Fe-enriched, sheared garnet peridotite. Surrounding unpatterned rock is pre-existing coarse garnet peridotite.

of peridotite at high pressure, but Cr accumulates as a residual element in the solid phases as the degree of melting increases. Ti and Fe may therefore be expected to be mobile relative to Cr in the presence of a small percentage of melt in the upper mantle.

Evolved Fe-Ti-rich liquid approximately intermediate in composition between ilmenite and clinopyroxene but possibly also rich in alkalis and H_2O is proposed as a possible medium for producing the metasomatism. Lamellar intergrowths of clinopyroxene and ilmenite are characteristic of kimberlites and have been interpreted as the products of "eutectic" crystallization of extremely Fe- and Ti-rich liquids on the basis of both geochemical and experimental evidence (MacGregor and Wittkopp, 1970; Gurney, et al., 1973; Wyatt, et al., 1975). The presence of lamellar clinopyroxene-ilmenite intergrowths at The Thumb indicates that such liquids were present during at least the final stages of crystallization of the megacrystalline rocks. The sheared garnet peridotites may have been produced by addition of several percent of such liquid to the coarse garnet peridotites, followed by subsolidus reequilibration and intense deformation. The common occurrence of disseminated to vein-like ilmenite in the sheared peridotites is consistent with this hypothesis.

The correlation between Ti-Fe-rich mineral chemistry and mosaic texture of olivine is explained as reflecting more intense deformation of the rocks along the margins of the mobile intrusive bodies. The apparent homogeneity of the minerals in many of the sheared peridotites (Table 2) indicates that the inferred metasomatic process was of such a nature as to effect complete reequilibration and homogenization of mineral compositions over distances ≤5 to 10 cm. Possibly this process was facilitated by the intense deformation.

Goetz (1975) and Mercier (1977) concluded that the mosaic-porphyroclastic texture displayed by peridotite xenoliths from kimberlites formed during an intense strain pulse of short duration. Mercier (1977) therefore suggested that the fabrics of the sheared lherzolites were produced by conduit formation associated with eruption of the host kimberlite. By analogy, the sheared peridotite xenoliths from The Thumb may also be the products of rapid, intense deformation (although dislocation densities have not been examined in these rocks).

But did this rapid deformation occur immediately before eruption of the minette or may it have occurred millions of years earlier? One might expect that the very fine-grained mosaic texture of the olivine in the sheared peridotites, with its very high surface energy compared with the alternative coarse texture (~0.2 vs. 2.0 mm average grain diameter), would be unstable over long periods of time at 1100°- 1200°C. Unfortunately, only qualitative reports of experimental data are available concerning the time required for grain growth to destroy fine-grained mosaic texture in olivine or quartz, and most of these experiments were not conducted under static conditions. This is an important consideration, however, because assumption of a short life expectancy for fine-grained mosaic olivine at high temperature implies that production of the sheared texture of the enriched xenoliths occurred immediately before or during eruption of the minette, and further implies that crystallization of the megacrystalline rocks and metasomatism of the sheared peridotites resulted from intrusion of the minette magma at depth.

Both sheared and coarse garnet peridotites from The Thumb have clinopyroxenes with low $^{87}Sr/^{86}Sr$ (~0.703) compared to the host minette (0.7064) and therefore cannot have equilibrated with the minette (Ehrenberg, 1976). Minerals of the megacrystalline rocks have not yet been analyzed for Sr and could have isotopic compositions very different from the peridotites. However, in the absence of isotopic data, discussion of the possible genetic relation between the megacrystalline group and the minette is premature.

Phlogopite Relations

Phlogopite in many of the peridotite and megacrystalline xenoliths appears to have been metasomatically introduced after crystallization of the anhydrous minerals because it occurs in veins, is interstitial to other minerals, or is localized around garnets and clinopyroxenes. However, because the phlogopite occurs in all three xenolith groups at The Thumb, it is not clear how its formation is related to the metasomatism which is suggested to have produced the sheared peridotites.

The textural evidence for secondary introduction of phlogopite is suggestive of interaction between the nodules and the phlogopite-rich minette host. However, coexisting phlogopites and clinopyroxenes in different xenoliths show approximately linear overall correlations of Fe/Mg and Ti/Cr ratios (Fig. 2). Similar correlations are observed between phlogopites and coexisting orthopyroxenes, garnets, and olivines. These relations suggest an approach to chemical equilibration between the micas and the coexisting anhydrous phases and therefore imply that formation of the phlogopite preceded incorporation of the xenoliths in the minette. A second argument against a late secondary origin is that the phlogopite is almost exclusively confined to rocks with equilibration temperatures <1100 C (temperatures by Wells' method). Similar relations are observed in the xenoliths from southern African kimberlites (Boyd and Nixon, 1974). The one exception to this rule at The Thumb is nodule 144THM (Table 3), which yields the highest calculated temperature of any of the samples (Fig. 6).

The phlogopite may reflect interaction between the xenoliths and the minette during an extended period preceding the eruption, but available data are insufficient to evaluate this possibility. For the present it may only be noted that alkaline volcanism is typically preceded by general alkali and H_2O metasomatism of the mantle regions sampled by the eruptions (Boettcher, et al., in press), and the Navajo volcanic field is no exception to this general pattern.

Comparison with Xenoliths from Southern African Kimberlites

The presence of the coarse, sheared, and megacrystalline nodule groups at The Thumb and the mineralogy of these rocks, with prominent garnet, phlogopite, and ilmenite, are strongly reminiscent of the xenolith suites described from kimberlites of several provinces, in particular the Cretaceous kimberlites of Lesotho studied by Boyd and Nixon (1974). In both The Thumb and the Lesotho xenolith suites, the coarse garnet peridotites are relatively depleted in their basaltic components, and the sheared peridotites are relatively rich in Fe and Ti. The megacrystalline rocks from The Thumb are very similar in mineral chemistry to the discrete nodules from the kimberlites. However, several major differences between the two suites may be noted. Coarse peridotites from The Thumb are not as depleted in their early-melting constituents as the coarse peridotites from Lesotho, and the sheared and coarse Lesotho peridotites do not overlap in their Cr contents as The Thumb peridotites do. Perhaps the most significant difference between these xenolith suites is that all three nodule types at The Thumb equilibrated at similar pressures, whereas in the Lesotho suite the sheared peridotites and discrete nodules equilibrated at much higher pressures than the coarse peridotites.

The megacrystalline rocks from The Thumb appear to be related to the discrete nodules characteristic of kimberlites because both groups have similar mineral chemistry and ultracoarse grain size. For example, megacrystalline clinopyroxenes show a range of 0.81 - 0.88 in Mg/(Mg+Fe), comparable to the variation in the discrete nodule group. Also, both groups include lamellar intergrowths of clinopyroxene and ilmenite. However, the two groups are very different in texture. The discrete nodules are euhedral single crystals believed to have formed as phenocrysts at 150-200 km depth. The megacrystalline rocks are pegmatite-like intergrowths which may have formed either as cumulates or by complete crystallization of pockets of magma at high pressure. Thus the textural difference may reflect variation in the time of sampling with respect to solidification of the magma: <u>before</u> in the case of the kimberlites and <u>after</u> in the case of the minette.

Pressure-temperature estimates for the Lesotho nodules define the well-known "kinked" geotherm of Boyd (1973), with the sheared peridotites and megacrysts along with the high P-T limb and the coarse peridotites along the low P-T limb. Green and Gueguen (1975) interpret the sheared peridotite nodules in the kimberlites as samples of hot diapiric masses of little-depleted mantle which intruded the overlying depleted mantle, represented by the coarse garnet peridotite nodules. Boyd and Nixon (1974) correlate the sheared peridotite with the asthenosphere and the coarse peridotite with the rigid lithosphere and view the discrete nodules as precipitates from liquids associated with the partially melted asthenosphere.

By analogy with the xenoliths from The Thumb, an alternative model for the Lesotho xenolith suite is that the sheared peridotites formed by extensive metasomatism of preexisting mantle of more depleted composition in response to intrusion of highly evolved liquids from which the discrete nodules were precipitated (Fig. 7B). Unlike The Thumb, however, the Lesotho kimberlite eruptions incorporated no unmetasomatized (i.e., coarse texture) xenoliths at the site of metasomatism, but entrained the partially metasomatized coarse peridotite group from the immediately overlying section. Gurney, et al., (1977) view the discrete nodules of the Monastery Mine as possible precipitates from the kimberlite, raising the possibility that the magma responsible for metasomatism and deformation of the sheared peridotite group is the host kimberlite.

<u>Relation of Xenoliths from The Thumb to Mantle Sampled by the Navajo Kimberlites</u>

Comprehensive investigations of individual kimberlite diatremes of the Navajo volcanic field by McGetchin and Silver (1970) and Smith and Levy (1976) have provided petrologic models for the upper mantle beneath the Colorado Plateau. Spinel peridotite and websterite xenoliths in the kimberlites are thought to represent the dominant lithologies of the partly hydrated uppermost mantle. Xenocrysts of pyropic garnet, pyroxenes, and olivine are believed to be completely fragmented samples of the garnetiferous mantle underlying the spinel peridotite (McGetchin, 1968), but intact garnet peridotite xenoliths are extremely rare.

Ca/(Ca+Mg) ratios of clinopyroxenes from both xenoliths and xenocrysts in the kimberlites are mainly in the range 0.46 to 0.52 (McGetchin, 1968; Smith and Levy, 1976), indicating temperatures of $\leq$900 C according to the En-Di solvus of Lindsley and Dixon (1976). If the geotherm below the Colorado Plateau is similar to the calculated continental shield geotherm of Clark and Ringwood (1964), as is tentatively suggested by the P-T estimates from The Thumb (Fig. 6B), then the low equilibration temperatures of the clinopyroxenes from the Navajo kimberlites imply that these eruptions sampled only a relatively shallow level of the mantle, at depths <100 km.

Smith and Levy (1976) suggested that the mantle regions sampled by the Navajo kimberlites were partly depleted in early-melting constituents long before the kimberlite eruptions. This is supported by the high Mg/(Mg+Fe) of the minerals from peridotites sampled by the kimberlites (Fig. 5) and the low TiO_2 contents ($\leq$0.2 wt.%) of pyropic garnet xenocrysts analyzed by McGetchin (1968) and Switzer (1977). The coarse-texture peridotites from The Thumb are similar in their mineral chemistry to the xenocrysts in the kimberlites (Fig. 5) except for effects attributable to higher pressures and temperatures of crystallization. Thus the depleted character of the Colorado Plateau upper mantle may extend to depths of at least 130 km. The Fe- and Ti-rich sheared peridotites and megacrystalline rocks at The Thumb have no compositional analogs in the Navajo kimberlites. Possibly the intrusive and metasomatic processes which affected the deeper mantle sampled by the minette did not extend to the shallowed levels from which it is suggested that the Navajo kimberlites were derived.

Relation between Minette and Kimberlite Magma Types

The minette of the Navajo volcanic field provides an important petrogenetic link between basaltic and kimberlitic volcanism in terms of its enrichment in LIL elements, phase compositions, and xenolith content. Although basic in composition, the minette has highly fractionated rare earth element abundances (Kay and Gast, 1970; Ehrenberg, 1978) which closely resemble those reported from kimberlites of regions other than the Colorado Plateau (Mitchell and Brunfelt, 1974; Frey, Ferguson, and Chappell, 1977). Certain kimberlites even duplicate the small negative Eu anomalies of the minettes (Mitchell and Brunfelt, 1974, Fig. 2).

The phase compositions in the minettes are also similar to those reported from kimberlites. Phlogopite in the minette is similar in composition to the average of phlogopite phenocrysts from kimberlite of Somerset Island (Clarke and Mitchell, 1975) and the rims of phlogopite phenocrysts in kimberlites reported by Boettcher, et. al. (in press). The diopsidic clinopyroxene of the minette is similar to the clinopyroxenes analyzed by Emeleus and Andrews (1975) and Dawson, et al. (1977), in the groundmass of kimberlites from Greenland and southern Africa, respectively. In common with many kimberlites, trachybasalt (the extrusive equivalent of the minette) at Washington Pass contains magnesian ilmenite (~5 wt.% MgO) in the groundmass (Ehrenberg, 1978).

A third significant parallel between minettes and kimberlites is the nature of the ultramafic xenolith suites which they contain. The xenloiths from The Thumb and possibly other localities, such as Black Rock Dike (Tables 2 and 3), are similar in many respects to the garnet peridotite and discrete nodule suites which have previously been found only in kimberlites.

Both minette and kimberlite may be interpreted as the products of very small percentages of melting of phlogopite-bearing garnet peridotite at depths of at least 130 km. The character and abundance of available volatiles and the composition of the parent rock may be the critical factors in determining the type of magma produced (Wyllie, 1977).

Conclusions

(1) Geochemical and textural relations of ultramafic xenoliths from The Thumb are interpreted as reflecting intrusion and metasomatism of previously depleted upper mantle at 130 km depth by Fe-Ti-rich liquids of deeper origin.

(2) P-T estimates for the xenoliths are interpreted as reflecting localized heating of the mantle at 130 km depth in response to the intrusive event. However, the P-T estimates for the least enriched rocks (peridotites with coarse texture) suggest that the Colorado Plateau geotherm may be similar to the calculated continental shield geotherm of Clark and Ringwood (1964).

(3) If fine-grained mosaic texture in olivine may be assumed to have short life expectancy under static conditions at 1100-1200°C in the upper mantle, then the strong correlation between the presence of mosaic olivine and Ti-Fe-rich mineral chemistry in the xenoliths implies that the inferred metasomatic/intrusive event occurred during formation and eruption of the minette magma.

(4) The Thumb's xenolith suite is analogous with the suites of coarse and sheared garnet peridotites and discrete nodules which have previously been reported only from kimberlites. This and other similarities indicate a close petrogenetic relationship between minette and kimberlite magma types.

(5) By analogy with the nodules from The Thumb, the Fe-Ti-rich sheared garnet peridotite group from Lesotho kimberlites is suggested to have formed by metasomatism of previously less enriched mantle in response to intrusion of magma (possibly related to the host kimberlite) from which the discrete nodule group crystallized.

(6) Peridotite sampled by the kimberlite pipes of the Navajo province lacks the variable enrichment in Fe and Ti of The Thumb's xenoliths and equilibrated at lower temperature, implying that the kimberlites were derived from a shallower level of the mantle which did not experience the intrusive and metasomatic processes affecting the deeper mantle sampled by The Thumb.

Acknowledgements. I thank K. D. Watson, W. G. Ernst, and A. L. Boettcher for their encouragement and advice during the course of this study. in addition to the above individuals, I thank F. R. Boyd, B. Leavy, and A. R. Basu for critically reviewing earlier drafts of this paper. Tom McGetchin initiated this study by suggesting that I investigate the minette volcanic centers. This work was supported in part by summer research positions at the Los Alamos Scientific Laboratory, by N. S. F. grant EAR 76-19067 to K. D. Watson, and by a Grant-in-Aid of Research from the Society of Sigma Xi. I am grateful to the Navajo Tribe for permission to conduct field studies and collect samples on their land.

References

Akella, J., Garnet pyroxene equilibria in the system $CaSiO_3$-$MgSiO_3Al_2O_3$ and in a natural mineral mixture, Amer. Mineral., 61, 589-598, 1976.

Baldridge, W. S., S. N. Ehrenberg, and T. R. McGetchin, Ultramafic xenolith suite from Ship Rock, N. Mex. (abstract), Eos Trans. AGU, 56, 464-465, 1975.

Boettcher, A. L., J. R. O'Neil, K. E. Windom, D. C. Stewart and H. G. Wilshire, Metasomatism of the basalts, Proceedings of the Second International Kimberlite Conference, in press.

Boyd, F. R., Quantitative electron probe analysis of pyroxenes, Carnegie Inst. Wash. Yearb., 66, 327-334, 1967.

Boyd, F. R., A pyroxene geotherm. Geochim. Cosmochim. Acta, 37, 2533-2546, 1973.

Boyd, F. R., and P. H. Nixon, Origins of the ultramafic nodules from some kimberlites of northern Lesotho and the Monastery Mine, South Africa, Phys. Chem. Earth, 9, 431-454, 1974.

Carmichael, I. S. E., The iron-titanium oxides of salic volcanic rocks and their associated ferro-magnesian silicates, Contr. Mineral. Petrol., 14, 36-64, 1967.

Carswell, D. A., Primary and secondary phlogopites and clinopyroxenes in garnet lherzolite xenoliths, Phys. Chem. Earth, 9, 417-494, 1974.

Clark, S. P., Jr., and A. E. Ringwood, Density distribution and constitution of the mantle, Rev. Geophys., 2, 35-88, 1964.

Clarke, D. B., and R. H. Mitchell, Mineralogy and petrology of the kimberlite from Somerset Island, N. W. T., Canada, Phys. Chem. Earth, 9, 123-135, 1974.

Dawson, J. B., J. V. Smith and R. L. Hervig, Late-state diopsides in kimberlite groundmass, Extended Abstracts, Second International Kimberlite Conference, Santa Fe, New Mexico, 1977.

Ehrenberg, S. N., Colorado Plateau garnet peridotite xenoliths: Strontium analyses of diopsides, Eos Trans. AGU, 57, 1026, 1976.

Ehrenberg, S. N., Petrology of potassic volcanic rocks and ultramafic xenoliths from the Navajo volcanic field, New Mexico and Arizona, Ph.D. dissertation, Univ. California, Los Angeles, 259, p.,1978.

Emeleus, C. H., and J. R. Andrews, Mineralogy and petrology of kimberlite dyke and sheet intrusions and included peridotite xenoliths from southwest Greenland, Phys. Chem. Earth, 9, 179-197, 1974.

Ernst, W. G., Petrochemical study of lherzolitic rocks from the western Alps, Contr. Mineral. Petrol., 19, in press.

Frey, F. A., J. Ferguson and B. W. Chappell, Petrogenesis of South African and Australian kimberlitic suites, Extended Abstracts, Second International Kimberlite Conference, Santa Fe, New Mexico, 1977.

Goetze, C., Sheared lherzolites: From the point of view of rock mechanics, Geology, 3, 172-173, 1975.

Green, H. W., II, and Y. Gueguen, Origin of kimberlite pipes by diapiric upwelling in the upper mantle, Nature, 249, 617-620, 1974.

Gurney, J. J., H. W. Fesq, and E. J. D. Kable, Clinopyroxene-ilmenite intergrowths from kimberlite: A reappraisal, in Lesotho Kimberlites, ed. P. H. Nixon, pp. 238-253, Lesotho National Development Corp. Maseru, Lesotho, 1973.

Gurney, J. J., W. R. O. Jakob and J. B. Dawson, Megacrysts from the Monastery Mine, Extended Abstracts, Second International Kimberlite Conference, Santa Fe, New Mexico, 1977.

Harte, B., Rock nomenclature with particular relation to deformation and recrystallization textures in olivine-bearing xenoliths, J. Geol., 85, 279-288, 1977.

Irving, A. J., Megacrysts from the Newer Basalts and other basaltic rocks of southeastern Australia, Geol. Soc. Amer. Bull., 85, 1503-1514, 1974.

Kay, R. W., and P. W. Gast, The rare earth content and origin of alkali-rich basalts, J. Geol., 81, 653-682, 1973.

Kelly, V. C., Monoclines of the Colorado Plateau, Geol. Soc. Amer. Bull., 66, 789-804, 1955.

Lindsley, D. H., and S. A. Dixon, Diopside-enstatite equilibria at 850°C to 1400°C, 5 to 35 kb, Amer. J. Sci., 276, 1285-1301, 1976.

MacGregor, I. D., and R. W. Wittkopp, Diopside-ilmenite xenoliths from the Monastery Mine, Orange Free State, South Africa, Geol. Soc. Amer. Abstracts, 2, 113, 1970.

Mercier, J. C. C., Peridotite xenoliths and the dynamics of kimberlite intrusion, Extended Abstracts, Second International Kimberlite Conference, Santa Fe, New Mexico, 1977.

McGetchin, T. R., The Moses Rock Dike: Geology, petrology and mode of emplacement of a kimberlite-bearing breccia pipe, San Juan County, Utah, Ph.D. dissertation, Calif. Inst. Tech., Pasadena, 405 p., 1968.

McGetchin, T. R., and L. T. Sliver, Compositional relations in minerals from kimberlite and related rocks in the Moses Rock dike, San Juan County, Utah, Amer. Mineral., 55, 1738-1771, 1970.

Mitchell, R. H., and A. O. Brunfelt, Rare earth element geochemistry of kimberlite, Phys. Chem. Earth, 9, 671-686, 1974.

Mori, T., and D. H. Green, Pyroxenes in the system $Mg_2Si_2O_6$-$CaMgSi_2O_6$ at high pressure, Earth Planet. Sci. Let., 26, 277-286, 1976.

Mysen, B. O., and A. L. Boettcher, Melting of hydrous mantle: II. Geochemistry of crystals and liquids formed by anatexis of mantle peridotite at high pressures and high temperatures as a function of controlled activities of water, hydrogen, and carbon dioxide, J. Petrol., 16, 547-593, 1975.

Ringwood, A. E., Composition and Petrology of the Earth's Mantle, McGraw-Hill, New York, 618 p., 1975.

Smith, D., and S. Levy, Petrology of the Green Knobs diatreme and implications for the upper mantle below the Colorado Plateau, Earth Planet. Sci. Let., 29, 107-125, 1976.

Switzer, G. S. Composition of garnet xenocrysts from three kimberlite pipes in Arizona and New Mexico, Smithsonian Contr. Earth Sci., 19, 1-21, 1977.

Wells, P. R. A., Pyroxene thermometry in simple and complex systems, Contr. Mineral. Petrol, 62, 129-139, 1977.

Williams, H., Pliocene volcanoes of the Navajo-Hopi country, Geol. Soc. Amer. Bull., 47, 76-104, 1936.

Wilshire, H. G., and E. D. Jackson, Problems in determining mantle geotherms from pyroxene compositions of ultramafic rocks, J. Geol., 83, 313-329, 1975.

Wilshire, H. G., and J. E. N. Pike, Upper mantle diapirism: Evidence from analogous features in alpine peridotite and ultramafic inclusions in basalt, Geology, 3, 467-470, 1975.

Wood, B. J., The solubility of alumina in orthopyroxene coexisting with garnet, Contr. Mineral. Petrol., 46, 1-15, 1974.

Wyatt, B., R. H. McCallister, F. R. Boyd and Y. Chashi, An experimentally produced clinopyroxene-ilmenite intergrowth. Carnegie Inst. Wash. Yearb., 74, 536-539, 1975.

Wyllie, P. J. Kimberlite magmas from the system peridotite-H_2O-CO_2, Extended Abstracts, Second International Kimberlite Conference, Santa Fe, New Mexico, 1977.

HYDROUS MINERALS AND CARBONATES IN PERIDOTITE INCLUSIONS FROM THE GREEN KNOBS AND BUELL PARK KIMBERLITIC DIATREMES ON THE COLORADO PLATEAU

Douglas Smith

Department of Geological Sciences, University of Texas at Austin, Austin, Texas 78712

Abstract. Over 90% of the peridotite inclusions at the Green Knobs and Buell Park diatremes contain hydrous phases which appear to have formed before the inclusions were incorporated in the kimberlitic eruptions, and a few percent contain magnesite of similar origin. Evidence that amphibole, chlorite, titanoclinohumite, and magnesite were formed before eruption includes observations that these phases were deformed together with anhydrous silicates, that they occur in apparent equilibrium metamorphic textures, and that they show systematic compositional variations together with anhydrous silicates. Evidence for the time of antigorite formation is ambiguous.

Amphibole (16-2 wt% Al_2O_3), chlorite (20-12%), and pyroxene (6-0.3%) become less aluminous with increased rock hydration, and orthopyroxene becomes less calcic. Titanoclinohumite contains the same TiO_2 content (5.5%) in different assemblages. Antigorite contains up to 5 weight % Al_2O_3 plus Cr_2O_3. Though compositional zoning and persistence of relict phases are common, local assemblages may have formed in equilibrium: magnesite-diopside and antigorite-enstatite are possible equilibrium pairs.

Hydration may have occurred over a range of water fugacities at temperatures below 700°C and depths from 45 to 60 plus km. Analysis of an Fe-Mg gradient in olivine indicates that some hydration preceded kimberlite emplacement by less than 25 million years. The fluid responsible for hydration was water-rich: constituents other than water and CO_2 were not added to the rocks. The hydrated peridotites likely represent continental mantle, not fragments of an oceanic slab. Intrusion of minette magma into hydrated peridotite may have caused dehydration reactions which triggered nonmagmatic kimberlitic eruptions.

Introduction

Most of the peridotite nodules included in kimberlitic tuff at the Green Knobs and Buell Park diatremes on the Colorado Plateau (Fig. 1) contain one or more of the following hydrous minerals: amphibole, chlorite, titanoclinohumite, and antigorite. A few percent contain magnesite. Peridotite-fluid reactions can occur during and after kimberlite emplacement, and hydrous minerals in peridotite inclusions in kimberlite may form in a number of ways (e.g. Carswell, 1975; Dawson and Smith, 1977). Evidence to be presented here indicates that amphibole, chlorite, titanoclinohumite, and likely part of the antigorite in peridotite inclusions from Green Knobs and Buell Park formed by peridotite-fluid reaction in the uppermost mantle before eruption began. The model of fluid-rock interaction has consequences for the generation of the kimberlitic diatremes and perhaps for the tectonic history of the Colorado Plateau.

The Buell Park diatreme has been described by Allen and Balk (1954), Schmitt et al. (1974), and Roden and Smith (1978). Green Knobs has been studied by Allen and Balk (1954) and Smith and Levy (1976). The kimberlitic tuffs at both localities appear to have formed from an erupting mixture of volatiles and physically disaggregated mantle and crustal rock, as outlined for other Plateau kimberlitic diatremes by McGetchin and Silver (1970) and McGetchin et al. (1973). There is no evidence that the kimberlitic tuffs were ever represented by a silicate melt.

Secondary Minerals in Peridotite Inclusions

Minerals which formed in the nodules during and after diatreme eruption are here termed secondary. In contrast, hydrous minerals and carbonates which appear to have formed in peridotite at depth before the peridotite was disaggregated and incorporated in the erupting breccia are called primary.

Alteration of crystal fragments in the kimberlitic tuffs is one guide to the secondary minerals. Olivine, pyroxene, chlorite, spinel, and many other minerals occur in angular crystal fragments in a fine-grained matrix in the diatremes; x-ray diffraction spectra of the maxtrix are attributed to serpentines, clays, chlorite, and talc (?). Spinel and pyroxene fragments are typically almost unaltered. At Green Knobs angular olivine fragments have been replaced by sheet silicates to a depth of as

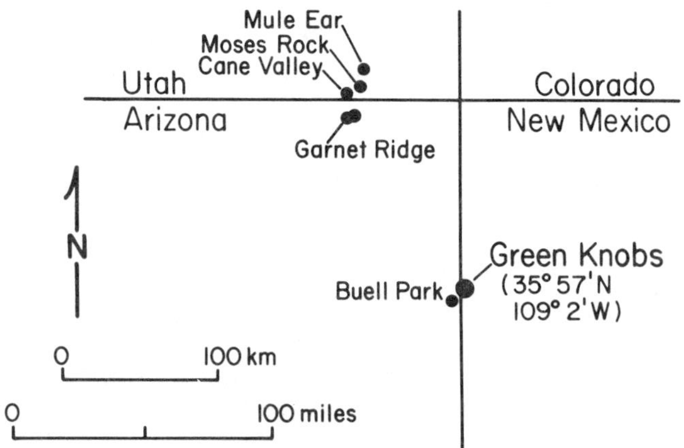

Fig. 1. Locations of the principal kimberlitic diatremes on the Colorado Plateau.

much as 0.4 mm; replacement is irregular, and some fragments appear to have been broken after the beginning of alteration. Olivine alteration at Buell Park is generally more severe. The alteration minerals are fine to medium-grained serpentines, minor talc (?), and an unidentified sheet silicate with a composition intermediate between serpentine and talc. Scarce carbonate occurs in the altered olivine.

Some hydrous minerals in peridotite nodules resemble those replacing the discrete olivine grains in the kimberlitic maxtrix. Such serpentine and associated sheet silicates occur in most nodules in sparse irregular veins along grain boundaries and transecting olivine crystals. In a few nodules from Buell Park these secondary minerals have almost completely replaced olivine, but pyroxenes and spinel are typically unaltered. Calcite and sparse fine-grained talc occur in some of the veins. The serpentine is typically very fine-grained and locally in fine cross-fibers, but in a few rocks veins contain blade-like crystals of antigorite. These antigorite crystals may be transitional to the antigorite of possible primary origin considered subsequently. With the exception of the bladed antigorite, all hydrous minerals described above are disregarded in further discussions.

Assemblages and Textures in Peridotite Inclusions with Primary Hydrous Phases and Carbonates

About 95% of the peridotite inclusions at Green Knobs contain hydrous minerals which are here called primary, because they formed before the peridotite was incorporated in the erupting breccia. Though the descriptions which follow are based mostly on nodules from Green Knobs, petrographic examination indicates that the suite from the Buell Park kimberlitic tuff is similar. Typical peridotite without hydrous phases is lherzolite with about 70% olivine (Fo 91), 20% orthopyroxene (En 92, 3-6% Al_2O_3), 8% diopside (Ca/(Ca+Mg) .48, 5-7% Al_2O_3) and 2% spinel (Al/(Al+Cr), 0.9-0.6) (Smith and Levy, 1976). Spinel-pyroxene clusters (Smith, 1977a) are common in relatively undeformed rocks. Websterites, typically rich in orthopyroxene, comprise about 3% of the peridotite inclusions. Peridotite textural types range from coarse to laminated-disrupted-mosaic-porphyroclastite, in the terminology of Harte (1976), but no correlation has been noted between deformation textures and extent of hydration or carbonation. Assemblages below are described in order of

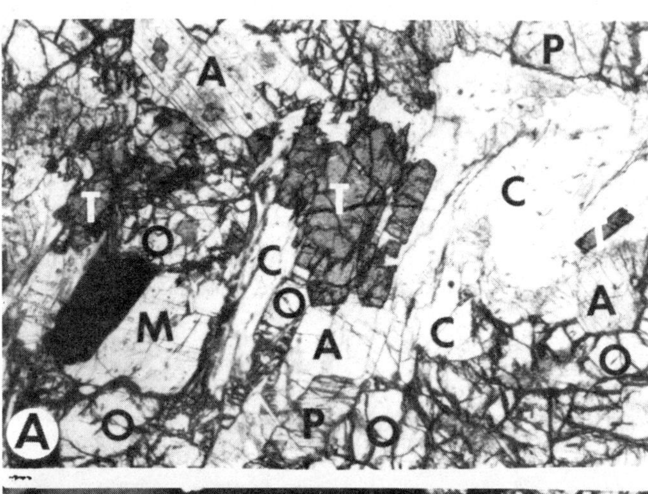

Fig. 2. Photomicrographs of relatively undeformed Group B rocks. (a) A typical cluster of chlorite crystals (C) with olivine (O), orthopyroxene (P) amphibole (A), magnesite (M), and unusually abundant, euhedral titanoclinohumite (T). The labelled olivine crystal in contact with the largest grain of titanoclinohumite appears to have crystallized with the hydrous phases. Analyzed rock N23 (Tables 1 and 2). Length of field, 3.1 mm. (b) An unusual occurrence of laths of chlorite (C) in a mosaic groundmass predominately of recrystallized olivine. Rock N155. Crossed nicols; length of field, 0.8 mm.

generally increasing abundance of hydrous minerals.

Group A: Assemblages with Some Aluminous Spinel Preserved

Rocks which contain at most several percent hydrous phases and yet which retain some aluminous spinel comprise less than five percent of the peridotite population. In the least hydrated rocks of this group, small idioblastic amphibole crystals occur within and at grain boundaries of pyroxenes and spinel. With increasing hydration chlorite is found at some spinel grain boundaries and amphibole is more abundant. Rims of spinel grains in contact with chlorite are opaque, reflecting a decrease in aluminum content. In extreme examples, some spinel grains have been reduced to shreds of opaque oxides surrounded by amphibole and chlorite; still other spinel grains in the same rocks are unaltered. In two rocks in this group, rims of garnet have developed about some of the spinel crystals (Smith and Levy, 1976). Fine lamellae and blades of pyroxene and spinel have exsolved from relict grains of both diopside and orthopyroxene, and the host pyroxenes are commonly slightly turbid. Small, clear, xenoblastic crystals of diopside have locally formed with chlorite, and some orthopyroxene crystals have clear rims. Antigorite in textures suggestive of a primary origin has not been observed in rocks of this group.

Group B: Assemblages with Amphibole and Chlorite and Without Aluminous Spinel

About 80% of the peridotite inclusions fall into this group. Most samples contain well over 70% combined relict pyroxene plus olivine.

Fig. 3. Photomicrograph of a cluster of deformed chlorite crystals in a Group B nodule. The cluster is bounded by orthopyroxene (P) and olivine (O). Note the kink bands in chlorite; they are particularly well-displayed in the crystal in the upper left. Crossed nicols; length of field, 1.2 mm.

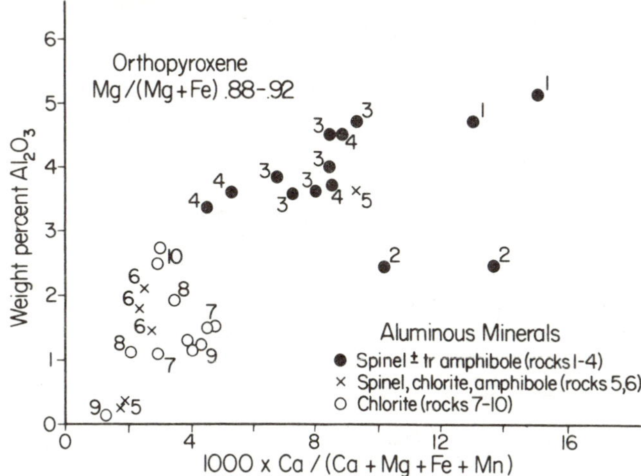

Fig. 4. Comparison of calcium and alumina contents of orthopyroxene crystals in rocks with different assemblages of aluminous minerals. Each point represents broad beam analyses of a crystal interior, except for the two low alumina points for rock 5; these two points represent clear volumes near a crystal rim. Note the ranges of composition for individual rocks. The following key relates the numbers plotted here to the rock numbers in tables and text: anhydrous assemblages - 2, N4; 3, N16; 4, N61; Group A - 1, N53; 5, N17; 6, N55; Group B - 7, N23; 9, N71; 10, N57; Group C - 8, N51.

Chlorite (up to 20%) typically occurs in clusters of crystals (Fig. 2a), some of which include residual oxides. In a few rocks chlorite also occurs in tabular crystals up to 2 mm long distributed with olivine and pyroxene crystals in apparent equilibrium metamorphic textures (Fig. 2b). Amphibole (up to 20%) forms crystals up to 3 mm in maximum dimension. Titanoclinohumite (up to 2%) occurs as single idioblastic crystals up to 1 mm long and as clusters of crystals, commonly within patches of chlorite (Fig. 2a); it also occurs as xenoblastic grains epitaxially intergrown with olivine. Antigorite (up to 5%) occurs in some rocks as sparse blades cutting pyroxene and olivine crystals and as rims about chlorite. Magnesite occurs in trace amounts in equant crystals up to 1 mm long in a few rocks (Fig. 2a). Relict pyroxene crystals are commonly slightly turbid, but small, clear crystals of diopside are locally present with the hydrous phases. In a few rocks small grains of recrystallized olivine also occur (Fig. 2b); most olivine grains, however, are relatively large with sparse kink bands, and they appear to predate hydration.

All these phases are found in peridotite nodules with little deformation recorded texturally and also in highly deformed rocks with flattened, strained pyroxene and olivine grains. Some chlorite, amphibole, and magnesite grains in

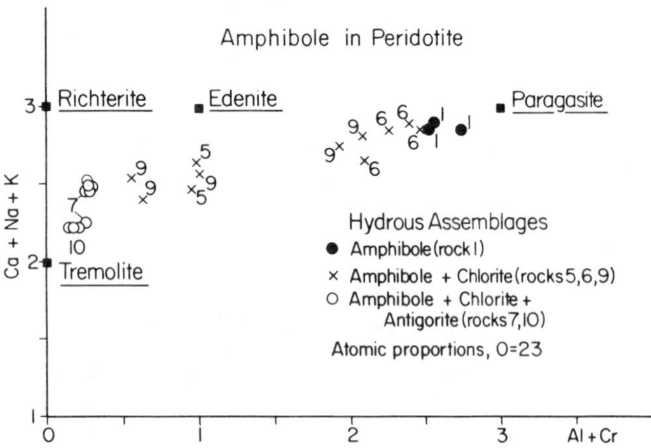

Fig. 5. Variations in amphibole composition with hydrous assemblage. Each point represents a small volume analyzed with a small-diameter beam, in contrast to most of the pyroxene analyses in Fig. 4. Note the compositional variation within single rocks. Key as in Fig. 4.

the deformed rocks are also strained (Fig. 3), and such textures confirm the pre-eruption origin of these minerals. Antigorite occurs both in undeformed blades at angles to deformational fabrics and in blades with irregular and sweeping extinction which appear to have been strained.

Group C: Rocks with abundant antigorite

These rocks comprise about 10% of the nodules; most of them contain 5-15% antigorite and at least 60% combined pyroxene plus olivine. Antigorite is relatively more abundant than amphibole and in typical samples amphibole is not present. The antigorite occurs in blades up to 3 mm long, which cut olivine and pyroxene and rim chlorite. Titanoclinohumite, chlorite, and amphibole occur in the textural settings described for Group B. Olivine crystals are typically relict, as judged from their relatively large size and kink bands, but apparently recrystallized olivine does occur. Most orthopyroxene and diopside crystals are clear, and some appear to be recrystallized. One rock (N15, Table 1) exceptionally rich in antigorite (65%) also contains large crystals of apparently relict olivine (33%), clear diopside (1%), and titanoclinohumite (trace).

Mineral Compositions

Mineral compositions reported here were obtained with an ARL EMX electron microprobe: corrections were made using the factors reported by Albee and Ray (1970).

Typical anhydrous phases in rocks of Group A are representative of those in peridotite inclusions without hydrous minerals (Smith and Levy, 1976): compositions of the inhomogeneous pyroxenes in an example (N53, Table 1) were determined by moving crystal interiors under a broad beam during electron probe analysis. Small recrystallized diopside grains in a similar rock (N17) contain much less alumina than turbid relict crystals (.41-.94% compared to 4.4%), and clear rims present on a few orthopyroxene crystals are much less aluminous and less calcic than the turbid crystal interiors (Rock 5, Fig. 4). Amphibole compositions in a rock without chlorite (N53, Table 1) approach ideal pargasite, with nearly complete occupancy of the "A" site (.96 Na+K) and up to 0.8 atoms of (Al+Cr) VI (23 oxygen basis). Where chlorite has developed, some amphibole is less aluminous (Fig. 5). Chlorite compositions fall near that of clinochlore (N55, Table 1).

Apparently residual enstatite and diopside crystals in Group B rocks are less aluminous than the pyroxenes in anhydrous lherzolite, and the orthopyroxene grains are also less calcic (Fig. 4 4). Amphiboles in some rocks are strongly zoned with broad ranges in composition; in an extreme example, amphibole with a range of 8 weight percent Al_2O_3 is present. In a few rocks, compositions are nearly uniform: in one such inclusion (N23, Table 1; Fig. 2a) the homogeneous amphibole is basically tremolite with limited richterite-edenite solid solution. In rocks with bladed antigorite, chlorite is distinctly less aluminous than the clinochlore in less hydrated rocks (Fig. 6).

The clear crystals of diopside and enstatite in Group C inclusions have relatively low aluminum contents (N15, Table 1); aluminum in diopside can be cast primarily in octahedral coordination as part of the jadeite molecule. Chlorite is restricted to a relatively low range of Al+Cr, and it occurs with aluminous antigorite. Antigorite in a chlorite-free rock is less aluminous (Fig. 6). Analyses of chlorite and antigorite in Group B and C inclusions consistently sum from 85 to 87 percent for all but one rock, even though repeated several times; the data suggest 13-15% H_2O, though antigorites typically contain only 12-13% (Whittaker and Wicks, 1970). Pyroxene and olivine analyses made during the same intervals yielded sums near 100%. Analyzed titanoclinohumite grains in Group B and C rocks are unzoned and contain 5.5 wt% TiO_2 (Table 1), though the bulk rock titanium contents (Table 2) are different.

Magnesite in all three inclusion groups is similar in composition, consisting of 93-96% magnesite, 4-7% siderite, and 0.2-0.8% calcite (mole proportions). The partition of Fe and Mg between magnesite and olivine in contact is systematic (Fig. 7). Dolomite in an ambiguous texture (primary or secondary ?) in one peridotite nodule has a composition near $Ca_{.53}Mg_{.46}Fe_{.015}$.

Grains in contact or in proximity were analyzed to test for exchange equilibrium. The results for Fe and Mg were broadly consistent for all

TABLE 1. Electron Probe Analyses of Minerals in Peridotite Inclusions.

Group	A	A	A	A	A	A	A
Rock	N53	N53	N53	N53	N53	N55	N55
Phase	Opx*	Cpx*	Ol	Sp*	Amph	Amph	Chlor
SiO2	54.3	52.6	na	nd	42.6	43.6	29.5
TiO2	.08	.22	na	nd	.28	.08	na
Al2O3	5.22	6.40	na	56.9	15.7	14.1	19.4
Cr2O3	.47	.87	na	11.9	.80	.75	.80
FeO**	6.07	2.20	9.45	12.2	3.91	4.10	3.01
MnO	.16	.10	na	.13	.06	.08	na
MgO	32.8	15.9	50.4	19.1	18.3	18.6	32.5
CaO	.77	20.2	.01	.02	12.0	12.2	.01
Na2O	.07	1.67	na	nd	3.49	3.37	na
K2O	nd	nd	na	nd	.03	.28	na
	100.0	100.2		100.2	97.2	97.2	85.2

Group	B	B	B	B	B	B	B	B
Rock	N23	N23	N23	N23	N23	N23	N23	N23
Phase	Opx	Cpx	Ol	Tclhm	Magne	Amph	Chlor	Antig
SiO2	57.3	54.7	40.4	35.6	na	55.6	31.8	39.9
TiO2	.06	.08	.01	5.55	na	.05	nd	na
Al2O3	1.47	1.73	nd	nd	na	1.64	12.4	3.87
Cr2O3	.29	1.24	nd	.02	na	.18	2.05	1.23
FeO**	6.24	2.03	11.6	11.7	5.28	2.85	4.28	4.74
MnO	.12	.06	.20	.18	na	.08	.01	na
MgO	35.6	16.8	49.1	46.2	42.0	23.2	34.2	37.2
CaO	.24	22.6	nd	.01	.17	9.23	na	.04
Na2O	nd	1.19	nd	na	na	3.83	na	na
K2O	na	na	na	na	na	.84	na	na
	101.3	100.4	101.2	99.2		97.4	84.8	87.0

Group	C	C	C	C	C	C	C	C
Rock	N51	N51	N15	N51	N51	N51	N51	N15
Phase	Opx	Cpx	Cpx	Ol	Tclhm	Chlor	Antig	Antig
SiO2	57.2	54.8	55.1	42.1	36.3	32.9	41.1	42.6
TiO2	.03	.02	na	nd	5.57	.02	.03	na
Al2O3	1.09	.33	1.89	nd	nd	12.3	2.82	1.45
Cr2O3	na	.18	1.44	na	na	1.95	.79	.24
FeO**	5.81	1.94	1.55	9.26	9.20	3.47	3.64	4.40
MnO	na	.08	na	.18	.19	nd	.04	na
NiO	na	na	na	.36	.29	.19	.16	na
MgO	35.1	17.4	16.4	49.1	46.5	34.9	37.5	40.2
CaO	.11	24.4	22.7	nd	.03	.04	.03	.03
Na2O	na	.59	1.35	na	na	na	na	na
	99.3	99.8	100.4	101.1	98.1	85.7	86.2	89.0

na Not analyzed nd Not detected

*Cr revised, analyses from Smith and Levy, 1976
**All Fe as FeO

samples, and data are given below for the ratio Fe/(Fe+Mg) in a number of grains in one rock characterized by relatively coarse-grained hydrous phases (N23, Table 1): diopside, .051-.065; calcic amphibole, .062-.070; magnesite, .066; chlorite, .066-.068; antigorite, .067-.069; orthopyroxene, .090; olivine, .115-.118; titano-clinohumite, .125. The data are for recrystallized diopside and olivine; values for relict diopside and olivine in the same rock are .07 and .10, respectively.

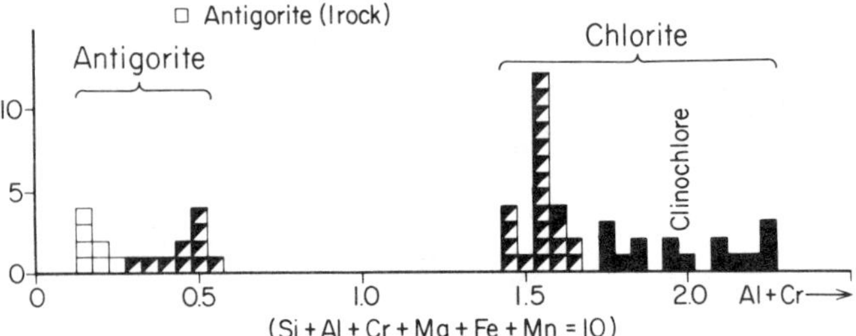

Fig. 6. Al plus Cr contents of chlorite and antigorite, according to the "primary" sheet silicates present. Each point represents a small volume analyzed with a small-diameter beam.

Interpretation of Hydrous Assemblages

On the basis of the anhydrous assemblages, Smith and Levy (1976) suggested that the peridotite nodules cooled to near or below 700°C at depths of 45-60 km before they were incorporated in the erupting breccia. These P-T conditions can be tested by comparison of the hydrate and carbonate assemblages with those in better-known metamorphic rocks and with experimental data. The assemblages may also provide information on hydrate and carbonate stabilities in a P-T range where data are limited, although compositional zoning and the persistence of relic phases and compositions show that disequilibrium was common.

Similar mineral assemblages are found in partly-hydrated peridotite masses with accessory chlorite exposed in many localities in western Europe (Rost 1961; Moore and Qvale, 1977) and in metamorphosed serpentinites, as summarized by Evans (1977). Assemblages in meta-serpentinites progress in the following order of decreasing metamorphic grade (Trommsdorff and Evans, 1974; Evans, 1977):

1. Forsterite, enstatite, diopside, MgAl spinel
2. Forsterite, enstatite, amphiboles, MgAl spinel
3. Forsterite, enstatite, tremolite, chlorite
4. Forsterite, magnesiocummingtonite, tremolite, chlorite
5. Forsterite, talc, tremolite, chlorite
6. Forsterite, antigorite, tremolite, chlorite
7. Forsterite, antigorite, diopside, chlorite

The Fe/Mg partition between these phases (Trommsdorff and Evans, 1974) is like that described above for phases in the nodules. The chlorite in the highest grade assemblages in metaserpentinite has 2.4 Al ions per 10 cations (Frost, 1975); for comparison, the most aluminous chlorites in Group A nodules have 2.3 (Al+Cr) ions. Magnesite is the predominant carbonate in metaserpentinites at temperatures above antigorite breakdown. Titanoclinohumite occurs in many of the antigorite-rich rocks.

The progression of assemblages with increasing hydration in the nodules is distinctly different, however, in that neither magnesiocummingtonite nor

TABLE 2. Bulk Compositions of Green Knobs Peridotite Inclusions.

Group Rock	(*) N61	(*) N16	A (*) N17	B N23	B N71	C N51	C N147
SiO_2	45.35	42.31	45.40	44.04	43.05	40.24	40.88
TiO_2	0.11	0.11	0.02	0.11	0.06	0.04	0.01
Al_2O_3	2.34	1.94	1.97	2.91	3.40	0.96	0.59
Cr_2O_3	0.42	0.38	0.45	0.34	0.43	0.33	0.35
Fe_2O_3	2.75	1.92	0.97	1.33	1.73	2.38	2.39
FeO	5.37	7.17	6.12	6.35	5.77	6.02	4.61
MnO	0.12	0.12	0.08	0.10	0.10	0.13	0.09
NiO	0.25	0.31	0.27	0.24	0.26	0.31	0.29
MgO	39.64	41.50	42.04	36.89	37.75	44.24	45.77
CaO	2.24	3.02	1.68	3.12	2.42	0.57	0.33
Na_2O	0.36	0.20	0.18	0.49	0.44	0.00	0.03
K_2O	0.01	0.01	0.01	0.05	0.03	0.00	0.00
H_2O^+	0.89	1.63	1.17	3.50	4.17	4.14	4.14
H_2O^-	0.09	0.13	0.09	0.10	0.13	0.15	0.30
CO_2	0.07	0.17	0.19	0.56	0.23	0.46	0.16
	100.01	100.92	100.64	100.13	99.97	99.97	99.93

Analyst: G. Karl Hoops, by modified rapid rock analysis. Ca, Na, K, Mn, Cr, Ni by atomic absorption.

* These three analyses were reported by Smith and Levy (1976).
The first two rocks in the table (N61, N16) contain no primary hydrous phases or carbonates.

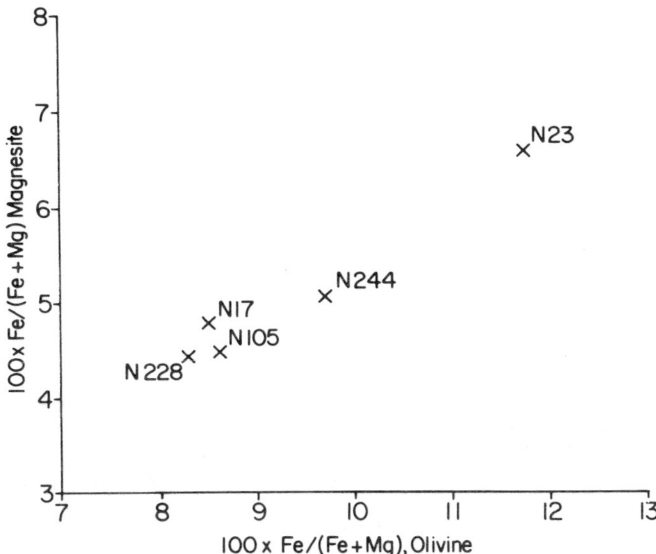

Fig. 7. Iron-magnesium partition between contacting olivine and magnesite crystals in five rocks.

anthophyllite occurs, and talc was seen only as a trace phase along apparently late fractures. Instead, enstatite occurs in contact with antigorite in rocks of Groups C and B. The absence of a low-Ca amphibole may reflect high pressure, since pure Mg-anthophyllite may be unstable above about 5-7.5 kb (Chernosky, 1976; Delany and Helgeson, 1978) Magnesiocummingtonite may be more stable than anthophyllite at high temperatures in the presence of tremolite (Rice et al., 1974), but a similar pressure limitation may occur. The field of talc plus forsterite, however, occupies a temperature interval of about 140° below the breakdown of enstatite to forsterite plus talc and above the upper stability limit of antigorite at low pressures (Evans, 1977). All nodule antigorite may be secondary, but the hypothesis that some antigorite is part of the primary hydration sequence is supported by the systematic changes in mineral composition from Groups A to C, by the absence of bladed antigorite from rocks without chlorite, and by some of the antigorite textures.

Two arguments suggest that enstate-antigorite could have been in equilibrium. First, talc and forsterite typically contain very little alumina plus chrome, whereas antigorite contains as much as 5.1 weight percent of the combined oxides in Group B nodules, and a minimum of 1.5 weight percent in Group C nodules. Chernosky (written communication, 1975) found that preliminary data indicated 3.7 wt% Al_2O_3 raises by about 100°C the metastable high-temperature breakdown of lizardite to forsterite-bearing assemblages. Evans et al. (1976), however, could detect no difference at 10 kb in the stability limits of synthetic antigorite and natural antigorite with about 1% Al_2O_3. Second, thermodynamic calculations support stability of antigorite-enstatite at high pressures. Hemley et al. (1977) calculated that the stability field of antigorite encroaches on that of talc plus forsterite with increasing water pressure. Delany and Helgeson (1978) calculated that the field of talc plus forsterite terminates at an invariant point at 16.7 kb, and that at higher pressures forsterite plus clinoenstatite plus water reacts directly to antigorite.

Magnesite occurs in proximity to recrystallized diopside in some nodules, and the systematic Fe-Mg partition between olivine and magnesite (Fig. 7) suggests that at least Fe-Mg exchange equilibrium was established between magnesite and one of the recrystallized silicates. Kushiro et al. (1975), however, indicate that the minimum pressure for stability of diopside-magnesite relative to enstatite-dolomite exceeds 30 kb at 700°C. Their definitive experiments were at and above 1000°C, but a thermodynamic calculation by Eggler et al. (1976) suggests that the projection of Kushiro et al. to lower temperatures is reasonable. Brey and Green (1978), however, projected the lower-temperature stability of diopside-magnesite to much lower pressures -- about 8 kb at 750°C: the occurrence in these nodules, presuming equilibrium, is consistent only with the projection of Brey and Green.

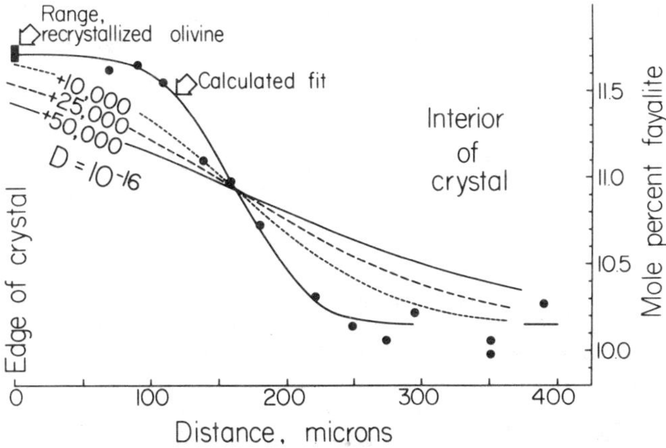

Fig. 8. Observed compositional zoning in an olivine crystal in rock N23, together with a calculated fit and curves indicating the decay of the fit with time for a diffusion coefficient of 10^{-16}. Times for the calculated fit to decay are in years. The fitted curve and the decay curves are from an equation relating compositional variations with time in a pair of semi-infinite solids, each initially homogeneous and in contact at a flat interface (e.g., Shewmon, 1963). The model for the fitted curve is not realistic for the actual formation of the gradient, and the decay curves do not accurately represent the crystal near its boundary. Exact decay solutions would yield shorter times for reduction of the compositional gradient.

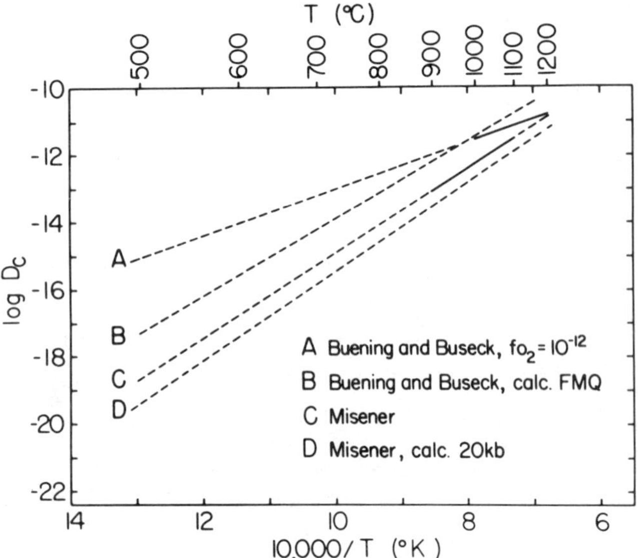

Fig. 9. Experimental determinations (solid lines) and extrapolated variations (dashed lines) of the diffusion coefficient (cm^2/sec) for Fe-Mg diffusion in olivine in the c-axis direction for a composition of 90% forsterite. The one-atmosphere data of Buening and Buseck (1973) have been extrapolated to conditions of the FMQ buffer (Wones and Gilbert, 1969) using their equation for variation with oxygen fugacity. The low-pressure data of Misener (1974) have been extrapolated to 20 kilobars according to his determination of the activation volume.

The above discussions indicate that the primary hydrate-carbonate assemblages may have formed in local equilibrium, and their stabilities can be used to infer conditions of volatile-solid reactions. Pargasite (Group A) and richterite-tremolite (Group B) may be stable to temperatures over 1000°C at 10-20 kb water pressure (e.g., Kushiro, 1970; Hariya and Terada, 1973). Clinochlore (Group A) breaks down to enstatite, forsterite, spinel, and water at 20 kb, 894°C (Staudigel and Schreyer, 1977). Group A hydration reactions could thus have occurred at temperatures up to almost 900°C at 20 kb. Antigorite, if pure and a primary phase, establishes limits for rocks of Groups B and C of 650°C, 15 kb water pressure (Evans et al., 1976); effects of Al plus Cr remain conjectural. The maximum temperature for stability of antigorite plus diopside is at least 25°C lower than antigorite breakdown at 10 kb (Trommsdorff and Evans, 1977). Titanoclinohumite is likely stabilized by titanium, and its stability is not well-defined (Merrill et al., 1972).

Hydration reactions in all groups may thus have occurred at 600-700°C at 45-60 km, values consistent with those inferred by Smith and Levy (1976) from the anhydrous assemblages; the variations from Groups A to C may reflect an increasing availability of water. Single thin sections of some Group A rocks contain both unaltered aluminous spinel and chlorite-amphibole clusters which replaced spinel: such variations suggest that hydration was limited by the supply of water, and water fugacity may have been controlled by local assemblages.

The volatiles which caused the reactions were likely mostly H_2O, even though magnesite was formed. At 2 kb fluid pressure of CO_2-H_2O mixes, magnesite can form at very low mole fractions of CO_2, while antigorite is stable only at mole fractions of H_2O greater than 0.8 (Trommsdorff and Evans, 1977). A comparison of the bulk compositions of nodules without primary hydrous phases with those of nodules in Groups A, B, and C (Table 2) provides no evidence for addition of alkalis, titanium, or other elements during hydration: sodium may actually have been lost during formation of Group C assemblages. Titanium and sodium concentrated in titanoclinohumite and amphibole, respectively, were probably drawn from diopside during reaction.

Timing of Hydration

Temperatures suitable for peridotite hydration in the uppermost mantle below the diatremes may have existed for over 1 billion years, since the last major plutonism in the central Plateau region probably occurred near 1430 m.y. (Silver, 1968), and the last major igneous activity may have been intrusion of diabase sills at about 1200 m.y. (Silver, 1960). The Buell Park and Green Knobs diatremes presumably erupted at about the same time, 25 m.y. ago (Roden, 1977).

Fe-Mg zoning in olivine crystals, however, can be used to place an approximate limit on the maximum age of the most recent hydration. In a Group B nodule (N23, Table 1, Fig. 2a), 5 recrystallized olivine grains intergrown with hydrous phases have 11.66-11.75 mole percent fayalite, whereas large, relict olivine crystals in the same thin section have from 10.26 to 11.20 mole percent. Olivine gains iron during formation of hydrous phases with lower Fe/Mg ratios. A compositional profile from the rim towards the interior of a relict grain in contact with amphibole is shown in Fig. 8. The profile may reflect growth of olivine (Fa 11.7) onto the rim of the relict crystal as well as Fe-Mg exchange and diffusion. If hydration ceased, diffusion would eliminate the compositional gradient. Fe-Mg diffusion rates depend on temperature, pressure, oxygen fugacity, composition, and crystallographic direction; two independent studies of the diffusion coefficient extrapolate to values in the range 10^{-14} to 10^{-16} at 700°C (Fig. 9). The solution used to model decay of the fitted curve (Fig. 8) contains the product of the diffusion coefficient (D) and time, and hence a change in one variable causes an inverse change in the other for the same solution. Assuming a D of 10^{-16}, the profile would be substantially changed in only 25,000 years. The assumed D is unlikely to be too large by more than a factor of 1000, and hence the hydration reactions and consequent partition of Fe into olivine almost certainly continued to or occurred at some time during the 25 m.y. preceding eruption. The uncertainties in both the extrapolation of the diffusion coefficient (Fig. 9) and in the nodule temperature history are such that it is not fruitful to attempt to set more precise limits on the possible decay time or to model the more complex processes which may have formed the gradient.

Discussion

The discrete crystals of some of the hydrous phases reported from Plateau diatremes are most likely fragments of hydrated peridotite inclusions which were distrupted by comminution during eruption. McGetchin et al. (1973) recognized such a source for chlorite in the Cane Valley diatreme (Fig. 1), but they believed that the phase formed by reaction of CO_2 and H_2O with aluminous clinopyroxene in spinel lherzolite, perhaps yielding carbonatite melt as an additional reaction product. Textural evidence in the nodules indicates that spinel was the most essential phase for chlorite formation and that no melt byproduct was formed. Aoki et al. (1972, 1976) and Aoki (1977) reported crystal fragments of tremolite containing richterite solid solution, of titanoclinohumite, and of titanochondrodite in the Buell Park diatreme, and they interpreted them to be crystallization products of kimberlite magma. Smith (1977b) argued that the fragments were from comminuted, solid peridotite, and the hydration assemblages reported here support this interpretation. McGetchin and Besancon (1973) have described carbonate grains apparently included in garnets from Red Mesa and Garnet Ridge, but the origin of these carbonates is unclear. Their analyzed magnesite grain has a Fe/(Fe+Mg) ratio of 0.19, and by analogy with fractionations measured in Green Knobs nodules and in experiments by Brey and Green (1978), the magnesite could not have been in equilibrium with either the clinopyroxene (Fe/(Fe+Mg), 0.08) included in the same garnet or with forsterite.

The demonstration that partly hydrated spinel peridotite exists at depth below the Colorado Plateau supports part of the hypothesis of McGetchin et al. (1973) that the Plateau eruptions began as solid-fluid systems without presence of a silicate melt. They hypothesized that volatiles accumulated along peridotite grain boundaries until the resulting low density made the system mechanically unstable. From evidence described above, it is likely that the volatiles, chiefly H_2O, reacted continuously to form hydrous phases, and that they did not accumulate in excess until a triggering event. Minettes and kimberlitic microbreccias erupted in the same region at approximately the same time on the Colorado Plateau (Naeser, 1971), and the finding that the eruptions actually overlapped in time in at least one center, Buell Park (Roden, 1977), suggests that minette magma provided the trigger. Heat from ascending minette magmas may have reversed the hydration reactions, releasing the volatiles needed to spur the kimberlitic eruption. The minette magmas were generated from below the relatively depleted garnet peridotite sampled by the ultramafic microbreccia (Smith and Levy, 1976; Ehrenberg, 1977).

Since the kimberlitic eruptions began in a garnet peridotite source (McGetchin and Silver, 1970; Smith and Levy, 1976; Switzer, 1977), the hypothesis requires that hydration also took place below the transition from spinel to garnet peridotite, near 60 km below Green Knobs and Buell Park at the time of eruption (Smith and Levy, 1976). The rare rocks of Group A with garnet rims about spinel must have cooled near the transition, and chlorite masses in several Group B nodules resemble pseudomorphs of garnet. Mercier (1976) reported a relict garnet in a peridotite inclusion with chlorite from the Moses Rock kimberlite (Fig. 1). McGetchin et al. (1970) have described titanoclinohumite inclusions in garnet fragments from Moses Rock.

Mercier (1976) suggested that the hydrated peridotite nodules in the diatremes represent fragments of ophiolite complexes subducted off the west coast and underthrust below the Colorado Plateau. Helmstaedt and Doig (1975) advocated a similar origin for relatively low-temperature eclogite inclusions from some of the northern Plateau diatremes. Hydrated peridotites may be an important component of the oceanic crust (e.g., Bonatti, 1976; Clague and Straley, 1977), and serpentine, chlorite, and tremolite-actinolite are among the hydrous phases observed in dredged peridotite samples (Prinz et al., 1976). Subducted oceanic crust should undergo progressive heating during its passage beneath the continent, but no nodule textures have been recognized which indicate progressive heating and metamorphism of hydrated peridotite. Such textures are well-documented in meta-serpentinites (Trommsdorff and Evans, 1974). Textures in these nodules instead indicate late hydration during or after cooling. Only the inclusions of titanoclinohumite in garnet reported by McGetchin et al. (1970) could be interpreted as prograde metamorphism of hydrated peridotite. Moreover, garnet peridotite, which is not recognized in the oceanic crust or uppermost oceanic mantle, has also been subject to late hydration, as discussed above. It is likely that the peridotite nodules represent cooled, hydrated continental mantle, sheared along deep fault zones but otherwise generally undisturbed since the Precambrian. The source of the fluid phase could have been from dewatering of a subducted slab at greater depth (e.g., Best, 1975).

Several arguments bear on the lateral extent of mantle affected by hydration reactions. All but one of the kimberlitic diatremes on the Colorado Plateau occur near or on major monoclines bordering uplifts; textures in many Green Knobs peridotite nodules suggest deformation shortly before eruption, perhaps as Laramide fault zones below the monoclines were reactivated at depth (Smith and Levy, 1976). These fault zones which localized eruptions may have done so because they acted as channel ways for fluid transport, perhaps begun after Laramide deformation. Such localized hydration along a potential transport system is easier to conceive than widespread hydration limited by diffusion rates. It is possible that large regions were affected, however, in a variant of the serpentization hypothesis discussed by Hess (1955) to explain uplift of the entire Colorado Plateau. Best (1974) described amphibole within peridotite inclusions in alkali basalt from near the western margin of the Plateau, and he suggested some of the amphibole may have formed by reaction of spinel, diopside, and an aqueous fluid. About half of the many peridotite nodules examined by Best (1974) contain textural evidence that amphibole or other hydrous phases were present but destroyed in melting and dehydration reactions, while in only a few percent of these nodules is amphibole preserved. Since the stability limits of chlorite and antigorite are lower than that of alkalic amphibole, it is unlikely that they would be preserved during transport in a basalt. Kimberlites, which erupt at lower temperatures, may provide a more complete representation of hydrous phases, and the confirmation of chlorite inclusions within diamond (Mitchell and Giardini, 1977) shows that the mantle chlorite described here is not unique. Julian (1970) found the 45-100 km depth range below the Plateau to be best modeled by peridotite with P-wave velocities near 7.9 km/sec. McGetchin and Silver (1972) noted that this value is low for upper mantle beneath stable shield areas, and they suggested the model P-wave velocity reflected the presence of an interstitial volatile-rich fluid in the upper mantle below the entire plateau. More detailed seismic studies, however, are necessary to evaluate the hypotheses discussed above for the extent of hydration.

Acknowledgments. The permission of the Navajo Tribe to conduct these studies is gratefully acknowledged: the cooperation of Martin Link, Director of Museum and Research of the Navajo Tribe, was particularly helpful. Susan Levy aided in parts of the electron probe analysis and in modal studies. I thank J. V. Chernosky for comments on serpentine stability and for permission to quote unpublished results. Constructive reviews by A. L. Boettcher, B. W. Evans, and D. H. Eggler improved an earlier version of this manuscript. The Geology Foundation of the University of Texas at Austin supported several aspects of this study; the bulk of support was furnished by The National Science Foundation, Earth Sciences Section, NSF Grants GA-41414 and EAR 76-12368.

References

Albee, A.L., and L. Ray, Correction factors for electron probe microanalysis of silicates, oxides, carbonates, phosphates, and sulfates, Anal. Chem., 42, 1408-1414, 1970.

Allen, J.E., and R. Balk, Mineral resources of Fort Defiance and Tohatchi quadrangles, Arizona and New Mexico, New Mexico Bur. Mines Miner. Res. Bull., 36, 140, 1954.

Aoki, K., R.V. Fodor, K. Keil, and E. Dowty, Tremolite with high richterite-molecule content in kimberlite from Buell Park, Arizona, Amer. Mineral., 57, 1889-1893, 1972.

Aoki, K., K. Fujino, and M. Akaogi, Titanochondrite and titanoclinohumite derived from the upper mantle in the Buell Park kimberlite, Arizona, U.S.A., Contr. Mineral. Petrol., 56, 243-253, 1976.

Aoki, K., Titanochondrodite and titanoclinohumite derived from the upper mantle in the Buell Park kimberlite, Arizona, U.S.A., A reply, Contr. Mineral. Petrol., 61, 217-218, 1977.

Best, M.G., Mantle-derived amphibole within inclusions in alkalic-basaltic lavas, Jour. Geophys. Res., 79, 2107-2113, 1974.

Best, M.G., Migration of hydrous fluids in the upper mantle and potassium variation in calc-alkalic rocks, Geology, 3, 429-432, 1975.

Bonatti, E., Serpentine protrusions in the oceanic crust, Earth Planet. Sci. Letters, 32, 107-113, 1976.

Brey, G.P., and D.H. Green, Carbon in the upper mantle and the genesis of kimberlites and olivine melilitites, Proceedings of the Second International Kimberlite Conference, in press.

Buening, D.K., and P.R. Buseck, Fe-Mg lattice diffusion in olivine, Jour. Geophys. Res., 78, 6852-6892, 1973.

Carswell, D.A., Primary and secondary phlogopites and clinopyroxenes in garnet lherzolite xenoliths, Phys. Chem. Earth, 9, 417-429, 1975.

Chernosky, G.V., Jr., The stability of anthophyllite, brucite, clinochrysotile, enstatite, forsterite, quartz, and talc, Amer. Mineral., 61, 1145-1155, 1978.

Clague, D.A., and P.F. Straley, Petrologic nature of the oceanic Moho, Geology, 5, 133-136, 1977.

Dawson, J.B., and J.V. Smith, The MARID (mica-amphibole-rutile-ilmentite-diopside) suite of xenoliths in kimberlite, Geochim. Cosmochim. Acta, 41, 309-323, 1977.

Delany, J.M., and H.C. Helgeson, Calculation of the thermodynamic consequences of dehydration in subducting oceanic crust to 100 kb and $>800°C$, Amer. J. Sci., 278, 638-686, 1978.

Eggler, D.H., I. Kushiro, and J.R. Holloway, Stability of carbonate minerals in a hydrous mantle, Carnegie Inst. Wash. Yb., 75, 631-636, 1976.

Ehrenberg, S.N., Garnet peridotite xenoliths in minette from the Navajo volcanic field, Extended Abstracts, Second International Kimberlite Conference, Santa Fe, New Mexico, 1977.

Evans, R.W., W. Johannes, H. Oterdoom, and V. Trommsdorff, Stability of chrysotile and antigorite in the serpentine multisystem, Schweiz. Mineral. Petrogr. Mitt., 56, 79-93, 1976.

Evans, B.W., Metamorphism of alpine peridotite and serpentinite, Ann. Rev. Earth Planet. Sci., 5, 397-447, 1977.

Frost, R.B., Contact metamorphism of serpentinite, chloritic blackwall, and rodingite at Paddy-Go-Easy-Pass, central Cascades, Washington, Jour. Petrol., 16, 272-313, 1975.

Hariya, Y., and S. Terada, Stability of richterite 50-tremolite 50 solid solution at high pressures and possible presence of sodium calcic amphibole in the upper mantle conditions, Earth Planet. Sci. Letters, 18, 72-76, 1973.

Harte, B., Rock nomenclature with particular relation to deformation and recrystallization textures in olivine-bearing xenoliths, J. Geol., 85, 279-288, 1977.

Helmstaedt, H., and R. Doig, Eclogite nodules from kimberlite pipes of the Colorado Plateau - samples of subducted Franciscan-type oceanic lithosphere, Phys. Chem. Earth, 9, 95-111, 1975.

Hemley, J.J., J.W. Montoya, D.R. Shaw, and R.W. Luce, Mineral equilibria in the MgO-SiO_2-H_2O system: II Talc-antigorite-forsterite-anthophyllite-enstatite stability relations and some geologic implications in the system, Amer. J. Sci., 277, 353-383, 1977.

Hess, H.H., Serpentines, orogeny, and epeirogeny, Geol. Soc. Amer. Sp. Pap., 62, 391-408, 1955.

Julian, B., Regional variations in upper mantle structure beneath North America, unpublished Ph.D. thesis, Calif. Inst. Tech., 208p., 1970.

Kushiro, I., Stability of amphibole and phlogopite in the upper mantle, Carnegie Inst. Wash. Yb., 68, 245-247, 1970.

Kushiro, I., H. Satake, and S. Akimoto, Carbonate-silicate reactions at high pressures and possible presence of dolomite and magnesite in the upper mantle, Earth Planet. Sci. Letters, 28, 116-120, 1975.

McGetchin, T.R., and L.T. Silver, Compositional relations in minerals from kimberlite and related rocks in the Moses Rock dike, San Juan County,Utah,Amer. Mineral., 55, 1738-1771,1970.

McGetchin, T.R., L.T. Silver, and A.A. Chodos, Titanoclinohumite: a possible mineralogical site for water in the upper mantle, Jour. Geophys. Res., 75, 255-259, 1970.

McGetchin, T.R., and L.T. Silver, A crustal-upper mantle model for the Colorado Plateau based on observation of crystalline rock fragments in the Moses Rock dike, Jour. Geophys. Res., 77, 7022-7037, 1972.

McGetchin, T.R., and J.R. Besancon, Carbonate inclusions in mantle-derived pyropes, Earth Planet. Sci. Letters, 18, 408-410, 1973.

McGetchin, T.R., Y.S. Nikhanj, and A.A. Chodos, Carbonatite-kimberlite relations in the Cane Valley diatreme, San Juan County, Utah, Jour. Geophys. Res., 78, 1854-1869, 1973.

Mercier, J.C., Single-pyroxene geothermometry and geobarometry, Amer. Mineral., 61, 603-615, 1976.

Merrill, R.B., J.K. Robertson, and P.J. Wyllie, Dehydration reaction of titanoclinohumite: reconnaissance to 30 kilobars, Earth Planet. Sci. Letters, 14, 259-262, 1972.

Misener, D.J., Cationic diffusion in olivine to 1400°C and 35 kb, in Geochemical Transport and Kinetics, Carnegie Inst. Wash. Pub., 634, 117-129, 1974.

Mitchell, R.S., and A.A. Giardini, Some mineral inclusions from African and Brazilian diamonds: their nature and significance, Amer. Mineral., 62, 756-762, 1977.

Moore, A.C., and H. Quale, Three varieties of alpine-type ultramafic rocks in the Norwegian Caledonides and Basal Gneiss Comples, Lithos, 10, 149-161, 1977.

Naeser, C.W., Geochronology of the Navajo-Hopi diatremes, Four Corners area, Jour. Geophys. Res. 76, 4978-4985, 1971.

Prinz, M., K. Keil, J.A. Green, A.M. Reid, E. Bonatti, and J. Honnorez, Ultramafic and mafic dredge samples from the equatorial mid-Atlantic ridge and fracture zones, Jour. Geophys. Res., 81, 4087-4103.

Rice, J.M., B.W. Evans, and V. Trommsdorff, Widespread occurrence of magnesiocummingtonite in ultramafic schists, Cima di Gagnone, Ticino, Switzerland, Contr. Mineral. Petrol., 43, 245-251, 1974.

Roden, M.F., Field geology and petrology of the minette diatreme at Buell Park, Apache County, Arizona, Extended Abstracts, Second International Kimberlite Conference, Santa Fe, New Mexico, 1977.

Roden, M.F., and D.Smith, Field geology, chemistry, and petrology of the Buell Park minette diatreme, Apache County, Arizona, Proceedings of the Second International Kimberlite Conference, in press.

Rost, F., Chlorit und granat in ultrabasischen gesteinen: Fortschr. Mineral., 39, 112-126, 1961.

Schmitt, H.H., G.A. Swann, and D. Smith, The Buell Park kimberlite pipe, northeastern Arizona, in Geology of Northern Arizona (Geol. Soc. Amer. Rocky Mtn. Section, Flagstaff, 1974), 672-698, 1974.

Shewmon, P.G., Diffusion in Solids, McGraw-Hill, New York, 203 p., 1963.

Silver, L.T., Age determinations on Precambrian diabase differentiates in the Sierra Ancha, Gila Country, Arizona, Geol. Soc. Amer. Bull., 71 1973-1974, 1960.

Silver, L.T., Precambrian batholiths of Arizona, Geol. Soc. Amer. Spec. Pap., 121, 558-559, 1968.

Smith, D., and S. Levy, Petrology of the Green Knobs diatreme and implications for the upper mantle below the Colorado Plateau, Earth Planet. Sci. Letters, 29, 107-125, 1976.

Smith, D., The origin and interpretation of spinel-pyroxene clusters in peridotite, Jour. Geol. 85, 474-482, 1977a.

Smith, D., Titanochondrodite and titanoclinohumite derived from the upper mantle in the Buell Park kimberlite, Arizona, U.S.A.: A discussion, Contr. Mineral. Petrol., 61, 213-215, 1977b.

Staudigel, H., and W. Schreyer, The upper thermal stability of clinochlore $Mg_5Al(AlSi_3O_{10})(OH)_8$, at 10-35 Kb P_{H_2O}, Contr. Mineral. Petrol., 61, 187-198, 1977.

Switzer, G.S., Composition of garnet xenocrysts from three kimberlite pipes in Arizona and New Mexico, Smithsonian Contr. Earth Sci., no. 19, 1-21, 1977.

Trommsdorff, V., and B.W. Evans, Alpine metamorphism of peridotitic rocks, Schweiz. Mineral. Petrog. Mitt., 54, 333-352, 1974.

Trommsdorff, V. and B.W. Evans, Antigorite-ophicarbonates: phase relations in a portion of the system $CaO-MgO-SiO_2-H_2O-CO_2$, Contr. Mineral. Petrol., 60, 39-56, 1977.

Whittaker, E.J.W., and F.J. Wicks, Chemical differences among the serpentine "polymorphs": a discussion, Amer. Mineral., 55, 1025-1047, 1970.

Wones, D.R., and M.C. Gilbert, The fayalite-magnetite-quartz assemblage between 600° and 800°C, Amer. Jour. Sci., 267-A, 480-488, 1969.

GARNET CLINOPYROXENITE - CHLORITE ECLOGITE TRANSITION IN A XENOLITH FROM MOSES ROCK: FURTHER EVIDENCE FOR METAMORPHOSED OPHIOLITES UNDER THE COLORADO PLATEAU

Herwart Helmstaedt and Daniel J. Schulze*

Department of Geological Sciences, Queen's University, Kingston, Ontario, Canada K7L 3N6

Abstract. Pyrope and calcic clinopyroxene in a garnet clinopyroxenite from Moses Rock have reacted to stably coexisting magnesian chlorite, less pyropic garnet, and omphacite, the assemblage of a chlorite eclogite. The hydration reaction can be balanced, provided the system is open to sodium and oxygen. Retrogression may have occurred when garnet clinopyroxenite in the upper mantle came into contact with subducted, sodium-rich ocean floor rocks now represented by xenoliths of Group C eclogites and jadeite rocks. Iron-magnesium distribution coefficients of the secondary garnet-omphacite pair and of garnet and clinopyroxene rims in Group C eclogite xenoliths are compatible with this interpretation and suggest that retrogressive metamorphism of the garnet clinopyroxenite and progressive metamorphism of the Group C eclogites took place at similar physical conditions prior to kimberlite eruption. The similarity of the ultramafic xenolith suite from the Colorado plateau kimberlites with rocks exposed in metamorphosed ophiolite complexes of high-pressure metamorphic belts leaves little doubt that a shallow slab of subducted oceanic lithosphere existed under the Colorado plateau at the time of kimberlite eruption.

Introduction

The Moses Rock diatreme, Utah, has yielded a variety of ultramafic xenoliths including eclogites, jadeite-rich clinopyroxenites, spinel websterites, spinel lherzolites, garnet lherzolite, and various hydrated peridotites (McGetchin, 1968; Gavasci and Helmstaedt, 1969; McGetchin and Silver, 1970, 1972; McGetchin et al., 1977). This paper describes a garnet clinopyroxenite xenolith (collected by D. J. Schulze in May, 1976) which deserves attention for the following reasons:

1. The xenolith is a garnet clinopyroxenite which has been retrogressed to an eclogitic assemblage coexisting stably with magnesian chlorite.
2. The formation of the eclogitic assemblage required a net addition of sodium indicating sodium metasomatism at depth prior to incorporation into the kimberlite.
3. The subsolidus history of this rock provides important constraints for models of the upper mantle beneath the Colorado plateau and supports the hypothesis that eclogite xenoliths from the Four Corners Kimberlites are fragments of subducted oceanic lithosphere (Helmstaedt and Doig, 1973, 1975; Råheim and Green, 1975).

Eclogite nomenclature used in this paper is based on a combination of the classifications by Coleman et al. (1965) and Smulikowski (1965, 1968). Coleman et al. (1965) referred to all garnet-clinopyroxene rocks as eclogites and divided them into three groups according to their geologic occurrence. Group A eclogites are characterized by highly pyropic garnets and calcic clinopyroxenes and are found in association with ultrabasic rocks and as inclusion in kimberlites. Group B eclogites occur as lenses in migmatitic and gneissic terrains. Their garnets are more iron-rich than those of Group A eclogites and the clinopyroxenes are more sodic (low-jadeite omphacite). Group C eclogites are found together with glaucophane schists and contain the most iron-rich garnets and the most jadeitic clinopyroxenes.

Smulikowski (1965, 1968) distinguished essentially the same groups, but he preferred the name 'garnet clinopyroxenite' or 'griquaite' for garnet-clinopyroxene rocks associated with ultrabasites (≈Group A of Coleman et al., 1965) and reserved the name 'eclogite' for rocks composed of pyrope-almandine garnet and clinopyroxenes with a significant jadeitic component (omphacite). He referred to Groups B and C of Coleman et al. (1965) as common and ophiolitic eclogites respectively. Pointing out

*present address: Geosciences Department, University of Texas at Dallas, Richardson, Texas 75080, U.S.A.

TABLE 1. Mineral Analyses.

Electron microprobe analyses, anhydrous, weight percent					
	1	2	3	4	5
SiO_2	41.93	55.1	40.16	56.26	29.62
TiO_2	0.10	0.12	0.07	0.17	0.05
Al_2O_3	23.83	3.44	22.72	9.75	20.16
FeO	9.23*	1.76*	15.53*	3.00*	4.46*
MnO	0.33	0.03	1.16	0.05	0.06
MgO	20.81	15.56	10.89	10.20	30.98
CaO	3.79	21.45	9.96	14.00	0.02
Na_2O	0.32	2.38	0.12	6.55	0.17
K_2O	0.01	0.01	0.02	0.01	0.02
Cr_2O_3	0.09	0.20	0.16	0.12	0.06
Total	100.44	100.05	85.60	100.11	100.79
Cation contents**					
Si	2.946	1.973	2.968	1.984	2.854
Ti	0.005	0.003	0.004	0.005	0.004
Al	1.973	0.145	1.979	0.405	2.289
Fe^{3+}	0.164	0.053	0.086	0.063	0.025
Fe^{2+}	0.379	0	0.873	0.026	0.334
Mn	0.02	0.001	0.073	0.001	0.005
Mg	2.179	0.83	1.199	0.536	4.448
Ca	0.285	0.823	0.789	0.529	0.002
Na	0.044	0.165	0.017	0.448	0.032
K	0.001	0	0.002	0	0.002
Cr	0.005	0.006	0.009	0.003	0.005
H	0	0	0	0	8
Mol % end members					
Pyr	73		40		
And	2		3		
Alm	17		30		
Spes	1		2		
Gros	7		24		
Jd^+		11		40	
Ac^+		5		5	
Aug^+		83		55	

1. Garnet. 2. Cpx - Primary assemblage
3. Garnet. 4. Cpx. 5. Chlorite - second. assemblage.
* Total Fe calculated as FeO.
**Based on 12 oxygens for garnet, 6 oxygens for Cpx, 14 oxygens for chlorite
\+ Calculated as by Essene and Fyfe (1967).

that kimberlites may contain inclusions of garnet clinopyroxenites and various eclogites, Smulikowski (1968, 1968a) did not establish a special group for xenolithic eclogites.

In the present paper we follow Smulikowski's usage of the names 'garnet clinopyroxenite' and 'eclogite', although we are aware that there is no sharp boundary between the two rocks. We replace Group A of Coleman _et al_. (1965) by garnet clinopyroxenite but retain Groups B and C for general discussions of eclogites. Ophiolite, as used in the present paper, refers to an assemblage of mafic and ultramafic rocks representing former oceanic crust and upper mantle (Coleman, 1977), and metaophiolite is the metamorphosed version thereof.

Garnet clinopyroxenite-eclogite transition

The xenolith has the approximate dimensions of 5 x 5 x 7.5 cm. It can be distinguished from other eclogite xenoliths at Moses Rock by its fine- to medium-grained, pistachio-green matrix of clinopyroxene and chlorite that is studded with subequant grains of brownish-pink garnet up to 0.8 cm in diameter and making up to 40% of the rock. Thin section and electron microprobe studies reveal that the rock is composed of two separate mineral assemblages. A primary pyrope-calcic clinopyroxene assemblage has altered to less pyropic garnet, omphacite, and magnesian chlorite (Table 1). The relatively coarse garnets of the original assemblage are separated from diopside by reaction rims of chlorite (Fig. 1a) which is intergrown with small euhedral grains of the less pyropic garnet and omphacite (Fig. 1b). Most small garnets form euhedral plates parallel to the basal plane of the chlorite. Some of the more equant euhedra are not confined to the basal plane of the chlorite and are included also within the omphacite. Omphacite is always smaller and more euhedral than the calcic clinopyroxene of the original assemblage. Most of the omphacite occurs within the chlorite reaction rims, the remainder forms overgrowth on the calcic clinopyroxene. In such cases, a sharp break in chemical composition exists between calcic clinopyroxene core and omphacite rim (Fig. 2). Rutile appears to have been stable with both assemblages. It is present as tiny inclusions in the original garnet, as separate grains among the calcic clinopyroxene grains, and in the chlorite reaction rims.

Garnets of both assemblages are homogeneous and show no compositional zoning. Their compositions are compared to those of other xenoliths and xenocrysts from the Moses Rock diatreme in Figure 3. Both garnet types are abundant in the varied garnet xenocryst suite. The pyrope-rich garnet (Py_{73}) is typical of garnet pyroxenites and resembles the pinkish xenocrysts described also by McGetchin and Silver (1970). Although similar in pyrope content, these pinkish garnets differ from the deep purple garnets of the garnet lherzolite assemblage by containing virtually no Cr_2O_3.

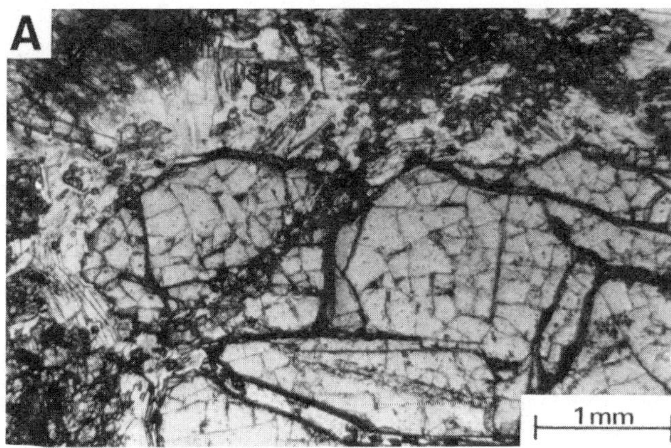

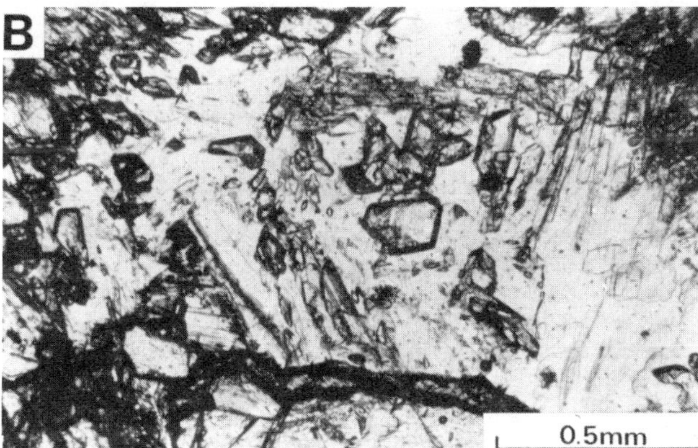

Fig. 1. A. Photomicrograph of pyropic garnet surrounded by reaction rim of chlorite with intergrown garnet and omphacite.
B. Enlarged part of reaction rim illustrating euhedral grains of secondary garnets.
Both photos plane polarized light.

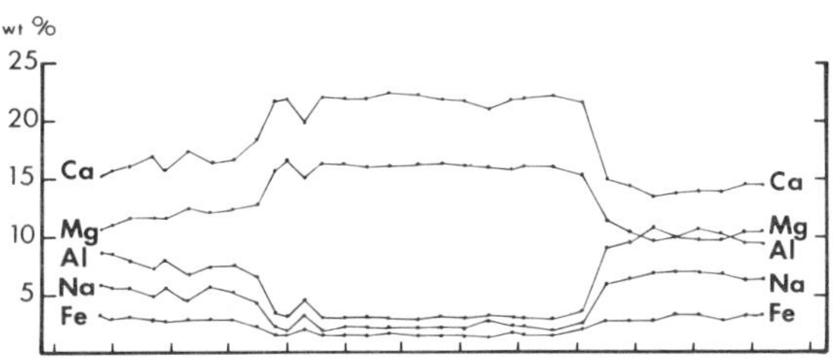

Fig. 2. Electron microprobe traverse across clinopyroxene of Moses Rock garnet pyroxenite xenolith showing contrast in major element oxide content between diopside and omphacite rim. Each dot represents a separate analysis. Horizontal scale divisions are 100 microns.

The pyrope content of the secondary garnets is much lower (Py_{40}), and the grossular content has increased to 24 mole percent. This composition corresponds to garnets of Group B eclogites.

The pyroxene of the primary assemblage is sodium-poor but near the sodium-rich end of Smulikowski's garnet pyroxenite group. Omphacite of the alteration assemblage plots on the edge of the pyroxene field from Group C eclogites (Fig. 4).

The two mineral assemblages can be related by the reaction

$$\text{pyrope} + \text{calcic clinopyroxene} + H_2O \rightleftarrows \text{pyrope-almandine} + \text{omphacite} + \text{chlorite} \qquad [1]$$

Several different versions of this reaction were modelled using the atomic proportions presented in Table 1. Reaction coefficients and residuals derived by a least square technique analogous to that described by Greenwood (1967) are summarized in Table 2. Attempts to balance for all major and minor components (reaction 1a) and for all major components except ferric iron (reaction 1c) yield very large residuals. The fit is better, but still far from satisfactory, when the system is left open to sodium (reactions 1b and 1d). However, a good fit is obtained when the reaction is balanced for total iron and all major components except sodium (reaction 1f). This has the effect of leaving the system open to

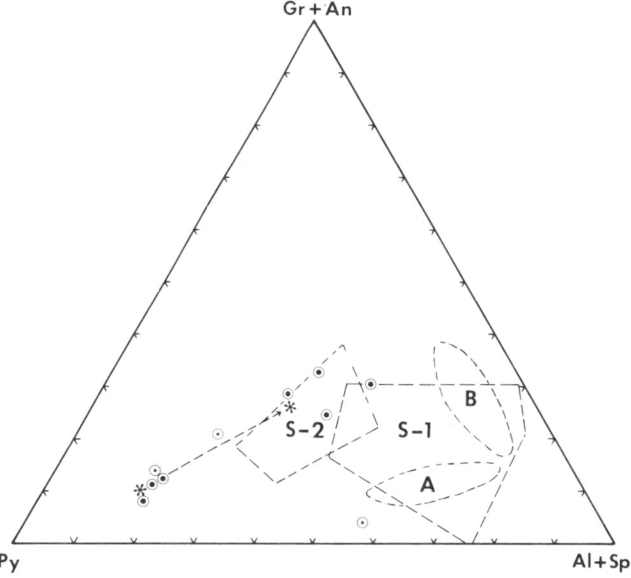

Fig. 3. Composition of garnets of garnet clinopyroxenite from Moses Rock (asterisks connected by dashed arrow) compared to that of garnet xenocrysts from Moses Rock (circles with small dots from McGetchin, 1968; circles with large dots, our data), garnets from eclogites of Colorado plateau kimberlites (Field A as compiled by Helmstaedt and Doig, 1975), garnets from eclogites of Group C eclogites (Field B after Coleman et al., 1965), and garnets from Sanbagawa belt eclogites (Field S-1 and S-2 as compiled by Ernst et al., 1970).

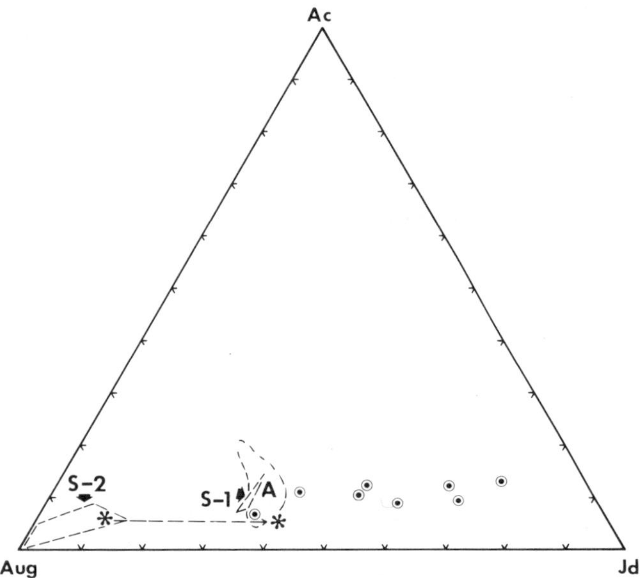

Fig. 4. Composition of clinopyroxenes of garnet clinopyroxenite from Moses Rock (asterisks connected by dashed arrow) compared to clinopyroxenes from eclogites from Colorado plateau kimberlites (circles with large dots, as compiled by Helmstaedt and Doig, 1975), from Group C eclogites (Field A, compiled by Coleman et al., 1965), and from Sanbagawa belt eclogites (Field S-1 and S-2 as compiled by Ernst et al., 1970).

oxygen and sodium. As balancing for total iron and the other major components including sodium (reactions 1g-h) yields unacceptably high residuals for many components, we conclude that reaction 1f is the best approximation of the 'real' reaction. The calculations show clearly that this reaction was not simply a hydration of the rock, but involved oxidation of iron, and depended on a net addition of sodium to proceed from left to right. The ΔV of this reaction is 16.5%. As the sodium content of the xenolith is relatively high compared to that of similarly ultrabasic garnet pyroxenites from surface occurrences (Table 3, see also Smulikowski, 1968), it is reasonable to conclude that some sodium metasomatism took place during re-equilibration.

The physical conditions of reequilibration cannot be estimated precisely. Values for Fe-Mg distribution coefficients (K_D) between garnet and clinopyroxene depend sensitively on the method of calculating the ferric iron content from the microprobe analyses. Assuming stoichiometry and calculating only the normal hypothetical end members, yielded K_D values of slightly above zero for the primary and of between 8.5 and 9 for the secondary garnet-clinopyroxene pairs (Helmstaedt and Schulze, 1977). If all minor components are considered as well (Table 1), no ferrous iron is available in the primary pyroxene, and a K_D value cannot be calculated for the primary pair. A K_D of 15 is obtained for the secondary pair. The relevant phase relationships and the K_D grid of Råheim and Green (1975) (Fig. 5) suggest that the primary pair was stable above 800°C at pressures between 15 and 20 Kb. The re-equilibration temperature may have been approximately 600°C or somewhat below, if pressures of between 12 and 15 Kb are assumed.

It is significant to note that chlorite is stable together with the eclogitic alteration assemblage. This confirms experimental studies by Staudigel and Schreyer (1977) (Fig. 5) and field evidence of Trommsdorff and Evans (1974) and Evans (1977) that magnesian chlorite is stable at high metamorphic grades and can exist in the eclogite facies.

Origin of eclogitic assemblage

The origin of eclogite nodules from kimberlites is a long debated problem. Nodules may be cognate (related to the origin of the kimberlite) or xenoliths (accidental inclusions from the vent walls). Xenoliths in turn may be igneous (crystallized from a basaltic melt under high P and T), metamorphic (recrystallized in the solid state from a basaltic or gabbroic protolith) or residual (remnants after extraction of a melt). In addition, they may have undergone extensive metamorphic recrystallization prior and during ascent with the kimberlite.

The non-diamondiferous Colorado plateau kimberlites are unique in containing xenoliths of relatively low-temperature eclogites with Group C affinities that have been interpreted as fragments of subducted oceanic lithosphere (Helmstaedt and Doig, 1973, 1975; Råheim and Green, 1975). The P-T path of these eclogites deduced from a decrease in K_D between cores and rims of garnets and clinopyroxenes is compatible with progressive metamorphism in a subduction zone (Råheim and Green, 1975; Krogh and Råheim, 1978) (Fig. 5).

A hypothesis for the origin of the eclogitic assemblage in the present xenolith must account for the fact that the metamorphic reaction producing it is retrogressive and involves hydration, oxidation, and sodium metasomatism. The primary assemblage has a low K_D indicating temperatures much higher than those reached by the Group C eclogites. The K_D of the secondary assemblage overlaps with that of the final stages of progressive metamorphism of the Group C eclogites for which values as low as 8.5 - 10 have been calculated (Råheim and Green, 1975; Helmstaedt and Schulze, unpublished data). Although a similarity in K_D does not prove a spatial association of the different xenolith types at depth, retrogressively metamorphosed garnet clinopyroxenites and progressively metamorphosed sodic eclogites are known to occur in close contact in metaophiolites of Japan (see

TABLE 2. Stoichiometric coefficients and residual components calculated from analyses of coexisting minerals (Table 1).

Reaction	Reactant coefficients: Garnet	Reactant coefficients: Calcic Cpx	Product coefficients: Garnet	Product coefficients: Omphacite	Product coefficients: Chlorite
1a	-7.20	-2.80	3.07	4.49	2.85
1b	-5.54	-4.46	1.70	6.91	2.27
1c	-7.24	-2.76	3.12	4.43	2.85
1d	-5.59	-4.41	1.74	6.84	2.28
1e	-7.57	-2.43	4.01	3.39	2.76
1f	-5.87	-4.13	2.18	6.24	2.31
1g	-7.42	-2.58	3.56	3.92	2.81
1h	-9.40	-0.60	5.07	2.30	3.17

Residual components

Reaction	SiO_2	TiO_2	Al_2O_3	Fe_2O_3	FeO	MnO	MgO	CaO	Na_2O	K_2O	Cr_2O_3	H_2O
1a	0.58*	0.00	0.10*	0.36*	-1.02*	-0.10*	-0.75*	-0.45*	-0.69*	0.00*	0.00*	-11.40
1b	-0.11*	-0.01	0.11*	0.25*	-0.32*	-0.03*	-0.07*	0.25*	-1.11	0.00*	0.00*	- 9.09
1c	0.60*	0.00	0.10*	0.36	-1.05*	-0.10	-0.72*	-0.48*	-0.68*	0.00	0.00	-11.40
1d	-0.09*	-0.01	0.11*	0.26	-0.34*	-0.03	-0.06*	0.23*	-1.10	0.00	0.00	- 9.12
1e	0.60*	0.00	-0.17*	0.37	-1.64	-0.16	-0.38	-0.81*	-0.47*	0.00	0.00	-11.02
1f	-0.02*	-0.01	0.02*	0.27**	-0.62**	-0.06	-0.01	0.04*	-0.98	0.00	0.00	- 9.23
1g	0.61*	0.00	-0.02	0.37**	-1.33**	-0.13	-0.54	-0.65*	-0.58*	0.00	0.00	-11.24
1h	0.24*	0.00	0.21*	0.46**	-1.88**	-0.20	-0.42	-2.05	-0.35*	0.00	-0.01	-12.682

* Components balanced in the least square fit.

** Balanced for total Fe. Residuals for total Fe are -0.07, -0.6, -1.06 for reactions 1f, 1g, 1h respectively.

next section of this paper). Miyashiro (1973) has suggested that such apparently mantle-derived garnet clinopyroxenites were caught up by solid intrusion of ultrabasic rocks into metabasites and then modified together with their country rocks by low T - High P metamorphism. We propose, therefore, that the xenolith may represent a mantle garnet clinopyroxenite which was retrogressively metamorphosed in the eclogite stability field when it came in contact with subducted oceanic lithosphere. Water necessary for the hydration was derived from dehydration reactions within the subducting slab and sodium became available from the spilitized ocean floor rocks carried down by the slab and now represented as xenoliths of Group C eclogites and jadeite pods. Sodium metasomatism of garnet clinopyroxenites should be considered as a viable mechanism for producing silica-poor eclogites typical of kimberlite inclusions. Some of the diamond-bearing eclogites from Yakutian kimberlites (Sobolev, 197) and an eclogite xenolith from Moses Rock found recently by F. Frey (Frey and Helmstaedt, unpublished data) contain garnets and pyroxenes almost identical to those of the secondary assemblage described here and may be the results of such process.

Constraints for upper mantle models

The xenolith and xenocryst suite from Moses Rock was used by McGetchin and Silver (1970, 1972) to establish a model of the upper mantle under the Colorado plateau. The kimberlite is thought to have been derived from physically disaggregated spinel and garnet peridotite in the mantle. The model assumes that the P-T conditions indicated by clinopyroxenes of the garnet peridotite assemblage (included in pyrope xenocrysts) corresponds to the depth at which the kimberlite originated (150 - 200 km).

Mercier (1976), also applying pyroxene geothermometry and barometry, obtained considerably different results. He observed that pyroxene xenocrysts plot along a rather well-defined oceanic geotherm, but that pyroxenes from spinel lherzolites yield a trend characterized by pressures and temperatures above this geotherm. Pyroxene inclusions in garnet from a hydrated garnet peridotite plot near the upper end of the oceanic geotherm in

TABLE 3. Whole rock chemical compositions of garnet clinopyroxenites (weight %).

	I	II	III	IV	V
SiO_2	43.30	44.07	46.7	46.9	46.98
TiO_2	0.27	0.25			1.22
Al_2O_3	17.35	16.04	11.8	12.1	15.26
Fe_2O_3	1.85	2.67	2.6	0.7	1.42
FeO	4.77	8.02	6.6	7.6	9.76
MnO	0.18	0.30			0.18
MgO	19.80	17.20	15.7	17.0	11.51
CaO	7.72	10.53	16.3	15.5	11.34
Na_2O	1.47	0.14	0.2	0.2	2.30
K_2O	0.04	0.01	0.1		0.03
Cr_2O_3		0.12			0.00
P_2O_5	0.04	0.01			
NiO		0.04			
H_2O^+					
H_2O^-	2.50	0.88			
CO_2					
Total	99.29	100.27	100	100	100.00

I Moses Rock Chlorite-bearing Garnet Clinopyroxenite, XRF analysis, F. Dunphy, analyst; FeO by C. Thornber.
II "Eclogite", Bellinzona, Switzerland (O'Hara and Mercy, 1967)
III "Garnet-diopside Eclogite" I Bessi District,
IV "Garnet-diopside Eclogite" II Japan (Banno & Yoshino, 1964)
V "Eclogite" (Griquaite), Kao Pipe, Lesotho (Nixon & Boyd, 1973), sample E 4.

the garnet peridotite field. However, the pyroxenes of the same rock outside the garnets have reequilibrated and plot on the spinel lherzolite trend. Pyroxenes in the kelyphitic rim around the garnet record intermediate P-T conditions. The highest pressures determined by Mercier (1976) are in the vicinity of 30 kb corresponding to a depth of less than 100 km. Mercier pointed out that these values correspond to conditions prior to the reequilibration caused by the massive hydration observed in most ultrabasic xenoliths. Mercier also emphasized that the depth at which the xenoliths were incorporated into the kimberlite must have been much shallower than inferred from geobarometry on anhydrous phases (see also Helmstaedt _et al._, 1972).

The xenolith described in the present paper confirms that the pyrope-bearing rocks have not escaped the hydration preceding incorporation in the kimberlite. Geothermometry and barometry on the metastably surviving minerals of the garnet peridotite assemblage do not help in portraying the P-T profile under the Colorado plateau at the time of kimberlite eruption. Conditions at that time can at best be inferred from P-T calibrations of the metamorphic assemblages, but even these may have survived metastably for some time prior to eruption. The xenoliths cannot be used to infer a depth of origin of the kimberlite greater than 100 km (see also Mercier, 1976), as compelling evidence for a depth greater than indicated by the reequilibration assemblages does not exist.

A realistic upper mantle model must consider the similarity of the xenoliths to rocks in surface geologic settings. A comparison of the Group C eclogite xenoliths from the Colorado plateau pipes with Franciscan eclogites first led to their interpretation as fragments of subducted oceanic lithosphere (Helmstaedt and Doig, 1973, 1975). Since then it has been demonstrated that the complex subsolidus history of the hydrated ultrabasic xenoliths is compatible with this hypothesis (Mercier, 1976). Equivalents for every major ultramafic xenolite type from the Colorado plateau kimberlite can be found in outcrops of metaophiolites in high-pressure metamorphic belts (Helmstaedt and Schulze, 1977; Smith, 1977, 1977a). A garnet clinopyroxenite-ecologite transition has not been documented previously in a single hand specimen. However, a similar reaction was inferred by Banno and Yoshino (1965) in metaophiolites of the Bessi district, Japan, where 'garnet-diopside eclogites' were transformed by high P-low T Sanbagawa metamorphism into hydrated 'garnet omphacite rocks'. Unaltered garnet clinopyroxenite (Table 3) consisting of pyropic garnet and diopside occurs in close contact with a hornblende-bearing eclogite consisting of less pyropic garnet and omphacite (S-2 eclogitic rocks of Ernst _et al._, 1970; Figs. 3 and 4 of present paper). The peridotite mass containing the garnet clinopyroxenites and their hydrous eclogitic derivatives is surrounded by metabasites containing lenses of eclogite with even more sodic omphacite and less pyropic garnets (S-1 eclogites of Ernst _et al._, 1970; Figs. 3 and 4). The field relationships in the Bessi district represent an example where high P, T garnet clinopyroxenites were retrogressed while juxtaposed basic volcanic rocks were progressively metamorphosed to high P-low T eclogites. As discussed in the previous section of this paper, we consider these field relationships as a possible model for the situation under the Colorado plateau.

The Colorado plateau xenoliths have counterparts also in the metaophiolites of the western Alps. Prograde Group C eclogites in

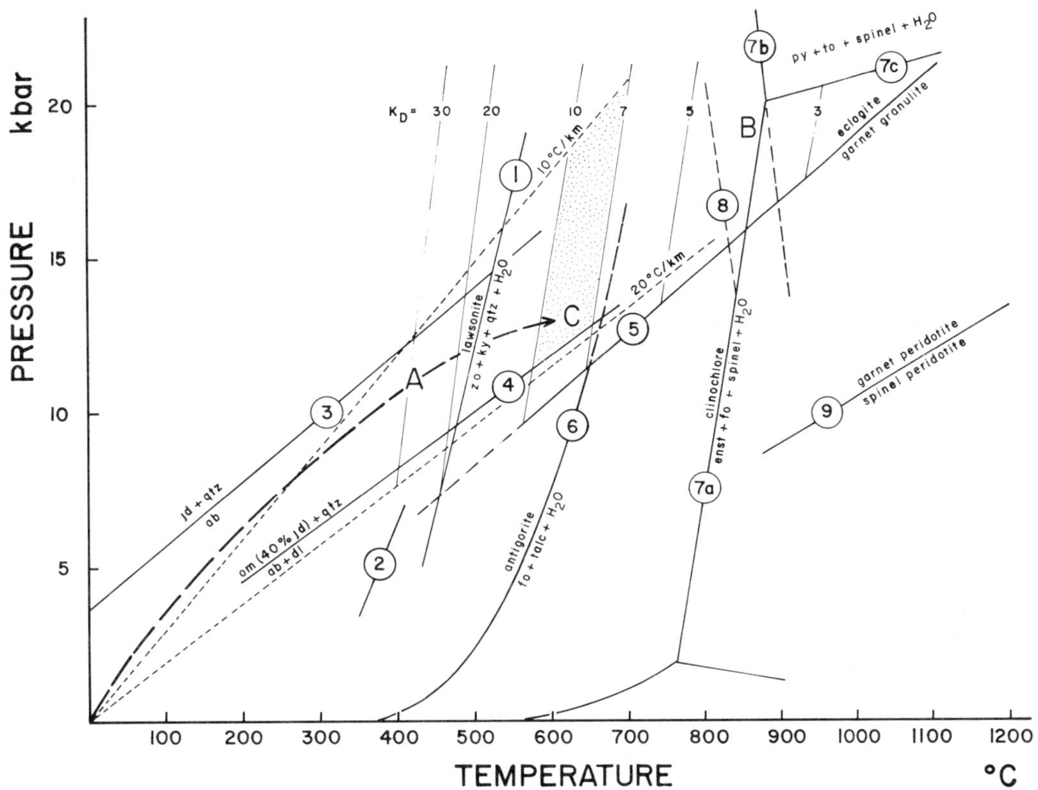

Fig. 5. Phase equilibrium curves pertaining to mineral assemblages of xenoliths from Colorado plateau kimberlites. Curve 1 (Newton and Kennedy, 1963); Curve 2 (same as Curve 1, but $P=P_{CO_2}$) (Nitsch, 1972); Curve 3 (Birch and Lecomte, 1960); Curve 4 (Kushiro, 1965); Curve 5 (Green and Ringwood, 1972); Curve 6 (Evans, 1977); Curves 7a-c (Staudinger and Schreyer, 1977); Curve 8 is an arbitrary shift to lower temperatures of the metastable extension of Curve 7b showing widening of stability field of Moses Rock pyrope, due to the lower MgO/FeO ratio in the pyrope than in chlorite; Curve 9 (Yoder, 1976).

$K_D = \frac{(FeO/MgO)Ga}{(FeO/MgO)CPX}$ as function of P and T after Råheim and Green, 1975.

A. probable stability field of lawsonite-bearing Group C ecologite xenoliths based on compositions of garnet and pyroxene cores
B. probable stability field of Moses Rock garnet clinopyroxenite assemblage
C. probable stability field of garnet and pyroxene rims in Group C eclogite xenoliths and eclogitic alteration assemblage of garnet clinopyroxenite

Heavy dashed arrow from origin to A and C is possible subduction path of the Group C eclogite xenoliths.
Dotted area is possible stability range of Moses Rock alteration assemblage.

association with hydrated ultrabasic rocks are known from the Pennine region (Bearth, 1965) and the Ligurian Alps (Cortesogno et al., 1977; Ernst, 1976). Titanoclinohumite, thought to be a major source of water in the mantle under the Colorado plateau, (McGetchin et al., 1970; Aoki et al., 1976) is common but volumetrically unimportant in many ophiolites of the western Alps (Moeckel, 1969; Trommsdorff and Evans, 1974; Evans and Trommsdorff, 1978). In the Lepontine Alps, eclogites, some of them kyanite-bearing, have been found in contact with garnet peridotite and garnet pyroxenite assemblages resembling those occurring in the Colorado plateau xenoliths (O'Hara and Mercy, 1966; Moeckel, 1969; Ernst, 1977) (Table 3). At Alpe Arami highly pyropic garnets show alteration to magnesian chlorite similar to that in the Moses Rock xenolith (Moeckel, 1969).

There is little doubt that the rock association represented by the ultramafic xenoliths from the Colorado plateau is typical of high P metamorphic ophiolites in orogenic belts. A comparison with surface occurrences leads thus to the recognition of the lithologic assemblage 'metaophiolite' under the Colorado plateau prior to the kimberlite eruption. However, the same comparison also points out the limitations of modelling the upper mantle on the basis of the xenoliths. Field relationships in exposed metaophiolites are extremely complex. The various xenoliths, often neatly separable on geothermometric and geobarometric plots, occur in close contact, and the degree of alteration may show strong variations over very short distances. Even in areas of abundant outcrops it is difficult and often impossible to establish the petrogenetic relationships between eclogites and peridotites. As there is no reason to assume that metaophiolites at depth are structurally less complex than those exposed in orogenic belts, it is concluded that geotherm models based on relatively few xenoliths and anhydrous xenocrysts yield a grossly oversimplified picture of the upper mantle stratigraphy and structure under the Colorado plateau.

Acknowledgements. We are thankful to D. M. Carmichael for stimulating discussions and help in balancing the reaction described in this paper.

This research was supported by National Research Council of Canada Grant No. A8375. Financial assistance for part of the cost to attend the Second International Kimberlite Conference to H. Helmstaedt from the Graduate Faculty of Queen's University and to D. J. Schulze from the Organizing Committee of the Kimberlite Conference is gratefully acknowledged.

References

Aoki, K., K. Fujino, and M. Akaogi, Titanochondrodite and titanoclinohumite derived from the upper mantle in the Buell Park kimberlite, Arizona, U.S.A., Contrib. Mineral. Petrol., 56, 243-253, 1976.

Banno, S., and G. Yoshino, Eclogite-bearing peridotite mass at Kigasiakaisi-Yama in the Bessi area, Central Sikoku, Japan, I.U.G.S. Upper Mantle Symposium, New Delhi, 1964, 150-160, Copenhagen, 1965.

Bearth, P., Zur Entstehung alpinotyper Eklogite, Schweiz. Mineral. Petrogr. Mitt., 45, 179-188, 1965.

Birch, F. and Lecomte, P., 1960, Temperature-pressure plane for albite composition. Amer. J. Sci., 258, 209-217.

Coleman, R. G., Ophiolites, ancient oceanic lithosphere? Springer-Verlag, New York, 1977.

Coleman, R. G., D.E. Lee, L.B. Beatty, and W.W. Brannock, Eclogites and eclogites: Their differences and similarities, Geol. Soc. Amer. Bull., 76, 483-508, 1965.

Cortesogno, L., W.G. Ernst, M. Galli, B. Messiga, G.H. Pedemonte, and G.B. Piccardo, Chemical petrology of eclogitic lenses in serpentinite, Gruppo di Voltri, Ligurian Alps, J. Geology, 85, 255-277, 1977.

Ernst, W.G., Mineral chemistry of eclogites and related rocks from the Voltri Group, Western Liguria, Italy, Schweiz. Mineral. petrograph. Mitt., 56, 293-343, 1976.

Ernst, W.G., Mineralogic study of eclogitic rocks from Alpe Arami, Lepontine Alps, Southern Switzerland, J. Petrology, 18, 371-398, 1977.

Ernst, W.G., Y. Seki, H. Onuki, and M.C. Gilbert, Comparative study of low-grade metamorphism in the California Coast Ranges and the outer metamorphic belt of Japan, Geol. Soc. Amer., Mem. 124, 1970.

Essene, E.J., and W.S. Fyfe, Omphacite in California metamorphic rocks, Contr. Mineral. Petrol., 15, 1-23, 1967.

Evans, B.W., Metamorphism of Alpine peridotite and serpentine, Ann. Rev. Earth Planet. Sci. 1977, 5, 397-447, 1977.

Evans, B.W., and V. Trommsdorff, Breakdown of titanoclinohumite in contact and regional metamorphism, Central Alps, EOS Trans. AGU, 59, 407, 1978.

Gavasci, A.T., and H. Helmstaedt, A pyroxene-rich garnet peridotite inclusion in an ultramafic breccia dike at Moses Rock, southeastern Utah, J. Geophys. Res., 74, 6691-6695, 1969.

Green, D.H., and A.E. Ringwood, A comparison of recent experimental data on the gabbro-garnet granulite-eclogite transition, J. Geol., 80, 277-288, 1972.

Greenwood, H.J., The n-dimensional tie-line problem, Geochim. Cosmochim. Acta, 31, 465-490, 1967.

Helmstaedt, H., O.L. Anderson, and A.T. Gavasci, Petrofabric studies of eclogite, spinel-websterite, and spinel-lherzolite xenoliths from kimberlite-bearing breccia pipes in southeastern Utah and northeastern Arizona, J. Geophys. Res., 77, 4350-4365, 1972.

Helmstaedt, H., and R. Doig, Eclogite nodules from kimberlite pipes of the Colorado Plateau - samples of subducted Franciscan-type oceanic lithosphere, Extended Abstracts, First International Kimberlite Conference, Capetown, South Africa, 171-172, 1973.

Helmstaedt, H., and R. Doig, Eclogite nodules from kimberlite pipes of the Colorado Plateau - samples of subducted Franciscan-type oceanic lithosphere, Physics and Chemistry Earth, 9, 95-111, 1975.

Helmstaedt, H., and D.J. Schulze, Type A - Type C eclogite transition in a xenolith from Moses Rock - further evidence for the presence of metamorphosed ophiolites beneath the Colorado plateau, Extended Abstracts, Second International Kimberlite Conference, Santa Fe, New Mexico, 1977.

Krogh, E.J., and A. Råheim, Temperature and pressure dependence of Fe-Mg Partitioning between garnet and phengite, with particular reference to eclogites, Contrib. Mineral. Petrol., 66, 75-80, 1978.

Kushiro, I., 1965, Clinopyroxene solid solutions at high pressure: the join diopside-albite. Carnegie Inst. Yearbook 64, 89-94.

McGetchin, T.R., The Moses Rock dike: Geology, petrology, and mode of emplacement of a kimberlite-bearing breccia dike, San Juan County, Utah, unpubl. Ph.D. thesis, Calif. Inst. Technol., Pasadena, Calif., 405 p., 1968.

McGetchin, T.R., and L.T. Silver, Compositional relations in minerals from kimberlite and related rocks from the Moses Rock dike, San Juan County, Utah, Amer. Mineral., 55, 1738-1771, 1970.

McGetchin, T.R., and L.T. Silver, A crustal-upper mantle model for the Colorado plateau based on observations of crystalline rock fragments in the Moses Rock dike, J. Geophys. Res., 77, 7022-7037, 1972.

McGetchin, T.R., L.T. Silver, and A.A. Chodos, Titanoclinohumite: a possible mineralogical site for water in the upper mantle, J. Geophys. Res., 75, 255-259, 1970.

McGetchin, T.R., D. Smith, S.N. Ehrenberg, M. Roden, and H.G. Wilshire, Navajo kimberlites and minettes guide, Second International Kimberlite Conference, Santa Fe, New Mexico, 1977.

Mercier, J.-C.C., Single-pyroxene geothermometry and geobarometry, Amer. Mineral., 61, 603-615, 1976.

Miyashiro, A., Metamorphism and metamorphic belts, George Allen & Unwin Ltd., London, 1973.

Moeckel, J.R., The structural petrology of the garnet-peridotite of Alpe Arami (Ticino, Switzerland), Leidse Geol. Meded., 42, 61-130, 1969.

Newton, M.S., and G.C. Kennedy, Jadeite, analcite, nepheline, and albite at high temperatures and pressures, Amer. J. Sci., 266, 728-735, 1968.

Nitsch, K.H., Das P-T-X_{CO_2} stabilitätsfeld von Lawsonit, Contr. Mineral. Petrol., 34, 116-137, 1972.

O'Hara, M.J. and E.L.P. Mercy, Garnet-peridotite and eclogite from Bellinzona, Switzerland, Earth and Planet. Sci. Letters, 1, 295-300, 1966.

Råheim, A. and D.H. Green, P, T paths of natural eclogites during metamorphism - a record of subduction, Lithos, 8, 317-328, 1975.

Smith, D., Titanochrondrodite and titano-clinohumite derived from the Upper Mantle in the Buell Park Kimberlite, Arizona, U.S.A., Discussion, Contrib. Mineral. Petrol., 61, 213-215, 1977.

Smith, D., The origin and interpretation of spinel-pyroxene clusters in peridotite, J. Geology, 85, 476-482, 1977a.

Sobolev, N.V., Deep-seated inclusions in kimberlites and the problem of the composition of the upper mantle, English Translation by D.A. Brown, edited by F.R. Boyd, AGU, 1977.

Smulikowski, K., Chemical differentiation of garnet and clinopyroxene in eclogites, Bull. Acad. Pol. Sc. ser. ged. geogr. XIII, 1, 11-18, 1965.

Smulikowski, K., Differentiation of eclogites and its possible causes, Lithos, 1, 89-101, 1968.

Smulikowski, K., Theoretical and geological arguments for eclogite occurrence in the upper mantle, Proc. 23rd Session, International Geological Cong., Prague, 1, 165-174, 1968a.

Staudigel, H. and W. Schreyer, The upper thermal stability of clinochlore $Mg_5Al[AlSi_3O_{10}](OH)_8$, at 10-35 kb P_{H_2O}, Contrib. Mineral. Petrol., 61, 187-198, 1977.

Trommsdorff, V. and B. Evans, Alpine metamorphism of peridotic rocks, Schweiz, Mineral. Petrograph. Mitt., 54, 333-352, 1974.

Yoder, H.S. Jr., 1976, Generation of basaltic magma, National Academy of Sciences, Washington, D.C., 265 pp.

V. XENOLITHS FROM BASALTS AND OTHER VOLCANICS

FRACTIONAL CRYSTALLIZATION IN THE MANTLE OF LATE-STAGE KIMBERLITIC LIQUIDS--EVIDENCE IN XENOLITHS FROM THE KIAMA AREA, N.S.W., AUSTRALIA

Suzanne Y. Wass

School of Earth Sciences, Macquarie University, North Ryde, N.S.W., 2113, Australia

Abstract. A unique suite of xenoliths rich in amphibole and apatite occurs in basanitic and nephelinitic dykes which outcrop over a restricted area near Kiama, N.S.W., Australia. Other minerals in these xenoliths include clinopyroxene, biotite, spineliferous titanomagnetite, magnesian ilmenite, olivine, rutile, accessory sulphides and primary calcite. Associated with these amphibole/apatite xenoliths are Cr-diopside lherzolite series and Al-augite series xenoliths, primary minerals of which show extensive replacement by carbonate and amphibole. Using textural and compositional criteria, the amphibole/apatite xenoliths are interpreted as representing fragments of a high-pressure differentiation sequence crystallized in the mantle from a fractionated magma of kimberlitic affinity, enriched in volatiles (including CO_2 and H_2O), alkalis and incompatible elements and depleted in Mg, Cr and Ni. The resultant upper mantle layered complex was then sampled by subsequent entrainment of fragments in later unrelated magmas. The altered Cr-diopside and Al-augite series xenoliths possibly represent metasomatic alternation of the inhomogeneous upper mantle wall-rock adjacent to the crystallizing kimberlitic magma.

Introduction and Occurrence

Amphibole/apatite-rich xenoliths comprise a distinct and unique suite restricted to nephelinitic and basanitic dykes and sills probably of Tertiary age outcropping around the Kiama area of southeastern New South Wales, Australia (Harper, 1915; Wilshire and Binns, 1961; Wass and Irving, 1976). Xenoliths described in this preliminary report were all collected from a single basanitic dyke exposed in the Bombo quarry, approximately 1 km north of Kiama. This dyke is approximately 0.6 m wide and is exposed over a length of about 200 m. Xenoliths constitute about 70% of the dyke and commonly are arranged in size-graded layers parallel to the dyke margin, with largest xenoliths (up to 30 cm across) occurring in the central layer. The outer margin is characterized by an abundance of small (about 1 to 2 cm across), fragmented xenoliths representing the entire mineralogical range shown by the xenoliths (Figure 1).

The xenolith assemblage of the Bombo dyke is dominated by the amphibole/apatite suite xenoliths. Volumetrically minor xenolith suites also recognized in this dyke include remnants of disaggregated and altered Cr-spinel-bearing, Cr-diopside lherzolites, Al-augite series xenoliths, and granofelses apparently of crustal origin.

Immediately northwest of this area lies the Southern Highlands alkali basaltic province, also Tertiary in age, which because of its similar basaltic types, was probably contemporaneous and genetically associated with the igneous episode that produced the basaltic host rocks containing the amphibole/apatite xenolith suite. The Southern Highlands basaltic flows, dykes and necks are rich in xenoliths and megacrysts of high-pressure origin (Wass and Irving, 1976), but these are confined to the Cr-diopside and Al-augite series, with no evidence of the amphibole/apatite assemblages. This (together with mineralogical and textural data discussed below) may indicate that ascending basaltic magmas extruded in geographically separated areas may have sampled discrete sections of an inhomogeneous mantle, as well as carrying cognate, high-pressure precipitates.

Mineralogy and Textures of Amphibole/Apatite Suite Xenoliths

The amphibole/apatite suite xenoliths show a wide range of textures and modes, within a restricted range of mineral assemblages, varying from monomineralic rocks to xenoliths layered in mineralogy and grainsize on various scales.

Mineralogy

As well as abundant amphibole and apatite, these xenoliths contain biotite, clinopyroxene, olivine, magnesian ilmenite, spineliferous

Fig. 1. Hand specimen from close to the margin of the dyke, showing the abundance and variety of xenoliths. The elongate, light-coloured xenolith centre left, is a crustal granofels. The remaining xenoliths belong to the amphibole/apatite suite. Scale is 2 cm long.

titanomagnetite, spinel, rare rutile and accessory sulphides (including chalcopyrite, pyrrhotite, pyrite and pentlandite) in varying proportions. Apatite commonly constitutes up to 20% of the mode (Figure 2, 3, 4, 5). Carbonate is widespread and some may be primary. All olivine is serpentinized or altered to carbonate. Discrete mineral assemblages commonly observed are: amphibole/clinopyroxene/apatite±ilmenite, spinel, biotite, olivine, carbonate; amphibole/biotite/apatite±ilmenite, spinel, carbonate; biotite/apatite±ilmenite, spinel; amphibole/apatite±carbonate, ilmenite, spinel; Fe-rich clinopyroxene/apatite/spinel± biotite; spinel/apatite±biotite, carbonate; and monomineralic rocks consisting of amphibole, biotite, apatite and spinel. Sulphides are ubiquitous accessories.

Textures

Grainsizes vary widely both within individual xenoliths and from xenolith to xenolith. Extremes in grainsizes are represented by large euhedral kaersutite crystals up to 5 cm in cross-sectional width, to grains less than 1 mm in mean diameter in some layers of banded xenoliths. Relict micro-structures including euhedral grain shapes, mineralogical layering and cumulus textures are consistent with a primary origin by precipitation from a melt.

Excellent examples of layering are evident in some of the Fe-rich clinopyroxene/apatite/spinel xenoliths. Layers consisting of approximately 50% Fe-rich clinopyroxene and 50% spinel, grade successively into a monomineralic spinel layer, then a spinel/apatite/biotite layer (Figure 4) culminating in a granular apatite layer. In the spinel-dominant layers, apatite crystals are commonly prismatic or acicular and elongate parallel to the mineralogical layering, in contrast to the equant grains with random orientations forming a mosaic microstructure in the apatite-dominant bands (Figure 4). No sharp boundaries between layers occur in this type of layered xenolith and individual layers are observed to vary from 1 mm to 5 cm wide. Some large spinel/apatite xenoliths (up to 12 cm across) show layering due to grainsize variation only (Figure 2). Other types of layered xenoliths are characterized by sharply-defined layers consisting of alternating bands of amphibole, clinopyroxene/amphibole and clinopyroxene/amphibole/biotite (Figure 6). The amphibole in the monomineralic layers commonly forms large crystals with a preferred orientation of the c-axes approximately parallel to the layering. This occurs on all scales, some layers being less than 0.5 cm wide, whereas others are up to 5 cm wide. Some xenoliths consisting of aggregates of subparallel amphibole crystals (Figure 7) are interpreted as broken fragments of such layers.

Xenoliths consisting dominantly of amphibole/clinopyroxene/apatite are characterized by euhedral grain shapes modified by mutual impingement during growth (Figure 5). The euhedral apatite prisms are comparable in grainsize to the associated silicate phases and may vary from 2 mm to 2 cm in length. In some xenoliths, oikocrysts of Fe-rich clinopyroxene enclose euhedral apatite and spinel grains, whereas in others, amphibole encloses euhedral clinopyroxene and altered olivine and possibly represents postcumulus growth. Carbonate occurs as rounded globules (Figure 6), either in grain boundary positions or as inclusions within grains (most commonly in amphibole). It is interpreted as being primary in origin, by analogy with textures in both experimental systems and natural occurrences as described by Gittins (1973).

Superimposed on these primary igneous textures in some xenoliths are the effects of recrystal-

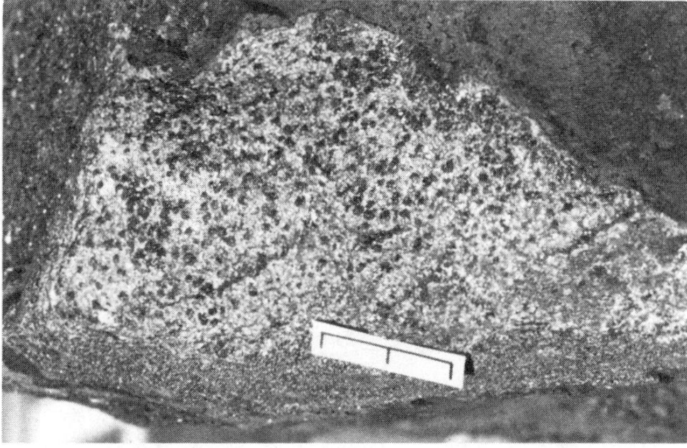

Fig. 2. Apatite/spinel xenolith with grainsize layering. The finer-grained layer starts abruptly near the scale-bar. Scale is 2 cm long.

lization and deformation. Two distinct effects are evident. The first occurs most commonly in the spinel/apatite/biotite, in the monomineralic spinel and apatite xenoliths, and in some amphibole-dominant xenoliths (Figures 3 and 8). A mosaic, interlocking aggregate dominated by smoothly-curved grain boundaries, triple junctions at approximately 120° and lack of inclusions is typical. Carbonate forms an integral part of this microfabric (Figure 3), indicating its primary nature. Such a microstructure suggests extensive grain-boundary adjustment at high but sub-solidus temperatures (Vernon, 1970) which may have occurred shortly after precipitation or significantly after precipitation by subsequent metamorphism. The other effect is solid-state deformation, evidenced by strongly-kinked micas and strained clinopyroxene with small, strain-free grains.

Mineralogy and Textures of Other Xenolith Suites

Three other xenolith suites are distinguishable in the Bombo dyke. 1. Granofelses of probable lower crustal origin show well-developed, high-grade, metamorphic textures and a mineralogy generally consistent with granitic composition.

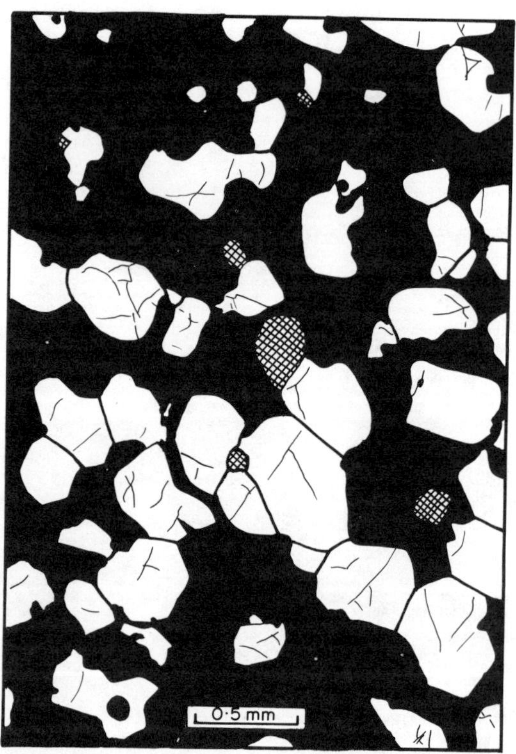

Fig. 3. Microstructure of spinel/apatite/calcite xenolith. Black areas represent spinel; clear grains with irregular fractures are apatite; cross-hatched areas represent calcite. Note that the carbonate forms part of the mosaic microfabric.

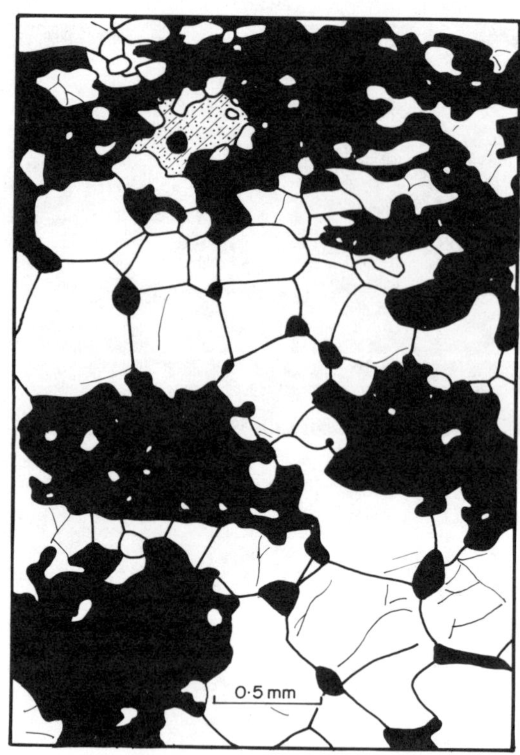

Fig. 4. A portion of a layered spinel/apatite/clinopyroxene/biotite xenolith. No clinopyroxene is observable in this section. Stippled areas with parallel lines represent biotite. Other symbols as for Figure 3.

2. Rare xenoliths containing amphibole have textures and mineralogy consistent with an origin as metasomatised Cr-diopside lherzolites, possibly representing mantle wall-rock altered by the magmatic fluid from which the amphibole/apatite suite rocks crystallized. Olivine (and orthopyroxene (?)) is pseudomorphed by calcite and layer silicates, Cr-diopside is extensively replaced by titaniferous pargasite and is crowded with exsolved Cr-spinel needles. Rare cores of Cr-spinel are surrounded by Ti- and Fe-rich spineliferous titanomagnetite. All the amphibole in these xenoliths is interpreted to be secondary on textural criteria because of its occurrence in veins cutting across primary lherzolite minerals, the patchy nature of its occurrence as an apparent alteration product in Cr-diopside grains, and its irregular grain shapes and occurrence in texturally discordant aggregates in the lherzolite, the primary phases of which show a typical metamorphic mosaic microfabric. The amphibole is compositionally equivalent to that in the amphibole/apatite xenoliths. 3. The remaining xenolith suite appears originally to have been normal, apatite-free, Al-augite series xenoliths (Wilshire and Shervais, 1975), consisting of clinopyroxene, spinel and olivine. Extensive alteration of

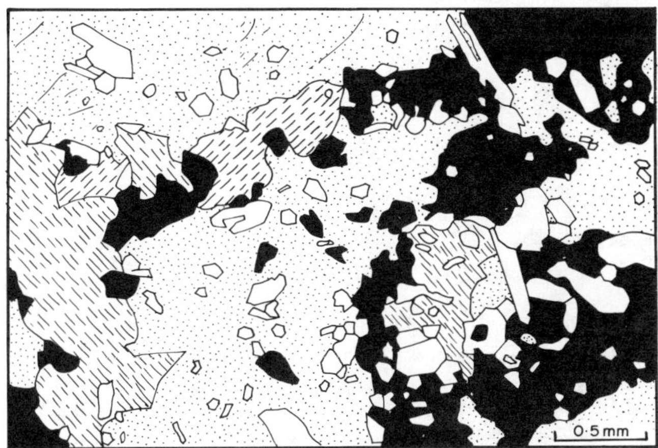

Fig. 5. Typical igneous texture of amphibole/apatite/clinopyroxene/spinel/ilmenite xenoliths showing euhedral crystal outlines and mutual impingement boundaries. Stippled areas represent amphiboles; areas of parallel lines represent clinopyroxene; clear grains represent apatite; black areas represent ilmenite and opaque spinel.

olivine to calcite, talc(?) and unidentified layer silicates and of the clinopyroxene to kaersutite and to the Fe-rich clinopyroxenes typical of some of the amphibole/apatite suite xenoliths has taken place. Altered clinopyroxenes usually show extensive exsolution of ilmenite and rutile needles, suggesting oxidizing conditions.

Chemical Composition of Minerals in Amphibole/Apatite Suite Xenoliths

Electron microprobe analyses were carried out using either the JEOL JXA5-A instrument at Melbourne University, with three computer-controlled crystal spectrometers using a correction program based on that of Mason et al. (1969), or the ETEC fully-automated instrument at Macquarie University, using a correction procedure based on the Bence and Albee (1968) method. At least 3 and usually 5 determinations were averaged for each analysis quoted. Representative microprobe analyses of the major minerals are given in Table 1.

Amphiboles in this xenolith suite vary in composition from titaniferous pargasite to titaniferous kaersutite using the terminology of Leake (1968). They are relatively rich in Al_2O_3 (12.2 to 15.2%) and have Mg values (Mg/(Mg±Fe) where Fe is total iron as Fe^{2+}) ranging from 50 to 73. NiO ranges from 0.15% to below detection limits, higher values correlating with the higher Mg ratios. Cr_2O_3 ranges up to 0.25% and MnO up to 0.30%, the latter increasing with decreasing Mg values. TiO_2 ranges from 4 to 6%.

Biotites are low in MnO, Cr_2O_3 and NiO (maximum values 0.15, 0.08 and 0.06%, respectively) and Mg varies from 50 to 66. TiO_2 lies within the range of 6.5 to 7.9%.

Clinopyroxenes show the greatest compositional variation. They are generally augites, titaniferous augites and salites with moderately high Al_2O_3 (4 to 8%) and TiO_2 from 1 to 3%. Cr_2O_3 varies from 0.8% for the most Mg-rich clinopyroxenes. Mg ranges from 50 to 78, the extremely Fe-rich compositions containing up to 15% FeO (total Fe as FeO). Calculation of an ideal structural formula indicates significant Fe^{3+} in these Fe-rich deep-green clinopyroxenes. Al is dominantly accommodated in the $CaAl_2SiO_6$ and $CaTiAl_2O_6$ molecules. Al^{iv}:Al^{vi} ratios are within the range for high-pressure clinopyroxenes precipitated from undersaturated magmas (Wass, in prep.).

Ilmenites are consistently low in Al_2O_3, NiO and Cr_2O_3; MnO ranges from about 0.5 to 1% TiO_2 is consistently about 50%.

Spinel is generally spineliferous titanomagnetite with Cr_2O_3 ranging from <0.1 to 7%, Al_2O_3 from 2.5 to 8.5%, MgO from 0.5 to 6%, MnO from 0.5 to 1.5% and NiO from 0.0 to 0.3%.

A typical apatite analysis is given in Table 1, but OH and minor and trace elements have not been determined. F is consistently about 1.3% and Cl approximately 0.4%.

Rutile analysed contains significant Fe and low totals suggest the possible presence of other elements. Reconnaissance analyses of carbonate suggest the presence of both calcite and dolomite.

Reverse zoning of amphiboles and clinopyroxenes in contact with the host magma is ubiquitous. The narrow outer rims are higher in MgO (and TiO_2) than the host grain, and the clinopyroxene rims are equivalent in composition to groundmass clinopyroxene grains in the host rock. This

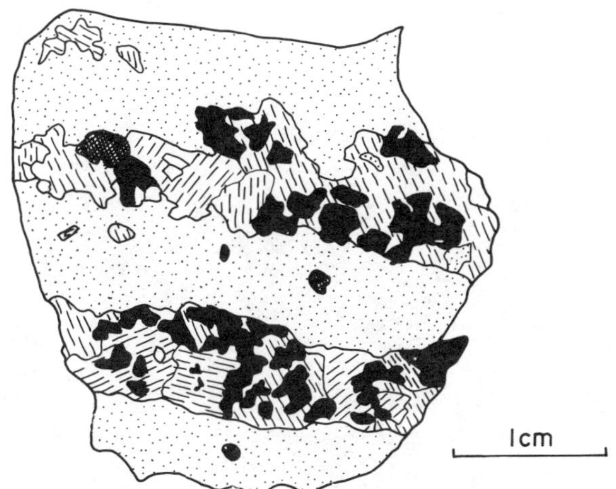

Fig. 6. Layered xenolith, symbols as in Figure 5, with cross-hatched areas representing calcite. Note the sharply-defined layer boundaries and the globular calcite. Amphibole crystals are generally one layer wide and are elongate with c-axes parallel to the layering.

reverse zoning is particularly pronounced around the margins of the green, Fe-rich clinopyroxenes.

A consistent pattern in the compositional differences of associated phases from discrete xenoliths is evident. Amphiboles, clinopyroxenes and biotites show mutually sympathetic variation in Mg value from xenolith to xenolith (Table 1). In some of the layered xenoliths, a gradational trend towards Fe-enrichment of constituent phases is observed in successive layers.

Comparisons and Discussion

The range of textural and chemical features of the amphibole/apatite xenolith suite is consistent with an origin as part of a fractional crystallization sequence. The lack of feldspar, the Mg-rich Al-poor nature of the ilmenite, and the low Al^{iv}: Al^{vi} ratios in the clinopyroxenes suggest high-pressure formation, probably at pressures in excess of 17 kb. The alteration of Cr-diopside series xenoliths to titaniferous pargasite (compositionally equivalent to that in some of the amphibole/apatite xenoliths) indicates metasomatic penetration of mantle wall-rock by a fluid phase associated with the formation of the amphibole/apatite suite, thus also supporting a mantle origin for this suite. The solid-state deformation effects observed in the amphibole/apatite suite, imply a significant time gap between their original precipitation and eventual entrainment in the host magma.

Compositionally similar clinopyroxenes and amphiboles in the Fen carbonatite complex in Norway were interpreted as cognate phenocrysts formed in subcrustal (upper mantle) magma chambers by Griffin and Taylor (1975). However, associated spinels and ilmenites have lower Al_2O_3 and NiO contents than those at Kiama. Potassic volcanics of South West Uganda and the

Fig. 7. Xenolith consisting of an aggregate of elongate amphibole prisms with minor ilmenite and calcite. Scale is 2 cm long.

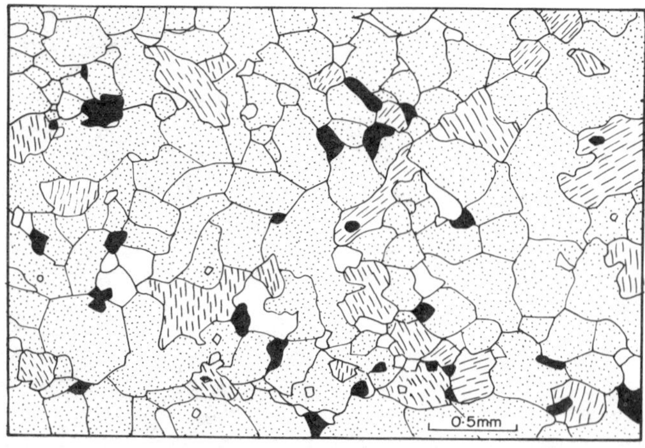

Fig. 8. Amphibole/apatite/clinopyroxene/spinel/ilmenite xenolith showing a recrystallized mosaic texture. Symbols as for Figure 5.

sodi-potassic volcanics of West Eifel, Germany (Lloyd and Bailey, 1975) contain nodules bearing similar Fe-rich clinopyroxenes, amphibole, dark mica and apatite, and are interpreted as products of light-element metasomatism of mantle material. Gittins et al. (1975) described a phenocryst assemblage in dykes of kimberlitic and carbonatitic affinity from Quebec, which closely resemble assemblages in the Kiama amphibole/apatite suite xenoliths, except for the absence of magnesian ilmenite. Associated spineliferous titanomagnetites are compositionally very similar to those in xenoliths from the Bombo dyke. Enrichment in Ti, P and Mn and depletion in Mg is characteristic of minerals in autoliths from kimberlitic pipes in Lesotho (Danchin et al., 1975). This parallels data from minerals in the Bombo dyke xenoliths except that the Lesotho spinels are richer in Cr_2O_3 and the ilmenites in MgO. The MARID suite of xenoliths (Dawson and Smith, 1977) shows similarities in overall assemblages, but the mineral compositions are distinct in geochemical detail, principally in their higher Mg, Cr and Ni concentrations, and in the occurrence of subcalcic potassic sodic richterites rather than titaniferous pargasites and kaersutites, and of phlogopites rather than biotites.

The amphiboles in the amphibole/apatite suite rocks at Kiama are compositionally similar to those in amphibole-bearing cumulate inclusions from Grand Canyon, Arizona, interpreted by Best (1975) as products of crystallization from a hydrous melt in the mantle. However, the Mg values given by Best are slightly higher, and no apatite or calcite is recorded in the cumulate assemblages, which also include garnet, olivine and orthopyroxenes as primary precipitates. The Bombo amphiboles can be closely matched compositionally with amphibole megacrysts from the Kakanui mineral breccia, New Zealand (Mason, 1966; and present author's unpublished data).

TABLE 1. Representative microprobe analyses of major minerals in amphibole/apatite suite xenoliths, Kiama, N.S.W.

Sample	K1023		K3			K13A1		K1041		K1025			K13b			K20		
No.	Amph.	Cpx.	Amph.	Cpx.	Ilmen.	Amph.	Cpx.	Amph.	Cpx.	Biot.	Amph.	Ilmen.	Biot.	Amph.	Cpx.	Cpx.	Spinel	†Apatite
SiO_2	41.8	48.3	38.8	47.5	0.10	39.7	46.2	40.4	50.7	35.5	39.0	n.d.	35.1	39.2	48.4	49.5	n.d.	0.27
TiO_2	4.76	2.36	6.11	2.10	51.2	4.07	2.61	5.48	0.69	6.94	4.82	51.9	7.90	5.20	1.46	1.01	20.9	n.d.
Al_2O_3	12.1	6.38	14.1	7.40	0.25	14.0	8.07	12.4	4.77	15.0	13.9	0.25	15.0	12.8	4.67	4.34	5.20	n.d.
Cr_2O_3	0.10	0.52	n.d.	n.d.	n.d.	n.d.	n.d.	0.25	0.20	0.06	0.01	0.08	n.d.	0.04	0.03	0.10	0.09	n.d.
FeO*	9.28	6.45	10.3	8.12	42.0	12.2	7.86	12.3	8.75	15.4	13.5	41.7	18.1	15.9	10.9	13.2	67.9	0.41
NiO	0.20	0.06	n.d.	n.d.	n.d.	n.d.	n.d.	0.03	n.d.	0.05	n.d.	n.d.	0.03	n.d.	n.d.	0.06	0.08	n.d.
MnO	0.08	0.12	n.d.	n.d.	0.57	0.13	0.16	0.20	0.20	0.09	0.12	0.53	0.12	0.20	0.29	0.33	0.80	0.06
MgO	13.7	12.8	12.6	11.0	6.20	12.0	11.2	11.3	11.3	12.8	10.4	5.20	10.5	9.20	11.3	8.90	0.52	0.23
CaO	12.0	22.3	12.3	22.6	0.20	12.2	23.3	11.5	21.6	n.d.	11.6	n.d.	n.d.	11.4	21.4	20.5	n.d.	53.2
Na_2O	2.46	0.78	2.07	1.44	n.d.	2.40	0.76	2.50	1.81	0.66	2.31	n.d.	0.58	2.32	1.11	1.86	n.d.	0.18
K_2O	1.75	n.d.	2.18	n.d.	n.d.	1.67	n.d.	2.11	n.d.	9.37	2.14	n.d.	8.54	1.95	n.d.	n.d.	n.d.	0.02
P_2O_5	n.d.	n.d.	n.d.	n.d.	n.d.	n.d.	n.d.	n.d.	n.d.	n.d.	n.d.	n.d.	n.d.	n.d.	n.d.	n.d.	n.d.	40.8
Total	98.2	100.1	98.5	100.2	100.5	98.4	100.2	98.3	100.0	95.9	97.8	99.7	95.9	98.2	99.6	99.8	95.5	96.9
100 MG / Mg+Fe*	72.5	78.0	68.6	70.7		63.7	71.8	62.1	69.7	59.7	57.9		50.8	50.8	64.9	54.6		

* Total Fe as Fe^{2+}: † includes F 1.3%, Cl 0.4%: n.d. = not detected.

Modes of Xenoliths

K1023: Cpx 50%, kaersutite 38%, apatite 12%, accessory spinel, ilmenite, carbonate.
K3: Kaersutite 98%, ilmenite 2%, cpx <1%, accessory sulphides.
K13A1: Cpx 42%, titaniferous pargasite 28%, apatite 20%, ilmenite and opaque spinel 10%, accessory sulphides, carbonate.
K1041: Kaersutite 50%, cpx 40%, apatite 4%, opaque spinel 4%, carbonate 2%, accessory sulphides, ilmenite.
K1025: Kaersutite 55%, biotite 35%, apatite 5%, ilmenite 5%, accessory carbonate, sulphides.
L13b: Kaersutite 35%, cpx 30%, biotite 12%, apatite 15%, spinel and ilmenite 8%, accessory carbonate, sulphides.
K20: Fe-rich cpx 50%, opaque spinel 30%, apatite 20%, accessory biotite, carbonate.
K1023-K3-K13A1-K1041-K1025-K13b-K20 represent xenoliths progressively enriched in Fe. Note the mutually sympathetic variation in 100 Mg/(Mg+Fe) of coexisting silicates.

Ilmenites from this locality are distinctly lower in TiO_2 and MgO and garnet also forms part of this megacryst assemblage. The Bombo amphiboles also overlap the compositional range of some amphiboles occurring as high-pressure megacrysts in alkali basaltic rocks (e.g. Irving, 1974), although the latter commonly have higher $Na_2O:K_2O$ ratios and lower Mg values.

The ilmenites analysed in the amphibole/apatite suite xenoliths at Kiama generally just overlap the lower part of the range of MgO cited for ilmenites of kimberlitic origin (Nixon and Boyd, 1973; Dawson and Smith, 1977; Mitchell, 1977). They are also similar to ilmenite megacrysts in basanitic lavas in NSW (Binns, 1969; Wass, 1971) and to an ilmenite megacryst from a nepheline benmoreite inferred to be of direct mantle origin from southeastern Queensland (Green et al., 1974). The ilmenite compositions quoted in Table 1 are closely comparable with those of ilmenites produced experimentally by crystallization of an olivine basanite composition at 30 kb and 27 kb, at 1050°C, and with 4.5% H_2O (Green and Sobolev, 1975). These synthetic ilmenites coexist with garnet.

Petrogenesis

The Kiama amphibole/apatite assemblages appear unique in their association of abundant apatite, magnesian ilmenite and calcite with the other minerals. They evidence the presence of mantle fluids low in SiO_2 and CaO, and rich in alkalis, Fe, volatiles (H_2O, CO_2, F, Cl) and incompatible elements. The mineralogical and grainsize layering and the sympathetic variation of Mg ratios of coexisting phases in the same xenolith or layer are consistent with an origin of these xenoliths as fragments of a crystalline body (or bodies) formed within the upper mantle by fractional crystallization of a highly undersaturated, hydrous, alkali-rich magma containing significant P and CO_2. The Fe-rich nature of the assemblages, as well as the abundance of hydrous phases and apatite suggests this magma may be residual after fractionation of a more primitive melt, resulting in the increased concentration of volatiles and incompatible elements, and in depletion of Mg, Cr and Ni. This prior fractionation may have occurred deeper in the mantle than the level of generation of the host basanite. The chemical characteristics of the postulated magma resemble those of kimberlitic magma which has undergone fractionation, resulting in a late-stage liquid depleted in MgO and enriched in H_2O, P and CO_2 relative to primitive kimberlite. The amphibole/apatite suite xenoliths may represent higher level, near-solidus equivalents of the MARID suite (Dawson and Smith, 1977). The inferred composition and fractionated nature of the magma crystallizing the amphibole/apatite xenoliths, the presence of inferred primary calcite, and the abundance of apatite strongly suggest a genetic relationship between kimberlitic and carbonatitic magmas at upper mantle depths, similar to that recorded from kimberlitic sills at Benfontein, South Africa by Dawson and Hawthorne (1973). The site of formation of the amphibole/apatite suite xenoliths is constrained to within the spinel lherzolite field in the upper mantle (by the interpretation of altered Cr-spinel, Cr-diopside lherzolite as metasomatised wall-rock) and probably to above the level of generation of basanitic magma.

The amphibole-apatite suite xenoliths are therefore interpreted as representing fragments of a high-pressure differentiation sequence crystallized in the mantle from a late-stage kimberlitic residuum and sampled by accidental entrainment in subsequent magmas generated in later, probably unrelated episodes of partial melting. Low-pressure equivalents of related kimberlitic magmas have not yet been found in southeastern NSW, although stream gravels in the area contain magnesian ilmenite, pyrope-almandine garnet and rare diamonds, which suggest the occurrence of kimberlitic rocks, and Ferguson et al. (1977) have reported rare kimberlitic diatremes 200 km to the west.

Acknowledgements. Valuable microprobe time was made available by the Geology Department, Melbourne University. The advice of R.H. Flood and S.E. Shaw on XRF and microprobe procedures at Macquarie University is acknowledged with thanks. R.H. Vernon and T.H. Green critically read the manuscript. Funding by a Macquarie University Research Grant made this study possible.

References

Bence, A. E., and A. L. Albee, Empirical correction factors for the electron microanalysis of silicates and oxides, J. Geol., 76, 382-403, 1968.

Best, M. G., Amphibole-bearing cumulate inclusions, Grand Canyon, Arizona and their bearing on silica-undersaturated hydrous magmas in the upper mantle, J. Petrology, 16, 212-236, 1975.

Binns, R. A., High-pressure megacrysts in basanitic lavas near Armidale, New South Wales, Amer. J. Sci. Schairer Vol., 267-A, 33-49, 1969.

Danchin, R. V., J. Ferguson, J. R. McIver, and P. H. Nixon, The composition of late-stage kimberlitic liquids as revealed by nucleated autoliths, Phys. Chem. of the Earth, 9, 235-246, 1975.

Dawson, J. B., and J. B. Hawthorne, Magmatic sedimentation and carbonatitic differentiation in kimberlite sills, Benfontein, South Africa, J. Geol. Soc. Lond., 129, 61-85, 1973.

Dawson, J. B., and J. V. Smith, The MARID (mica-amphibole-rutile-ilmenite-diopside) suite of xenoliths in kimberlite, Geochim. Cosmochim. Acta, 41, 309-323, 1977.

Ferguson, J., D. J. Ellis, and R. N. England, Unique spinel-garnet lherzolite inclusion in kimberlite from Australia, Geology 5, 278-280, 1977.

Gittins, J., The significance of some porphyritic textures in carbonatites, Can. Mineralogist, 12, 226-228, 1973.

Gittins, J., R. H. Hewins, and A. F. Laurin, Kimberlitic-carbonatitic dykes of the Saguenay River Valley, Quebec, Canada, Phys. Chem. of the Earth, 9, 137-148, 1975.

Green, D. H., A. D. Edgar, P. Beasley, E. Kiss, and N. G. Ware, An upper mantle source for some hawaiites, mugearites and benmoreites, Contrib. Mineral. Petrol, 48, 33-43, 1974.

Green, D. H., and N. V. Sobolev, Coexisting garnets and ilmenites synthesized at high pressures from pyrolite and olivine basanite and their significance for kimberlitic assemblages, Contrib. Mineral. Petrol., 50, 217-229, 1975.

Griffin, W. L., and P. N. Taylor, The Fen Damkjernite: petrology of a "central complex kimberlite", Phys. Chem. of the Earth, 9, 163-177, 1975.

Harper, L. F., Geology and mineral resources of the Southern Coalfield, Mem. Geol. Surv. N. S. W., 7, 276-405, 1915.

Irving, A. J., Megacrysts from the Newer Basalts and other basaltic rocks of southeastern Australia, Neues Jahrb. Mineral. Abhandl., 120, 147-167, 1974.

Leake, B. E., A catalog of analysed calciferous and subcalciferous amphiboles together with their nomenclature and associated minerals, Geol. Soc. Amer. Spec. Paper, 98, 210 pp., 1968.

Lloyd, F. E., and D. K. Bailey, Light element metasomatism of the continental mantle: the evidence and the consequences, Phys. Chem. of the Earth, 9, 389-416, 1975.

Mason, B., Pyrope, augite and hornblende from Kakanui, New Zealand, N. Z. J. Geol. Geophys., 9, 474-480, 1966.

Mason, P. K., M. T. Frost, and S. J. B. Reed, B.M.-I.C.-N.P.L. computer program for calculating correlations in quantitative X-ray microanalysis, Nat. Phys. Lab. IMS Report 1, 1969.

Mitchell, R. H., Geochemistry of magnesian ilmenites from kimberlites in South Africa and Lesotho, Lithos, 10, 29-30, 1977.

Nixon, P. H., and F. R. Boyd, Deep seated nodules, in Lesotho kimberlites, edited by P. H. Nixon, pp. 106-109, Lesotho Nat. Develop. Corp., 1973.

Vernon, R. H., Comparative grain-boundary studies of some basic and ultrabasic granulites, nodules and cumulates, Scot. J. Geol., 6, 337-351, 1970.

Wass, S. Y., Basaltic igneous activity Southern Highlands, N. S. W., unpublished Ph.D. thesis, University of Sydney, NSW, 1971.

Wass, S. Y., Geochemistry of clinopyroxenes of diverse origins in alkali basaltic rocks from Southern Highlands (NSW) and the Massif Central (France), in prep.

Wass, S. Y., and A. J. Irving, XENMEG: a catalogue of occurrences of xenoliths and megacrysts in volcanic rocks of eastern Australia, Australian Museum Sydney, 441 pp. 1976.

Wilshire, H. G., and R. A. Binns, Basic and ultrabasic xenoliths from volcanic rocks of New South Wales, J. Petrology, 2, 185-208, 1975.

Wilshire, H. G., and J. W. Shervais, Al-augite and Cr-diopside ultramafic xenoliths in basaltic rocks from the western United States, Phys. Chem. of the Earth, 9, 257-272, 1975.

MANTLE XENOLITHS FROM SOUTHEASTERN NEW ENGLAND

Brian D. Leavy
O. Don Hermes

Department of Geology, University of Rhode Island,
Kingston, Rhode Island

Abstract. Newly exposed mafic dikes that intrude granites in southwestern Rhode Island (USA) contain a suite of ultramafic xenoliths and megacrysts. This is the first reported occurrence of nodule-bearing rocks from southeastern New England. Ultramafic xenoliths in these dikes include spinel lherzolite, spinel harzburgite, and wehrlite. These have been mechanically transported upward in a lamprophyric matrix which crystallized relatively rapidly.

The mineral assemblages, textures, major element abundances, and mineral chemistry exhibited by these nodules indicate that they were derived from the upper mantle. Coincidence of the orientation of the Westerly dikes with Mesozoic lamprophyres in central New England suggests that the genesis of the host dikes and the mobilization of the included xenoliths was related to a major tectonic event which affected this part of New England during the post-Paleozoic.

Introduction

Dikes that contain ultramafic xenoliths recently have been exposed in roadcuts near Westerly, Rhode Island (Leavy and Hermes, 1977 a, b; 1978). Xenoliths contained in these dikes include spinel lherzolite, spinel harzburgite, and wehrlite, all of which are types of rocks that commonly are interpreted as having been derived from the upper mantle (White, 1966; Frey and Green, 1974; Frey and Prinz, 1978). Also present are megacrysts which include olivine, clinopyroxene, orthopyroxene, spinel, and fragments of the enclosing granites. To our knowledge, this is the first documented occurrence of such rocks in southeastern New England.

The nodule-bearing dikes described here are petrologically distinct from the diabase dikes that are common to this region. McHone (1977, 1978 a, b) has reported a number of alkalic lamprophyre dikes in central New England, including a camptonite from southern Vermont which contains nodules of undetermined origin. Other rocks in the New England area that may be similar to these are the highly altered ultramafic dikes reported from Syracuse, New York (Maynard and Ploger 1946; Apfel *et al.*, 1951; Hogeboom, 1957), and nodule-bearing kimberlites near Ithaca, New York (Schulze *et al.*, 1978).

Two of the dikes described here crop out in a roadcut in Ashaway, Rhode Island (Fig. 1). An offshoot of one of these appears to be the "mafic dike" described by Feininger (1964, 1965); he did not report the presence of xenoliths, although olivine megacrysts were observed. A third dike is exposed in a quarry within the town of Westerly and was described by Moore (1967) as a "porphyritic limburgite dike." Quinn (1971) classified both of these dikes as lamprophyres, "possibly monchiquite." The present study emphasizes the spinel lherzolites of the Ashaway outcrops, due to the relatively fresh nature of the host dikes and the abundance of the contained nodules.

Geologic Setting

The dike at Westerly intrudes Westerly Granite and Ordovician or older metavolcanic rocks that are intruded by the granite (Moore, 1967). The Ashaway dikes cross-cut both the Narragansett Pier Granite and a shallow, southerly dipping dike of Westerly Granite (Fig. 1). At present, absolute ages of these country rocks have not been convincingly documented by radiometric methods. Kocis *et al.* (1978) report concordant U-Pb ages of 275 m.y. on separated monazite from outcrops of Narragansett Pier Granite approximately 45 km to the east of Westerly. Earlier Rb-Sr and K-Ar ages on the Westerly Granite yield ages that range from 208-299 m.y., but some of these may represent ages that were reset due to a regional "disturbance" during the Permian period (Zartman, *et al.*, 1970). Regardless of the uncertainties in the absolute ages of these granites, they are recognized as belonging to the youngest plutonic igneous event in southern New England. Because these granites are cut by the lamprophyres, the dikes must post-date, or at least be contemporaneous with the most recent episode of plutonism in this region.

The dikes exhibit nearly vertical dips and strike N25°E, following a prominent regional

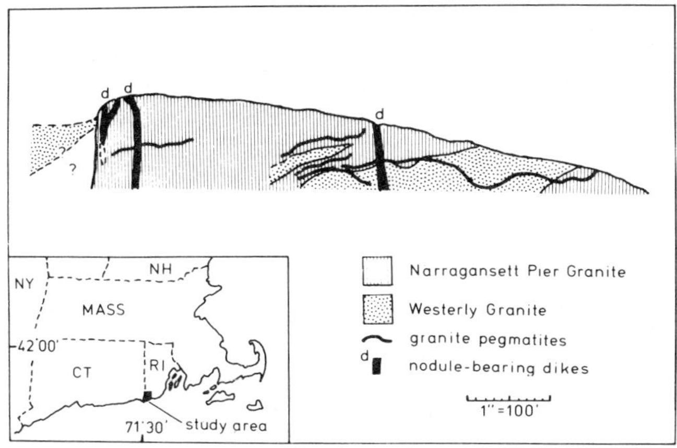

Figure 1. Cross section of roadcut in Ashaway, R.I., showing relationship of Westerly Granite, Narragansett Pier Granite, and the nodule-bearing dikes. Section is projected to a plane parallel to the face of the roadcut. Contacts of the Westerly Granite that is exposed in a quarry at the left of the figure are obscured by soil.

joint and fault trend. This trend is well displayed by the attitudes of Mesozoic diabases in the region (King, 1961; May, 1971; Pierce, 1976, 1978), and is also followed by Mesozoic lamprophyres of central New England (McHone, 1978 a, b). The general similarity of dike orientations and compositions of the Westerly dikes compared to other lamprophyric rocks in New England suggests that they may be correlative in time. Generally massive granite enclosing the dikes is closely jointed and locally brecciated for about 2 m on either side of the dikes. Veins of carbonate and stringers of dike material occur several centimeters from the dike-granite contact. Carbonate veins also commonly cut the dikes parallel to the contacts, and less commonly, at low angles. The calcite, and hydrous phases within the dike matrix, appear to be related to late stage, but primary, magmatic processes. The fine scale jointing, marginal brecciation, and probable presence of a volatile-rich phase in the dike magma are compatible with a relatively rapid and forceful mode of emplacement.

Mineralogy and Petrology

Host dikes

Mineralogically, the host dikes are lamprophyres, probably a variety of monchiquite. Alteration is common, but not pervasive. Primary euhedral phases include titaniferous augite (Wo:En:Fs= 54:32:14), kaersutite, phlogopite (Mg:Fe=2.12:1), titanomagnetite, and apatite (Table 2). The pyroxene, opaques, and apatite are enclosed in a matrix of secondary or late-stage minerals, predominantly chlorite, with minor calcite and serpentine. Neither fresh nor relict feldspars have been observed. Kaersutite and phlogopite are present in this groundmass, but are restricted mainly to ocelli similar to those described by McHone (1977), intergrown with stilbite and/or analcite, or calcite. Ocelli that lack amphibole or mica are also present. Clearly, a CO_2- and H_2O-rich fluid phase was associated with the dike magma during emplacement, and the patchy distribution of early and late stage matrix minerals indicates that there may have been some segregation of volatiles, through some mechanical or immiscible process, during the later stages of crystallization.

The grain size of the groundmass is variable, grading (sometimes abruptly) from aphanitic at dike margins to grains approximately 3 mm in longest dimension near the dike centers. Locally common flow layering, parallel to the dike-granite contact, is composed of alternating bands of fine grained dike groundmass and layers that are rich in megacrysts. This suggests that the dikes were emplaced in several pulses. The larger crystals of titanaugite display sector zoning similar to that found in augite from quenched alkali basalts and volatile-rich basaltic rocks (Nakamura, 1973; Leung, 1974).

Crustal xenoliths and xenocrysts

Locally abundant in the dikes are rock and mineral fragments that appear to have been stoped from the surrounding granites. These fragments include discrete grains of feldspars, biotite, and quartz, and larger rock fragments containing these minerals. These fragments generally are quite angular and free of marginal reaction features, attesting to the forceful and rapid mode of dike emplacement. So far, we have not observed any high grade metamorphic rocks in the xenolith suite, which might be representative of lower crustal rocks.

Megacrysts

Megacrysts in the Westerly dikes are of two general types: 1) cognate or xenocrystic single crystal nodules, and 2) disaggregated crystals or crystal fragments of the xenolithic nodules. Of the first type, the most common mineral is euhedral olivine, Fo_{89} (Table 1), which ranges in size from several millimeters up to 1 cm in longest dimension. The euhedral to subhedral shape and absence of marginal reaction with the groundmass suggests that these are cognate. Commonly these olivines have skeletal cores somewhat similar to quench textures of olivines from quickly-cooled basalts. Secondary or late stage processes have replaced many olivine megacrysts with bowlingite, calcite, or serpentine.
Less common are large clinopyroxenes (Wo:En:Fs= 44:45:11) that are low in Cr and high in Ti, Al, and Fe (Table 2). These display marginal zoning, or altered rims surrounded by overgrowths of clinopyroxene which has a composition that resembles that of the host dike pyroxene.

TABLE 1. Whole Rock Major Element Compositions in Westerly Dike Rocks and Xenoliths

	Dike Rocks		Lherzolites		Harzburgites	
Oxide (wt.%)	N3A	Monch.†	LN-4	World Ranges§	N-1	World Ranges§
SiO_2	32.04	39.00	42.15	42.69-45.59	45.20	42.60-47.52
TiO_2	4.99	3.60	0.07	0.00- 0.28	0.10	0.01- 0.05
Al_2O_3	6.30	11.72	1.87	0.47- 4.00	1.52	0.37- 1.93
Cr_2O_3	0.10	n.d.	n.d.	0.26- 0.48	0.11	0.04- 0.38
Fe_2O_3	4.79	4.11	1.40	0.38- 2.54	1.24	1.60- 2.37
FeO	11.10	8.19	6.60	5.89- 9.27	7.17	3.50- 6.21
MnO	0.43	0.16	0.11	0.08- 0.15	0.12	0.10- 0.21
MgO	17.09	12.24	40.27	37-40-45.32	39.85	41.14-48.20
CaO	10.03	11.80	2.75	0.57- 3.50	2.04	0.23- 0.91
Na_2O	0.70	3.04	0.16	0.05- 0.38	0.08	0.04- 0.05
K_2O	2.55	0.43	0.13	0.00- 0.10	0.09	0.01- 0.05
NiO	n.d.	n.d.	n.d.	0.22- 0.34	0.40	0.17- 0.42
P_2O_5	2.11	1.03	0.02	0.00- 0.08	0.01	0.00- 0.06
CO_2	7.50	n.d.	1.54	-	1.41	-
H_2O	0.27	1.30	2.92	-	0.65	-
Total	100.00	99.92	99.99	-	99.99	-
ΣFe=FeO	15.41	11.89	7.86	-	8.29	-
Mg/Mg+ΣFe	0.53	0.51	0.84	-	0.83	-
norms						
c	2.39	-	0.00	-	0.88	-
or	15.08	-	0.79	-	0.55	-
ab	5.92	-	1.13	-	0.70	-
an	0.00	-	4.93	-	1.10	-
wo	0.00	-	3.89	-	0.00	-
en	13.03	-	15.17	-	34.31	-
fs	3.03	-	1.87	-	4.43	-
fo	17.97	-	59.99	-	45.50	-
fa	4.61	-	7.90	-	6.47	-
mt	6.94	-	1.33	-	1.80	-
cm	0.00	-	0.00	-	0.17	-
il	9.47	-	0.31	-	0.20	-
ap	4.99	-	0.49	-	0.03	-
cc	12.94	-	0.00	-	3.22	-
mg	3.47	-	0.00	-	0.00	-
sal	23.39	-	7.28	-	3.24	-
fem	76.45	-	90.46	-	96.12	-
Total	99.84	-	97.74	-	99.36	-

† monchiquite, Hopi Buttes, Arizona (Williams, 1936).

§ worldwide ranges for spinel lherzolites and spinel harzburgites compiled from data of Donaldson, 1978; Frey and Green, 1978; Suwa, et al., 1975; Frey and Green, 1974; Kuno and Aoki, 1970.

n.d. not determined

A striking feature of these megacrysts is the presence of iron sulfide blebs which lie in a grid-like pattern along crystal planes.

Megacrysts of the second type are present as anhedral single grains or small aggregates; compositions are similar to those minerals of the same type found within the lherzolite and harzburgite nodules. Olivine is altered in much the same manner as the cognate megacrysts, but may be distinguished by its anhedral shape, higher Mg content (Fo_{92}), and the deeply embayed margins of many grains. Clinopyroxenes and orthopyroxenes commonly are overgrown by clinopyoxene similar to that of the dike matrix, and brown spinel is rimmed by Cr-magnetite.

Ultramafic xenoliths

The ultramafic xenoliths are subrounded (Fig. 2) and up to 8 cm in longest dimension. Based on the textural classification of Boullier and Nicolas (1975), at least two major textural types are present: 1) coarse granular (tabular) and 2) tectonite (mosaic). Both of these varieties display metamorphic textures common to xenoliths of presumed mantle origin (Pike and Schwarzman, 1977). For a given rock type, mineral proportions are fairly constant from nodule to nodule, and although mineral distribution within single xenoliths may be patchy, no distinct layering has been observed. Likewise, distinct cumulate igneous textures have not been recognized.

Granular textural types include spinel lherzolite and spinel harzburgite. These consist of orthopyroxene (Wo:En:Fs=1:90:9) ± chrome diopside (Wo:En:Fs=46:49:5); olivine (Fo=91-92); and Cr-rich spinel (Table 3). Pargasitic amphibole is present in some lherzolites, but it is not common. The fabrics of the lherzolite xenoliths range from nearly equidimensional silicates with interstitial spinels to tabular, strain-free silicate grains with stringers or elongated clusters of spinel parallel to the

TABLE 2. Mineral Compositions in Westerly Dike Rocks and Megacrysts

	dike minerals				megacrysts		
	cpx	phlog	amph	opaq	ol	cpx†	cpx§
Oxide (wt.%)	RB1C	RB5B	RB5B	RB1C	RB1A	LNA-2	LNA-2
SiO_2	39.30	38.97	37.94	0.35	41.11	48.77	43.46
TiO_2	6.98	4.50	7.50	21.97	0.35	1.66	4.67
Al_2O_3	11.50	13.07	13.86	5.19	0.00	8.26	8.11
Cr_2O_3	0.01	0.05	0.01	0.06	0.00	0.05	0.25
FeO*	8.07	10.88	10.65	67.17	9.67	6.09	6.71
MnO	0.10	0.16	0.20	1.44	0.12	0.09	0.04
MgO	9.96	19.99	12.20	1.68	48.53	14.04	10.57
CaO	23.04	0.19	12.12	0.15	0.24	19.23	24.04
Na_2O	0.61	0.65	1.68	0.00	0.01	1.25	0.39
K_2O	0.07	8.81	0.00	0.03	0.00	0.00	0.00
NiO	0.00	0.00	0.00	0.00	0.00	0.00	0.00
P_2O_5	n.d.	1.19	n.d.	n.d.	n.d.	n.d.	n.d.
Total	99.64	98.46	96.16	98.04	99.44	99.44	98.24
Mg/Mg+ΣFe	0.55	0.65	0.53	0.02	0.83	0.70	0.61
cations							
Si	1.526	6.951	5.628	0.028	1.010	1.791	1.661
Ti	0.196	0.603	0.835	0.023	0.000	0.046	1.134
Al	0.503	2.751	2.419	0.093	0.001	0.358	0.366
Cr	0.000	0.000	0.000	0.001	0.000	0.001	0.008
Fe	0.253	1.623	1.271	1.118	0.200	0.187	0.215
Mn	0.002	0.019	0.019	0.033	0.001	0.003	0.001
Mg	0.563	5.310	2.695	0.034	1.750	0.768	0.602
Ca	0.950	0.007	1.925	0.001	0.000	0.757	0.985
Na	0.046	0.218	0.594	0.000	0.000	0.089	0.029
K	0.000	2.001	0.315	0.000	0.000	0.000	0.000
Ni	0.000	0.000	0.000	0.000	0.000	0.000	0.000
P	-	0.520	-	-	-	-	-
Total	4.041	20.003	15.701	1.581	2.980	4.000	4.001
# oxygens	6	24	23	3	4	6	6

† core
§ rim
* ΣFe=FeO; data from electron microprobe analyses.
n.d. not determined

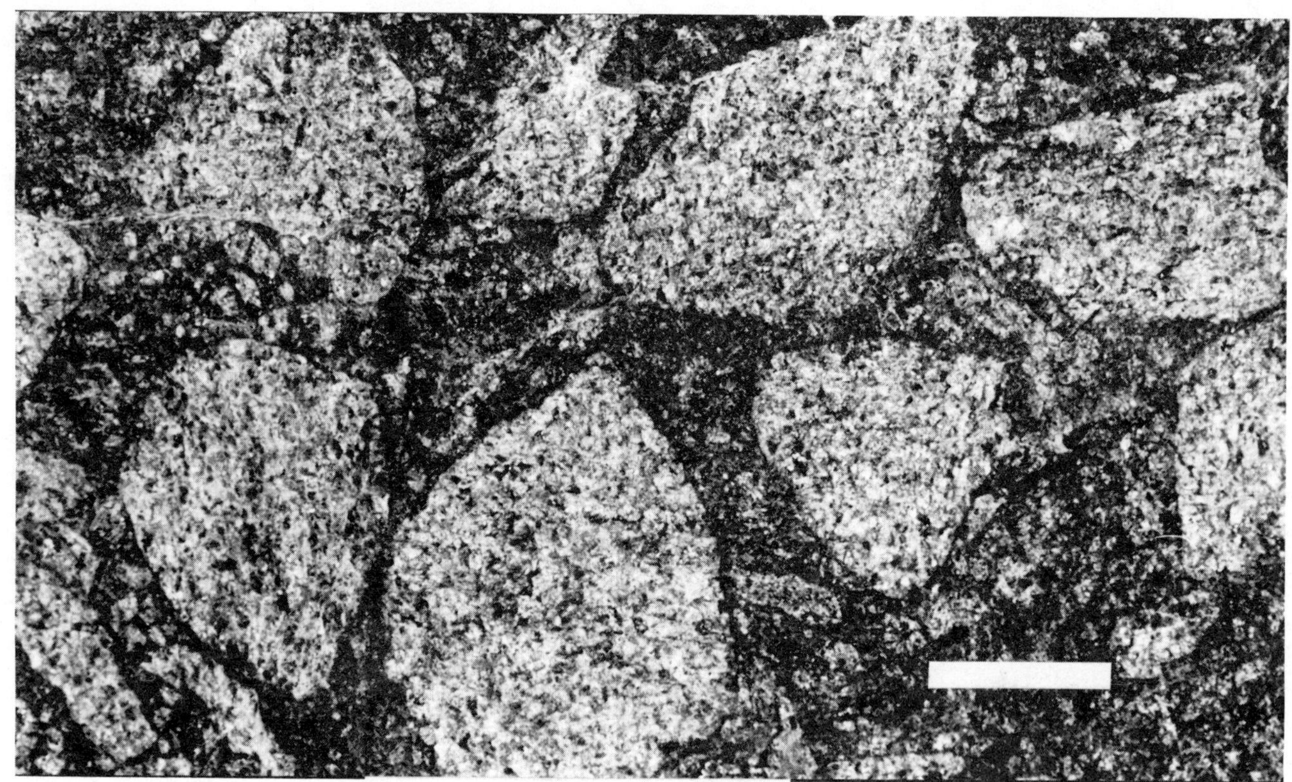

Figure 2. Portion of Sample RB1 from Ashaway, R.I., showing ultramafic nodules. Bar is 1 cm.

foliation. No foliation is evident in the harzburgites.

The tectonite type includes only wehrlites. In these xenoliths, large (up to 1 cm) strained grains of olivine and clinopyroxene are surrounded by a finer grained matrix of the same minerals (Table 3). The larger grains display undulatory extinction and commonly have ragged edges; matrix minerals are equigranular and granoblastic. Grain margins in the wehrlites are coated with a thin layer of a red mineral, probably hematite, which imparts a reddish color to the entire xenolith.

At the margins and along fractures in the lherzolites and harzburgites, xenolith minerals show reaction with the dike groundmass. Olivine is altered in a manner similar to the megacrystic olivine, and pyroxenes are overgrown by dike clinopyroxene. Pargasite has reacted to form amphibole closer in composition to that of the host dike, notably becoming richer in titanium. Brown spinel is rimmed by Cr-magnetite. These effects clearly indicate disequilibrium and support the view that these are foreign xenoliths and are not cognate with the host dike.

Whole-rock Chemistry

The whole-rock major element chemistry for one host dike sample and two xenolith types are given in Table 1. Due to extensive alteration and the abundance of megacrysts, it is difficult to estimate the initial chemical composition of the host dikes. Because mineral proportions in the dikes are highly variable, the chemistry will depart from the given analysis where ocelli dominate over the pyroxene-rich matrix. Despite these difficulties, it appears that the composition is comparable to that of some lamprophyres (Table 1). The relatively low Al_2O_3, SiO_2, and Na_2O may be attributed to the absence of feldspars and/or feldspathoids; the correspondingly higher TiO_2, MgO, and FeO values may reflect megacrysts of mafic minerals included in the dike matrix. The high K_2O may be due to the abundance of phlogopite.

Compared to the xenoliths, the dikes are lower in SiO_2 and MgO, and higher in TiO_2, Al_2O_3, ΣFe=FeO, CaO, Na_2O, K_2O, and P_2O_5. Representative analysis of the granular xenoliths are comparable to similar xenolith types worldwide (Table 1). The high CaO content of the harzburgite is most likely due to the presence of host dike-derived carbonate; this is supported by the correspondingly high CO_2 values.

CaO and Al_2O_3 are higher in the lherzolites than in the harzburgites as might be expected for the less refractory nature of the clinopyroxene-bearing assemblage, as suggested by Frey and Prinz (1978). However, such differences are not striking, probably due to the fact that clinopyroxene abundance in the lherzolites is low; the name lherzolite was assigned to these rocks following the usage of Boyd (1973).

Mineral chemistry

Representative microprobe analyses of minerals from the host dikes and the contained nodules are presented in Tables 2 and 3.

Host dikes. All mafic minerals comprising the dike matrix are rich in Ti. Clinopyroxene is a high Al titanaugite that exhibits sector zoning. In some cases this pyroxene is overgrown by kaersutite which has a composition similar to that of the discrete grains of amphibole in the groundmass and ocelli. Similar amphibole, although with generally higher alkali content, is commonly reported as an upper mantle phase (Mason, 1968; Varne, 1970; Francis, 1976) Phlogopite has appreciable Na_2O and P_2O_5. Fine grained, commonly euhedral opaques are homogeneous titanomagnetite.

Xenoliths. The chemistry of the xenolith phases is distinctly different from that of the host dike minerals. Mineral compositions are generally homogeneous within grains, from grain to grain, and in many cases between rock types. Olivine in the lherzolite is slightly more Mg-rich (Fo_{92}) than that of the harzburgite (Fo_{91}); olivine in the wehrlite is the most Fe-rich observed, and exhibits two distinct populations: 1) porphyroclasts (Fo_{82}), and 2) recrysallized matrix grains (Fo_{81}). Similar populations are noted for the wehrlitic clinopyroxenes, porphyroclasts with Wo:En:Fs=38:52:10 versus recrystallized matrix grains with Wo:En:Fs=42:47:11. Spinel shows

TABLE 3. Mineral Compositions in Westerly Xenoliths

	lherzolite					harzburgite			wehrlite			
	cpx	opx	ol	amph	sp	opx	ol	sp	cpx†	cpx§	ol†	ol§
Oxide (wt.%)	RB1C	RB5B	RB1C-1	LN-4	RB1C	RB1M	RB1M	RB1M	RB4A	RB4A	RB4A	RB4A
SiO_2	52.49	55.64	40.49	42.07	0.00	55.97	44.67	0.00	45.54	47.17	35.58	38.82
TiO_2	0.30	0.07	0.03	2.17	0.18	0.08	0.00	0.03	1.71	2.48	0.04	0.03
Al_2O_3	5.46	3.46	0.03	12.71	47.83∇	2.57	0.02	47.05∇	6.63	4.98	0.00	0.00
Cr_2O_3	0.89	0.29	0.03	2.22	20.35∇	0.52	0.00	19.25∇	0.21	0.03	0.02	0.01
FeO*	2.32	5.43	8.89	3.56	11.99	4.75	7.98	11.75	6.29	7.11	17.82	17.78
MnO	0.11	0.12	0.16	0.21	0.18	0.11	0.12	0.10	0.13	0.11	0.21	0.20
MgO	15.78	34.16	50.01	18.07	18.93	34.78	47.09	20.93	18.63	15.17	45.50	43.18
CaO	20.37	0.63	0.04	11.20	0.02	0.64	0.04	0.00	18.94	22.28	0.11	0.09
Na_2O	1.54	0.07	0.02	3.65	0.04	0.06	0.02	0.07	1.69	0.64	0.01	0.01
K_2O	0.00	0.00	0.32	0.14	0.00	0.00	0.00	0.00	0.03	0.03	0.00	0.00
NiO	0.00	0.00	0.27	0.00	0.29	0.00	0.00	0.21	0.00	0.00	0.02	0.08
Total	99.26	99.87	100.29	96.00	99.81	99.48	99.94	99.33	99.80	100.00	99.31	100.20
Mg/Mg+ΣFe	0.87	0.86	0.85	0.84	0.61	0.88	0.86	0.64	0.75	0.68	0.72	0.71
cations												
Si	1.913	1.901	0.983	6.111	0.000	0.941	0.991	0.000	1.699	1.808	0.921	0.978
Ti	0.003	0.002	0.000	0.234	0.005	0.000	0.000	0.001	0.047	0.044	0.001	0.001
Al	0.228	0.163	0.000	2.173	1.494∇	0.089	0.000	1.469∇	0.291	0.266	0.000	0.000
Cr	0.021	0.006	0.000	0.249	0.425∇	0.013	0.000	0.402∇	0.003	0.004	0.000	0.000
Fe	0.078	0.167	1.169	0.430	0.265	0.143	0.166	0.259	0.196	0.192	0.387	0.329
Mn	0.002	0.004	0.001	0.019	0.003	0.001	0.002	0.001	0.003	0.003	0.005	0.003
Mg	0.859	1.743	1.863	3.914	0.748	1.791	1.849	0.827	1.036	0.863	1.758	1.653
Ca	0.807	0.022	0.001	1.741	0.000	0.022	0.000	0.000	0.756	0.771	0.002	0.002
Na	0.077	0.004	0.000	1.028	0.000	0.001	0.000	0.000	0.121	0.103	0.000	0.002
K	0.000	0.000	0.000	0.019	0.000	0.000	0.000	0.000	0.000	0.000	0.000	0.000
Ni	0.000	0.000	0.000	0.000	0.008	0.000	0.000	0.004	0.000	0.000	0.000	0.001
Total	3.985	4.012	3.014	15.925	2.948	4.000	3.008	2.963	4.155	4.055	3.073	3.017
# oxygens	6	6	4	23	4	6	4	4	6	6	4	4

† porphyroclast
§ matrix
∇ Variable; above values are averages.
* ΣFe=FeO; data from electron microprobe analyses.

the greatest range of heterogeneity, exhibiting grain-to-grain ranges of 11-35% Cr_2O_3 and 20-35% Al_2O_3. As there is no obvious correlation between these variations and any textural features in the spinels or their associated xenoliths, it is not possible to speculate on the cause of the variation at this time.

Discussion

A cognate origin of the xenoliths seems unlikely due to the disequilibrium between the xenoliths and the host dikes, as well as to their different mineral assemblages and mineral compositions. A cumulative origin at a shallow depth is discounted by the high pressure mineral assemblages, chemical homogeneity of phases, and absence of cumulate textures. The possibility remains that they initially may have been formed as deep-seated cumulate rocks and were subsequently metamorphosed to produce the observed fabrics. Based on the mineral assemblages and field relationships, it seems certain that the xenoliths, and therefore the host dike magma, were generated within the upper mantle (White, 1966; Frey and Prinz, 1978). At present our data does not permit us to determine whether the xenoliths represent mantle residue that is directly related to the genesis of the host dike magma, or instead were included accidentally during its ascent.

Using the pyroxene geothermometer of Wood and Banno (1973), calculated temperatures of final equilibration of the lherzolites are estimated to lie in the range 1000-1200°C. While at present there is no generally accepted method for precise determination of the pressures of equilibration for the spinel lherzolite assemblage, the absence of garnet and plagioclase places these xenoliths in the range of 9-25 kb (Frey and Prinz, 1978). Temperatures of 1000-1200°C are high for these depths in the upper mantle, but this may reflect a local tectonic disturbance of the geothermal gradient. One possibility is the influence of epeirogenic doming following a tectonic episode, here most likely the Alleghany orogeny. Such a relationship of similar rocks has been suggested by Irving (1976) and Ferguson *et al.* (1977). Dikes similar to the lamprophyres at Westerly are associated with major granitic intrusions elsewhere, such as the appinite-Newer Granites association of Northern Ireland and Scotland (Carmichael *et al.*, 1974, p. 584), and the lamprophyres associated with Mesozoic granites in Vermont and southern Quebec (McHone, 1977, 1978 a, b). This sort of cogenetic relationship may exist between the Westerly dikes and the Narragansett Pier and Westerly granites.

The trend of the Westerly dikes is coincident with that of regional lineaments and Mesozoic dikes of the region (King, 1961; McHone, 1978 a, b; Pierce, 1978). May (1971) has suggested that this trend reflects a deep-seated stress field associated with the last opening of the Atlantic Ocean. Possibly, such a stress field could have mobilized upper mantle material as represented by the Westerly nodule suite. This is supported by the apparently strict association of ultramafic xenoliths with areas of mantle upwelling, or Wilson-Morgan hotspots (Basu, 1975). These nodules thus are thought to represent relics of such activity during the late stages of continental rifting during the last opening of the Atlantic Ocean.

Acknowledgements. Part of this work was supported through a grant to BDL from Sigma Xi. Microprobe work was funded through an NSF Grant (EAR 75-21795) to ODH. The JEOL microprobe facility used was supported by an NSF Grant (OCE 75-13295) to Drs. H. Sigurdsson and J-G. Schilling of the Graduate School of Oceanography, University of Rhode Island.

References

Apfel, E.T., Maynard, J.E., and Ploger, L.W. (1951) Possible diatremes in Syracuse, New York: *Geol. Soc. Amer. Bull. 62*; 1421.

Basu, A.R. (1975) Hotspots, mantle plumes, and a model for the origin of ultramafic xenoliths in alkali basalts: *E. Planet. Sci. Lett. 28*; 261-274.

Boullier, A.M., and Nicolas, A. (1975) Classification of textures and fabrics of peridotite xenoliths from South African kimberlites: *Phys. Chem. Earth 9*; 467-475.

Boyd, F.R. (1973) A pyroxene geotherm: *Geochim. et Cosmochim. Acta 37*; 2533-2546.

Carmichael, I.S.E., Turner, F.J., and Verhoogen, J. (1974) *Igneous Petrology*: McGraw-Hill, New York; 739 p.

Donaldson, C.H. (1978) Petrology of the uppermost upper mantle deduced from spinel lherzolite and harzburgite at Carlton Hill, Derbyshire: *Contr. Min. Petrol. 65*; 363-377.

Feininger, T. (1964) Petrology of the Ashaway and Voluntown quadrangles, Connecticut-Rhode Island: unpub. Ph.D. thesis, Brown University.

________ (1965) Bedrock geologic map of the Ashaway quadrangle, Connecticut-Rhode Island: U.S.G.S. map GQ-403.

Ferguson, J., Ellis, D.J., and England, R.N. (1977) Unique spinel-garnet lherzolite from Australia: *Geology 5*,5; 278-280.

Francis, D.M. (1976) The origin of amphibole in lherzolite xenoliths from Nunivak Island, Alaska: *Jour. Petrol. 17*,3; 357-378.

Frey, F.A., and Green, D.H. (1974) The mineralogy, geochemistry, and origin of lherzolite inclusions in Victorian basanites: *Geochim. et Cosmochim. Acta 38*; 1023-1059.

________ and Prinz, M. (1978) Ultramafic inclusions from San Carlos, Arizona: Petrologic and geochemical data bearing on their petrogenesis: *E. Planet. Sci. Lett. 38*; 29-176.

Hogeboom, W.L. (1957) The Petrology of the Green Street peridotite dikes in Syracuse, New York: unpub. M.S. thesis, Syracuse University.

Irving, A.J. (1976) On the validity of paleogeotherms determined from xenolith suites in basalts and kimberlites: Amer. Min. 61, 7&8; 638-642.

King, P.B. (1961) Systematic pattern of Triassic dikes in the Appalachian region: U.S.G.S. Prof. Paper 424-B; B93-B95.

Kocis, D.E., Hermes, O.D., and Cain, J.A. (1978) Petrologic comparison of the Narragansett Pier Granite, Rhode Island (abstr.): Geol. Soc. Amer. Abstr. w/Prog. 10; 71.

Kuno, H. and Aoki, K. (1970) Chemistry of ultramafic nodules and their bearing on the origin of basalt magmas: Phy. E. Planet. Int. 3; 273-301.

Leavy, B.D., and Hermes, O.D. (1977a) Lherzolite nodules in dikes from southeastern New England (abstr.): Geol. Soc. Amer. Abstr. w/Prog. 9; 293.

__________ (1977b) Mantle xenoliths from southeastern New England (abstr.): Extended Abstracts, Second International Kimberlite Conference, Santa Fe, New Mexico.

__________ (1978) Petrology of mantle-derived mafic dikes from southwestern Rhode Island (abstr.): Geol. Soc. Amer. Abstr. w/Prog. 10; 73.

Leung, I.S. (1974) Sector-zoned titanaugites: Morphology, crystal chemistry, and growth: Amer. Min. 59; 127-138.

Mason, B. (1968) Kaersutite from San Carlos, Arizona, with comments on the petrogenesis of this mineral: Min. Mag. 36; 997-1002.

May, P.R. (1971) Pattern of Triassic-Jurassic diabase dikes around the North Atlantic in the context of predrift positions of the continents: Geol. Soc. Amer. Bull. 82; 1285-1292.

Maynard, J.E., and Ploger, L.W. (1946) An unrecorded dike in Syracuse, New York: Amer. Min. 31; 200-201.

McHone, G. (1977) Dike types and distribution in central New England (abstr.): Geol. Soc. Amer. Abstr. w/Prog. 9; 300.

__________ (1978a) Petrogenesis of lamprophyre dikes and related White Mountain-Monteregian intrusive complexes (abstr.): Geol. Soc. Amer. Abstr. w/Prog. 10; 75.

__________ (1978b) Distribution, orientations, and ages of mafic dikes in central New England: Geol. Soc. Amer. Bull. , in press.

Moore, G.E. (1967) Bedrock geologic map of the Watch Hill quadrangle, Washington County, Rhode Island, and New London County, Connecticut: U.S.G.S. map GQ-655.

Nakamura, Y. (1973) Origin of sector-zonimg in igneous clinopyroxenes: Amer. Min. 58; 986-990.

Pierce, T.A. (1976) Petrology of dolerite and metadolerite dikes of southeastern New England: unpub. M.S. thesis, University of Rhode Island.

__________ (1978) Petrology and geochemistry of diabase and metadiabase dikes in southeastern New England (abstr.): Geol. Soc. Amer. Abstr. w/Prog. 10; 80.

Pike, J.E.N., and Schwarzman, E.C. (1977) Classification of textures in ultramafic xenoliths: Jour. Geol. 85; 49-61.

Quinn, A.W. (1971) Bedrock Geology of Rhode Island: U.S.G.S. Bull. 1295; 68 p.

Schulze, D.J., Helmstaedt, H., and Cassie, R.M. (1978) Pyroxene-ilmenite intergrowths in garnet pyroxenite xenoliths from a New York kimberlite amd Arizona latites: Amer. Min. 63; 258-265.

Suwa, K., Yusa, Y., and Kishida, N. (1975) Petrology of peridotite nodules from Ndonyuo Olnchoro, Samburu District, central Kenya: Phys. Chem. Earth 9; 273-286.

Varne, R. (1970) Hornblende lherzolite and the upper mantle: Contr. Min. Petrol. 27; 45-51.

White, R.W. (1966) Ultramafic inclusions in basaltic rocks from Hawaii: Contr. Min. Petrol. 12; 245-314.

Williams, H. (1936) Pliocene volcanoes of the Navajo-Hopi country: Geol. Soc. Amer. Bull. 47; 111-171.

Wood, B.J., and Banno, S. (1973) Garnet-orthopyroxene and orthopyroxene-clinopyroxene relationships in simple and complex systems: Contr. Min. Petrol. 42; 109-124.

Zartman, R.E., Hurley, P.M., Krueger, H.W., and Giletti, B.J. (1970) A Permian disturbance of K-Ar radiometric ages in New England: its occurence and causes: Geol. Soc. Amer. Bull. 87; 3359-3374.

MAJOR TRACE ELEMENTS OF AL-AUGITES AND CR-DIOPSIDES FROM ULTRAMAFIC NODULES IN EUROPEAN ALKALI BASALTS

Emile Jagoutz, Volker Lorenz* and Heinrich Wänke

Max-Planck-Institut für Chemie, Saarstraße 23, 65 Mainz, F. R. Germany

*Institut für Geowissenschaften, J.-Gutenberg-Universität, 65 Mainz

Abstract. Clinopyroxenes from ultramafic inclusions (spinel lherzolites and wehrlites) included in European Tertiary basalts (Massif Central, France; Dreiser Weiher and Vogelsberg, Germany and Kapfenstein, Austria) were analyzed for major elements by microprobe and trace elements by neutron activation. The REE pattern of the clinopyroxenes reflect the bulk composition of the inclusion and from the combination of major and trace elements a similar complex history arises for the inclusions as for those described by Frey and Green, 1974 and Frey and Prinz, 1978. It is shown that Yb and Sc increase with the jadeite and jadeite+Tschermaks component, resp., in the clinopyroxene and it is concluded that the uptake into clinopyroxene of such trace elements is a function of major element composition and probably pressure and temperature.

Introduction

Ultramafic inclusions are very common in alkali basalts from all over the world (e.g., Forbes and Kuno, 1966). The different types were discussed by White (1966) in a study on Hawaiian ultramafic inclusions. He suggested the wehrlitic nodules to be cumulates and the lherzolitic (spinel lherzolites) nodules to be fragments of the upper mantle unrelated to the host basalt. Already in 1935 Ernst reached similar conclusions based on textural evidence for spinel lherzolites from the Westberg, Lower Saxony, Germany. Frechen (1948, 1963) described the ultramafic inclusions from the Dreiser Weiher, Germany, and suggested a cognate origin. Aoki, et al. (1968), arguing from the major element chemistry of the clinopyroxenes and Paul (1972) based on Sr-isotopes also considered the wehrlitic nodules as cumulates and the lherzolites as xenoliths from the upper mantle. The major element chemistry and mineralogy of ultramafic inclusions from France were studied by Hutchison, et al. (1975). Data on the rare earth geochemistry of Eruopean ultramafic nodules are so far very scarce (e.g., Philpotts, et al., 1972). Frey and Green (1974), however, show for Australian nodules and Frey and Prinz (1978) for nodules from San Carlos, Arizona, that a combination of trace and major element chemistry and microprobe data may yield important information on the composition and processes of the upper mantle.

In the present study we concentrated on the major and trace element composition of the clinopyroxenes which can be used as indicators of the geochemical environment since they are the host for the major part of the REE content in spinel lherzolites and wehrlites (see Frey and Green, 1974).

Experimental Methods

Samples from four areas in Europe (Dreiser Weiher and Vogelsberg, Germany, Kapfenstein, south-east Austria, and Massif Central, France) were used in the present study. Their localities, petrographic types, mineralogy and textures are described in the Appendix.

None of the samples included in this paper contained amphibole, garnet or apatite. Some nodules contain veins or patches partly filled with Na-rich brown glass. Since the trace element study was intended to be done on clean mineral separates these veins may cause contamination. Mineral separates were handpicked using a binocular microscope.

Consequently, the separated clinopyroxenes were crushed in an agate mortar and sieved.

The fraction between 0.5 and 1 mm (for $C_1 > 0.2$ mm) was etched for 5 mins in cold 5% HF, washed in H_2O and then etched again for 10 mins in 2n HNO_3 in an ultrasonic bath. These treated clinopyroxenes were handpicked again in alcohol to exclude mineral inclusions (mainly oxides) and dried at 110°C. In order to get clean mineral separates all these precautions are considered necessary as shown by the following test: 100 mg of supposedly

TABLE 1. Major and Trace Elements of Al-Augites from Wehrlites

	W1	W2	W3	W23	W29	E1
SiO_2	51.83	52.04	51.83	53.39	50.39	53.66
TiO_2	.69	.55	.70	.45	1.12	.37
Al_2O_3	4.87	4.64	4.80	4.37	6.11	4.00
Cr_2O_3*	.93	1.09	1.03	1.21	.39	1.03
FeO	4.70	4.70	4.61	3.95	5.02	4.31
MnO*	.124	.119	.117	.116	.099	.06
MgO	15.88	16.26	16.28	17.47	15.09	17.49
CaO	19.65	19.18	19.67	18.33	20.94	18.36
Na_2O*	.89	.96	.93	.97	.92	.86
Total	99.65	99.53	99.96	100.25	100.07	100.14
Si	1.896	1.902	1.889	1.922	1.846	1.935
Al	.104	.098	.111	.077	.154	.065
Al	.106	.102	.096	.108	.111	.105
Ti	.019	.015	.019	.012	.031	.010
Cr	.027	.031	.030	.034	.011	.029
Fe^{2+}	.144	.144	.140	.119	.154	.130
Mn	.004	.004	.004	.004	.003	.002
Mg	.865	.885	.884	.937	.823	.940
Ca	.770	.751	.768	.707	.821	.709
Na	.063	.068	.066	.068	.065	.060
Total	3.998	4.000	4.007	3.988	4.019	3.985
$\frac{Mg}{Mg+Fe}$	.8575	.8604	.8628	.8874	.8426	
Sc	41	51	41	44.5	63	
Co	31	32	40	32	29	
Ni	nd	nd	nd	365	250	
Hf	1.7	1.72	1.66	.6	1.6	
La	2.5	2.56	2.1	2.3	3.14	
Ce	nd	nd	nd	12.5	14	
Sm	1.65	2.4	1.70	1.74	2.8	
Eu	.57	.8	.64	.6	.8	
Tb	nd	nd	nd	.22	nd	
Dy	1.5	1.6	1.2	1.5	1.98	
Yb	.42	.53	.43	.54	.6	
Lu	.078	.063	.069	.058	.07	

*Analyzed by NAA

pure handpicked olivine mineral separate was neutron irradiated in the reactor. The then pulverized sample was treated with cold 3n HCl for 1 hour. The γ-spectra of the dissolved and undissolved portions of the sample were measured on a Ge(Li) detector. In the dissolved portion La was enriched by a factor of 8 and Eu by a factor of 4 relative to the undissolved portion. This result shows that part of the REE in ultramafic nodules sits on grain boundaries and suggests that the major part of the observed REE "in" olivine originates from grain surfaces.

It proved to be very difficult to obtain glassfree clinopyroxene separates from some glass-rich wehrlites. For the sample E_1 glass contaminated clinopyroxenes were eliminated by the following method: single clinopyroxene grains were irradiated and immediately analyzed for Na and Mn. Sodium is strongly enriched in the glass, Mn partitiones nearly equally between liquid and crystals. Thus increasing Na/Mn correlates with increasing contamination by glass and consequently, only those grains with the lowest Na/Mn were used for further measurements.

The samples (50-100 mg) were analyzed by instrumental neutron activation analysis for Sc, Cr, Mn, Co, Ni, Na, K, La, Ce, Sm, Eu, Dy, Yb, Lu and Hf. The technique used is described by Wänke, et al. (1973). Major elements from the same mineral separates (analyzed by NAA for trace elements) were determined with a KEVEX energy dispersive system attached to an ARL SEMQ38 microprobe, using the commercially available software (Magic V) from KEVEX. The geochemical standards from the Smithsonian Institute were used.

Experimental Results

A. Wehrlitic nodules. Only samples from the Dreiser Weiher (Westeifel/Germany) were available for this study. The results are given in Table 1. As already shown by Aoki, et al. (1968), the wehrlite clinopyroxenes (Al-augites) differ in composition from the clinopyroxenes (Cr-diopside) of lherzolitic inclusions in having higher CaO, FeO, MnO and TiO_2 and lower MgO and Cr_2O_3 (see also Fig. 2).

From the samples analyzed it appears that Al^{VI} and Na in the Al-augites is constant (Table 1 and Fig. 1), but Al^{IV} varies from 0.007 to 0.016. This is due to an increase in the component $CaTiAl_2O_6$. Simultaneously Mg/Mg+Fe and Cr decrease, whereas the totals of the structural formula increase (Table 1), suggesting an increase in Fe^{3+}. A similar trend can be seen from the data of Becker (1978). The concentrations of the incompatible elements increase slightly, Co remains nearly constant and Ni decreases. The Al-augites show a fractionation of LREE to HREE (Fig. 2).

B. Lherzolitic Nodules. The analyses of clinopyroxene separates from 20 spinel lherzolites (see Appendix) are given in Table 2. These Cr-diopsides show a wider range in trace element composition than the Al-augites from the wehrlitic suite and partly overlap with them (Fig. 2). The Al-content of the Cr-diopsides in the octahedral site is similar to that of Al-augites for one group and higher for a second group (Fig. 1). For both groups Al^{VI} is higher than the additional jadeite component (Fig. 1) and vice versa for the Na-Al^{VI} (jadeite) part of Fig. 1. These two groups are also distinguishable in Yb- and not quite as clear as Sc-contents (Figs. 1, 3, 4). Cr_2O_3 shows no systematic correlation with Mg/Mg+Fe (Fig. 2).

The LREE contents scatter over a wide range

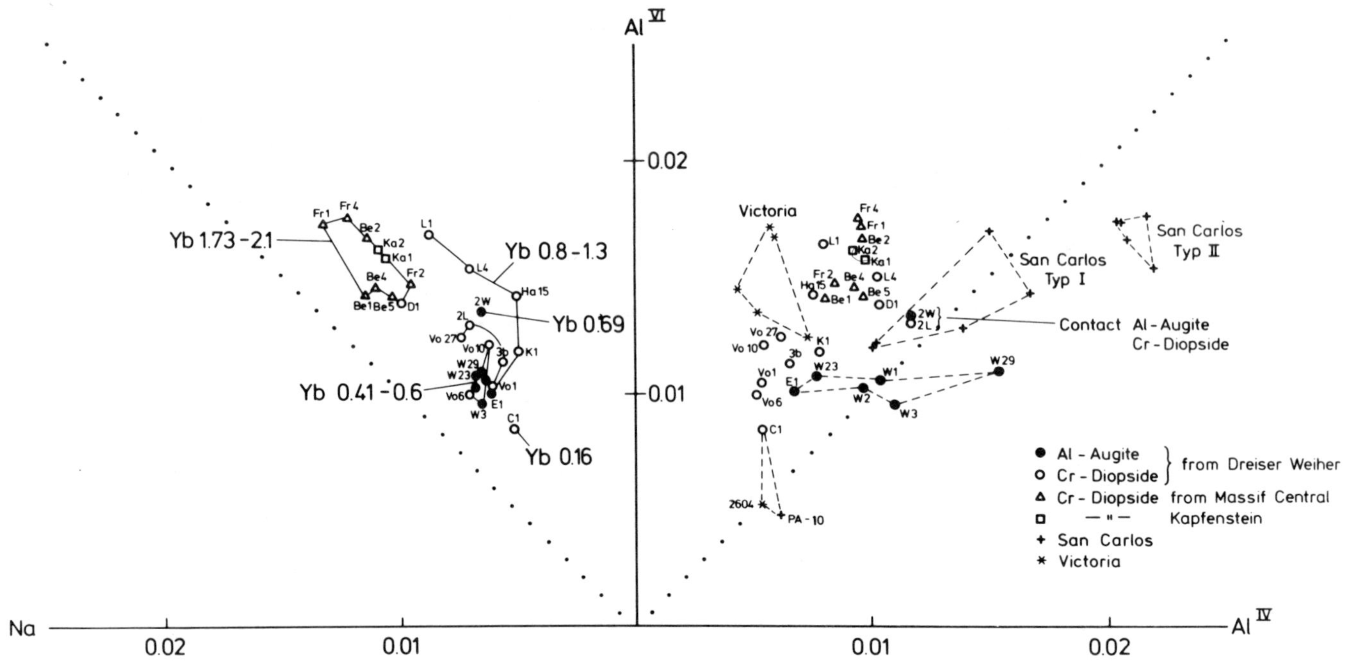

Fig. 1. Relative proportions of Al in the tetrahedral and in the octehedral site and Na in the M_1 position of Al-augites and Cr-diopsides. The lines with slope 1 and -1 represent the lines for the Tschermaks and jadeite component, respectively. Symbol o for Cr-diopsides from the Dreiser Weiher also includes those from Vogelsberg.

and show, independent of regional occurrences and the above described groupings either enrichment or depletion compared to the HREE (Fig. 2 and 5).

The Cr-diopside from the very refractory lherzolite (or harzburgite) C_1 from the Dreiser Weiher falls outside these groups by having very low Yb (Fig. 3) and high LREE contents (Table 2). It is very similar in this respect to the samples 2604 of Frey and Green (1974) and PA-10 of Frey and Prinz (1978) (see Fig. 1). These similarities are expressed further by the bulk composition of the nodule in having low CaO (0.52%), low Lu (0.035 ppm) and high La/Yb (10.6).

C. The Composite Nodule (composite wehrlite = CW, and composite lherzolite = CL from Dreiser Weiher). In this nodule lherzolite and wehrlite portions are in a planar contact to each other. Clinopyroxenes from either side of the contact (~1 cm distance) were analyzed by microprobe and INAA (instrumental neutron activation analysis) and microprobe analyses alone on a thin section across the contact. The Mg/(Mg+Fe) is very similar on both sides of the contact (Table 2) and very uniform throughout the nodule. The lherzolite part (CL) shows the lowest Mg/Mg+Fe of the Dreiser Weiher lherzolites for all phases, except for spinel (ol 88.77, opx 88.34, cpx 87.08, sp 69.48). Also in contrast to other lherzolites olivine has a higher Mg/Mg+Fe than orthopyroxene and

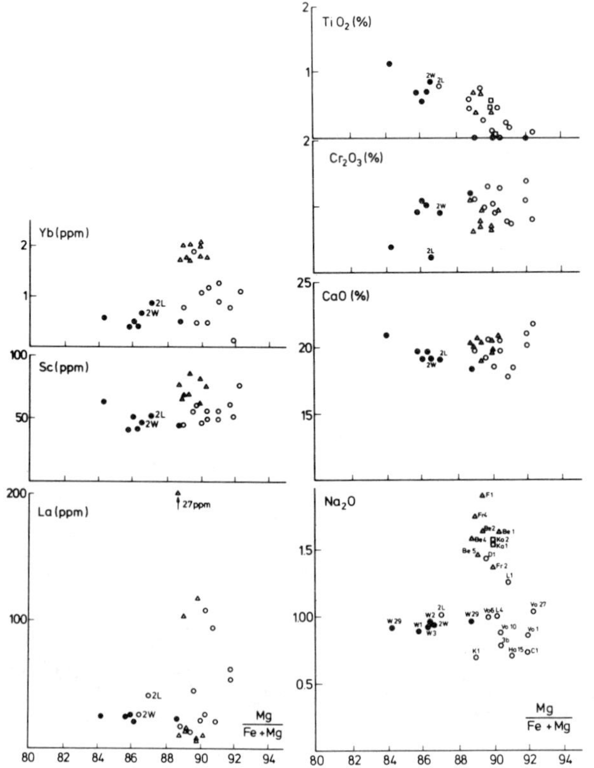

Fig. 2. Mg/Mg+Fe versus various oxides and elements in clinopyroxenes from European ultramafic nodules. Symbols are as in Fig. 1.

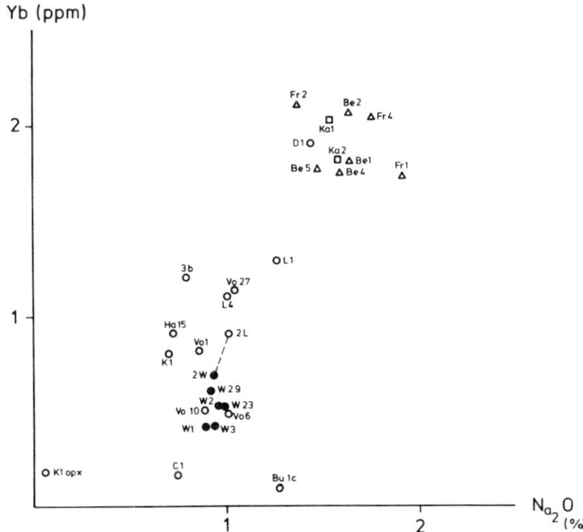

Fig. 3. Yb versus Na_2O for clinopyroxenes from European ultramafic nodules. Symbols are as in Fig. 1.

clinopyroxene which means a relatively lower temperature of equilibration (see e.g., Mori and Green, 1978). In contrast to the uniformity of M^{2+} ions, M^{3+} ions change rapidly over a short distance across the contact. A similar behavior of M^{2+} and M^{3+} ions was found by Wilshire and Shervais (1975).

Discussion and Conclusion

The main subject of investigation in this study was the major and trace element chemistry of clinopyroxenes from ultramafic nodules from

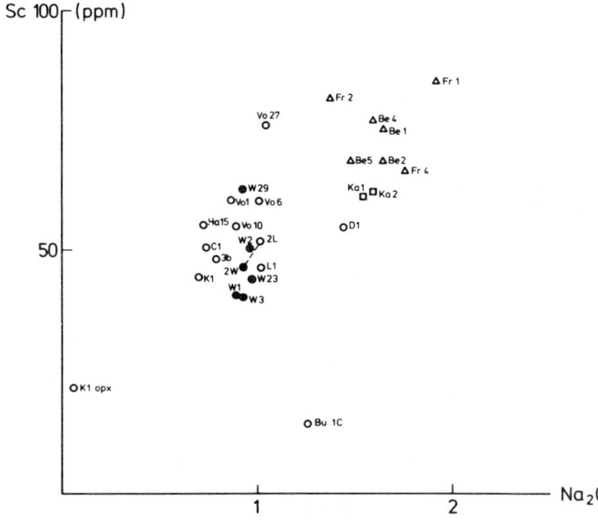

Fig. 4. Sc versus Na_2O for clinopyroxenes from European ultramafic nodules. Symbols are as in Fig. 1.

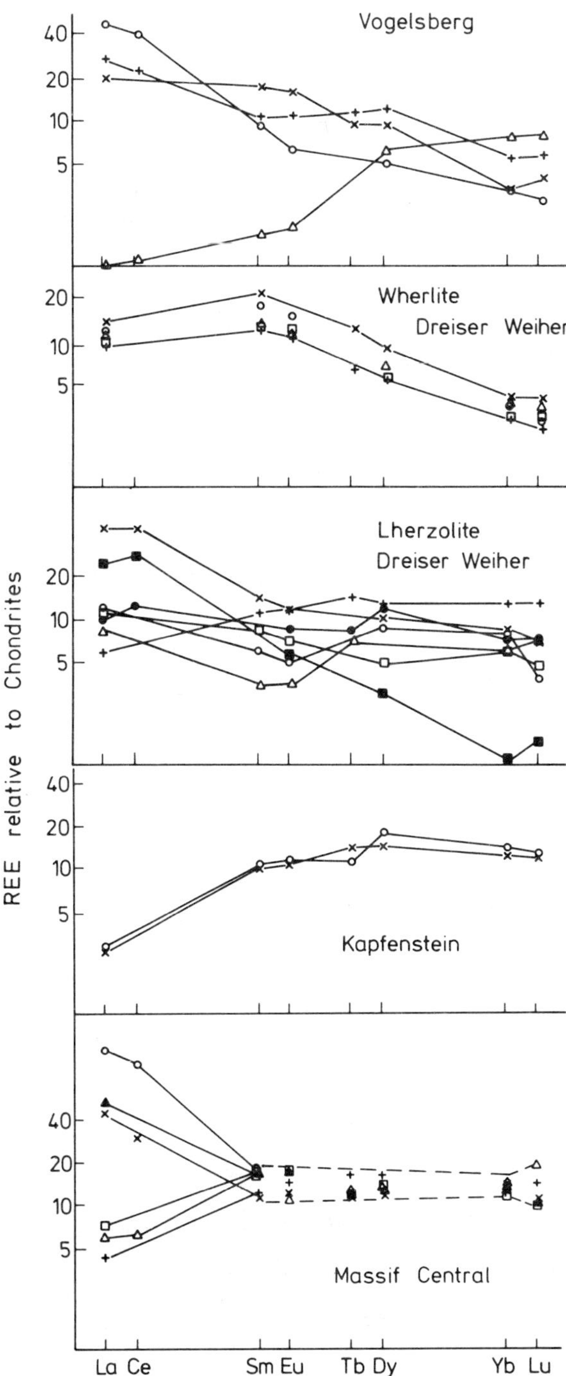

Fig. 5. REE patterns of clinopyroxenes from European ultramafic nodules.

four localities in Europe. Since the major part of the REE in ultramafic inclusions resides in clinopyroxene (e.g., Frey and Green, 1974) its composition reflects--in equilibrium--the composition of bulk nodule. Moreover, the very clean mineral separates used here have the advantage that contamination by REE

TABLE 2. Major and Trace Elements of Cr-Diopsides from Lherzolites

	HA15	L4	L1	D1	C1	3b	K1	Klopx	2L	2W	Vo27	Vo10
SiO_2	53.79	52.52	53.36	52.68	54.22	52.91	53.15	54.55	52.19	51.87	53.03	54.28
TiO_2	.16	.18	.23	.27	-	-	-	-	.78	.86	.07	-
Al_2O_3	5.16	5.94	5.86	5.71	3.26	4.15	4.61	4.38	5.80	5.89	4.32	4.14
Cr_2O_3*	.75	.92	.77	1.00	1.39	1.29	1.13	.84	.92	.23	.81	1.03
FeO	3.13	3.32	3.09	3.47	2.81	3.16	3.74	5.77	4.32	4.52	2.32	3.16
MnO*	.113	.10	.10	.14	.082	.094	.097	.10	.115	.115	.08	.087
MgO	17.94	17.04	17.12	16.63	17.85	16.67	16.92	32.43	16.35	16.24	15.50	16.59
CaO	18.44	18.48	17.77	19.17	20.15	19.71	19.69	1.30	19.03	19.19	21.76	20.52
Na_2O*	.72	1.01	1.26	1.44	.74	.79	.7	.08	1.01	.93	1.04	.89
Total	100.20	99.68	99.56	100.45	100.50	98.77	100.03	100.14	100.51	99.84	98.93	100.69
Si	1.924	1.897	1.920	1.896	1.946	1.934	1.921	1.897	1.882	1.883	1.938	1.946
Al	.076	.103	.080	.104	.054	.066	.079	.103	.117	.117	.062	.054
Al	.142	.150	.168	.138	.084	.113	.118	.076	.129	.136	.124	.121
Ti	.004	.005	.006	.007	-	-	-	-	.021	.023	.002	-
Cr	.021	.026	.022	.028	.039	.037	.032	.023	.026	.007	.023	.029
Fe	.093	.100	.093	.104	.084	.097	.113	.168	.130	.137	.071	.095
Mn	.003	.003	.003	.004	.002	.003	.003	.003	.004	.004	.002	.003
Mg	.956	.917	.918	.892	.954	.908	.911	1.680	.876	.878	.844	.886
Ca	.707	.715	.685	.739	.775	.772	.762	.048	.735	.746	.852	.788
Na	.050	.071	.088	.100	.051	.056	.049	.006	.071	.065	.074	.062
Total	3.976	3.987	3.983	4.012	3.989	3.986	3.988	4.004	3.991	3.996	3.992	3.984
$\frac{Mg}{Mg\ Fe}$	.9108	.9014	.908	.8952	.9188	.9038	.8896		.8709	.8649	.9225	.9034
Sc	55.5	46.4	49	55	51	49	45		52	47	76.3	55.3
Co	31.5	33.2	28	34	28	30	28		35	30	22.6	34.8
Ni	500	500	332	500	380	nd	nd		300	270	nd	310
Hf	0.25	.5	.35	.58	nd	nd	nd		1.8	1.1	nd	1.16
La	2.42	2.22	9.5	1.3	5.40	2.7	1.7		4.15	2.64	.23	10.8
Ce	7.00	7.1	24	nd	15	nd	nd		nd	nd	nd	23
Sm	1.2	1.14	1.9	1.5	1.10	.83	.4		2.56	2.30	.22	1.26
Eu	0.4	.46	.63	.6	.30	.27	.15		.75	.75	.1	.34
Tb	.23	.3	.24	nd	nd	nd	nd		.41	.3	nd	nd
Dy	1.00	2.6	2.31	2.9	.57	2.0	nd		nd	nd	1.4	1.16
Yb	.9	1.1	1.29	1.9	.16	1.2	.80		.9	.69	1.13	.5
Lu	.12	.17	.16	.3	.03	.09	.10		.2	.14	.18	.064

sitting on grain boundaries is excluded. A further aim for the work on such mineral separates was to evaluate whether the partitioning of trace elements into clinopyroxene depends on its major element composition.

Ultramafic inclusions may originate as cumulates from basaltic compositions or may be residues from a partial melting process in the upper mantle. Wehrlitic inclusions with Al-augite are generally thought to represent cumulates either cognate or accidental (with or without metamorphic features). This origin (as cumulates) is most likely for the wehrlites from Dreiser Weiher presented in this paper since, with decreasing Mg/(Mg+Fe), TiO_2, CaO and REE increase and Cr_2O_3 decreases--all features consistent with an origin as cumulates from a liquid. The (calculated) REE pattern (using partition coefficients from Consolmagno and Drake (1977)) for a possible coexisting liquid is very similar to that of the host basinite (Philpotts, et al., 1972).

There is no clearcut boundary between wehrlite and some types of spinel lherzolites (Fig. 1), neither in mineralogy, texturally and chemistry so that for some lherzolites a cumulative origin is conceivable.

Composite nodules of lherzolite + wehrlite (like the sample (CW+CL)) could possibly illuminate the genetic relationship, but an interpretation is very difficult. Our sample (CW+CL) is metamorphically overprinted. While the M^{2+} ions apparently are equilibrated throughout the nodule, the M^{3+} ions vary rapidly within

TABLE 2. Cont'd.

	Vo6	Vo1	Ka2	Ka1	Fr4	Fr2	Fr1	Be5	Be4	Be2	Be1
SiO_2	54.21	53.89	52.90	52.94	52.24	53.45	52.60	52.98	52.96	52.33	53.04
TiO_2	.12	–	.46	.56	.69	.34	.66	.38	.58	.75	.46
Al_2O_3	3.56	3.72	5.99	6.03	6.28	5.45	6.28	5.64	5.63	6.13	5.23
Cr_2O_3*	1.3	1.09	.69	.65	.62	.70	.789	.945	1.093	.705	.947
FeO	3.37	2.64	3.11	3.06	3.19	3.27	3.24	3.27	3.41	3.09	2.88
MnO*	.096	.085	.093	.089	.085	.086	.104	.083	.099	.079	.079
MgO	16.51	16.74	15.50	15.28	14.31	16.35	15.19	15.49	15.00	14.43	14.98
CaO	20.57	21.01	19.83	20.49	20.01	19.64	18.91	20.67	20.34	20.34	20.85
Na_2O*	1.006	.86	1.58	1.54	1.752	1.37	1.91	1.47	1.59	1.64	1.64
Total	100.74	100.03	100.15	100.63	99.17	100.65	99.68	100.92	100.7	99.49	100.1
Si	1.948	1.946	1.907	1.902	1.905	1.916	1.904	1.903	1.907	1.903	1.919
Al	.052	.055	.093	.098	.095	.084	.096	.097	.093	.097	.081
Al	.099	.104	.162	.158	.175	.146	.172	.142	.146	.166	.142
Ti	.003	–	.012	.015	.019	.009	.018	.010	.016	.021	.013
Cr	.037	.031	.020	.018	.018	.020	.023	.027	.031	.020	.027
Fe	.101	.080	.094	.092	.097	.097	.098	.098	.103	.094	.087
Mn	.003	.003	.003	.003	.003	.003	.003	.003	.003	.003	.003
Mg	.884	.900	.832	.818	.777	.873	.819	.829	.805	.782	.807
Ca	.792	.812	.767	.789	.781	.754	.733	.795	.784	.792	.808
Na	.070	.060	.110	.107	.124	.095	.134	.102	.111	.115	.115
Total	3.989	3.991	3.991	4.000	3.994	3.997	4.000	4.006	3.999	3.993	4.002
$\frac{Mg}{Mg\ Fe}$	.8972	.9187	.8988	.8990	.8888	.889	.893	.8904	.8868	.8927	.9026
Sc	60.8	60.8	62.3	61.4	66.5	82	85.22	68.8	77	69	75.4
Co	34.7	35	24.2	25.2	24.1	23	26	21	23	21	20.5
Ni	nd	730	230	340	470	nd	410	300	300	300	230
Hf	2.03	.65	1.25	.96	1.2	nd	1.22	nd	nd	nd	nd
La	4.5	6.19	.59	.65	1.00	11.7	1.6	10	27.6	1.33	1
Ce	nd	13	nd	nd	nd	nd	nd	17	55	3.6	nd
Sm	2.35	1.42	1.43	1.46	1.67	2.51	2.26	1.53	2.36	1.49	1.2
Eu	.85	.59	.59	.61	.76	.97	.94	.65	.93	.63	.56
Tb	.33	.4	.49	.39	.57	nd		.41	.38	.52	.42
Dy	2.1	2.8	3.3	4.1	3.7	2.6	3.1	3.25	2.6	3.9	3.2
Yb	.5	.82	1.81	2.02	2.03	2.1	1.73	1.77	1.74	2.05	1.8
Lu	.092	.13	.28	.29	.32	nd	.42	.25	.22	.31	.22

*Analyzed by NAA

a short distance from the contact. This may be due to very different diffusion rates and thus the M^{3+} ions may have preserved a pre-metamorphic history of the inclusion.

Geochemical and mineralogical evidence (for arguments see e.g., Frey and Prinz, 1978) suggests that most lherzolites represent residua from various degrees of partial melting. Such a simple story becomes complicated by the fact that the supposedly most residual nodules (i.e., nodules with the highest Mg/Mg+Fe, and lowest CaO and Al_2O_3) show a strong enrichment in LREE and generally an enrichment in large ion lithophile elements (Frey and Green, 1972 and Frey and Prinz, 1978). Frey and Prinz (1978) suggest the following as an origin of these LREE enriched nodules: after a partial melting event, this depleted mantle material was contaminated (and equilibrated at subsolidues temperatures) with an ascending liquid, rich in incompatible elements and with a high La/Yb as a result from small degree of partial melting in the garnet peridotite stability field. Because of the common occurrence of such nodules, similar, so far poorly understood processes must go on worldwide.

The clinopyroxenes from European lherzolite nodules fall into two groups with respect to their jadeite and Tschermaks component (Fig. 1) and Yb and Sc (Fig. 3 and 4). They are similar in CaO, TiO_2 and vary irregularly in abundances of LREE. All of our samples from France and Austria fall into the Al^{VI}, Na, Yb and Sc rich group, the samples from Germany form two groups. This may be a sampling problem on our part or, what we favor, reflect the sampling from different areas in the upper mantle (different P, T conditions and maybe chemical compositions) by the volcanic processes. The sampling of different portions of the upper mantle under different physical conditions is also suggested by the formation of separate groups by the Victorian and San Carlos lherzolites with respect to the Tschermaks molecule and jadeite component.

One also may say that the jadeite component (presented as Na_2O in place of jadeite in Figs. 3 and 4) correlates positively with the Yb and Sc content of the Cr-diopsides. The Yb-Na_2O correlation (Fig. 3) roughly goes through the origin whereas the Sc axis (Fig. 4) is cut at values greater than zero. The Yb content in the Cr-diopsides does not correlate with the Tschermaks component (Al^{IV} corrected for Cr and Ti) whereas Sc increases with Al^{IV} (both not shown in a figure here). We therefore conclude that the Yb content in clinopyroxenes is a function of the jadeite component and that Yb is six-fold coordinated in the M_2 position. Part if not all Sc goes on the same place, but because of the additional dependence on the Tschermaks molecule it may also replace Al in the four-fold coordination. Both the Tschermaks and jadeite component in clinopyroxene are a function of pressure, temperature and composition and thus we conclude that the uptake of trace elements like Yb and Sc is also a function of P, T and composition. These preliminary results suggest a compositional dependence of Henry's Law constant, different for different M^{3+} ions. Thus, as Wood and Fraser (1976) suggest, the study of both major and trace elements may lead us very much further in our understanding of the behavior of trace elements. Orthopyroxene K1 falls on the correlation trends for the clinopyroxenes which suggests similar dependences on major element chemistry.

The Cr-diopside from a garnet lherzolite from Bulfontein, Bu 1c, (Fig. 3 and 4, own unpublished results) has very much lower Yb and Sc, but higher Na_2O than those from spinel lherzolites. As it is known from other studies (e.g., Philpotts et al., 1972) the presence of garnet changes the partitioning of HREE and M^{3+} ions drastically.

We plan to extend our studies to the other coexisting minerals and into garnet peridotites. Thus, we should get a better understanding on the behavior of trace elements as a function of major elements and hopefully physical conditions.

Acknowledgements. We would like to thank Dr. G. Brey for fruitful discussions and his help with the manuscript. F. Kunstler prepared the thin section, J. Huth and B. Spettel helped to carry out microprobe and neutron activation analyses. We extend our thanks to all of them.

Appendix

All ultramafic inclusions investigated were brought to light by the tertiary (Massif Central, Vogelsberg, Kapfenstein) and quarternary (Dreiser Weiher) basaltic volcanism in Europe. Only a brief description of texture and the petrography of the nodules is given here.

Dreiser Weiher (Nodules included in basanitic tuffs).

A. Lherzolites:

- D_1 Lherzolite, tabular equigranular texture. Some porphyroclastic olivines up to 5 mm. Redbrown spinel, orthopyroxene and clinopyroxene. Fine partly deformed exsolution lamellae in clinopyroxene. Veins with some glass.
- Ha_{15} Pyroxenite (<10% recrystallized olivine ~0.5 mm). Kink bands are very rare; orthopyroxene and dominantly clinopyroxene. No spinel was found.
- L_1 Lherzolite, recrystallized with abundant clinopyroxene and orthopyroxene. Olivine <1 mm. Some foliation, spinel.
- L_4 Lherzolite, tabular equigranular texture. Brown spinel, clinopyroxene and orthopyroxene.
- 3b Lherzolite, mainly recrystallized with some olivine up to 5 mm. Kink band boundaries are common, foliation, redbrown spinel, clinopyroxene and orthopyroxene with few exsolution lamellae.
- K_1 Lherzolite, recrystallized with porphyroclastic olivine relicts up to 10 mm. Redbrown spinel, in some cases associated with clinopyroxene and orthopyroxene.
- C_1 Harzburgite, tabular recrystallized, foliation, red spinel, orthopyroxene and very little clinopyroxene.

B. Wehrlites:

- W_1 Wehrlite with inhomogeneous distribution of clinopyroxene. In the dunitic part triple point junctions and kink bands are common. In some parts poikilitic clinopyroxene is dominant. Clinopyroxene shows wavy extinction. Glass-filled veins are very common. Some exsolution lamellae of orthopyroxene.
- W_2 Wehrlite with little recrystallized olivine. Some exsolution lamellae in the clinopyroxene.
- W_3 This olivine rich wehrlite is partly recrystallized and the olivine shows undulose extinction. On the irregularly

shaped grain borders fine grain recrystallization appear. Brown spinel and some glass.

W_{23} Poikilitic clinopyroxene with some olivine and spinel. Clinopyroxenes are up to some cm in dimension, veins with some glass.

W_{29} Clinopyroxenite. Clinopyroxene is dark and up to 3 cm. Some olivines are present and no spinel was found. Larger olivine crystals show kink band boundaries. Veins up to 5 mm are very common and partly filled with brown glass.

E_1 Clinopyroxenite with some orthopyroxene neoblasts, but mainly clinopyroxene and few olivine. Recrystallization very common. Thin glass-filled veins and some spinel.

Vogelsberg (Nodules included in olivine-nephelinite-lava-lake).

Vo_1 Lherzolite, granoblastic olivine ~4 mm, kink band boundaries. Slight foliation, rich in orthopyroxene, clinopyroxene less abundant, some recrystallization. Red brown spinel.

Vo_6 Granoblastic lherzolite, olivine ~5 mm, kink band boundaries common, recrystallization common, orthopyroxene, clinopyroxene and spinel.

Vo_{27} Lherzolite granoblastic, olivine ~5 mm, kink band boundaries very common, grey-brown spinel, few recrystallization, fine spinel exsolutions in the pyroxene.

Vo_{10} Lherzolite granoblastic, olivine ~5 mm, kink band boundaries very common, pronounced foliation, red brown spinel.

Massif Central (Nodules included in basanitic tuffs).

Landoz

Fr_1 Granoblastic lherzolite, olivine ~5 mm, foliation, recrystallization common, some tabular neoblasts, brown spinel.

Fr_2 Like Fr_1, but clinopyroxene appear more curvilinear, clinopyroxene and orthopyroxene have reaction rims, spinel.

Chauderolles

Fr_4 Granoblastic lherzolite, olivine ~4 mm, some kink band boundaries, orthopyroxene and clinopyroxene show exsolution lamellae, grey-brown spinel.

Beyzac

Be_1 Lherzolite recrystallized, kink band boundaries common, spinel, clinopyroxene and orthopyroxene wavy extinction, exsolution lamellae in orthopyroxene and clinopyroxene, glass-filled veins

Be_2 Lherzolite, mosaic recrystallization, orthopyroxene and clinopyroxene, abundant spinel, glass-filled veins.

Be_4 Similar to Be_1.

Be_5 Granular lherzolite with some recrystallization, olivine, orthopyroxene, clinopyroxene, abundant vermicular spinel, clinopyroxene have some exsolution lamellae, veins with glass containing small crystals.

Kapfenstein (Nodules included in tuffs).

Ka_1 Lherzolite, granoblastic, olivine ~3 mm, kink band boundaries abundant, grey-brown spinel, some recrystallization, exsolution lamellae in orthopyroxene, clinopyroxene.

Ka_2 Lherzolite, granoblastic, olivine up to 6 mm, kink band boundaries, grey-brown spinel, fine exsolution lamellae in orthopyroxene, glass-filled veins, clinopyroxene.

Dreiser Weiher (Composite nodule)

CL Lherzolitic part, olivine ~3 mm, partly recrystallized, red spinel, cpx, opx.

CW Wehrlitic part, olivine in poikilitic clinopyroxene; foliation plane in both parts of the nodule about normal to the planar contact.

References

Aoki K. and I. Kushiro, Some clinopyroxenes from ultramafic inclusions in Dreiser Weiher, Eifel, Contrib. Mineral. Petrol., 18, 32, 1968.

Becker, H. J., Pyroxenites and Hornblendites from the Maar-type volcanoes of the Westeifel, Federal Republic of Germany, Contr. Mineral. Petrol., 65, 45, 1977.

Consolmagno, G. J. and M. J. Drake, Composition and evolution of the eucrite parent body: Evidence from rare earth elements. Geochim. Cosmochim. Acta, in press, 1977.

Ernst, T. G., Olivinknollen der Basalte als Bruchstücke alter Olivinfelse. Nachr. Akad. Wiss. Göttingen, Math. Phys. Kl, 1, 147 (1935).

Frechen, J., Die Genese der Olivinausscheidungen vom Dreiser Weiher und Finkenberg (Siebengebirge), N. Jb. Mineral. Geol. Paleontol. 79, 317, 1948.

Frechen, J., Kristallisation, Mineralbestand, Mineralchemismus und Förderfolge der Mafitite vom Dreiser Weiher in der Eifel. N. Jb. Mineral. Mh. 9-10, 205, 1963.

Frey, F. A. and D. H. Green, The mineralogy, geochemistry and origin of lherzolite inclusions in Victorian basanites. Geochim. Cosmochim. Acta 38, 1023, 1974.

Frey, F. A. and M. Prinz, Ultramafic inclusions from San Carlos, Arizona: Petrologic and geochemical data bearing on their petrogenesis. Earth Planet. Sci. Lett. 38, 129, 1978.

Forbes, R. B. and H. Kuno, The regional petrology of peridotite inclusions and basaltic host rocks. Proc. 22nd Intern. Geol. Congr., Upper Mantle Symp. New Delhi, 161, 1965.

Hutchison, R., A. L. Chambers, D. K. Paul, and P. G. Harris, Chemical variation among French ultramafic xenoliths evidence for a heterogeneous upper mantle. Mineral. Mag. 40, 153, 1975.

Mori T. and D. H. Green, Laboratory duplication of phase equilibria observed in natural garnet lherzolites, J. Geol. 86, 83, 1978.

Paul, D. K., Strontium isotope studies on ultramafic inclusions from Dreiser Weiher, Eifer, Germany, Contrib. Mineral. Petrol. 34, 22, 1972.

Philpotts, J. A., C. C. Schnetzler and H. H. Thomas, Petrogenetic implications of some new geochemical data on eclogitic and ultrabasic inclusions, Geochim. Cosmochim. Acta 36, 1131, 1972.

White, R. W., Ultramafic inclusions in basaltic rocks from Hawaii, Contrib. Mineral. Petrol. 12, 245, 1966.

Wilshire, H. G. and J. W. Shervais, Al-augite and Cr-diopside ultramafic xenoliths in basaltic rocks from western United States, Phys. Chem. Earth 9, 257, 1975.

Wänke, H., H. Baddenhausen, G. Dreibus, E. Jagoutz, H. Kruse, H. Palme, B. Spettel, and F. Teschke, Multielement analyses of Apollo 15, 16 and 17 samples and the bulk composition of the moon, Proc. Lunar Sci. Conf. 4th, 1461, 1973.

Wood, B. J. and Fraser, D. G., Elementary thermodynamics for geologists. Oxford Press, 1976.

GEOCHEMISTRY OF ULTRAMAFIC XENOLITHS FROM SAN QUINTIN, BAJA CALIFORNIA

Asish R. Basu

Department of Geology and Geophysics, University of Minnesota, Minneapolis, Minnesota 55455

Abstract. The major element chemistry, the abundances of K, Rb, Sr, Ba, and the $^{87}Sr/^{86}Sr$ ratios of separated silicate minerals in a suite of ultramafic xenoliths and plagioclase megacryst in alkalic basalts from San Quintin, Baja California, are reported. Data also include the trace elements and $^{87}Sr/^{86}Sr$ ratios in two host alkalic basalts. Although the xenoliths are enriched in modal diopsides, the trace element abundances are extremely low in all the silicate phases. For example, the diopsides contain as low as 10.7 ppm K, 0.01 ppm Rb, 1.7 ppm Sr and 0.6 ppm Ba. The diopsides from the pyroxenite layers usually show slightly higher abundances, such as 107 ppm K, 0.07 ppm Rb, 22 ppm Sr and 33 ppm Ba.

In four lherzolites, the coexisting olivines, orthopyroxenes, and clinopyroxenes were analyzed after separation by hand-picking. In three of these samples, each mineral fraction was analyzed twice - without washing and after washing in 2N cold HCl for three minutes. In the acid washed minerals (AW), K, Rb, Sr, and Ba abundances are reduced by about half from the unwashed abundances. In addition, $^{87}Sr/^{86}Sr$ ratios are also lowered by the acid washing. In one of these samples, the acid washed minerals and the acid washed whole rock reveal an isochron of $T = 3.41 \pm 0.3$ (2σ) and $I = 0.70057 \pm 0.0004$ (2σ). In the other three samples the coexisting silicates show clear disequilibrium in their $^{87}Sr/^{86}Sr$ ratios.

The clinopyroxenes from the different xenoliths show the lowest and the most variable $^{87}Sr/^{86}Sr$ ratios of all the silicates, from .70196±11 to .70445±5. Thus the data indicate that at least part of the vertical mantle profile beneath San Quintin, represented by the xenoliths studied here, is heterogeneous in its Sr-isotopic ratios. This heterogeneity must reflect complex processes in the mantle, such as partial melting, removal of melt, cumulus processes, recrystalization, etc.

The extreme depletion of the trace elements in the silicate phases of the lherzolites, as reported here, contrasts remarkably with the diopside-rich nature of the xenoliths. No simple scheme of partial melting of these xenoliths can produce any normal basalt with its appropriate trace elemental abundances and ratios.

(Present address: Department of Geological Sciences, University of Rochester, Rochester, New York 14627)

Introduction

The San Quintin volcanic field is located approximately 260 kilometers south of the United States border along the west coast of Baja California in Mexico. It consists of a dozen basaltic cinder cones together with their associated lava flows. The series of eruptions probably commenced in late Pleistocene and continued to Recent in age; the last volcanic activity may even be historic (Woodford, 1928).

Mafic and ultramafic inclusions occur abundantly mostly in the northern part of the field in both cones and flows as blocks and cores of some bombs in agglutinate and agglomerate beds and lava flows. The host lavas are mostly nepheline normative olivine basalts (Bacon and Carmichael, 1973). The xenoliths are predominantly therzolites with some less abundant dunites, pyroxenites, gabbros and green spinel bearing granulites. The therzolites consist of varying proportions of olivine, enstatite, green chrome diopside and chromite spinels. The largest xenoliths may be as large as 40 cms in their greatest dimension. The larger xenoliths occasionally show subangular to rounded margins; the smaller xenoliths show a tendency to be ellipsoidal.

The modal proportions of olivine, orthopyroxene and clinopyroxene of the xenoliths (when plotted in a conventional triangular plot for ultramafic rocks) show that a great majority of the ultramafic xenolith population in San Quintin are therzolites, followed by dunites, websterites and harzburgites. Deformational features are commonly observed in these xenoliths. The typical therzolites show evidence of deformation in the hand specimen by flattened and stretched minerals, resulting in a strongly foliated aspect. A porphyroclastic texture is common in the lherzolites and is defined by elongated or flattened olivine, orthopyroxene, clinopyroxene and spinel porphyroclasts in a fine grained equigran-

TABLE 1. Major Element Composition of the Minerals.

	SQ 2-82				SQ 2-13			
	OL	OPX	CPX	SP	OL	OPX	CPX	SP
SiO_2	41.84	53.45	52.10		41.29	54.04	55.04	
TiO_2		0.11	.43	.09		0.13	.37	.08
Al_2O_3		6.13	6.42	56.98		5.00	4.72	56.08
Cr_2O_3		0.54	.80	9.85		0.56	.55	8.91
FeO (total)	9.83	6.43	2.83	10.94	10.83	6.70	2.96	10.91
MnO	0.12	0.14	.08	.14	0.14	0.12	.08	.13
MgO	48.86	31.81	16.03	19.76	46.57	32.05	15.70	22.79
NiO	0.32		.02	.38	0.39		.02	.39
CaO		1.05	20.11			1.08	19.69	
Na_2O			.99				1.39	
K_2O								
Total	100.96	99.67	99.80	98.13	99.21	99.68	100.58	99.29
				Atomic Proportions				
Oxygen	4	6	6	4	4	6	6	4
Si	1.014	1.858	1.872		1.022	1.880	1.965	
Ti		0.002	.012	.002		.003	.011	.002
Al		0.251	.272	1.772		.204	.197	1.724
Cr		0.014	.023	.205		.015	.017	.184
Fe	0.199	0.187	.085	.242	0.224	.194	.088	.238
Mn	0.002	0.004	.002	.003	0.002	.003	.002	.003
Mg	1.765	1.649	.859	.777	1.718	1.662	.834	.886
Ni	0.005		.001	.008	0.007		.001	.008
Ca		0.039	.774			.040	.753	
Na			.069				.096	
K								
Total	2.984	4.003	3.969	3.009	2.973	4.000	3.968	3.045
Mg/Mg+Fex100	89.9	89.8	91.0	77.9*	88.5	89.5	90.4	87.3*

*for spinels $(Mg/Mg+Fe^{2+}) \times 100$

TABLE 1 contd.

	SQ 2-41				SQ 2-67			
	OL	OPX	CPX	SP	OL	OPX	CPX	SP
SiO_2	41.55	54.95	51.11		41.54	53.94	51.97	
TiO_2		0.13	0.43	0.12		0.11	0.34	.09
Al_2O_3		4.18	7.18	57.72		5.11	5.60	54.34
Cr_2O_3		0.48	0.94	10.56		0.58	.82	13.03
FeO (total)	9.82	6.46	2.73	10.24	9.02	5.79	2.34	10.25
MnO	0.13	0.12	0.08	0.12	0.11	.13	.08	.16
MgO	47.64	33.50	15.00	20.88	49.45	33.37	15.07	19.97
NiO	0.31		0.02	0.35	0.33		.04	.34
CaO		0.84	20.34			0.57	20.34	
Na_2O			1.40				1.24	
K_2O								
Total	99.44	100.66	99.23	99.99	100.45	99.61	97.82	98.18
				Atomic Proportions				
Oxygen	4	6	6	4	4	6	6	4
Si	1.022	1.889	1.848		1.009	1.869	1.910	
Ti		0.003	0.012	0.002		.002	.009	.002

TABLE 1 contd.

	SQ 2-41				SQ 2-67			
	OL	OPX	CPX	SP	OL	OPX	CPX	SP
Al		0.168	0.306	1.758		.208	.241	1.703
Cr		0.013	0.027	0.216		.015	.024	.274
Fe	0.201	0.186	0.083	0.221	0.183	.167	.072	.228
Mn	0.002	0.003	0.002	0.003	0.002	.003	.002	.003
Mg	1.746	1.716	0.809	0.804	1.790	1.724	.822	.792
Ni	0.005		0.000	0.007	0.005		.001	.007
Ca		0.030	0.788			.020	.797	
Na			0.098				.088	
K								
Total	2.976	4.007	3.973	3.011	2.989	4.009	3.957	3.009
(Mg/Mg+Fe)x100	89.7	90.2	90.7	80.5*	90.7	91.2	92.0	79.4*

TABLE 1 contd.

	SQ 1-7-28 cpx	SQ 2-68 pxite layer + glass	SQ 2-30 cpxite layer	SQ 1-8-5 Black cpxite	SQ 2-81 Black cpxite	SQ 2-36 Opxite	SQ 1 - 6 - 1 OL	Plag	CPX	SP	SQ 3-1 plag
SiO_2	51.8	49.86	52.06	51.39	52.95	54.18	40.63	44.11	53.06	0.03	57.3
TiO_2	.53	1.12	.47	0.21	.21	.07			.16		
Al_2O_3	6.72	8.12	5 82	5.25	3.74	3.27		34.65	2.68	61.84	25.8
Cr_2O_3	.97	.00	.97	0.24	.24	.42		0.00	.37	0.16	
FeO (total)		2.86	5.18	2.33	5.56	6.37	7.52	20.56	5.58	20.60	0.20
MnO	.09	.18	.08	0.18	.20	.16	.28		.17	0.23	
MgO	15.90	15.60	15.57	14.24	15.36	32.26	39.20	0.05	15.70	15.76	
NiO	.08		.02		.03		.14		.05	0.13	
CaO	20.57	18.76	20.01	21.66	20.47	.66	.03	20.23	22.32	0.05	7.69
Na_2O	1.88	1.09	1.48	0.52	.30			0.25	0.31		6.47
K_2O											1.22
Total	100.78	99.91	98.86	99.25	99.86	98.53	100.84	99.29	100.42	98.80	98.68
Atomic Proportions											
Oxygen	6		6	6	6	6	4	8	6	4	32
Si	1.827		1.883	1.893	1.937	1.912	1.033	2.058	1.937	0.001	10.439
Ti	.014		.013	0.006	.006	.001			.004		
Al	.283		.248	0.128	.123	.135		1.906	0.115	1.941	5.540
Cr	.027		.028	0.007	.007	.011			0.011	0.003	
Fe	.085		.070	0.171	.195	.222	.437		0.170	0.440	.029
Mn	.003		.002	0.006	.006	.004	.006		0.005	0.005	
Mg	.846		.839	0.782	.837	1.697	1.486	.003	0.854	0.610	
Ni	.002		.001	0.000	.001		.003		0.001	0.003	
Ca	.787		.775	0.855	.802	.023	.001	1.011	0.873	0.001	1.500
Na	.130		.104	0.037	.021			0.022	.022		2.285
K											0.278
Total	4.004		3.966	3.984	3.973	4.005	2.967	5.000	3.995	3.004	20.071
(Mg/Mg+Fe)x100	90.9		92.3	82.1	81.1	88.4	77.3	An mol % 97.9	83.4	58.1	

Note: OL = olivine, opx = orthopyroxene, cpx = clinopyroxene, sp = spinel, and Plag = plagioclase

ular groundmass consisting of these four minerals. Spinel and orthopyroxene often clearly show lineation in the deformed specimens The textures and the deformational features of these xenoliths have been reported elsewhere (Basu, 1975a; Basu, 1977a,b).

The purpose of this paper is to report the major element chemistry and the trace element chemistry of K, Rb, Sr, Ba and the isotopic composition of Sr of separated minerals in a suite of ultramafic xenoliths from San Quintin. The data also include the trace element abundances and the $^{87}Sr/^{86}Sr$ ratios of two alkalic basalts - the hosts for the xenoliths, and of one megacryst of plagioclase feldspar.

Analytical Method

The analyses for the major elements were made with an ARL electron microprobe at the U.C. Davis Geology Department. Some of the analyses were performed with the automated microprobe. There was good agreement between the results obtained from the manually operated probe and the automated probe system. Mineral grains were separated by hand picking. To test homogeneity of the minerals, more than one grain was analyzed in several instances from the same rock. To check compositional zoning, the same grain was tested at several different points. No significant compositional zoning of one single grain or chemical variation from one grain to another in the same rock was noticeable in the case of olivine and spinels. However, many of the pyroxenes show inhomogeneities in their major element composition which are due to the exsolution of the pyroxenes in one another, and also due to the exsolution of $MgAl_2O_4$ spinels in the orthopyroxenes.

The sample no. SQ2-68 of a very fine grained pyroxenite layer with glass was analyzed using a defocussed electron beam.

For trace element analysis grains of olivine, orthopyroxene and clinopyroxene were separated by hand picking under a binocular microscope. Grains with clear surfaces and free from apparent inclusions and exsolution lamellae were chosen for analysis. The process was repeated several times after grinding the minerals into a finer grain size. The separates were then washed with distilled acetone dried and digested after further grinding in a boron carbide martar. The abundances of K, Rb, Sr, and Ba were determined by the isotope dilution method. Low-level blanks were essential for the determination of the extremely low concentrations of these elements. The average blanks for the entire chemical procedure were 20 ng of K, 0.1 ng of Rb, 1 ng of Sr and 6 ng of Ba. These blank corrections were applied for the data in Table 2.

A different set of analyses was made on the same mineral separates after washing in 2N cold HCl for three minutes followed by washing in distilled water and acetone, drying and digestion. This washing experiment was designed to remove any grain boundary contamination of the minerals by the host basalts or by other possible sources.

The mass spectrometric analyses were carried out at the University of Minnesota in the laboratory of Dr. V. R. Murthy on a 30-cm radius, single focusing mass spectrometer using a mini-computer based system with on-line data reduction and magnetic field switching capabilities. Operational procedures and statistical aspects of the data collection are described in Murthy et al (1971).

Major element chemistry

Electron microprobe analyses of the mineral phases in the xenoliths under discussion are given in Table 2. Although space does not permit extensive discussion of the major element chemistry of the coexisting minerals in the ultramafic xenolith suite from San Quintin, a brief summary of the salient aspects of the chemistry is in order. The following summary is based on 696 mineral analyses from 200 spinel lherzolite xenoliths by the electron microprobe (Basu, 1975b).

$\frac{Mg X 100}{Mg+Fe}$ in olivine (88.5-91.8); in orthopyroxene (86.5-92.0) and in spinel (68.5-88.5)

$\frac{Cr X 100}{Cr+Al}$ in spinel varies from (8.6-54.5)

Al_2O_3 Wt % in orthopyroxene (1.98-7.71); in clinopyroxene (2.6-8.98)

Na_2O Wt % in clinopyroxene (0.40-1.88)

Cr_2O_3 Wt % in clinopyroxene (0.27-1.48)

The Al_2O_3 content in the coexisting pyroxenes show a positive correlation. The Al_2O_3 contents of clinopyroxenes vary inversely with CaO contents whereas Na_2O and Cr_2O_3 show a positive correlation. Na_2O and Al_2O_3 also show a positive correlation in clinopyroxenes. The Cr_2O_3 content of the orthopyroxenes correlates negatively with the Al_2O_3 content of the coexisting clinopyroxene. The Cr_2O_3 content of the clinopyroxenes correlates negatively with the Al_2O_3 content of the ortho and clinopyroxenes. TiO_2 shows positive correlation in coexisting pyroxenes.

The variation in the distribution coefficients of Mg - Fe and Cr - Al partitioning between the various mineral pairs in the xenoliths are considered to be caused by one or more of the following: temperature variation, pressure variation, and bulk compositional variation, (Basu, 75a, 77a; MacGregor and Basu, 1976).

K, Rb, Sr, Ba contents and $^{87}Sr/^{86}Sr$ ratios in silicate minerals of the xenoliths

The trace-element contents and the $^{87}Sr/^{86}Sr$ ratios of the minerals and the rocks, under discussion, are given in Table 2. Some of the analyses have been reported elsewhere (Basu and Murthy, 1977a, b).

TABLE 2. Trace element concentrations and the isotopic composition of Sr in the minerals of the xenoliths and in the host lava.

Sample	Mineral	K ppm	Rb ppm	Sr ppm	Ba ppm	$^{87}Rb/^{86}Sr$	$^{87}Sr/^{86}Sr$
	OL	13.6	.042	.380	.424	.3195	.70709±25
2-82	OL(AW)	4.6	.018	.233	.114	.2233	-
Spinel	OPX	33.3	.059	.556	.280	.3068	.70680±18
Lherzolite	OPX(AW)	22.4	.034	.322	.172	.3053	-
	CPX	15.5	.049	1.951	1.490	.0726	.70340±12
	CPX	12.39	.038	1.703	.650	.0645	.70261±8
	OL	42.6	.100	2.643	16.00	.1094	.70723±5
2-13	OL(AW)	17.5	.044	1.301	9.71	.0978	.70555±12
Spinel	OPX	52.2	.110	3.552	13.02	.0895	.70765±10
Lherzolite	OPX(AW)	28.7	.083	2.976	9.98	.0806	.70717±14
	CPX	50.9	.090	2.857	9.6	.0911	.70528±12
	CPX(AW)	23.9	.050	2.136	7.07	.0677	.70484±10
	OL	39.3	.063	4.32	13.0	.0422	.70682±11
	OL(AW)	21.2	.035	.953	8.48	.1062	.70585±18
2-41	OPX	204.4	.186	5.74	8.82	.1094	.70702±7
Spinel	OPX(AW)	43.6	.057	1.51	6.34	.1091	.70579±7
Lherzolite	CPX	92.2	.123	11.17	12.72	.0318	.70545±12
	CPX(AW)	38.7	.054	5.295	12.58	.0295	.70196±11
	Whole Rock (AW)	33.6	.043	1.591	8.10	.0781	.70452±8
2-67	OL(AW)	22.6	.050	.189	.321	.7648	.70644±16
Spinel	OPX(AW)	4.00	.003	.338	.827	.0256	.70302±13
Lherzolite	CPX(AW)	10.70	.011	7.353	.597	.0043	.70230±6
1-7-28 Spinel Lherzolite	Cr-Diopside (AW)	63.14	.059	5.643	.732	.0302	.70275±9
2-68 Pxite layer in Spinel Lherzolite	CPX + glass (AW)	43.70	.050	32.83	.666	.0044	.70303±8
2-30 Cpxite layer	Cr-Diopside (AW)	107.9	.073	21.99	1.005	.0096	.70295±7
1-8-5 Black Cpxite (augite)	CPX (AW)	64.38	.049	18.707	-	.0076	.70413±8
2-81 Black Cpxite (augite)	CPX (AW)	31.61	.049	22.144	33.58	.0064	.70445±5
2-36 Sp-Lh W/ Opxite layer	OPX (AW)	94.96	.070	.449	.242	.4541	-
1-6-1	Plag (An_{98})	36.80	.042	342.32	16.65	.0004	.70407±7

TABLE 2 (continued)

Sample	Mineral	K ppm	Rb ppm	Sr ppm	Ba ppm	$^{87}Rb/^{86}Sr$	$^{87}Sr/^{86}Sr$
Plag - Peridotite	(AW) CPX(AW)	18.03	.046	13.04	4.03	.0102	.70425±13
3-1 Plag Megacryst Andesine	Plag (AW)	10591	12.776	2503	348.7	.0148	.70317±9
1-2-6 Host Basalt	Whole Rock (AW)	15390	35.96	611.7	380.6	.1699	.70311±8
SQB Host Basalt	Whole Rock (AW)	13853	26.58	560.7	315.8	.1371	.70314±7

Note: AW = washed in cold 2N HCl for three minutes.

The data in Table 2 include the abundances of K, Rb, Sr, Ba and the isotopic composition of Sr in the coexisting olivines, orthopyroxenes and clinopyroxenes of four lherzolites. In three of these lherzolites, each mineral fraction was analyzed twice - without washing and after washing the minerals in cold 2N HCl for three minutes. In the acid washed minerals (AW), K, Rb, Sr, and Ba abundances are reduced by about half from the unwashed abundances. In addition, $^{87}Sr/^{86}Sr$ ratios of the minerals are also considerably lowered by the acid washing experiment. In one of these lherzolites (no SQ 2-41), the analytical data for the acid washed minerals and the whole rock define a straight line relationship on a $^{87}Rb/^{86}Sr$ versus $^{87}Sr/^{86}Sr$ diagram. The slope corresponds to an age of 3.4±0.3 AE (2σ) and an initial of $^{87}Sr/^{86}Sr$ of 0.70057±0.0004 (2σ). In the other three samples, the coexisting silicates show clear disequilibrium in their $^{87}Sr/^{86}Sr$ ratios both in the acid washed and the unwashed minerals. In the case of the plagioclase peridotite, the coexisting clinopyroxene and plagioclase show essentially the same $^{87}Sr/^{86}Sr$ ratios within the (2σ) mean of each other.

The abundances of K, Rb, Sr, and Ba are very low in all the silicate minerals of the peridotite xenoliths. Although the lowest contents of K, Rb, Sr, and Ba are found in acid washed olivines and orthopyroxenes of the lherzolites, this generality does not always hold among the coexisting minerals of any individual xenolith. For example, in specimen no. SQ-2-67, the orthopyroxenes have 4.0 ppm K, while the olivines and clinopyroxenes have 22.6 and 10.7 ppm K, respectively.

K and Sr contents of the clinopyroxenes from the various types of rocks show a good positive correlation. However, the most striking trace element characteristics of the clinopyroxenes from the lherzolites is their low Sr contents. The range of Sr contents in these pyroxenes, from 1.7 ppm to 7.3 ppm, is possibly the lowest ever reported from similar pyroxenes. The higher contents of Sr, from 13 ppm to 32.8, are found in pyroxenes from layers within therzolites or in black clinopyroxenites, all of which have different major element chemistry compared to the lower Sr-bearing chrome-diopsides.

Of all the minerals olivines and orthopyroxenes, invariably, show higher $^{87}Sr/^{86}Sr$, and Rb/Sr ratios than the coexisting clinopyroxenes. The clinopyroxenes from the different xenoliths show the lowest Rb/Sr and $^{87}Sr/^{86}Sr$ ratios. Most of the chrome-diopsides from the xenoliths show $^{87}Sr/^{86}Sr$ ratios lower than 0.7030, whereas the iron-rich pyroxenes show much higher $^{87}Sr/^{86}Sr$ ratios.

Discussion

A significant aspect of the geochemistry of the constituent minerals in the xenoliths, reported here, is the contrasting nature between the major elements and the minor elements. This contrast is revealed by the general enrichment of Ca, Al, Fe, and Na in the whole xenolith (as indicated by the major element chemistry of the minerals and the generally higher modal abundances of diopside in the xenolith) and by the extreme depletion of the trace elements, K. Rb, Sr, and Ba in the minerals of the xenoliths. This apparent dichotomy between the major and trace elements might be explained in the following way. It is often argued that the large ion lithophile elements have no structural sites in the anhydrous silicate minerals. However, it is well documented now that some mantle-derived clinopyroxenes have K contents ranging between 400-800 ppm and more

(Shimizu, 1975; Sobolev, 1974; Basu and Murthy, 1977d). The high-K pyroxenes are considered to be stable in the upper mantle at pressures of the order of 50-150 kb (Sobolev, 1974). Along with K, the clinopyroxenes from the San Quintin xenoliths also show extreme depletion in Rb, Sr, and Ba. Therefore, a reasonable explanation is that if these elements were present in the silicate minerals of the xenoliths - particularly in clinopyroxenes, a subsequent partial melting episode had removed these elements from the xenoliths. However, the major element considerations require that the degree of this partial melting must have been very small. Yet another possibility is that the trace elements resided in a hydrous phase like amphibole in these xenoliths which was subsequently removed by partial melting.

The ubiquitous disequilibrium in $^{87}Sr/^{86}Sr$ ratios among the coexisting silicate minerals of the xenoliths may also be explained due to the growth of radiogenic Sr since an episode of partial melting. The Rb-Sr data of sample SQ-2-41 has been interpreted (Basu & Murthy 1977a) in that way. One difficulty of such an interpretation is that Sr should have equilibrium distribution ratios among the silicate minerals at the time of partial melting. Assuming that Sr behaves like Ca, the distribution of Sr among the acid washed olivine, orthopyroxene and clinopyroxene in specimen no. SQ 2-41 does not suggest an equilibrium distribution. However, at the low concentration levels of Sr in the sample SQ 2-41, it is not at all certain whether Sr should behave like Ca. It is quite possible that at low concentrations the trace element distributions among coexisting minerals are controlled by defect structures in the crystal lattice.

The acid washing experiment of the three lherzolites in Table 2 also bear an important implication regarding the residence sites of the trace elements. It is clear from this data that a substantial amount of these elements are along the grain boundaries or cracks of the silicate minerals, which are easily removed by mild acid washing. Although the extremely high concentrations of the LIL elements in the host basalt make it a possible source of contamination, their low $^{87}Sr/^{86}Sr$ ratio of 0.70311 (Table 2) rules out such a possibility. This is because the contaminating source must have a $^{87}Sr/^{86}Sr$ ratio greater than 0.70765, the highest measured ratio in the unwashed orthopyroxene in sample SQ 2-13 (Table 2). The very close proximity of San Quintin volcanic field to the Pacific Ocean points towards a possible sea water contamination, since sea water has the appropriate $^{87}Sr/^{86}Sr$ ratio of 0.7090. However, low K/Rb ratios of the minerals reported here as compared to the high K/Rb ratios in sea water and the lack of visible alteration, such as serpentinization, of the minerals in the optical and in the optical and in the scanning electron microscope rule out sea water as a contaminant. In view of this situation, I am inclined to believe that the contamination of the xenoliths along the grain boundaries took place in the upper mantle by a vapor phase prior to the inclusion of the xenoliths by the host basalts. In this respect it is interesting to note the increasing evidence of mantle metasometism in kimberlite nodules (Boyd, personal communication). Such metasomatic processes might have affected the San Quintin xenoliths by vapor phase reactions resulting in minor element deposition on grain boundary surfaces. These processes might also be a factor in causing some of the variations in major and minor element concentrations observed in the silicate minerals of the xenoliths.

All the green chrome diopsides from the xenoliths, except in specimen SQ 2-13, show low $^{87}Sr/^{86}Sr$ ratios, usually less than 0.7030. The iron-rich black clinopyroxenes show a much higher ratio, greater than 0.7040. Therefore, these two groups of pyroxenes are not genetically related. It is probable that the black clinopyroxenites formed as deep-seated cumulates in alkali basaltic magma chambers having high $^{87}Sr/^{86}Sr$ ratios.

Specimen no. SQ 2-13, with the lowest Mg/Mg+Fe ratios among all the xenoliths studied, show the highest $^{87}Sr/^{86}Sr$ (0.70484) bearing clinopyroxene in the entire suite. This data indicate that this xenolith had a different genetic history as compared to the low $^{87}Sr/^{86}Sr$ (in clinopyroxene) bearing xenoliths. A cumulate mode of origin in an alkali basaltic magma is preferable for this xenolith. The differences in Rb/Sr ratios and in the $^{87}Sr/^{86}Sr$ ratios in the minerals of this xenolith, however, do not define a clear isochron relationship. Rb-Sr evolution in this xenolith and in xenolith no. Sq 2-67 is possibly disturbed.

The extremely fine grained whole rock layer (SQ 2-68), consisting of clinopyroxene and glass, possibly represents a product of partial melting of the host therzolite under upper mantle conditions. It is interesting to note that this layer has slightly higher K content and much higher Sr content than the average chrome-diopside in the xenoliths. The $^{87}Sr/^{86}Sr$ ratio of this layer (0.70303) may represent the same ratio in the residual diopside of the host rock. Because of the extremely deformed nature of the host rock (fine grain size) and of the low modal proportions of the pyroxenes, it was not possible to separate enough mineral grains from the host rock for analysis.

In the plagioclase peridotite, no. 1-6-1, both plagioclase and clinopyroxene show the same $^{87}Sr/^{86}Sr$ ratios. Possibly, this peridotite formed by crystal settling in an alkali basalt magma chamber in a relatively shallower upper mantle region within the stability field of plagioclase periodtite. The K-poor (36.8 ppm), anorthite-rich composition of the plagioclase in this rock is noteworthy.

The plagioclase megacryst of andesine composition, specimen no. 3-1, show essentially the same $^{87}Sr/^{86}Sr$ ratio as its host basalt. This plagioclase is, therefore, a true phenocryst. On this basis the partition coefficients between plagioclase/basalt for K, Rb, Sr, and Ba are 0.688, 0.355, 4.09, and 0.916, respectively. These estimates are two to ten times higher than the values commonly used. However, the andesine magacryst is likely to be a high pressure phenocryst, and the above distribution coefficients may be valid only at high pressures.

Conclusion

This study has demonstrated that some mantle-derived spinel lherzolites, although enriched in modal amounts of diopside and in the major elements like Ca. Al, Fe, and Na, are extremely depleted in their trace element contents of K, Rb, Sr, and Ba. It is possible to generate "basaltic liquids" by partially melting these xenoliths, but these liquids will resemble basalt only in major element composition. The trace element abundances and ratios of K, Rb, Sr, and Ba in these liquids will not resemble any natural basalt. Therefore, other phases, like amphibole and phlogopite, must be taken into consideration as reservoirs of trace elements in the source regions of basalt in the upper mantle. The role of kaersutitic amphibole in the origin of mid ocean ridge basalts has been documented from this point of view (Basu and Murthy, 1977c).

The present study has also shown that, in terms of $^{87}Sr/^{86}Sr$ ratios, the coexisting silicate minerals of the xenoliths are in disequilibrium. If these xenoliths represent a vertical cross-section of part of the upper mantle, then that part of the upper mantle is heterogeneous in Sr-isotopic composition in the scale range of a few millimeters. This heterogeneity reflects complex processes - such as partial melting, removal of melt, cumulus processes and recrystallization - in the upper mantle regions from where the xenoliths are derived.

Appendix: Sample Petrology

In this section a brief description of the samples analyzed is given.

SQ 2-82 It is a coarse grained, 12 cms in long dimension, spinel lherzolite xenolith. Average grains size is 4 mm. The sample consists of 79% olivine, 10% orthopyroxene, 10% clinopyroxene, and 1% chromite spinel. The pyroxenes show very little exsolution. Texture is porphyroclastic but shows high degree of recovery and recrystallization.

SQ 2-13 This is a typically porphyroclastic spinel lherzolite with all the four minerals forming porphyroclasts in a finer grained mosaic of the same minerals. The groundmass average grain size is 1 mm, while the porphyroclasts are 8 mm X 3 mm in average size. The pyroxene porphyroclasts contain abundant exsolution lamellae of each other and of spinel. The groundmass minerals are free of any exsolution. The modal proportions of the minerals are: olivine - 41%, orthopyroxene - 30%, clinopyroxene - 28%, spinel 1%. This xenolith is 9 cms in long dimension.

SQ 2-41 This xenolith is much larger in size, 25 cms in long dimension, and resembles the sample SQ 2-13 in texture. It consists of 52% olivine, 26% orthopyroxene, 20% clinopyroxene, and 2% spinel. The geochemical data of this xenolith has already been published (Basu and Murthy, 1977a).

SQ 2-67 This is a fist-size xenolith, 8 cms in long dimension showing porphyroclastic texture. Average grain size of the porphyroclasts is 9 mm X 4 mm. Average groundmass grain size is 3 mm. The pyroxene porphyoclasts contain common, coarse, exsolution of each other along with some fine tablets of spinel. Modal proportions of the minerals are olivine 53%, orthopyroxene 20%, clinopyroxene 25%, and spinel 2%.

SQ 1-7-28 This is a porphyroclastic spinel lherzolite xenolith. Texturally, it is very similar to SQ 2-67 but much larger in size, measuring 30 cms in the long dimension. Modal mineralogy is olivine - 55%, orthopyroxene - 20%, clinopyroxene - 23%, and spinel - 2%.

SQ 2-68 This is an extremely deformed spinel lherzolite specimen. Although still retaining some of the porphyroclasts of orthopyroxene and clinopyroxene, the average grain size of the groundmass of this sample is 0.3 mm. Of special interest in this sample is the 4 cms broad pyroxenite layer containing glass. The pyroxenite layer itself is fine grained, 0.5 mm in average grain size, and is parallel to the foliation of the host rock. The pyroxenite and glass layer has most probably formed due to in situ partial melting under upper mantle conditions. This xenolith measures 10 cms in the long dimension including the pyroxenite layer. The host lherzolite has the following modal abundances: olivine - 60%, orthopyroxene - 20%, clinopyroxene - 17%, and spinel - 3%. The pyroxenite layer consists of 85% clinopyroxene and 15% (glass & spinel).

SQ 2-30 This is a fist-size, fine grained lherzolite xenolith (average grain size 2 mm) with a 1 cm thick clinopyroxenite layer (average grain size 3 mm).

SQ 1-3-5 and SQ 2-81 These two samples are large (12 cms in long dimension) polycrystalline, monomineralic xenoliths. They consist of coarse aggregates of pitch-black clinopyroxene grains of 4 mm average grain size.

SQ 2-36 This sample is again a composite xenolith consisting of a 5 cms thick coarse (5 mm grain size) polycrystalline layer of orthopyroxenite in a spinel lherzolite host xenolith. The specimen size is 15 cms in long dimension. The orthopyroxenite layer shows boudin structures suggesting plastic deformation.

SQ 1-6-1 It is a rather rare variety of plagioclase peridotite xenolith (measuring 8 cms in the long dimension) among the xenolith suite in San Quintin. The sample shows evidence of plastic deformation (boudin structure in plagioclase layers) and consists of iron-rich olivine (10%), clinopyroxene (57%), orthopyroxene (3%), plagioclase (28%), and spinel (2%). The plagioclase is of composition An_{98}, and shows a reaction relationship to olivine to form orthopyroxene and clinopyroxene and spinel.

SQ 3-1 It is a clear, glassy, unzoned, single crystal of plagioclase measuring 3 cms in the long dimension. The plagioclase has the composition - $Ab_{56.2}$ $An_{36.9}$ $Or_{6.9}$ and may be termed andesine. It was found embedded in the alkalic basalt SQB.

SQ 1-2-6 and SQB These two specimens are typical representatives of the alkalic basalt flows in San Quintin which have enclosed the numerous ultramafic xenoliths. These are typical holocrystalline alkali olivine basalts containing phenocrysts of olivine and plagioclase in a groundmass of olivine, augite, feldspar, and iron-titanium oxide minerals.

Acknowledgment: This research has been supported by grants from the Geological Society of America, and by NSF grant DES-75-14806 to V. Rama Murthy. The author is grateful to Dr. V. Rama Murthy for introducing him to mass spectrometry and Dr. F. R. Boyd for his suggestions which improved the manuscript.

References

Bacon, C. R. and Carmichael, I.S.E., 1973, Stages in the P-T path of ascending basalt magma: An example from San Quintin, Baja California, Contr. Min. Pet. 41: 2-11.

Basu, A. R., 1975a, Hot Spots, mantle plumes and a model for the origin of ultramafic xenoliths in alkali basalts, Earth Planet. Sc. Lett., 28, 261-274.

Basu, A. R., 197b, Petrogenesis of the ultramafic xenoliths from San Quintin, Baja California, Ph.D. thesis, Univ. of California, 229 p.

Basu, A. R., 1977a, Olivine-spinel equilibria in ultramafic xenoliths from San Quintin, Baja California, Earth Planet. Sci. Lett., 33, 443-448.

Basu, A. R., 1977b, Textures, microstructures and deformation of the ultramafic xenoliths from San Quintin, Baja California, Tectonophysics, 43.

Basu, A. R. and Murthy, V. R., 1977a, Ancient lithospheric lherzolite xenolith in alkali basalt from Baja California, Earth Planet. Sci. Lett., 35, 239-246.

Basu, A. R. and Murthy, V. R., 1977b, Trace elements and Sr-isotopic geochemistry of the constituent minerals in ultramafic xenoliths from San Quintin, Baja California, Extended Abstr., 2nd Int'l Kimberlite Conf., Santa Fe, N.M.

Basu, A. R. and Murthy, V. R., 1977c, Kaersutites, suboceanic low-velocity zone, and the origin of mid-oceanic ridge basalts, Geology, 5, 365-368.

Basu, A. R. and Murthy, V. R., 1977d, Significance of amphibole and phlogopite in basalt genesis - trace elemental and Sr-isotopic evidence, Extended Abstr., Int'l Confer. Exp. Trace Element Geochemistry, Sedona, Arizona, 2-4.

Murthy, V. R., Evensen, N. M., Jahn, B. and Coscio, M. R., 1971, Rb-Sr ages and elemental abundances of K, Rb, Sr, and Ba in samples from the ocean of storms: Geochim. Cosmochim. Acta, 35, 1139-1153.

Shimizu, N., 1975, Geochemistry of ultramafic inclusions from Salt Lake Crater, Hawaii, and from southern African Kimberlites, Physics, Chem. Earth, vol. 9, 655-669.

Sobolev, N. V., 1974, Deep-seated inclusions in Kimberlites and the problem of the composition of the upper mantle, English Translation, Am Geophys. Union, Washington, D.C., 1977, 279 p.

Woodford, A. O., 1928, The San Quintin volcanic field, Lower California, Am. J. Sci., 15, 337-345.

GARNET BEARING LHERZOLITES AND DISCRETE NODULE SUITES FROM THE MALAITA ALNOITE, SOLOMON ISLANDS, S.W. PACIFIC, AND THEIR BEARING ON OCEANIC MANTLE COMPOSITION AND GEOTHERM

Peter H. Nixon

Department of Geology, P.O. Box 4820, University, Papua New Guinea

F. R. Boyd

Geophysical Laboratory, Carnegie Institution, 2801 Upton Street, N.W., Washington, D.C. 20008

Abstract. A suite of xenoliths of deep-seated origin, similar to those found in continental kimberlites, occur in a phlogopite-melilite-olivine-pyroxene (alnoite) breccia, 33.9 m.y. old, in an oceanic environment. The coarse lherzolites have modal proportions of olivine 50-75, orthopyroxene 5-45, clinopyroxene 0-20, garnet 0-15 and spinel 0-10. Some specimens show slight deformation. There are small, but significant variations in mineral composition indicating depletion in the nodules of deepest origin (low Na, Ti, Fe/Mg and Al/Cr). Pyrope-rich garnet is restricted to nodules of deeper origin, and Al-rich spinel to the nodules of shallower origin. The discrete nodule (mega-cryst) suite includes pyrope-rich garnet (single specimens up to 8.2 kb), augite, augite-ilmenite lamellar intergrowths, subcalcic diopside and bronzite. A pyroxene geotherm, calculated for these garnet-bearing nodules, has a much higher gradient than those previously estimated for continental kimberlite nodules. A high-temperature inflection in the geotherm can be taken as evidence for a lithosphere thickness of 110 km. At depths less than the inflection there is a lower horizon of lherzolite (source of Alite volcanics ?) overlain by less depleted lherzolites and an upper horizon of harzburgites (source of oceanic crust?). The alnoite appears to have been emplaced near the edge of a thick-ened portion of the Pacific Plate (the Ontong Java Plateau).

Introduction

The interest in the alnoite volcanic centre at Malaita stems from its deep-seated origin in an oceanic environment. The xenolith suites described here, normally characterize kimber-lites that occur in contental cratonic settings, but even in those, a single pipe rarely displays the range of rock types found in Malaita.

The island of Malaita is geologically anomal-ous compared with the remainder of the Solomon Islands, and has been placed apart in the "Pacific province" by Coleman (1965, 1968). In the north it has a basal succession of at least 700 m of marine basalts, pillow lavas and pyroclastic rocks (Alite volcanics, Fig. 1) overlaid and partly intruded into 1500 m of Upper Cretaceous (Coleman, 1968) mudstones, limestones, chalk and siltstones (The Malaita group) according to Rickwood (1957). The very early Oligocene age of the Malaita alnoite (33.9 m.y. Davis, 1977) determined from a zircon in gravels indicates a penecontemporaneous origin with the associated sedimentary rocks.

The aims of the present investigation are the determination of upper mantle stratigraphy and temperature distribution at the time of emplacement from a study of the deep seated inclusions and comparison of the xenoliths with those found in continental kimberlites.

The Malaita Alnoite

The discovery of boulders of ultrabasic rocks in streams in 1951 (Rickwood, 1957) and coarse ilmenite and garnet gravels in Kwai Harbour in 1956 by J. C. Grover, then Chief Government Geologist, were the prelude to further explora-tion that led to the discovery in 1962 of alnoite and ankaratrite outcrops (Fig. 1) by mining company geologist E. Gerryts and Survey geologist R. B. M. Thompson (Gerryts, 1965). There were only a limited number of samples collected, and four of these were subsequently described by Allen and Deans (1965) whose report inspired the present investigation. Difficulties relat-ing to access (Hackman, 1968; Bradshaw, 1968) have prevented adequate sampling until this investigation.

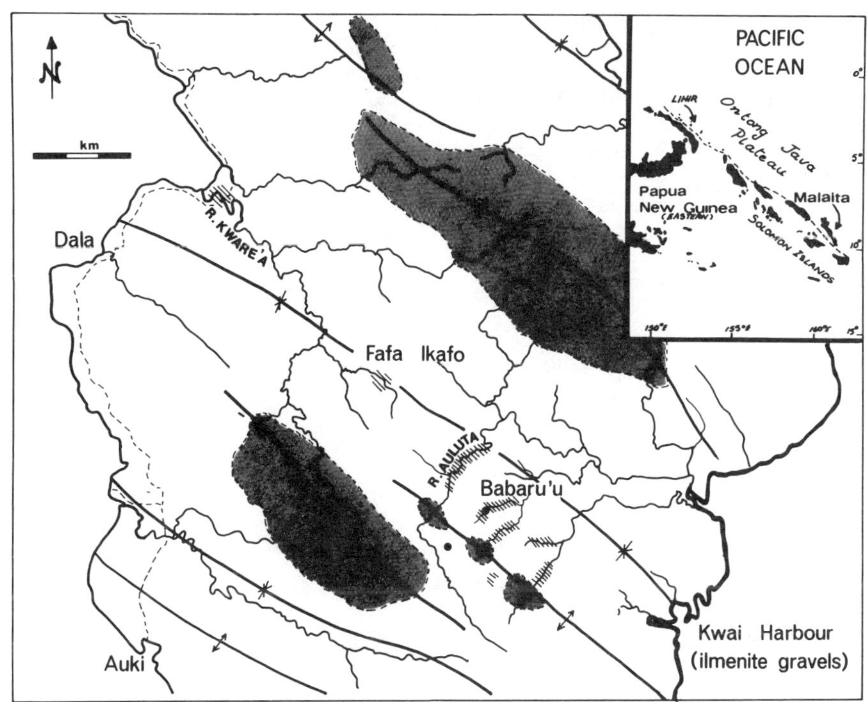

Fig. 1. Geology of part of northern Malaita, Solomon Islands, from Rickwood (1957) showing the Alite Volcanics (shaded) overlain by Upper Tertiary Malaita limestones and mudstones (plain) folded into NW trending anticlines and synclines. The alnoite at Babaru'u on the Hauhaumela stream and the dot indicating an ankaratrite 2 km to the SW, together with 'shows' of coarse ilmenite gravels (diagonal lines) were discovered by Survey and exploration geologists (see text) and figured by Allen and Deans (1965). The position of the Fafa Ikafo gravels has been relocated slightly to the south of that shown by these authors. The inset shows the Melanesian Archipelago with a line drawn near the periphery of the Ontong Java Plateau portion of the Pacific Plate. This line bounds the alkaline volcanic rocks of the Lihir Group (Johnson et al., 1976) and Malaita 1000 km to the SE where it coincides with the margin of the "Pacific Province" of Coleman (1968).

Alnoite breccia is well exposed for over 100 metres at Babaru'u along the Hauhaumela stream and a small tributary. A steeply dipping contact against weathered mudstones along the southeast margin, together with the rounded nature of xenoliths in the breccia, suggests a pipe-like intrusion. It would appear that the outcrop owes its presence to the erosion of the sediments along the crest of an anticline mapped by Hackman (1968).

Much of the breccia is soft, bluish-green and superficially resembles kimberlite "blue ground" but there are fresh, dark grey, harder aphanitic varieties forming rounded patches and inclusions. There are also tuffs containing altered alnoite lapillae, and autoliths with cores of mineral or rock fragments. The proportion of xenoliths varies, but is generally greater than 10%; country rock limestones, mudstones and basalts have been noted. Rounded, weathered ultrabasic nodules together with discrete nodules of ilmenite, dark pyroxene, and red-brown garnet are conspicuous. This suite and coarse flakes of mica form a gravel zone immediately overlying the solid rock, and can be traced in the soils of the surrounding area. The widespread disposition of ilmenite-pyrope gravels as far afield as the Fafa Ikafo Valley and the mouth of the river Kware'a (Fig. 1) suggest either the presence of a very large volcanic centre at Babaru'u, or several intrusions striking in a NW-SE direction. The ankaratrite to the southwest of Babaru'u (Fig. 1; Gerryts, 1968) was not visited, although ilmenite probably shed from it was collected in the Auluta river.

The case for use of the term "alnoite" has been argued by Allen and Deans (1965) notwithstanding the presence of the suite of "kimberlitic" inclusions. More recently, Marchand (1970) and Kresten (1976) have described inclusions of deep-seated origin in alnoites at Ile Bizard, Quebec and Alno, Sweden, respectively, and it is clear that both alnoites and kimberlites sample portions of the outer layers of the earth from great depths. Comparisons of the groundmass petrography are particularly difficult where alteration has destroyed the primary minerals and, if the presence of melilite cannot be proved, the rock

TABLE 1. Modal Composition of Ultrabasic Nodules from the Malaita Alnoite. The figures are only approximate because of the coarse grain size and small nodule size. A comparison with modes of other garnet-bearing nodules from kimberlites from southern Africa is included. (1) Thaba Putsoa, Lesotho (Boyd and Nixon, 1972); (2) Matsoku, Lesotho (Cox, et al., 1973); and (3) Bultfontein, South Africa (Chen, 1971). The range of modal olivine for the Malaita nodules does not include the abnormally low value of 27 for PHN 3568.

PHN	Olivine*	Orthopyroxene	Clinopyroxene	Garnet**	Spinel
3533	70	15	12		3
3535	58	14	22		6
3536***	60	10	12	18	
3537	40	15	40		5
3538	72	20	8	tr	
3539	64	15	6	10	5
3540	50	18	25		7
3549B	70	23	7		
3553B	55	30	10		5
3567***	53	10	21	12	4
3568	27	37	32	tr	4

* "Olivine" consists mainly of alteration products.
** "Garnet" includes kelyphitic alteration.
*** amphibole present.
tr = trace.

		Olivine	Orthopyroxene	Clinopyroxene	Garnet	Spinel or Chromite
Malaita		50-72	10-37	6-40	0-18	0-7
Southern Africa	1	50-75	5-45	0-20	0-15	0-10
	2	47-75	20-50	0-5	0-11	
	3	59	33	2	6	

Very small nodules for which modes are meaningless are tentatively classified as follows:

3532J	garnet lherzolite	3550A	garnet pyroxenite
3534	amphibole-bearing lherzolite	3550B	gt-opx-amph chip
3542A	spinel wehrlite	3551A	spinel griquaite
3542B	spinel lherzolite	3553A	spinel pyroxenite

greatly resembles kimberlite, particularly if MgO is high and $K_2O >> Na_2O$ (Nixon, Mitchell and Rogers, in prep.).

In thin section the proportion of olivine (Fo_{85}) microphenocrysts, some with cruciform twins, varies between 25-35%. The median size is 0.2 mm. Stumpy crystals or anhedral plates of melilite (akermanite-soda melilite) measure up to 1 mm and make up a third or more of the rock (Nixon, Mitchell and Rogers, in prep.). Primary salites occur together with xenocrysts of augite and bronzite that are fragments of much larger discrete nodules (see below) and commonly exhibit reaction rims in which olivine has formed. The groundmass consists of euhedral perovskite, magnetite, some apatite, melanite, K-poor nepheline and glass with late-formed poecilitic phlogopite. The breccias are cemented by varying amounts of chalcedony, zeolite (including gonnardite) and calcite mottled with MnO_2 in which coarse (>1 cm) phlogopite is embedded.

Ultrabasic Nodules

These are rounded and commonly a few cm in diameter; there is one exceptional specimen of 30 cm. Unless they are encased in the indurated variety of alnoite they weather pale brown and are very friable. The olivine is almost completely altered in contrast to the associated minerals. It can be seen from approximate modes (Table 1),

TABLE 2. Chemical Compositions of Minerals in Ultrabasic Nodules from the Malaita Alnoite, wt. %.

PHN	3532J				3533			3534		3535		
	gt	cpx	opx	ol	cpx	opx	sp	cpx	opx	cpx	opx	sp
SiO_2	42.46	52.58	56.30	40.52	52.59	56.46	0.08	51.76	55.85	51.34	55.11	0.04
TiO_2	0.14	0.48	0.16	0.01	0.17	0.04	0.25	0.46	0.16	0.49	0.11	0.11
Al_2O_3	23.26	4.08	2.71	0.01	4.23	3.25	34.61	4.29	2.91	5.96	4.88	55.25
Cr_2O_3	0.82	0.70	0.32	0.04	1.26	0.65	31.45	1.07	0.50	0.70	0.43	9.56
Fe_2O_3	-	-	-	-	-	-	5.38	-	-	-	-	4.61
FeO	8.87	2.54	6.37	10.13	2.30	5.13	9.31	2.71	6.36	2.45	6.34	8.25
MnO	0.40	0.10	0.16	0.14	0.11	0.15	0.18	0.10	0.15	0.11	0.19	0.16
MgO	19.98	15.95	34.79	49.80	16.23	35.28	18.27	15.67	34.35	15.07	34.08	20.75
CaO	5.20	21.28	0.50	0.04	20.53	0.64	0.01	20.67	0.53	21.50	0.44	0.01
Na_2O	0.01*	1.53	0.08	n.d.	1.68	0.10	NiO 0.25	1.75	0.07	1.54	0.05	NiO 0.42
Total	101.14	99.24	101.39	100.69	99.10	101.70	99.79	98.48	100.88	99.16	101.63	99.16
Number of cations for n oxygens												
n =	12	6	6	4	6	6	4	6	6	6	6	4
Si	2.991	1.921	1.920	0.988	1.921	1.911	0.002	1.910	1.915	1.879	1.875	0.001
Ti	0.007	0.013	0.004	0.000	0.005	0.001	0.005	0.013	0.004	0.013	0.003	0.002
Al	1.931	0.176	0.109	0.000	0.182	0.130	1.162	0.187	0.118	0.257	0.196	1.704
Cr	0.046	0.020	0.009	0.001	0.036	0.017	0.708	0.031	0.014	0.020	0.012	0.198
Fe^{3+}	-	-	-	-	-	-	0.115	-	-	-	-	0.091
Fe^{2+}	0.523	0.078	0.182	0.207	0.070	0.145	0.222	0.084	0.182	0.075	0.180	0.181
Mn	0.024	0.003	0.005	0.003	0.003	0.004	0.004	0.003	0.004	0.003	0.005	0.004
Mg	2.098	0.869	1.769	1.811	0.884	1.780	0.776	0.862	1.756	0.822	1.729	0.810
Ca	0.392	0.833	0.018	0.001	0.804	0.023	-	0.817	0.019	0.843	0.016	-
Na	0.001	0.108	0.005	n.d.	0.119	0.007	Ni 0.006	0.125	0.005	0.109	0.003	Ni 0.009
Total	8.013	4.021	4.021	3.011	4.024	4.018	3.000	4.032	4.017	4.021	4.019	3.000
Ca	13.0	46.8	0.9		45.7	1.2		1.0	46.4	48.5	0.8	
Mg	69.7	48.8	89.9		50.3	91.3		89.7	48.9	47.2	89.8	
Fe	17.3	4.4	9.2		4.0	7.5		9.3	4.7	4.3	9.4	
$Mg/(Mg+Fe^{2+})$%	80.1	91.8	90.7	89.8	92.6	92.5	69.7	91.2	90.6	91.6	90.6	81.8
Ca/(Ca+Mg)%	-	48.9	-	-	47.6	-	-	48.7	-	50.6	-	-

* detectability limit is 0.03 for most elements

(continued)

Table 2, page 2

PHN	3536			3537			3538			3539			
	gt	cpx	opx	cpx	opx	sp	gt	cpx	opx	gt	cpx	opx	sp
SiO_2	42.74	53.03	55.96	51.85	55.19	0.04	42.81	53.01	56.81	42.93	53.50	56.77	0.17
TiO_2	0.14	0.44	0.16	0.42	0.11	0.12	0.00	0.00	0.00	0.02	0.00	0.01	0.09
Al_2O_3	22.98	4.22	2.71	5.77	4.57	54.31	20.77	2.26	2.45	20.11	3.00	2.93	24.59
Cr_2O_3	1.45	1.05	0.45	0.80	0.44	11.91	4.61	0.99	0.86	5.46	1.40	0.95	43.39
Fe_2O_3	-	-	-	-	-	3.34	-	-	-	-	-	-	3.99
FeO	8.35	2.65	6.22	2.19	6.09	8.17	6.05	2.56	5.15	5.99	2.57	5.20	10.67
MnO	0.43	0.09	0.14	0.11	0.17	0.17	0.32	0.14	0.15	0.32	0.13	0.14	0.23
MgO	20.38	15.80	33.38	15.18	32.73	20.73	20.52	19.56	33.69	20.23	18.88	33.71	16.29
CaO	5.07	20.72	0.53	21.47	0.43	0.01	6.75	20.52	1.22	6.82	19.79	1.19	0.02
Na_2O	0.02*	1.70	0.09	1.60	0.05	NiO 0.42	0.00	0.28	0.02	0.01	0.73	0.07	NiO 0.24
Total	101.56	99.70	99.64	99.39	99.78	99.22	101.83	99.32	100.35	101.89	100.00	100.97	99.68
Number of cations for n oxygens													
n =	12	6	6	6	6	4	12	6	6	12	6	6	4
Si	2.995	1.927	1.938	1.890	1.907	0.001	3.005	1.929	1.948	3.018	1.930	1.936	0.005
Ti	0.007	0.012	0.004	0.012	0.003	0.002	0.000	0.000	0.000	0.001	0.000	0.000	0.002
Al	1.898	0.181	0.111	0.248	0.186	1.681	1.718	0.097	0.099	1.666	0.128	0.118	0.868
Cr	0.080	0.030	0.012	0.023	0.012	0.247	0.256	0.028	0.023	0.304	0.040	0.026	1.027
Fe^{3+}	-	-	-	-	-	0.066	-	-	-	-	-	-	0.090
Fe^{2+}	0.489	0.081	0.180	0.067	0.176	0.179	0.355	0.078	0.148	0.352	0.078	0.148	0.267
Mn	0.026	0.003	0.004	0.003	0.005	0.004	0.019	0.004	0.004	0.019	0.004	0.004	0.006
Mg	2.130	0.856	1.724	0.825	1.686	0.811	2.149	1.061	1.723	2.121	1.016	1.714	0.728
Ca	0.381	0.807	0.020	0.839	0.016	-	0.508	0.800	0.045	0.514	0.765	0.043	0.001
Na	0.003	0.120	0.006	0.113	0.003	Ni 0.009	0.000	0.020	0.001	0.001	0.051	0.005	Ni 0.006
Total	8.009	4.017	3.999	4.020	3.994	3.000	8.010	4.017	3.991	7.996	4.012	3.994	3.000
Ca	12.7	46.3	1.0	48.5	0.8		16.9	41.3	2.3	17.2	41.2	2.3	
Mg	71.0	49.1	89.6	47.7	89.8		71.3	54.7	90.0	71.0	54.6	89.9	
Fe	16.3	4.6	9.4	3.9	9.4		4.0	7.7	7.7	11.8	4.2	7.8	
$Mg/(Mg+Fe^{2+})\%$	81.3	91.4	90.5	92.5	90.5	81.9	85.8	93.2	92.1	85.8	92.9	92.0	73.1
$Ca/(Ca+Mg)\%$	-	48.5	-	50.4	-	-	-	43.0	-	-	43.0	-	-

* detectability limit is 0.03 for most elements

(continued)

Table 2, page 3

PHN	3540			3542A			3542B				3549B	
	cpx	opx	sp	cpx	ol	sp	cpx	opx	ol	sp	cpx	opx
SiO_2	51.48	54.74	0.04	54.14	40.54	0.09	53.19	54.44	40.96	0.06	53.61	55.93
TiO_2	0.50	0.11	0.13	0.06	0.00	0.08	0.52	0.13	0.00	0.15	0.01	0.00
Al_2O_3	6.11	4.13	56.36	4.14	0.01	50.68	6.26	4.75	0.01	58.58	2.24	2.38
Cr_2O_3	0.65	0.29	9.11	0.81	0.04	16.68	0.70	0.39	0.05	9.11	0.99	0.84
Fe_2O_3	-	-	3.88	-	-	4.06	-	-	-	3.01	-	-
FeO	2.51	6.50	8.28	2.32	9.00	9.12	2.52	6.28	7.86	9.19	2.44	5.00
MnO	0.11	0.18	0.16	0.12	0.17	0.25	0.13	0.18	0.14	0.18	0.12	0.13
MgO	15.01	32.52	20.88	16.72	50.36	20.18	15.17	34.06	51.48	21.05	18.88	33.43
CaO	21.22	0.42	0.01*	22.22	0.05	0.02	21.39	0.50	0.06	0.01	20.92	1.20
Na_2O	1.70	0.05	NiO 0.43	0.79	n.d.	NiO 0.42	1.53	0.05	n.d.	NiO 0.46	0.28	0.01
Total	99.29	98.94	99.28	101.32	100.17	101.58	101.41	100.78	100.56	101.80	99.49	98.92

Number of cations for n oxygens

n =	6	6	4	6	4	4	6	6	4	4	6	6
Si	1.881	1.911	0.001	1.931	0.989	0.002	1.896	1.870	0.990	0.002	1.945	1.945
Ti	0.014	0.003	0.003	0.002	0.000	0.002	0.014	0.003	0.000	0.003	0.000	0.000
Al	0.263	0.170	1.729	0.174	0.000	1.565	0.263	0.192	0.000	1.750	0.096	0.098
Cr	0.019	0.008	0.188	0.023	0.001	0.346	0.020	0.011	0.001	0.183	0.028	0.023
Fe^{3+}	-	-	0.076	-	-	0.080	-	-	-	0.058	-	-
Fe^{2+}	0.077	0.190	0.180	0.069	0.184	0.200	0.075	0.180	0.159	0.195	0.074	0.145
Mn	0.003	0.005	0.004	0.004	0.004	0.006	0.004	0.005	0.003	0.004	0.004	0.004
Mg	0.818	1.693	0.810	0.889	1.832	0.789	0.806	1.744	1.855	0.796	1.021	1.734
Ca	0.831	0.016	0.000	0.849	0.001	0.001	0.817	0.018	0.002	0.000	0.813	0.045
Na	0.120	0.003	Ni 0.009	0.055	n.d.	Ni 0.009	0.106	0.003	n.d.	Ni 0.009	0.020	0.001
Total	4.026	3.999	3.000	3.996	3.011	3.000	4.001	4.026	3.010	3.000	4.001	3.995
Ca	48.2	0.8		47.0			48.1	0.9			42.6	2.3
Mg	47.4	89.2		49.2			47.5	89.8			53.5	90.1
Fe	4.4	10.0		3.8			4.4	9.3			3.9	7.6
Mg/(Mg+ Fe^{2+})%	91.4	89.9	81.1	92.8	90.9	79.8	91.5	90.6	92.1	80.3	93.2	92.3
Ca/(Ca+Mg)%	50.4	-	-	48.9	-	-	50.3	-	-	-	44.3	-

* detectability limit is 0.03 for most elements

(continued)

Table 2, page 4

PHN	3550A			3550B			3551A			3553A		
	gt	cpx	opx	amph	gt	opx	gt	cpx	sp	cpx	opx	sp
SiO_2	42.44	53.11	56.39	44.21	42.96	55.58	42.69	53.17	0.03	51.65	54.62	0.04
TiO_2	0.09	0.48	0.13	2.63	0.07	0.09	0.11	0.37	0.10	0.39	0.11	0.11
Al_2O_3	23.24	5.04	2.54	14.21	23.15	2.61	23.51	5.02	64.06	5.90	4.94	55.31
Cr_2O_3	0.65	0.53	0.25	1.07	1.48	0.48	0.23	0.16	3.04	0.75	0.46	10.38
Fe_2O_3	-	-	-	K_2O 1.23	-	-	-	-	2.42	-	-	3.53
FeO	9.26	2.65	6.75	4.57	8.16	5.92	6.75	2.05	6.61	2.33	6.25	8.23
MnO	0.41	0.07	0.13	0.07	0.41	0.15	0.21	0.03	0.07	0.09	0.18	0.16
MgO	19.79	15.31	33.25	17.92	20.58	33.37	19.98	15.89	22.80	15.04	32.68	20.76
CaO	5.11	20.61	0.46	9.42	5.01	0.48	6.41	22.25	0.00	21.21	0.63	0.01
Na_2O	0.01*	1.77	0.06	3.30	0.01	0.09	0.00	1.16	NiO 0.61	1.63	0.08	NiO 0.36
Total	101.00	99.57	99.96	98.63	101.83	98.77	99.89	100.10	99.74	98.99	99.95	98.89
Number of cations for n oxygens												
n =	12	6	6	23	12	6	12	6	4	6	6	4
Si	2.997	1.927	1.949	6.222	2.998	1.940	3.016	1.918	0.001	1.890	1.888	0.001
Ti	0.005	0.013	0.003	0.276	0.004	0.002	0.006	0.010	0.002	0.011	0.003	0.002
Al	1.934	0.216	0.103	2.354	1.904	0.107	1.957	0.213	1.889	0.254	0.201	1.709
Cr	0.036	0.015	0.007	0.115	0.082	0.013	0.013	0.005	0.060	0.022	0.013	0.215
Fe^{3+}	-	-	-	K 0.219	-	-	-	-	0.046	-	-	0.070
Fe^{2+}	0.547	0.080	0.195	0.537	0.476	0.173	0.399	0.062	0.138	0.071	0.181	0.180
Mn	0.025	0.002	0.004	0.008	0.024	0.004	0.013	0.001	0.001	0.003	0.005	0.004
Mg	2.083	0.828	1.713	3.759	2.141	1.736	2.104	0.854	0.851	0.821	1.685	0.811
Ca	0.387	0.801	0.017	1.421	0.375	0.018	0.485	0.860	0.000	0.832	0.023	0.000
Na	0.001	0.124	0.004	0.896	0.001	0.006	0.000	0.081	Ni 0.012	0.116	0.005	Ni 0.008
Total	8.015	4.006	3.995	15.807	8.005	3.999	7.993	4.004	3.000	4.020	4.004	3.000
Ca	12.8	46.9	0.9	24.9	12.5	0.9	16.2	48.4		48.3	1.2	
Mg	69.1	48.4	89.0	65.7	71.6	90.1	70.4	48.1		47.6	89.2	
Fe	18.1	4.7	10.1	9.4	15.9	9.0	13.3	3.5		4.1	9.6	
$Mg/(Mg+Fe^{2+})$%	79.2	91.2	89.8	29.0	81.8	90.9	84.1	93.3	86.0	92.0	90.3	76.4
Ca/(Ca+Mg)%	-	49.2	-	27.4	-	-	-	50.2	-	50.3	-	-

* detectability limit is 0.03 for most elements

(continued)

Table 2, page 5

PHN	3553B			3567					3568			
	cpx	opx	sp	gt	cpx	opx	ol	sp	cpx	opx	ol	sp
SiO_2	51.88	54.75	0.05	42.24	52.59	55.20	40.14	0.13	52.23	54.91	40.16	0.05
TiO_2	0.48	0.13	0.15	0.09	0.64	0.16	0.00	0.35	0.51	0.13	0.00	0.08
Al_2O_3	5.68	4.50	54.14	23.11	4.28	2.77	0.01	40.51	5.60	4.66	0.01	54.52
Cr_2O_3	0.71	0.40	11.01	0.93	0.68	0.37	0.03	24.02	0.77	0.44	0.04	10.87
Fe_2O_3	-	-	3.72	-	-	-	-	6.12	-	-	-	4.00
FeO	2.45	6.39	8.54	8.70	2.78	6.74	10.63	12.98	2.38	6.94	10.29	8.59
MnO	0.09	0.18	0.16	0.49	0.09	0.18	0.18	0.36	0.06	0.18	0.15	0.15
MgO	15.00	32.55	20.38	19.02	16.29	34.54	49.63	16.75	14.92	32.73	49.93	20.48
CaO	21.12	0.43	0.01*	5.49	21.90	0.46	0.04	0.02	20.76	0.43	0.04	0.01
Na_2O	1.62	0.06	NiO 0.38	0.04	1.02	0.04	n.d.	NiO 0.38	1.80	0.09	n.d.	NiO 0.37
Total	99.03	99.39	98.54	100.11	100.27	100.46	100.66	101.62	99.03	100.51	100.62	99.12
Number of cations for n oxygens												
n =	6	6	4	12	6	6	4	4	6	6	4	4
Si	1.897	1.902	0.001	3.007	1.905	1.906	0.983	0.004	1.907	1.893	0.982	0.001
Ti	0.013	0.003	0.003	0.005	0.017	0.004	0.000	0.007	0.014	0.003	0.000	0.002
Al	0.245	0.184	1.688	1.939	0.183	0.113	0.000	1.324	0.241	0.189	0.000	1.689
Cr	0.021	0.011	0.230	0.052	0.019	0.010	0.001	0.527	0.022	0.012	0.001	0.226
Fe^{3+}	-	-	0.074	-	-	-	-	0.128	-	-	-	0.079
Fe^{2+}	0.075	0.186	0.189	0.518	0.084	0.195	0.218	0.301	0.073	0.200	0.210	0.189
Mn	0.003	0.005	0.004	0.030	0.003	0.005	0.004	0.008	0.002	0.005	0.003	0.003
Mg	0.818	1.686	0.803	2.019	0.880	1.778	1.811	0.692	0.812	1.682	1.820	0.803
Ca	0.828	0.016	0.000	0.419	0.850	0.017	0.001	0.001	0.812	0.016	0.001	0.000
Na	0.115	0.004	Ni 0.008	0.006	0.072	0.003	n.d.	Ni 0.008	0.127	0.006	n.d.	Ni 0.008
Total	4.015	3.997	3.000	7.995	4.013	4.031	3.018	3.000	4.010	4.006	3.017	3.000
Ca	48.1	0.8		14.2	46.9	0.9			47.9	0.8		
Mg	47.5	89.4		68.3	48.5	89.4			47.9	88.7		
Fe	4.4	9.8		17.5	4.6	9.8			4.2	10.5		
$Mg/(Mg+Fe^{2+})$%	91.6	90.1	81.0	79.6	91.3	90.1	89.3	69.7	91.8	89.4	89.6	81.0
Ca/(Ca+Mg)%	50.3	-	-	-	49.1	-	-	-	50.0	-	-	-

* detectability limit is 0.03 for most elements

determined by limited point counting and visual estimation, that most of the peridotites are lherzolites (with pyroxene-rich varieties) as has been found for nodules of upper-mantle origin elsewhere; e.g., in an oceanic area, from Hawaiian basaltic rocks (White, 1966) and in a continental area, from South African kimberlites (Table 1, lower portion). Thus, olivine averages 56% and the pyroxenes are next in abundance, although diopside outweighs enstatite in many nodules. It is notable that garnet coexists with primary Cr-Al spinel in three specimens.

Under the microscope the lherzolites have a coarse texture (Harte, 1977) usually with little or no deformation, and with well defined triple junctions. A slight foliation due to orientation of tabular olivines and orthopyroxenes (cf. Boullier and Nicolas, 1973) was seen in 3539, and is emphasized by concordant late-formed calcite stringers. Some shattering of the pyroxenes was observed.

Subrounded olivine, 2-5 mm across, with fine grained opaque inclusions, may show faint kink bands in thin section. It is traversed by serpentine stringers, but most commonly is completely pseudomorphed by serpentine and calcite with original crystals outlined by concentrations of limonite. There is a little veining by natrolite, chalcedony and later stage calcite. Five determinations showed a mean composition of Fo_{90} (Table 3).

Rounded, pink-purple garnet, average diameter 2 mm, maximum 1 cm, was observed in half the 26 specimens examined. In specimen 3567, lenticular kelyphitised garnets up to 4 mm long, define a crude lineation. Under the microscope, garnet is associated with the pyroxenes, which it sometimes encloses together with occasional altered olivine. The analyses in Table 2 indicate about 70% pyrope content, and variable amounts of chromium with the highest values occurring in specimens 3538 and 3539 (Cr_2O_3, 4.61 and 5.46 wt. % respectively). The garnet from a garnet-spinel chip (3551B) is pale pink and fine grained (0.3-1.0 mm) although compositionally very similar to those from the lherzolites with which it is identified.

The clinopyroxene varies from the characteristic bright emerald green typical of "chromium diopsides" to dull grey green. In thin section it is interstitial to the other primary minerals, and has the characteristic lobate appearance seen elsewhere in mantle-derived lherzolite nodules. The grains are finer (1 mm) than most other primary constituents and have a slight turbid appearance due to alteration. The mineral is not deformed except in specimens 3538, 3539 and 3568 where slight shattering was observed. Fine exsolution lamellae (?opx) were detected in three nodules. The clinopyroxenes are classified as diopsides and show only a small range in Mg/(Mg + Fe) % from 91.2 to 93.2 (Table 2). The most magnesium-rich specimens are also the least calcic, i.e., Ca/(Ca + Mg) % = 43-44 (3538, 3539, and 3549B), and most depleted in Na_2O and TiO_2. The range from calcic to subcalcic diopsides (Fig. 2) is sufficient to form a basis for geotherm calculations (see below). The level of Al_2O_3 (2.24-6.86 wt. %, Table 2, and unpublished analyses) plus minor Cr_2O_3 (0.53-1.40 wt. %) indicates high proportions of Ca-Tschermaks component combined with appreciable jadeite/kosmochlor (ureyite) components.

Pale green-yellow orthopyroxene has faint pink pleochroism in thick sections. The grain size averages 1-2 mm (maximum 5 mm) with little evidence of strain, except in specimen 3538 where the orthopyroxene grains are shattered and cemented by haematite-strained serpentine (Fig. 3a). Very narrow (100) exsolution lamellae of clinopyroxene were occasionally seen. At the periphery of the nodules in contact with the alnoite, the orthopyroxene is replaced by oriented granules of olivine similar to those seen in discrete pyroxene nodules. The cores of some orthopyroxene grains contain symplectite-like patches of yellow to red-brown spinel.

The orthopyroxenes have restricted ranges of chemical composition (Mg/(Mg + Fe) % = 89.4-92.3, and are mostly enstatites with appreciable Al_2O_3 (2.38-4.94 wt. %) and Cr_2O_3 (0.25-0.86 wt. %) and with a trace of TiO_2 (<0.03-0.16 wt. %). The Al + Cr can be accommodated in the structure equally in the 4 and 6 coordinated sites, with only a small surplus to balance the minor amounts of Na present. The CaO varies from 0.42-1.22 wt. %, the higher values (in specimens 3538, 3539 and 3549B) reflecting increased mutual solubility with diopside at higher temperatures.

Smokey-brown amphibole was observed in one small garnet-bearing inclusion (3550B, Table 2) and also found in washed concentrates of three other nodules (Table 4; J. V. Smith, personal communication). They are similar in composition to paragasitic amphiboles observed in symplectites from ultrabasic nodules by Boyd (1971) and Dawson and Smith (1975).

Spinel is a common mineral in the ultrabasic nodules. It occurs as coarse (up to 3 mm) rounded primary grains, as interstitial blebs and as secondary skeletal (symplectite) forms in orthopyroxene. The primary spinels (Table 2) are all Ti poor, and are of two types: olive green in nodules lacking garnet that is Mg rich and more aluminous than any hitherto recorded in ultrabasic nodules from alnoites and kimberlites (Fig. 4); and, reddish brown spinel, richer in Cr, which does coexist with garnet and is similar to that observed by Ferguson, et al. (1977), in a kimberlite nodule from New South Wales. These are transitional to the chromites that occur in some depleted garnet lherzolites (e.g., Boyd and Nixon, 1972).

A small garnet-clinopyroxene (griquaite) nodule, 3551A, with granoblastic texture and

Table 3. Chemical Compositions of Minerals in the Discrete Nodule (Megacryst) Suite from the Malita Alnoite, wt. %.

	Subcalcic Diopsides										Augites	
PHN	3532/E3	3532/E2	3532/E4	3521/B1	3508/D1	3520	3521/B3	3521/B2	3521/B4	3508/B	3521/A1	3521/A3
SiO_2	54.50	54.30	54.16	53.11	52.08	52.50	53.57	53.55	52.78	52.22	54.06	53.54
TiO_2	0.32	0.32	0.44	0.57	0.59	0.75	0.71	0.77	0.92	1.02	0.92	0.87
Al_2O_3	5.04	5.02	5.09	5.14	5.22	5.41	5.28	5.24	5.35	5.08	5.13	4.84
Cr_2O_3	0.82	0.78	0.40	0.29	0.27	0.18	0.14	0.06	0.03*	0.03	0.03	0.02
Fe_2O_3	-	-	-	-	-	-	-	-	-	-	-	-
FeO	4.07	4.10	5.44	5.43	5.38	5.93	5.69	5.96	6.48	6.88	8.83	9.12
MnO	0.12	0.14	0.12	0.14	0.14	0.13	0.13	0.14	0.13	0.15	0.18	0.18
MgO	20.07	19.65	18.63	18.14	18.00	17.04	17.18	17.14	16.15	15.41	13.15	12.93
CaO	14.63	15.09	14.66	14.99	15.09	15.13	15.75	15.70	15.56	15.94	15.02	15.10
Na_2O	1.44	1.38	1.58	1.70	1.81	2.06	1.86	1.91	2.13	2.41	3.43	3.49
Total	101.01	100.78	100.52	99.51	98.58	99.13	100.31	100.47	99.53	99.14	100.75	100.09
Numbers of cations for n oxygens												
n=	6	6	6	6	6	6	6	6	6	6	6	6
Si	1.925	1.925	1.933	1.921	1.906	1.914	1.926	1.925	1.922	1.919	1.964	1.964
Ti	0.009	0.009	0.012	0.016	0.016	0.021	0.019	0.021	0.025	0.028	0.025	0.024
Al	0.210	0.210	0.214	0.219	0.225	0.232	0.224	0.222	0.230	0.220	0.220	0.209
Cr	0.023	0.022	0.011	0.008	0.008	0.005	0.004	0.002	0.001	0.001	0.001	0.001
Fe^{3+}	-	-	-	-	-	-	-	-	-	-	-	-
Fe^{2+}	0.120	0.122	0.162	0.164	0.165	0.181	0.171	0.179	0.197	0.211	0.268	0.280
Mn	0.004	0.004	0.004	0.004	0.004	0.004	0.004	0.004	0.004	0.005	0.006	0.006
Mg	1.057	1.039	0.991	0.978	0.982	0.926	0.921	0.919	0.877	0.844	0.712	0.707
Ca	0.554	0.573	0.561	0.581	0.592	0.591	0.607	0.605	0.607	0.627	0.585	0.593
Na	0.099	0.095	0.109	0.119	0.128	0.146	0.130	0.133	0.150	0.172	0.242	0.248
Total	4.001	3.999	3.997	4.010	4.026	4.020	4.006	4.010	4.013	4.027	4.023	4.032
Ca	32.0	33.1	32.7	33.7	34.0	34.8	35.7	35.5	36.1	37.3	37.4	37.6
Mg	61.1	59.9	57.8	56.8	56.5	54.6	54.2	54.0	52.2	50.1	45.5	44.7
Fe	6.9	7.0	9.5	9.5	9.5	10.6	10.1	10.5	11.7	12.6	17.1	17.7
$Mg/(Mg+Fe^{2+})$%	89.8	89.5	85.9	85.6	85.6	83.7	84.3	83.7	81.6	80.0	72.6	71.7
Ca/(Ca+Mg)%	34.4	35.6	36.1	37.3	37.6	39.0	39.7	39.7	40.9	42.6	45.1	45.6

* detectability limit is 0.03 for most elements.

(continued)

Table 3, Page 2

PHN	Augites			Bronzites						Pyropes		
	3521/A2	3519	3521/A4	3542N	3532/K1	3532/K4	3532/K5	3556/3A	3532/K3	3511	3512	3514
SiO_2	54.00	52.40	53.05	54.11	55.46	55.46	54.86	54.16	55.06	42.82	42.47	42.55
TiO_2	0.88	0.82	0.88	0.36	0.34	0.33	0.30	0.33	0.29	0.54	0.59	0.58
Al_2O_3	4.82	4.34	5.02	3.29	3.22	3.39	3.46	3.50	3.61	23.13	23.14	23.07
Cr_2O_3	0.02*	0.02	0.02	0.04	0.03	0.06	0.07	0.04	0.16	0.10	0.08	0.17
Fe_2O_3	-	-	-	-	-	-	-	-	-	-	-	-
FeO	9.05	9.61	9.68	9.76	9.82	9.14	9.14	9.43	8.20	10.67	10.73	10.14
MnO	0.18	0.18	0.16	0.18	0.21	0.16	0.17	0.17	0.18	0.33	0.33	0.32
MgO	12.86	12.46	11.76	31.00	30.91	31.27	31.08	29.51	31.73	19.43	19.09	19.69
CaO	15.23	15.00	15.09	1.38	1.33	1.42	1.42	1.38	1.54	4.64	4.60	4.75
Na_2O	3.52	3.63	3.86	0.26	0.29	0.28	0.28	0.29	0.29	0.04	0.05	0.05
Total	100.56	98.46	99.52	100.38	101.61	101.51	100.78	98.81	101.06	101.70	101.08	101.32
Numbers of cations for n oxygens												
n=	6	6	6	6	6	6	6	6	6	12	12	12
Si	1.970	1.964	1.965	1.900	1.920	1.916	1.911	1.925	1.905	3.011	3.006	3.000
Ti	0.024	0.023	0.025	0.010	0.009	0.009	0.008	0.009	0.008	0.029	0.031	0.031
Al	0.207	0.192	0.219	0.136	0.131	0.138	0.142	0.147	0.147	1.917	1.931	1.917
Cr	0.001	0.001	0.001	0.001	0.001	0.002	0.002	0.001	0.004	0.006	0.004	0.009
Fe^{3+}	-	-	-	-	-	-	-	-	-	-	-	-
Fe^{2+}	0.276	0.301	0.300	0.287	0.284	0.264	0.266	0.280	0.237	0.627	0.635	0.598
Mn	0.006	0.006	0.005	0.005	0.006	0.006	0.005	0.005	0.005	0.020	0.020	0.019
Mg	0.699	0.696	0.649	1.623	1.595	1.611	1.614	1.564	1.637	2.037	2.015	2.070
Ca	0.595	0.602	0.599	0.052	0.049	0.053	0.053	0.053	0.057	0.350	0.349	0.359
Na	0.249	0.264	0.277	0.018	0.019	0.019	0.019	0.020	0.019	0.005	0.007	0.007
Total	4.027	4.049	4.040	4.032	4.020	4.018	4.020	4.004	4.019	8.002	7.998	8.010
Ca	37.9	37.7	38.7	2.6	2.6	2.7	2.7	2.8	3.0	11.6	11.6	11.9
Mg	44.5	43.5	41.9	82.7	82.7	83.6	83.5	82.4	84.7	67.6	67.2	68.3
Fe	17.6	18.8	19.4	14.6	14.7	13.7	13.8	14.8	12.3	20.8	21.2	19.8
$Mg/(Mg+Fe^{2+})$%	71.7	69.8	68.4	85.0	84.7	85.9	85.8	84.8	87.3	76.5	76.0	77.6
Ca/(Ca+Mg)%	46.0	46.4	48.0	-	-	-	-	-	-	-	-	-

* detectability limit is 0.03 for most elements.

(continued)

Table 3, Page 3

PHN	Pyrope with inclusion 3516		Lamellar intergrowths 3522		3523	3532/C1		3532/C2		3532/C3	3532/C4	3532/C6
	py	cpx	cpx	ilm	ilm	cpx	ilm	cpx	ilm	cpx	ilm	cpx
SiO_2	41.91	53.43	52.86	0.00	0.00	53.91	0.00	53.63	0.00	53.97	0.00	53.73
TiO_2	0.55	0.84	0.95	51.06	50.73	0.84	51.24	0.81	50.84	0.80	50.75	0.72
Al_2O_3	22.71	5.27	4.83	0.74	0.70	5.00	0.99	4.94	1.05	4.98	0.73	4.97
Cr_2O_3	0.08	0.05	0.00*	0.00	0.00	0.02	0.03	0.03	0.03	0.02	0.03	0.02
Fe_2O_3	-	-	-	9.70	10.08	-	10.17	-	9.85	-	10.05	-
FeO	10.86	5.97	7.91	13.67	32.97	7.58	30.89	8.11	31.36	8.17	31.89	7.57
MnO	0.32	0.13	0.18	0.27	0.27	0.16	0.26	0.16	0.27	0.14	0.29	0.14
MgO	19.00	16.51	14.91	7.78	6.91	14.93	8.35	14.23	7.88	14.07	7.51	14.68
CaO	4.48	15.73	15.67	0.04	0.03	15.54	0.03	15.55	0.03	15.66	0.05	15.88
Na_2O	0.04	1.96	2.76	-	-	2.48	-	2.74	-	2.82	-	2.48
NiO			-	0.05	0.02	-	0.00	-	0.00	-	0.00	-
Total	99.95	99.89	100.07	101.31	101.71	100.46	101.96	100.20	101.31	100.63	101.30	100.19
Numbers of cations for n oxygens												
n=	12	6	6	3	3	6	3	6	3	6	3	6
Si	3.004	1.932	1.932	0.000	0.000	1.952	0.000	1.954	0.000	1.958	0.000	1.953
Ti	0.030	0.023	0.026	0.905	0.902	0.023	0.897	0.022	0.898	0.022	0.901	0.020
Al	1.919	0.225	0.208	0.021	0.019	0.213	0.027	0.212	0.029	0.213	0.020	0.213
Cr	0.005	0.001	0.000	0.000	0.000	0.001	0.001	0.001	0.001	0.001	0.001	0.001
Fe^{3+}	-	-	-	0.172	0.179	-	0.178	-	0.174	-	0.178	-
Fe^{2+}	0.651	0.180	0.242	0.623	0.651	0.229	0.601	0.247	0.616	0.248	0.629	0.230
Mn	0.019	0.004	0.006	0.005	0.005	0.005	0.005	0.005	0.005	0.004	0.006	0.004
Mg	2.031	0.890	0.813	0.273	0.243	0.806	0.290	0.773	0.276	0.761	0.264	0.795
Ca	0.344	0.609	0.614	0.001	0.001	0.603	0.001	0.607	0.001	0.609	0.001	0.618
Na	0.006	0.137	0.196	-	-	0.174	-	0.194	-	0.198	-	0.175
Ni			-	0.001	0.000	-	0.000	-	0.000	-	0.00C	-
Total	8.009	4.001	4.037	2.000	2.000	4.006	2.000	4.015	2.000	4.014	2.000	4.009
Ca	11.4	36.3	36.8			36.8		37.3		37.6		37.6
Mg	67.1	53.0	48.7			49.2		47.5		47.1		48.4
Fe	21.5	10.7	14.5			14.0		15.2		15.3		14.0
il				63.5	66.2		61.3		62.9		64.0	
gk				27.8	24.7		29.6		28.2		26.9	
hm				8.7	9.1		9.1		8.9		9.1	
Mg/(Mg+ Fe^{2+})%	75.7	83.1	77.1	30.5	27.2	77.8	32.5	75.8	30.9	74.4	29.6	77.6
Ca/(Ca+Mg)%	-	40.6	43.0	-	-	42.8	-	44.0	-	44.4	-	43.7

* detectability limit is 0103 for most elements.

(continued)

Table 3, Page 4

PHN	Lamellar intergrowths						Ilmenites				
	3532/C5		3532/C7		3543B		3532/A1	3532/A2	3543C	3556/5A	3556/5B
	cpx	ilm	cpx	ilm	cpx	ilm					
SiO_2	53.73	0.00	53.98	0.00	53.77	0.00	0.00	0.00	0.00	0.00	0.00
TiO_2	0.79	52.23	0.77	50.60	0.88	51.10	52.51	50.47	50.00	52.09	49.49
Al_2O_3	5.04	0.97	4.90	0.93	5.23	0.90	0.65	0.57	0.73	0.49	0.41
Cr_2O_3	0.02*	0.03	0.03	0.03	0.03	0.03	0.00	0.00	0.04	0.00	0.00
Fe_2O_3	-	8.67	-	10.26	-	9.53	9.36	10.12	10.12	10.07	10.41
FeO	7.52	30.93	7.90	30.68	7.84	31.02	29.31	33.88	32.65	28.91	36.55
MnO	0.14	0.25	0.15	0.25	0.15	0.25	0.35	0.30	0.29	0.52	0.31
MgO	14.77	8.82	14.98	8.15	14.49	8.20	9.80	6.27	6.70	9.71	4.27
CaO	15.90	0.05	15.29	0.03	15.61	0.05	0.06	0.02	0.06	0.07	0.02
Na_2O	2.49	-	2.52	-	2.80	-	-	-	-	-	-
NiO	-	0.00	-	0.00	-	0.00	0.01	0.00	0.00	0.00	0.00
Total	100.40	101.95	100.52	100.93	100.80	101.08	102.05	101.63	100.59	101.86	101.46
Numbers of cations for n oxygens											
n=	6	3	6	3	3	3	3	3	3	3	3
Si	1.948	0.000	1.955	0.000	1.945	0.000	0.000	0.000	0.000	0.000	0.000
Ti	0.022	0.910	0.021	0.895	0.024	0.904	0.910	0.901	0.897	0.906	0.899
Al	0.215	0.027	0.209	0.026	0.223	0.025	0.018	0.016	0.021	0.013	0.012
Cr	0.001	0.001	0.001	0.001	0.001	0.001	0.000	0.000	0.001	0.000	0.000
Fe^{3+}	-	0.151	-	0.182	-	0.168	0.162	0.181	0.182	0.175	0.189
Fe^{2+}	0.228	0.600	0.239	0.604	0.237	0.609	0.565	0.673	0.652	0.559	0.739
Mn	0.004	0.005	0.005	0.005	0.005	0.005	0.007	0.006	0.006	0.010	0.006
Mg	0.799	0.305	0.809	0.286	0.781	0.287	0.337	0.222	0.239	0.335	0.154
Ca	0.618	0.001	0.593	0.001	0.605	0.001	0.001	0.001	0.002	0.002	0.001
Na	0.175	-	0.177	-	0.196	-	-	-	-	-	-
Ni	-	0.000	-	0.000	-	0.000	0.000	0.000	0.000	0.000	0.000
Total	4.010	2.000	4.009	2.000	4.017	2.000	2.000	2.000	2.000	2.000	2.000
Ca	37.6		36.1		37.3						
Mg	48.5		49.3		48.1						
Fe	13.9		14.6		14.6						
il		61.2		61.5		62.1	57.4	68.3	66.4	57.0	74.8
gk		31.1		29.2		29.3	34.3	22.5	24.3	34.1	15.6
hm		7.7		9.3		8.6	8.3	9.2	9.3	8.9	9.6
Mg/(Mg+ Fe^{2+})%	77.8	33.7	77.2	32.1	76.7	32.0	37.3	37.4	26.8	37.4	17.2
Ca(Ca+Mg)%	43.6	-	42.3	-	43.6	-	-	-	-	-	-

* detectability limit is 0.03 for most elements.

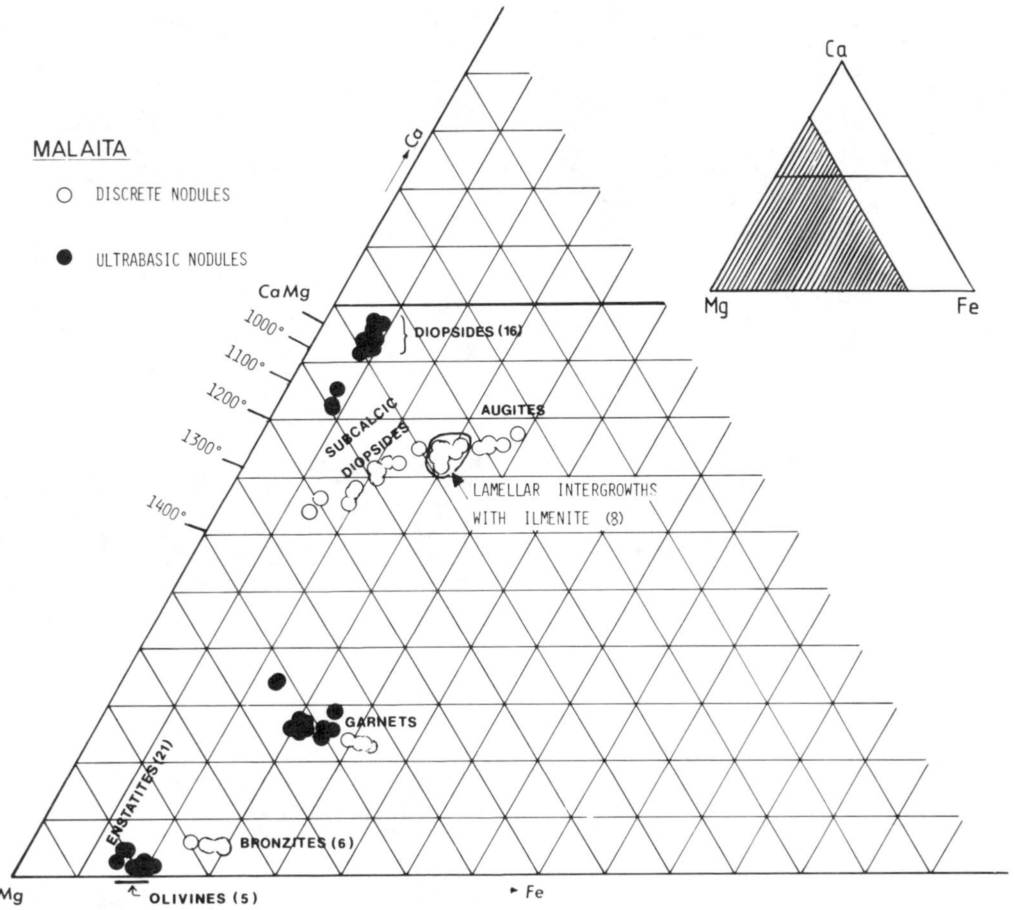

Fig. 2. Compositions of minerals from the lherzolitic and discrete nodule suites from the alnoitic pipe, Babaru'u, Malaita, Solomon Islands, S.W. Pacific.

subordinate primary spinel is included in Table 2, but it may not be a modal variant of the ultrabasic nodule suite because of the low Cr_2O_3 values. It resembles recrystallized nodules originating by unmixing of clinopyroxene, described by Shervais, et al. (1973).

Discrete Nodule (Megacryst) Suite

Large, rounded, single crystals of garnet, clinopyroxene, orthopyroxene and ilmenite that are found in kimberlite are believed to be consanguineous (Nixon and Boyd, 1973). Granular aggregates of two or more of these phases and lamellar intergrowths (particularly of pyroxene and ilmenite) have been found at many localities but are less common than the monomineralic discrete nodules. A similar suite from Malaita is dominated by particularly large, smooth, reddish-brown garnets, e.g., specimens 3511 and 3512 that weigh 8.2 and 3.3 kg respectively (Fig. 3b). Specimen 3516 is an 8 cm nodule containing inclusions of subcalcic diopside. Although heavily fractured, the garnet nodules contain clear zones of gem-quality garnet up to one cm in size. They are pyrope-rich (py 67-68, Table 3, Fig. 2) with minor amounts of TiO_2 and Cr_2O_3, and with detectable Na_2O.

The clinopyroxenes are dark green, fractured, rounded (where unbroken) and are several cm in diameter (Fig. 3c), with pale grey alteration skins. Small fragments can be divided into two color groups when examined under a strong light or microscope: greyish-green augites ($n > 1.695$) and much paler subcalcic diopsides ($n < 1.695$; lower 2V +).

The ranges of composition of the discrete augites and subcalcic diopsides (Table 3, Figs. 2 and 5) are distinct but the gap between them is bridged by analyses of clinopyroxenes from lamellar intergrowths. Within the subcalcic group the least calcic specimens are the richest in MgO and Cr_2O_3, and poorest in Na_2O and TiO_2, as was found for the diopsides from the ultrabasic nodules. The discrete augites have compositions that extend these trends with the possible exception of TiO_2 (Fig. 5). The clinopyroxene inclusions in the discrete pyrope crystal, 3516, clearly belong to the subcalcic group (compare with analysis 3521/B2 in Table 3).

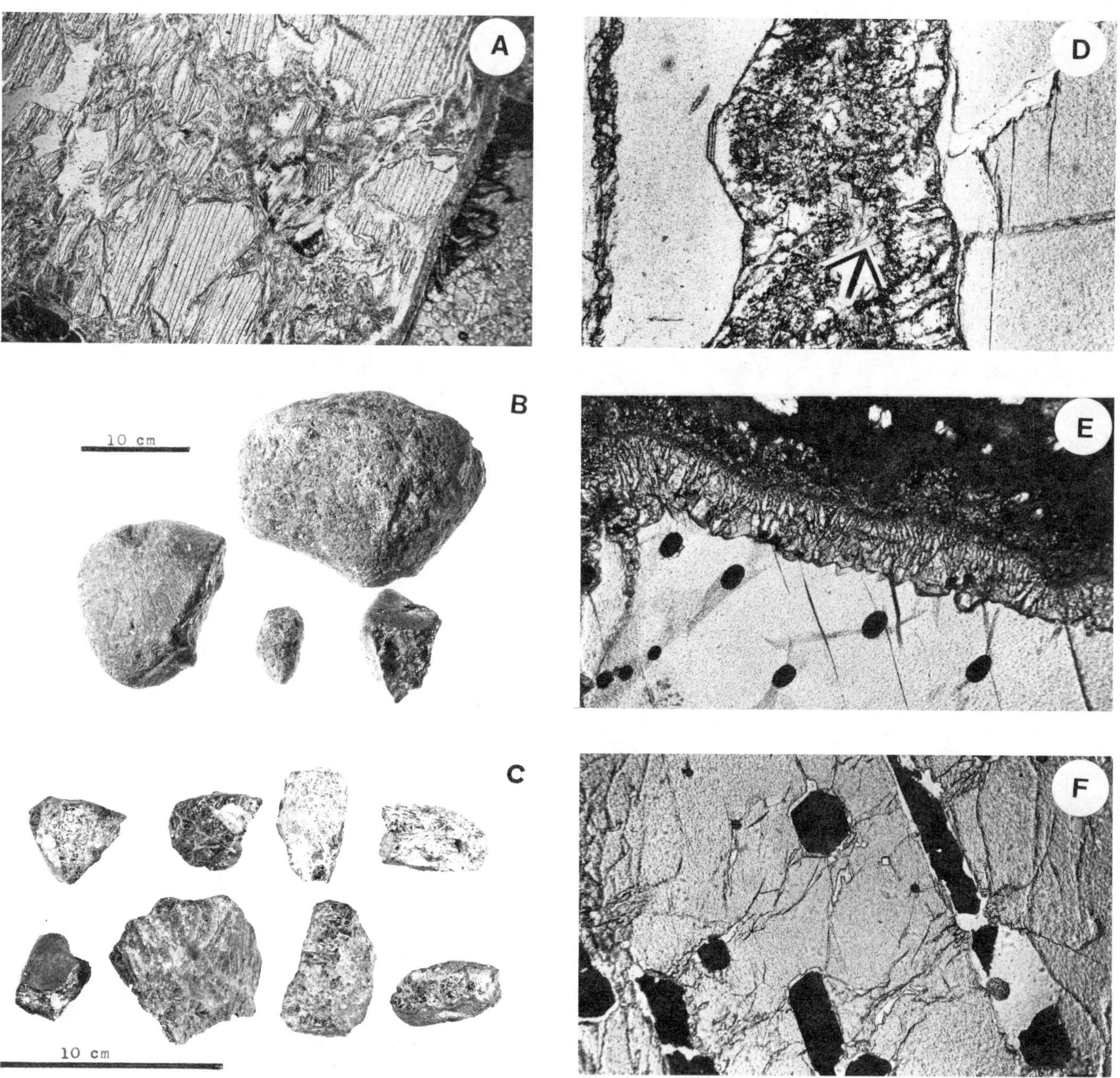

Fig. 3.

(a) Shattered and recemented orthopyroxene in lherzolite xenolith PHN 3538, ordinary light, width of section is 2 mm.

(b) Discrete garnet nodules from the Malaita alnoite. The largest specimen, PHN 3511 weighs 8.2 kg.

(c) Fragments of discrete pyroxene nodules from the Malaita alnoite. The four specimens on the left are augites and those on the right are subcalcic diopsides.

(d) Alteration along crack in discrete augite PHN 3521/A2 showing development of kaersutite (arrowed). Ordinary light, width of alteration zone is 1 mm.

(e) Bronzite megacryst with aligned sulphide blebs and reaction rims against alnoite. Ordinary light, width of alteration zone is 1 mm.

(f) Clinopyroxene-ilmenite lamellar intergrowth. Small black spots are sulphide blebs. Ordinary light, width of section in 2 mm.

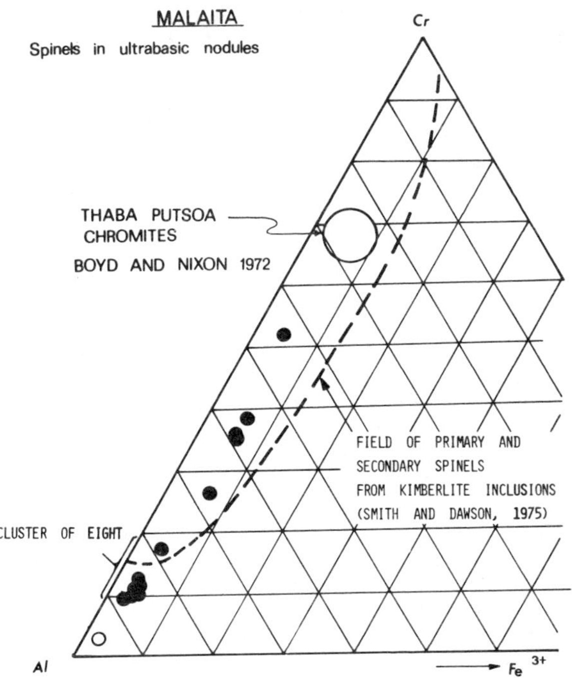

Fig. 4. Plots of spinel compositions, in terms of Cr, Al and Fe^{3+}, from Malaita nodules. Note the highly aluminous nature of those from seven spinel lherzolites and one spinel wehrlite (cluster of eight), and from garnet-pyroxene specimen PHN 3551A (open circle). Four spinels, more highly chromiferous, are closer in composition to chromites in garnet-bearing nodules such as those from Thaba Putsoa, Lesotho (Boyd and Nixon, 1972). The dashed line encloses published analyses of primary and secondary (symplectite) spinels from deep seated xenoliths (Smith and Dawson, 1975). Points are plotted for analyses in Table 2 and unpublished results.

A two-fold division of megacrysts from the Sloan kimberlite pipes, Colorado, has been delimited by Eggler and McCallum (1976). Nevertheless the Sloan megacrysts differ from the Malaita suite. Cr-rich diopsides from Sloan are compositionally similar to those from ultrabasic nodules, although much coarser grained. The Cr_2O_3 values in both Malaita groups are within the Cr-poor range at Sloan. The Malaita augite group appears to have no counterpart in kimberlite nodule suites, but there are similarities with some clinopyroxene megacrysts from deep-seated, garnet-bearing alkaline tuffs, e.g., at Kakanui, New Zealand (Mason, 1968; Dickey, 1968; Mason and Allen, 1972), Elie Ness, Scotland (Chapman, 1976) and kimberlite with alnoitic affinities, Ile Bizard, Quebec (Marchand, 1970). Table 4 illustrates comparable levels of TiO_2, Cr_2O_3, and MgO/FeO and the fact that these augites are mildly subcalcic. Surface alteration of discrete pyroxenes is ascribed to reaction with the alnoite liquid. Secondary amphibole (kaersutite) has formed along cracks in discrete augite 3521/A2 (Fig. 3d, Table 4). Kaersutite is a secondary mineral in augite megacrysts in basanites, New South Wales (Binns, 1969) and in kimberlite with alnoitic affinities, Ile Bizard, Quebec (Marchand, 1970).

The discrete orthopyroxene nodules are coarse, cleaved, brown crystals of bronzite composition (av. En_{86}, Table 3, Fig. 2). They are sufficiently magnesian and calcic to have equilibrated with the subcalcic diopsides rather than the augites. With increasing calcic nature, i.e., increasing equilibration temperature, Mg/(Mg + Fe) and Cr_2O_3 wt. % tend to increase, whereas TiO_2 decreases. There is no evidence that there is an orthopyroxene or pigeonite group that coexisted with the augites. Aligned blebs of sulfide were observed in the discrete orthopyroxenes (Fig. 3e).

The ilmenite nodules are usually <1 cm in diameter and are most conspicuous in enriched river gravels. They are rounded, with smooth

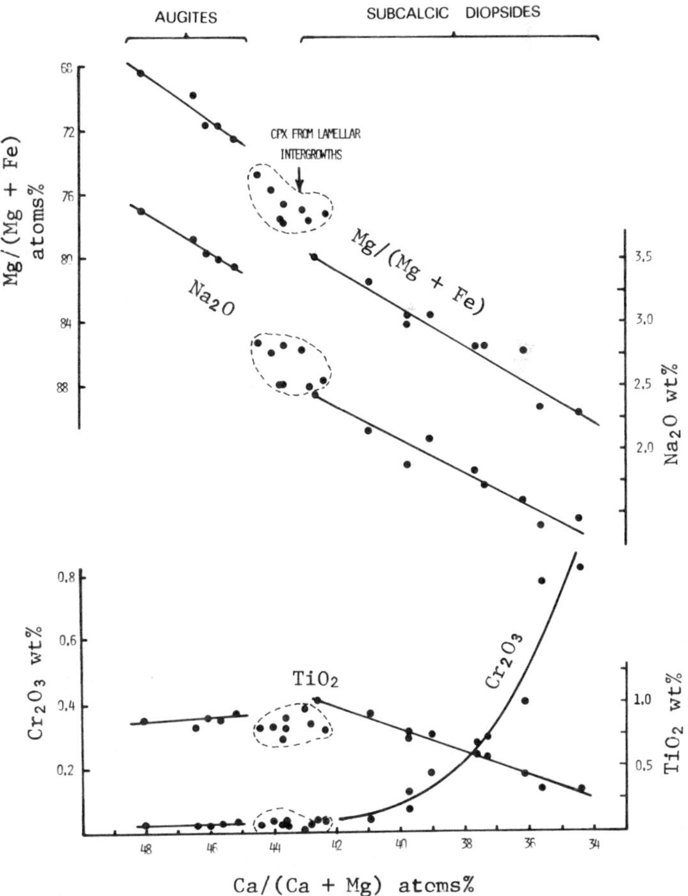

Fig. 5. Chemical variation within the discrete clinopyroxene group of the Malaita alnoite, showing the division into augites and subcalcic diopsides.

TABLE 4. **Comparisons of Compositions of Discrete Augite Megacrysts and Secondary Amphibole from Malaita with Those from Other Deep-Seated Volcanic Intrusions.**

	1	2	3	4	5	6
SiO_2	52.40-54.06	49.18	50.73	51.30	40.95	40.81
TiO_2	0.82- 0.92	0.79	0.74	0.64	6.08	3.15
Al_2O_3	4.34- 5.13	8.29	7.86	4.72	12.30	14.54
Cr_2O_3	0.02- 0.03	0.10	-	-	0.08	-
Fe_2O_3	-	2.98	3.69	-	-	-
FeO	8.83- 9.68*	3.88	3.45	5.19	9.61	9.91
MnO	0.16- 0.18	0.16	0.13	-	0.11	-
MgO	11.76-13.15	16.30	16.65	17.09	13.46	13.72
CaO	15.00-15.23	15.96	15.82	18.85	11.03	12.13
Na_2O	3.43- 3.86	1.29	1.27	1.33	2.94	2.12
K_2O	-	0.00	0.00	0.00	1.36	2.05
	- -	98.93	100.34	99.12	97.92	98.43

* Total Fe as FeO.

1. Augite, range in compositions (Table 3), Babaru'u, Malaita.
2. Augite megacryst from Elie Ness tuff (Chapman, 1976).
3. Augite megacryst from Kakanui; Trace element determinations include Cr, 785 ppm by wt. (Mason and Allen, 1973).
4. Augite megacryst from Kimberlite with alnoitic affinities, Ile Bizard (Marchand, 1970).
5. Amphibole alteration of augite PHN 3521/A2, Babaru'u, Malaita.
6. Amphibole, probably formed from augite, kimberlite with alnoitic affinities, Ile Bizard (Marchand, 1970).

to dimpled surfaces occasionally peppered with secondary perovskite granules. Fracture varies from conchoidal to hackly and granular, indicating that both single crystals and aggregates of partially annealed grains (0.5 mm) are present. Voids lined with a soft, amorphous, brown mineral occur. Both physically and chemically (Table 3), the ilmenites resemble those from kimberlites although $MgTiO_3$ contents are in some cases lower (Fig. 6). Calculated values

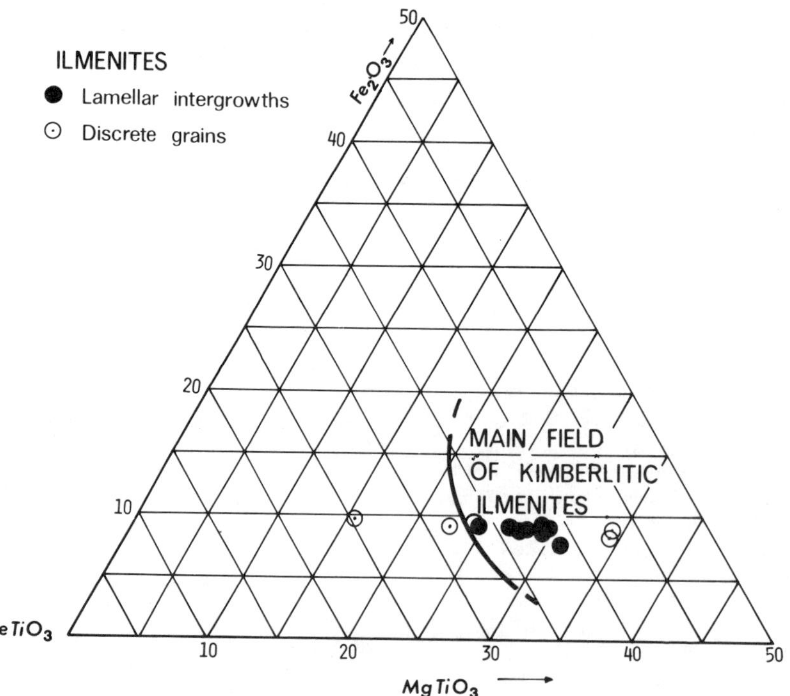

Fig. 6. Compositions of Malaita ilmenites compared with the field of kimberlite ilmenites from Boyd and Nixon (1973), Haggerty (1975) and other sources.

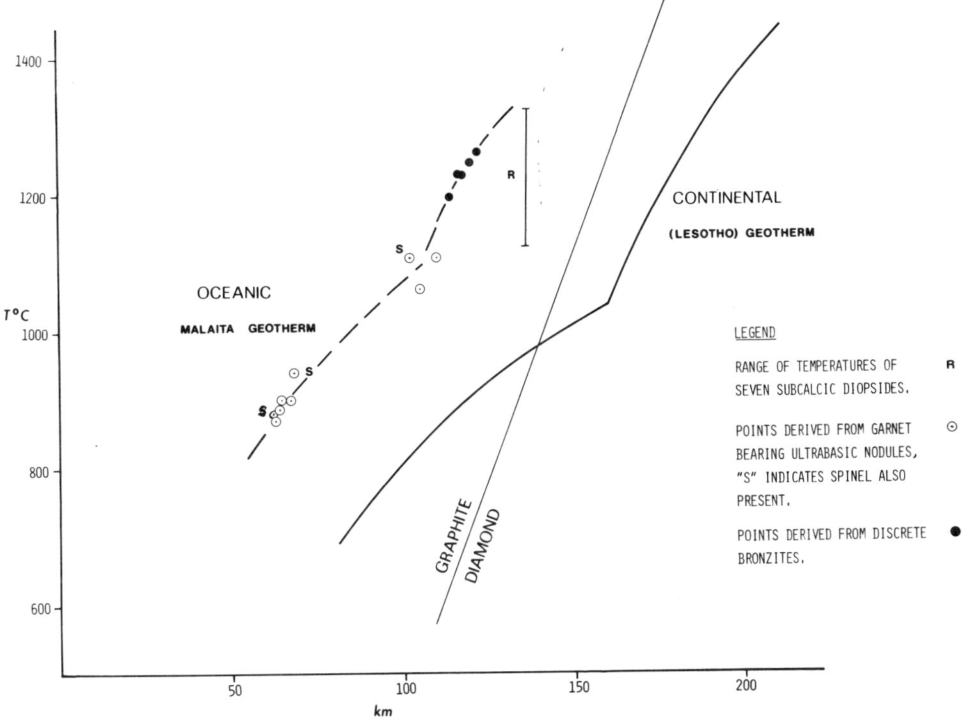

Fig. 7. Estimates of temperature and depth of equilibration for garnet lherzolite and bronzite discrete nodules from the Malaita alnoite (oceanic) compared with the Lesotho kimberlite (continental) geotherm. Three small weathered lherzolites did not contain fresh garnet (see text). The graphite/diamond equilibrium boundary is from Kennedy and Kennedy (1976).

of Fe_2O_3 for the analyzed ilmenites are also low and lie within the range 8-10%. The following minor oxides in the Malaita ilmenites were normally below probe detection limits, in contrast to kimberlitic ilmenites, for which representative ranges in wt. % are given in parentheses: SiO_2 (0.05-0.4), Cr_2O_3 (<0.03-1) and NiO (0.05-0.2).

The Malaita ilmenite-clinopyroxene nodules are commonly several cm in maximum dimension and consist of oriented lamellae or rods of ilmenite in clinopyroxene (Fig. 3f; also Allen and Deans, 1965). They appear similar to those found in many kimberlites (e.g., Williams, 1932; Gurney, et al., 1973). Individual lamellae are about 0.4-0.7 mm thick with clinopyroxene predominating. Ilmenite analyses from the intergrowths are similar to those of monomineralic nodules (Table 3), with the exception of a slight enrichment in Al in the lamellar ilmenites. The clinopyroxenes are similar in composition to those in lamellar intergrowths from kimberlite. They are relatively poor in Cr and rich in Fe compared with diopsides from lherzolites and subcalcic discrete nodules (Fig. 2; Boyd, 1971; Gurney, et al., 1973).

Zircon is possibly a member of the discrete nodule suite as it has been found in southern Africa to occur with ilmenite as inclusions and intergrowths. Only rounded single crystals were observed on Malaita and the largest (7 mm) from the Fafa Ikafo gravels was found to have an age of 33.9 m.y. (G. L. Davis, 1977).

The Geotherm

A plot of estimated equilibration temperatures and depths for the garnet lherzolites (Fig. 7) was constructed using the "raw Al_2O_3" method described by Boyd (1973) and Boyd and Nixon (1973). This method involves temperature determinations from a comparison of natural diopside compositions with the experimental diopside-enstatite solvus at 30 kbar (Davis and Boyd, 1966), and pressure determinations from a comparison of Al_2O_3 contents of natural enstatites with equilibria determined for the system enstatite-pyrope (MacGregor, 1973). Improvements in the synthetic phase relations have been made (e.g., Lindsley and Dixon, 1976; Akella, 1976) and alternative methods proposed (e.g., Wells, 1977; Wood, 1974). The primary purpose in constructing the plot in Fig. 6, however, is comparison with earlier geotherm calculations and it is a convenience to retain the original method. Results obtained by the "raw Al_2O_3" method are consistent with occurrences of diamond and graphite in lherzolites

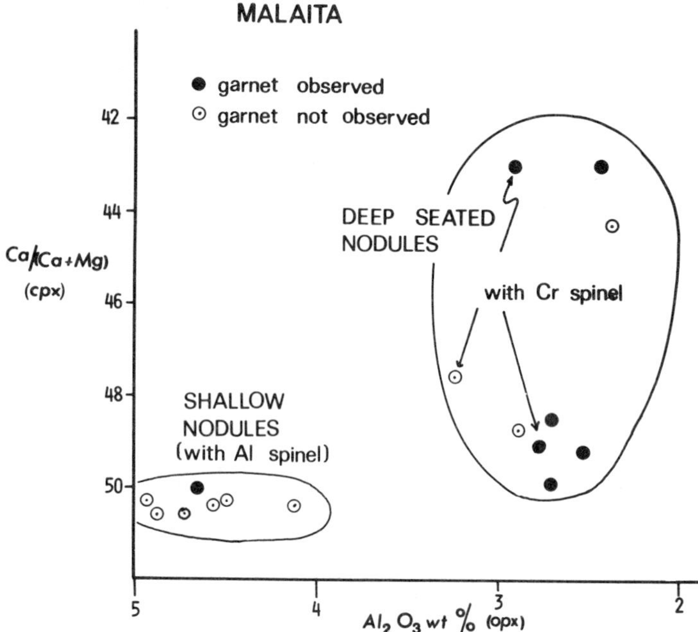

Fig. 8. Division of the Malaita lherzolites into two groups using Ca/(Ca + Mg) % cpx (temperature variable) and Al_2O_3 wt. % opx (pressure variable) in order to identify deep-seated xenoliths believed to have equilibrated in the garnet stability field although garnet was not always observed (possibly due to alteration of small size of the nodule). Note that Cr, Al-spinel was observed in two nodules from the deep-seated group but that in the garnet-free shallow lherzolites, with one exception, the spinel is highly aluminous.

and it, therefore, seems improbable that they are in gross error.

It has been suggested that solution of Al_2O_3 in pyroxenes in a Tschermak couple reduces the solid solution between diopside and enstatite (O'Hara, 1967). It can be seen from experimental data (Boyd, 1970) that this effect is sufficiently small to be negligible for Al_2O_3 contents up to 3 wt. %. The Al_2O_3 contents of diopsides in garnet lherzolite nodules from kimberlites are normally less than 3 wt. % and the atomic proportions of Al + Cr and Na are usually about equal (Boyd, 1970). The clinopyroxenes in the Malaita lherzolites, however, contain up to ~6 wt. % Al_2O_3. At such high concentrations the effect of Al on the diopside solvus might be significant. Nevertheless, Wells (1977) found no consistent relationship between the diopside-enstatite miscibility gap and Al_2O_3 concentrations in synthetic pyroxenes of up to 12 wt. %.

Equilibration pressures for spinel lherzolites cannot at present be estimated (e.g., Herzberg and Chapman, 1976) and a group of low-temperature (shallow) lherzolites (Fig. 8) were hence excluded from geotherm calculations. These lherzolites contain Al-rich spinels, all with one exception, lacking garnet. Six garnet-bearing lherzolites appear to form a coherent group with three additional nodules in which garnet was not observed (Fig. 8), possibly due to small grain size or the altered nature of the xenoliths. Garnet was, nevertheless, assumed to be present in these. Three nodules, including one without observed garnet also contained spinel. These spinels are Cr-rich (Fig. 4), however, and do not appear to be a reaction product of garnet; cf. Thaba Putsoa chromite lherzolites (Boyd and Nixon, 1972). These three nodules of uncertain origin plot close to the geotherm formed by points for unambiguous garnet lherzolites (Fig. 7).

Equilibration conditions can also be estimated for the discrete pyroxenes if the bronzite, subcalcic diopside, and pyrope are consanguineous (Nixon and Boyd, 1973). The temperature range of equilibration of the subcalcic diopsides is considerably higher than that indicated by the lherzolites--1120° to 1320°C (Fig. 7).

Both temperature and pressure can be estimated for bronzites, assuming they crystallized as part of the assemblage opx-cpx-gt. Inasmuch as the pressure effect on the enstatite solvus is uncertain, it is necessary to calibrate Ca/(Ca + Mg) for the natural orthopyroxenes against Ca/(Ca + Mg) for coexisting clinopyroxenes. Curves constructed for pyroxenes in lherzolites from Malaita and Lesotho (Fig. 9) differ and it is clear that each area must be treated separately. As a test, it should be seen that the range of equilibration temperatures estimated for the discrete bronzites, is similar to that of the discrete subcalcic diopsides for a particular occurrence. The reason for the difference in the Malaita and Lesotho calibration curves is probably due to the effect of pressure on the solubility of diopside in enstatite. Increasing pressure decreases Ca/(Ca + Mg) (Akella, 1974). The Malaita enstatites have higher values of Ca/(Ca + Mg) for given equilibration temperatures than do enstatites from Lesotho because the former have equilibrated in a shallower depth range. Differences in minor element composition may also affect the calibration curves.

The Malaita geotherm gives an estimate of a palaeo-temperature distribution in an oceanic lithosphere, and reflects the high heat flow, relative to continental (shield) areas where garnet-bearing nodule data have previously been obtained. Other oceanic estimates can be derived from garnet pyroxenite nodules from Salt Lake Crater, Hawaii (Pleistocene-Holocene) described by Beeson and Jackson (1970). If temperature-depth estimates are made for their calculated analyses of unexsolved, deep-seated minerals, they are found to plot at higher temperatures than points for the Malaita suite.

It is possible that diapiric upwelling takes

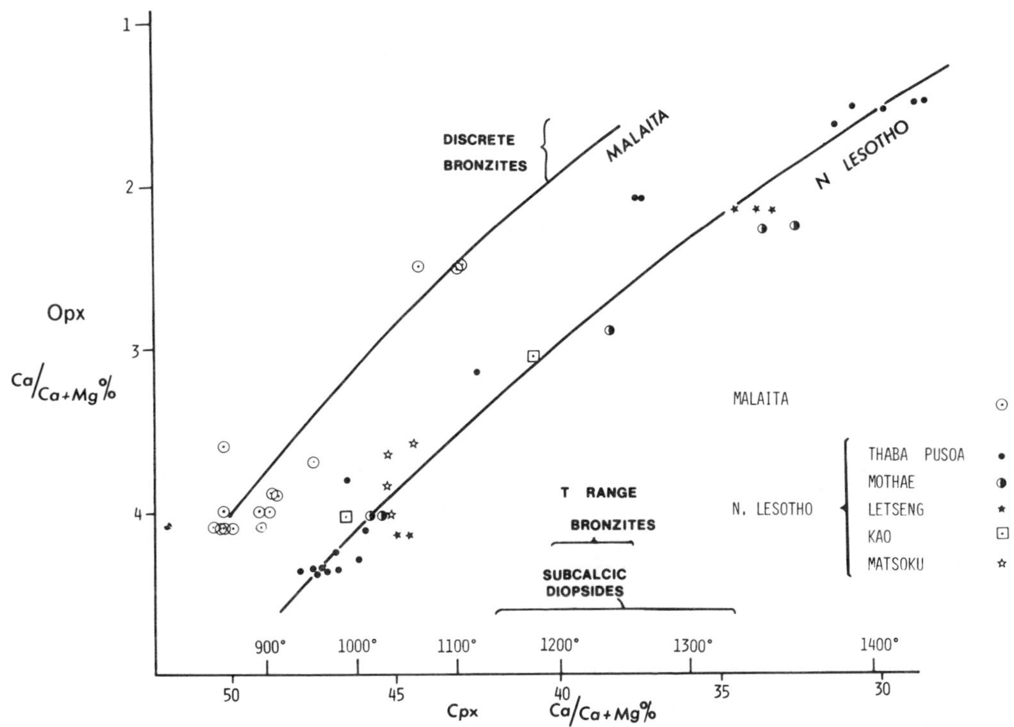

Fig. 9. Calibration curves for coexisting opx and cpx in lherzolites from the Malaita alnoite and Lesotho kimberlites used for determining equilibration temperatures of discrete bronzite nodules. The temperature ranges of the Malaita discrete bronzites and subcalcic diopsides are compared at the bottom centre. See text for discussion of the difference between these curves.

place in the asthenosphere (Green and Gueguen, 1974) with kimberlite eruption being initiated where a diapir impinges on the base of the lithosphere (Boyd, 1976). The evidence from Malaita (Fig. 7) suggests that diapiric movement followed by alnoitic intrusion took place from the base of a lithosphere having a thickness of 110 km. Although the estimated lithospheric thickness is less than under continents where kimberlites occur, it is, nevertheless, unusually thick for an oceanic region. The role played by this thickened portion of the Pacific Plate (Ontong Java Plateau) in the emplacement of the Malaita alnoite is discussed by Nixon and Coleman (in press).

The shallower depth of alnoite formation compared with that of kimberlite may be a factor in melilite crystallization as this mineral is rare or absent in kimberlite; however, host rock composition may also be important, and, although both alnoite and kimberlite are ultrabasic and enriched in alkalis, volatiles and similar trace elements, small increases in Na_2O content might govern the appearance of melilite (F. E. Hough, in Velde and Yoder, 1976). It is perhaps significant that the Malaita ankaratrite, described by Allen and Deans (1965), that lacks melilite is also very low in Na_2O.

The Oceanic Mantle

There is little evidence of a variation of deformation with depth of origin as was found

TABLE 5. Average Chemical Values of Minerals in Malaita Garnet Lherzolite Nodules Illustrating Greater Depletion in the More Deep-Seated Rocks. Deep-seated noudles are PHN 3538, 3539, and 3549B; shallow-seated rocks are all other garnet lherzolites listed in Table 2.

	Mg/(Mg+Fe)%			Na_2O wt. %			TiO_2 wt. %			Cr_2O_3 wt. %		
	opx	cpx	gt	opx	cpx	gt	opx	cpx	gt	opx	cpx	gt
Shallow-seated	90.5	91.7	81.0	0.06	1.58	<0.03	0.12	0.45	0.11	0.41	0.81	1.23
Deep-seated	92.1	93.1	85.8	0.03	0.43	<0.03	<0.03	<0.03	<0.03	0.88	1.13	5.04

MALAITA OCEANIC MODEL

Depth, km	Succession		Spinel/Garnet
0	Basaltic crust		
20			
	Harzburgite (very depleted)	LITHOSPHERE	Cr spinel
40			
			Al spinel
60	Lherzolite		Garnet
80			Cr Al spinel & Cr garnet
	Depleted lherzolite		
100			
120	(ilmenite-rich zone) Zone rich in large crystals of garnet and pyroxene.	LVZ	Garnet
140			

Fig. 10. Stratigraphic succession through the upper mantle in Melanesia, S.W. Pacific largely based on the study of inclusions from the Malaita alnoite, Solomon Islands. Depths are obtained by the pyroxene geotherm method described in the text. The nature of the harzburgite mantle immediately underlying the crust is derived from studies of obducted ophiolites described by Thompson (1965) and Davies (1971).

for Lesotho nodules (Nixon, et al., 1973). Too few samples were obtained to confirm whether the foliation observed in specimen 3539 might be confined to nodules of more deep-seated origin. Shattering of pyroxenes and recementation was observed in the nodules of deepest origin, and also in one nodule of shallower origin (3568, previously described). This shattering is interpreted as being a product of explosive eruption of the host alnoite.

The following features of the paleo-lithosphere are believed to have existed on the basis of studies of the lherzolite nodules: (1) the modal mineralogy was similar to that observed elsewhere, but with possibly greater amounts of diopside; (2) aluminous spinel was present at intermediate depths (70 km); (3) pargasitic amphibole occurred at depths down to at least 70 km; (4) phlogopite was surprisingly sparse; and (5) graphite would have been stable rather than diamond in the whole of the mantle section penetrated by the alnoite (down to 135 km).

Bulk analyses of the nodules were not feasible due to their highly altered nature. Microprobe analyses of minerals (Table 2) illustrate a chemical variation (summarized in Table 5) which shows that the three nodules of deepest origin are relatively depleted in Fe, Na, Ti and enriched in Cr relative to Al. A nodule originating at slightly shallower depth (Table 2, 3533), has mineral compositions intermediate between those of deep and shallow origin. The base of the lithosphere would thus appear to be more depleted than the portion immediately above it. The Cretaceous-Eocene succession of Alite volcanics may have originated in this deeper zone ($\simeq$ 100 km). The uppermost lithosphere is not represented by inclusions, but it is known from upthrust wedges elsewhere in Melanesia to consist of harzburgite (Thompson, 1965; and Davies, 1971) that may have been depleted during initial formation of the overlying crust (Coleman, 1971). It is suggested on the basis of limited samples that the central zone of the lithosphere may be the least depleted (Fig. 10).

The Mantle Role of Discrete Nodules (Megacrysts)

Large chemical variations within this group, particularly of $Mg/(Mg + Fe^{2+})$, suggest that igneous fractionation has taken place. Nixon and Boyd (1973) have previously suggested that this fractionation occurred in the asthenosphere while liquids seeped upwards through a crystal mush following a P-T path along the geotherm. The discrete nodule suite crystallized in contact with the rising liquid which became enriched in Fe, Na, and Ti (cf. "zone refining," Harris, 1957). Ilmenite and ilmenite intergrowths crystallized in the upper part of the Low Velocity Zone (Fig. 10).

The lack of evidence in the Malaita suite for lherzolites that equilibrated at depths equivalent to the discrete nodules suggests that the upper asthenosphere consisted mainly of coarse crystals of which the discrete nodules are fragments, i.e., essentially of garnet and pyroxenes with some ilmenite. These silicates are more ferriferous than similar species in the lherzolites. If olivine were present it must have been comminuted or have been altered. The discrete augite megacrysts have a similar fractionation trend to that of the subcalcic diopsides (Fig. 5) but they have probably equilibrated in the lithosphere. They are comparable to megacrysts found in alkaline basalts that are believed to have crystallized from evolving, undersaturated magmas (Binns, et al., 1970; Wilkinson, 1975; Chapman, 1976). The extent of alnoite-ankaratrite differentiation is not known at Malaita, hence it is not yet possible to support the idea of cognate augites crystallizing from these magmas over a range of P-T conditions. The augites could be relicts within the mantle of an earlier partial melting episode involving fractionation of a liquid, similar to the process suggested for the discrete subcalcic diopsides.

Acknowledgements. Mr. Selwyn Wandili, a chief and landowner in Kawara'ae district, Malaita, facilited communication with the local people so that specimens could be collected. Mr. Wally Bush and other residents of Auki gave friendly assistance and advice. In 1976, Dr. R. B. Thompson, then Chief Government Geologist, kindly supplied a small heavy mineral concentrate collected on an earlier expedition to the area. Helpful correspondence and advice were provided by Dr. T. Deans.

References

Akella, J., Garnet pyroxene equilibria in the system $CaSiO_3$-$MgSiO_3$-Al_2O_3 and in a natural mixture, Am. Mineral., 61, 589-598, 1976.

Akella, J., Solubility of Al_2O_3 in orthopyroxene coexisting with garnet and clinopyroxene for compositions on the diopside-pyrope join in the system $CaSiO_3$-$MgSiO_3$-Al_2O_3, Carnegie Inst. Washington Year Book 73, 273-278, 1974.

Allen, J. B. and T. Deans, Ultrabasic eruptives with alnoitic-kimberlitic affinities from Malaita, Solomon Islands, Mineral. Mag., 34, 16-34, 1965.

Beeson, M. H. and E. D. Jackson, Origin of the garnet pyroxeneite xenoliths at Salt Lake Crater, Oahu, Mineral. Soc. Amer. Spec. Pap. 3, 95-112, 1970.

Binns, R. A., High-pressure megacrysts in basanitic lavas near Armidale, New South Wales, Amer. Jl. Sci., 267-A, 33-49, 1969.

Binns, R. A., M. B. Duggan, and J. F. G. Wilkinson, High pressure megacrysts in alkaline lavas from north-east New South Wales, Amer. Jl. Sci., 269, 132-68, 1970.

Boullier, A. M. and A. Nicolas, Texture and fabric of peridotite nodules from kimberlite at Mothae, Thaba Putsoa, and Kimberley, in Lesotho Kimberlites (editor P. H. Nixon), pp. 57-66. Lesotho National Development Corp., Maseru, Lesotho, 1973.

Boyd, F. R., Enstatite-ilmenite and diopside-ilmenite intergrowths from the Monastery Mine, Carnegie Inst. Washington Year Book 70, 134-138, 1971.

Boyd, F. R., Pargasite-spinel peridotite xenolith from the Wesselton Mine, Carnegie Inst. Washington Year Book 70, 138-142, 1971.

Boyd, F. R., Garnet peridotites and the system $CaSiO_3$-$MgSiO_3$-Al_2O_3, Mineral. Soc. Amer. Spec. Pap. 3, 1970.

Boyd, F. R., A pyroxene geotherm, Geochim. Cosmochim. Acta, 37, 2533-2546, 1973.

Boyd, F. R., Inflected and non-inflected geotherms, Carnegie Inst. Washington, Year Book 75, 521-523-1976.

Boyd, F. R. and P. H. Nixon, Ultramafic nodules from the Thaba Putsoa kimberlite pipe, Carnegie Inst. Washington Year Book 71, 362-373, 1972.

Boyd, F. R. and P. H. Nixon, Origin of the ilmenite-silicate nodules in kimberlites from Lesotho and South Africa, in Lesotho Kimberlites (editor P. H. Nixon), pp. 254-268. Lesotho National Development Corp., Maseru, Lesotho, 1973.

Bradshaw, N., Petrographic examination of seven ultrabasic eruptive rocks and one limestone from Malaita, British Solomon Islands, Spec. Report No. 255, Overseas Division, Inst. Geol. Sci., London, 1968.

Chapman, N. A., Inclusions and megacrysts from undersaturated tuffs and basanites, East Fife, Scotland, Jl. Petrol. 17, 472-498, 1976.

Chen, J., Petrology and chemistry of garnet lherzolite nodules in kimberlite from South Africa, Amer. Min., 56, 2898-2110, 1971.

Coleman, P. J., Stratigraphical and Structural Notes on the British Solomon Islands with reference to the first geological map, 1962, British Solomon Islands Geol. Rec., II, 1959-1962, 17-31, 1965.

Coleman, P. J., Upper Cretaceous Deep water pelagic sediments from Northern Malaita, British Solomon Islands Geol. Rec., III, 1963-67, 53-58, 1968.

Coleman, R. G., Plate tectonic emplacement of upper mantle peridotites along continental edges, Jl. Geophys. Res., 76, 1212-1222, 1971.

Cox, K. G., J. J. Gurney, and B. Harte, Xenoliths from the Matsoku pipe, in Lesotho Kimberlites,

(P. H. Nixon, ed.) Lesotho National Development Corp. Maseru, Lesotho, pp. 76-100, 1973.

Davies, H. L., Peridotite-gabbro-basalt complex in Eastern Papua: an overthrust plate of oceanic mantle and curst, Bull. 128, Bureau Mineral Resources, Canberra, 1971.

Davis, B. T. C. and F. R. Boyd, The join $Mg_2Si_2O_6$-$CaMgSi_2O_6$ at 30 kilobars pressure and its application to pyroxene from kimberlites, Jl. Geophys. Research, 71, 3567-3576, 1966.

Davis, G. L., The ages and uranium contents of zircons from kimberlites and associated rocks, Extended Abstracts, Second International Kimberlite Conference, Santa Fe, New Mexico, 1977.

Dawson, J. B. and J. V. Smith, Chromite-silicate intergrowths in upper-mantle peridotites, Phys. Chem. Earth, 9, 339-350, 1975.

Dickey, J. S., Eclogitic and other inclusions in the mineral breccia member of the Deborah Volcanic Formation at Kakanui, New Zealand, Amer. Mineral, 53, 1304-19, 1968.

Eggler, D. H. and M. E. McCallum, A geotherm from megacrysts in the Sloan kimberlite pipes, Colorado, Carnegie Inst. Washington, Year Book, 538-541, 1976.

Ferguson, J., D. J. Ellis, and R. N. England, Unique spinel-garnet lherzolite inclusions in kimberlite from Australia, Geology, 5, 278-280, 1977.

Gerryts, E., A visit to Malaita Island, 1962, British Solomon Islands, Geol. Rec., II, 1959-62, 141-142, 1965.

Green, H. W. and Y. Gueguen, Origin of kimberlite pipes by diapiric upwelling in the upper mantle, Nature, 249, 617-620, 1974.

Gurney, J. J., H. W. Fesq and E. J. D. Kable, Clinopyroxene-ilmenite intergrowths from kimberlite: A reappraisal, in Lesotho Kimberlites (editor P. H. Nixon), pp. 238-253. Lesotho National Development Corp., Maseru, Lesotho, 1973.

Hackman, B. D., Observations on folding in the Oligocene-Miocene limestones of Central Kwara'ae, Malaita, The British Solomon Islands Geol. Rec., III, 1963-67, 47-50, 1968.

Haggerty, S. E., The chemistry and genesis of opaque minerals in kimberlites, Phys. Chem. Earth, 9, 295-308, 1975.

Harris, P. G., Zone refining and the origin of potassic basalts, Geochim. Cosmochim. Acta, 12, 195-208.

Harte, B., Rock nomenclature with particular relation to deformation and recrystallization textures in olivine-bearing xenoliths, Jl. Geol., Sci., 85, 279-288, 1977.

Herzberg, C. and N. A., Chapman, Clinopyroxene geothermometry of spinel-lherzolites, Amer. Mineral., 61, 626-637, 1976.

Johnson, R. W., D. A. Wallace and D. J. Ellis, Felspathoid-bearing potassic rocks and associated types from volcanic islands off the coast of New Ireland, Papua New Guinea: a preliminary account of geology and petrology, in Volcanism in Australia, R. A. Johnson (Editor), Elsevier, Amsterdam, 1976.

Kennedy, C. S. and G. C. Kennedy, The equilibrium boundary between graphite and diamond, Jl. Geophys. Res., 81, 2467-2470, 1976.

Kresten, P., Chrome pyrope from the Alnö complex, Geol. Føren. Stockholm Forhand, 98, 179-180, 1976.

Lindlsey, D. H. and S. A. Dixon, Diopside-enstatite equilibria at 850° to 1400°C, 5 to 35 kbar, Amer. Jour. Sci., 276, 1285-1301, 1976.

MacGregor, I. D., The system MgO-Al_2O_3-SiO_2: Solubility of Al_2O_3 in enstatite for spinel and garnet peridotite compositions, Amer. Mineral., 59, 110-119.

Marchand, M., Ultramafic nodules from Ile Bizard, Quebec, M. Sc. thesis, Univ. McGill, 1970.

Mason, B., Eclogite xenoliths from volcanic breccia at Kakanui, New Zealand, Contrib. Mineral. Petrol. 19, 316-327, 1968.

Mason, B. and R. O. Allen, Minor and trace elements in augite, hornblende, and pyrope megacrysts from Kakanui, New Zealand, N. Z. Jl. Geol. Geophys. 16, 935-17, 1972.

Nixon, P. H. and F. R. Boyd, The discrete nodule association in kimberlites from northern Lesotho, in Lesotho Kimberlites (editor, P. H. Nixon), pp. 67-75. Lesotho National Development Corp., Maseru, Lesotho, 1973.

Nixon, P. H., F. R. Boyd and A. M. Boullier, The evidence of kimberlite and its inclusions on the constitution of the outer part of the earth, in Lesotho Kimberlites (editor, P. H. Nixon), pp. 312-318. Lesotho National Development Corp., Maseru, Lesotho, 1973.

Nixon, P. H. and P. J. Coleman, Garnet-bearing lherzolites and discrete nodule suites from the Malaita alnoite, Solomon Islands, and their bearing on the nature and origin of the Ontong Java Plateau, Bull. Aust. Soc. Geophysicists (Special Issue 2nd SW Pacific Earth Sci. Seminar)(in press).

O'Hara, M. J., Mineral parageneses in ultrabasic rocks, in Ultramafic and Related Rocks, P. J. Wyllie, ed., Wiley and Sons, N. Y., 1967.

Rickwood, F. K., Geology of the Island of Malaita, in Marshall, C.E., et al., University of Sydney (1957), Colon. Geol. Min. Res., 6, 300-305, 1957.

Shervais, J. W., H. G. Wilshire and E. C. Schwarzman, Garnet clinopyroxenite xenolith from Dish Hill, California, Earth Planet Sci. Lett. 19, 120-130, 1973.

Smith, J. V. and J. B. Dawson, Chemistry of Ti-poor spinels, ilmenites and rutiles from peridotite and eclogite xenoliths, Phys. Chem. Earth, 9, 309-322, 1975.

Thompson, R. B., Ultrabasic rocks in the Solomons-a preliminary statement, British Solomon Islands Geol. Rec., II, 1959-62, 33-34, 1965.

Velde, D. and H. S. Yoder, The chemical composition of the melilite-bearing eruptive rocks, Carnegie Inst. Washington Year Book 75, 574-580, 1976.

Wells, P. R. A., Pyroxene thermometry in simple and complex systems, Contr. Mineral. Petrol., 62, 129-139, 1977.

White, R. W., Ultramafic inclusions in basaltic rocks from Hawaii, Contrib. Mineral. Petrol., 12, 245-314, 1966.

Wilkinson, J. F. G., Ultramafic inclusions and high pressure megacrysts from a nephelinite sill, Nandewar Mountains, north-eastern New South Wales, and their bearing on the origin of certain ultramafic inclusions in alkaline volcanic rocks, Contrib. Mineral. Petrol. 51, 235-262, 1975.

Williams, A. F., The genesis of the diamond, 2 vols. E. Benn Ltd., London, 1932.

Wood, B. J., The solubility of alumina in orthopyroxene coexisting with garnet, Contr. Mineral. Petrol., 46, 1-15, 1974.

Author Index

The authors are indexed by the volume in which their papers appear. Volume I: Kimberlites, Diatremes, and Diamonds: their Geology, Petrology and Geochemistry is indicated by (I); Volume II: The Mantle Sample: Inclusions in Kimberlites and Other Volcanics is indicated by (II).